SURFACE AND SUBSURFACE HYDROLOGY

Proceedings of the Fort Collins Third International Hydrology Symposium, on Theoretical and Applied Hydrology, held at Colorado State University, Fort Collins, Colorado, USA, July 27–29, 1977.

Edited by

Hubert J. Morel–Seytoux
(general editor)

Jose D. Salas,
Thomas G. Sanders
and Roger E. Smith
(technical editors)

Water Resources Publications

P. O. Box 303

Fort Collins, Colorado 80522, USA

205925

For information and correspondence:

WATER RESOURCES PUBLICATIONS
P.O. Box 2841
Littleton, Colorado 80161 USA

SURFACE AND SUBSURFACE HYDROLOGY

ISBN: 0-918334-28-4

U.S. Library of Congress Catalog Card Number 78-68496

This publication is printed and bound by LithoCrafters, Chelsea, Michigan, U.S.A.

GENERAL TABLE OF CONTENTS

PREFACE

This book SURFACE AND SUBSURFACE HYDROLOGY combines the submitted papers and the discussions of the Third International Hydrology Symposium held at Colorado State University, Fort Collins, Colorado, USA, June 27-29, 1977. This Symposium was sponsored by the following professional organizations: the Hydrology Section of the American Geophysical Union, the U.S. National Committee for Scientific Hydrology, the International Association of Hydrological Sciences (I.A.H.S.), the U.S. National Committee for I.A.H.S., and the International Water Resources Association. The National Science Foundation, the United Nations Educational Scientific and Cultural Organization, and Colorado State University provided financial support for the Symposium. These sponsorships are acknowledged.

The original, pre-symposium theme for the Symposium was: Theoretical Hydrology. Because the authors of submitted papers had very different concepts of the scope of theoretical hydrology, the submitted papers represented, in the view of the Planning and Organizing Committees of the Symposium, both the theoretical (or more analytical) hydrology and the applied (e.g., experimental or field) hydrology. It was then necessary, by accepting both types of papers, to change the original Symposium theme of Theoretical Hydrology to the de facto theme: "Theoretical and Applied Hydrology." With this transformed objective, the papers have been accepted, presented and discussed at the Symposium. Furthermore, the decision was made by the Planning and Organizing Committees, as well as by the Director of the Symposium, to present to the public the proceedings of the Symposium under this transformed or de facto theme of "Theoretical and Applied Hydrology."

Most hydrologists have been or are very much connected either with solving the current hydrologic problems related to water resources development, conservation and control, or they perform research activities and investigations that are closely related to applied hydrologic and water resources problems. Symposium, by the material presented and discussed, has demonstrated the severe difficulty hydrologists have to detach themselves from practical problems in order to treat the basic problems by theoretical approaches, or in other words conduct studies in hydrology in the way geologists, meteorologists and oceanographers sometimes treat the basic problems of the earth crust, atmosphere and oceans. The Third Fort Collins International Symposium of 1977 has demonstrated that the qualitative jump in treating the basic hydrologic problems at the highest theoretical levels has yet to be undertaken. If the Fort Collins Symposium of July 27-29, 1977 has effectively contributed to both aspects, (1) displaying the current state of theoretical levels of problem treatment in hydrology and (2) providing the impetus for the future to attempt this jump in a sequence of whatever logical steps are feasible, then the holding of the Symposium was justified beyond the mere presentation of interesting contributions of applied and combined applied-theoretical types of papers.

The second part of the Symposium contributions is presented in another book MODELING HYDROLOGIC PROCESSES, being published simultaneously with this book SURFACE AND SUBSURFACE HYDROLOGY. The separation of the material in two books, under two different titles, was a natural result of the pre-symposium division of material in sessions, with sessions beginning with a presentation of general reports, followed by additional interventions by the authors of papers, and finally by a general discussion of papers from the audience. The session materials have become the section materials in the books. Some overlap between books was unavoidable because proceeding otherwise would have required the rewriting and reorganization of several general reports.

Editors have made their best efforts to polish the language and the form of presentation of most papers, as well as to check the validity of some statements and results, by corresponding with the authors. Four papers are written in French, as they were submitted and presented to the Symposium. The summaries in English, with mathematical, tabular and graphical presentations are expected to help English language readers to grasp also the basic contributions by these papers.

Fort Collins, Colorado
November, 1978

Vujica Yevjevich
Chairman, Symposium Planning Committee

Hubert J. Morel-Seytoux
Chairman, Symposium Organizing Committee and Symposium Director

MAR DEL PLATA ACTION PLAN

By

Yahia Abdel Mageed
Banquet Speaker, Tuesday Evening
June 28, 1977

Mr. President,
Distinguished delegates,
Ladies and Gentlemen,

It is a great pleasure and privilege for me to speak to you this evening and to express my appreciation of the good work being done by the Organizing Committee of this Symposium with active assistance and involvement of the Colorado State University in cooperation with other United States and international organizations interested in various aspects of water resources. I am, of course, referring to the valuable symposia which you have organized on floods and droughts and on the most elusive problem of "decision-making with inadequate hydrologic data" in 1972 and the earlier symposium on the "Application of Mathematics in Hydrology" in 1967. The proceedings of these symposia which you publish from time to time are of lasting value to researchers in these important subjects.

Historically, starting from the dawn of human civilizations which were born and which flourished in some of the celebrated river valleys, the developments in the field of hydrology have always influenced, and been influenced by, water development activities of the times in which they occurred. Nevertheless, the new technologies which emerged in the period after the Industrial Revolution have contributed to significant advances in hydrology and water resources development in recent times. In the period after the Second World War, and particularly in the developing countries, the potential of water development as a potent instrument for social and economic change has received wide recognition. Water development was accorded priority in national development plans in many countries after independence. While this is so, it appears that we seem to be far away from meeting the minimum needs of a majority of mankind, even in respect of food and fiber. A quarter of the world's population is starving and another quarter is undernourished. Two thirds of mankind has no reasonable access to safe or adequate water supply. Several countries are burdened with an energy crisis while, ironically, more than 95 percent of hydroelectric potential remains untapped in some regions. In spite of all the advances in science and technology, there is as yet no reasonable degree of immunity from the fury of floods or death and devastation by droughts. Over a greater part of the globe, we merely react to these events after they occur, rather than take action in time to mitigate the damage or even forecast their occurrence. In some regions of the world, only two to three percent of the potential water resources are utilized at present, while, in other regions, increasing pollution is rapidly making water a critically rare commodity. All this emphasizes the need for a critical evaluation of the present water situation in the world and for taking urgent action to redress and remedy the situation.

The United Nations Water Conference was held at Mar del Plata in this world situation of a growing complexity of water problems both in the developed as well as in the developing countries. This was also the first occasion when the range and complexity of the problems in the water sector were considered in depth in a comprehensive and integrated manner in an international forum. Further, the Conference addressed itself to decision-makers at the policy-making level in all countries of the world. The principal outcome of the Conference is what has now come to be known as the Mar del Plata Action Plan, which covers a wide spectrum of activities in the field of water resources development and management such as assessment of water resources; water use and efficiency in such aspects as drinking water supply and agriculture; environment, health and pollution control; planning, management and institutional aspects; education, training and research; and regional and international cooperation. I have chosen to speak to you today on the Mar del Plata Action Plan, because I believe there is an intrinsic connection between this Action Plan and many of the problems in the field of theoretical hydrology you have chosen to discuss in your symposium.

With this in mind, it will now be appropriate to take a somewhat closer look at the principal components of the Action Plan.

Principal Components of the Action Plan

Assessment of water resources. The first area that the Plan deals with is assessment of water resources. The Plan notes that, in most countries, there are serious inadequacies in the availability of data on water resources, particularly regarding ground water and water quality. The processing and compilation of data have also been seriously neglected. Adequate and reliable data is of great importance not only in the field of design and development but also to all aspects of theoretical as well as applied hydrology. Investigations and experiments using inaccurate data will obviously lead to inaccurate conclusions. To improve the present situation, the Plan calls for the promotion, review and coordination of network densities, strengthening of the mechanisms for data collection, processing and publication and the monitoring of water quality. It also calls for appropriation of increased financial resources for activities related to assessment, the strengthening of the mechanisms for data collection, processing and publication and the monitoring of water quality. It also calls for appropriation of increased financial resources for activities related to assessment, the strengthening of training programmes and the expansion of international assistance for the assessment of surface and ground waters, as well as snow and ice. The Plan recommends further that the activities of the International Hydrological Programme and the Operational Hydrological Programme should be keyed to the targets set by the United Nations Water Conference.

Water Use and Efficiency

The second important area dealt with by the Plan relates to the use of water for different purposes such as drinking water supplies, agriculture, industry and hydropower generation.

In the field of cummunity water supply and waste disposal, the Plan endorsed the call of the United Nations Conference on Human Settlements: HABITAT, held in Vancouver in 1976 for the provision of ample and safe drinking water supply to all the population of the world by 1990, if possible, and in order to achieve this goal, adopted a comprehensive Plan of Action addressed to governments as well as international organizations.

One of the important features of this Plan of Action is that it points out the imperative need for a multidisciplinary approach to this problem and urges that rural water supplies and sanitation should form part of integrated rural development projects. Assigning priority in joint programmes of water resources development for agriculture and livestock was identified as one of the key strategies for the provision of drinking water supplies in rural areas.

The Plan also recommends that the period 1980-1990 should be designated the International Drinking Water Supply and Sanitation Decade and should be devoted to implementing national plans in this sector. This proposal has tremendous potentialities for dynamic action at the national and international levels on a vital issue which concerns the daily lives of every man, woman and child in all countries, irrespective of political, economic, social or cultural differences. Serious attention should be given to a reorientation of our basic approaches to problems in the domain of theoretical hydrology to help in the solution of such practical problems as the provision of drinking water supplies.

With regard to agricultural water use, the Plan contains specific provisions designed to realize the targets set by the World Food Conference of 1974, with related recommendations concerning a phased programme for financing, training, extension and research, promotion of national advisory services, and needed international support. Toward that goal, a 15-year global programme was outlined calling for improvement and new irrigation development in addition to drainage over extensive areas.

The Plan emphasizes the need for preparing phased programmes of financial requirements of countries and urges that these be available within two years for presentation to the appropriate intergovernmental bodies. Also within two years, a report is to be prepared on world training and research facilities and activities. Such a report would serve as a basis for the establishment of new facilities and programmes, as and where appropriate.

This recommendation on research on water use for agriculture is of interest to the intensification of research on such problems as the determination of water requirements for different crops, studies on potential evaporation and evapotranspiration rates from different soil and crop covers, salt tolerant or drought resistant crop varieties, soil and water conservation, soil erosion and sedimentation, etc.

Regarding industrial water use, there is need for more systematic and comprehensive study, both quantitatively and qualitatively, including such aspects as input and output quality, level of treatment where required, reuse and recycling. The Plan notes that these matters may be crucial to the attainment of industrialization targets in the developing countries in accordance with the Lima Declaration and the Plan of Action evolved at the second General Conference of the United Nations Industrial Organization, in 1975.

Environment, Health and Pollution Control

Large scale water development projects have important environmental repercussions of a physical, chemical, biological, social and economic nature. The Plan urges that these should be evaluated and taken into consideration. Safeguarding public health must receive paramount consideration. Water pollution from domestic and industrial effluents and the agricultural use of chemical fertilisers and pesticides is on the increase. Control measures regarding the discharge of urban, industrial and minimg effluents are inadequate. The Plan calls for increased emphasis on environmental protection of ecosystems, including pollution control, sound methods of waste management and adequate safeguards to ensure public health.

Planning, Management and Institutional Aspects

The Action Plan underlines the importance of the formulation of an adequate and sound policy for the development of water resources and the imperative need for making institutional arrangements for the planning and management of resources. Countries were called upon to formulate and keep under review a general statement of policy in relation to the use, management and conservation of water and to ensure real coordination between all bodies responsible for the investigation, development and management of water resources. In this connection, the establishment of efficient water authorities was suggested to provide for proper coordination.

In its recommendations on policy and planning, the Plan emphasizes the need for formulating master plans for countries and river basins so as to provide a long term perspective for planning, including resource conservation and using modern planning tools of systems analysis and mathematical modelling. Although advances in these concepts have been made over the years, there has not been much progress in their application to major river basin developments, particularly in the developing countries. The difficulties in applying these concepts arise not only because of the lack of adequate hydrological data, but also because of deficiencies of all other required data such as those on topographic and land use, agricultural and industrial potential and the needs and possibilities of hydropower generation and inland navigation. Above all, there is an urgent need to transfer knowledge required for planning to the countries concerned. As a result, the Action Plan calls for the building-up of indigenuous technological capability at the national and regional levels, with priorities accorded to the technologies of low capital cost and the use of local raw materials and resources.

Education, Training and Research

The Plan, therefore, attaches considerable importance to the elements of education, training and research. Countries were urged to undertake manpower surveys to determine national needs for administrative, scientific and technical manpower in the water resources sector, in the context of the needs for overall national development. Such manpower surveys would provide the background for undertaking training programmes, not on an ad hoc basis as at present, but on a more systematic basis for all categories and disciplines and for all levels of personnel.

International Cooperation

In recognition of the vast dimensions of the problem, the Plan calls for a greater measure of international cooperation including an expansion of international technical and advisory services on the part of the United Nations, its specialized agencies and other international organizations, so as to assist countries in various aspects of water development.

The Relevance of the Action Plan to the Symposium

The questions you have taken up at this symposium are of great importance and significance to a further deepening of our understanding of the fundamental processes in the domain of hydrology.

Broadly speaking, hydrology is common to the entire gamut of water resources technology. From this point of view, all the five topics you have chosen are of great relevance to the entire Action Plan.

A consideration of the temporal distribution of energy is of interest to problems associated with convective rain, snow melt or geothermal water flow. One of the important components of the Action Plan is a call for the preparation of an inventory of mineral and thermal waters with a view to studying and developing their industrial potential as well as their use as spas. The relation of water quality and water quantity is becoming increasingly complex in view of uncontrolled and unregulated discharge of industrial, agricultural and domestic effluents into natural water sources, whether surface or subsurface. The Action Plan contains important components which deal with the study of this problem and has brought up the necessity for a greater exchange of information amongst the countries with regard to the various remedial approaches they may now be adopting. The Plan also calls for a greater use of the techniques of structural models for a study of hydrologic processes in planning the development of both surface as well as ground waters and also in the matter of their conjunctive utilization. The study of climatic changes is still in infancy and so are our attempts at weather modification. All these have important repercussions on our ability to deal with the problems of minimizing losses either due to floods or droughts. The unprecedented and catastrophic drought which has produced devastating consequences in the Sahel in Africa in recent years, and which has resulted in a number of deaths and agony to those who have survived, has brought up the importance of intensifying our efforts to gain a greater insight into the fundamental processes which cause drought.

<u>Closer integration between scientific theory and development policy</u>. I would suggest that we now consider the nature of theoretical investigations into the fundamental hydrologic processes and their relevance to the solution of pressing problems in the field of development and management of water resources. This has relevance to the broader perspective which should govern the relation between theoretical research and the problems of practical development. Research into problems of hydrology in recent times has proceeded under the influence of scientific and technological developments in such related fields as remote sensing, rendered possible by satellite technology, or computerized analysis of intricate mathematical problems which could not be attempted in earlier times before the availability of facilities of methods of electronic digital computerization. The impact of satellite or computer technologies or the development of instrumentation of a sophistication not possible earlier is undoubtedly desirable and is resulting in a qualitative leap in the analytical techniques and methodologies for investigation. Nevertheless, our understanding of the fundamental natural processes leaves much to be desired. The type of basic discoveries in the realms of hydraulics or hydro-dynamics that took place from the 18th to the 20th Centuries, is not continuing today. The advances in computational facilities and the facilities for the formulation of techniques like mathematical models enable us to attempt mathematical solutions of complex multivariable functions. While this is so, we are in a situation where facilities for mechanical or electronic computation far exceed our insight and experimental investigations into the fundamental processes of hydrological phenomena like the physical mechanisms of rainfall or snowmelt, evaporation, transpiration, infiltration and seepage or nature of flow through porous media, etc. This gives rise to a tendency to work out their interrelationships in a formalistic way with the aid of sophisticated analytical and computational tools or models. While the resulting solutions are of great help in solving immediate practical problems, it should be recognized that these are essentially empirical solutions which do not negate the need for rational and physical solutions. There is undoubtedly a need for greater attention to problems of fundamental research, not under the impact of scientific and technological advances in other disciplines, nor in isolation, but under the impact of the problems specifically in the domain of the development and management of water resources. Herein lies the need for a greater understanding and appreciation of the elements of the Mar del Plata Action Plan on the part of those who conduct research in theoretical hydrology. I have every confidence that a greater interaction between the research worker on the one hand and the policy and decision-maker on the other, will

to more fruitful cooperation in the future for the benefit of both and, in a broader sense, for the benefit of manking.

It is precisely in this context that the Action Plan has drawn attention to further research on such areas as weather modification; climatology and agro-climatology; weather forecasting; effects of climatic changes on water availability; geohydrology including artificial replenishment of aquifers; conservation and reservoir operation; physical modelling; application of techniques of systems analysis to problems of water planning and management including aspects of water quality and pollution control; desalination including treatment of brackish waters; waste treatment; flood and drought control; and many other related problems. It calls upon countries to formulate a clear policy for research, within the overall framework of a national science policy, to forge a closer relationship between research and development, to ensure that duplication and overlapping are minimised and that the results of long- and short-term basic research are used in the solution of specific problems. The Plan recommends that international organizations should assist in conducting a review and evaluation of the research work accomplished so far with a view to outlining the directions of future research, in facilitating exchange of information and experience and dissemination of research results, in conducting global studies of environmental trends, in the standardisation of research methodologies and in investigation of the potential of new technologies such as satellite and computer technologies for the solution of water problems.

At the international level, by virtue of their respective responsibilities, a number of agencies in the family of the United Nations and a number of international non-governmental organizations concern themselves with different aspects of water resources research. There is a great need to coordinate the work of all the numerous organizations in their activities related to problems of research in water resources technology. It is only then that a significant breakthrough will be possible in charting new directions in the domain of theoretical hydrology, in giving it a new orientation and a new sense of purpose which stems from a close relation to applied hydrology and to enable it to make important contributions to intricate problems encountered in the development and management of water resources, as called for by the Mar del Plata Action Plan.

SECTION I

CATCHMENT HYDROLOGY

Section Chairman, David A. Woolhiser
Research Hydraulic Engineer
Agricultural Research Service
U.S.D.A., Fort Collins, Colo.

Rapporteur, Donald E. Overton
Associate Professor of
Environmental Engineering
University of Tennessee
Knoxville, Tennessee

SECTION I

CATCHMENT HYDROLOGY

SECTION I

CATCHMENT HYDROLOGY GENERAL REPORT

By

Donald E. Overton
Associate Professor of Environmental Engineering
The University of Tennessee
Knoxville, Tennessee

Introduction

Land use activities are continuing with an increasing worldwide intensity. Assessment of the environmental impact of these activities is of great concern both before and after the fact. Urbanization, agricultural practices, coal strip mining, and logging operations are examples of land use activities that have allegedly contributed to flooding and stream water quality degradation. In order to develop defensible environmental impact statements associated with these activities, it is essential that the most scientifically based methodology be applied to the problems. Since little hydrologic data are available on catchments, it is becoming widely accepted that mathematical modeling is the only available means of making reliable predictions of the effects of land use changes on runoff quantity and quality.

If the focus of the investigator is primarily on stormwater pollution it must be carefully understood that the transport mechanism for water pollutants is the water itself. Therefore, one will not be able to generate stormwater pollutional loadings without first gaining a fundamental knowledge of the rainfall-runoff process.

Mathematical Modeling of Catchment Hydrology

Since no process can be completely observed, any mathematical expression of a process will involve some element of uncertainty. Hence, any mathematical model will be conceptual to some extent and the reliability of the model will be based upon the extent to which it can be or has been verified. Model verification is a function of the data available to test scientifically the model and the resources available (time, manpower, and money) to perform the scientific tests. Since time, manpower, and money always have finite limits, decisions must be made by the modelers as to the degree of complexity the model is to have, and the extent of the verification tests that are to be performed.

The initial task of the modeler then is to make decisions as to which to use or to build, how to verify it, and how to determine its statistical reliability in application, e.g., feasibility, planning, design, or management. This decision-making process is initiated by clearly formulating the *objective* of the modeling endeavor and placing it in the context of the available resources on the project for fulfilling the objective.

If the initial model form does not achieve the intended objective, then it simply becomes a matter of revising the model and repeating the experimental verifications until the project objective is met. Hence, mathematical modeling is by its nature *heuristic* and *iterative*. The choice of model revisions as well as the initial model structure will also be heavily affected by the range of choice of modeling concepts available to the modeler, and by the skill which the modeler has or can develop in applying them.

Figure 1 is a schematic representation of the modeling process. The modeling process is not new but is nothing more than a modern expression of the classical scientific thought processes involved in the design of an experiment.

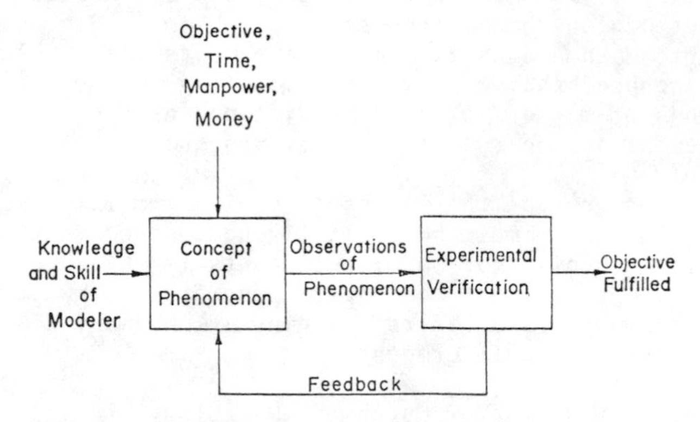

Fig. 1. The modeling process (after Overton and Meadows [1])

Classification of Systems

The distinction between *linear* and *nonlinear* systems is of paramount importance in understanding the mechanism of mathematical modeling. A linear system is defined mathematically by a linear differential equation, the principle of superposition applies and system response is only a function of the system itself and the input intensity. An example of a nonlinear system representation is the equation of gradually varied open channel flow. It is well known that the real world systems are very nonlinear, but linear representations have often been made because of lack of knowledge of the system or because of the pressures exerted by the resource constraint.

System *memory* is the length of time in the past over which the input affects the present state. If stormwater from a basin today is affected by the stormwater flow yesterday, the system (basin) is said to have a finite memory. If it is not affected at all, the system has no memory; and, if it is affected by storm flows since the beginning of the world, the system is said to have an infinite memory. Memory of surface water flow systems is mostly a function of antecedent moisture conditions.

A *time-invariant* system is one in which the input-output relation is not dependent upon the time at which the input is applied to the system. Stormwater flow systems are both time-variant and time-invariant depending upon size (acreage) and land use. The sediment load can also induce time-variance since channel roughness is directly affected.

A *lumped* variable or parameter system is one in which the variations in space either do not exist or have been ignored. The input is said to be lumped if rainfall into a system model is considered to be spatially uniform. Lumped systems are represented by ordinary differential equations and distributed systems are represented by partial differential equations.

A system is said to be *stochastic* if for a given input there is an element of chance or probability associated with obtaining a certain output. A *deterministic* system has no element of chance or probability associated with it, hence for a given input a completely predictable output results for given initial and/or boundary values. A *purely random* process is a system with no deterministic component, and output is completely given to chance. A parametric or conceptual model does have a stochastic component since there will always be errors in verifying it on real data. A "black box" model relates input to output by an arbitrary function, and therefore has no inherent physical significance.

The Modeling Process

Mathematical models of catchment hydrology are needed in land use planning if the consequences of development strategies on the water resource are to be evaluated. Even where actual data collected under land use conditions similar to that being proposed are available, differences in site characteristics will tend to invalidate results that are simply transferred. Further, in the typical situation, very little if any data are available for directly assessing the consequences of alternate development strategies. As a result mathematical models must be employed in the planning process.

These models are needed to account for effects of various site characteristics and to simulate the consequences of alternate development schemes.

There are two conceptual modeling approaches presented in this session. An approach often employed in urban planning has been termed *deterministic* modeling or system simulation. These models have a theoretical structure based upon physical laws and measures of initial and boundary conditions and input. When conditions are adequately described, the output from such a model should be known with a high degree of certainty. In reality, however, because of the complexity of the stormwater flow process, the number of physical measures required would make a complete model intractable. Simplifications and approximations must therefore be made. Since there are always a number of unknown model coefficients or parameters that cannot be directly or easily measured, it is required that the model be verified. This means that the results from usable deterministic models must be verified by being checked against real watershed data wherever such a model is to be applied.

The second conceptual approach is *parametric modeling*. In this case, the models are somewhat less rigorously developed and generally simpler in approach. Model parameters are not necessarily defined as measurable physical entities although they are generally rational. Parameters for these models are usually determined by fitting the model to hydrologic data with an optimization technique.

The two modeling approaches thus appear to be similar and indeed for some subcomponent models, the differences are relatively minor. One practical difference between the two approaches often lies in the number of coefficients or parameters involved. The typical deterministic model has more processes included and thus more coefficients to be determined. Because of the inherent interactions among processes in nature, these coefficients become very difficult to determine in an optimum sense. Because of interactions within the model a range of values for various coefficients may all yield similar results. Hence, without rigorous model verification, the output from a deterministic model is suspect. The parameters in a parametric model on the other hand, are determined by optimization (objective best fit criteria).

Both modeling approaches require data before the model can be employed. The significant difference lies where the data must be located. For the deterministic model, the data should be available at the site of the application. For the parametric model, this latter requirement can be avoided by employing a two-step approach. The model can be fitted to data at locations where it is available in order to obtain optimum model parameters. These parameter values can presumably then be correlated with the physical characteristics of the catchment or watershed. When this is done over a geographic area, the model is said to be regionalized. Once regional relationships between the site characteristics and model parameters are developed, it then becomes possible to measure the site characteristics at locations where water resource data are unavailable and to predict reliably the model parameters and hence make scientifically based predictions.

There are advantages to both modeling approaches. If observed rainfall and storm hydrographs are available, then both approaches can be employed in the development of the stormwater system process models.

Deterministic Models

The problem with the complete simulation model lies in the data required to define many of the boundary conditions. Since many of these measures are not available at the application site (since they necessitate research-level data gathering procedures), the results obtained from research studies in other cities and even other countries are often used to assign boundary condition values to the area of study. The simulated results obtained from such models at an application site, even when a degree of verification is achieved, are still suspect because of bias incorporated by the modeler and because boundary values determined at other locations are not necessarily applicable.

The components of stormwater system modeling for which assignment of boundary values is most difficult in the simulation models are those that quantify nonpoint sources of streamflow quality and quantity in urban and rural subareas. For example, a most difficult problem is quantifying the change in streamflow quantity and quality resulting from building a subdivision on an agricultural area. If, as in the typical simulation approach, estimates of a water quality constituent are considered only after urbanization occurs and these estimates are based upon boundary values measured in other cities, then the results of the modelers and the inappropriate data are incorporated into the findings.

To be objective, the urban planning models must be capable of reliably estimating the streamflow quality and quantity as it was prior to urbanization, and as it will be under alternate development strategies. Only when this is done can the impact of urbanization be evaluated. For example, simply identifying that pollution is associated with urbanization is not sufficient since rural streams and even streams in pristine areas are also polluted using accepted quality criteria (at least during flood periods). Urban planning should aim only at minimizing the impact of development upon water quantity and quality. Any schemes to improve water quality or reduce flooding beyond that of a natural stream can hardly be economically or environmentally justified.

Parametric Models

Parametric modeling lies between deterministic and stochastic (purely random) modeling. In essence, the parametric approach strives for the definition of functional relations between hydrologic and geometric and land use characteristics of a catchment or watershed. This modeling approach involves the model formulation, data collection, data processing, model evaluation by optimization, regionalization of model parameters, and prediction of stormwater flows from ungauged watersheds.

The stochastic scale of modeling approaches is shown in Fig. 2. If the modeler has almost no information on cause-

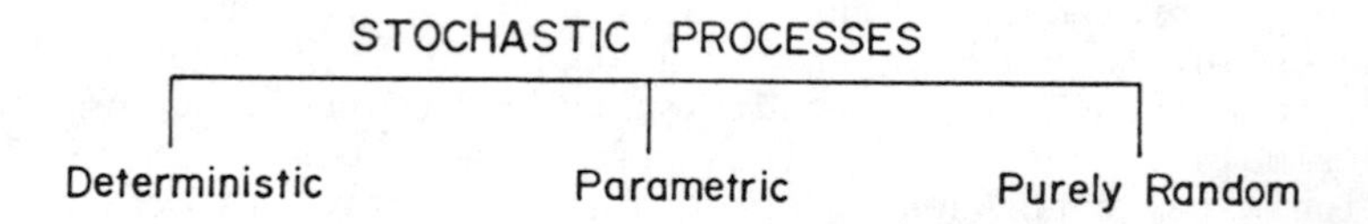

Fig. 2. Stochastic scale of stormwater models (after Overton and Meadows [1])

effect, then the stormwater process must be regarded as purely random. In this instance, the runoff process must be treated as being based entirely on chance. If the modeler has much information on cause-effect, then the process may be treated as deterministic. Seldom, however, are we on the extremes of the scale, because we usually have some notion of the cause-effect of the process and seldom do we have enough of the required information for a rigorous deterministic treatment of the process. The parametric approach is a compromise between the two extremes, but it also involves an effort to improve our notions of the process. *Understanding of the process, in the opinion of this reporter, is something we must take on faith.*

Another distinction among the modeling approaches can be made not only on the amount of, but also upon the type of information available. We must have boundary conditions and initial values in order to utilize the deterministic approach, but we do not need historical observed rainfall and runoff (hydrographs) data. In the parametric approach, it is imperative that we have observed time series of runoff, but it is not necessary that we have rainfall records. However, in stochastic modeling, since there is little, if any cause-effect concepts built into the model, analysis of the time series must be done only during periods of constant land use.

Figure 3 indicates the steps involved in parametric modeling; it must be emphasized that the process is *heuristic* and *iterative*. If, after parameter optimization, it is concluded that the model has done a poor job of fitting the data,

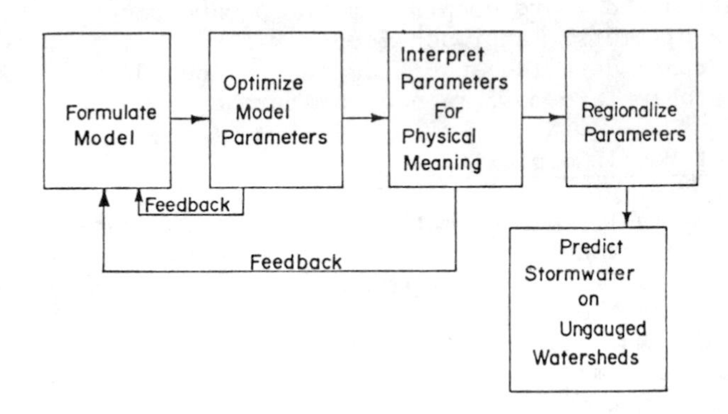

Fig. 3. Steps in parametric modeling (after Overton and Meadows [1])

then, adjustment of the model structure can be made and the experiment repeated. Further, it could be concluded that even though the model does a good job of fitting the data, little physical interpretation can be placed on the optimized model parameters. At this point, the model is a "black box" and any attempt to regionalize the parameters would be futile. Therefore, another adjustment of the model structure would be necessary and the experiment would be repeated. A sensitivity analysis would be included in the parameter interpretation. Not all of the parametric models utilize an objective best fit criteria in optimization; some have "eyeball" fits.

Linkage Between Parametric and Deterministic Models

Parametric and deterministic stormwater models can be and should be complimentary. Parametric models have the capability of assessing the gross effects of land use on runoff, but they do not have the capability of assessing the sensitivity of internal distributions of land use on runoff. Deterministic models are distributed system representations and can simulate the transport mechanism at the source of runoff production to the basin outlet.

Hence, if the hydrologic and physical data are available for the study site, both modeling approaches can be effectively utilized. The parametric model can be utilized as the most scientific basis for predicting basin storm hydrographs on a regional basis, and the deterministic model can be utilized to investigate various land use scenarios on runoff and to simulate transport mechanisms including quality constituents. For the most reliable results, the two model types should be correlated. Further, as the deterministic approach provides information and improvements for the parametric approach, so will the parametric approach feedback information to indicate where further detailed specification is needed and where areas of the problem are most in need of further study.

Choice of Model Complexity

If a highly complex mathematical representation of the system under study is made, either parametric or deterministic, then the risk of not representing the system will be minimized but the difficulty of obtaining a solution will be maximized. Much data will be required, programming effort and computer time will be large, and the general complexity of the mathematical handling may even render the problem formulation intractable. Further, the resource constraints of time, money, and manpower may be exceeded. Hence, the modeler must determine the proper degree of complexity of the mathematical model such that the best problem solution will result and the effort will meet the project constraints. Conversely, if a greatly simplified mathematical model is selected or developed, the risk of not representing the system will be maximized but the difficulty in obtaining a solution will be minimized. The main point here is that the modeler must make a decision from the range of choice of models available or from the models which could be built. But, as pointed out in Fig. 1 refinements in the model can be made by the modeler and indeed this is usually true.

Figure 4 is called a "trade-off diagram" because it illustrates the consequences of the decision of how complex the model should be. If after preliminary verification, it is determined that the initially chosen model is either too complex or not complex enough, then the modeler may move along the abscissa scale in Fig. 4 and experiment with another degree of complexity. This modeling effort should continue until the project objective is attained within the resource constraint.

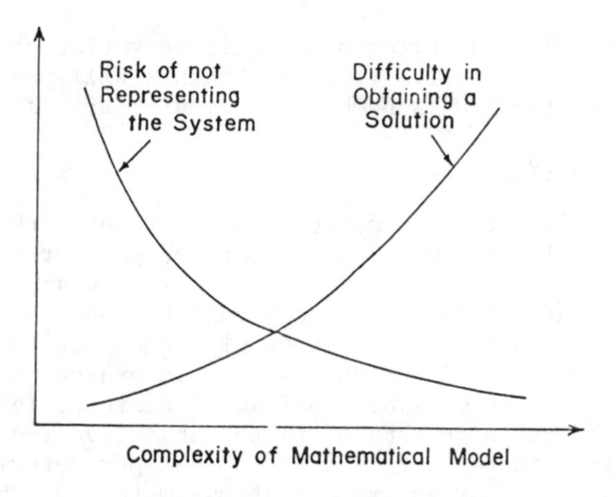

Fig. 4. The trade-off diagram
(after Overton and Meadows [1])

Model Optimization

Since parametric models are conceptual, a set of unknown coefficients or parameters will appear in the mathematical formulation. The parameter values in the model are experimentally determined in the verification procedure. Intuitively, the proper coefficient values would produce the best fit or linkage between storm rainfall (input) and the stormwater hydrograph (output). There is an instinctive temptation, which has appeared in modeling literature, to derive model parameters from observed storms by the trial-and-error "eyeballing" best fit procedures. There are certain distinct and far reaching disadvantages associated with this approach to model verification. They are:

1. If the model is of average complexity, about four or five parameters, then there are a very large number if not an infinite set of coefficients which will produce essentially the same fit. Hence, a large *operational bias* is induced into the modeling process and attributing physical significance to and regionalizing the model parameters may be precluded.

2. If the goodness of fit between the model and the observed stormwater hydrograph is not quantified, the "eyeballing" technique itself induces another operational bias and the same negative effects as above will result.

3. The trial-and-error process is very time consuming and inefficient. Time constraints will permit but a relatively few number of computer trials.

Sensitivity Analysis

Model verification is not complete without a thorough sensitivity analysis. Once the calibrated parameters are arrived at by a best fit procedure, sensitivity analysis proceeds by holding all parameters constant but one, and perturbating that one such that variation of the objective function (measure of fit between the observed storm hydrograph and the fitted model) can be examined. If small perturbations of the parameter produce large changes in the objective function, the system is said to be sensitive to that parameter. This gives a measure of how accurate that parameter must be estimated if the model is to be used in prediction. If the objective function is not sensitive to the perturbated parameter, then the parameter need not be accurately estimated in prediction. If the system is extremely insensitive to the perturbated parameter, the parameter and its associated system component may be redundant and could be deleted from the model.

Each refinement to a model adds more parameters with a marginally diminishing net gain in accuracy. The added parameters will be less sensitive than the original ones. Yet, even though the model will appear more realistic, the fitted parameters will represent less and less by their numerical values. This is a natural consequence of moving to the right along the abscissa scale in the trade-off diagram in Fig. 4. Further, small errors in the hydrologic data may generate large errors in some of the less sensitive parameters. Apparently the residual errors in the computed values may be small, notwithstanding that some of the more insensitive parameters have large errors in them.

It should be recognized that the goodness of fit of the model to the observed data, in the components as well as in any overall model, is a function of the structure of the mathematical model, the accuracy of the data used, the method of fitting the model to the data, and the criteria used for goodness of fit, i.e., the objective function. The first two are somewhat related to the closeness of fit, but the last two may be independently important. Any fitting method makes assumptions about the model fitted primarily because the two must be mathematically compatible. The shape of the response surface will determine to a great extent the viability of the optimization. Therefore, the last two criteria mentioned above are interrelated, because the response surface is the mathematical statement of the test of the goodness of fit.

SUMMARY OF PAPERS AT FIRST SESSION "CATCHMENT HYDROLOGY"

There are fourteen (14) papers in this session and they combine to form a wide range of topics on *Catchment Hydrology*. I have grouped the papers under the following topics:

A. Linear and Nonlinear Surface Water Response

1. Stormwater

a) Ramaseshan, S., "Linearly Structured Models in Hydrology"

b) Gupta, V. L. and S. M. Afaq Moin, "Nonlinearities of Hydrologic Response"

c) Singh, B. P. and B. S. Mathur, "A Conceptual Model for the Binomially Distributed Runoff Elements of a Natural Catchment"

d) Muslu, Y., "A Method of Computing Volume of Storm Water Retention Basins"

2. Snowmelt

e) Singh, R. and B.S. Mathur, "A Linear Distributed Model for a Partially Snowbound Catchment"

f) Martinec, J., "Snowmelt Hydrographs from Spatially Varied Input"

B. Optimized Parametric Models

g) Van der Beken, "A Monthly Water Balance Model Applied to Two Different Watersheds"

h) Seth, S. M., "Conceptual Rainfall Runoff Model Based on Retention Concept"

i) Wheater, H. S., "A Physically Realistic Formulation for Conceptual Model Development"

C. Linkage Between Parametric and Deterministic Models

j) Kadoya, M. and A. Fukushima, "Concentration Time of Flood Runoff in Smaller River Basins"

k) Yen, B. C., Y. Shen and V. T. Chow, "Time of Peak Surface Runoff from Rainstorms"

D. Subsurface Stormwater Response

l) Beven, K., "Experiments with a Finite Element Model of Hillslope Hydrology - the Effect of Topography"

m) Beven, K. and M. J. Kirkby, "Considerations in the Development and Validation of a Simple Physically-Based, Variable Contributing Area Model of Catchment Hydrology"

n) Lynch, J. A., E. S. Corbett and W. E. Sopper, "Effects of Antecedent Soil Moisture on Stormflow Volumes and Timing."

DISCUSSION OF PAPERS

These papers will be reported on by topics.

A. 1. Linear Stormwater Models

The four papers under this topic provide an interesting contrast of *linear* vs. *nonlinear* and *lumped* vs. *distributed* models. Ramaseshan has written an overview of linearly structured models and defines such as "... those that are defined by a set of linear algebraic equations..." This is in contrast to the definition offered by this reporter, i.e., linear differential equation. Generally, both definitions are consistent, however, there are occasions where they are not compatible. A nonlinear differential equation could yield a linear algebraic solution both from finite and infinitesimal calculus. He has also classified linear models into deterministic, statistical and *stochastic* categories and his distinction is whether or not the probablistic nature of the process and presence of errors are taken into consideration. I fully agree with the definition that a stochastic model takes into consideration the probabilities of the process and that statistical models consider the modeling errors, such as in regression; but, the deterministic models discussed by Ramaseshan do not appear to have been derived from the laws of motion. Hence, I would classify them (as well as his statistical models) as *parametric* models. If parametric models (or any models) are evaluated by optimization, I find it difficult to classify them as deterministic, for if they are deterministic they would be *simulation* models and nothing would be left to chance. Further, errors associated with parametric models are considered by evaluating the errors between the model and the data resulting from the optimization. Thus, Ramaseshan has provided what I think to be an interesting contrast in model classification with the definitional scheme proposed by this reporter. However, any classification of models would be, by its very nature, arbitrary to some extent and from my point of view classification schemes would be useful only if they assist the analyst in solving a problem or fulfilling an objective. On the last point, I believe Ramaseshan and I are together for as he states in closing, "Since mathematical models of similar structure may be assumed or derived from different considerations, they are to be classified according to the procedure used in fitting the model"; and, "... the suggested classification is viable ... and may serve in understanding the limitations and advantages of fitted models." Since model structure should be compatible with objective fitting, my views and those of the author appear compatible.

Gupta and Afaq Moin have presented a nonlinear model with four parameters which relates storage to outflow, unsteadiness, and excess rainfall intensity. If the term associated with excess rainfall is dropped, their model would be the same as proposed by Prasad [2]. They used a central difference scheme to integrate the differential equation. Two of the parameters were satisfactorily evaluated by analyzing peak flows from observed storm events for several experimental watersheds, but improvement in the systematic evaluation of the remaining two parameters was deemed necessary by the authors. Gupta and Afaq Moin have shown that their model can satisfactorily reproduce 28 storm events from small experimental watersheds in the S.W. United States, and that many of the disadvantages associated with the linearity of the unit hydrograph approach are rectified by their model. However, there are a couple of problems yet to cope with. The rainfall excess hydrograph was "estimated" by them but they were not specific as to how this was done. Hence, errors associated with this estimation would be dumped into estimated hydrograph parameters. Further, although there were attempts made by the authors to investigate the physical significance of the parameters, it is hard for me to understand it other than in very general terms. I could possibly see this significance if an explicit soil loss component were built into the model and all resulting parameters were optimized on a regional basis and then related to watershed characteristics. This is what has been done with the TVA stormwater model [1]. Yet, the authors have performed an effective sensitivity analysis of the hydrograph parameters and have demonstrated how model output is affected by each one.

Singh and Mathur have developed a model of a cascade of linear channels allowing for distributed inputs for the Ajai Basin in Bihar, India. The local input into each of the linear channels of the cascade is assumed to be distributed binomally which was introduced to account for storage effects in the basin. The authors describe their model to have two parameters: (1) response subarea and (2) translation time of the linear channels. The natural catchment was divided and adjustments were made to obtain the "response subareas". Translation time was considered to be the time to peak of the IUH previously derived from basin hydrologic records. The authors have shown that their model is capable of effectively fitting observed storm hydrographs from the basin. Hence, their model appears to have fulfilled their immediate objectives. It was not at all clear to me how they determined the rainfall excess hyetograph. Hence, errors associated with this estimation would be dumped into the estimation of the hydrograph parameters. Further, the sensitivity of the two hydrograph parameters on stormwater response has not been explicitly demonstrated, and this information should be useful in future modeling experiments on this basin. If a representative sample of stormwater from basins in their region could be obtained, and if a mathematical component for soil loss could be incorporated into their model, the model could be optimized and the parameters regionalized such as has been done with the TVA stormwater model [1]. However, the main consideration would be that the objective of the authors' endeavor has been fulfilled within their project constraints. This appears to be the case.

Muslu has developed a method of computing volume of stormwater retention by offsetting the S-curve a distance equal to the duration of the design storm. His approach is similar to the classical unit hydrograph model except that the peak flow for the resulting design hydrograph was defined by the *rational* method. In some manner, Muslu has derived a maximum storage. The end result of his work is a ready method for finding the storages during various storms by means of a single S-curve plotted for a reference storm.

Recognizing that the runoff process is nonlinear thereby producing a family of S-curves associated with the generating rain excess intensity, the author has not indicated which S-curve he is using. Hence, an average S-curve probably would be used. Further, time of concentration used in the author's paper is also considered a watershed constant. With such fundamental averages being used, I question whether or not the rigor proposed by the author is justified. I agree that methodology in this area should be directed at a search for maximizing detention storage for a given return period, but I must confess that I became somewhat lost in the intricate detail of this paper and I would appreciate a future simplification of the method.

2. Linear Snowmelt Models

The paper by Singh and Mathur is an application of the previously reported model by Singh and Mathur to the Beas River which originates in the Northwestern Himalayas. About 30% of the catchment is under permanent snow cover. The snowmelt analysis is based on the energy budget approach and the melt from the delineated zones is assumed to be binomially distributed. As with the previous paper, the authors have shown close agreement between the observed and model hydrographs, and they offer this as proof of the validity of their model. The same considerations which were discussed for the previous paper are reinforced here, namely that (1) errors in snowmelt simulation and the binomial assumption would be dumped into the estimation of hydrograph shape parameters, and (2) the sensitivity of the two hydrograph parameters to snowmelt be useful information for any future simulations. Furthermore, since this is a linear system representation of a nonlinear system, hydrograph parameters would most likely vary from storm to storm, as has been demonstrated in a number of studies [1]. However, it should be kept in mind that this model has solved a problem for the authors, and we should keep watch for further developments which they may report.

Martinec offers another approach to snowmelt as contrasted with the previous model. The author has developed a *lumped linear* model for the Dischma Basin in the Swiss Alps. Snowmelt is generated by an experimental equation [3] derived by the author. The degree-day factor is increased stepwise in relation to the current density of snow. Since melt takes place on top of the snowpack, the daily meltwater volume is proportional to the current snow-covered area. Discharge from the basin is composed of a linear sum of the immediate runoff (meltwater routed to the outfall) and recession flow. Hence, the model is not applicable to storm conditions (this was true of the previous paper). Martinec found that this linear system model was a very accurate representation of the basin snowmelt response process. This could be due to the fact that the melt process is much less dynamic than stormwater response, since the absolute value of the energy (solar) and its variability in time is much less than that generated by rainfall. The author has provided an error analysis in the melt simulation component, but none in the runoff component. In contrasting Martinec's model with the previous model, the Swiss Basin has a much more extensive snow cover which perhaps could explain the need for the distributed model used by Singh and Mathur.

B. Optimized Parametric Models

The papers presented by Van der Beken and by Seth are quite similar in approach although they are attacking slightly different problems with different model structures.

Seth has developed a model which is based upon retention of rainwater in the soil and is depleted by deep seepage and evapotranspiration. The components of his model are *retention capacity*, *direct contribution*, *leakage*, *evapotranspiration capacity* and *storage and translation* effects of the *direct runoff*. The model was applied to an experimental catchment at Wallingford, England. His model contains 12 parameters and was optimized by Rosenbrock's [14] method, which he found to be quite effective. He used four years of data and split the record into dry and wet seasons. Split sample testing and sensitivity analysis were also used to judge the performance of his model. Optimized model results were checked with observed hydrograph and soil moisture data. His study showed that using a relatively simple model gave good performance in fitting observed data. It would be of interest if Seth's model could be regionalized in England or in India such as has been done by TVA [1].

Van der Beken has developed a model for *monthly water balance* and applied it to two watersheds in Belgium. Although pattern search was used to optimize his model rather than Rosenbrock's method he has approached his modelling problem much the same way as Seth has. The same basic components have been built in by both investigators except that the time intervals are different (3 hours compared to 1 month). Optimized results shown by Van der Beken indicate consistency and reliability of the model. For example, average actual ET in the sandy watershed calculated for 6 years came out to be 62% of evaporation while in the loamy region it was 86%. In both cases, total streamflow volumes were simulated with a relative error of less than 3%.

Wheater has reported an eleven (11) parameter model for several agricultural catchments which are tributaries of the River Severn near Gloucester, England. The catchments have extensive tile drainage installed at 2 meters depth. Rainfall is added to an overland flow reservoir from which infiltration into the surface soil zone is calculated by Holtan's equation. Excess rainfall is routed through a linear reservoir as overland flow. Any initial soil moisture deficit is apportioned to the topsoil as a negative storage subject to a maximum allowable deficit. If a greater deficit exists, it is taken up by the subsoil. Runoff is only permitted from positive storage. The residual infiltration term in the Holtan equation passes as seepage to the subsoil. Lateral flow from the topsoil is calculated using the linear reservoir, as is also the subsoil response generating baseflow. The combined outflows of throughflow, overland flow and baseflow are routed through the channel as another linear reservoir. Finally, a parameter accounts for poor drainage in local areas with no direct drainage outlet. The model was optimized using a multivariate stochastic search technique on 15 storms. The results of this model application were compared with the results of a 4-parameter "black box" model optimized on the same 15 storms.

The results of these optimizations combined with a split data test showed that both models performed about equally well in fitting and in prediction but the "black box" performed better in system identification, thereby indicating that problems associated with system identification increase with model size - an observation previously mentioned.

With enough data, "black box" models could be regionalized, however, parametric models offer the advantage that the parameters - we hope - have physical significance and thereby are more amenable to estimation in situations of land use change. And, there is some hope that moving ahead with the parametric model will lead us to determinism. Wheater states that this is an end of parametric models. Yet, adding more parameters produces more difficulty in obtaining a solution with a diminishing net gain in accuracy. Thus, which direction do we head in? My solution to the apparent dilemma is - concentrate on fulfilling an objective, i.e., solving the engineering problem and not worry too much about determinism. Besides, my solution will guarantee a good nite's sleep and no ulcers.

In all three of these modeling experiments this reporter finds that a systematic evaluation of the models has been carried out and that after all of the computations have been completed, the "proof of the pudding" is to convince oneself that the results look right. In the words of a famous *hydrologic philosopher*, "it has to make me feel warm inside". Evidently these three investigators have that feeling about these projects. It looks good to me, too.

C. Linkage of Models

The two papers that I have grouped under this topic basically pursue the same problem but with different approaches. Both are attempting to relate *hydrologic response time* to catchment characteristics and input. I call this a *linkage of parametric* and *deterministic* models because the basic model form is derived from the equations of motion (i.e., kinematic wave approximations) or from idealized physical catchment experiments. The intent is, however, to optimize the parameters in these models for response time for more complicated geometric and rainstorm conditions and under field conditions. When this latter experiment has been extensively accomplished, we would be in a position to draw an inference as to the surface drainage process under field conditions. We would have also developed a *similarity criteria* or a *regionalization scheme* for hydrologic response.

Kadoya and Fukushima have taken the time of concentration (t_c) relation derived from the kinematic wave equations and optimized the model on four basins. The model says that t_c is equal to a basin coefficient times storm rain excess intensity raised to a power. As long as catchment land use is stable, the coefficient should be constant. For surface water Manning-turbulent flow, the exponent should be equal to -0.4. Their experimental results indicate that the relation applies to their basins and the exponent was very nearly equal to -0.4 on all four basins. At this point, the optimized coefficients can be correlated to basin characteristics since the nonlinearities of the system have been filtered out. The authors have related the basin coefficient (which I call lag modulus [1] to length of the main stream and channel slope. With enough basins, many more catchment characteristics could be correlated to lag modulus (approximately one physical characteristic per two catchments) and a multivariate technique such as principal components analysis [1] could be applied to (a) filter out redundant characteristics, (b) produce the most meaningful regression model, and (c) assist in drawing an inference as to the drainage process.

Yen et al. appear to be attempting to solve the same problem but with their well known Watershed Experimentation System (WES). WES is 40 feet square and 2000 experiments were performed on four types of surfaces including moving rainstorms. Rather than *time of concentration*, they were concerned with *time to peak* (t_p). They related t_p to area, length of channel, slope, rain intensity and duration, and longitudinal velocity of rainstorm movement. Their experiments have provided basic information on rainfall-runoff relationships and a mechanism for drawing an inference as to the response mechanisms of prototype catchments. It would be interesting to see components analysis applied to their data as well.

D. Subsurface Stormwater Response

The last three papers in this session have been submitted by Beven, Beven and Kirkby, and Lynch et al. The first two papers are two different approaches to essentially the same problem. The paper by Beven is a deterministic approach to subsurface stormwater response and the paper by Beven and Kirkby is a parametric approach.

The deterministic model utilizes the finite element solution of Darcy's law combined with the equation of continuity. Simulations were made for various hillslope topography including slope convergence/divergence, and slope convexity/concavity. Beven's results showed that for 2 pulse rainstorms the resulting hydrograph response from the hillslope into the channel is highly nonlinear during the wetting period. Further, hydrograph response was very sensitive to slope convergence as well as the initial conditions. Slope convergence considerably increased peak flow and total volume of flow. Beven's results reinforce the concept that hillslope discharge depends directly, but in a complex manner on upslope extent of the saturated zone or "variable source area". Once wetted up, the hillslope could produce peak flows within minutes after the end of the rainstorm but these peaks are low relative to the generating rain intensities. Results such as these could significantly contribute to a linkage between parametric and deterministic catchment models.

The *parametric* model presented by Beven and Kirkby, by contrast, could provide a simulation tool for drawing an inference as to the source of subsurface stormwater flow. The model provides a balance between being physically based yet relatively simple mathematically.

Lynch, Corbet and Sopper have performed an outdoor physical simulation of rainstorms on a small forested watershed using a sprinkler irrigation system. A 1.5 inch storm over 6 hours was replicated eight times at various antecedent soil moisture (ASM) contents. Hydrograph properties which they define as *quickflow, delayed flow, total storm runoff*, and *maximum peakflow* were found to strongly correlate with antecedent soil moisture. Further, the percentage of rainfall converted to total storm runoff ranged from 12 to 86 percent, while antecedent soil moisture ranged from 87 to 99 percent of field capacity. The authors have concluded that (1) these relationships only apply to the range of antecedent soil moisture conditions included in their study, and (2) additional information is needed to assess the effects of seasonal changes. Since the storm duration and intensity were held constant, it would be interesting to see if the unit response function significantly varied from storm to storm. Such variation is usually attributed to rainfall excess intensity, but if it varies with ASM, then the nonlinear response would be partially attributed to ASM. It would also be interesting to vary rainfall intensity while holding ASM constant. This would permit an investigation of the nonlinearity of system response as affected by input intensity. Forested watersheds tend towards linearity due to the damping effect of the forest cover. In conducting future simulated rain events, I believe it to be important to (1) simulate at durations and intensities which are representative of actual frequencies of the area, and (2) specify those frequencies. Finally, I would very much like to see the TVA stormwater model [1] applied to this catchment since it has two responses (quick and delayed) built into it. It would also permit a ready and efficient means of answering the questions posed in this report as to the nonlinear nature of this catchment. It would also provide another sample of catchment hydrology for which the TVA model would have been applied to and thereby expand our generalities drawn in *catchment hydrology*.

Concluding Remarks

It has been my pleasure to have had the opportunity to report on these papers. I was stimulated by the diversity of the papers on the topic of *Catchment Hydrology*. Also, I am looking forward to reading forthcoming results in the open literature which I am sure will be reported by these authors on such interesting and important problems.

References

[1] Overton, D. E. and Meadows, M. E., 1976. "Stormwater Modeling," Academic Press, Inc., 358 p.

[2] Prasad, R., 1966. "Nonlinear Simulation of a Regional Hydrologic System," J. of Hydr. Div., ASCE, pp. 201-205.

[3] Martinec, J., 1976. "Snow and Ice," Ch. IV, In: Facets of Hydrology, ed. by J. C. Rodda, John Wiley & Sons Lts., pp. 85-118.

[4] Rosenbrock, H. H., 1960. "An Automatic Method of Finding the Greatest or Least Value of a Function," The Computer Journal, 3:175-184.

CONSIDERATIONS IN THE DEVELOPMENT AND VALIDATION OF A SIMPLE PHYSICALLY-BASED, VARIABLE CONTRIBUTING AREA MODEL OF CATCHMENT HYDROLOGY

By

K. J. Beven
Institute of Hydrology, Maclean Building
Crowmarsh Gifford, Wallingford
Oxon, England

and

M. J. Kirkby
School of Geography, University of Leeds
Leeds LS2 9JT, England

Abstract

A hydrologic forecasting model is presented that attempts to combine the important distributed effects of channel network topology and dynamic contributing area with the advantages of simple lumped parameter catchment models. The model parameters are physically based in that they may be determined directly by measurement and the model may be used at ungauged sites. The model provides satisfactory simulations of catchment discharge and the pattern of spreading contributing area, and should be capable of further improvement.

Introduction

There is an undoubted need for a simple, physically-based hydrological forecasting model for medium sized catchments. In particular, a model with parameters that are directly measurable for a given catchment would obviate the need to derive parameter values for unguaged catchments from regional statistical generalizations (as per Nash, 1960; James, 1972; Institute of Hydrology, 1975). However, there is an immediate dilemma facing attempts to formulate such a physically-based model. Every catchment is an exceedingly complex open system with component processes and state variables that may change rapidly over space and time. Even if the processes operating were fully understood then an impossibly large number of parameters would be necessary to model the response of the spatially structured system in any but the crudest detail (Stephenson and Freeze, 1974). On the other hand exceedingly simple models with only one or two parameters can provide a good empirical fit to the response of a particular catchment (Nash, 1957; Dooge, 1959; Lambert, 1969). This paper presents a model for humid temperate areas that attempts to combine the advantages of simple lumped parameter catchment models with the important distributed effects of variable contributing areas and flow routing through the channel network, while retaining the possibility of deriving parameters by direct measurement within the catchment under study.

Variable Contributing Area Concepts

Runoff may occur in a uniform catchment in at least four major ways:

a. Rainfall intensity exceeds soil infiltration or storage capacity resulting in overland flow all over the catchment. This is the classical version of Horton's (1933) model and is thought to have considerable relevance in areas of low vegetation cover and high rainfall intensities. However, in humid temperate areas with a vegetation cover the measured infiltration capacities of soils are generally high in comparison with normal rainfall intensities (Kirkby, 1969; Freeze, 1972). In this case the Horton model of catchment response is not applicable.

b. Rainfall intensity exceeds soil infiltration or storage capacity on a variable area of near-saturated soils. This is the basis for Betson's (1964) partial area conceptual model of catchment response in which it is recognized that the spatially variable nature of infiltration capacities and differences in moisture status at the soil surface caused by downslope flow of water will result in some parts of the catchment being far more likely to produce infiltration excess overland flow than others. Engman and Rogowski (1974) have produced a relatively simple physically-based model based on this concept.

c. Rain falling on stream channels and completely saturated soil. Where the latter are adjacent to stream channels (as is common) this source of overland flow contributes directly to the storm hydrograph (Dunne and Black, 1970). The zone of soil saturation may extend completely from bedrock or may build up above a relatively impermeable layer within the soil.

d. Downslope lateral flow of saturated or unsaturated soil water. Most of this flow will be within the soil ('Subsurface stormflow') but it may locally exceed the soil storage capacity and return to flow over the surface at much higher velocities ('Return flow', Dunne and Black, 1970). Subsurface flow velocities are commonly too slow to contribute appreciably to the peak of the storm hydrographs although in volume terms subsurface flow may dominate the overall response of the catchment in providing the hydrograph tail and low flows.

In small humid temperature catchments mechanisms 2 and 3 would appear to be the critical sources of storm flow, with subsurface flow making a highly significant contribution in setting up the soil water conditions prior to a further storm rainfall. The nature of these processes are thought to explain the observed non-linearity of catchment response and any simple physically-based hydrograph model must reflect this general conceptual knowledge of the mechanisms involved.

However, a choice is available between an infiltration rate approach to the prediction of overland flow, as in the model of Engman and Rogowski (1974), and a soil storage based approach in which the infiltration rate is essentially considered to be non-limiting such that the prediction of overland flow occurs when storage capacity is exceeded. The latter approach has been adopted here both because it would appear to be more physically realistic in British catchments and because it has operational advantages with respect to moisture accounting.

Considerations in Model Development

A number of physically-based deterministic models of the variable contributing area concept of catchment response are reported in the literature (Calver, Kirkby and Weyman, 1972; Freeze, 1972; Hewlett and Troendle, 1975). These models, of varying degrees of sophistication and methodological rigor, have been essentially based on distributed moisture accounting for soil elements within segments of hillslope. The data and computing requirements of these models are, however, so great as to restrict their practical application to research projects (Stephenson and Freeze, 1974) where economic criteria are less dominant. In formulating a simpler model, the present study attempts to integrate the important distributed effects described above with a simple lumped model of the average response of soil water storage in the catchment, while taking into account all information about a given catchment that is available within the pertaining physical and economic constraints. Hydrological forecasting on small and medium sized catchments will normally be allocated only limited funds and the need for simple operational models will undoubtedly continue. Simplicity is not held to be a virtue in itself but is a pragmatic response to a desire to produce a model that is capable of being applied operationally whilst reflecting the current state of knowledge of hydrological processes.

Kirkby (1975) provides an analysis of simple lumped storage models which forms a basis for the present study. The effect of combining several stores of differing time constants in a series chain is particularly noted. The slowest responding store must be most accurately modelled in terms of its non-linearity, because the outflow function is less sensitive to the *form* of the faster responding stores provided an appropriate time parameter is used. A linear approximation may commonly be sufficient.

The effect of the channel network probably becomes important for catchments larger than about 10 km^2 where the time constants of the network compare with those of the infiltration phase. However, following similar arguments it is suggested that a linear network model may be sufficient in catchments of less than 1000 km^2 in which routing time becomes more comparable to the subsurface response time. It is also suggested that the important effects of the channel link frequency distribution of a given network on the form of the outflow hydrograph should be taken into account (Kirkby, 1976). In fact it is convenient to use the channel network to subdivide the catchment so that areas of markedly different hydrologic characteristics can be modelled separately. This preliminary study, however, concentrates on modelling the response of one catchment sub-unit.

The exact structure of the model must necessarily reflect the types of hydrologic characteristics that are quick, convenient and economic to measure for a particular catchment. These include the topographic structure of the catchment together with infiltration rates, overland and channel flow velocities, a small number of discharge measurements and some simple measurements of the soil hydrologic characteristics. There will be a number of ways of interpreting such, essentially crude, measurements in terms of simple storage elements consistent with the discussion above. A preliminary attempt has been described in Beven and Kirkby (1976) and the present model formulation (Fig. 1) represents a further stage in the continuing development of the concepts.

Modelling a Variable Contributing Area

In attempting to model the response of a catchment in simple terms, the dynamic variation of moisture over the catchment must be grossly simplified. For a given average level of moisture storage there will be a wide range of possible spatial distributions, even assuming spatially uniform rainfall conditions. Thus, in attempting to integrate the distributed effects of a variable contributing area with a lumped model of average soil water response described above, some assumptions must be made about the magnitude and duration of rainfall inputs. The simplest is to assume a time-independent steady state at a rainfall rate $\bar{i}$.

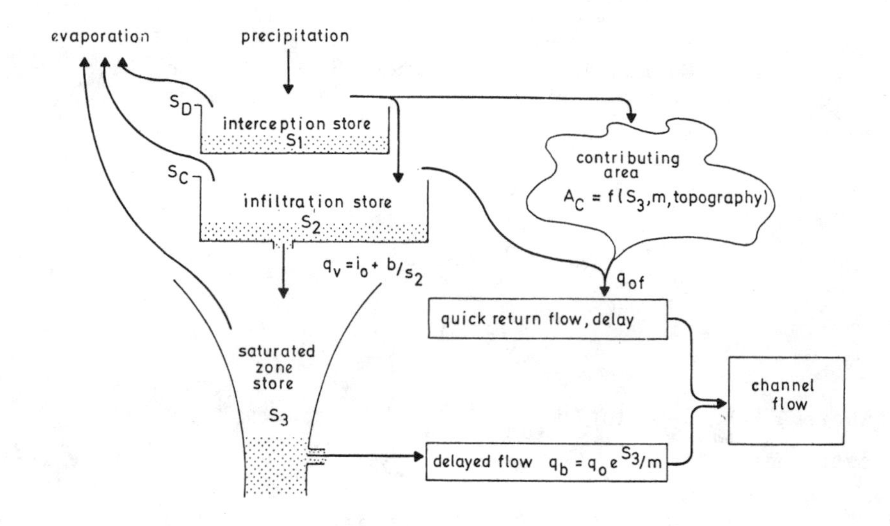

Fig. 1. Schematic representation of the model

For a point in the catchment let the area drained per unit contour length be a, with a local slope angle, B, and a subsurface soil water for which transmissivity at a point may be approximated by an exponential function of the form

$$q = K_o \, e^{S/m} \tan B \tag{1}$$

where q is the flow downslope under an assumed hydraulic gradient due to gravity alone and $K_o \tan B$ is the flow when the storage, S, in rainfall equivalent units (e.g., mm) is zero. Note that for this exponential relationship, providing the value of the constant K_o is chosen to scale the values of q correctly, it is necessary only to specify *relative* changes in the values of subsurface storage, S.

Then under steady state conditions

$$q = \bar{i} \, a \, K_o \, e^{S/m} \tan B$$

or

$$S = m \log_e \left(\frac{\bar{i}}{K_o} \; \frac{a}{\tan B}\right) \tag{2}$$

The saturated area may then be defined as the area for which $S > S_T$

or

$$\frac{a}{\tan B} > \frac{K_o}{\bar{i}} e^{S_T/m} \qquad (3)$$

for some local maximum storage values S_T. Over the whole catchment of area A, for constant K_o and m, mean storage is given by

$$\bar{S} = \frac{1}{A} \int S \, dA$$

$$= \frac{1}{A} \int m \log_e \left(\frac{\bar{i}}{K_o} \frac{a}{\tan B}\right) dA$$

$$= \frac{1}{A} \int m \log_e \left(\frac{\bar{i}}{K_o}\right) dA = m\lambda \qquad (4)$$

where $\lambda = \frac{1}{A} \int \log_e \left(\frac{a}{\tan B}\right) dA$ is a constant of the catchment.

Combining Eqs. (3) and (4), the saturated area A_c is that for which

$$\log_e \left(\frac{a}{\tan B}\right) > \frac{S_T}{m} - \frac{\bar{S}}{m} + \lambda \qquad (5)$$

The spatial distribution of $\log_e \left(\frac{a}{\tan B}\right)$ can be readily mapped for a particular catchment unit. Then if S_T can be assumed spatially constant, the parameters, λ, S_T and m relate the topographic structure of the catchment and average soil moisture storage to saturated area. In a subsequent rainfall the area obtained in this way may be closely identified with the concept of contributing area so that overland flow may be estimated as

$$q_{of} = iA_c \qquad (6)$$

where i is an instantaneous rainfall intensity and A_c is the saturated area calculated from Eq. (5). Obviously this procedure does not allow for the mechanism of return flow described in section 2 but does introduce a major source of non-linearity into the model. An example of the relationship for a 0.23 km^2 headwater area of the Crimple Beck catchment, near Harrogate, Yorkshire, England, is given in Fig. 2. Such curves do not provide a direct relationship between topography, average storage and contributing area since this also depends on the soil parameters. It should however provide the basis for the form of the relationship.

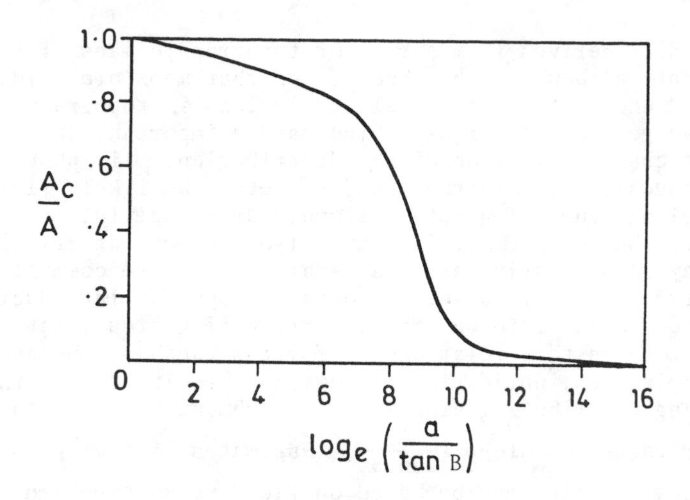

Fig. 2. The relationship between saturated contributing area and values of $\log_e (\frac{a}{\tan B})$ (values of a in metres), for the 0.23 km^2 Lanshaw catchment unit of the Crimple Beck catchment, near Harrogate, Yorkshire, England

Sensitivity analysis of earlier model formulations showed that some form of overland flow routing component in the model would be necessary on all but the smallest catchment units. This was backed up by field measurements of flow velocities. Considering the stated aims of this modelling exercise the use of kinematic or other relatively complex methods of flow routing could not be justified, even if it is accepted that they can provide an adequate representation of the patterns of overland flow in reality. Accordingly, a simplified procedure was used based on a constant average overland flow velocity and a *variable* histogram relating calculated contributing area with distance from the catchment outlet. This procedure provides an easily computed non-linear delay function.

Model Calibration

The parameters of the model consist of the $\log_e (\frac{a}{\tan B})$ distribution for the area and the constants S_D, S_C, i_o, B, q_o, λ, m, and S_T as defined in Fig. 1 and section 4. All the parameters are physically-based in the sense that they may be determined by measurements made in the field, so that the model may be used at sites without a continuous discharge record. The $\log_e (\frac{a}{\tan B})$ distribution and the constant λ

are readily derived from survey or topographic maps of the catchment, although it has been found that maps are seldom detailed enough to define small variations in topography, such as convergent hollow areas, which may be important in the correct characterization of the distribution. Air photographs or an investigation in the field to determine likely flow lines along lines of greatest slope, may be helpful in this respect. Dunne et al. (1975) have also reviewed other evidence that may be suggestive as regards areas that are commonly saturated. There is also, of course, scope for introducing impervious areas into the final contributing area relationship. Figure 2 shows this relationship for the Lanshaw sub-catchment of Crimple Beck, while Fig. 3 shows the predicted pattern of spreading contributing area in the catchment based on the calculated values of $\log_e (\frac{a}{\tan B})$. Estimates of average overland flow velocity may be based on field measurements using visible dye as a tracer.

A value of the soil parameter m is obtained by assuming that it is constant throughout the area and equal to the equivalent parameter for the lumped exponential subsurface storage component. An estimate may then be derived from the storage/discharge relationship during a recession period, preferably during winter when the effects of evapotranspiration are minimized. Preliminary analyses have shown that this may be a useful approximation. Remembering that soil storage values need only be specified in relative units, if

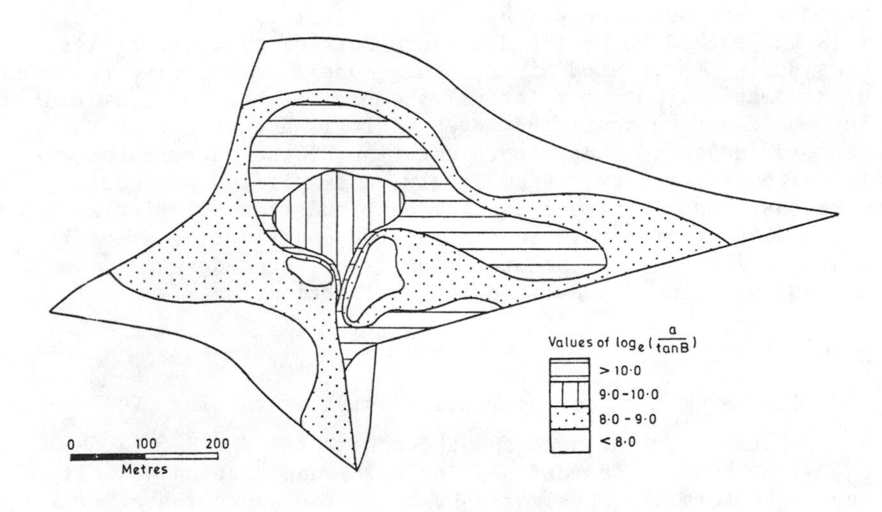

Fig. 3. The distribution of $\log_e (\frac{a}{\tan B})$ values in the Lanshaw catchment

the average maximum storage capacity of the soil profile, S_T, is taken as zero, then all values of subsurface storage S_3 may be taken as representing amounts of soil moisture deficit (negative values) or surplus relative to this value. On the basis of measured or estimated soil moisture deficits (below saturation), the value of q_0 can be calculated by substituting known discharge values into the subsurface storage equation.

The final soil parameters associated with the interception and subsurface storage components are measured in the field using a sprinkling infiltrometer apparatus. Model calibration is completed by estimates of the initial storage values at the start of a simulation. If conditions are dry then the upper storage elements may be taken as empty, while an estimate of average soil moisture deficit provides a starting value for the subsurface store. Further details of model calibration are given in Beven and Kirkby (1976).

Considerations in Model Validation

Since all models are simplifications of our perception of the real world, they cannot hope to reproduce the behavior of the prototype system in all details. Thus there can be no absolute validation of any model and the term can only be applied in a relative sense with reference to some specified criteria of comparison between the observed operation of the real system and the model simulation. In addition, practical considerations may result in a model being accepted as sufficiently accurate for a given purpose without being accepted as a validated representation of the real system (Kisiel, 1971).

Models that are to any degree physically-based must satisfy further validation criteria with respect to the dynamic change in the internal state variables of the model in comparison with the catchment system. In this particular lumped model, changes in soil moisture status and surface saturation cannot be predicted in any but a broad generalized way. However, a comparison of the predicted spread of saturated area (Fig. 3) with observed saturated area (Fig. 4) for the Lanshaw catchment unit does give some support for the underlying topographically based contributing area concepts incorporated in the model. The broad pattern of change is very similar. Some of the anomalies can be explained by distinct soil differences in the catchment. A similar degree of conformity has been shown in other small upland catchments (Kirkby and Weyman, 1974), although it is expected that sensitivity of contributing area response may occur only in a certain range of topography and soil parameters within a climatic regime (Kirkby et al., 1976).

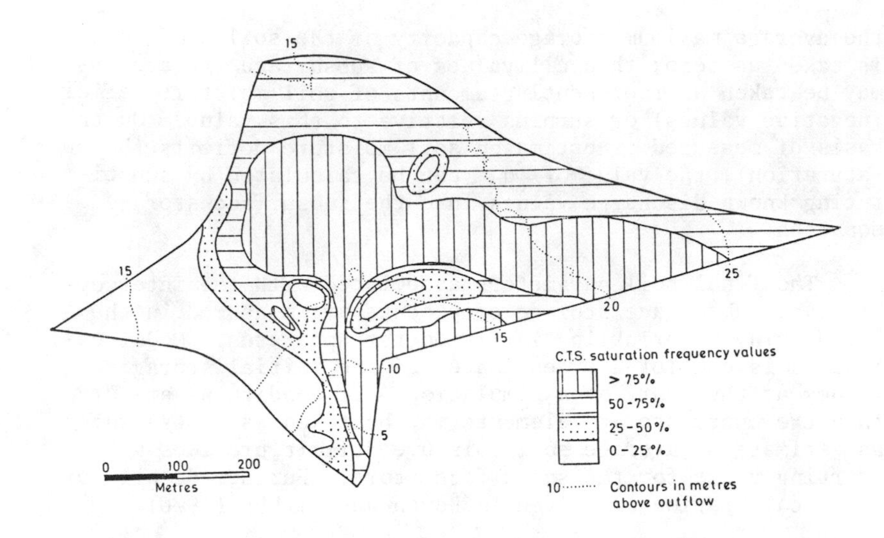

Fig. 4. Frequency of surface saturation in the Lanshaw catchment, measured using crest saturation tubes (CST's)

A comparison of observed and predicted discharges reveals that the model can provide a reasonably accurate representation of the catchment response under both winter and summer conditions (Fig. 5). It should be noted that these simulations have used only measured and estimated parameters with the catchment being treated as essentially ungauged. Sensitivity analysis reveals that the simulations could be slightly improved by optimization of the parameter values, and some seasonal variation in both optimum and field measured values is apparent.

In conclusion, the model can provide simulations of catchment behavior and outflow that may be sufficiently accurate for many purposes, particularly on ungauged catchments. Evidence of inconsistency in the model residuals and seasonality in the parameter values may result from either model or data errors but suggests that further improvement will be possible. The model has the advantage of simplicity while retaining at least some of the distributed effects apparent in catchment behavior, and focusing attention quite clearly on the structure and nature of response of the catchment.

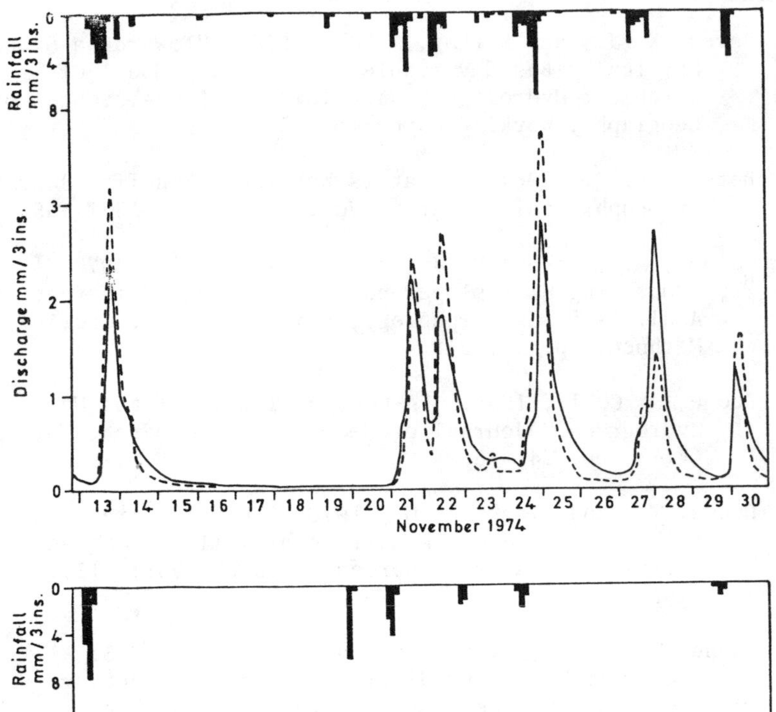

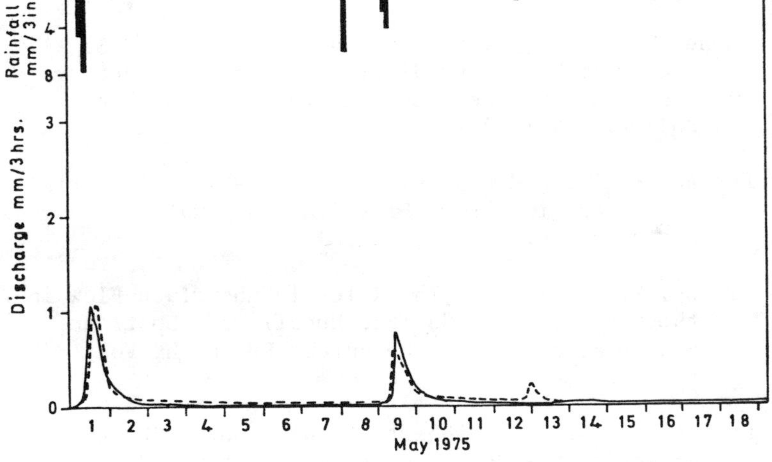

Fig. 5. Observed (solid line) and predicted (broken line) discharge from the Lanshaw catchment over winter and summer periods

References

[1] Beven, K. J., and Kirkby, M. J., 1976. "Towards a Simple Physically Based Variable Contributing Model of Catchment Hydrology," University of Leeds, School of Geography, Working Paper No. 154, 26 p.

[2] Betson, R. P., 1964. "What is Watershed Runoff?" Journal of Geophysical Research, Vol. 69(8), pp. 1541-1552.

[3] Calver, A., Kirkby, M. J., and Weyman, D. R., 1972. "Modelling Hill-slope and Channel Flows," in Spatial Analysis in Geomorphology, edited by R. J. Chorley, Methuen, pp. 197-218.

[4] Dooge, J. C. I., 1959. "A General Theory of the Unit Hydrograph," Journal of Geophysical Research, Vol. 64(1), pp. 241-256.

[5] Dunne, T., and Black, R. G., 1970. "Partial Area Contributions to Storm Runoff in a Small New England Watershed," Water Resources Research, Vol. 6(5), p. 1296.

[6] Dunne, T., Moore, T. R., and Taylor, C. H., 1975. "Recognition and Prediction of Runoff-Producing Zones in Humid Regions," Hydrological Sciences Bulletin, Vol. 20(3), p. 305.

[7] Engman, E. T., and Rogowski, A. S., 1974. "A Partial Area Model for Storm Flow Routing," Water Resources Research, Vol. 10(3), p. 464.

[8] Freeze, R. A., 1972. "The Role of Subsurface Flow in the Generation of Surface Runoff, 2. Upstream Source Areas," Water Resources Research, Vol. 8(5), p. 1272.

[9] Hewlett, J. D., and Hibbert, A. R., 1967. "Factors Affecting the Response of Small Watersheds to Precipitation in Humid Areas," in International Symposium on Forest Hydrology, edited by W. E. Sopper and E. W. Lull, Pergamon Press, pp. 275-290.

[10] Hewlett, J. D., and Troendle, C. A., 1975. "Non-Point and Diffused Water Sources: A Variable Source Area Problem," in Symposium on Watershed Management, American Society of Civil Engineers, pp. 21-46.

[11] Horton, R. E., 1933. "The Role of Infiltration in the Hydrological Cycle," Transactions of the American Geophysical Union, Vol. 14, pp. 446-460.

[12] Institute of Hydrology, 1975. "Flood Studies Report," Natural Environment Research Council, London.

[13] James, L. D., 1972. "Hydrologic Modelling, Parameter Estimation and Watershed Characteristics," Journal of Hydrology, Vol. 17, pp. 283-307.

[14] Kirkby, M. J., 1969. "Infiltration, Throughflow and Overland Flow," in Water, Earth and Man, edited by R. J. Chorley, Methuen, pp. 215-228.

[15] Kirkby, M. J., 1975. "Hydrograph Modelling Strategies," Chapter 3 in Processes in Physical and Human Geography, edited by Peel, Chisholm and Haggett, Heinemann.

[16] Kirkby, M. J., 1976. "Tests of the Random Network Model and Its Application to Basin Hydrology," Earth Surface Processes, Vol. 1(3), Wiley, pp. 197-212.

[17] Kirkby, M. J., and Weyman, D. R., 1974. "Measurement of Contributing Area in Very Small Drainage Basins," Seminar Paper Series B, No. 3, Department of Geography, University of Bristol.

[18] Kirkby, M. J., Callen, J., Weyman, D. R., and Wood, J., 1976. "Measurement and Modelling of Dynamic Contributing Areas in Very Small Catchments," University of Leeds, School of Geography, Working Paper No. 167, 65 p.

[19] Kisiel, C. C., 1971. "Efficiency of Parameter and State Estimation Methods in Relation to Models of Lumped and Distributed Hydrologic Systems," in V. Yevjevich (Editor), Systems Approach to Hydrology, Water Resource Publications, Fort Collins, Colorado.

[20] Lambert, A. O., 1969. "A Comprehensive Rainfall-Runoff Model for Upland Catchment," Journal of the Institute of Water Engineers, Vol. 23, pp. 231-238.

[21] Nash, J. E., 1957. "The Form of the Instantaneous Unit Hydrograph," IASH Publication No. 45, Vol. 3, pp. 114-121.

[22] Nash, J. E., 1960. "A Unit Hydrograph Study with Particular Reference to British Catchments," Proceedings of the Institute of Civil Engineers, Vol. 17, pp. 249-282.

[23] Stephenson, G. R., and Freeze, R. A., 1974. "Mathematical Simulation of Subsurface Flow Contributions to Snowmelt Runoff, Reynold's Creek Watershed, Idaho," Water Resources Research, Vol. 10(2), pp. 284-298.

[24] Wooding, R. A., 1965. "A Hydraulic Model for the Catchment-Stream Problem: I Kinematic Wave Theory," Journal of Hydrology, Vol. 3, pp. 254-267; "II Numerical Solutions," Journal of Hydrology, Vol. 3, pp. 268-282; "III Comparison with Runoff Observations," Journal of Hydrology, Vol. 4, pp. 21-37.

EXPERIMENTS WITH A FINITE ELEMENT MODEL OF HILLSLOPE HYDROLOGY - THE EFFECT OF TOPOGRAPHY

By

K. J. Beven
Institute of Hydrology
Wallingford, England

Abstract

Physically-based simulation models may be used to explore the implications of specific assumptions about the nature of a real world system and then to make predictions of the behavior of that system under a set of naturally occurring conditions. It is important that understanding generated by the former use should be gained before predictive use of the system model. This paper describes and uses a finite element model of transient, partially saturated water flow within a hillslope soil mantle overlying an impermeable bedrock to investigate the effects of topography on the hillslope hydrograph. The simulations produce peak flows which are generally low but may occur within minutes of the end of a pulse rainstorm. The response of all the slopes is non-linear in both peak flow and total flow volume. With soil parameters held constant, slope convergence increases the rate of discharge from a hillslope and the propensity towards saturation at the base of the slope and the subsequent generation of overland flow. The influence of topography on the initial moisture conditions within the soil mantle serves to increase these effects on concave slopes.

Introduction

There are two major aims of simulation models. The first is to explore the implications of certain specific assumptions about the nature of a real world system; the second is to predict the behavior of that real system under a set of naturally occurring conditions. This paper is concerned with the former objective and, in particular, with a deterministic physically-based model of the dynamic response of partially saturated water flow on hillslopes. The intention is to promote understanding of the response of the system model to different boundary conditions and to varied external influences imposed upon it. There is an underlying implication that, by analogy, one is also extending understanding of the real world prototype. However, without supporting evidence, such conclusions must be made with care, as demonstrated by Freeze (1975) for the specific case of subsurface water flow. All simulation models are essentially simplifications of reality and, as such, will be subject to uncertainty and error. Ultimately it is the working of the model that we understand but it is essential that such understanding should precede predictive use of such models.

The work presented in this paper complements the earlier work of Freeze (1972a, 1972b) who used coupled finite difference models of two dimensional subsurface flow and one dimensional overland and channel flows to explore primarily the effects of changing soil hydraulic conductivity on the predicted hillslope hydrograph. He concluded that only soils with the highest possible saturated hydraulic conductivities are likely to contribute subsurface storm flow sufficiently quickly to be termed 'storm flow'. On the other hand, soils with very low conductivities are more likely to produce complete saturation of the soil profile at the base of the slope thereby generating overland flow. Freeze suggests that for any slope/soil system the mechanism of quick response subsurface storm flow (Hursh, 1944; Whipkey, 1965, 1967) can only be supported above a certain threshold value of conductivity and that there is a range of conductivities for which the response is highly lagged. Freeze's work does support the concept of a dynamic contributing area of overland flow close to the channel as observed in the field by Dunne and Black (1970a, 1970b).

A previous study (Beven, 1977) has shown that other soil parameters and the topographic configuration of the slope can affect the predicted hillslope hydrograph at least as much as variations in hydraulic conductivity. This paper explores further the effect of slope topography.

Model Assumptions

(a) It is assumed that the hillslope is underlain by a shallow impermeable bedrock, and that the upslope boundary at the watershed is one of zero flow.

(b) The hillslope is assumed to be externally coupled to the channel at its base such that the solution of the former does not depend on the depth of flow in the channel. This restriction precludes the simulation of bank storage effects that may be important in many natural locations but may be valid in steep upland catchments with incised channels such as the system described by Weyman (1970, 1973). This is not a theoretical limitation (Freeze, 1972b) but represents a convenient subdivision of the problem.

(c) It is further assumed that within a given hillslope segment, downslope variations of soil characteristics are much greater than variations across the slope, so that, if changes in slope width are continuous and gradual, the system may be averaged across the slope into a two dimensional representation.

(d) Flow is assumed to be laminar and of low velocity in an isothermal and nondeformable porous medium.

(e) Only single phase flow of water, of constant unit density and viscosity, in response to hydraulic pressure gradient is considered, the effect of air and water vapor flows, temperature, osmotic and other forces being neglected.

Within the foregoing constraints an equation of flow can be derived by combining an equation of continuity with Darcy's law. Thus

$$BC(\Psi)\frac{\partial\Psi}{\partial t} - \frac{\partial}{\partial x}\{BK_{x'x}(\Psi)\frac{\partial\Psi}{\partial x} + BK_{x'z}(\Psi)\frac{\partial\Psi}{\partial x} + BK_{x'z}(\Psi)\}$$

$$- \frac{\partial}{\partial z}\{BK_{x'z}(\Psi)\frac{\partial\Psi}{\partial x} + BK_{z'z}(\Psi)\frac{\partial\Psi}{\partial x} + BK_{z'z}(\Psi)\} = 0 \qquad (1)$$

where Ψ is the capillary potential in the soil,

$C(\Psi) = \frac{d\Theta}{d\Psi}$ is the specific moisture capacity of the soil,

Θ is soil moisture content by volume,

B is the width of the slope,

$[K(\Psi)]$ defines the effective hydraulic conductivity tensor which will vary with Ψ in some complex manner,

x and z are the global horizontal and vertical axes,

x' and z' are the local principal axes of conductivity of the soil, and

t is time.

At saturation $C(\Psi) = 0$ the term involving the time derivative disappears and the hydraulic conductivity components will be approximately constant. In both cases the boundary conditions are:

$\Psi = \Psi_o$ at $t = 0$ (2a)

$\Psi = 0$ on seepage faces (2b)

$\frac{\partial\Psi}{\partial l} = 0$ at no flow boundaries where l is the direction of the outward normal to the boundary (BC and CD of Fig. 1) (2c)

$K(\Psi)\frac{\partial\Psi}{\partial l} = -q$ at boundaries across which there is a specified flux per unit width, q (such as rainfall across AB of Fig. 1). (2d)

The Finite Element Method

The finite element method (Zienkiewicz, 1971) provides a powerful tool for obtaining approximate numerical solutions to partial differential equations. The method is based on the choice of a functional that, when minimized over the complete flow domain, is a solution to the flow equation, in this case Eq. (1). The most general method for deriving such functionals is the Galerkin method, which is used here in conjunction with a finite element discretisation of linear triangles. Here, a backward difference formulation of the time derivative has been used to step the solution through time.

At each time step a set of non-linear simultaneous equations in the nodal values of Ψ at the end of the time increment are obtained. Estimates of the mid-time step values of the equation coefficient are calculated and the complete system is solved by a Gauss elimination algorithm. The solution is then iterated, improved values of the non-linear coefficients being calculated at each iteration until the calculated nodal values of Ψ at the end of the time step converge. This process leads to a fully implicit solution scheme and convergence is very rapid at most time steps. An occasional tendency towards (stable) oscillation was observed but this could be avoided by reducing the time step at the expense of increased computational effort.

The finite element method has a number of advantages over the alternative finite difference method for problems encountered in hillslope hydrology. In particular:

(a) The method can more easily approximate complex geometrics such as the natural curvature of hillslopes and the variable depths of profile horizons within the soil;

(b) the method can more easily cope with boundary conditions described by a differential equation (conditions 2c and 2d);

(c) arbitrary variations in hydraulic conductivity and anisotropy within the flow region are handled easily; and

(d) the method allows a great deal of flexibility in varying the closeness of the mesh of elements in regions of the flow domain where rapid or important changes are expected to occur.

Applications of the finite element method to partially saturated flow problems include Bruch and Zylovski (1973),

Neuman (1973) and Neuman et al. (1975). The finite element solution is coupled to a one-dimensional downslope kinematic solution for overland flow through the mechanism of infiltration in a similar manner to that used by Freeze (1972a, 1972b) and Smith and Woolhiser (1971). More details of the model and the procedure used to calculate hillslope discharge are given in Beven (1977).

The Model Parameters

The parameters required by the model fall into three groups: topographic parameters, soil parameters and initial conditions. In addition precipitation and evaporation data may be required at each time step.

For the purposes of this investigation the principal axes of the soil were assumed to be coincident with the global horizontal and vertical axes throughout the simulations. The remaining soil parameters are concerned with defining the characteristic relationships between capillary potential, hydraulic conductivity and soil moisture content. Single valued relationships are used, based on primary wetting curves, although there is no inherent reason why adequately defined hysteretic relationships could not be used in the model. The functional equations relating Ψ, $K(\Psi)$ and $C(\Psi)$ take a form modified from that suggested by Taylor and Luthin (1969). Thus

$$K(\Psi) = \frac{K_S}{1 + A_K(-\Psi)^3} \tag{3a}$$

$$C(\Psi) = \frac{\Theta_S A_\Theta}{2(-\Psi)^{\frac{1}{2}} (1 + A_\Theta(-\Psi)^{\frac{1}{2}})^2} \tag{3b}$$

where K_S is the saturated hydraulic conductivity of the soil,

Θ_S is the saturated moisture content of the soil,

A_K and A_Θ are equation coefficients (see Beven, 1977).

The soil parameters chosen for the present simulations are $K_S = 0.02$ cm/min, $\Theta_S = 0.4$, $A_K = 0.0008$, and $A_\Theta = 0.05$.

The choice of a pattern of initial moisture content for the slope is arbitrary. Here the initial condition is that of complete gravity drainage defined by $\Psi + z = 0$ everywhere on the slope. This condition is easily defined for all the slopes under consideration and gives a common initial discharge ($Q = 0$) to facilitate comparison between the simulations. However, such a condition is unlikely to be approached on real slopes since it presupposes an extremely long period of drainage and disregards the effect of evaporation on the pattern of water content on the slope.

The Hillslopes Simulated

The initial application of the model is to a hypothetical straight hillslope, 100 metres long and 1 metre wide with a constant surface slope of 1 in 5. The isotropic homogeneous soil mantle is 1 metre thick throughout on an impermeable base. The finite element discretisation used with 141 nodal points and 204 elements is illustrated in Fig. 1. To ensure reasonable accuracy in the calculation of discharge at the base of the slope it is necessary to use a fine discretisation close to the seepage face. However, the rapid increase in computer run times associated with an increasing number of nodes necessitates a relatively coarse discretisation of the upper slope, which must reduce the overall accuracy of the simulations.

The development of further hillslope topologies is based on two criteria, slope convergence/divergence and slope convexity/concavity. The degree of convergence and convexity of the particular slopes used here are defined by the following relationships. For a point at distance x_i metres from the base of the slope at which the height of the base of the soil profile is z_i metres, and the width is B_i metres. Then:
For a convergent slope

$$B_i = 101 - 10^{(100-x_i)/50} \tag{4a}$$

For a divergent slope

$$B_i = 10^{(100-x_i)/50} \tag{4b}$$

For a convex slope

$$z_i = 20 - 0.002\ (100-x_i)^2 \tag{4c}$$

For a concave slope

$$z_i = 0.002\ x_i^2 \tag{4d}$$

For all the slopes the length of slope, depth of the soil profile and height of the divide are kept constant. A total of seven slope configurations are simulated as follows:

a. Constant slope, constant width.
b. Constant slope, divergent.
c. Constant slope, convergent.

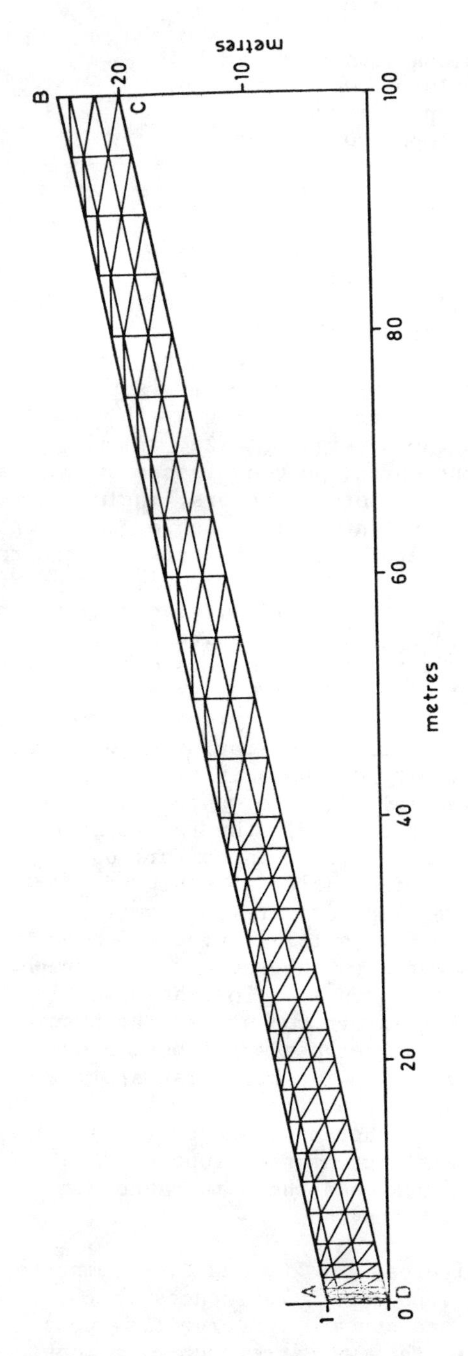

Fig. 1. Finite element discretisation of a hypothetical straight slope with 141 nodes and 204 elements. Vertical exaggeration of soil profile only: x 5

d. Convex slope, constant width.
e. Concave slope, constant width.
f. Convex slope, divergent.
g. Concave slope, convergent.

The Results

All the simulated hillslopes were subjected to 2 pulse rainstorms, each of 600 minutes at a rate of 0.01 cm/min, with a period of 1400 minutes without rainfall between the storms. The calculated outflow hydrographs for the different slopes are shown in Fig. 2. These hydrographs show that the response of all the slopes to rainfall is highly non-linear during the wetting up period covered by the two storms. The first three hydrographs (Figs. 2A, 2B, 2C) show clearly the dominating influence of slope convergence on the calculated outflows, even when all other factors, including the initial conditions remain constant. Slope convergence considerably increases both the magnitude of the peak and the total volume of flow. The response of the divergent slope however, differs little from that of constant width, with only slightly lower peak flows and recession limbs resulting from the relatively lower rainfall input and flow from upslope.

It appears that the convexity/concavity of the slope is effective primarily through the influence of the topography on the complete gravity drainage initial condition. This leads to higher initial moisture contents within the soil at the base of the concave slope. The resulting higher unsaturated hydraulic conductivity, (in comparison with the drier convex slope) are sufficient to give higher velocities of flow on the concave slope despite the lower slope gradient at its base. Consequently, a larger proportion of the concave slope contributes water to the outflow hydrograph. The joint effect of these influences is illustrated in Figs. 2D and 2E. The higher initial moisture content at the base of the concave slope leads to higher peaks for both storms and a higher recession discharge following the first storm. Outflow from the convex slope after the first storm is very low but leaves the slope with much higher moisture contents prior to the second storm. The effect of the slope gradient is then apparent in the delayed peak and the maintenance of the recession limb of the second hydrograph.

The final figures, 2F, 2G, and 3, attempt to further illustrate these topographic influences in the simulation of convex divergent and concave convergent slopes. The slope configurations are the two extreme combinations of topography and this is reflected in the outflow hydrographs. They are, however, the types of slope that make up many headwater catchment areas. Physically-based simulation also allows

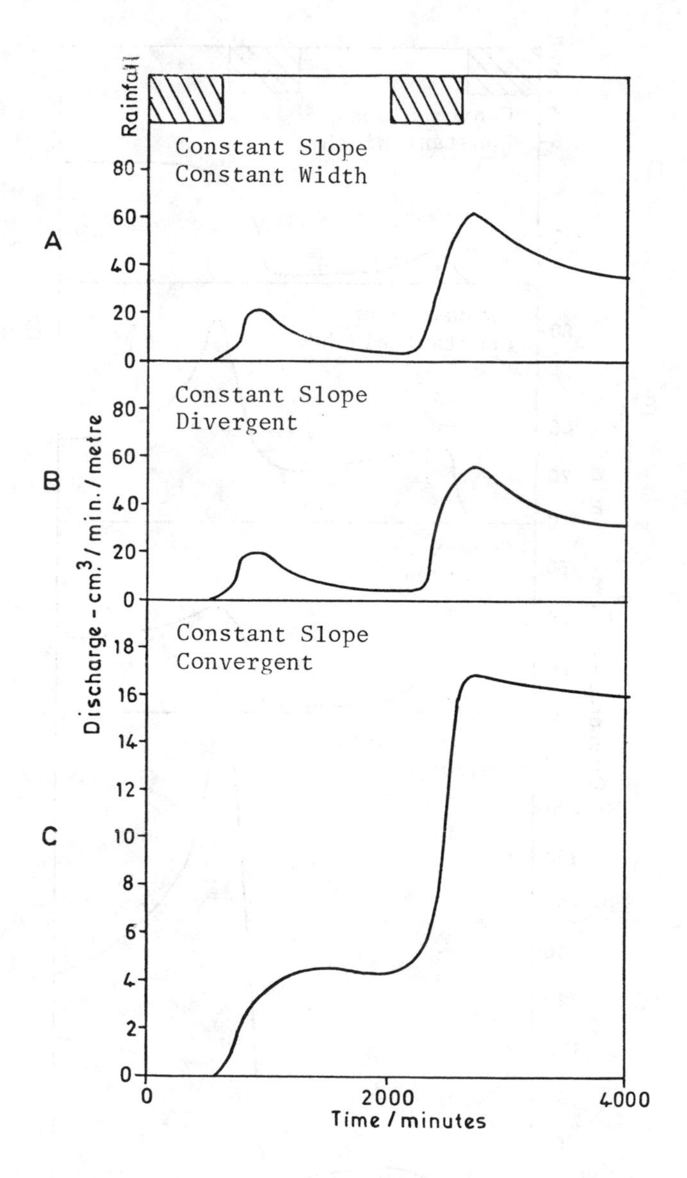

Fig. 2. Simulated hillslope discharge hydrographs (flow per unit width of slope) resulting from rainstorms of 0.01 cm/min for 600 minutes.

A. Constant slope of constant width
B. Constant slope, divergent
C. Constant slope, convergent
D. Convex slope of constant width
E. Concave slope of constant width
F. Convex divergent slope
G. Concave convergent slope

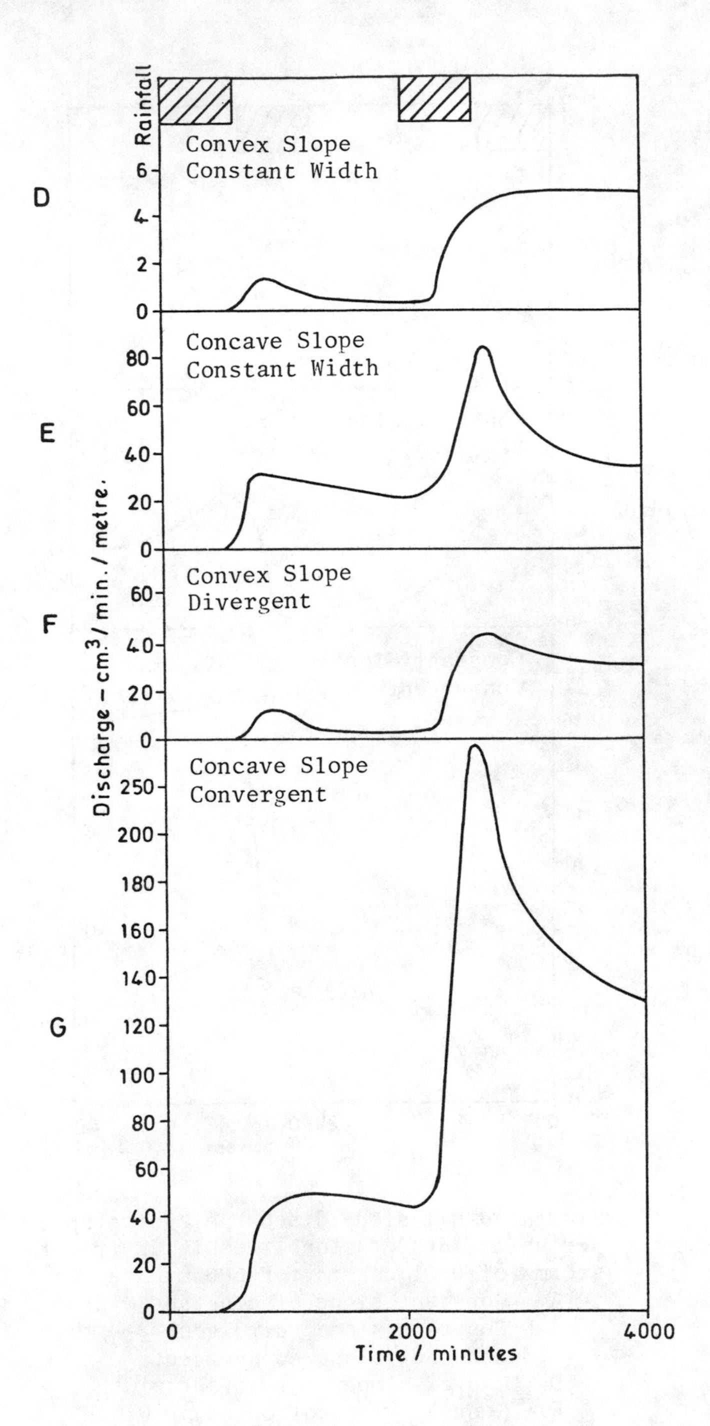

Fig. 2. Continued

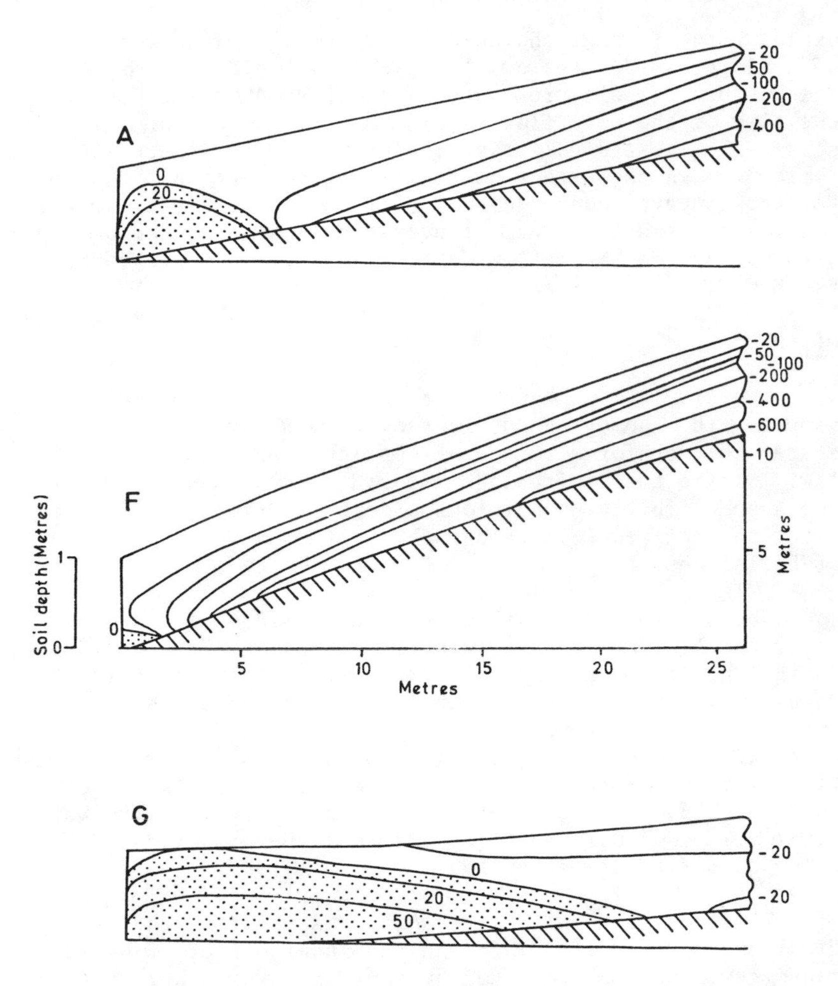

Fig. 3. Simulated pattern of capillary potential, ψ, at the end of the second rainstorm (t = 2600 minutes). Saturated zone shaded.

A. Constant slope of constant width
F. Convex divergent slope
G. Concave convergent slope

exploration of other characteristics of the dynamic response of a slope besides the outflow hydrograph. Figure 3 shows the pattern of moisture content on slopes A, F and G at about the time of the peak flow resulting from the second rainstorm. The most significant feature of these figures is the difference in the extent of the saturated zone on the slopes. For the concave, convergent slope saturation extends to the surface and small amounts of overland flow generated on the area of surface saturation contributed to the hydrograph of Fig. 2G.

Discussion

The simulations have dealt only with hypothetical hillslopes with a homogeneous, isotropic, non-hysteretic soil system that conforms to the assumptions inherent in Darcy's law. Within the limitations imposed by such assumptions and the assumed initial conditions some suggestions may be made about the influence of topography on the hydrologic response of hillslopes.

These simulations reinforce the findings of the field experiments of Hewlett and Hibbert (1963) and Weyman (1973) in that hillslope discharge is shown to depend directly, but in a complex manner, on the upslope extent of the saturated zone or 'variable source area'. The simulated hillslopes, once wetted up by the first rainstorm, could all produce peak flows within minutes of the end of a pulse rainstorm, but these peak flows were generally low in comparison with the total volume of delayed flow that would be produced in an extended period of recession.

The present results give further support to the earlier findings of Beven (1977) for the dominant effect of hillslope convergence on the predicted hydrograph, over other topographic parameters. Convergent slopes produce both higher peak discharge rates and total flow volumes during the simulated period. It should be noted however, that this result only holds for outflow per unit width of slope. Considering the two extreme cases, the peak flow per unit width of slope. It may be that the *overall* contribution of subsurface stormflow divergent and straight slopes may exceed that of the convergent areas, albeit relatively delayed. There are however further complications, primarily the threshold effects introduced by the increased tendency towards the generation of overland flow on the concave, convergent slope (Kirkby and Chorley, 1967; Kirkby, 1969) a feature that is also supported by field evidence (Dunne and Black, 1970; Kirkby and Weyman, 1972).

This study must conclude by emphasising the simplified picture of reality and limited range of conditions upon which these suggestions are based, a point that has been stressed by Hewlett (1974). There are undoubtedly many complicating factors, some of small scale, which may be important on real slopes and that may be very difficult to simulate. In particular the three-dimensional effects associated with spatial variability in soils, especially local zones of rapid seepage, and the dynamic extension of areas of channelled surface flow, have not been considered. Such factors may mask or reinforce the influences of topography suggested here. The ability of simulation techniques to isolate the effects of different external parameters can be misleading as well as advantageous. It is hoped that the present study may suggest some broad outlines upon which further consideration of the effects of topography may be based.

References

[1] Beven, K. J., 1977. "Hillslope Hydrographs by the Finite Element Method," Earth Surface Processes, Vol. 2, in press.

[2] Bruch, J. C. and Zylovski, G., 1973. "A Finite Element Weighted Residual Solution to One-Dimensional Field Problems," International Journal of Numerical Methods in Engineering, Vol. 6, p. 577.

[3] Dunne, T. and Black, R. G., 1970a. "Experimental Investigation of Runoff Production in Permeable Soils," Water Resources Research, Vol. 6, pp. 478-490.

[4] Dunne, T. and Black, R. G., 1970b. "Partial Area Contributions to Storm Runoff in a Small New England Watershed," Water Resources Research, Vol. 6, p. 1296.

[5] Freeze, R. A., 1972a. "The Role of Subsurface Flow in the Generation of Surface Runoff, 1, Base Flow Contributions to Channel Flow," Water Resources Research, Vol. 8, pp. 609-623.

[6] Freeze, R. A., 1972b. "The Role of Subsurface Flow in Generating Surface Runoff, 2, Upstream Sources Areas," Water Resources Research, Vol. 8, p. 1272.

[7] Freeze, R. A., 1975. "A Stochastic-Conceptual Analysis of One-Dimensional Groundwater Flow in Non-Uniform Homogeneous Media," Water Resources Research, Vol. 11, pp. 725-741.

[8] Hewlett, J. D., 1974. "Comments on letters relating to 'Role of Subsurface Flow in Generating Surface runoff, 2, Upstream Source Areas', by R. Allen Freeze, Water Resources Research, Vol. 10, pp. 605-607.

[9] Hewlett, J. D., and Hibbert, A. R., 1963. "Moisture and Energy Conditions Within a Sloping Soil Mass During Drainage," Journal of Geophysical Research, Vol. 68, pp. 1081-1087.

[10] Hursh, C. R., 1944. "Appendix B - Report of Sub-Committee on Subsurface Flow," Transactions of the American Geophysical Union, Vol. 25, pp. 743-746.

[11] Kirkby, M. J., 1969. "Infiltration, Through Flow and Overland Flow," in: Chorley, R. J., Water, Earth and Man, Methuen, London, pp. 215-228.

[12] Kirkby, M. J. and Chorley, R. J., 1967. "Throughflow, Overland Flow and Erosion," Bulletin International Association of Scientific Hydrology, Vol. 12, No. 3, pp. 5-21.

[13] Kirkby, M. J., and Weyman, D. R., 1972. "Measurements of Contributing Area in Very Small Drainage Basins," University of Bristol, Department of Geography Seminar Paper, Series B, No. 3.

[14] Neuman, S. P., 1973. "Saturated - Unsaturated Seepage by Finite Elements," Journal of the Hydraulic Division, Proceedings of the American Society of Civil Engineers, Vol. 99, No. HY12, p. 2233.

[15] Neuman, S. P., Feddes, R. A. and Bresler, E., 1975. "Finite Element Analysis of Two-Dimensional Flow in Soils Considering Water Uptake by Roots," I Thoery, Soil Science Society of America, Proceedings, Vol. 39, p. 231.

[16] Ragan, R. M., 1967. "An Experimental Investigation of Partial Area Contributions," International Association of Scientific Hydrology Publications, No. 76, General Assembly of Berne, Vol. 2, p. 241.

[17] Smith, R. E. and Woolhiser, D. A., 1971. "Overland Flow on an Infiltrating Surface," Water Resources Research, Vol. 7, No. 4, p. 899.

[18] Taylor, G. S. and Luthin, J. N., 1969. "Computer Methods for Transient Analysis of Water Table Aquifers, Water Resources Research, Vol. 5, No. 1, p. 144.

[19] Weyman, D. R., 1970. "Throughflow on Hillslopes and Its Relation to the Stream Hydrograph," Bulletin of the International Association of Scientific Hydrology, Vol. 15, No. 3, p. 25.

[20] Weyman, D. R., 1973. "Measurements of the Downslope Flow of Water in a Soil," Journal of Hydrology, Vol. 20, pp. 267-288.

[21] Whipkey, R. Z., 1967. "Theory and Mechanics of Subsurface Stormflow," in International Symposium on Forest Hydrology, Sopper, W. E. and Lull, H. W., eds, Oxford, Pergammon.

[22] Whipkey, R. Z., 1969. "Storm Runoff from Forested Catchments by Subsurface Routes," in Floods and Their Computation, Symposium of Leningrad, International Association of Scientific Hydrology, Publication No. 85, pp. 773-779.

[23] Zienkiewicz, O. C., 1971. "The Finite Element Method in Engineering," London, McGraw-Hill.

NONLINEARITIES OF HYDROLOGIC RESPONSE

By

Vulli L. Gupta
Research Professor, Water Resources Center
Desert Research Institute, and
Professor of Civil Engineering
University of Nevada, Reno, Nevada

and

Syed M. Afaq Moin
Graduate Research Fellow
Water Resources Center, Desert Research Institute
Reno, Nevada

Abstract

A methodology, using finite difference calculus, is developed and tested for analyzing the drainage basin response in terms of nonlinear storage effects, unsteady flow effects, and the variation of rainfall excess intensity with time during a storm event. Physical significance of the four parameters characterizing the nonlinear response equation was investigated. The methodology was applied to data comprising 28 storm events for watersheds located at McCredie, Missouri; Chickasha, Oklahoma; and Tombstone, Arizona. Principal findings of the study are (i) storage-discharge relationship is nonlinear in contrast to the implications of the unit hydrograph theory, (ii) as can be evidenced from the unsatisfactory attempts to reproduce observed dimensionless hydrographs, trial and error approaches to estimate the parameter, 'K_2' warrant alternative procedures to be considered, (iii) conceptually, relatively rural drainage basins are likely to be characterized by small values for parameters, N, and large values of parameters, 'K_1' and 'K_2'. Parameter, 'K_3', descriptive of the temporal variation of rainfall excess intensity, was found to have limited effect upon the profiles of rising and recession limbs of the hydrographs.

Introduction

The translation of rainfall excess into surface runoff is traditionally conducted by one of the two methods, namely, adoption of unit hydrograph, and storage routing of instantaneous unit hydrograph. In spite of the over-simplified assumptions, unit hydrograph theory has and will continue to enjoy a considerable degree of popularity in the field of hydrologic engineering. Recent studies by Amorocho (1963), and Prasad (1966, 1976), however, have raised several questions concerning the validity of assumptions characteristic of the unit hydrograph theory. Whether or not basin response is a linear or a nonlinear process was investigated by several researchers (Chow, 1964; Kulaindaswamy, 1964; and Prasad, 1966) and some of their findings include supporting evidence that drainage basins seldom conform to the linearity hypothesis.

The three major assumptions of unit hydrograph theory are:

1. Hydrologic response of a drainage basin can be considered as a lumped-parameter system rather than a distributed parameter system.

2. Surface runoff hydrograph pattern is dependent only upon the storm duration and is independent of the temporal pattern of the rainfall excess. This assumption advocates temporal invariance of the cause and effect phenomenon in surface water hydrology.

3. Flow ordinates of surface runoff hydrographs having a common time base are in linear proportion to the total volume of rainfall excess. This assumption permits the applicability of the principles of proportionality and superposition.

The implications of the foregoing assumptions are to treat the drainage basin as a linear system with steady state and spatially invariant response properties. The analysis, naturally, would be more complex if in fact the basin behaves as a non-linear system with unsteady state and spatially variant response properties.

This paper documents an effort to describe drainage basin response in terms of nonlinear and unsteady state properties. The scope and objectives of the study reported herein were limited to the following:

1. development of an equation for the temporal change of surface runoff in the context of non-linear and unsteady properties of basin response,

2. devising a solution methodology for the foregoing equation using finite difference schemes, and

3. illustration of the methodology with the aid of selected watershed runoff data.

Linearity Versus Nonlinearity

The structure of the relationship between basin storage and surface runoff, generally, forms the basis for discerning the response as linear or nonlinear. A generalized relationship may be formulated as:

$$S = KQ^{m} + K' \tag{1}$$

where S = basin storage; Q = volumetric rate of surface runoff; K = storage coefficient, and m, K' are constants. If $m = 1$, and $K' = 0$ in Eq. 1, the generalized relationship reduces to that of a linear system. Contrasting features of linear and nonlinear response behavior of drainage basins and their physical concepts can be inferred from Fig. 1 wherein different control configurations are illustrated. For instance, linear storage reservoir implies a rectangular reservoir wherein the storage is proportional to flow depth. Similarly, a linear outlet control refers to a situation where the rate of flow is directly proportional to flow depth, i.e.,

$$S = Ay \tag{2}$$

and

$$Q = By \tag{3}$$

elimination of 'y' in Eqs. 2 and 3, leads to a linear storage situation given by,

$$S = \frac{A}{B} Q = KQ \tag{4}$$

In contrast, nonlinear controls can be represented by power functions as indicated below, and as illustrated in Fig. 1.

$$S = A^{1} y^{k} \tag{5}$$

$$Q = B^{1} y^{1} \tag{6}$$

where A^{1} = storage at unit flow depth; k = slope of the curve relating 'log S' with 'log y'; B^{1} = flow rate at unit flow depth; 1 = slope of the curve relating 'log Q' with 'log y'. Elimination of 'y' from Eqs. 5 and 6, leads to the storage-discharge relationship as:

$$S = K_{1} Q^{m} \tag{7}$$

where

$$K_{1} = \frac{A^{1}}{(B^{1})^{m}} \tag{8}$$

and

$$m = \frac{k}{1}$$

Nonlinear Response Equation

In the preceding formulations, unsteady flow effects were not considered. More important, basin storage is not

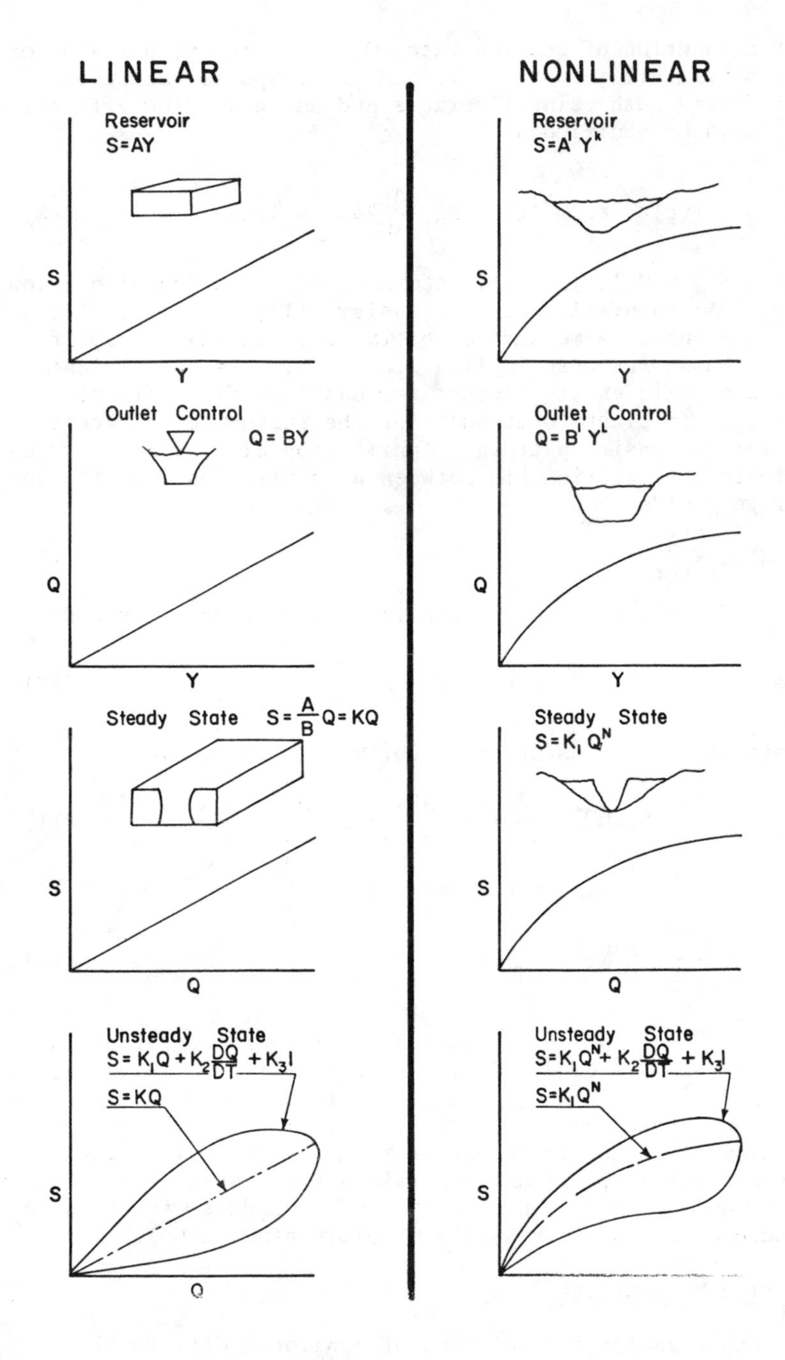

Fig. 1. Schematic of linear and nonlinear response

only a function of outflow rate, Q, and time, t; but also of excess rainfall intensity, I, and its temporal variation. Considering both rainfall excess and unsteady flow effects, Eq. 7 can be modified as

$$S(Q,I,t) = K_1 Q^N(t) + K_2 \frac{dQ(t)}{dt} + K_3 I(t) \tag{8}$$

In Eq. 8, $S(Q,I,t)$ = basin storage as a function of outflow rate, 'Q'; rainfall excess intensity, 'I'; and time, 't'; N = exponent as a measure of basin nonlinearity, i.e., if N = 1, basin response implies linearity; K = coefficient; K_2 = a coefficient to account for unsteady flow effects; K_3 = a coefficient to account for the influence of excess rainfall intensity pattern. Application of continuity principle to the relationship between precipitation, runoff, and storage yields:

$$I - Q = \frac{ds}{dt} \tag{9}$$

and differentiating Eq. 8 with respect to time, t, yields:

$$\frac{ds}{dt} = K_1 N Q^{N-1} \frac{dQ}{dt} + K_2 \frac{d^2Q}{dt^2} + K_3 \frac{dI}{dt} \tag{10}$$

Substituting for ds/dt from Eq. 9, Eq. 10 becomes

$$I - Q = K_1 N Q^{N-1} \frac{dQ}{dt} + K_2 \frac{d^2A}{dt} + K_3 \frac{dI}{dt} \tag{11}$$

and upon rearranging the terms becomes,

$$K_2 \frac{d^2Q}{dt^2} + K_1 N Q^{N-1} \frac{dQ}{dt} + Q = I - K_3 \frac{dI}{dt} \tag{12}$$

Eq. 12 is descriptive of the nonlinear character of basin response, as a second order, nonlinear differential equation. Prasad (1966, 1976) modelled the nonlinear response, and the functional relationship developed therein is similar to Eq. 12, but disregards the influence of rainfall intensity patterns on basin storage. If uniform rainfall intensity were to be a criterion, 'dI/dt' vanishes and Eq. 12 reduces to the same structure as that of Prasad's model (Prasad, 1966).

Solution Methodology

Exact analytical solution of nonlinear differential equations such as Eq. 12 is not feasible. Therefore, graphical, numerical and analog methods are commonly used. An analog assemblage coupled with the rationale of successive approximation was utilized by Prasad (1976). A procedure is described herein whereby Eq. 12 can be solved using finite difference techniques. The end product of this solution technique is the evaluation of runoff Q, at successive time intervals. Input information consists of rainfall excess hyetograph, basin and storm characteristics, K_1, K_2, K_3 and the exponent N, using a forward difference scheme, the time rate of change of rainfall excess intensity can be described as:

$$\frac{dI}{dt} = \frac{I(J+1) - I(J)}{2(\Delta t)} \tag{13}$$

Similarly, central difference schemes can be used to describe the variation of flow rate with respect to time and the resulting relationships are given below:

$$\frac{dQ}{dt} = \frac{Q(J+1) - Q(J-1)}{2(\Delta t)} \tag{14}$$

and

$$\frac{d^2Q}{dt^2} = Q(J+1) - \frac{2Q(J) + Q(J-1)}{(\Delta t)^2} \tag{15}$$

Eq. 12 can be rewritten utilizing the above finite difference schemes,

$$\frac{K_2[Q(J+1) - 2Q(J) + Q(J-1)]}{(\Delta t)^2} + \frac{K_1NQ^{N-1}(J)\ [Q(J+1) - Q(J-1)]}{2(\Delta t)}$$

$$+ Q(J) = I(J+1) - \frac{K_3[I(J+1) - I(J)]}{\Delta t} \tag{16}$$

Rearranging the terms in Eq. 16 and solving for Q(J+1),

$$Q(J+1) = \frac{I(J+1) - \dfrac{K_3[I(J+1) - I(J)]}{\Delta t} - Q(J)\ [1 - \dfrac{2K_2}{\Delta t}] - Q(J-1)\left[\dfrac{K_2}{(\Delta t)^2} - \dfrac{K_1NQ^{N-1}(J)}{2(\Delta t)}\right]}{\dfrac{K_2}{(\Delta t)^2} + \dfrac{K_1NQ^{N-1}(J)}{2(\Delta t)}} \tag{17}$$

A computer program was developed to handle the solution methodology.

Parameter Evaluation

In the methodology for describing the nonlinear behavior of basins, four coefficients or parameters, K_1, K_2, K_3 and N are introduced. The following steps describe a suggested procedure for evaluating these parameters.

1. Separation of the hydrograph into components of base flow, interflow and direct surface runoff by utilizing the generally known graphical separation procedure (Butler, 1957; Chow, 1964).

2. A systematic evaluation of K_1, K_3, and N without an excessive trial and error exercise as given below. Recalling Eq. 8, basin storage, S, can be described as

$$S = K_1Q^N + K_2\frac{dQ}{dt} + K_3I \tag{8}$$

at peak discharge, dQ/dt vanishes. Storage at peak rate timing, S_p, becomes:

$$S_p = K_1Q_p^N + K_3I_p \tag{18}$$

continuity principle applied to flow volumes yields

$$S_p = \int_0^{T_p} (I-Q)dt \tag{19}$$

wherein the value of the integral can be evaluated by planimetering the hydrograph area beginning from the time of rise and ending at the time to peak. From Eqs. 18 and 19, it follows,

$$K_1Q_p^N + K_3I_p = \int_0^{T_p} (I-Q)dt \tag{20}$$

in which 'Q_p' and 'I_p' are known for a given hydrograph and storm description. Repeating the process for a given basin over a number of storm events, values of K_1, K_3, and N, can be estimated using least squares analysis.

3. Direct evaluation of 'K_2' is not possible without recourse to trial and error approach. Prasad (1966) employed trial and error approaches to evaluate both the parameters, 'K_1' and 'K_2'. In this study, trial finite differences are employed.

Parameter Sensitivity

Parameter sensitivity was studied by successively varying one each of the parameters K_1, K_2, K_3 and N as illustrated in Figs. 2 thru 5, for a specific hydrograph simulation. Nominal parameter values were 50, 80, 0.20, and 1.20, respectively.

As shown in Fig. 2, increases in N resulted in a) increased peak discharge, b) shift of the time to peak

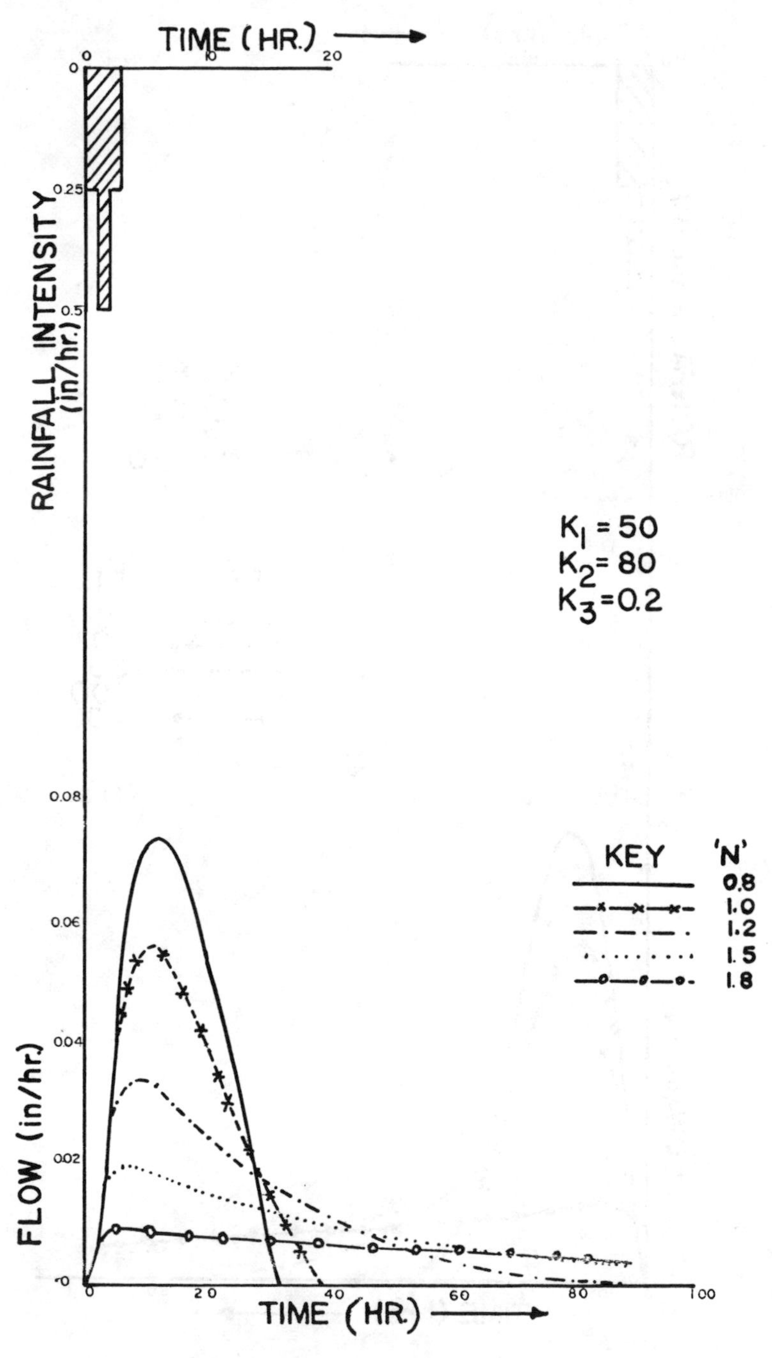

Fig. 2. Illustration of the effects due to changes in parameter, 'N'

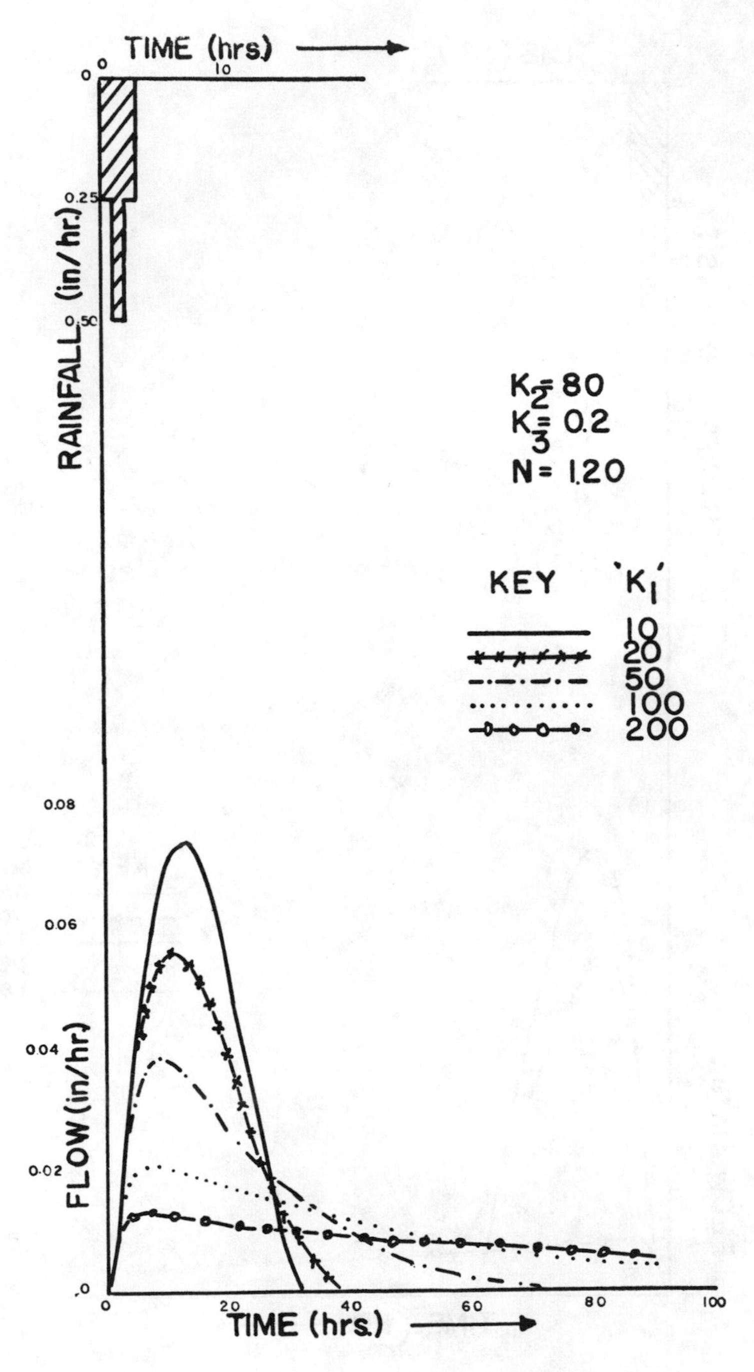

Fig. 3. Illustration of the effects due to changes in parameter, 'K_1'

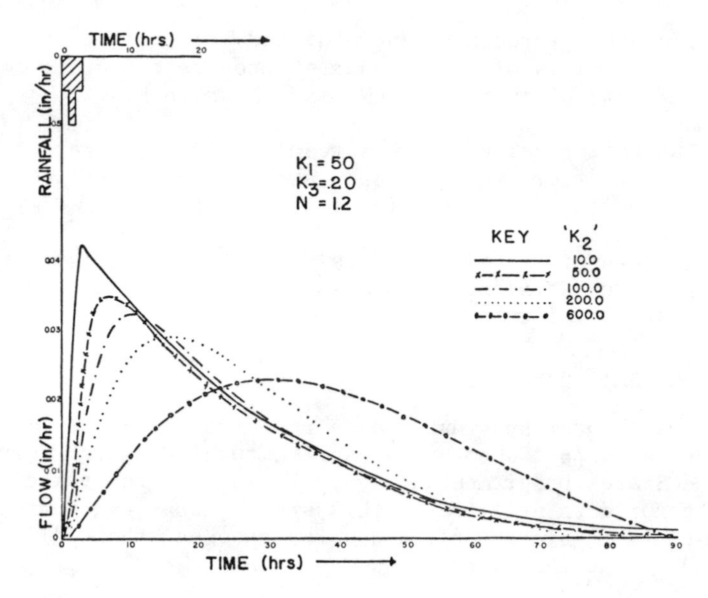

Fig. 4. Illustration of the effects due to changes in parameter, 'K_2'

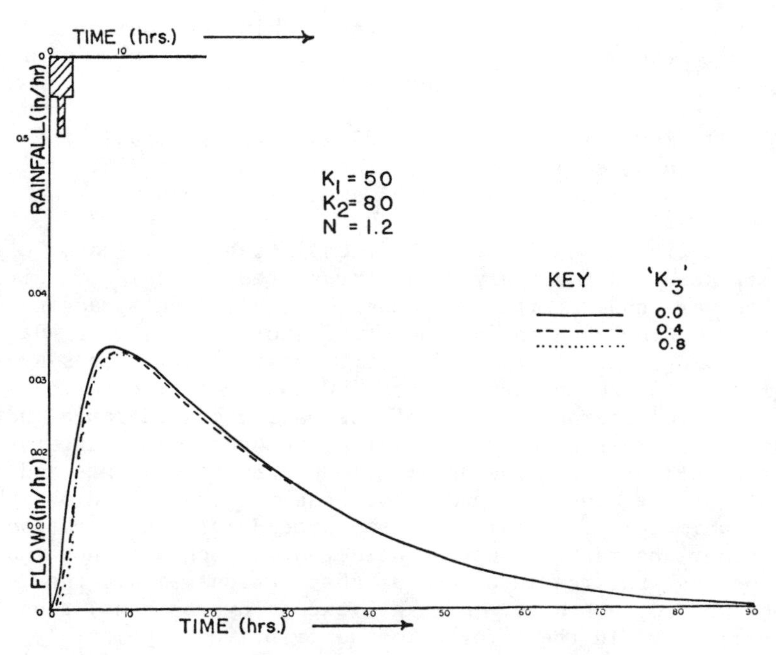

Fig. 5. Illustration of the effects due to changes in parameter, 'K_3'

away from the hydrograph centroid, and c) steeper recession. Increasing values of K_1 produces opposite effects from the above effects of increasing N, as shown in Fig. 3.

Increasing values of K_2 resulted in a) lower flood peaks, b) larger times to peak, and c) larger values of time of concentration, as Fig. 4 illustrates.

K_3 cannot exceed 1, and is theoretically related to rainfall excess intensity. As shown in Fig. 5, this parameter had relatively little effect.

Data Analysis and Results

Small watershed runoff data resources collected and periodically published by the Agricultural Research Service, United States Department of Agriculture, are the principal source for data utilized in this study (Hobbs and Burford, 1970). There are three reasons for choosing the small watershed data, as indicated below:

(1) availability of precipitation hyetograph and the consequent outflow hydrograph information, on a short and systematic time base,

(2) the desirability of investigating whether or not the solution methodology reported herein is applicable to small watershed phenomenon, and

(3) investigating the validity of the generally assumed linearity hypothesis underlying the hydrologic response of small watersheds.

Location of basins and the identification of storm events selected for analysis are presented in Table 1. Five basins were chosen with the number of storm events ranging from 4 to 8 on each basin. Observed storm events were analyzed to obtain an estimated profile of rainfall excess hyetograph. In the absence of detailed data relative to the infiltration characteristics of the basin, considerable judgment was exercised in hypothesizing the hyetograph patterns of rainfall excess. The data included 28 storm events and in each instance, the area under the observed runoff hydrograph was planimetered. Trial and error procedures, infiltration and other abstractions were considered in such a manner that the hyetograph area in excess of abstractions equals the runoff hydrograph area. Rainfall excess hydrographs, thus developed, yield the information descriptive of temporal variation of rainfall excess intensity, I. This information serves as one of the input elements of the framework warranted by Eq. 17.

TABLE 1. — SMALL WATERSHED DATA SELECTED FOR ANALYSIS

Watershed Features			Rainfall Features			Runoff Features	
Location	Drainage Area, acres (square kilometers)	A.R.S. Identification Number	Event Date	Duration hours	Total Depth, inches (cms)	Total Depth. inches (cm)	Peak Rate of Runoff, inches per hour (cm/hf)
(1)	(2)	(3)	(4)	(5)	(6)	(7)	(8)
McCredie, Missouri Watershed Number W-1	154(0.623)	25.1	July 3, 1941	2.08	1.67(4.24)	0.87(2.21)	0.93(2.36)
			June 10-11, 1942	1.08	1.36(3.45)	0.89(2.26)	0.92(2.34)
			June 8, 1943	3.50	1.09(2.77)	0.94(2.39)	0.67(1.70)
			May 14-15, 1945	2.00	1.19(3.02)	0.93(2.36)	0.82(2.08)
			May 1-2, 1948	2.00	1.78(4.52)	0.22(0.56)	0.20(0.51)
			July 22, 1948	3.94	1.23(3.12)	0.75(1.91)	0.39(0.99)
			Sept. 12-13, 1949	0.75	1.09(2.77)	0.81(2.06)	0.55(1.39)
			June 29-30, 1957	2.75	1.55(3.94)	1.19(3.02)	1.04(2.64)
Chickasha, Oklahoma; Watershed Number 111	16,634(67.32)	69.10	Sept. 15, 1962	5.65	2.20(5.59)	0.14(0.36)	0.05(0.13)
			April 26-27, 1963	8.00	1.47(3.73)	0.04(0.10)	0.04(0.10)
			May 9-10, 1964	3.40	2.17(5.51)	0.13(0.32)	0.04(0.102)
			May 10-11, 1964	0.75	1.27 (3.22)	0.15(0.38)	0.05(0.102)
			June 11, 1964	2.50	1.64(4.16)	0.04(0.10)	0.02(0.127)
			Nov. 3-4, 1964	6.70	1.67(4.24)	0.05(0.13)	0.02(0.051)
			May 9-10, 1965	2.50	1.01(2.56)	0.03(0.08)	0.01(0.026)

TABLE 1.— SMALL WATERSHED DATA SELECTED FOR ANALYSIS (Continued)

Watershed Features			Rainfall Features			Runoff Features	
Location (1)	Drainage Area, acres (square kilometers) (2)	A.R.S. Identification Number (3)	Event Date (4)	Duration hours (5)	Total Depth inches (cms) (6)	Total Depth inches (cm) (7)	Peak Rate of Runoff, inches per hour (cm/hf) (8)
Chickasha, Oklahoma;' Watershed Number 131	25,660(103.84)	69.11	Sept. 15, 1962	4.52	1.65(4.19)	0.02(0.05)	0.004(0.053)
			April 26-27, 1963	13.00	2.23(5.66)	0.06(0.15)	0.007(0.104)
			May 10-11, 1964	9.33	1.35(3.43)	0.10(0.25)	0.017(0.264)
			Nov. 3-4, 1964	5.75	1.97(5.00)	0.10(0.25)	0.004(0.064)
			May 9-10, 1965	2.15	2.44(6.20)	0.10(0.25)	0.018(0.224)
Tombstone, Arizona; Walnut Gulch Watershed Number W-6	23,500(95-10)	63.06	July 25-26, 1962	3.25	1.80(4.57)	0.07(0.18)	0.075(0.175)
			Sept. 4, 1962	0.90	1.00(2.54)	0.03(0.08)	0.046(0.107)
			Aug. 19, 1963	1.00	1.47(3.73)	0.11(0.28)	0.084(0.178)
			Aug 25-26, 1963	1.05	0.77(1.96)	0.08(0.20)	0.067(0.163)
Tombstone, Arizona; Walnut Gulch Watershed Number W-8	3,830(15.50)	63.08	Aug. 19, 1963	1.03	1.47(3.73)	0.14(0.36)	0.156(0.396)
			July 22, 1964	0.50	1.94(4.93)	0.27(0.69)	1.108(2.814)
			Sept. 9-10, 1964	9.75	1.14(2.89)	0.19(0.48)	0.172(0.437)
			Sept. 11, 1964	2.83	1.02(2.59)	0.15(0.38)	0.496(1.260)

Evaluation of 'K_1' and 'N', in accordance with Eqs. 18 through 20, was possible by logarithmically regressing the hydrograph recession area with peak rate of runoff, for each basin. It is realized that the number of storm events included in the analysis represent a very small sample. A listing of the planimetered areas for all the 28 storm events along with the principal findings of multiple regression analysis are presented in Table 2. Best estimates of 'K_1' and 'N' obtained through regression analysis are listed in Table 2. Multiple regression coefficients were found to range from a low value of 0.676 to a high value of 0.998. Unexplained variability thus was found to range from a low of 0.4 percent to a high of 54.3 percent. Exponent 'N', for the five basins, was found to range from a low value of 0.775 to a high value of 1.21 thereby implying nonlinearity of basin response characteristics.

Estimation of 'K_2' was done by trial and error procedures. Trial values for 'K_2' ranging from 0.15 to 11.25 in approximate increments of 0.2 were introduced into the computerized solution of Eq. 17. For each trial value of 'K_2', surface runoff hydrograph ordinates were calculated with efforts to select the appropriate value of 'K_2'. A comparison of the observed dimensionless hydrographs and the computed dimensionless hydrographs was conducted and the sample results are illustrated in Figs. 6 through 10. General inspection of these figures indicate (a) wide variance between the observed and computed results, and (b) strong sensitivity of the computed hydrograph profiles to changes in 'K_2'. Causative factors for the disparity of results are speculated to include the following:

1. inadequacy of the graphical procedures adopted for hydrograph separation into various components including the surface runoff,

2. evaluation of 'K_1' and 'N' by regression analysis implies that the parameters can at best be considered as best possible estimates based on very few number of storm events per basin,

3. whereas Eq. 17 considers the variation of rainfall intensity with respect to time, the methodology used herein sets 'K_3' to be zero, and

4. adoption of trial and procedures to evaluate 'K_2' can at best be considered as an approximate rather than exact solution.

TABLE 2. ESTIMATION OF 'K_1' and 'N' by REGRESSION ANALYSIS

WATERSHED IDENTIFICATION			HYDROGRAPH FEATURES		REGRESSION EQUATION Log X = Log K_1 + N Log Q			
LOCATION	A.R.S. NUMBER	STORM DATE	RECESSION AREA OR STORAGE 'S', inches (cm)	PEAK RATE OF RUNOFF, 'Q', inches per hour (cm/hr	K_1	N	MULTIPLE REGRESSION COEFFICIENT 'R'	STANDARD ERROR OF ESTIMATE
(1)	(2)	(3)	(4)	(5)	(6)	(7)	(8)	(9)
McCredie,	25.1.	July 3, 1941	0.31(0.79)	0.93 (2.36)				
Missouri,		June 10-11,1942	0.41(1.04)	0.92(2.34)				
Watershed		June 8, 1943	0.31(0.79)	0.67(1.70)				
Number W-1		May 14-15,1945	0.29(0.74)	0.82(2.08)	0.385	1.163	0.676	0.218
		May 1-2, 1948	0.07(0.18)	0.20(0.51)				
		July 22, 1948	0.12(0.30)	0.39(0.99)				
		Sept.12-13,1949	0.04(0.10)	0.55(1.39)				
		June 29-30,1957	0.17(0.43)	1.04(2.64)				
Chickasha,	69.10	Sept. 15, 1962	0.040(0.101)	0.05(0.127)				
Oklahoma,		April 26-27,1963	0.180(0.457)	0.04(0.102)				
Watershed		May 9-10,1964	0.281(0714)	0.04(0.102)				
Number 111		May 10-11,1964	0.292(0.742)	0.05(0.127)	2,490	1.142	0.998	0.017
		June 11, 1964	0.097(0.246)	0.02(0.051)				
		Nov. 3-4,1964	0.109(0.277)	0.02(0.051)				
		May 9-10, 1965	0.203(0.516)	0.01 (0.026)				
Chickasha,	69.11	Sept. 15, 1962	0.04(0.102)	0.021(0.053)				
Oklahoma,		April 26-27,1963	0.17(0.432)	0.041(0.104)				
Watershed		May 10-11,1964	0.29(0.737)	0.104(0.264)	4.651	1,210	0.876	0.168
Number 131		Nov. 3-4, 1964	0.04(0.102)	0.025(0.064)				
		May 9-10,1965	0.20(0.508)	0.088(0.224)				

TABLE 2. ESTIMATION OF 'K_1' and 'N' by REGRESSION ANALYSIS (Continued)

WATERSHED IDENTIFICATION			HYDROGRAPH FEATURES		REGRESSION EQUATION $\text{Log } X = \text{Log } K_1 + N \text{ Log } Q$			
LOCATION	A.R.S. NUMBER	STORM DATE	RECESSION AREA OR STORAGE, 'S', inches (cm)	PEAK RATE OF RUNOFF, 'Q', inches per hour (cm/hr)	K_1	N	MULTIPLE REGRESSION COEFFICIENT 'R'	STANDARD ERROR OF ESTIMATE
(1)	(2)	(3)	(4)	(5)	(6)	(7)	(8)	(9)
Tombstone, Arizona, Walnut Gulch Watershed Number W-6	63.06	July 25-26,1962	0.028(0.071)	0.069(0.175)	0.389	0.936	0.838	0.066
		Sept.4,1962	0.018(0.046)	0.042(0.107)				
		Aug. 19,1963	0.038(0.097)	0.070(0.178)				
		Aug. 25-26,1963	0.027(0.069)	0.064(0.163)				
Tombstone, Arizona, Walnut Gulch Watershed Number W-8	63.08	Aug.19,1963	0.031(0.079)	0.156(0.396)	0.184	0.775	0.987	0.034
		July 22,1964	0.114(0.289)	1.108(2.814)				
		Sept.9-10,1964	0.047(0.119)	0.172(0.437)				
		Sept.11,1964	0.067(0.170)	0.496(1.260)				

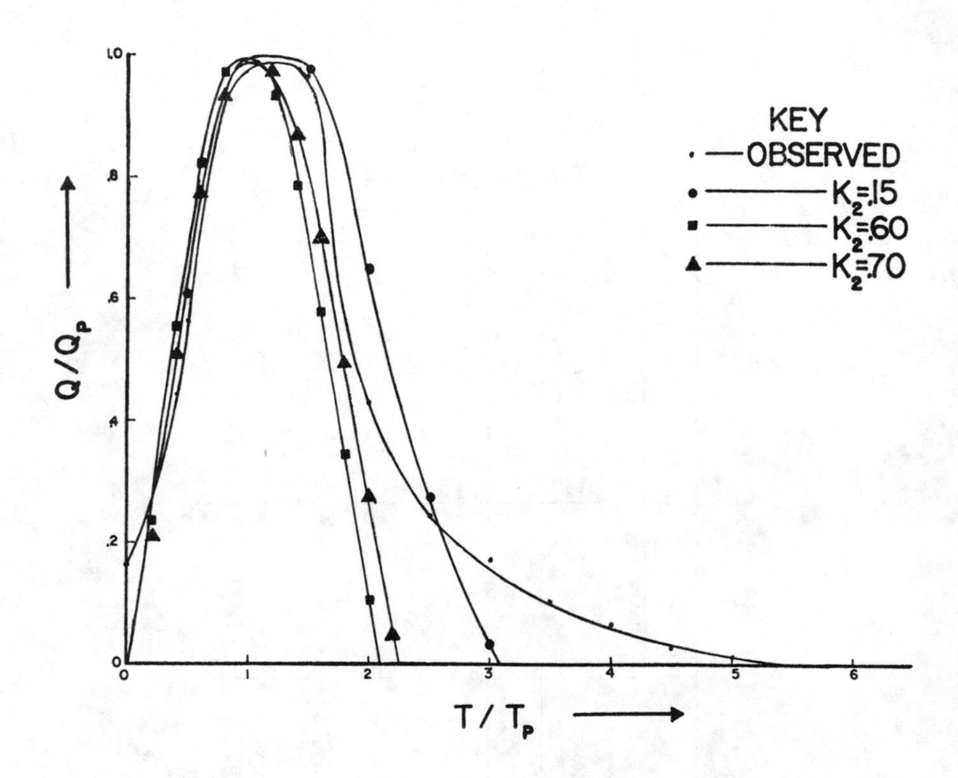

Fig. 6. Observed and computed dimensionless hydrographs, Event: June 10-11, 1942, McCredie, Missouri, Watershed W-1

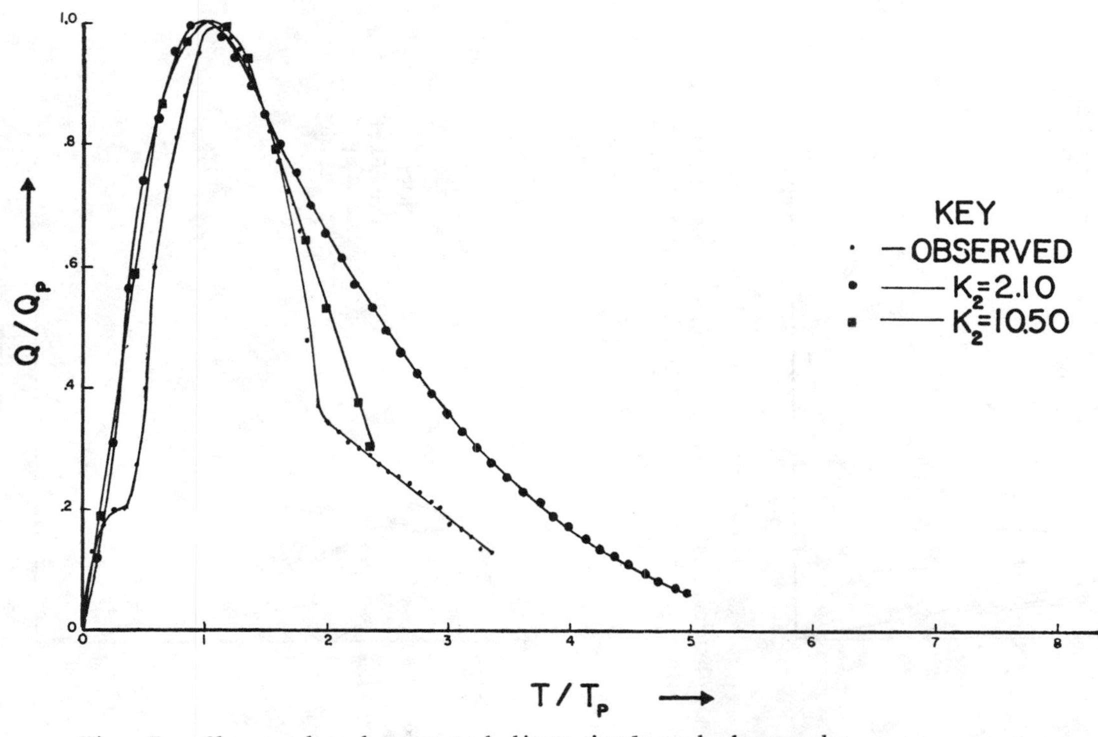

Fig. 7. Observed and computed dimensionless hydrographs,
Event: May 10-11, 1964, Chickasha, Oklahoma, Watershed No. 131

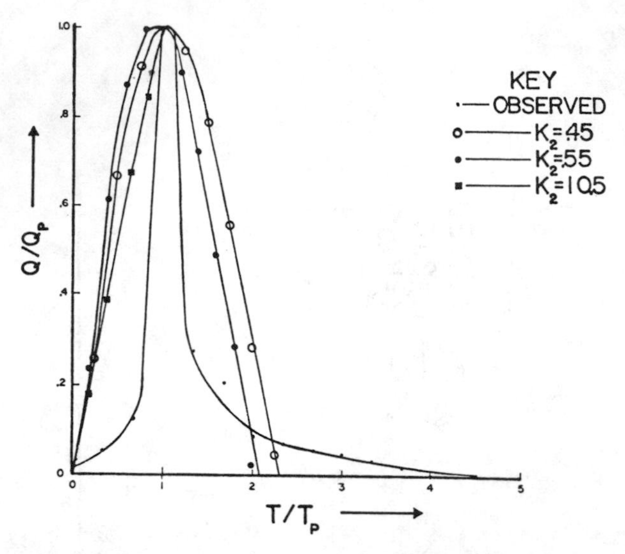

Fig. 8. Observed and computed dimensionless hydrographs, Event: Sept. 9-10, 1964, Tombstone, Arizona, Watershed No. 063

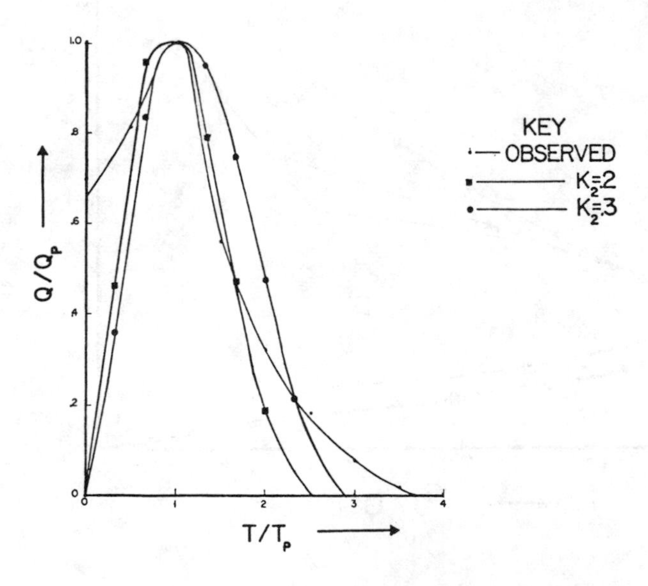

Fig. 9. Observed and computed dimensionless hydrographs, Event: July 25-26, 1962, Tombstone, Arizona, Watershed No. W-6

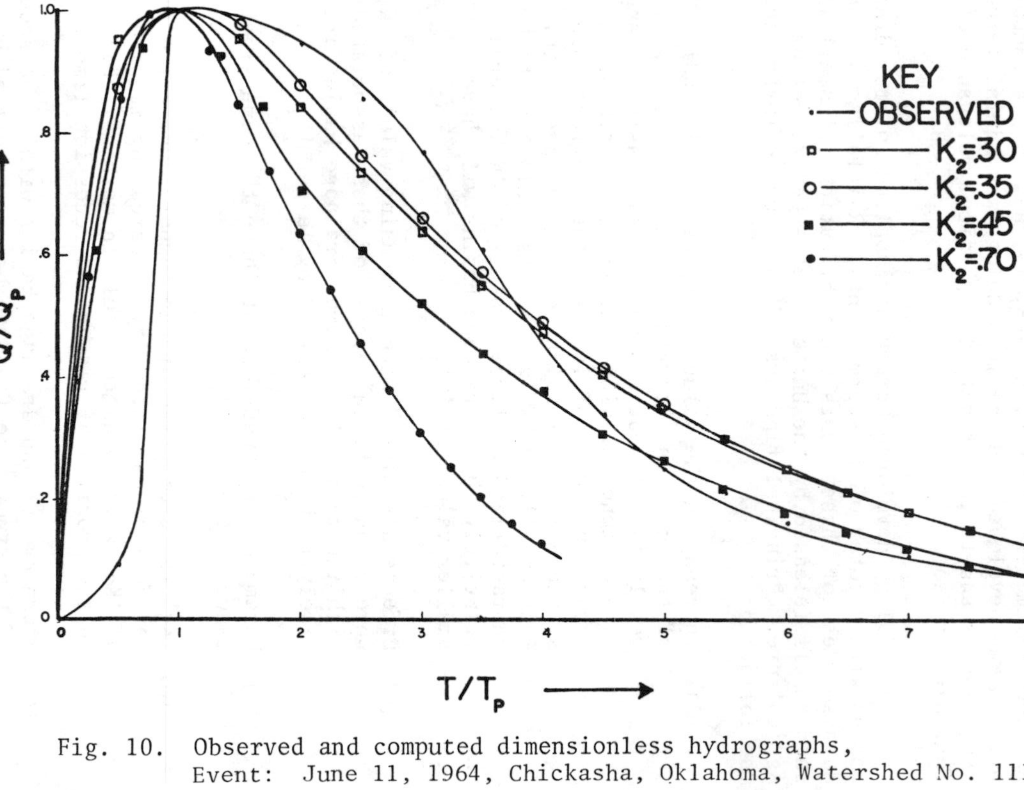

Fig. 10. Observed and computed dimensionless hydrographs,
Event: June 11, 1964, Chickasha, Oklahoma, Watershed No. 111

Summary and Conclusions

Event-based surface runoff hydrographs can be modelled, in contrast to unit hydrograph theory, to reflect the effects of (a) nonlinear basin response, and (b) variation of rainfall intensity with respect to time. Utilizing the tools of finite difference calculus, a methodology was developed to evaluate the four parameters, namely, K_1, K_2, K_3, and N. A computer program was developed and the methodology was administered to known events and to investigate the physical relevance of the four parameters.

Methodology developed herein was applied to four drainage basins with a total of 28 storm events. Rainfall and runoff data sources for these events are publications of small watershed runoff released by the United States Department of Agriculture. Principal findings of the study efforts include the following:

(a) Increase in the value of exponent, 'N', indicative of the nonlinear storage response effects, would result in (i) peak discharge, (ii) basin lag and time of concentration, and (iii) steppening of recession limb profile. All these effects are symptomatic of urbanization of drainage basin. Consequently, urbanized basins would tend to be characterized by larger values of 'N', whereas relatively rural drainage basins will have smaller values of 'N'.

(b) Effects associated with increasing values of 'K_1' were found to be in reverse of those associated with increasing 'N'. Rural drainage basins are likely to be characterized by larger 'K_1' values.

(c) Parameter 'K_2' was found to have the same significance as that of 'K_1'.

(d) 'K_3' is descriptive of the temporal variation of rainfall excess intensity. Increasing magnitudes of 'K_3' were found to result in marginal changes in the profiles of rising and recession limbs.

(e) Storm events and drainage basins within the scope of the study were found to be characterized by 'N' values ranging from 0.775 to 1.210. Basin response, consequently, can be tractable as nonlinear in contrast to the linear hypothesis implied by unit hydrograph theory.

(f) Methodology developed in the study administers trial and error procedures for the estimation of

'K_2' parameter. It was found that such an approach is very cumbersome and dissatisfactory. This is evidenced from attempts to reproduce observed hydrographs of the storm events.

Acknowledgments

The authors wish to express their appreciation to Douglas Trodeau and Russel McMullen for their cooperation in the conduct of the study. Thanks are extended to Alfred Cunningham, John Fordham and Karen Fallon for their review and suggestions.

The work reported herein is a conceptual outgrowth of projects supported in part by the United States Department of Interior, Office of Water Resources Research as authorized under the Water Resources Research Act of 1964 (P.L. 88-379), and in part by funds provided by the Desert Research Institute, University of Nevada System, Reno, Nevada.

References

[1] Ames, W. F., 1964. "Nonlinear Problems of Engineering," Academic Press, New York, N. Y.

[2] Amorocho, J., 1963. "Measures of the Linearity of Hydrologic Systems," Journal of Geophysical Research, Vol. 68, No. 8, pp. 2237-2249.

[3] Butler, S. S., 1957. "Engineering Hydrology," Prentice-Hall, Inc., Englewood Cliffs, N. J., pp. 227-229.

[4] Chow, V. T., 1964. "Handbook of Applied Hydrology," McGraw-Hill Book Company, Inc., New York, N. Y.

[5] Cosgriff, R. L., 1958. "Nonlinear Control Systems," McGraw-Hill Book Company, Inc., New York, N.Y.

[6] Edson, C. G., 1951. "Parameters for Relating Unit Hydrographs to Watershed Characteristics," Trans. of the American Geophysical Union, Vol. 32, pp. 591-596.

[7] Kulaindaswamy, V. C., 1964. "A Basic Study of the Rainfall Excess-Surface Runoff Relationship in a Basin System," Ph.D. dissertation, University of Illinois, Urbana, Ill.

[8] Minshall, N. E., 1960. "Predicting Storm Runoff on Small Experimental Watersheds," Journal of Hydraulics Division, ASCE, Vol. 86, No. HY8, pp. 17-38.

[9] Prasad, R., 196b. "Nonlinear Simulation of a Regional Hydrologic System," Ph.D. dissertation, University of Illinois, Urbana, Ill., 138 p.

[10] Prasad, R., 1976. "Nonlinear Simulation of a Regional Hydrologic System," Journal of Hydraulics Division, ASCE, pp. 201-225.

[11] Singh, K. P., 1962. "A Nonlinear Approach to the Instantaneous Unit Hydrograph Theory," Ph.D. dissertation, University of Illinois, Urbana, Ill.

[12] Hobbs, H. W., and Burford, J. B., 1970. "Hydrologic Data for Experimental Agricultural Watersheds in the United States," Agricultural Research Service, U.S. Department of Agriculture, Misc. Publ. Nos. 994, 1070, 1164, 1194, 1216.

CONCENTRATION TIME OF FLOOD RUNOFF IN SMALLER RIVER BASINS

By

Mutsumi Kadoya, Professor of Hydrology
Disaster Prevention Research Institute
Kyoto University, Kyoto, Japan

and

Akira Fukushima, Assistant Professor of
Agricultural Engineering, Shimane University
Matsue, Japan

Abstract

A study was undertaken to develop a formula for estimating the concentration time of flood runoff in a watershed from the theoretical and practical viewpoint. The concentration time is composed of travel times of rainwater input from the remotest slope and in channels. But it is easily shown by applying the kinematic wave theory that their functional expressions are similar to each other, which gives us a useful clue for the practical treatment of concentration time. In this paper, a functional expression of the concentration time is discussed first theoretically and then substantially using hydrological data in mountainous river basins to propose a practical formula. With some discussion of its form, the formula is modified to extend its applicability to developed watersheds.

Introduction

The rational formula is often used yet to estimate the flood peak discharge at the outlet of small or medium river basins. The formula has been regarded as a typical empirical formulae and criticized as inadequate by several hydrologists (Chow, 1964). The formula is nevertheless useful if the river channel systems have no flood control facilities and if the river basin is so small that the catchment dynamics allows lumped evaluation. A practical merit of the formula is that it is often simple and easy to correlate rainfall intensity with flood peak discharge for the desired frequency.

The rational formula involves some hypotheses to be satisfied in order to use it rationally. The most difficult but essential problem along the hypotheses is to estimate the concentration time of rainwater in a watershed. This paper discusses the concentration time from both the theoretical and practical viewpoint.

Conception of Concentration Time

The concentration time, which is a key for the rational use of the rational formula, is generally defined as the time required for a rainwater input pulse to propagate from the remotest part of the watershed to the point being considered. In order to get a functional expression for the concentration time, we shall adopt the following assumptions: (1) The hyetograph of the effective rainfall intensity, the duration of which is longer than the concentration time, is the same everywhere over the watershed. (2) The peak of hyetograph of the effective rainfall intensity is not extremely sharp so that the average intensity involving the peak can be used for evaluating the catchment dynamics. (3) The kinematic runoff model is applicable to evaluate the propagation of rainwater disturbance.

The fundamental equation of the kinematic flow of rainwater is given as follows: For the flow on a slope;

$$h = kq^{p}, \quad \partial h/\partial t + \partial q/\partial x = r_e \tag{1}$$

where h and q are the depth and the discharge of rainwater flow, respectively, t is the time, x is the distance, r_e is the effective rainfall intensity, and k and p are the parameters depending on the type of flow, for example, as follows:

$$\text{Interflow of Darcy type;} \quad k = \lambda/k_d s, \quad p = 1 \tag{2}$$

$$\text{Overland flow of Manning type;} \quad k = (N/\sqrt{s})^{p}, \quad p = 3/5 \tag{3}$$

where λ and k_d are the effective porosity and the permeability of the upper soil layer, respectively; s is the gradient of the slope, and N is the equivalent roughness. For the flow in a channel;

$$F = KQ^{P}, \quad \partial F/\partial t + \partial Q/\partial x = q \tag{4}$$

where F and Q are the sectional area and the discharge of channel flow, respectively, q is the inflow and K and P are the parameters.

Applying the characteristic theory, we get

$$dx/dt = (pk)^{-} q^{1-p} \tag{5}$$

$$r_e\, dt = pkq^{p-1} dq, \quad r_e\, dx = dq \tag{6}$$

and

$$dx/dt = (PK)^{-1} Q^{1-P} \tag{7}$$

$$dt = PKQ^{p-1}dQ, \quad q\,dx = dQ \tag{8}$$

By using Eqs. (6) and (8) for a constant rate of the effective rainfall, we get the fundamental expression for the concentration time, t_p, as follows:

$$\left.\begin{aligned} t_p &= t_s + t_c, \\ t_s &= kb^p r_e^{p-1}, \\ t_c &= \sum_i [Kq^{-1}(Q^P - Q_u^P)]_i = \sum_i [KLY(r_e A)^{P-1}]_i, \\ Y &= [1 - (A_u/A)^P]/[1 - (A_u/A)] \end{aligned}\right\} \tag{9}$$

where t_s is the propagation time of rainwater traveling on a slope situated in the remotest part of a watershed in the sense of catchment dynamics, and t_c is the propagation time in river channels in the lower side of the slope; b is the slope length; Q_u and Q are the discharges at the upper and lower ends of a reach of river channel, respectively, A_u and A are the catchment areas at the upper and lower ends of the reach, respectively, and L is its reach length.

The above expression for the concentration time is easy to understand if we consider a model of watershed such as Fig. 1(a) in the practical sense. We have only to set up i in Eq. (9) for the route to be maximum in the propagation time.

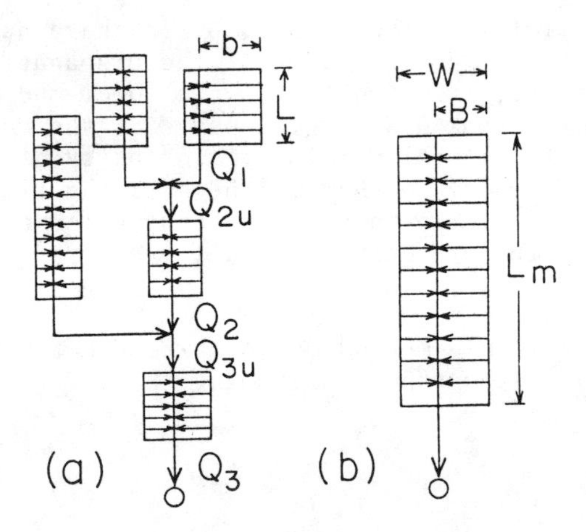

Fig. 1. Models of a watershed

It is not always easy to evaluate the effect of the propagation time in channels on the concentration time in general, because it depends upon the definition of river channel networks and upon the land surface condition. However, if we choose a sub-basin attached to a channel of the stream order u-2 or u-3 as a unit-basin in a river basin of the order u to construct a watershed model for runoff analysis, the effect of channels on the concentration time may be evaluated as 10 ~ 20% in a natural river basin and as 30 ~ 40% in an urban area. That is, it may be considered that the concentration time is mainly governed by the propagation time on a slope in the remotest part of watershed from the viewpoint of engineering practice.

Moreover, the above equation introduced for a constant rate of effective rainfall can also be approximately used for the effective rainfall intensity averaged over the duration of the concentration time as discussed below.

Estimation of Concentration Time Through Hyetograph

In order to simplify the discussion, we shall assume that there exists a simple model such as Fig. 1(b) to be equivalent to a detail model such as Fig. 1(a) to obtain simulated hydrographs at the outlet of the watershed by the technique of runoff analysis. Moreover, we assume that the effect of the channel on the concentration time can be combined with that of the slope, for example, by slight modification of the slope length or a parameter so far as the flood peak is discussed.

We can then refer to the studies by Ishihara et al. (1959) and Ishihara (1961). They showed that the propagation time of rainwater disturbance from the top to the lower end of a slope is minimum when the peak discharge appears at the end of the slope, and that the effective rainfall intensity at the arrival time, t_2, is equal to that at the starting time, t_1. This result suggests to us that the concentration time can be estimated through the effective rainfall hyetograph as seen in Fig. 2.

The above mentioned method may be extended to the observed rainfall hyetograph by assuming that

$$r_e = f\ r \qquad (10)$$

or

$$r_e = r - f_c \qquad (11)$$

where r is the observed rainfall intensity, f is the runoff coefficient and f_c is the temporal average loss rate. Every assumption is questionable in the theoretical sense, but

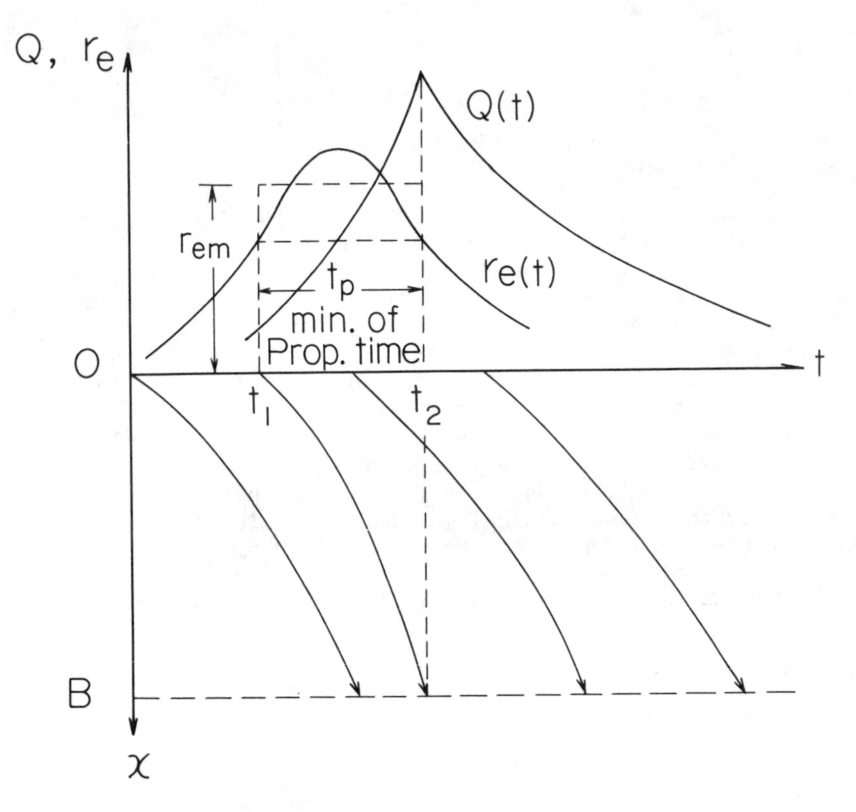

Fig. 2. A definition of the concentration time

has a practical merit to make possible the direct estimation of the concentration time through observed hyetographs.

Concentration Time in Mountain River Basin

In order to apply the previous equations for concentration times to a mountain or hill river basin in a natural state, we shall assume a simple model such as Fig. 1(b) and employ Hack's law for the relation between the main channel length, L_m, and the catchment area, A, (Hack, 1957). Then we have

$$B = \varepsilon W, \qquad W = A/L_m, \qquad L_m = u A^v \tag{12}$$

where ε is the ratio of the larger side slope length, B, to the watershed width, W, and u and v are constants.

Substitution of Eq. (12) into Eq. (9) yields

$$\left.\begin{aligned} t_p &= t_s + t_c \\ t_s &= k\ B^p\ r_e^{p-1} = k(\ /u)^p\ A^{(1-v)p}\ r_e^{p-1} \\ t_c &= K\ L_m\ (r_e A)^{P-1} = K\ uA^{v+P-1}\ r_e^{P-1} \end{aligned}\right\} \quad (13)$$

We shall obtain definite values for parameters in the above equation to estimate the practical form of the concentration time. Assuming the Manning's law for overland flow near the peak of flood runoff, we may put $p = 0.6$. It is difficult to define the value of equivalent roughness clearly, but the value of k may be put as $k = 1.1 \sim 1.4$ in the metric system. The value of P is regarded as $P = 0.65 \sim 0.75$ or 0.7 depending on the sectional form of a river channel. The parameter K has also a range $K = 0.8 \sim 1.2$ in the metric system. The Hack's constants u and v may be put as $u = 1.35 \sim 1.45$ and $v = 0.6 \sim 0.65$ in the km system, if we refer to a topographical map of the scale 1/25000. The ratio of the slope length may be $\varepsilon = 0.5 \sim 0.6$.

Substituting these values into Eq. (13), we get

$$\begin{aligned} t_p &= t_s + t_c \\ t_s &= (250 \sim 350) A^{(0.21 \sim 0.24)}\ r_e^{-0.4} \\ t_c &= (30 \sim 50) A^{(0.25 \sim 0.40)}\ r_e^{-(0.25 \sim 0.35)} \end{aligned} \quad (14)$$

where t is shown in minutes, A in km^2 and r_e in mm/hr.

The effect of river channel on the concentration time is evaluated as follows:

$$t_c/t_p = (t_c/t_s)/(1 + t_c/t_s) \quad (15)$$

$$t_c/t_s = (K/k)\ u^{1+p}\ \varepsilon^{-p}\ A^{P-(1-v)(1+p)}\ r_e^{P-p} \quad (16)$$

We can find the channel effect roughly by putting $p=0.6$, $P=0.7$, $K/k=0.75$, $\varepsilon=0.55$ and the Hack's constants $u=1.45$ and $v=0.62$ in the km system in Eq. (16), as follows:

$$t_c/t_s = 0.14\ A^{0.1}\ r_e^{0.1} \quad (17)$$

Eq. (17) tells us that the larger catchment area becomes and the more intense rain falls, the larger effect the channel has on the concentration time. But the extent of the channel effect is evaluated as $t_c/t_p = 0.15 \sim 0.21$ for $1 \le A \le 10$ km^2 and $10 \le r_e \le 50$ mm/hr and as $t_c/t_p = 0.28$ for $A=1000$ km^2 and $r_e = 30$ mm/hr. This fact suggests to us that the concentration time is mainly governed by the propagation time on the slope.

Figure 3 shows examples of the concentration time in mountain and hill river basins. The concentration time was estimated through a hyetograph of observed rainfall by applying the method shown previously. The effective rainfall intensity was obtained from the observed peak discharge using the rational formula inversely. The reason for the scatter of the plotted data is not clear. It may be due to the extreme areal distribution of rainfall, the radical change of rainfall intensity near the peak of the hyetograph for which all slopes of the watershed cannot be mobilized to form the runoff peak, or the heterogeneity in surface conditions of the watershed. The plotted data, however, show that there exists a functional relation between the concentration time and the effective rainfall intensity in each watershed.

Inserting an average line roughly into the data plotted on logarithmic paper such as Fig. 3, the concentration times corresponding to $r_e = 3$ and 30 mm/hr have been estimated in

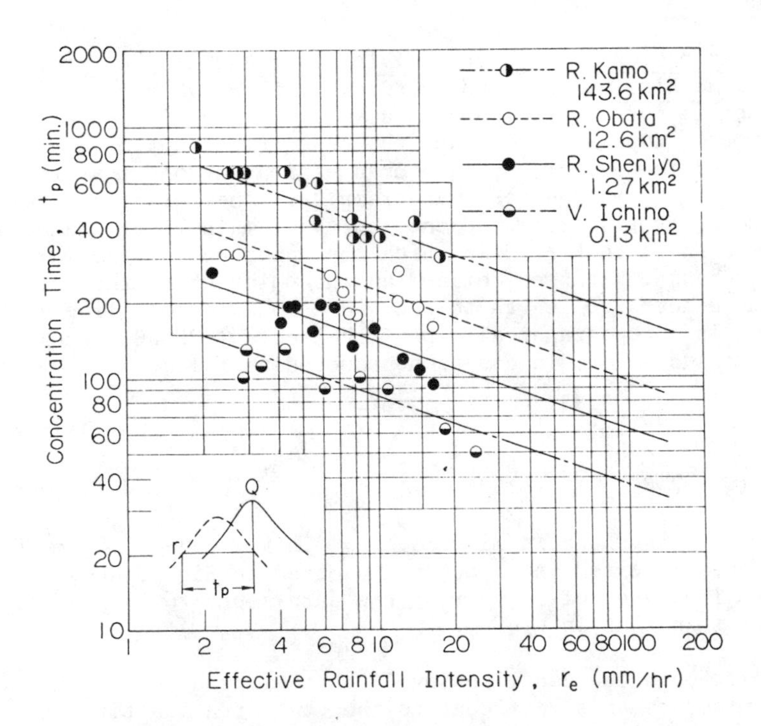

Fig. 3. Relations between the effective rainfall intensity and the concentration time in mountain and hill river basins

every watershed and then plotted for their catchment area to obtain Fig. 4. Figure 4 shows that there clearly exists a functional relation between the concentration time and the catchment area.

We then propose a formula for estimating the concentration time in a mountain or hill river basin as follows:

$$\left.\begin{aligned} t_p &= C\, A^{0.22}\, r_e^{-0.35} \\ C &= 290, \quad \text{for a mountain or hill river basin} \end{aligned}\right\} \qquad (18)$$

where t_p is shown in minutes, A in km^2 and r_e in mm/hr.

It may be said that Eq. (18) is a simple expression of Eq. (14), because a value estimated through Eq. (18) is always contained within the range defined by Eq. (14). The lines inserted into the data plotted in Figs. 3 and 4 are the ones obtained by Eq. (18).

Additional Discussion on Concentration Time in Mountain River Basin

Expression by main channel length. An example for validating Hack's law is shown in Fig. 6. Although a river channel is not always found in a small watershed when we refer to a topographical map, the concentration time can be expressed in term of the main channel length so far as the Hack's law is recognized. An example of the relation between the concentration time and the length of main river channel, L_m, is shown in Fig. 5 in which the upper end of the main river channel has been defined as the point where the ratio of the penetrated length to the width in a contour line (drawn in a topographical map of the scale 1/25000) becomes unity. And a formula has been proposed as follows:

$$t_p = 250\, L_m^{0.35}\, r_e^{-0.35} \qquad (19)$$

or

$$t_p = 240\, L_m^{0.37}\, r_e^{-0.35} \qquad (20)$$

where t_p is shown in minutes, L_m in km and r_e in mm/hr. The lines inserted into the data plotted in Fig. 5 show Eq. (19). The range of scatter of the data seems to be slightly larger than that in Fig. 4.

Although the contribution of channel gradients is hardly recognized to be significant to the concentration time except in a few special cases from the theoretical viewpoint, we have also tried to express the concentration time in terms of the main channel gradient, S_c, as in previous work (Schultz and Lopez, 1974) to get Eq. (21) in place of Eq. (19) or (20).

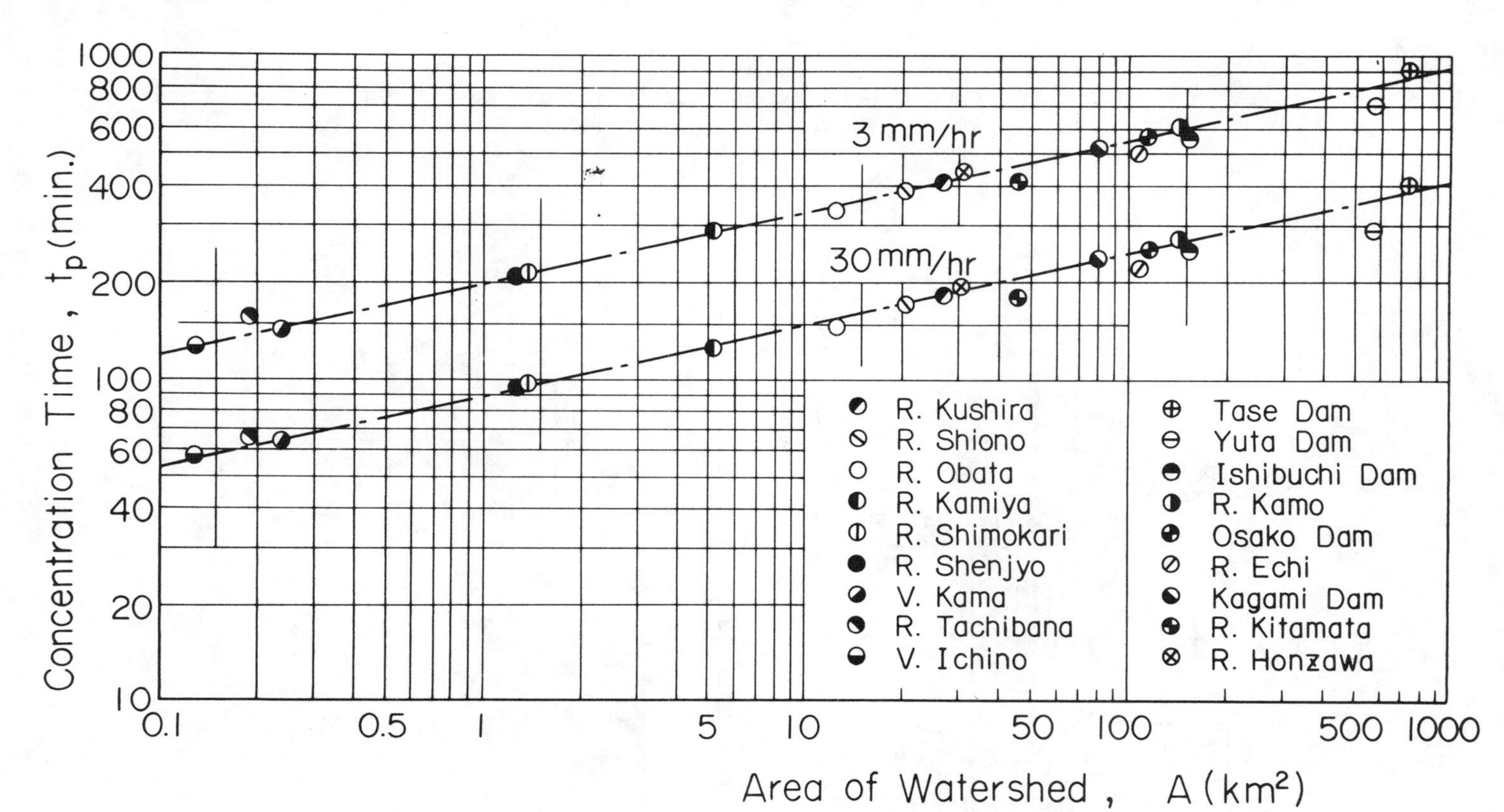

Fig. 4. Relations between the concentration time and the area of mountain or hill river basin

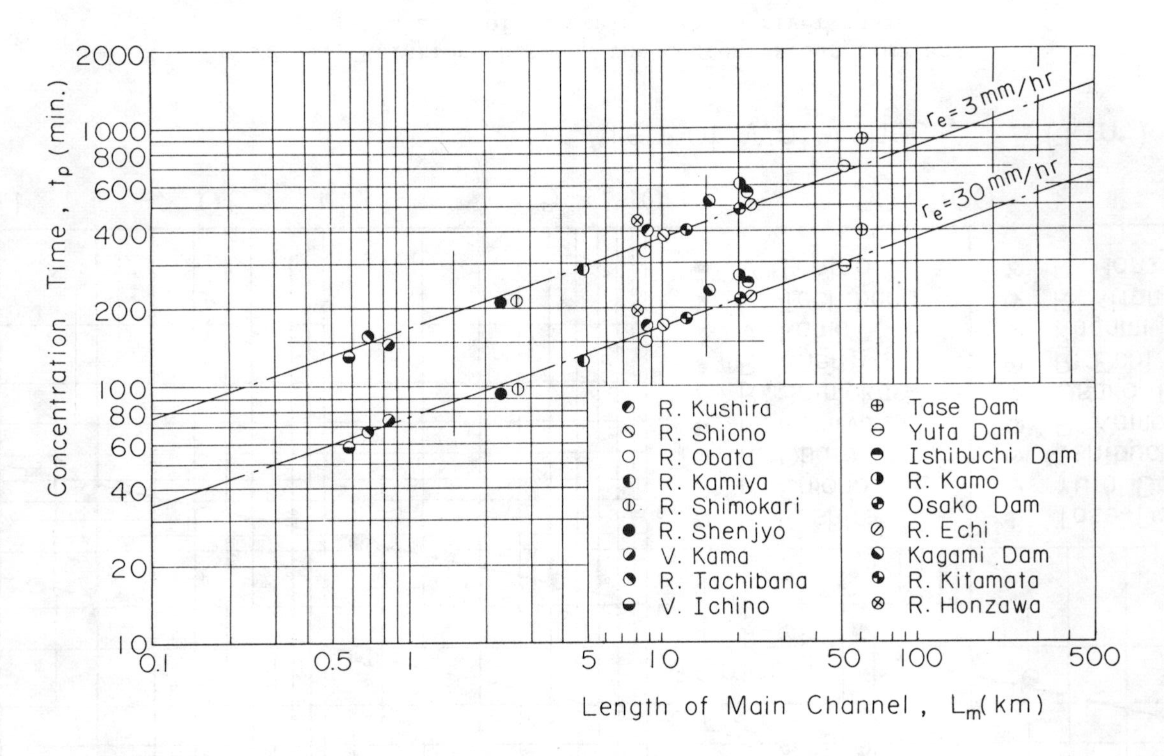

Fig. 5. Relations between the concentration time and the main river channel length in mountainous and hilly watersheds

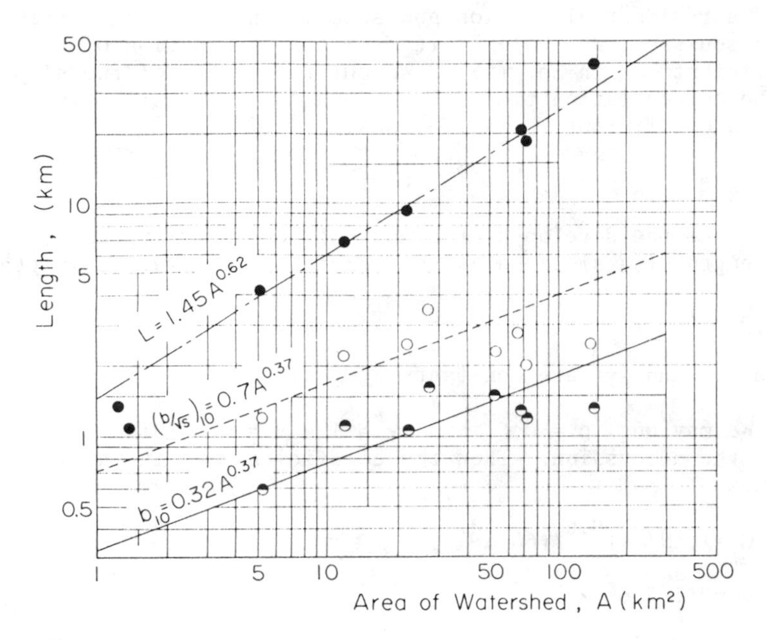

Fig. 6. Geomorphological characters of watersheds

The actual data can well be explained by Eq. (21), but they show a range of scatter larger than that in Fig. 5.

$$t_p = 170\ (L_m/\sqrt{S_c})^{0.32}\ r_e^{-0.35} \tag{21}$$

Remotest slope in a distributed model. When a distributed watershed model is used, the slope used to compute the concentration time should be the largest in the slope length, b, or the characteristic length, $b/\sqrt{s}$, in the uppermost reach. The most remote slope would control the concentration time under the condition that the area of the slope is not extremely small. Examples of possible relations of b and $b/\sqrt{s}$ to the catchment area are shown in Fig. 6. For example, the slope length corresponding to the probability of 10% exceedance in area has been taken instead of the largest slope length in a distributed model of a watershed. Figure 6 suggests to us that there exists a relation similar to Eq. (16) in the slope length assumed to be the remotest slope, allowing use of expression of Eq. (23) (below). There may nevertheless be an upper limit in the slope length for a large watershed.

Concentration Time in Developed Watershed

Here, we deal with the concentration time in a watershed, the surface of which has been developed to form a grazing area, a residential section and so on. In order to shorten the discussion, we shall start from the examination of the coefficient C in Eq. (18). We put the weight of the slope on the concentration time as β_s and change it slightly using a coefficient β as follows:

$$t_s = \beta_s t_p = \beta\, r_e^{-(p-0.35)}\, t_p \tag{22}$$

Since the developed area may not be extremely large, the length of a slope being the remotest slope may be written as

$$b = \delta A^{\gamma} \tag{23}$$

where δ and γ are constants.

We may put $p = 0.6$ and $\gamma = 1 - v = 0.37$ for simplifying the discussion. Then the coefficient C in Eq. (18) becomes

$$\left.\begin{aligned} C &= (\delta N/\sqrt{s})^{0.6}/(60\beta\alpha^{0.4}) \\ \alpha &= (1/3.6) \times 10^{-6} \end{aligned}\right\} \tag{24}$$

Eq. (24) gives important information for evaluating changes in the concentration time by watershed developments. For example, when a natural watershed is changed to a grazing or golf field, the equivalent roughness may be changed from $N_b = 0.8 \sim 1.2$ to $N_a = 0.4 \sim 0.6$ to yield slight change in β even if the values δ and s are not changed. For the overall urbanization of a watershed, we may assume the following changes (Kadoya, 1972). Before urbanization; $N_b = 0.8 \sim 1.2$, $s_b = 0.2 \sim 0.5$, $\beta_b = 0.80 \sim 0.85$. After urbanization; $N_a = 0.005 \sim 0.015$, $s_a = 0.01 \sim 0.02$, $\beta_a = 0.4 \sim 0.6$. In these expressions, subscripts b and a are used for parameters of before and after states of development, respectively.

Summarizing the above discussions regarding the first proposal, we propose a practical formula for estimating the concentration time in a river basin as follows:

$$\left.\begin{aligned} t_p &= CA^{0.22}\, r_e^{-0.35} \\ C &= 290, \quad \text{for a mountain river basin} \\ C &= 190 \sim 210, \quad \text{for a grazing or golf field} \\ C &= 60 \sim 90, \quad \text{for an urban area} \end{aligned}\right. \tag{25}$$

where t_p is shown in minutes, A in km^2 and r_e in mm/hr.

Figure 7 shows a few examples of the concentration time in a grazing area (C=190) and urban areas (C=65).

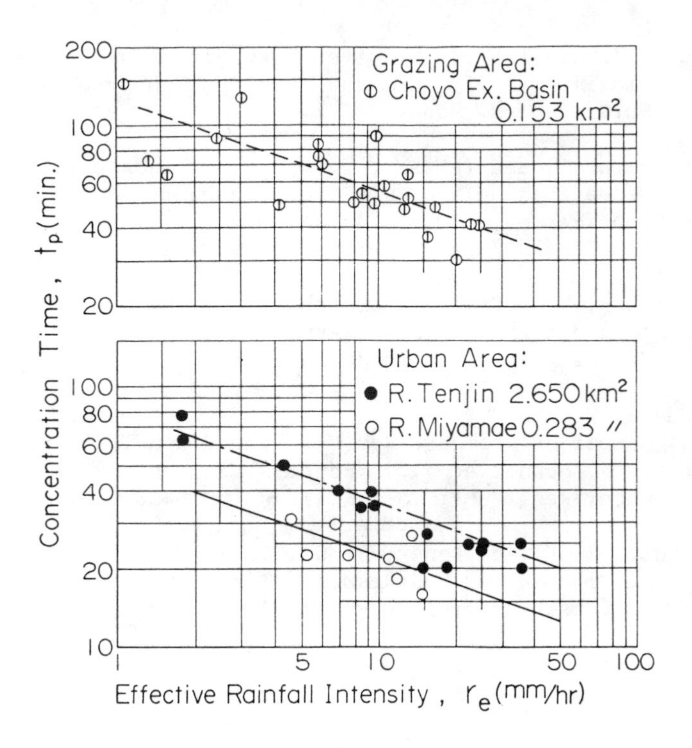

Fig. 7. Examples of the concentration time in grazing and urban areas

Conclusion

The concentration time in a river may be used as a clue for solving several problems in catchment dynamics. For example, the rational formula can be used rationally by considering both relationships of the effective rainfall intensity to its duration and to the concentration time. If the kinematic runoff model is used for runoff analysis, a good first approximation of the equivalent roughness is found by decomposing the concentration time.

This paper has dealt with concentration time simply, but theoretically and practically. Because the concentration time is a lumped expression of complicated runoff phenomena, there remain a few problems to be studied in the future.

References

[1] Chow, V. T., 1964. "Runoff." In: V. T. Chow (Editor), *Handbook of Applied Hydrology*. McGraw Hill, New York, 14; 1-54.

[2] Hack, J. T., 1957. "Studies of Longitudinal Stream Profiles in Virginia and Maryland," U.S. Geol. Surv. Profes. Paper 294-B.

[3] Ishihara, T. and Takasao, T., 1959. "Fundamental researches on the unit hydrograph method and its application," Trans. Japan Soc. Civil Eng., 60, 3:1-34 (In Japanese with English abstract).

[4] Ishihara, Y., 1961. "Geomorphological effect on flood peak discharge," Proc. Ann. Meet. Japan Soc. Civil Eng. II: 101-102 (In Japanese).

[5] Kadoya, M., 1972. "Predictive Study on Urbanizing Effect of Drainage Basin of Flood Runoff," In: E. F. Schulz et al., (Editors) *Floods and Droughts*, Water Res. Publ. Fort Collins, pp. 436-449.

[6] Schulz, E. F. and Lopez, O. G., 1974. "Determination of Urban Watershed Response Time," Colorado State University Hydrology Paper 71.

EFFECTS OF ANTECEDENT SOIL MOISTURE ON STORMFLOW VOLUMES AND TIMING

By

James A. Lynch, Assistant Professor of Forest Hydrology
School of Forest Resources
The Pennsylvania State University
University Park, PA 16802

Edward S. Corbett, Principal Hydrologist
U.S.D.A. - Forest Service
Northeastern Forest Experiment Station
Pennington, New Jersey 08534

and

William E. Sopper, Professor of Forest Hydrology
Institute for Research on Land and Water Resources
The Pennsylvania State University
University Park, PA 16802

Abstract

The effects of antecedent soil moisture content on stormflow amounts (quickflow, delayed flow, total storm runoff, and maximum peakflow) and timing (time to maximum peakflow, time of recession, and duration of quickflow) were studied by applying 38.1 mm of simulated rainfall at an intensity of 6.1 mm per hour to a 7.93 ha forested watershed. An irrigation system, which consisted of 8 kilometers of aluminum and plastic pipe and utilized 520 rotating sprinklers mounted on 1.5 m risers to distribute the water pumped from a nearby lake, was used to apply the simulated rainstorms. A 38.1 mm simulated rainstorm was replicated eight times at various antecedent soil moisture contents.

Quickflow, total storm runoff, maximum peakflow, time of recession, and duration of quickflow resulting from these artificial rainstorms were positively correlated with increasing antecedent soil moisture. Delayed flow and time to maximum peakflow were, for the most part, unaffected by changes in antecedent soil moisture. The percentage of rainfall converted to total storm runoff ranged from 12 to 86 percent, while antecedent soil moisture ranged from 87 to 99 percent of field capacity.

Introduction

One of the most important and fundamental needs in watershed research is to fully understand watershed behavior--that is to understand the entire hydrologic system which affects the disposition of precipitation within the drainage basin and its conversion to streamflow. This conversion is the result of complex interactions between hydrologic processes and numerous climatic and physiographic factors. One of these factors is antecedent soil moisture (ASM).

Although the importance of ASM in controlling storm-runoff relations has been realized, its actual affects on storm hydrograph characteristics and components have not been quantitatively defined. The primary problem encountered when utilizing natural rainfall-runoff events to evaluate the effects of ASM on stormflows is the variability of meteorological and antecedent watershed conditions. This variability makes it highly improbable that a precipitation event will be replicated on an experimental watershed at the same intensity and duration over a range of antecedent conditions. Physical laboratory models and computer models enable one to achieve the necessary replications. In addition, the utility of these models allows one to hold certain factors constant while at the same time varying others. However, one of the major disadvantages is the difficulty of transferring the results of model studies to natural watersheds (Black, 1970; Graveto and Eagleson, 1970; Roberts and Klingeman, 1970). Although both of these approaches have unquestionable value, their usefulness still depends upon a thorough knowledge of the hydrologic processes which can be extrapolated to the physical laboratory model or used to select parameters and linkages for conceptual models to characterize watershed behavior and response to rainfall.

The purpose of this study was to quantitatively evaluate the effects of antecedent soil moisture on stormflow amounts (quickflow, delayed flow, total storm runoff and maximum peakflow) and timing (time to maximum peakflow, time of recession and duration of quickflow). This was accomplished using an irrigation system installed on a 7.93 ha forested watershed in central Pennsylvania. The watershed is essentially elliptical, with symmetrical sides facing north and south. Average slideslope steepness is 30 percent. The rainfall simulation facility allows for control of watershed conditions prior to the application of simulated rainfall as well as rainfall intensity, duration, and consequently, amount.

The Rainfall Simulation System

The system consisted of approximately 8 kilometers of aluminum and plastic pipe. Two heavy-duty centrifugal pumps mounted in tandem supplied water from a nearby lake. Five-hundred and twenty rotating sprinklers with 4 mm nozzles mounted on 1.5 m risers distributed the water over the basin. The sprinklers were located on the watershed in a 12 m grid spacing. Flow control valves on each sprinkler allowed a known quantity of simulated rainfall to be applied once the system was pressurized. The flow control valves had a maximum application rate of 6.1 mm per hour. For this study the irrigation system covered 7.59 ha of the 7.93 ha watershed; only the extreme upper portion of the rear half and several small areas on the ridge top were not irrigated.

The watershed was divided into sub-basins by three streamflow gaging stations on the perennial stream channel. Another streamflow gaging station monitored intermittent streamflow from a secondary channel in the upper portion of the watershed. The hydrologic status of the watershed was monitored at 31 soil moisture sites, 15 tensiometer sites, 52 shallow groundwater observation wells and 3 deep groundwater observation wells. Simulated and natural rainfall were measured by an extensive network of 8 recording, 40 standard, and 17 trough-type raingages.

The general shape of the hydrograph of a single storm occurring over the drainage area is shown in Fig. 1, which also shows the components of a storm hydrograph as used in this report. Separation of quickflow from delayed flow was based on the technique described by Hewlett and Hibbert (1967). A line was projected from the beginning of each storm hydrograph at a slope of 0.547 liters per second per square kilometer per hour until it intersected the recession side of the hydrograph. However, in our analysis, base flow volumes resulting from antecedent storms were excluded from delayed flow. This was accomplished by extending a horizontal line of constant discharge equal to antecedent flow rate from the beginning of storm runoff until it intersected the recession limb of the storm hydrograph.

Calibration of the Rainfall Simulation Facility

In order to quantitatively evaluate the effects of ASM on storm hydrographs resulting from simulated rainfall events, it was necessary to determine what, if any, difference in the storm hydrographs could be attributed solely to variations in the operation of the simulated rainfall facility.

The calibration involved the replication of a 38.1 mm simulated rainstorm at low antecedent soil moisture and a 25.4 mm simulated rainstorm at high antecedent moisture conditions. The two 38.1 mm storms were applied during August, while the 25.4 mm simulated rainstorm was replicated in October.

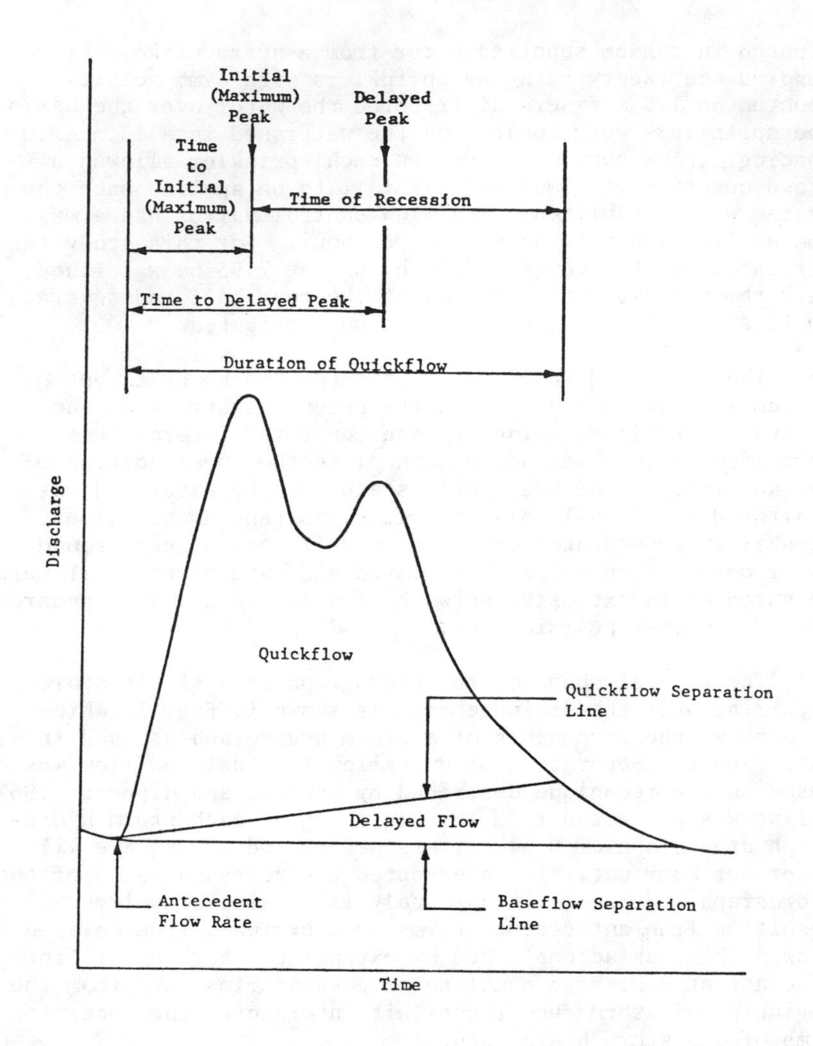

Fig. 1. Components and characteristics of a storm hydrograph resulting from a single storm occurring on the experimental watershed

The hydrographs resulting from the 38.1 mm rainstorm replications were almost identical with respect to rising limbs, peakflows, recession limbs, and storm runoff. Maximum peakflows (0.099 and 0.098 m^3s^{-1}/km^2), both of which occurred within minutes after cessation of rainfall, differed by only one percent. The times of recession and durations of stormflow differed by only 23 and 19 minutes, respectively. Total storm runoff from the replicated storms measured 4.5 and 4.4 mm, a variation of less than 2.2 percent.

The results of the 25.4 mm replications were very similar to those obtained from the 38.1 mm replications. The rising limbs, peakflows, and most of the recession limbs were almost identical for both storms. The only major discrepancy occurred during the first 13 hours of the second storm recession. During this period, a delayed peak occurred that was not evident following the first storm. Based on shallow groundwater well observations, the source of stormflow for this delayed peak was attributed to slightly wetter soil in the valley floor above the perennial stream at the time of the second storm. In terms of volume, the first simulated rainstorm produced a total storm runoff of 19.6 mm compared to 20.6 mm for the second rainstorm. The difference, 1.0 mm, represents only five percent of the total storm discharge.

Based on the results of both the 38.1 and 25.4 mm storm replications, very little difference in the size and shape of the storm hydrographs could be attributed to variations in the operation of the simulated rainfall facility.

Procedure

Soil moisture determination. Soils are mostly shaly silt loams and silt loams ranging in depth from 0.3 to 0.9 m on the upper side slopes and ridge top to 0.9 to 2.7 m on the lower side slopes and channel area. Soil infiltration and percolation rates are greater than the maximum expected rainfall intensity for the area. Field capacity for the entire soil mantle (average depth 1.42 m) has been estimated at 319.8 mm of water.

Using the neutron scattering technique, soil moisture was measured at 31 sites to monitor moisture fluctuations in the soil mantle. Both surface and depth probes were used. Thirteen of the soil moisture sites were located on a transect perpendicular to the stream channel at a point midway between two streamflow gaging stations. Each of the remaining soil moisture sites represented a physiographic unit of the watershed.

Effects of antecedent soil moisture on storm hydrographs. In order to quantitatively evaluate the effects of ASM on stormflow volumes and timing, eight 38.1 m simulated rainstorms were applied to the entire watershed under various ASM conditions. Rainfall intensity for all storms was 6.1 mm per hour; rainfall duration was 6.25 hours. The storms were applied between August 1 and September 19, 1974 under as similar climatic conditions as possible.

Quickflow, delayed flow, total storm runoff, maximum peakflow, time to maximum peakflow, time of recession and duration of quickflow from each simulated rainstorm were compared with ASM using correlation analysis. In addition, empirical relationships between the individual storm hydrograph parameters and ASM were derived using regression analysis.

Results and Discussion

Antecedent soil moisture measurements. Antecedent soil moisture on a total watershed basis reached its lowest level (278.1 mm) on 8/1/74 and progressively increased to its highest level (316.2 mm) on 9/19/74 (Table 1). A comparison of the watershed response to the simulated rainfall applications at the lowest and highest antecedent moisture contents is illustrated in Fig. 2. The relatively small difference in ASM (38.1 mm) between the applications of simulated rainfall under "wet" and "dry" conditions was due to the high ASM on 8/1/74. The high ASM on 8/1/74 was the result of 292.1 mm of rainfall (both natural and artificial) that fell on the watershed during the two months preceding this rainstorm. The abundance of rainfall during this period continually replenished soil moisture, but did not appreciably affect streamflow because of available moisture storage within the soil mantle. Consequently, streamflow prior to the 8/1/74 rainstorm was seasonally low at 0.0012 m^3s^{-1}/km^2. Although streamflows below 0.001 m^3s^{-1}/km^2 are not uncommon on the

Table 1. Stormflow volumes for eight 38.1-mm simulated rainfall events at different antecedent soil moisture contents

Date	Antecedent Soil Moisture	Total Storm Runoff	Quick-flow	Delayed Flow	Maximum Peakflow
	mm				(m^3s^{-1}/km^2)
8/01/74	278.1	4.55	2.64	1.90	0.099
8/07/74	287.5	12.19	8.36	3.84	0.201
8/14/74	290.1	15.52	11.94	3.58	0.253
8/19/74	295.1	15.90	12.27	3.63	0.239
8/23/74	304.8	20.75	17.02	3.73	0.304
8/27/74	310.4	23.34	19.58	3.76	0.328
9/16/74	313.2	29.74	25.78	3.96	0.526
9/19/74	316.2	32.66	28.88	3.78	0.628

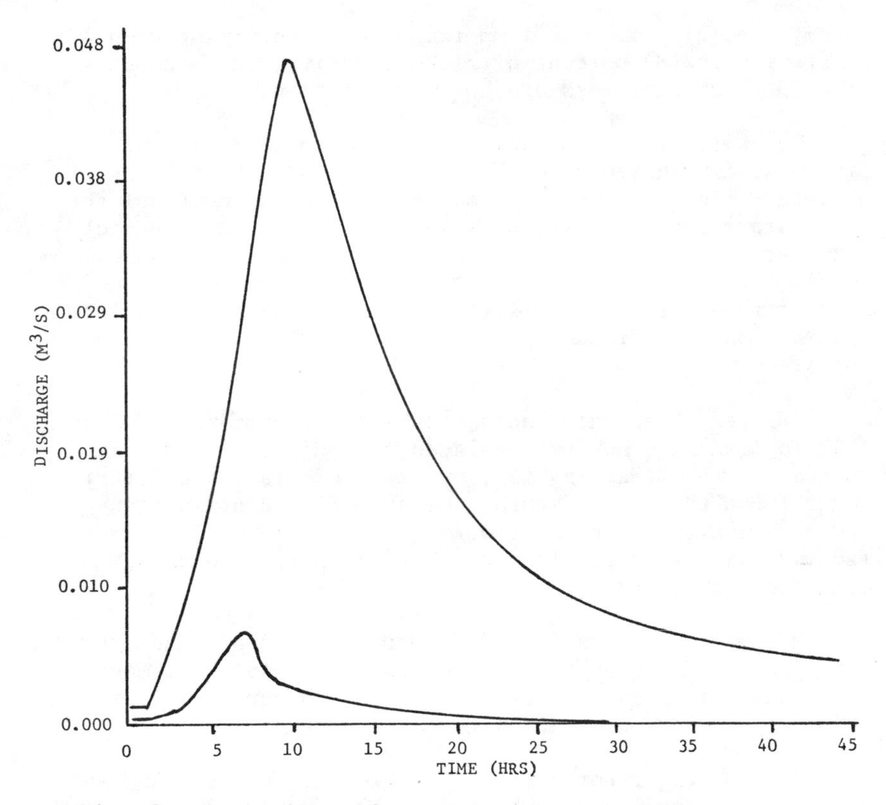

Fig. 2. Comparison of storm hydrographs from 38.1 mm simulated rainstorms at the lowest and highest measured antecedent moisture content

watershed, they occur only during long rainless periods with high evapotranspiration rates. Streamflow recession during such periods is very slow. As a result, soil moisture can decrease substantially with little change in stream discharge.

Stormflow volumes. Quickflow, delayed flow, total storm runoff and maximum peakflows from eight simulated 38.1 mm rainstorms at different ASM contents are given in Table 1. It is evident that ASM conditions of the basin were very important in determining the magnitude of storm discharge. Total storm runoff ranged from 4.55 to 32.66 mm. This increase in the efficiency of the watershed in converting precipitation to streamflow occurred over a relatively small range in ASM. The amount of moisture stored in the soil mantle at field capacity was estimated to be 319.8 mm. Based on this estimate, ASM on 8/1/74 and 9/19/74 represented 87 and 99 percent, respectively, of the total moisture stored at field capacity. This range represents a span of only 12 percent of the available moisture below field capacity. Storm runoff resulting from 38.1 mm rainstorms, occurring when ASM is at or below 87 percent of field capacity, can account for only 12 percent or less of the precipitation.

Quickflow volumes ranged from 2.64 to 28.88 mm and progressively increased with increasing ASM. Regardless of ASM, quickflow accounted for most (58 to 88 percent) of the total storm runoff from each simulated rainstorm. In fact, with the exception of the 8/1/74 and 8/7/74 storms, quickflow consistently accounted for 77 to 88 percent of storm runoff. Seven percent of the 8/1/74 rainstorm occurred as quickflow while 76 percent of the 9/19/74 rainfall was converted into quickflow.

Delayed flow contributions to storm runoff ranged from 1.90 to 3.96 mm. However, delayed flow did not progressively increase with increasing ASM, but remained fairly constant for all but the 8/1/74 storm. Delayed flow contributions to storm discharge in terms of percent of total runoff decreased from 42 percent at the lowest ASM to 12 percent when ASM was near field capacity.

Maximum peakflows ranged from 0.099 to 0.628 m^3s^{-1}/km^2. Although peakflows generally increased with ASM, the most substantial increases in terms of volume occurred when soil moisture contents approached field capacity.

Simple linear correlation coefficients between ASM and the various stormflow parameters ranged from 0.687 for delayed flow to 0.976 for total storm runoff. Linear correlation coefficients between ASM and quickflow and maximum peakflow were 0.975 and 0.903, respectively. All of the correlation coefficients, with the exception of delayed flow, were significant at the one percent level of confidence.

A good linear relationship existed between ASM and quickflow and total storm runoff, whereas no relationship was found between ASM and delayed flow. Almost 95 percent of the variation in quickflow and total storm runoff can be explained by the simple linear relationship $Y = b_o + b_1X$, where Y = quickflow or total storm runoff and X = ASM. These relationships apply only to the range of ASM conditions included in this study. However, it should be noted that the increase in ASM within this range resulted in a 74 percent increase in total storm runoff from a 38.1 mm rainfall.

Despite the good linear association between maximum peakflows and ASM, the mathematical function that best described this relationship was curvilinear. This curvilinear relationship is represented by the model $\ln Y = b_o + b_1X$, where Y = maximum peakflow and X = ASM. Using this model, approximately

89 percent of the variation in peak discharge could be explained on the basis of the soil moisture content prior to rainfall. A simple linear relationship could explain approximately 81 percent.

Stormflow timing. Time to maximum peakflow, time of recession and duration of quickflow are presented in Table 2. Maximum peakflows for storms applied at moisture contents of 304.8 mm or less occurred several minutes after cessation of rainfall. At ASM contents greater than 304.8 mm, maximum storm discharge occurred several hours after cessation of rainfall. This shift in the time of occurrence of peak discharge was caused by a change in the source-area controlling maximum peakflow as well as a decrease in the time for subsurface stormflow to reach the stream channel. Maximum peakflows at the lower ASM contents were regulated for the most part by the location and extent of the saturated soil zone along both the perennial and intermittent stream channel. This in turn determined the amount of direct channel interception of rainfall and consequently the magnitude and timing of the maximum peak. At higher ASM contents, the major source of stormflow contributing to peakflows was from subsurface storm water from the side slopes.

The duration of quickflow from the 8/1/74 simulated rainstorm lasted 25.43 hours. On 8/7/74 the duration of quickflow was 39.08 and on 8/27/74 storm runoff continued for 50.72

Table 2. Stormflow timing parameters for eight 38.1-mm simulated rainfall events at different antecedent soil moisture contents

Date	Antecedent Soil Moisture	Time to Maximum Peakflow	Time of Recession	Duration of Quickflow
	(mm)	-----	hrs.	-----
8/01/74	278.1	6.38	19.05	25.43
8/07/74	287.5	6.45	32.63	39.08
8/14/74	290.1	6.35	36.13	42.48
8/19/74	295.1	6.37	37.42	43.79
8/23/74	304.8	9.87	37.58	47.45
8/27/74	310.4	9.12	41.60	50.72
9/14/74	313.2	8.82	40.38	49.20
9/19/74	316.2	8.93	40.35	49.28

hours. Thereafter, the duration of quickflow decreased to 49.20 and 49.28 hours for the 9/16/74 and 9/19/74 storms, respectively.

The time of recession of the storm applied at the lowest ASM was only 19.05 hours. At the next lowest ASM, storm recession lasted 32.63 hours. Gradual increases in the length of storm recession occurred for each of the next four storms. The longest time of recession followed the 8/27/74 storm and was 41.60 hours. Thereafter, the duration of storm recession decreased to 40.38 and 40.35 hours for the 9/16/74 and 9/19/74 storms, respectively.

The linear correlation coefficient between ASM and time to maximum peakflow was 0.848. Although this correlation coefficient, which is significant at the one percent confidence level, indicated that ASM could explain 72 percent of the variation in the time to peakflow using a simple linear relationship, such a relationship is unrealistic in that it assumes a gradual transition over the range of ASM studied. This gradual transition was not supported by the data. In addition, no curvilinear relationship was evident between ASM and time to peakflow.

Despite the relatively good linear associations between ASM and time of recession and duration of quickflow ($r = 0.852$ and 0.905, respectively), a curvilinear function best described these relationships. This curvilinear function, a second-degree parabola, is represented by the model $Y = b_o + b_1X + b_2X^2$, where Y = time of recession or duration of quickflow and X = ASM. Using this model, 97 and 90 percent of the variation in the duration of quickflow and time of recession, respectively, could be explained on the basis of variations in ASM conditions. These relationships were significantly better than straight-line linear relationships.

It should be emphasized that these curvilinear relationships apply only to the range of ASM studied. Both the duration of quickflow and the time of recession would become fairly constant as ASM approached field capacity as indicated by the data. This would not occur if the models are extended to or beyond field capacity.

Summary and Conclusions

Quickflow volume, total storm runoff, and maximum peakflow generally increased with increasing antecedent soil moisture, while delayed flow and time to maximum peakflow were, for the most part, unaffected. Time of recession and duration of quickflow progressively increased throughout the lower antecedent moisture contents, leveling-off as antecedent soil moisture approached field capacity. The changes in the size and shape of the storm hydrographs occurred over a relatively small range in antecedent soil moisture conditions. This range represented a span of only 12 percent of the available moisture below field capacity. Consequently, it appears that specifying moisture contents of 20, 50 and 80 percent of field capacity to represent low, average, and high antecedent moisture conditions of a drainage basin, so common in stormflow modeling, is too insensitive to provide an adequate description of the affects of antecedent soil moisture on storm runoff.

Although significant linear or curvilinear relationships were found between antecedent soil moisture and quickflow volume, total storm runoff, maximum peakflow, time of recession and duration of quickflow, these relationships only apply to the range of antecedent soil moisture conditions included in this study.

Additional information is needed to assess the effects of seasonal changes on the relationships between these stormflow volumes and timing parameters and antecedent soil moisture.

Acknowledgment

Financial support was provided by the United States Department of the Interior, Office of Water Research and Technology; as authorized under the Water Resources Research Act of 1964; the Pennsylvania Science and Engineering Foundation of the Pennsylvania Department of Commerce; the Pennsylvania Department of Environmental Resources; and National Science Foundation Grant GA-22860.

References

[1] Black, P. E., 1970. "Runoff from Watershed Models," Water Resour. Res., 6(2):465-478.

[2] Graveto, V. M. and Eagleson, P. S., 1970. "Scale Effects in the Physical Modeling of Surface Runoff," Hydrodynamics Laboratory Report No. 120, Mass. Inst. of Tech., Cambridge, Mass., 328 p.

[3] Hewlett, J. D. and Hibbert, A. R., 1967. "Factors Affecting the Response of Small Watersheds to Precipitation in Humid Areas," Internatl. Symposium on Forest Hydrology, pp. 275-290, Pergamon Press, N. Y.

[4] Roberts, M. C. and Klingeman, P. C., 1970. "The Influence of Land Form and Precipitation Parameters on Flood Hydrographs," Jour. of Hydrology, 11(4):393-411.

SNOWMELT HYDROGRAPHS FROM SPATIALLY VARIED INPUT

By

J. Martinec
Federal Institute for Snow and Avalanche Research
7260 Weissfluhjoch/Davos, Switzerland

Abstract

The spatial and temporal variability of the meltwater production in mountain watersheds results from two main reasons: 1) The energy input varies according to the elevation and exposure of the different parts of the basin, apart from short-term and seasonal changes. 2) The meltwater production is related to the areal extent of the snow cover which gradually decreases during the snowmelt season.

The altitude-dependent temperature is taken into account by dividing the watershed into several elevation bands. The snow coverage is determined by monitoring the watershed from aeroplanes and satellites. In a rugged terrain the snow cover is dispersed into scattered patches which necessitate a computerized evaluation of the imagery.

In the proposed model the transformation of the meltwater input into the discharge from the basin is based only on the recession coefficient which is however constantly adjusted by a relation to the current intensity of runoff. This relation is characteristic for a given watershed although it can vary slightly from year to year.

In the basin studied the changing energy input and the areal extent of the snow cover in different parts of the basin were found to be more important than the spatial variation of areas contributing to the meltwater production in computing the daily snowmelt runoff. The runoff mechanism was revealed by tracing and dating techniques using environmental tritium. A major part of meltwater infiltrates and immediately stimulates the outflow from the subsurface storage. This phenomenon seems to be characteristic for several basins of Central Europe, enabling the complex snowmelt-runoff process to be simulated by a relatively simple model.

Introduction

Factors involved in computations of the snowmelt runoff are so numerous that it is practically impossible to take them all into account. Some of them are not regularly measured or are available only from experimental sites. A snowmelt-runoff model to be used in normal watersheds should cope with this siruation. It is therefore necessary to decide about factors to be considered and details to be omitted. The areal extent of the seasonal snow cover appears to be an important variable in mountain watersheds with a considerable elevation range.

Varied Energy Input and Changing Snow Cover

Although a complicated energy balance governs the snowmelt, the temperature along with appropriate degree-day ratios can give satisfactory results. Figure 1 shows the varying positive degree days as measured by an automatic meteorological station situated at 2370 m a.s.l. in the Dischma basin in the Swiss Alps. The degree-day factor is increased stepwise in relation to the current density of snow according to an experimentally derived equation (Martinec, 1976, pp. 90-91)

$$a = 1,1 \frac{\rho_s}{\rho_w} \tag{1}$$

where a is the degree-day factor $[cm^oC^{-1}.d^{-1}]$, and ρ_s, ρ_w are the density of snow and water, respectively.

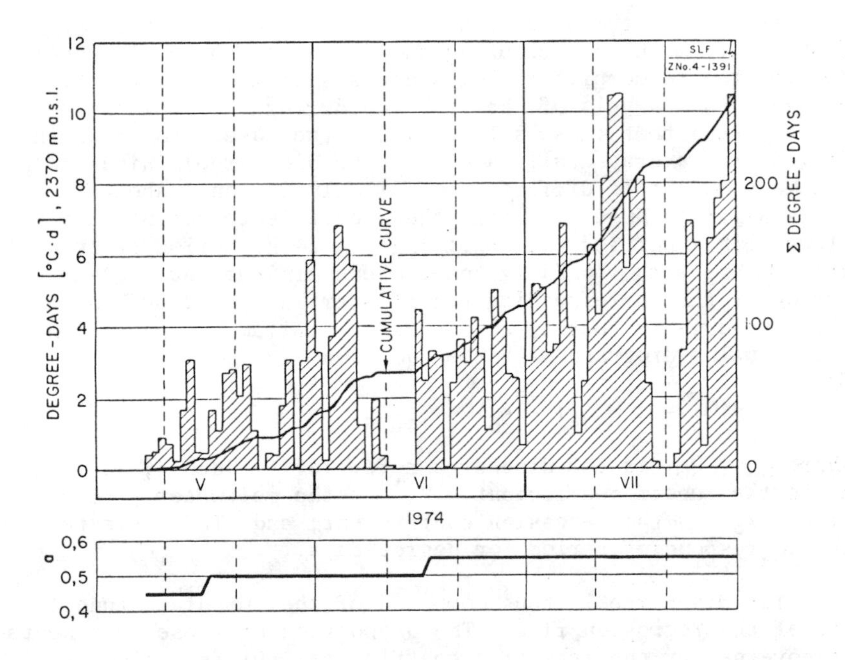

Fig. 1. Daily degree-day numbers in the Dischma basin at 2370 m a.s.l.

Since the melting takes place on top of the snowpack, the daily meltwater volume is proportional to the current snow-covered area. While additional melting at the boundaries of scattered snow patches might be neglected, it is not advisable to take average values of temperature and of the snow coverage for the whole basin. A hypothetical example in Table 1 illustrates the serious errors which can result. If the snowline is set at 50% snow coverage, the area below is interpreted as snow-free (I) and the area above as totally snow-covered (II). This gives a lower meltwater volume than the total of the three elevation bands. On the other hand, if average values are applied for the whole basin, the obtained meltwater volume appears to be too high.

Figure 2 shows an orthophoto of the Dischma basin divided into three elevation bands with an altitude interval of 500 m. The gradual decrease of the snow coverage is plotted separately for each zone in Fig. 3. The variable temporal distribution of the energy input in the respective years can result in different depletion curves and in a considerable time shifting as is illustrated by Fig. 4. Deviations from the usually S-shaped form of the curve (Leaf, 1967) can also occur.

Components of the Snowmelt Hydrograph

In view of the size of Dischma, 43,3 km^2, the varying distance from the contributing snowfields to the outlet would be expected to complicate the synthesis of the hydrograph. However, an analysis of the time lag during the snowmelt period shows that this is luckily not the case. As shown in Fig. 5, the average daily hours of the discharge minima and maxima show little difference until July and only about 2-hours delay in August. It is thus possible to compute daily flows with a unified time shifting: A 24-hr period starting at 6 00 hrs with the daily temperature minimum and a 24-hr period of the corresponding runoff starting at 12 00 hrs when the discharge is roughly at the daily minimum. Figure 6 shows the actual hydrograph with the equation

$$Q_5 = c\ M_5\ (1 - k_5) + Q_4\ k_5 \qquad (2)$$

where Q_4, Q_5 is the discharge referring to June 4, June 5; c is the runoff coefficient; M_5 is the meltwater production; k_5 is the recession coefficient; and T_5 indicates the corresponding period for degree-days.

The discharge is thus composed of the immediate runoff and of the recession flow. The proportion of these components is governed by the recession coefficient. Different concentrations of the environmental tritium in the precipitation,

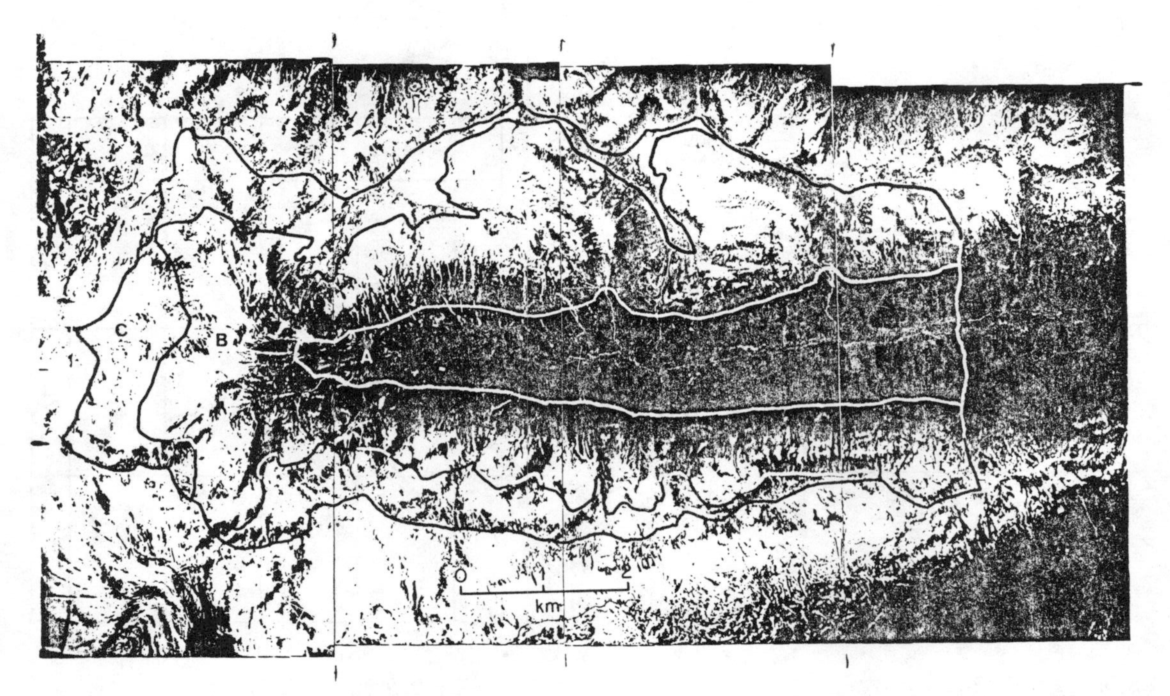

Fig. 2. Orthophoto showing the snow cover in Dischma on 5 July 1974 and the boundaries of the elevation bands A, B, C

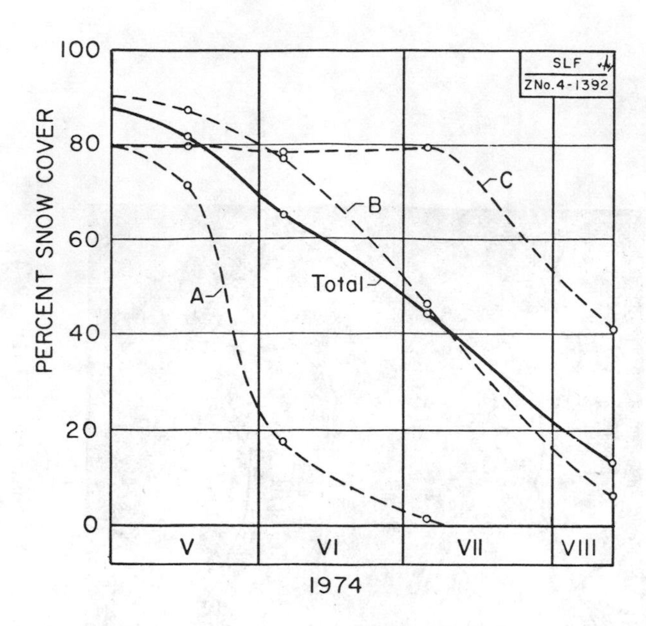

Fig. 3. Depletion curves of the snow-covered area for the zones A,B,C and for the whole watershed

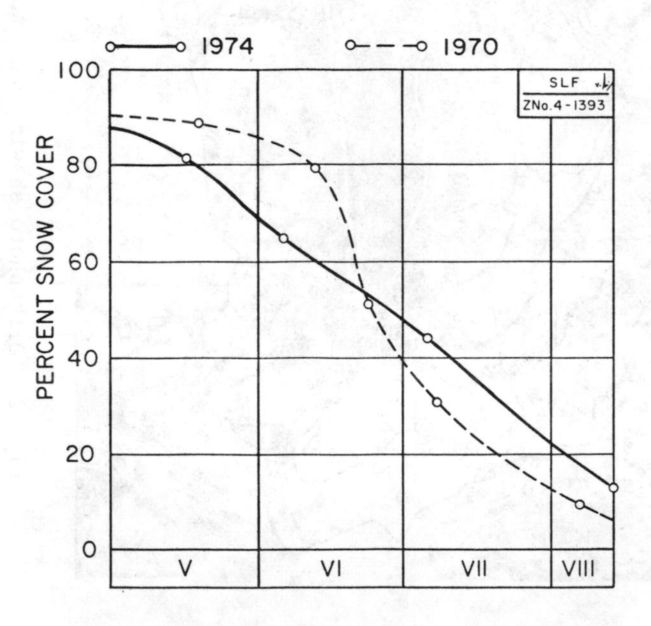

Fig. 4. Comparison of depletion curves of the snow-covered area in 1970 and 1974

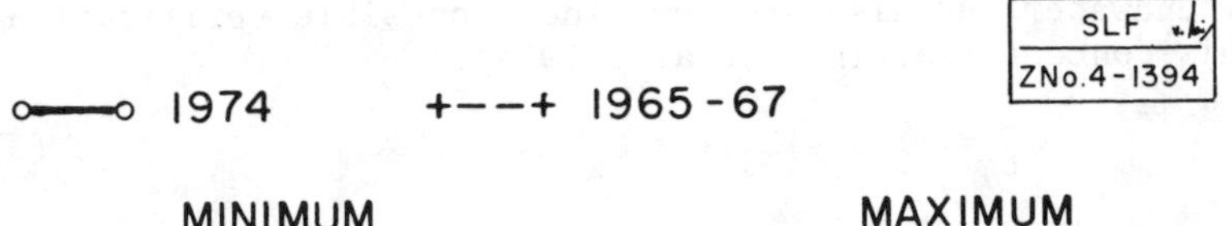

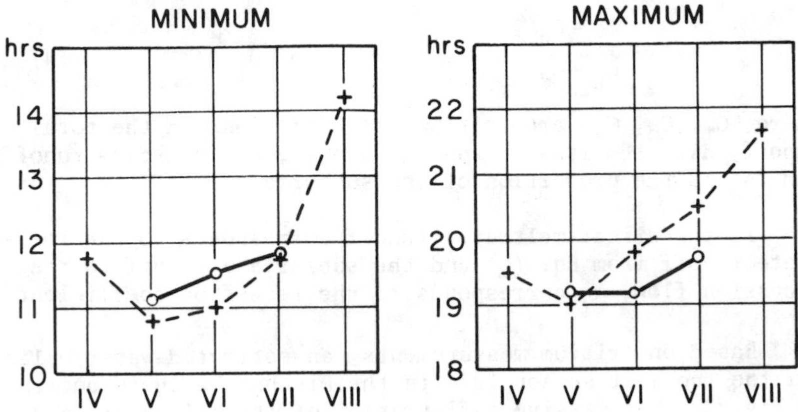

Fig. 5. The time-shifting of the daily lows and peaks of the snowmelt hydrograph, average hours for the respective months

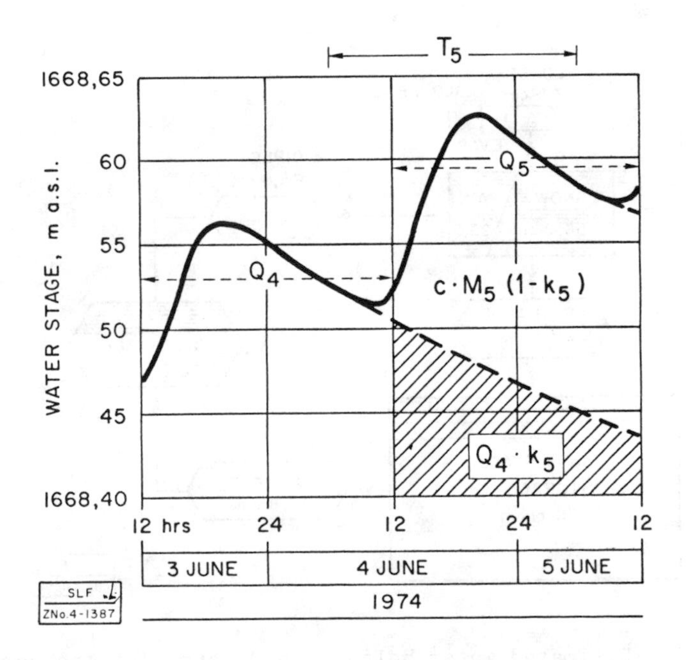

Fig. 6. Example of a snowmelt-runoff hydrograph

groundwater and discharge provide a possible verification for this concept (Martinec et al., 1974):

$$C_t = (1 - s)C_d + s\,C_s \tag{3}$$

$$s = \frac{C_t - C_d}{C_s - C_d} \tag{4}$$

where C_t, C_d, C_s are tritium concentrations in the total runoff, direct meltwater runoff (snow) and subsurface runoff, and s is the proportion of the subsurface runoff.

If the direct meltwater runoff corresponds to the immediate runoff from Eq. (2) and the subsurface runoff to the recession flow, s corresponds to the recession coefficient k.

Based on tritium measurements, an estimated water balance for the snowmelt season 1974 in the Dischma basin is outlined in Fig. 7. The massive infiltration of the meltwater leads quickly to an increased outflow from the subsurface storage, so that this old water attains a major proportion in the discharge. This runoff mechanism is a possible explanation of the relative uniformity of the time lag (Fig. 5) in spite of

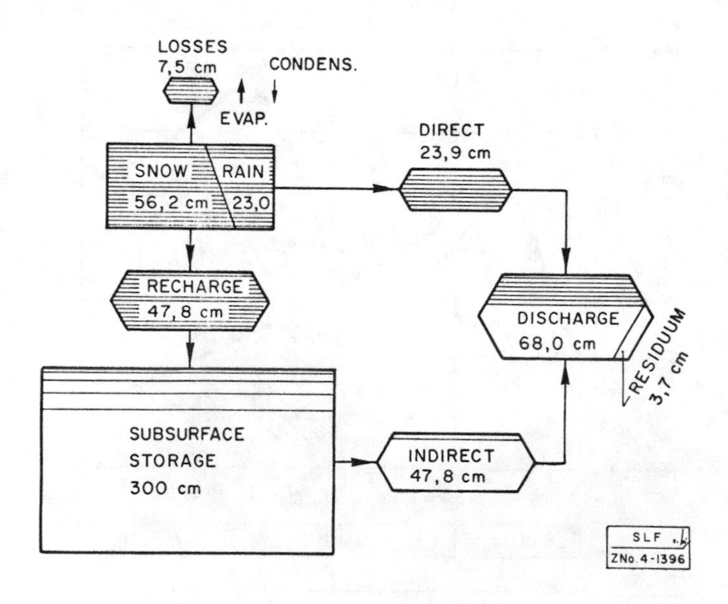

Fig. 7. Estimated water balance in Dischma for the period 8 May-30 July 1974 in terms of runoff depths. The residuum 3,7 cm at the end follows as the recession flow

the varying spatial distribution of the meltwater production and of different percolation velocities in the snowpack (de Quervain, 1948).

Spatial and Temporal Distribution of Meltwater Production

Recalling the example in Table 1, it is advisable to compute the snowmelt runoff in Dischma by determining the meltwater production separately for each elevation band. In view of the relatively rapid and uniform watershed response, due to the runoff mechanism just explained, the total daily discharge is obtained simply by adding up the respective contributions and transforming the total into the output:

Table 1. Meltwater Production from Different Assessments of the Snow Coverage

Zone	Area m^2	Snow Coverage	Degree-days °C.d	Degree-day factor cm $°C^{-1}.d^{-1}$	Meltwater Production m^3
A	10.10^6	0,25	7,5	0,5	93 750
B	20.10^6	0,5	4,5	0,5	225 000
C	10.10^6	0,75	1,5	0,5	56 250
Total	40.10^6				375 000
I	20.10^6	0	6,75	0,5	0
II	20.10^6	1,0	2,25	0,5	225 000
Total	40.10^6				225 000
Entire basin	40.10^6	0,5	4,5	0,5	450 000

$$Q_n = c_n \Big\{ [a_n(T_n+\Delta T_A) \cdot S_{An} + P_{An}] \frac{A_A \cdot 10^{-2}}{86400} + [a_n(T_n+\Delta T_B) \cdot S_{Bn} + P_{Bn}] \frac{A_B \cdot 10^{-2}}{86400} + [a_n(T_n+\Delta T_C) \cdot S_{Cn} + P_{Cn}] \frac{A_C \cdot 10^{-2}}{86400} \Big\} (1-k) + Q_{n-1} \cdot k \quad (5)$$

where Q is the average daily discharge $[m^3 s^{-1}]$

c_n is the runoff coefficient

a_n is the degree-day factor $[cm.°C^{-1}.d^{-1}]$

T_n is the measured number of degree-days [°C.d]

ΔT is the correction by the temperature lapse rate [°C.d]

S is the snow coverage

P is the precipitation contributing to runoff [cm]

A is the area [m^2]

k is the recession coefficient

n is an index referring to the sequence of days

A,B,C as index refers to the three elevation bands.

Fig. 8 illustrates substantial differences in the spatial and temporal distribution of the computed snowmelt depths in two subsequent years.

Seasonal Snowmelt Hydrograph

For day-to-day computations of the snowmelt runoff by Eq. (5), T, ΔT, S, P, A are measured, extrapolated or known, a is determined by Eq. (1), c has to be estimated according to the assessed losses. The recession coefficient k was found to be a function of the current discharge:

$$k_n = b.Q_{n-1}^{y} \qquad (6)$$

When the snow cover has disappeared, any future daily discharge Q_n of the recession flow can be computed from the starting discharge Q_o by the equation

$$Q_n = b^{\sum_{j=0}^{n-1} \frac{n!}{j!(n-j)!} y^{n-1-j}} \cdot Q_o^{\sum_{j=0}^{n-1} \frac{n!}{j!(n-j)!} y^{n-j} + 1} \qquad (7)$$

For the Dischma basin, b = 0,85 and y = - 0,086, thus

$$k_n = 0,85\ Q_{n-1}^{-0,086} \qquad (8)$$

The snowmelt hydrograph computed for the season 1974 is compared with the measured runoff in Fig. 9. The outlined snowmelt-runoff model is based on the changing areal extent of the seasonal snow cover. The progress in the application of satellites to the snow areal mapping (Rango, 1975) gives some hopes to obtain this information in a more efficient way than by aeroplanes.

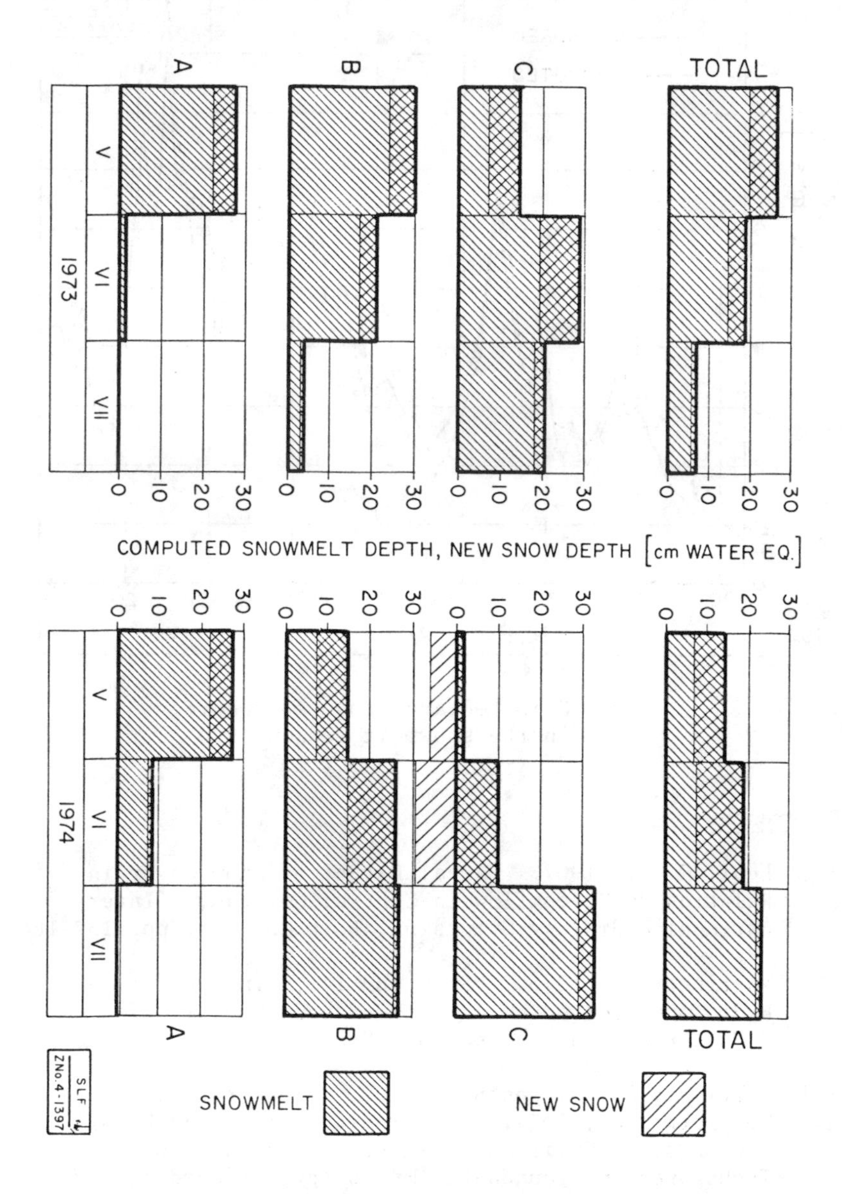

Fig. 8. Computed snowmelt depths for the three elevation bands. In May and June 1974, there was more new snow than snowmelt in the zone C

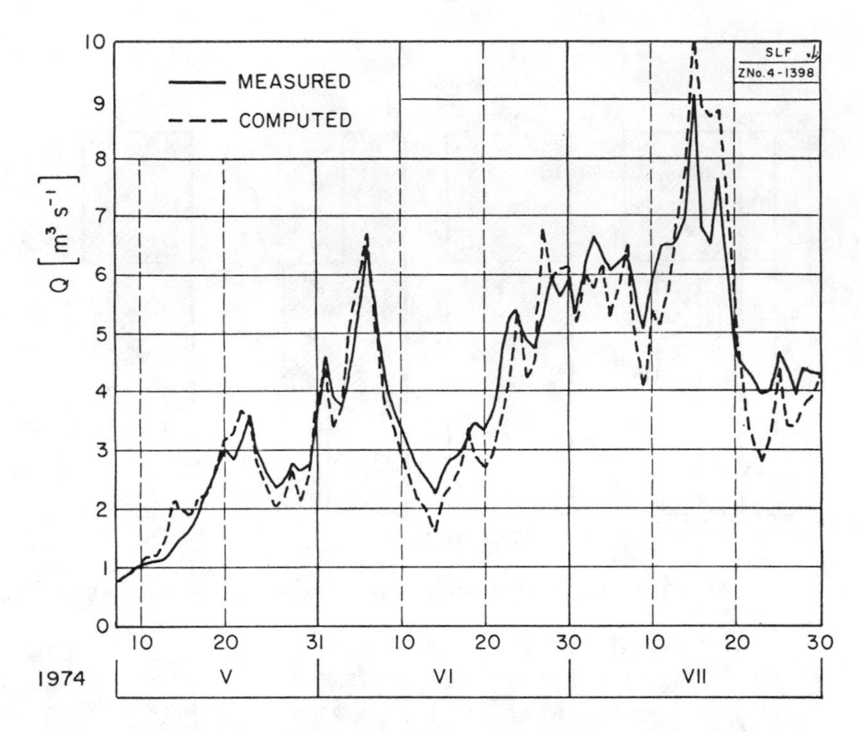

Fig. 9. Computed and measured runoff in the snowmelt season 1974

References

[1] Leaf, C. F., 1967. "Areal Extent of Snow Cover in Relation to Streamflow in Central Colorado," International Hydrology Symposium, Fort Collins, pp. 157-164.

[2] Martinec, J., 1976. "Snow and Ice," in: Facets of Hydrology, edited by J. C. Rodda, Chapter IV, John Wiley & Sons Ltd., London-New York-Sydney-Toronto, pp. 85-118.

[3] Martinec, J., Siegenthaler, U., Oeschger, H., and Tongiorgi, E., 1974. "New Insights into the Runoff Mechanism by Environmental Isotopes," in: Isotope Techniques in Groundwater Hydrology, Proceedings of a Symposium in Vienna, International Atomic Energy Agency, Vienna, Vol. I, pp. 129-143.

[4] de Quervain, M. R., 1948. "Ueber den Abbau der Alpinen Schneedecke," in: Snow and Ice, Vol. II, Proceedings IUGG General Assembly in Oslo, IAHS Publ. No. 30, pp. 55-68.

[5] Rango, A., 1975. "An Overview of the Applications Systems Verification Test on Snowcover Mapping," in: Operational Applications of Satellite Snowcover Observations, edited by A. Rango, Proceedings of a Workshop at South Lake Tahoe, California, NASA, Washington, D.C., pp. 1-12.

A METHOD OF COMPUTING VOLUME OF STORM WATER RETENTION BASINS

By

Yilmaz Muslu
Professor, Division of Environmental Science
and Technology, Civil Engineering Faculty
Technical University of Istanbul
Istanbul, Turkey

Abstract

Although storm water retention basins are becoming a more frequent part of combined and separate storm water collection system, their design can be made with many simplifications and there remains a need to develop a simple but reliable method which considers as many factors effecting the capacity of the basins as possible. The proposed method is based on the basin area and unit hydrograph S-curve of the watershed and enables determination of the required volume of the basin with respect to the physical and hydrological characteristics of the watershed.

The method developed in this study is then applied to 6 different dimensionless S-curves including a linear relationship also, and the storms which require the maximum storage have been computed. By means of these results the effect of neglecting an area to be measured by a planimeter on the total volume stored has been investigated and an equation derived to determine the design storm for the capacity of the retention basin.

Introduction

Retention basins are important elements of storm or combined sewerage systems. They store a part of the storm water from intense rains, and discharge the stored water gradually into downstream channels, pumping stations, or treatment plants. Hence, retention basins constructed at the end of a drainage area reduce the peak rate of runoff in the main sewer. In this way smaller diameter more economical sewers are possible. Retention basins reduce the wastewater load flowing into a river, and therefore are very advantageous from the sanitary point of view. Retention basins may be emptied by means of pumps where there is no sufficient slope.

Inflow hydrographs or mass curves may be used to solve various storage problems such as the design of water supply reservoirs, dams, and retention basins. The capacity of water supply reservoirs, for example, can be determined according to the variation of water consumption of a community with respect to time. In this case inflow is constant, outflow fluctuates. However, difficulties are encountered in determination of required capacity of retention basins because the inflow hydrograph changes from one storm to the other, and depends upon the shape of the watershed. It is necessary to compute many inflow hydrographs for various storm durations and find the storage corresponding to each storm. The maximum storage computed gives the required capacity of the retention basin. This is a long and tedious procedure. For this reason, a number of methods have appeared in the literature which have presented alternative solutions for the problem with many simplifications (Müller, 1939; Müller-Neuhaus, 1947; Randolph, 1958; and Yrjanainen and Warren, 1973). They are far from satisfactory. A method considering the shape of the S-curve of the watershed may be used for solving practical problems. This method has been developed and its application on six dimensionless S-curves is shown here.

A Method for Computation of Retention Basin Capacity

The inflow hydrograph has been considered for a storm with uniform intensity. Hence the rate of runoff at a moment t is directly obtained as the difference between the ordinates of the two S-curves shifted horizontally by a distance equal to storm duration t_d (see Fig. 1).

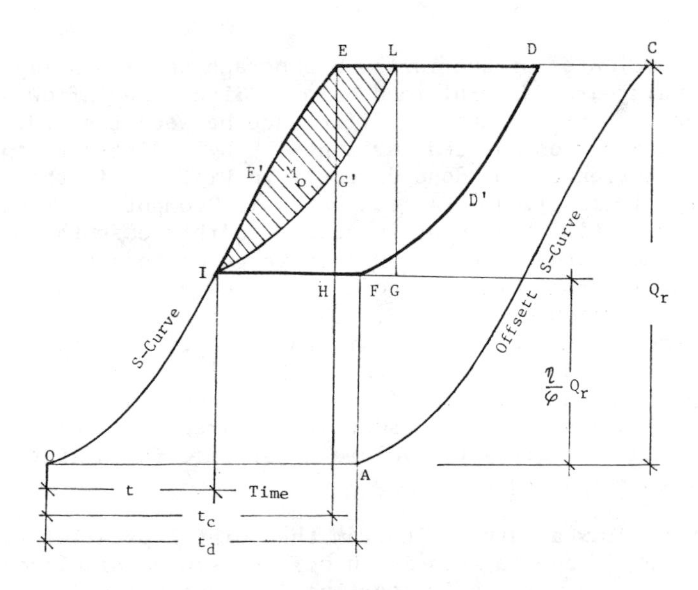

Fig. 1. The general case in determination of the water stored in the basin during a storm of duration t_d

Let us express in general the storm intensity as $i = i_r \cdot \phi(t_d)$ for a selected frequency in which i_r is the intensity of a reference storm in l/sec/ha for a certain duration $(t_d)_r$ and ϕ is a function of storm duration t_d. If the S-curve is plotted once for a storm i_r, there is no need to plot the S-curves for other storms. In order to get the S-curves for other storms, it is enough to multiply the ordinates of the S-curve corresponding to the storm i_r by ϕ or to use a discharge scale increased by $1/\phi$. i_r will be called as reference storm.

Let the S-curve resulting from the storm of uniform intensity i_r and duration $(t_d)_r$ be plotted as shown in Fig. 1 for a location where the retention basin will be constructed. Let us express the outflow from the basin as $Q_c = \eta \cdot Q_r$ where

$$Q_r = i_r \sum_{n=1}^{n=n} C_n A_n \tag{1}$$

Q_r is the maximum ordinate of the original S-curve in ℓ/sec, A_n is the surface area of the various parts of the drainage basin in ha, η is a dimensionless quantity, and C_n is the runoff coefficient. The storage for a storm with duration t_d is obtained by subtracting the value

$$\overline{FA} = \frac{\eta}{\phi} \cdot Q_r \tag{2}$$

from the ordinates of the inflow hydrograph and measuring the corresponding area as explained below. Since the inflow hydrograph is obtained as the difference between the ordinates of the two S-curves shifted horizontally by a distance t_d, this subtraction can be done directly in Fig. 1. If the storm intensity i is given, $\overline{FA}$ can be easily computed. There is no storage till the point I because in this case the outflow is greater than the inflow. If we go on to subtract the value $\overline{FA}$ from the ordinates of the inflow hydrograph, we obtain the horizontal line $\overline{IF}$. The area above this line denotes the volume to be stored in the retention basin. Thereafter we plot the curve FD'D which is obtained by shifting the S-curve vertically by a distance $\overline{FA}$ and horizontally by a distance t_d, so that the area IFD'DEE'I multiplied by ϕ gives the volume of storage for a storm with the duration t_d.

Let us draw a curve IG'L from the point I parallel to the curve FD'D. Of course this is an offset S-curve displaced horizontally and vertically starting from the point I. If we denote the hatched area with M_o, so the area IFD'DEE'I becomes equal to the sum of the parallelogram IFD'DLG'I and M_o. From Fig. 1 we can write:

$$\overline{IF} = t_d - t; \qquad \overline{HE} = (1 - \frac{\eta}{\phi}) \cdot Q_r$$

Hence the storage becomes:

$$V = (\overline{IF} \cdot \overline{HE} + M_o) \cdot \phi = (t_d - t) Q_r \cdot (1 - \frac{\eta}{\phi}) \phi + M_o \cdot \phi \tag{3}$$

or

$$V = (t_d - t)\ Q_r\ (\phi - \eta) + M_o \cdot \phi \tag{4}$$

Dividing both sides of Eq. 4 by Q_r and substituting

$$\frac{V}{Q_r} = B \quad \text{and} \quad \frac{M_o}{Q_r} = N_o$$

the following equation is obtained:

$$B = \frac{V}{Q_r} = (t_d - t)\ (\phi - \eta) + N_o \cdot \phi \tag{5}$$

$$B = \phi [t_d (1 - \frac{\eta}{\phi}) - t(1 - \frac{\eta}{\phi}) + N_o] \tag{6a}$$

$$B = \phi [(t_d - t)\ (1 - \frac{\eta}{\phi}) + N_o] \tag{6b}$$

The following two special cases will be discussed:

Special Case I. Let the point D of the curve FD'D coincide with the end point of the S-curve OIE ($t_d = t_k$ in Fig. 2).

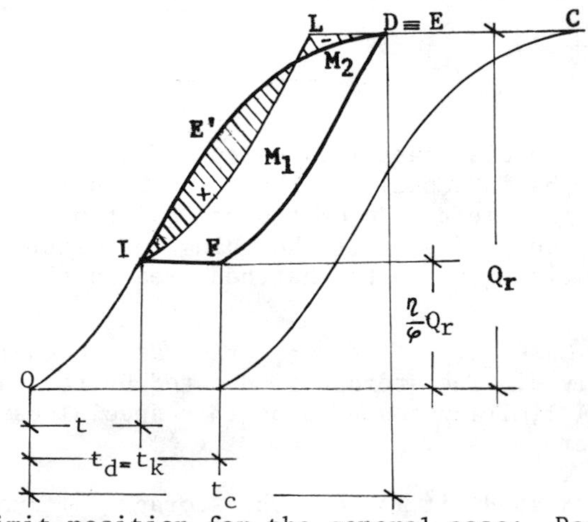

Fig. 2. Limit position for the general case: Point D coincides with the end point of the S-curve OIE

This is a limiting case and even here Eq. 3 can be applied, that is

$$V = IFD'DLI + M_1 - M_2 \tag{7}$$

where $M_o = M_1 - M_2$.

But if the case seen in Fig. 3 is under consideration ($t_d<t_k$) Eq. 3 cannot be applied because here the storage corresponds to the hatched area. If the area shown in black in Fig. 3 is subtracted from the area M_2, which has a negative sign, then Eq. 3 can be applied. Thus, it can be stated that if $t_d<t_k$, the volume of storage will be directly found by measuring the hatched area in Fig. 3.

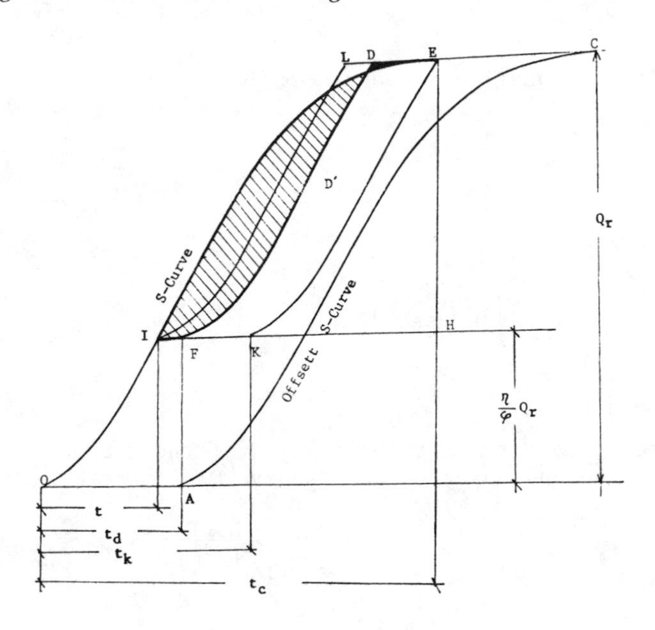

Fig. 3. Special case I in calculation of retention basin capacity: The point D is located on the left side of the end point of the S-curve ($t_d<t_k$). The volume of water stored is equal to the hatched area in this case

Special Case II. If $t_d < t$, the case shown in Fig. 4 will be involved. The storage is denoted by the hatched area in Fig. 4. A limiting position of this special case is shown in Fig. 5 where $V = 0$.

From this study it is seen the storage is found in general by applying Eq. 3 and Eq. 6a or Eq. 6b and, in special cases, by measuring the hatched areas shown in Figs. 3 and 4.

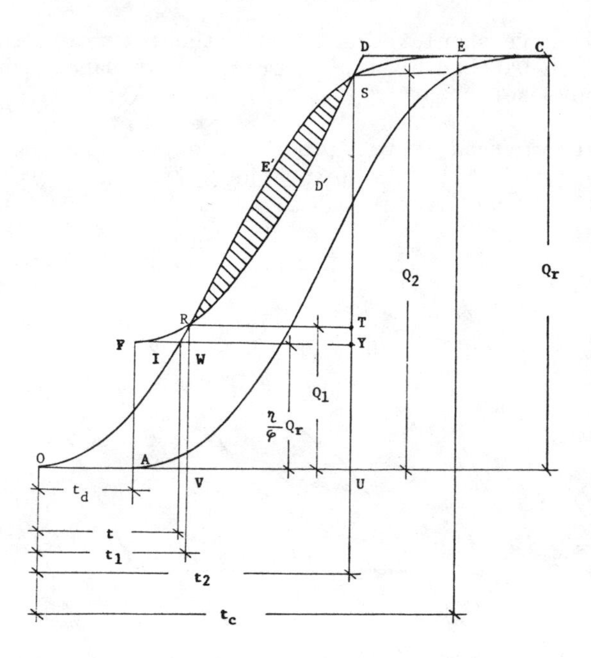

Fig. 4. Special case II in calculation of retention basin capacity: The point F is located on the left side of the S-curve. The volume of water stored is obtained by measuring the hatched area

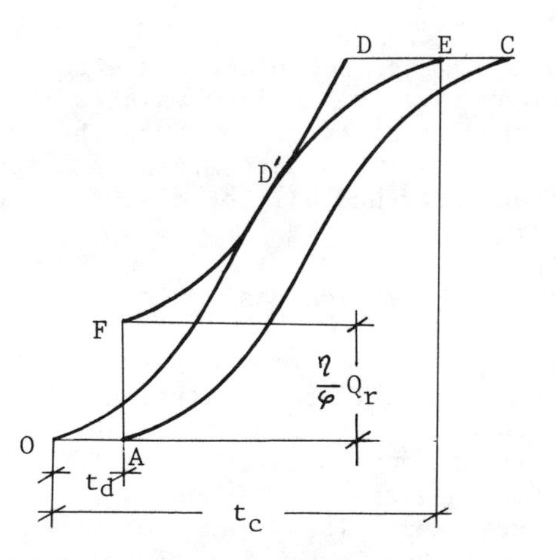

Fig. 5. Limiting position for the special case II. The greatest ordinate of the runoff hydrograph is equal to the retention basin outflow, hence the required capacity of the retention basin is zero

The S-curve should be shifted without turning in order to plot these hatched areas. A transparent paper can be used for this purpose.

A better method is to plot the S-curve on a template as shown in Fig. 6. With its help, the shifted S-curves can be drawn easily in any place.

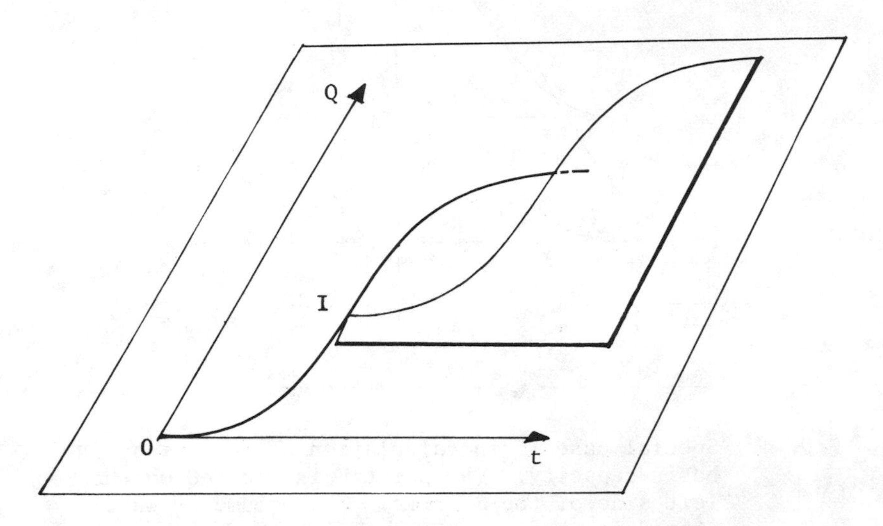

Fig. 6. Plotting the S-curve on a cardboard and cutting it away in order to have its form

The value of B in Eq. 6 depends upon the S-curve and the storm duration t_d. The maximum value of B gives the required capacity of the retention basin. How the required capacity of the retention basin is determined as a function of the S-curve of the watershed will be discussed below and applications illustrated.

Determination of Retention Basin Capacity

M_o or N_o is a function of storm duration t_d for a certain value of η. If the analytical expression of the S-curve is known, the storm duration which gives the required capacity of the retention basin can be determined from $(dB/dt_d) = 0$, in a similar way as explained above. However, in general the S-curve is a line, for which an analytical expression cannot be given. Therefore, a graphical approach to the problem is necessary in this case. The values η/ϕ and t can be taken from the S-curve for the point where the retention basin will be constructed. The storm duration t_d corresponding to these values, and the expression (t_d-t) $[1-(\eta/\phi)]$ in Eq. 6b can then be easily computed. For example if $\eta=0,4$; $\eta/\phi = 0.82$, from $\phi = 24/(t_d+9)$, $t_d = 40,5$ min is found and if $t = 30,5$ min, $(t_d-t)[1-(\eta/\phi)] = 108$ sec is obtained. By adding N_o to the expression $(t_d-f)(1 - \eta/\phi)$, we determine the function $B/\phi=(t_d-t)[1-(\eta/\phi)] + N_o$ for various storm durations t_d (Fig. 7). When t_d changes, the value of this function also changes. However, ϕ is inversely proportional to t_d. Hence the product $(B/\phi)\phi$ which is equal to B, becomes maximum for a certain value of t_d. We therefore find the point of the curve $B/\phi = f(t_d)$ shown in Fig. 7, for which the value B becomes maximum. The function $B/\phi = f(t_d)$ plotted for constant values of B can be used in order to find the value B for any point on curve $B = f(t_d)$. For this purpose the straight lines for B = 50,100,150,... seconds corresponding to the function

$$\frac{B}{\phi} = \frac{50(t_d+9)}{24}, \quad \frac{B}{\phi} = \frac{100(t_d+9)}{24}, \ldots \tag{8}$$

Have been shown in Fig. 7. These straight lines cross the axis t_d at the point M for which $t_d = -9$ min. For $t_d=15$ min, we obtain $\phi=1$ and $B/\phi=B$. If such a diagram, plotted on a transparent paper as explained above, is placed on the curve $B/\phi = f(t_d)$ in such a way that the axes on both figures coincide, the point of the curve which is tangent to the uppermost line of the diagram gives the maximum B. For example, the maximum value of B is 85 seconds in Fig. 7. For $\phi = 24/(t_d+9)$, the tangent drawn from the point M to the curve $B/\phi = f(t_d)$ gives the required value of B.

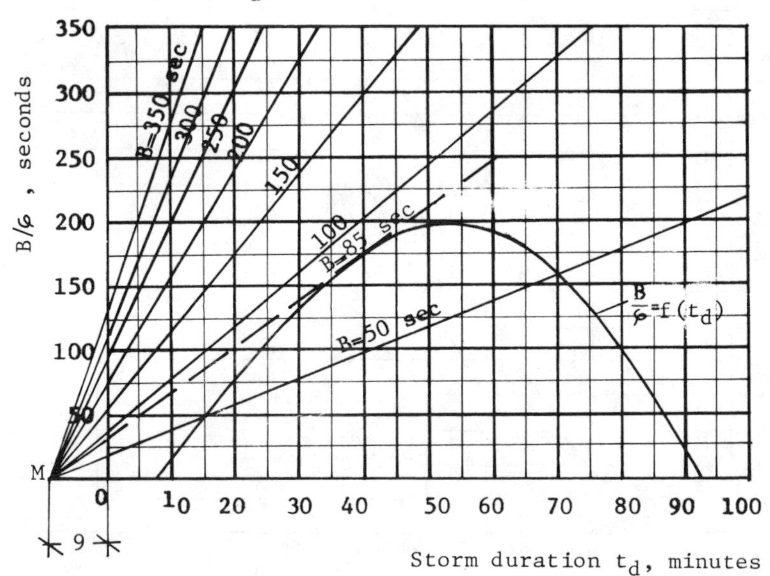

Fig. 7. Determination of the point of the curve of $B/\phi=f(t_d)$, where the value of B becomes maximum

Research on various urban areas have shown that S-curves in drainage areas are in between the boundaries seen in Fig. 8. Forcheimer (1930) gave the formula

$$h(t) = H \cdot \left(\frac{t}{t_r}\right)^m \cdot e^{m\left(1 - \frac{t}{t_r}\right)} \tag{9}$$

where $h(t)$ is the water depth at any moment t, and H is the maximum water depth at the moment t_r. Later Schoenefeldt (1937) accepted this formula in his study on runoff hydrograph of a storm sewerage system [11]. Randolf (1958) also used the same expression in his doctoral thesis on determination of retention basin capacity [9]. When Eq. 9 is applied to urban drainage, a linear relationship is assumed between water depth and discharge. Hence Eq. 9 becomes:

$$Q(t) = Q_r \left(\frac{t}{t_c}\right)^m \cdot e^{m\left(1 - \frac{t}{t_c}\right)} \tag{10}$$

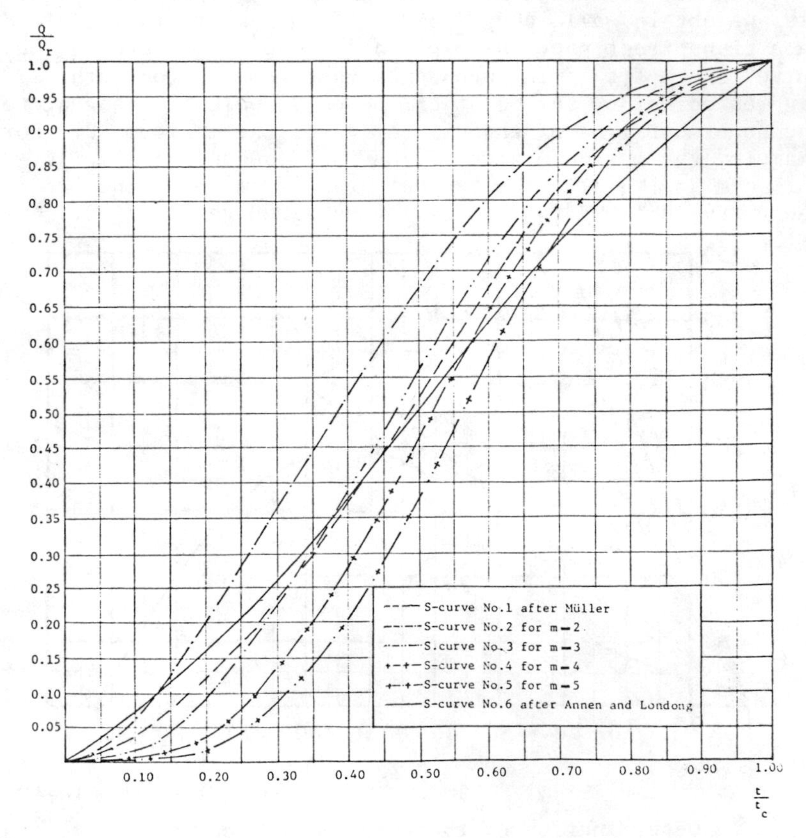

Fig. 8. Variation of S-curves in urban areas

where t_c denotes time of concentration and m is a constant which changes between m=2 and m=5. Q/Q_r is computed from Eq. 10 for various values of m and the dimensionless S-curves are plotted in Fig. 8 by means of these values. The dimensionless S-curves by Müller (1939), Anen and Londong (1960) are also shown in Fig. 8 for the purpose of comparison [1]. It is seen from Fig. 8 that the S-curves for m=2 and m=5 bound the curves given by Müller, Annen and Londong.

The graphical method developed in this paper has been applied to the dimensionless S-curves 1 to 5 shown in Fig. 8. For this purpose the curves were plotted on a template. With these templates, the shifted S-curves could be drawn easily in any place. Then the hatched areas M_o were measured by a planimeter and the B/ϕ values determined. For the given values of t_c and η, the function of B/ϕ has been plotted for various storm durations t_d (see Fig. 9 prepared for

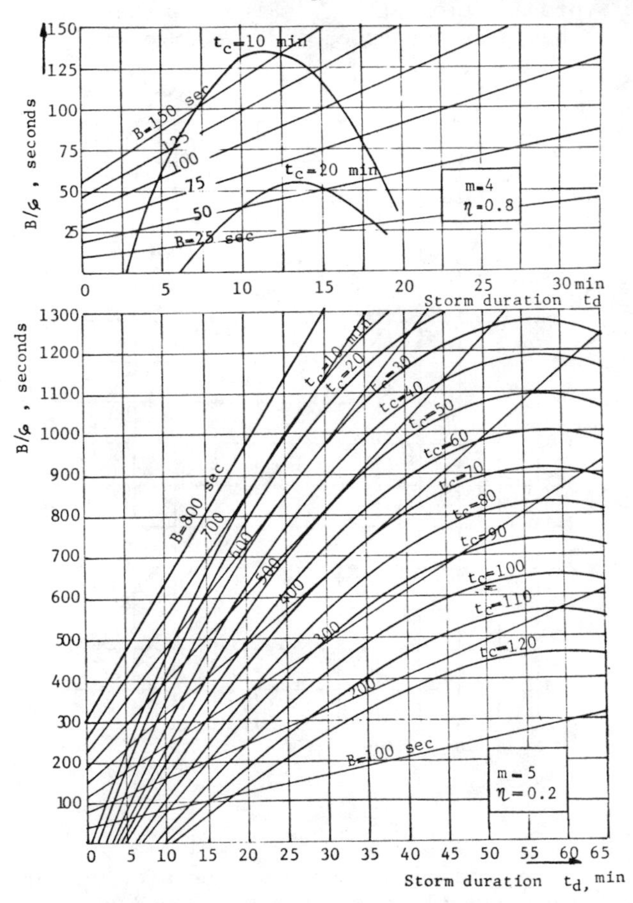

Fig. 9. Investigation of storm duration $(t_d)_e$ for which the volume of water stored in the basin becomes maximum

$\eta = 0.8$ and $\eta = 0.2$ corresponding to S-curve No. 4 and No. 5 respectively, as an example. Complete data tables and graphs are available in Muslu, 1966). The tangent drawn from the point M to the curve corresponding to a certain concentration time t_c gives the value B and the storm duration $(t_d)_e$ for the required capacity of retention basin. Some of the functions of $B = f(\eta, t_c)$ for different dimensionless S-curves are plotted in Figs. 10, 11, 12, 13 and 14, as an example.

The retention basin capacities and the storm durations $(t_d)_e$ computed in this way are obtained without neglecting any value. However their determination requires the measuring of many areas by planimeter. Therefore the number of areas to be measured should be minimized. For this purpose, the effect of the function of N_o upon the volume of water stored in the basin has been studied and a simple procedure developed to obtain the storm duration $(t_d)_e$. Firstly, the function of N_o in Eq. 6a has been neglected and the error made by this

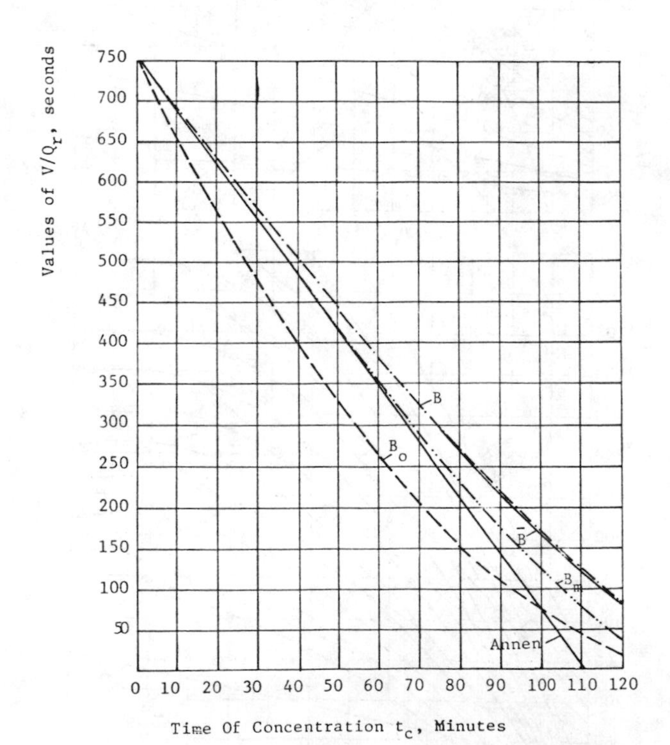

Fig. 10. The variation of the values of V/Q_r for dimensionless S-curve No. 1 under various assumptions, when $\eta = 0.2$

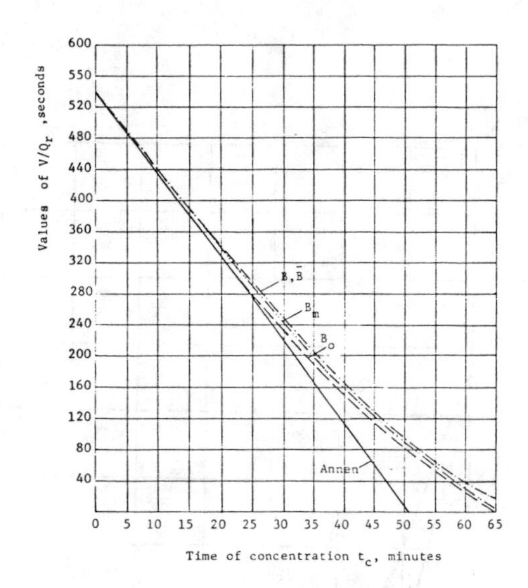

Fig. 11. The variation of the values of V/Q_r for dimensionless S-curve No. 2 under various assumptions, when $\eta = 0.4$

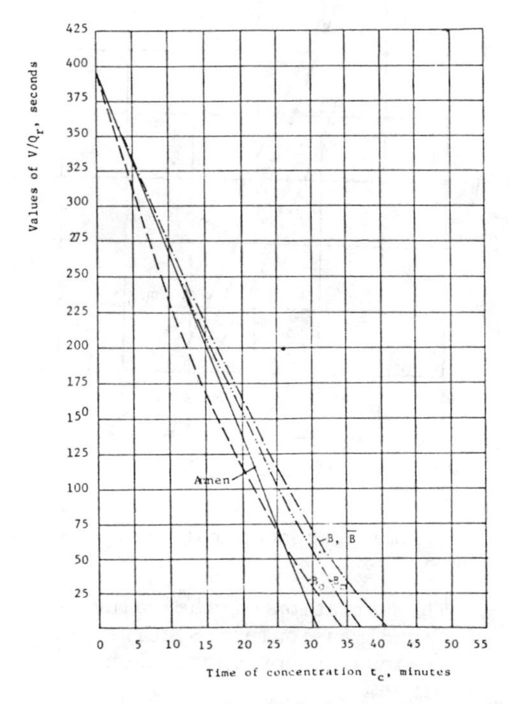

Fig. 12. The variation of the values of V/Q_r for dimensionless S-curve No. 3 under various assumptions, when $\eta = 0.6$

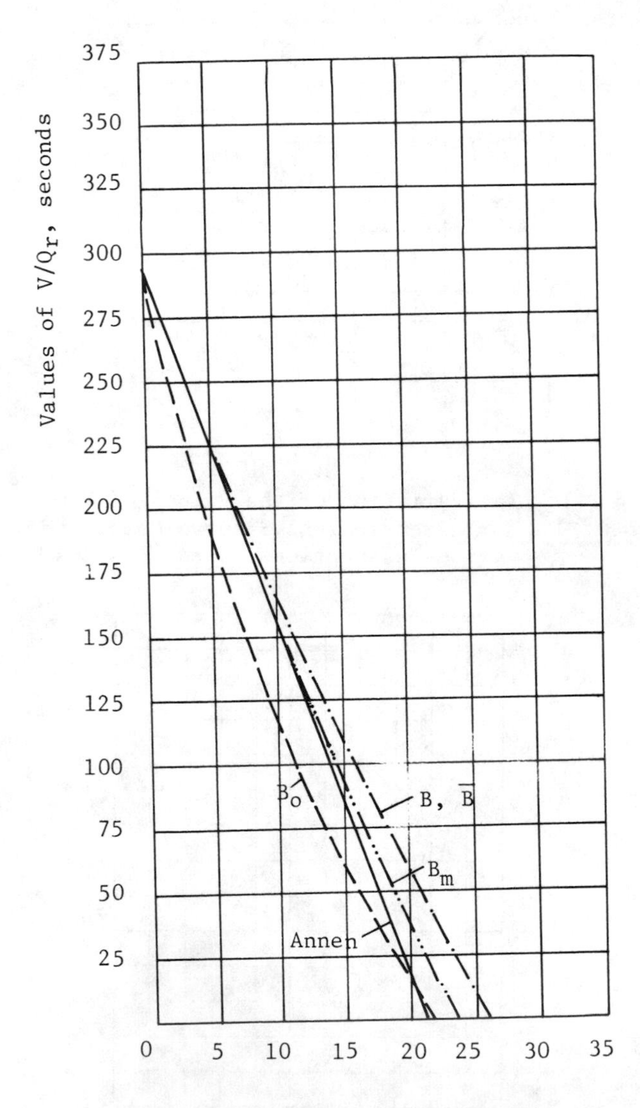

Fig. 13. The variation of the values of V/Q_r for dimensionless S-curve No. 4 under various assumptions, when $\eta = 0.8$

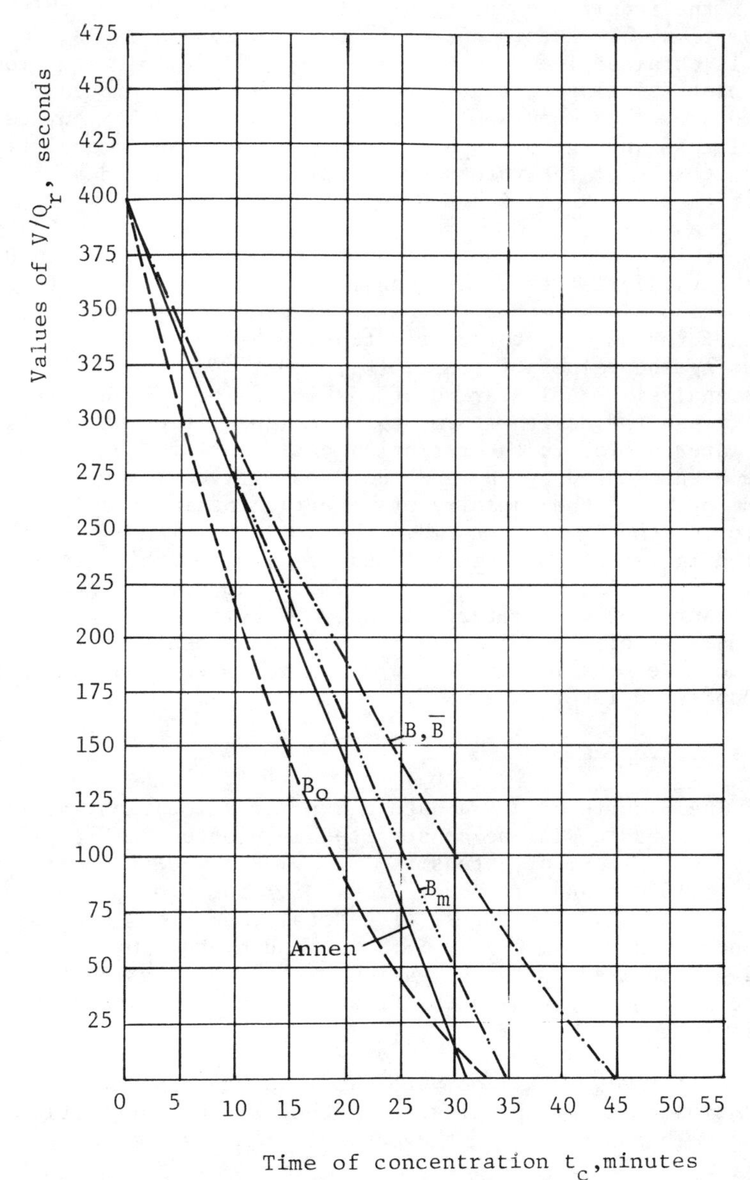

Fig. 14. The variation of the values of V/Q_r for dimensionless S-curve No. 5 under various assumptions, when $\eta = 0.6$

assumption has been found. B_o denotes the quantity which gives the required capacity of retention basin under this assumption. The storm duration $(t_d)_o$ which makes B_o maximum is obtained from the equation of $(dB_o/dt_d) = 0$ or from the plots similar to that shown in Fig. 9. The procedure outlined above has been applied to the dimensionless S-curves No. 1 to 5 in Fig. 8 graphically and S-curve No. 6 analytically. Some of the values of B_o are also plotted as a function of t_c in Figs. 10-14.

It is seen from these figures that the quantity of B_o is always smaller than the quantity of B because of the assumption of $N_o = 0$. Further, the problem of the determination of the design storm for retention basin has been studied by using the values of $(t_d)_o$, $(t_d)_e$, B_o and B obtained in this analysis. A linear relationship has been found between $(t_d)_o$ and the design storm $\overline{t}_d$. The quantities which give the water stored in the retention basin for $\overline{t}_d$ and $(t_d)_o$ have been denoted by $\overline{B}$ and B_m, respectively. Here the storm by which the capacity of retention basin is computed, is termed the "design storm". The volume of runoff to be stored in retention basin becomes maximum for this storm. $(t_d)_e$ is replaced by $\overline{t}_d$ to minimize the number of areas to be measured by planimeter. Ideally $\overline{t}_d$ is equal to $(t_d)_e$. The details of these computations and the results of the study on the determination of design storm for retention basins will be published later on.

Illustrative Example

The S-curve of a drainage basin for a location where the retention basin will be constructed is plotted in Fig. 15 for a reference storm of intensity i_r = 178 l/sec/ha, duration $(t_d)_r$ = 15 min and frequency f = 0.2. The time of concentration is t_c = 44.1 minutes. The total impervious area of the drainage basin is $\Sigma C_n A_n$ = 33.2 ha and hence, Q_r = 178 x 33.2 = 5920 l/sec. The dimensionless coordinates of the S-curve have been given in the first and second columns of Table 1.

At first we shall compute the required capacity of the retention basin by applying the method without neglecting any value. The outflow from the basin is Q_{out} = 2368 l/sec. Thus,

$$\eta = Q_{out}/Q_r = 2368/5920 = 0.4$$

The storm duration t_d has been found using the values t and Q/Q_r given in Column 1 and Column 2 of Table 1 as explained above. The values of t_d calculated in this way are given in Column 3 of Table 1. The V/ϕ values, i.e., the

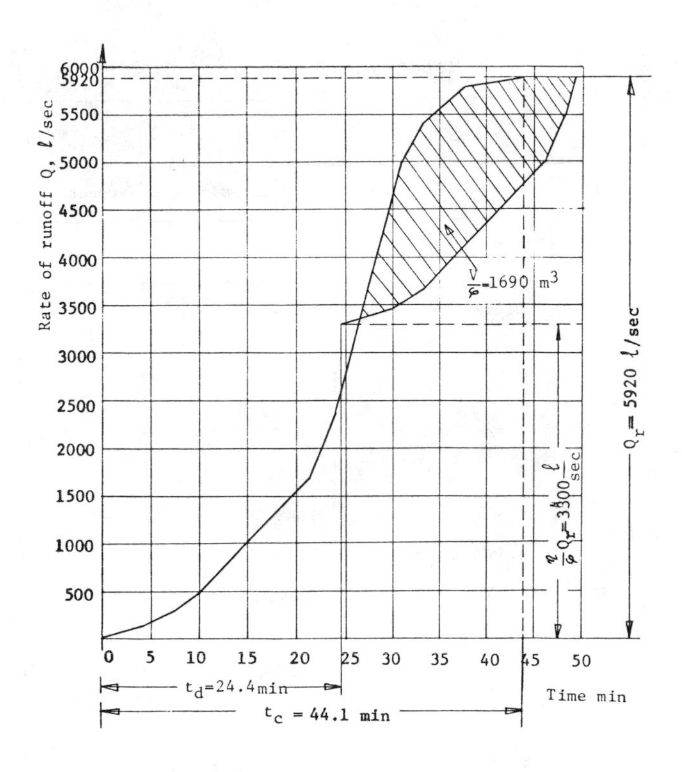

Fig. 15. Determination of the required capacity of retention basin in the illustrative example according to the design storm

volume of storage V for this storm duration t_d divided by ϕ has been measured with a planimeter and written in Column 4 of Table 1. Then $V/\phi = f(t_d)$ is plotted in Fig. 16 and the required capacity of the basin is obtained as 1360 m by drawing a straight line from the point M, which is tangent to the V/ϕ-curve.

This value obtained is exactly consistent with rational unit hydrograph theory. It has been necessary to measure 8 different areas in order to get this result.

Summary and Conclusion

The problem of computing the capacity of the retention basin has been studied for half a century. Those who studied this problem either followed an excessively simplified approach or used runoff hydrographs for various storms to find the maximum storage. In each hydrograph the volume of the storage has to be determined, which is a very cumbersome and time consuming procedure.

Table 1. Design of Retention Basin in Illustrative Example

t	$\frac{\eta}{\varphi} = \frac{Q}{Q_r}$	t_d	$\frac{V}{\varphi}$	$(t_d - t)$	$B_o/\varphi = (t_d - t)(1 - \frac{\eta}{\varphi})$
minutes	-	minutes	m^3	minutes	Seconds
1	2	3	4	5	6
3.9	0.022	-	-	-	-
7.8	0.048	-	-	-	-
10.1	0.081	-	-	-	-
18.8	0.228	4.7	157.5	-	-
21.3	0.282	7.9	521.0	-	-
23.9	0.395	14.7	1185.0	-	-
25.1	0.466	19.0	1570.0	-	-
27.1	0.600	27.0	1810.0	-	-
27.7	0.648	29.8	-	2.1	44
28.3	0.695	32.7	1485.0	4.4	84
29.6	0.760	36.7	-	7.1	102
30.8	0.825	40.5	965.0	9.7	102
32.2	0.870	43.2	-	11.0	86
33.6	0.915	45.8	428.0	12.2	62
35.9	0.947	47.8	-	11.9	38
38.1	0.981	49.8	-	11.7	13
41.1	0.991	50.5	-	9.4	5
44.1	1.000	51.0	-	6.9	0

In this paper a new method has been developed to find the storages during various storms by means of a single S-curve plotted for a reference storm i_r. The numerical example given above illustrated this method.

This method has been applied to six S-curves expressed in dimensionless variables, and in each case the storm which gives the maximum storage has been computed. The computed capacities of the retention basins are dependent on the assumptions implied in the unit hydrograph theory. In addition, a procedure has been proposed which minimizes the number of areas necessary to measure. In this method results of sufficient accuracy can be found by measuring only one area.

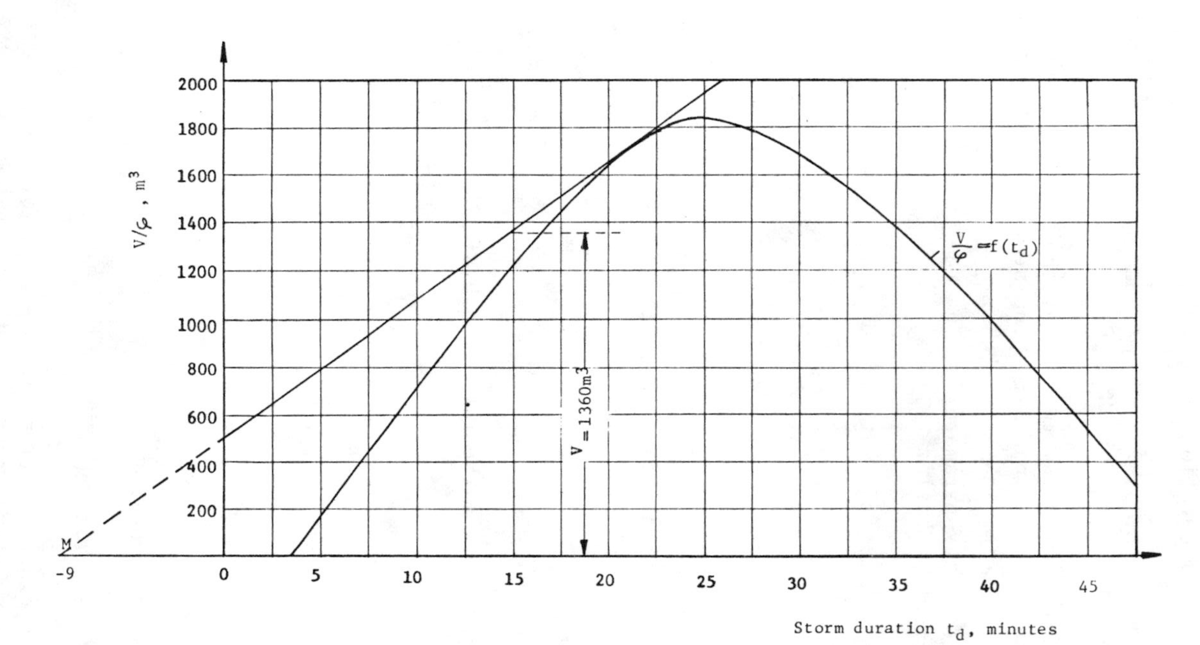

Fig. 16. Determination of the exact value of the required capacity of the retention basin in the illustrative example

Notations

A	Surface area of the drainage basin;
B_m	A parameter giving the volume of the basin for a storm of duration $(t_d)_o$;
B_o	A parameter giving the maximum value of B under the assumption of $N_o = 0$;
$\bar{B}$	A capacity parameter giving the volume of water stored in the basin for the storm of duration $\bar{t}_d$;
$B = \frac{V}{Q_r}$	A capacity parameter giving the exact volume of retention basin;
C	Runoff coefficient;
i	Storm intensity;
i_r	intensity of the reference storm;
m	A constant in Eq. 10;
M_o	The hatched area in Fig. 1;
N_o	M_o/Q_r;
Q	Rate of runoff
Q_c	Outflow from the retention basin;
Q_r	$\sum_{n=1}^{n=n} C_n i_r A_n$;
t_c	Time of concentration;
t	Time;
$(t_d)_o$	Storm duration for which B_o becomes maximum;
$\bar{t}_d$	Design storm;
t_d	Storm duration;
$(t_d)_e$	The exact duration of the storm for which the volume of water stored in the basin becomes maximum;

$(t_d)_r$ Duration of the reference storm;

V Volume of retention basin;

$\bar{V}$ Volume of retention basin for storm of duration $\bar{t}_d$;

η Q_c/Q_r ;

ϕ A function expressing storm intensity in term of duration.

References

[1] Annen, G. and Londong, D., 1960. "Vergleichender Beitrag zur Bemessungverfahren von Rückhaltebecken," Technische-Wissenschaftliche Mitteilungen Emschergenossenschaft-Lippeeverband, H. 3.

[2] Forchheimer, Ph., 1930. "Hydraulik," Teubner, Berlin and Leibzig.

[3] Imhoff, K., 1972. "Taschenbuch der Stadtentwässerung, R. Oldenbourg Verlag.

[4] Lautrich, R., 1956. "Die Graphische Ermittlung des Speicherraums fur Hochbehälter, Sicher und Regenausgleichbecken und die Grenzwerte der Verzogerung," Wasser und Boden, 344.

[5] Müller, G., 1939. "Regenwasseraufhaltebecken in Städtischen Entwasserungsnetzen," Beihefte zum Ges. Ing., Reihe II, H. 11-19.

[6] Müller-Newhaus, G., 1947. "Die Ermittlung von Regenwasserabflussen," Ges. Ing., 143.

[7] Muslu, Y., 1966. "Meskûn Bölge Hidrologisinde Hesap Yagmurunun ve Akim Dalgasinin Tespiti Üzerine Bir Arastirma," Doctoral Thesis in Turkish.

[8] Ordon, C. J., 1974. "Volume of Storm Water Retention Basins," Journal of the Environmental Engineering Division, ASCE, Vol. 100, No. EE5.

[9] Randolf, R., 1958. "Zur Berechnung von Regenwasseraufhaltebecken in Städtischen Entwässerungsnetzen," Dis. T.H. Dresden.

[10] Reinhold, F., 1940. "Regenspenden in Deutschland," Archiv für Wasserwirtschaft, Nr. 56.

[11] Schoenefeldt, O., 1937. "Die Abflusswelle bei der Stadtentwässerung," Gesundheits Ingenieure H. 25, 402.

[12] Yrjanainen, G. and Warren, A. W., 1973. "A Simple Method for Retention Basin Design, Water and Sewage Works," Vol. 120, No. 12.

LINEARLY STRUCTURED MODELS IN HYDROLOGY

by

S. Ramaseshan
Department of Civil Engineering
Indian Institute of Technology
KANPUR, U.P. 208016, INDIA

Abstract

Several hydrologic models dealing with basin runoff, channel flow and groundwater flow consist of a set of linear algebraic equations in the time domain or in a transformed domain. They are defined to be linear in structure.

Mathematical models of hydrologic systems are assumed empirically or derived from a conceptual approach to the physical nature of the process. They are represented by conceptual models, differential and convolution equations, differential equations and discrete kernals, time series models and state variable models. The models may also take into consideration the presence of errors and the probabilistic nature of the process. When the models have a linear structure, the methods of identification, analysis or detection for the several models are generally similar.

The models may be classified into deterministic models when randomness of variables and presence of errors are not taken into consideration, statistical models when data and modelling errors are taken into consideration and essentially a regression procedure is used; and stochastic models which explicitly take into consideration the sequential probabilistic nature of the process. As similar models may be obtained from different approaches, the classification is to be based on the procedure used in fitting the model.

The suggested classification is illustrated with examples of linearly structured hydrologic models. It seems to be viable and meaningful and it may help in a good understanding of the limitations and advantages of the fitted model.

Introduction

Hydrologic systems are sequential dynamic physical systems with water as the throughput. They are distributed in space and their characteristics vary in time. Hence they are to be represented by a set of partial differential equations (Freeze and Harlan, 1969). Yet, from considerations of simplicity, convenience and data availability, they are often represented by lumped system models and sets of differential, integral, difference or algebraic equations (Amorocho, 1970; Eagleson, 1969). A number of such models are represented or approximated by linear algebraic equations and are solved by methods of linear algebra. They are referred to herein as linearly structured models. These models are generally linear in the statistical regression sense, and may or may not be linear in the system sense (Clarke, 1973).

Linearly structured models have been studied quite extensively with reference to their analysis and identification (Astrom and Eykhoff, 1971; Box and Jenkins, 1970; Dooge, 1973; Eagleson, 1969; Leondes, 1976) and sensitivity (Rao and Usul, 1973). They are also useful in detection of inputs to the hydrologic system (Ramaseshan et al., 1973). Linearly structured models have also been used quite widely in hydrology and so are considered here in detail.

Analogous Nature of Hydrologic Models

Hydrologic models are either assumed arbitrarily or derived conceptually from deterministic, statistical or stochastic considerations. Quite often several models are similar in mathematical structure and form, though divergent in origin (Yevjevich, 1974). There is a correspondence between the different representations of a hydrologic system. Delleur and Rao (1971), Eagleson et al. (1965), O'Connor (1976) and Ramaseshan et al. (1973) consider the correspondence between continuous and discrete models; Chow (1964), Chow and Kulandaiswamy (1971), Dooge (1967) and Ramaseshan et al. (1973) consider the correspondence between conceptual and system models; and Chow et al. (1975) indicate the correspondence between input-output and state variable representations. Furthermore, Delleur and Rao (1973), Dooge (1972), Dooge (1973), Klemes and Boruvka (1975), Quimpo (1973), Spolia and Chander (1974) and Venetis (1969) consider the correspondence between deterministic system models and time series models. Ramaseshan et al. (1973) and O'Connor (1976) consider the correspondence between discrete, conceptual and time series models. It is very desirable to keep in the mind the analogous nature of models assumed or derived from different considerations while analysing data or interpreting results.

Classification of Hydrologic Models

When different approaches lead to similar models, they may provide a comparative insight to the process. Yet what distinguishes them is the method of analysis used in modelling the process using available data. It is suggested that when the formulation and fitting of a model are based on different approaches, it is the fitting procedure that should be used to classify the model. This paper deals with linearly structured models in hydrology and their classification into deterministic, statistical and stochastic models.

Deterministic System Models

These generally deal with transient and "short" duration hydrologic events. They do not take into consideration the stochastic nature of the processes and further assume that the data and model are free of errors. They may treat time (or space, whichever is the independent variable) as a continuous variable and so involve differential and integral equations, or as a discrete variable and so may involve difference, discrete and/or algebraic equations. They may be represented by input-output or transfer function models (Amorocho, 1970; Eagleson, 1969) that relate the outputs of a system to the present and past inputs to the system; or state variable models (Chow et al., 1975; Duong et al., 1975; Muzik, 1974) that characterize the state of the system by so-called "state variables", say, storages and their derivatives, and relate the outputs and rate of change of state variables to the present and past values of the states and outputs.

Deterministic linear models in hydrology. A number of linearly structured deterministic models have been used in hydrology. For example, consider an initially relaxed linear system with input R, output Q and unit response function U which are vector functions of time. For discrete data,

$$Q = [U]\ R \tag{1}$$

where [ ] indicates a matrix and [U] is a lower half Toeplitz matrix, LHTM (Ramaseshan et al., 1973) with U as the first column. It can also be represented in the alternative form

$$Q = [R]\ U \tag{1a}$$

where [R] is a LHTM with R as the first column. This model has been used in hydrology for basins (Chow, 1964), for channels (Dooge and Harley, 1967; Eagleson, 1969; Keefer and McQuivey, 1974; Sauer, 1973) and for groundwater systems (Moench and Kisiel, 1969; Morel-Seytoux and Daly, 1975).

Amorocho and Brandstetter (1967) and Bidwell (1971) considered the nonlinear functional representation of the basin

$$Q = [X]\ U \tag{2}$$

where U is a Mx1 column vector of M unknown ordinates of a nonlinear system response function, and [X] is a NxM rectangular matrix of past system inputs and their products. This is also linear in structure and Eq. (1a) is a special case of Eq. (2).

Chow et al. (1975) and Leondes (1976) give the basic form of the state variable model. They are considered in greater detail later.

Equations (1a) and (2) are sets of linear equations. When the number of unknowns in the system response vector U are equal to the number of independent equations in the set, it may be assumed that there are no errors and the system response vector U can be estimated by the relationships

$$U = [R]^{-1} Q \quad \text{or} \qquad (3)$$
$$U = [X]^{-1} Q$$

Equation (3) can be solved by direct elimination or synthetic division (Eagleson et al., 1965) or equivalently by z-transform (Kulandaiswamy, 1966; Choube, 1971; Delleur and Rao, 1971) or other orthogonal transforms including polynomials (Papazafiriou, 1975); Fourier series (O'Donnell, 1960); Fourier and Fast Fourier transforms (Blank and Delleur, 1968), Laplace transform (Diskin, 1964); Laguerne or Meixner transform (Dooge, 1964; Amorocho and Brandstetter, 1967); Walsh functions (Emsellem and De Marsily, 1971); etc. Since they are all mathematically equivalent, the results should be independent of the specific procedure except for numerical approximations, rounding off, and errors in the different procedures. If high frequency or random data errors are suspected, they may be reduced by using smoothing and filtering techniques in the analysis (Delleur and Rao, 1974; Kavvas and Schultz, 1973; Kishi, 1971).

Statistical Models in Hydrology

In general errors are always present in data, in modelling and in numerical analysis, and they lead to large fluctuations and instabilities in the solution of Eq. (3). It is hence necessary to take into consideration these errors by formulating the statistical model, say,

$$Q = [R]\, U + \varepsilon \quad \text{or}$$
$$= [X]\, U + \varepsilon \qquad (4)$$

where ε is the error vector. This model may be assumed empirically or else derived on a conceptual basis from the physical nature of the process (Clarke, 1973; Chow et al., 1975).

The system response vector from Eq. (3) has a large number of parameters which may vary from one data set to another for the same system. To facilitate study of the time-variant and quasilinear behavior of the system and from considerations of parsimony in parameters, some assumptions about the system can be made. These in turn lead to differential or difference equations of finite small order or a finite small number of state variables. Thus Chow and Kulandaiswamy (1971) assumed a finite order differential equation and Ramaseshan et al. (1973) assumed a finite order difference equation for the basin system, and Ramaseshan (1976) assumed a finite order difference equation for a channel on the basis of Muskingam equation. More specifically conceptual models of various types may be assumed for the system in terms of a small number of basic components and parameters (Chow, 1964) and they correspond essentially to fitting specified functional forms to the unit response function.

The statistical state variable representation of a system is given by Chow et al. (1975), Kashyap and Rao (1973), Leondes (1976) and Lettenmaier and Burges (1976). When they involve essentially regression analysis, they are to be considered as statistical models.

Parameters of statistical models can be estimated from historical data by statistical procedures including method of moments, maximum likelihood method and the method of least squares (Chow et al., 1975; Fahlbrush and Muir, 1973; Yevjevich, 1972; Yevjevich, 1976) and Bayesian methods (Duckstein and Davis, 1976); or programming techniques. The method of least squares deals with the minimization of sum of squared errors,

$$S_e^2 = \varepsilon^T \cdot \varepsilon \tag{5}$$

where superscript T stands for the transpose. For a linearly structured model with independent or correlated random errors, standard procedures are available for parameter estimation (Box and Jenkins, 1970; Chow et al., 1975; Leondes, 1976). Method of least squares has been used for modelling hydrologic basins by Bidwell (1971), Chow et al. (1975), Papazafiriou (1975), Kashyap and Rao (1973), Singh (1976) and Snyder (1955) and for channel routing using Muskingum model by Ramaseshan (1976).

Parameter estimation may be based on other criteria concerning errors: for example, the minimization of

i. sum of absolute errors, $S_1 = \Sigma|\varepsilon|$ or

ii. maximum absolute error, $\varepsilon_{max} = \max_i |\varepsilon_i|$ (6)

Additional constraints based on conservation of matter, shape or functional form of the unit system response, etc., may also be specified. For linearly structured models, if these constraints are also linear with respect to parameters, they lead to a linear programming problem (Chow et al., 1975; Deininger, 1969; Eagleson et al., 1965; Neuman and DeMarsily, 1976). If the constraints are nonlinear and/or the model is not linearly structured, both the method of least squares (Eq. 5) and the programming models (Eq. 6) lead to nonlinear programming problems (Clarke, 1973; Dawdy and O'Donnell, 1965; Diskin and Boneh, 1973; Diskin and Boneh, 1975; Ramaseshan and Anant, 1974; Ramaseshan et al., 1973).

Stochastic Models in Hydrology

Stochastic models take into consideration the stochastic (i.e., sequential and probabilistic) nature of the process and hence data and modelling errors as well. They use explicitly the auto and cross-correlation structure of the data, if any, in the modelling of the process. They include:

i. time series models (Box and Jenkins, 1970; Clarke, 1973; Kisiel, 1969; Yevjevich, 1976);

ii. queuing, inventory and storage models (Lloyd, 1967; Moran, 1959); and

iii. other stochastic process models including linear and nonlinear system models (Bayazit, 1965; Quimpo, 1971; Klemes and Boruvka, 1975; Yevjevich, 1972a), and state variable models (Chow et al., 1975; Leondes, 1976).

Stochastic modelling involves the estimation and use of the stochastic characteristics of the process rather than regression analysis. Thus the use of Wiener-Hopf equations, Yule Walker relationships, univariate and multi-variate time series modelling using auto and cross-correlograms and spectral analysis, transition probability models, etc., (Kisiel, 1969; Shen, 1976; Yevjevich, 1972a) are examples of stochastic models which may be linearly structured.

The classification between statistical and stochastic models may be slightly ambiguous and in some cases the method of analysis may be considered as statistical or stochastic, e.g., decision theory applications (Duckstein and Davis, 1976; Thomas-Fiering model (Thomas and Fiering, 1962) and stochastic state variable equations. In such cases it does not generally matter as to how the model is classified. But in other cases the differences may be important and it is necessary to keep in mind the procedure used so that one is aware of the limitations of the model. Thus according to this classification Eagleson et al. (1965), Kashyap and Rao (1973), Ramaseshan et al. (1973), Szollisi-Nagi (1976) and Thomas and Fiering (1962) use statistical models while Quimpo (1973), Klemes and Boruvka (1975) and Young and Jettmar (1976) use stochastic models.

Conclusion

Linearly structured models are those that are defined by a set of linear algebraic equations in time domain or a transformed domain. They constitute an important group of hydrologic models. Since mathematical models of similar structure may be assumed or derived from different considerations, they are to be classified according to the procedure used in fitting the model.

This paper describes their classification into deterministic system models, statistical models or stochastic models according as the system is assumed respectively to be purely deterministic; to be modelled by statistical and regression techniques; or to explicitly take into consideration the stochastic nature of the process. The suggested classification is illustrated with linearly structured hydrologic models. It is suggested that the classification may serve in understanding the limitations and advantages of fitted models.

References

[1] Amorocho, J., 1970. "The Use of Input-Output (Transfer Function) Models in Hydrology," in: Systems Analysis of Hydrologic Problems, Utah Water Research Laboratory, Utah State Univ., Logan, pp. 231-248.

[2] Amorocho, J. and Brandstetter, A., 1967. "The Representation of Storm Precipitation Field Near Ground Level," Journal of Geophysical Research, 22(4):1145-1164.

[3] Astrom, K. J. and Eykhoff, P., 1971. "System Identification - A Survey," Automatica, 7(2): 123-162.

[4] Bayazit, M., 1965. "A Numerical Method to Determine the Unit Hydrograph," Bull. Tech. Univ., Istanbul, 18(1): 127-143.

[5] Bidwell, V. J., 1971. "Regression Analysis of Nonlinear Catchment Systems," Water Resources Research, 7:1118-1126.

[6] Blank, D. and Delleur, J. W., 1968. "A Program for Estimating Runoff from Indiana Watersheds - Part I," Tech. Report No. 4, Water Resources Research Center, Purdue Univ., Lafayette, Indiana.

[7] Box, G.E.P. and Jenkins, G. M., 1970. "Time Series Analysis, Forecasting and Control," Holden Day, San Francisco, CA.

[8] Clarke, R. T., 1973. "Mathematical Models in Hydrology," FAO, Rome.

[9] Choube, U. C., 1971. "Inverse Analysis of Rainfall Excess-Surface Runoff Process," M. Tech. Thesis, Dept. of Civil Engrg., Indian Institute of Technology, Kanpur, India.

[10] Chow, V. T. (Ed.), 1964. "Handbook of Applied Hydrology," McGraw-Hill, New York.

[11] Chow, V. T. (Ed.), 1969. "The Progress of Hydrology," Dept. of Civil Engrg., Univ. of Illinois at UC, Urbana.

[12] Chow, V. T. and Kulandaiswamy, V. C., 1971. "General Hydrologic System Model," Journal of the Hydraulics Division, ASCE, 97(HY6): 791-804.

[13] Chow, V. T., Kim, D. H., Maidment, D. R. and Ula, T. A., 1975. "A Scheme for Stochastic State Variable Water Resources System Optimisation," Res. Rept. No. 105, Water Resources Center, Univ. of Illinois at UC, Urbana.

[14] Dawdy, D. R. and O'Donnel, T., 1965. "Mathematical Models of Catchment Behaviour," Journal of the Hydraulics Division, ASCE, 91 (HY4): 123-137.

[15] Deininger, R. A., 1969. "Linear Programming for Hydrologic Analysis," Water Resources Research, 5(5):1105-1109.

[16] Delleur, J. W. and Rao, R. A., 1971. "Linear Systems Analysis in Hydrology - The Transform Approach, The Kernel Oscillations and The Effect of Noise," paper presented at the First U.S.-Japan Bilateral Seminar in Hydrology, Honolulu, Jan. 1971, published in the book Systems Approach to Hydrology WRP, Fort Collins, Colo., pp. 116-138.

[17] Delleur, J. W. and Rao, R. A., 1973. "Some Extensions of Linear System Analysis in Hydrology," in: Schulz et al., Floods and Droughts, 2nd International Symposium, Fort Collins, Colo., Sept. 11-13, 1972.

[18] Delleur, J. W. and Rao, R. A., 1971. "Characteristics and Filtering of Noise in Linear Hydrologic Systems," Mathematical Models in Hydrology Symposium, Vol. 2, Warsaw, July 1971, pp. 570-579.

[19] Diskin, M. H., 1964. "A Basic Study of the Linearity of the Rainfall-Runoff Process in Watersheds," Ph.D. Thesis, Univ. of Illinois, Urbana.

[20] Diskin, M. H. and Boneh, A., 1973. "Determination of Optimal Kernals for Second Order Stationary Surface Runoff Systems," Water Resources Research, 9(2):311-325.

[21] Diskin, M. H. and Boneh, A., 1975. "Determination of an Optimal IUH for Linear Time Invariant Systems from Multi-storm Records," Journal of Hydrology, 24 (1/2):57-76.

[22] Dooge, J.C.I., 1964. "Analysis of Linear Systems by Means of Laguerre Functions," SIAM Journal on Control, Series A, 2(3): 396-408.

[23] Dooge, J.C.I., 1967. "The Hydrologic System as a Closed System," Proc. Intl. Hydrology Symposium, Fort Collins, 2:98-115.

[24] Dooge, J.C.I., 1972. "Mathematical Models of Hydrologic Systems," Proc. Symp. Modelling Techniques and Water Resources Systems, Environment Canada, 1:171-189.

[25] Dooge, J.C.I., 1973. "Linear Theory of Hydrologic Systems," USDA, ARS Tech. Bull. 1468.

[26] Dooge, J.C.I. and Harley, B. M., 1967. "Linear Routing in Uniform Open Channels," Proc. Intl. Hydrology Symp., Fort Collins, 1:57-63.

[27] Duckstein, L. and Davis, D. R., 1976. "Application of Statistical Decision Theory," in: Shen (ed.), Institute on Stochastic Approaches to Water Resources, Vol. I, Colorado State University, Fort Collins, 1975.

[28] Duong, N., Wynn, C. B., and Johnson, G. R., 1975. "Modern Control Concepts in Hydrology," IEEE Trans. on Systems, Man and Cybernetics, SMC-5(1):46-53.

[29] Eagleson, P. S., 1969. "Deterministic Linear Hydrologic Systems," The Progress of Hydrology, Proceedings of 1st International Seminar for Hydrology Professors, Vol. 1, July 13-25, 1969, pp. 400-419.

[30] Eagleson, P.S., Mejia, R., and March, F., 1965. "The Computation of Optimum Realizable Unit Hydrographs from Rainfall and Runoff Data," MIT Hydrodynamic Lab. Rept. No. 84.

[31] Emsellem, Y. and DeMarsily, G., 1971. "An Automatic Solution for the Inverse Problem," Water Resources Research, 7(5): 1264-1283.

[32] Fahlbrush, F. and Muir, T. C., 1973. "Experience with a Monthly Rainfall-Runoff Transfer Function Model," in: Decisions with Inadequate Hydrologic Data, Water Resources Publications, Fort Collins, pp. 121-131.

[33] Freeze, R. A. and Harlan, R. L., 1969. "Blue Print for a Physically Based Digitally Simulated Hydrologic Response Model," Journal of Hydrology, 9:237-258.

[34] Kashyap, R. L. and Rao, R. A., 1973. "Real Time Recursive Prediction of River Flows," Automatica, 9:175-183.

[35] Kavvas, M. L. and Schulz, E. F., 1973. "Removal of Unit Hydrograph Oscillations by Filtering," in: Schulz et al., Floods and Droughts, 2nd International Symposium, Fort Collins, Colo., Sept. 11-13, 1972, pp. 167-174.

[36] Keefer, T. N. and McQuivey, 1974. "Multiple Linearization Flow Routing Model," Journal of Hydraulics Division, ASCE, 100 (HY7): pp. 1031-1046.

[37] Kishi, T., 1971. "Effect of an Error in Discharge Measurements on the Detection Process in Runoff System Analysis," paper presented at the First US-Japan Bilateral Seminar in Hydrology, Honolulu, Jan. 1971, published in the book Systems Approach to Hydrology WRP, Fort Collins, Colo., pp. 143-162.

[38] Kisiel, C. C., 1969. "Time Series Analysis of Hydrologic Data," in: Chow, V. T. (Ed.), Advances in Hydroscience, Academic Press, 5: 1-119.

[39] Klemes, V. and Boruvka, L., 1975. "Output from a Cascade of Discrete Linear Reservoirs with Stochastic Input," Journal of Hydrology, 27(1): 1-13.

[40] Kulandaiswamy, V. C., 1966. "Derivation of the Instantaneous Unit Hydrograph Using Z-Transform," Irrigation and Power, 23(3): 251-255.

[41] Leondes, C. T. (Ed.), 1976. "Control and Dynamical Systems," Academic Press, Vol. 12.

[42] Lettenmaier, D. P. and Burges, S. J., 1976. "Use of Estimation Techniques in Water Resources System Modelling," Water Resources Research, 12(1): 83-99.

[43] Lloyd, E. H., 1967. "Stochastic Reservoir Theory," in: Chow, V. T. (Ed.), Advances in Hydroscience, Academic Press, 4: 281-339.

[44] Moench, A. and Kisiel, C. C., 1969. "The Convolution Relation as Applied to Estimating Recharge From an Ephemeral Stream," paper presented at Annual Meeting, Amer. Geophys. Union, Washington, D.C.

[45] Moran, P.A.P., 1969. "The Theory of Storage," Methuen, London.

[46] Morel-Seytoux, H. J., and Daly, C. J., 1975. "A Discrete Kernel Generator for Stream Aquifer Studies," Water Resources Research, 11(2): 253-260.

[47] Muzik, I., 1974. "State Variable Model of Overland Flow," Journal of Hydrology, 22(3/4): 347-364.

[48] Neuman, S. P. and DeMarsily, G., 1976. "Identification of Linear System Response by Parametric Programming," Water Resources Research, 12(2): 253-262.

[49] O'Connor, K. M., 1976. "A Discrete Linear Cascade Model for Hydrology," Journal of Hydrology, 29(3/4): 203-242.

[50] O'Donnell, T., 1960. "Instantaneous Unit Hydrograph Derivation by Harmonic Analysis," IASH Publ. No. 51, pp. 546-557.

[51] Papazafiriou, Z. G., 1975. "Polynomial Approximation of Kernels of Closed Linear Hydrologic Systems," Journal of Hydrology, 27: 319-321.

[52] Quimpo, R. G., 1971. "Kernels of Stochastic Linear Hydrologic Systems," paper presented at the First US - Japan Bilateral Seminar in Hydrology, Honolulu, Jan. 1971, published in the book Systems Approach to Hydrology WRP, Fort Collins, Colo., pp. 163-177.

[53] Quimpo, R. G., 1973. "Link Between Stochastic and Parametric Hydrology," Journal of the Hydraulics Division, ASCE, 99 (HY3): pp. 461-470.

[54] Ramaseshan, S., 1976. Lecture Notes - "Advanced Summer School on Systems Analysis of Hydrologic Problems," Dept. of Civil Engrg., Indian Institute of Tech., Kanpur, India.

[55] Ramaseshan, S. and Anant, R. S., 1971. "A Multiple Input System Model for the Hydrologic Basin," Mathematical Models in Hydrology Symposium, Vol. 2, Warsaw, July 1971, pp. 540-547.

[56] Ramaseshan, S., Choube, U. C. and Sharma, A. K., 1973. "A Discrete Linear Model for the Hydrologic Basin," in: Chow, V. T., Scallany, S. C., Krizek, R. J., and Preul, H. C. (Eds.), Water for Human Environment, IWRA, Champaign, 4:367-376.

[57] Rao, R. A. and Usul, N., 1973. "Sensitivity Analysis of Linear Lumped System Models of the Rainfall-Runoff Process," Purdue Univ., Water Resources and Hydromechanics Lab Tech. Rept. No. 43, Lafayette, Indiana.

[58] Rodda, J. C. (Ed.), 1974. "Proc. Intl. Symposium on Mathematical Models," Warsaw, Poland, 1971, IASH Publ. No. 101.

[59] Sauer, V. B., 1973. "Unit Response Method of Open Channel Flow Routing," Journal of Hydraulics Division, ASCE, 99(HY1) 179-193.

[60] Shen, H. W. (Ed.), 1976. "Stochastic Approaches to Water Resources," P.O. Box 606, Fort Collins, Colo.

[61] Schulz, E. F., Koelzer, V. A., and Mahmood, K. (Eds.), 1973. "Floods and Droughts," Water Resources Publications, Fort Collins, CO, pp. 167-174.

[62] Singh, K. P., 1976. "Unit Hydrographs - A Comparative Study," Water Resources Bulletin, 12(2): 381-392.

[63] Snyder, W. H., 1955. "Hydrograph Analysis by the Method of Least Squares," Proc. ASCE, 81:793.

[64] Spolia, S. K. and Chander, S., 1974. "Modelling of Surface Runoff Systems by ARMA Model," Journal of Hydrology, 22: 317-332.

[65] Szollisi-Nagi, A., 1976. "Adaptive Identification and Prediction Algorithm for the Real Time Forecasting of Hydrological Time Series," Hydrological Sciences Bulletin, 21(1): 163-176.

[66] Thomas, H. A., Jr. and Fiering, M. B., 1962. "Mathematical Synthesis of Streamflow Sequences for the Analysis of River Basins by Simulation," in: Maass et al., Design of Water Resource System, Harvard University Press.

[67] Venetis, C., 1969. "The IUH of the Muskingum Channel Reach," Journal of Hydrology, 7: 444-447.

[68] Yevjevich, V. (ed.), 1971. "Systems Approach to Hydrology," Water Resources Publications, Fort Collins, Colorado.

[69] Yevjevich, V., 1971a. "The Structure of Inputs and Outputs of Hydrologic Systems," paper presented at the First U.S.-Japan Bilateral Seminar in Hydrology, Honolulu, Jan. 1971, published in the book Systems Approach to Hydrology WRP, Fort Collins, Colorado, pp. 5-71.

[70] Yevjevich, V., 1972. "Probability and Statistics in Hydrology," Water Resources Publications, Fort Collins, Colorado.

[71] Yevjevich, V., 1972a. "Stochastic Processes in Hydrology," Water Resources Publications, Fort Collins, Colorado.

[72] Yevjevich, V., 1974. "Determinism and Stochasticity in Hydrology," Journal of Hydrology, 22:225-238.

[73] Yevjevich, V., 1976. "General Overview of Application of Stochastic Methods to Water Resources Problems," in Shen (ed.), Institute on Stochastic Approaches to Water Resources, Vol. I, Colorado State University, Fort Collins, pp. 1.1 - 1.46.

[74] Young, G. K. and Jettmar, R. U., 1976. "Modelling Monthly Hydrologic Persistence," Water Resources Research, 12(5): 829-835.

CONCEPTUAL RAINFALL RUNOFF MODEL
BASED ON RETENTION CONCEPT

By

S. M. Seth
Reader in Hydrology
International Hydrology Course
University of Roorkee
Roorkee-247672, India

Abstract

This paper describes the development of a rainfall runoff model based on retention concept of dividing the rainfall into runoff and non-runoff (retention). The components of the model for simulating retention capacity, evapotranspiration capacity, direct contribution, leakage, and storage and translation effects were tested using the data from Grendon Underwood experimental catchment of the Institute of Hydrology, Wallingford, England. The data consisted of three-hourly values of rainfall, runoff and potential evaporation for a four year period. Some data of soil moisture available at weekly intervals, has also been used in the study. The Rosenbrock technique of optimisation was used for estimation of parameters. Split sample testing and sensitivity analysis were also used to judge the performance of the model.

The simple model developed, gave a good performance in fitting data records extending over a year. The moisture retained in the soil as given by model showed a good correlation with the observed values. A seasonal character is imparted to optimised parameter values obtained by fitting a particular season's data and this gave a poor reconstruction of catchment runoff in another season. It was found that at least one full year's data is necessary for proper estimation of model parameters. Only then can the controlling influence of one season on the other prevent apparent over estimation or under estimation of parameters.

Introduction

The problem of determining runoff from rainfall is one of the basic problems of scientific hydrology. The relationship between rainfall and runoff is needed not only for flood forecasting, but also for predicting the possible effects of water resources development works or other changes in the catchments on the flow regime. The hydrologic environment is in a state of non-equilibrium and there is need for better analysis procedures. This is particularly important for small catchments in view of their large numbers and highly sensitive response characteristics due to very short lag times. The rainfall runoff process of a catchment is a very complicated phenomena. In recent years, the application of systems concept and widespread availability of computers has given a great impetus to development of conceptual modelling techniques in hydrology. It provides a better estimation of design flood than is possible by the use of short term runoff data or by using rainfall data with arbitrary assumptions for abstractions. Moreover, the model can incorporate the present state of understanding of physical processes and it can be developed for regional application by relating model parameters to physiographic and climatic characteristics of the region.

Conceptual rainfall runoff models. The conceptual model simulates the rainfall runoff process by means of some mathematical representations or concepts. The structure of a particular model depends on the ideas or physical concepts used to build its different components, the method of linking together of these components, the objective criteria used to test the performance of the model and the type of data used. Various theoretical conceptual models have been hypothesized to describe the conversion of effective rainfall to direct runoff. Virtually all these models treat the catchment system as some combination of simulated components to describe the storage, attenuation, and translation effects. Total response models have also been developed which simulate the complete rainfall runoff process. The Standford watershed model developed by Crawford et al. (1966) represents a complex model structure which simulates the physical processes of the catchment system by means of conceptual components. It is based on the traditional infiltration concept of considering total runoff as being made up of contributions in the form of surface runoff, interflow and groundwater flow. Most other models have been developed on similar lines and employ a simplified structure or a smaller number of parameters. Another category of total response models are those which are based on retention concept of dividing rainfall into two parts - runoff and non-runoff (retention). Studies of Kohler et al. (1962), Bell (1967), and Mandeville et al. (1970) fall under this category.

Optimisation of conceptual models. Optimisation is the collective process of finding out the set of values of model parameters within constraints (if any) for the particular model structure which will reduce the difference between observed and calculated runoff to a minimum, using rainfall and potential evaporation as input data. The optimisation technique developed by Rosenbrock (1960) has been extensively used for optimisation of catchment models. Dawdy et al. (1965) made the first study of its kind using Rosenbrock optimisation technique with a conceptual model of rainfall runoff process.

Ibbitt (1970) compared different optimisation techniques in order to evaluate their performance in fitting conceptual catchment models and found that Rosenbrock technique with some modifications is the most effective technique.

The models developed by different researchers range from simple to very complex in structure. Nearly all of them try to simulate complex antecedent and storm period conditions using different structures of model components and different interlinking. These aspects of models have been discussed in detail elsewhere by the author, (Seth, 1972). The data required for a complex model is generally not available, and hence it is desirable to develop the model structure beginning with a simple structure and then make it complex to the desired level consistent with data availability. These ideas formed the basis for the study presented in this paper which involved development of a simple model of rainfall runoff process based on retention concept.

Description of the Model

In the traditional approach of infiltration theory, the rate of generation of surface runoff is assumed to be equal to the excess of rainfall intensity over the infiltration capacity of the soil. Some portion of the infiltrated water from present or earlier storms reappears as surface flow in the form of interflow and baseflow contributions. Both the interflow and baseflow involve moisture movement through soil and its complex physical processes. The phenomenon becomes more complicated due to spatial variability of catchment properties. This causes the surface flow and subsurface flow to change from one phase to the other and vice versa, at different times and locations. As both interflow and baseflow constitute reappearance of some portion of the infiltered water, the various assumptions in infiltration calculations regarding time distribution, spatial variability, recovery rates, etc., are carried into interflow and baseflow computations. Basically, the surface runoff, interflow and baseflow concepts are equivalent to the routing of total rainfall excess through three parallel reservoirs with different response characteristics. The retention concept of dividing rainfall into runoff and non-runoff parts is based on the physical processes of soil moisture retention. The use of retention concept in rainfall runoff models was advocated in an earlier study (Bell, 1967). The basic idea of retention concept is to separate rainfall into two parts; one which will become runoff and another which will not become runoff. The non-runoff part contributes to the retention store of the catchment, which is depleted only by evapotranspiration and leakage out of the catchment. The runoff storage constitutes the water present in natural catchment (above surface and below surface) which will appear at the catchment outlet after passing through different flow paths. The different components of the model structure based on retention concept were tested in the present study using data from Grendon Underwood clay catchment (area 19 sq km) of Institute of Hydrology, England.

Retention storage S. Retention storage S simulates the volume of water present in a natural watershed which will not become runoff. To simulate the high rates of retention during early periods of a storm and the slow rates during later periods, and the effect of evaporation processes on retention characteristics, the retention store has been specified by two layers of fixed storage volumes. The maximum storage volume V of retention store for the catchment is subdivided into V_1 for the upper layer and V_2 for the lower layer. The retention storage at any time in the upper layer is represented by S_1 and that in the lower layer by S_2. The total retention storage is $S = S_1 + S_2$, which was specified as depth of water in mm over the catchment.

Retention capacity Y. Retention capacity Y (mm per unit time interval) is the rate at which retention store can be filled and its maximum value Y_m corresponds to zero retention storage, i.e., $S_1 = 0$, $S_2 = 0$. The maximum retention capacity Y_m depends on catchment characteristics only. The retention capacity Y at any time will depend on the moisture status, i.e., values of S_1 and S_2. To simulate higher retention capacity in the beginning of a storm after long dry period, as well as to simulate reduction of these rates for any immediately following storm, Y_m was considered in two parts Y_1 and Y_2, where Y_1 represents maximum retention capacity for upper layer and Y_2 represents the maximum retention capacity for the lower layer and $Y_1 >> Y_2$. To retain simplicity of the model structure the following linear relationship was used as also shown in Fig. 1.

$$Y = Y_1[1 - (S_1/V_1)] + Y_2[1 - (S_2/V_2)]$$

Actual retention AY. The actual retention AY in any time interval will depend on the moisture supply G available and value of retention capacity Y for that interval. The simplest form of relationship between AY, G and Y is as below:

$$AY = Y, \quad CQ = G - AY, \quad \text{if } Y < G$$

$$AY = G, \quad CQ = 0 \quad , \quad \text{if } Y \geq G$$

where CQ represents the contribution to runoff storage during that time interval.

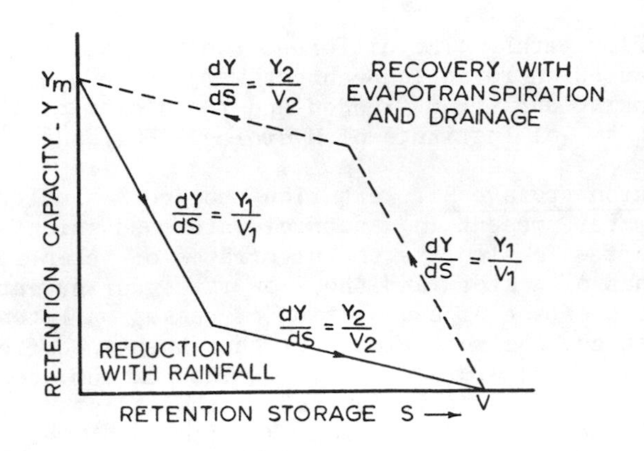

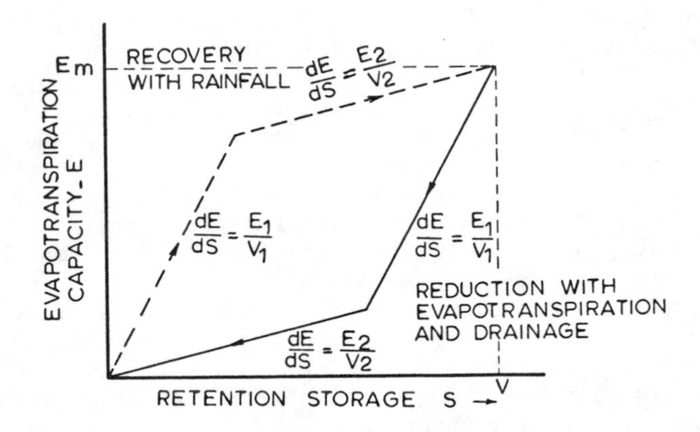

Fig. 1. Structures of functions for retention capacity and evapotranspiration capacity

Evapotranspiration capacity E. Evapotranspiration capacity E (mm per unit time interval) is the rate at which evapotranspiration can take place and its maximum value E_m represents the condition $S_1 = V_1$ and $S_2 = V_2$, i.e., when the retention storage is maximum and the available potential for evaporation is very high. Thus, E_m depends on the catchment characteristics only. To simulate the effects of reduction in evapotranspiration due to reduction in moisture contents of the soil, the maximum evapotranspiration capacity. E_m is considered in two parts, E_1 and E_2 for the upper and lower layer respectively with $E_1 > E_2$. The following linear relationship was used for evapotranspiration capacity E, as also shown in Fig. 1.

$$E = E_1(S_1/V_1) + E_2(S_2/V_2)$$

As $E_1 > E_2$, the same amount of moisture change in upper or lower layer will have a correspondingly bigger or smaller effect respectively on E.

Actual evapotranspiration AE. The actual evapotranspiration AE in any time interval will depend on the potential evaporation G_e which is to be satisfied and the evapotranspiration capacity E for that time interval. The simplest form of relationship between AE, G_e and E is as below:

$$AE = E, \quad \text{if} \quad E < G_e$$

$$AE = G_e, \quad \text{if} \quad E \geq G_e$$

Other components. In addition to these main components various alternative structures of direct contribution, leakage and evapotranspiration were tested. The direct contribution component simulates areal variations in retention capacity, which can cause a contribution to runoff storage even when moisture supply available is less than average retention capacity Y for the whole catchment. The leakage component simulates leakage out of the catchment which may cause significant recovery in retention capacity.

Any water that becomes runoff storage was routed through runoff routing components to simulate the storage and delay effects of catchment response. In the retention approach, the runoff storage constitutes all the water which in the traditional infiltration approach comes as overland flow, interflow and groundwater flow. Different alternatives were tried using linear reservoir, linear channel and nonlinearised (piecewise linear) reservoir in various combinations. The nonlinearised reservoir simulates the effects of relatively higher delay times for flow contributions which travel through subsurface paths. The linear reservoir and linear channel simulate the effects of delay and storage in overland flow path and channel flow.

Special care was taken in formulating the structure of these components so as to have good physical correspondence with Grendon Underwood catchment and to avoid interdependent parameters which cause difficulty in optimisation of parameters. These model components in different combinations were then tested for the performance, reliability and efficiency in the simulation of rainfall runoff process of a small clay catchment using typical data sets.

Data Used for Study

The data record used for testing the model consisted of 3 hourly values of rainfall, runoff and potential evaporation expressed in mm of depth over the Grendon Underwood catchment. It is a small, well instrumented, experimental clay catchment, 19 sq km in area, on the river Ray, which is a tributary of the river Thames, and it is maintained by Institute of Hydrology, Wallingford. In addition to above mentioned data which was available for a 4-year period from November 1963 to November 1967, the soil moisture data was also available in the form of depth of water in mm in the top 100 cm of soil at approximately weekly intervals for some period.

For the purpose of the present study, various portions of available data were selected which represented different seasonal and antecedent conditions. The relevant particulars of these data sets, nine in number, have been given elsewhere (Seth, 1972). The data set seven extending over 1496 three hourly intervals constituted a typical dry season record, beginning with a very wet period and followed by a long dry period. The data set eight with 1064 three hourly values of rainfall, runoff and potential evaporation constituted a period of typically wet season which followed the period covered in data set seven. The combination of these two data sets constituted data set nine which represented almost a year's data. The final tests of the model were made on these three typical data records.

Testing Criteria

The objective criterion for the evaluation of optimised parameters, as well as component performance, was the minimisation of the sums of squares of the difference between the observed and calculated values of runoff. The rainfall and potential evaporation data were used as input. The objective function F adopted for the study can be written as

$$F = \sum_{I=1}^{M} [Q(I) - QC(I)]^2$$

where $Q(I)$ and $QC(I)$ are observed and calculated values of runoff respectively in the Ith interval and M = number of data pieces or three hourly time intervals. The Rosenbrock technique of optimisation was modified and used for evaluation of parameters by optimisation. The values of upper and lower constraints for the parameters and the starting values of the parameters were determined from the physiographic characteristics of the catchment. Predictive testing on a split sample basis was used as a criterion for testing the performance of the model. The performance was judged by the ability of the model to fit the first part of the data record and then predict the second (immediately following) part. Sensitivity tests for testing the sensitivity of model response (i.e., change in value of objective function F after optimisation run) due to change in any one of the model parameters from its optimised value, were also used to judge the performance of model components. The product moment correlation coefficient, RM, was also evaluated for judging the model performance, which measures the degree of association between observed and calculated values of runoff.

The computer program for optimisation and testing the performance of different model components was written in Fortran IV and was run on ICL 1906 A computer of Regional Computer Centre, Manchester (U.K.). To save computer time a limit of 200 was placed in the number of function evaluations.

Discussion of Results

After preliminary tests with different data sets and various alternative model components, a 12 parameter model was evolved (Seth, 1972). This model consisted of components for simulating retention capacity, evapotranspiration capacity, direct contribution, leakage and catchment storage and lag effect. The model was tested with data sets 7, 8 and 9. This involved: (i) Optimisation of typical wet season data set eight, (ii) optimisation of typical dry season data set seven and prediction of data set eight, and (iii) optimisation of typical annual data set nine (combination of 7 and 8).

The optimisation run with data set eight, gave $F = 6.624$ mm^2 and $RM = 0.961$. The totals of rainfall, runoff and potential evaporation for this data set extending over 133 days were 344.47 mm, 144.28 mm and 116.11 mm respectively. The total calculated runoff was within 3.5% of the total observed runoff. Thus the degree of fitting achieved was very good.

The optimisation run with data set seven also gave quite encouraging results as indicated by $F = 4.59$ mm^2 and $RM = 0.968$. The totals for rainfall, runoff and potential evaporation for this period of 187 days were 390.26 mm, 112.56 mm and 539.05 mm respectively. The total calculated runoff was within 1.1% of the total observed runoff. However, the results for prediction of immediately following data set eight were not good. The value of $F = 48.74$ mm^2 and $RM = 0.85$ indicate this. Moreover, the total calculated runoff was 25% more than the total observed runoff. Figure 2 shows a typical hydrograph from data set 8, for both optimisation run and prediction run. It clearly shows model's capability to fit in optimisation with somewhat poorer prediction of the hydrograph.

The optimisation run using data set nine extending over 2560 three hourly time intervals of 320 days data, also gave equally good fit as shown by $F = 14.07$ mm^2 and $RM = 0.96$.

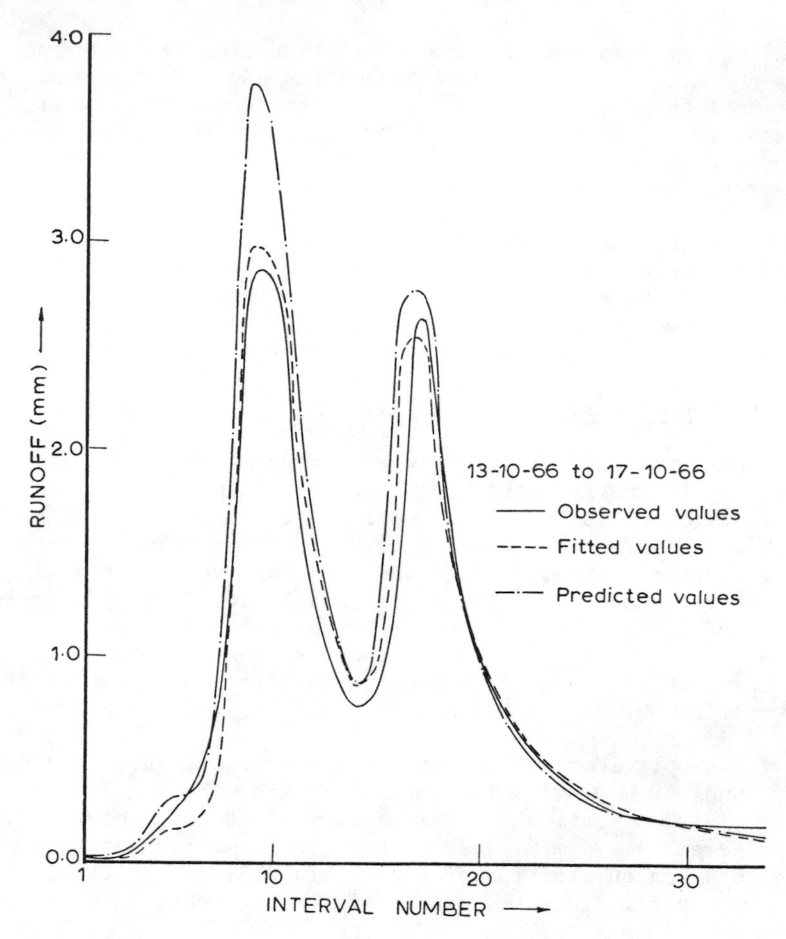

Fig. 2. Observed hydrograph and corresponding fitted and predicted hydrographs from data set eight. Interval no. refers to number of 3-hour time intervals

The total observed runoff was within 6% of the total observed runoff.

Thus the model developed proved to be quite versatile in fitting the data of dry season or wet season or full year but gave poorer prediction of following wet season data when optimisation was done on dry season data. Figure 3 shows a plot of observed values of soil moisture and moisture retained in soil as given by model. The observed soil moisture values and model values for optimisation run with data set nine follow almost a similar trend. However, for dry season record (data set seven optimisation run) alone the model values are higher than those for combined dry season wet season record of data

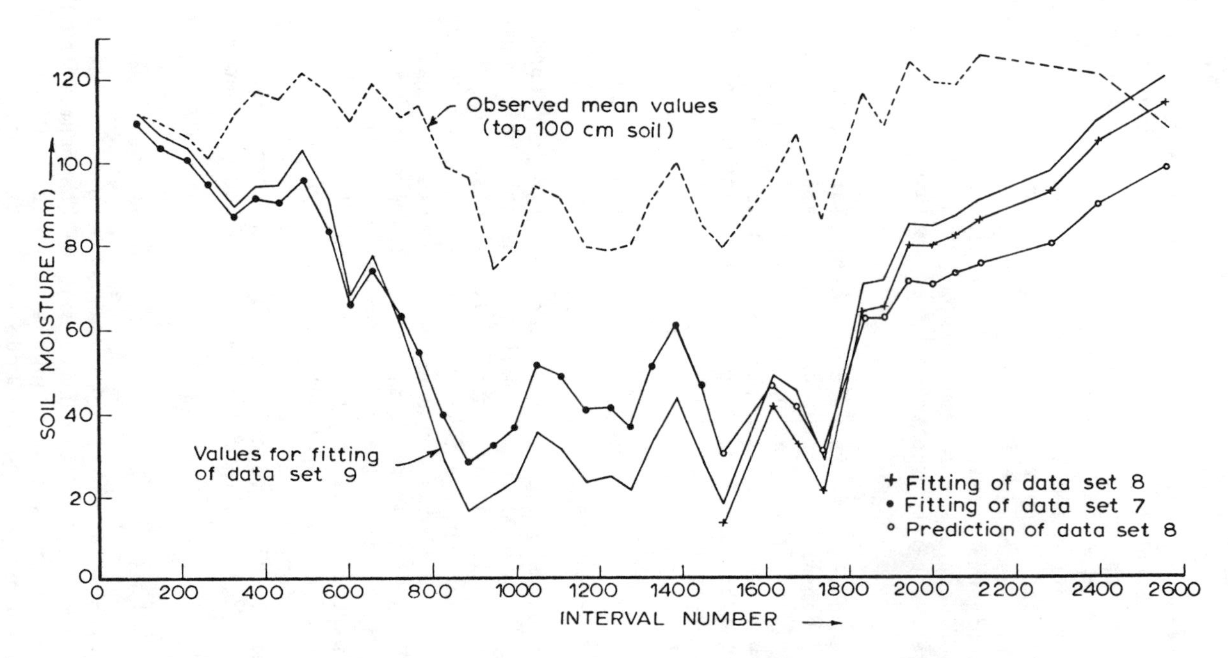

Fig. 3. Observed soil moisture data and corresponding values of soil moisture retained as given by model

set nine. This clearly shows that though the runoff values fitted well by using objective function F based on runoff values, the soil moisture was not reduced to the correct level by the evapotranspiration process and this resulted in poor prediction of the following wet season record.

The response sensitivity analysis was also done for these tests, which also supported the above findings (Seth, 1972). It was observed that from use of a particular season's data (dry or wet season) for an optimisation run, a seasonal character is introduced in the estimated parameters values and this results in poor reconstruction of runoff in another season. There is always an apparent overestimation or underestimation of some parameters when dry season or wet season data is used separately in comparison to the optimised values of parameters obtained when a combined wet and dry season record is used.

When a year's data is used for optimisation, the dry season data acts as a control on those parameters which have more sensitivity in wet season's conditions alone and vice versa. Hence, the optimised parameters obtained by a complete year's data should give a better reconstruction of the following year's runoff. However, optimisation using a year's record that does not include certain extreme events cannot give good prediction of those events.

Conclusions

This study shows that using the retention concept a simple model can be developed which gives good performance in fitting and prediction of runoff. It has also been shown that to take into account the full range of storm and antecedent conditions the optimisation run should extend over data for a few years. In view of the computation involved for such cases, the decision will depend on overall economic feasibility for a particular case. The Rosenbrock technique proved to be quite effective in optimisation of parameters. However, there is need for further research in this direction. The development of objective criteria to include soil moisture data also in addition to runoff may give good parameter estimates with limited data.

Acknowledgments

The author thankfully acknowledges help and guidance from Prof. D. M. McDowell of the University of Manchester in this study. The author is also thankful to Prof. T. O'Donnell for suggesting the use of Grendon Underwood data, which was supplied by the Institute of Hydrology, Wallingford, England.

References

[1] Bell, F. C., 1967. "An Alternative Physical Approach to Watershed Analysis and Streamflow Estimation," Proceedings, International Hydrology Symposium, Colorado State University, Fort Collins, Colorado, 1: 86-93.

[2] Crawford, N. H. and Linsley, R. K., 1966. "Digital Simulation in Hydrology: Stanford Watershed Model IV," Technical Report No. 39, Department of Civil Engineering, Stanford University.

[3] Dawdy, D. R. and O'Donnell, T., 1965. "Mathematical Models of Catchment Behaviour," Journal of the Hydraulics Division, ASCE, No. HY4, 91:123-237.

[4] Ibbitt, R. P., 1970. "Systematic Parameter Fitting for Conceptual Models of Catchment Hydrology," Ph.D. Thesis, Imperial College, University of London, London (U.K.).

[5] Kohler, M. A. and Richards, M. M., 1962. "Multi Capacity Basin Accounting for Predicting Runoff from Storm Precipitation," Journal of Geophysical Research, 13: 5187-5197.

[6] Mandeville, A. N., O'Connell, P. E., Sutcliff, J. V., and Nash, J. E., 1970. "River Flow Forecasting Through Conceptual Models, Part III - The River Ray Catchment at Grendon Underwood," Journal of Hydrology, 11:109-129.

[7] Rosenbrock, H. H., 1960. "An Automatic Method of Finding the Greatest or Least Value of a Function," The Computer Journal, 3:175-184.

[8] Seth, S. M., 1972. "Conceptual Modelling of the Rainfall Runoff Process of a Small Clay Catchment," Ph.D. Thesis, University of Manchester, Manchester (U.K.).

A CONCEPTUAL MODEL FOR THE BINOMIALLY DISTRIBUTED RUNOFF ELEMENTS OF A NATURAL CATCHMENT

By

B. P. Singh
Assistant Engineer, Advance Planning
Bihar Irrigation Department
Patna (Bihar), India

and

B. S. Mathur
Reader in Hydrology, School of Hydrology
University of Roorkee
Roorkee-247672, India

Abstract

In this paper, a two parameter linear hydrologic model is proposed to evaluate the response of a natural catchment. The catchment is conceptually represented by a cascade of linear channels. One parameter of the model is translation time of linear channels which is estimated from the response due to excitation of the system by an instantaneous input. The other parameter of the model is the response subarea. The input to the hydrologic model is considered as distributed and it is assumed to enter its corresponding linear channel at its upper bound, along with the input translated down to it by the earlier system. To account for the storage effects of the catchment, the runoff generated on each day is assumed to be distributed into runoff elements which are considered to be binomially distributed. Net response of this linear system is obtained by superimposing the runoff elements of each day. The model has been developed and tested for the Ajay Basin in Bihar in India.

Introduction

A linear distributed hydrologic model can mathematically be expressed by a second order partial differential equation. Mathematical solution to such a problem would require data pertaining to initial and boundary conditions. Rarely can such data be found in the field. Therefore, solutions to such a system are approximated with the help of conceptual identities like linear reservoirs (Dooge, 1959) and linear channels (Mathur, 1977).

In this paper, a conceptual model has been developed utilising the properties of linear channels arranged in series.

The input to the system is considered as distributed. Basically it is a two parameter model. The first parameter is 'response subarea' and the other is 'translation time' of the linear channels of the cascade which is kept uniform throughout. The two parameters are interdependent. The natural catchment is divided and adjustments are made to obtain 'response subareas'. Runoff from each response subarea is the distributed input entering at the upper bound of a linear channel along with the translated input from the upstream elements. The two input components help in maintaining the continuity of the hydrograph.

In order to account for the storage effects of the catchment, the runoff generated on each subarea on any day is assumed to be divided into runoff elements which are binomally distributed and reach the outlet on subsequent days. Proper initial time lags for these runoff elements are accounted for by the lower bound translation values for each subarea of the system.

Formulation of the Model

The response of a linear hydrologic system for a lumped input is given by the convolution integral:

$$Q(t) = \int_0^t I(\tau) \cdot U(t-\tau)d\tau \tag{1}$$

where $I(\tau)$ = instantaneous excitation at time τ
$U(t-\tau)$ = kernel function characterising the hydrologic system.

In terms of area function, the response of the same hydrologic system may be expressed as:

$$Q(t) = \int_0^t I(\tau) \frac{dA}{d\tau} d\tau \tag{2}$$

where $\frac{dA}{d\tau}$ = time response of catchment area characterising the system.

Comparison of the two equations would give the following relationship

$$\frac{dA}{d\tau} = U(t-\tau) \tag{3}$$

Therefore, if the inpulse response function of a natural catchment is known the area may be divided into response subareas which may be conceptualised through a serial system of linear channel. However, such a system will be a pure translatory scheme. To account for the storage effects of the drainage basin, an other concept has to be divised. Writing Eq. (2) in discritized form;

$$Q(t) = \sum_{\tau=1}^{t} \frac{\Delta A}{\Delta \tau} \cdot I(\tau) \cdot \Delta\tau \quad , \tag{4}$$

$$Q(t) = C_{1,t} \cdot A_{(1)} I'_1 + C_{2,t} A_{(2)} I'_2 + \ldots . + C_{\tau,t} \cdot A_{(\tau)} \, I'(\tau) \quad ,$$

$$Q(t) = \sum_{\tau=1}^{t} C_{\tau,t} \cdot A_{(\tau)} I'(\tau) \tag{4a}$$

where $C_{\tau,t}$ = coefficient of runoff distribution for τ^{th} subarea for instant t. It must account for the distribution of input in time.

$A(\tau)$ = response subarea, obtained by dividing the catchment in accordance with the proportionate distribution scheme proposed in Eq. (3).

In Eq. (4) the total runoff generated over a response subarea, i.e., $A_{\tau} I'(\tau)$ is distributed over successive time periods into runoff elements whose magnitude will be characterised by the coefficient $C_{\tau,t}$ which is a function of time. In this study these coefficients are assumed to be binomially distributed. Thus, the value of the coefficient may be computed by the following relation.

$$C_{\tau,t} = \frac{(t-1)!}{(\tau-1)! \, (t-\tau)!} \, \omega^{\tau} \, (1-\omega)^{t-\tau} \tag{5}$$

The magnitude of travel coefficient ω will depend upon the physiographic characteristics of the catchment. Accordingly, the runoff elements from each response subarea may be computed over subsequent periods. Being a linear system the runoff elements may be superimposed to compute total response of the system. The mechanics of the process is explained in Table 1.

Application of the Model on Ajai River Catchment

The proposed response model has been tested on the Ajai Basin in Bihar. The catchment area is approximately 1000 sq miles. Daily rainfall data for a few intense storms and corresponding daily discharges at the outlet located at Sikatia are available. The Instantaneous Unit Hydrograph for the catchment was derived from the observed records of storm dated October 15, 1971, using Nash's model (Nash, 1958, 1960). The ordinates of IUH and correspondingly the magnitudes of response subareas are given in Table 2. The peak of IUH appears at 0.5 day from the beginning. Therefore, the unit translation

Table 1 Scheme for the Binomially Distributed Runoff Elements for the Computation of Flood Hydrograph

Response Subarea No.	Area of the Response Subarea (acres)	Rainfall Excess in		Generated Runoff Col. 2x4 (cusees)	Runoff Elements Distributed over Subsequent Days and Reaching the Outlet Point						
		Time days	inches		1	2	3	4	5	6	7
1	2	3	4	5	6	7	8	9	10	11	12
1.	A	1	I_{a1}	R_{a1}	$C_{11}R_{a1}$	$C_{12}R_{a1}$	$C_{13}R_{a1}$	$C_{14}R_{a1}$	-	-	-
		2	I_{a2}	R_{a2}	-	$C_{11}R_{a,2}$	$C_{12}R_{a,2}$	$C_{13}R_{a,2}$	$C_{14}R_{a,2}$	-	-
		3	I_{a3}	R_{a3}	-	-	$C_{11}R_{a,3}$	$C_{12}R_{a3}$	$C_{13}R_{a3}$	$C_{14}R_{a3}$	-
2.	B	1	I_{b1}	R_{b1}	-	$C_{22}R_{b1}$	$C_{23}Rb_1$	$C_{24}R_{b1}$	-	-	-
		2	I_{b2}	R_{b2}	-	-	$C_{22}Rb_2$	$C_{23}Rb_2$	$C_{24}R_{b2}$	-	-
		3	I_{b3}	R_{b3}	-	-	-	$C_{22}R_{b3}$	$C_{23}R_{b3}$	$C_{24}R_{b3}$	-
3.	C	1	I_{c1}	R_{c1}	-	-	$C_{33}R_{c1}$	$C_{34}R_{c1}$	$C_{35}R_{c1}$	-	-
		2	I_{c2}	R_{c2}	-	-	-	$C_{33}R_{c2}$	$C_{34}R_{c2}$	$C_{35}R_{c2}$	-
		3	I_{c3}	R_{c3}	-	-	-	-	$C_{33}R_{c3}$	$C_{34}R_{c3}$	-

Table 2 Ordinates of Instantaneous Unit Hydrograph derived from the storm registered on 15.10. 1971 and correspondingly the response subarea distribution

Time (t) (days)	Ordinates of IUH $\times 10^3$	IUH Ordinates Expressed in %	Response Subarea (in 1000 acres)
0.5	19.8	39.45	250
1.0	14.8	29.47	187
1.5	8.3	16.57	105
2.0	4.1	8.25	52
2.5	1.6	3.25	21
3.0	0.87	1.73	11
3.5	0.38	0.76	4.8
4.0	0.16	0.32	2.0
4.5	0.06	0.13	0.8
5.0	0.03	0.06	0.4

for all the linear channels in the proposed cascade model is uniformly kept as 0.5 day. Velocities are calculated from different parts of the basin using empirical equations and the isochromes are drawn. Readjustments of the areas are carried out by using the proportionate scheme given in Eq. (3). Structure of the proposed cascade of linear channels with its distributed inputs is shown in Fig. 1.

Thiessen's polygon method has been used for considering spatial uniformity of precipitation. The weighted effective depth of precipitation (P_i) is computed using equations

$$p_i = P_i \; [p_{ave}/(\sum_{i=1}^{n} P_i A_i / \sum_{i=1}^{n} A_i)] \tag{6}$$

where P_i = gross precipitation
A_i = Thiessen polygon area
p_{ave} = average depth of precipitation over the catchment.

The runoff from each response subarea is computed using the genetic runoff formulae

$$Q(t) = \int_0^t C_{\tau,t} \; I(\tau) \cdot \frac{dA}{d\tau} \cdot d\tau \tag{7}$$

The runoff elements are thus considered to be binomially distributed over subsequent days and follow the runoff scheme as given in Table 1. The rainfall excesses as computed by Eq. (6) for the storms recorded on 3.6.1971, 9.7.1971, and 14.10.1971

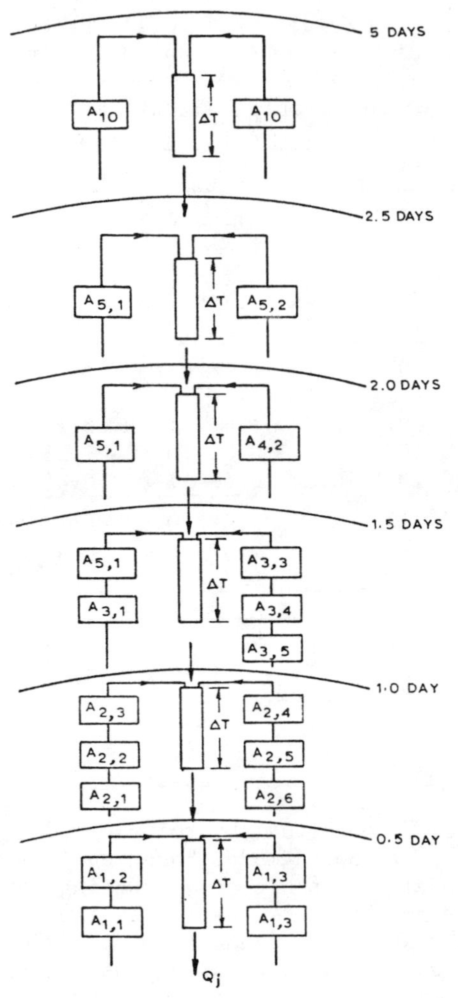

Fig. 1. Cascade of Linear Channels Representing Ajai Basin in Bihar (India)

and 11.10.1973 are given in Table 3. The runoffs are computed as detailed in previous sections and the base flows were added to them. A comparison of computed and observed hydrographs is shown in Figs. 2 and 3. A close agreement has been found between the two.

Table 3 Computed Values of Rainfall Excess for Different Storms Over the Ajai Basin in Bihar (India)

Storm	Date	Rainfall Excess on Response Subarea (inches)						
		1	2	3	4	5	6	7 to 10
June 3-7 1971	3.6.1971	-	0.2	1.16	1.03	0.82	1.6	-
	4.6.1971	1.9	1.48	1.08	0.2	0.23	0.03	0.53
	5.6.1971	0.25	0.27	0.19	-	-	-	-
	6.6.1971	0.49	0.84	0.17	1.91	1.98	1.75	2.38
	7.6.1971	0.03	0.06	0.26	0.32	0.26	0.46	-
July 9-13 1971	9.7.1971	0.16	0.18	-	-	-	-	-
	10.7.1971	0.39	-	-	-	-	-	-
	11.7.1971	0.07	1.66	1.89	0.93	0.74	1.5	-
	12.7.1971	-	0.25	0.60	1.25	1.25	1.32	1.17
	13.7.1971	0.02	0.69	1.00	0.40	0.36	0.37	0.35
October 15-16 1971	15.10.1971	-	0.3	0.52	0.15	0.04	0.32	-
	16.10.1971	1.19	0.97	0.78	0.44	0.60	0.04	1.43
October 11-12 1973	11.10.1973	0.49	0.53	0.56	0.97	1.02	0.27	2.12
	12.10.1973	2.17	2.30	1.87	1.44	1.66	1.25	2.39
	13.10.1973	-	0.51	0.61	-	-	-	-

Conclusions

The proposed conceptual hydrologic model is basically a two parameter distributed model based on the linearity principle. The model is capable of accounting for the spatial non-uniformity of the precipitation as input to the model is considered as distributed and not lumped. Runoffs from each response subarea are computed using the genetic runoff formula and are considered to be binomially distributed over subsequent days. The assumption of linearity enables superposition of runoff elements of each day to compute total response of the system.

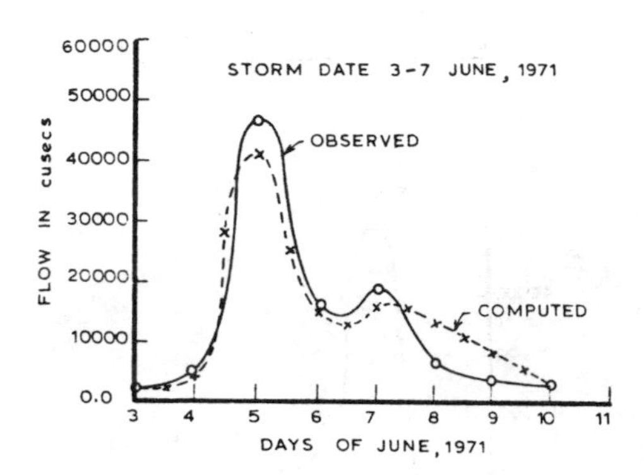

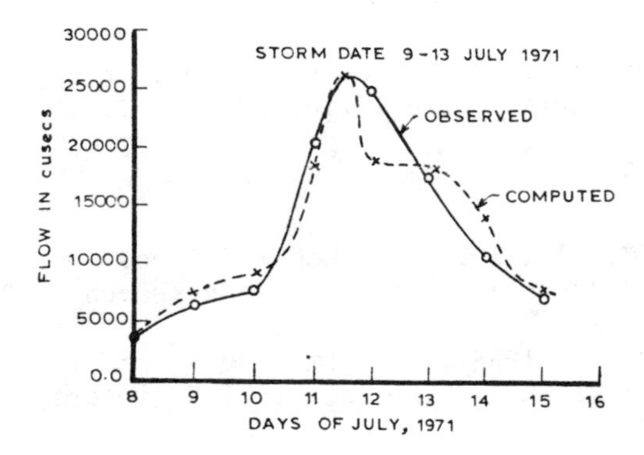

Fig. 2. Comparison of computed and observed hydrographs for the storms dated 3.6.1971 and 9.7.1973

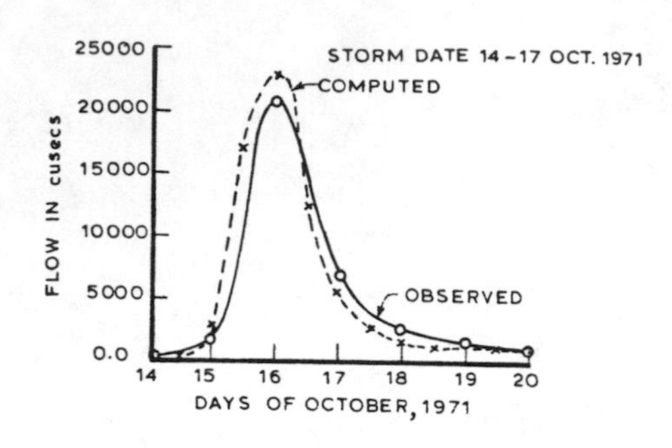

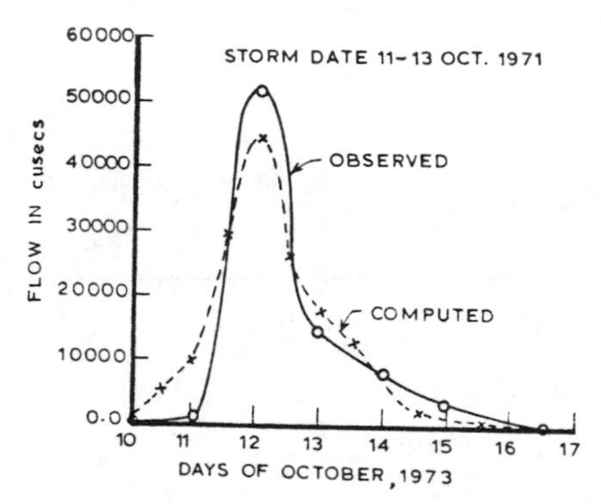

Fig. 3. Comparison of computed and observed hydrographs for the storms dated 14.10.1971 and 11.10.1973

References

[1] Dooge, J.C.I., 1959. "A General Theory of Unit Hydrographs," Journal of Geophysical Research 64(2),241-256.

[2] Nash, J. E., 1958. "Determining Runoff from Rainfall," Proceedings Institution of Civil Engineers, Vol. 10, pp. 163-184.

[3] Nash, J. E., 1960. "A Unit Hydrograph Study with Particular Reference to British Catchments," Proceedings of Institution of Civil Engineers, No. 17.

[4] Mokliak, V. I. and Tchkushkina, T. A. "Calculation of Flood Discharges by Means of Unit Hydrograph," Proceedings of IASH Symposia on "Floods and Their Computations," Leningrad.

[5] Mathur, B. S., 1972. "Runoff Hydrographs for Uneven Spatial Distribution of Rainfall," Ph.D. dissertation, Indian Institute of Technology, New Delhi.

A LINEAR DISTRIBUTED MODEL FOR A PARTLY SNOWBOUND CATCHMENT

By

Ranjodh Singh
Executive Engineer
H. P. State Electricity Board
Catholic Club
Simla-1 (H.P.) India

and

B. S. Mathur
Reader in Hydrology, University of Roorkee
Roorkee (U.P.) India 247 672

Abstract

The complex rainfall-runoff relationship is further complicated if the catchment is partly under permanent snow cover. In this paper, a new mathematical model is proposed for the Beas River which originates from the northwestern Himalayas. Substantial part (around 30%) of the Beas catchment is under permanent snow cover.

In the proposed model, for the rain affected area, a series arrangement of linear channels has been used. The input to the system is considered as distributed. The model is a two parameter model. The first parameter is the translation time of the linear channels which is kept uniform for all linear channels of the system. The second parameter is the response areas whose distributed input is fed to the channel components of the system.

The snowmelt analysis is based on the energy budget approach. The snowmelt runoff from different zones has been considered as binomially distributed.

The observed and computed runoffs have been compared for different months of a year. The close agreement between the two demonstrates the utility of the proposed mathematical model.

Introduction

A large number of rivers in northern India originate from the Himalayan ranges. Precipitation in their catchments is generally in the form of rain in the lower reaches and snow at higher elevations. These rivers receive part of their runoff from the melting of the snow over a considerable period of the year.

The process whereby rainfall excess, generated over a catchment, is transformed into runoff observed at the outlet is very complex and cannot be studied by direct application of physical laws. However, the process can be studied scientifically through a conceptual model. An attempt is made in this study to evolve a suitable hydrologic model utilising the information from the existing models (Dooge, 1959; Mokliak et al, 1967; Mathur, 1972).

Presence of permanent snowcover in the catchment greatly influences the runoff characteristics of the area under certain basic conditions of air temperature and rainfall. The complexity of the snowmelt process coupled with a general lack of systematic and reliable data about the glaciers necessitates the use of a rather simplified approach to estimate snowmelt. Therefore, a simpler model for snowmelt relationships is used in which snowmelt is considered to be a function of meteorologic parameters. Time lags for snowmelt transformation could be properly accounted for by utilising the properties of travel coefficients which are binomially distributed in time.

In the proposed model, the effect of rain and snow are studied separately. The approach is linear so the two effects are superimposed to develop the response.

Estimation of Peak Flood by a Series System of Linear Channels

A linear distributed model based upon the series arrangement of linear channels having uniform translation is proposed by Mathur (Mathur, 1972). A catchment translates and attenuates the input hydrograph, whereas a linear channel purely translates the input. To incorporate the net effect the input to the proposed system is considered as distributed. As indicated in Fig. 1, input to a linear channel is considered as comprised of two parts. Each linear channel receives the primary response of input component (Q_j) from the earlier channels of the system. It is assumed to enter at the upper bound of the channel along with the secondary response of the input corresponding to area to which it represents. The scheme is indicated in Fig. 1(b).

Since the input is considered as distributed each linear channel of the series network has two parameters viz., the translation time and the response subarea. Both parameters are interdependent. The unit translation is adopted such that sufficient number of linear channels are produced for properly defining the shape of the hydrograph. For simplicity the time step of input hyetograph may be chosen as the unit translation.

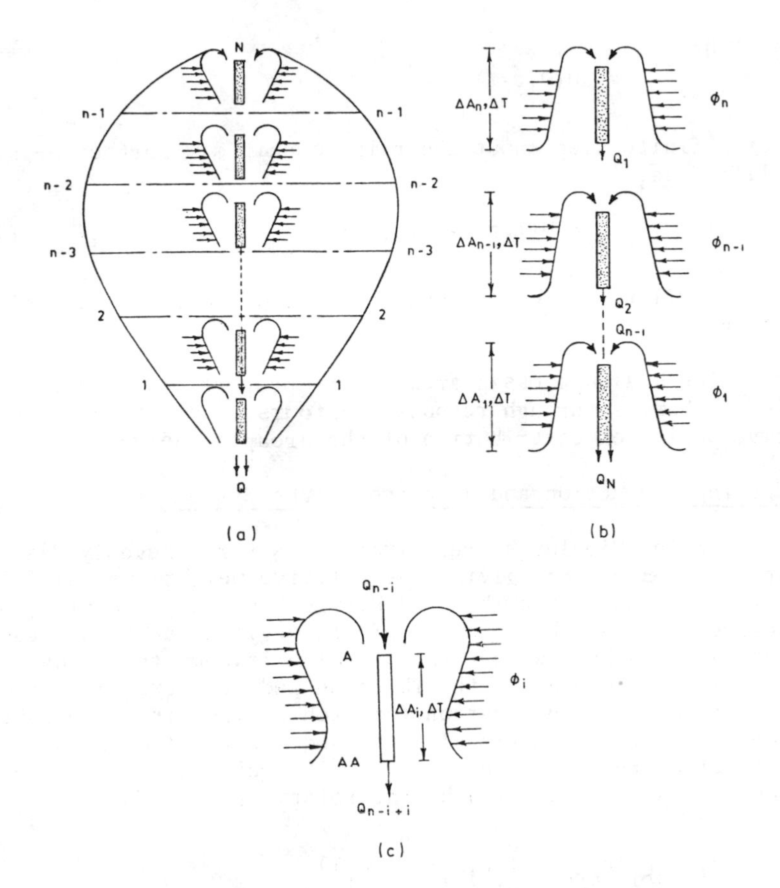

Fig. 1. Distributed input to a series network of linear channels

The response subarea is defined as part of drainage basis area which contributes to the upper bound of the linear channel marked by the net translation of the subsystem. The response of any linear hydrologic system is given by the concolution integral:

$$Q_t = \int_0^t i(\tau)\ U(0,t-\tau)d\tau \tag{1}$$

and the response from any subarea can be obtained using Clark's time-area method (e.g., Viessman et al., 1977) and would be given by

$$Q_t = \int_0^t i(\tau) \cdot \frac{dA_{t-\tau}}{d\tau} \cdot d\tau \tag{2}$$

Comparing Eqs. (1) and (2)

$$\frac{dA_{t-\tau}}{d\tau} = U(0,t-\tau) \tag{3}$$

For a finite step input the relationship may further be simplified as:

$$A_t - A_{t-1} = U(1,t) - U(1,t-1) \tag{4}$$

where $U(1,t)$ is unit graph ordinate of unit time duration at instant t.

The drainage basin area may be proportionally divided into subareas through response contours, indicating the effective period of contribution of the area to the gauge.

The Input Function and Response of the System

In the Himalayas precipitation is very unevenly distributed in space. For giving the relative weights to rainfall at different gauges the Thiessen Polygon method is adopted, but before doing so the point rainfalls registered at each gauge are adjusted by a weight factor which is computed on an average basis by comparing the Thiessen and isohyetal methods. The effective rain duration for each polygon is established keeping in view the translations and the total duration. The effective precipitation (p_i) at ith rain gauge over the subarea A_i is computed by the relations

$$p_i = P_i [p_{ave} / (\sum_{i=1}^{n} P_i A_i / \sum_{i=1}^{n} A_i)] \tag{5}$$

where P_i = gross rainfall recorded over the area, and

p_{ave} = average storm depth over the basin.

The response of the system may be computed using the genetic principle of runoff.

$$Q = \int_0^t i(\tau)\, dA_{t-\tau} \cdot d\tau \tag{6}$$

Where $i(\tau)$ refers to both components of input to the linear channel and $dA_{t-\tau}$ is the subarea whose distributed input is fed to the channel.

Estimation of Melt from Snowbound Areas

Computation of snowmelt runoff is a different problem as it depends upon the available heat for melting and would therefore depend upon adequate prediction of energy supplies. The main sources of heat energy for snowmelt are shortwave radiation, terrestrial radiation, convection condensation, conduction of heat from the ground and heat content of rainfall. The dominant source of heat being the solar radiations. The following typical (e.g., Viessman et al., 1977) relationships have been used for the computation of snowmelt using meteorologic parameters:

$$M = [2.29 + (1.33 + 0.21\ KW + 0.13R)\ (T-TF)] \qquad (7)$$

For a partly forested basin, total basin melt for a snowpeak at 0°C with 3% water content is computed by the relation:

$$M = C[0.1\ K'S(1-F)(1+A) + 0.21KW(0.22T'+0.78TD')+1.33T'F] \qquad (8)$$

Where M = snowmelt in mm,
C = coefficient for undetermined variation in snowmelt,
K = 1.0 (unforested basin) to 0.3 (densely forested) and gives convective condensation melt,
K' = shortwave radiation melt factor (0.9-1.1),
T = air temperature in °C at 3 m above the snow,
T' = difference in air temperature at 3 m and at snow surface,
W = wind speed in Km/hr at 15 m above snow,
R = rainfall in mm per day,
TF = snowpeak temperature,
S = solar radiation at the snow surface,
F = average forest canopy cover (expressed as fraction),
TD'= difference in due point temperature (in °C) at 3 m and at snow surface in °C.

The Runoff Mechanics of Snowmelt

The area under permanent snow cover is subdivided into elevation zones. Snowmelt from each zone is computed using Eq. (7) or Eq. (8). The runoff generated on each day in each zone is distributed binomially over the subsequent days. Proper time lags for the distribution of melt runoff are assigned. If the time taken by the melt in reaching the outlet be τ, the melt appears at the gauge from the instant τ and onwards. If the area of subzone be A_1, and M_1 is the melt on any day, total runoff volume is given by

$$q_1 = A_1 M_1 \qquad (9)$$

The total melt given by Eq. (9) will be distributed binomially i.e., during interval τ and $\tau + \Delta\tau$ the contribution at time $(t \geq \tau)$ is $r_{\tau}^{(\tau)}\ q_1$. Similarly during the interval $\tau + \Delta\tau$ and $\tau + 2\Delta\tau$ the contribution from the zone, will be $r_{\tau+1}^{(\tau)} \cdot q_1$ and so on, thus:

$$\sum_{t=\tau}^{\infty} r_t^{(\tau)} = 1 \tag{10}$$

The superscript of the coefficient r indicates the subarea and the suffix stands for the time when the melt (M) generated over the subzone would reach the outlet. Snowmelt from other subzones are also subjected to similar distributions. The coefficient r is computed by the relation given below:

$$r_t^{(\tau)} = \frac{(t-1)!}{(\tau-1)!\ (t-\tau)!}\ \omega_\tau\ (1-\omega)^{(t-\tau)} \tag{11}$$

The travel coefficient ω has the value in a range of 0.8 to 0.9 for the Himalayan region.

Application to Beas Catchment

The proposed model has been applied to compute peak floods and the snowmelt runoff for the catchment of Beas River in the Himalayas. Nearly 1400 sq km of the catchment is comprised of high mountain ranges above 4270 m and is under permanent snow cover. The main discharge measuring site is located at Larji.

The daily rainfall data for the eight rain gauging stations in and around the catchment is available. Meteorological data needed in Eqs. (7) and (8) has been interpolated from the nearby meteorologic stations which are subjected to the same meteorologic disturbances as the Beas catchment. For temperature a lapse rate of 1°C per 200 m as recommended by Hill has been adopted. The following relationship has been used for estimating dewpoint temperature from observed saturation vapour pressure data (e_s):

$$e_s = 6.11 \times 10^{at(a+b)}$$

where $a = 7.5$, $b = 277.3$ (for water),
$a = 9.5$, $b = 265.3$ (for ice).

Records of the observatories at Gilgit, Simla, Gulmarg Kargil and Len have been used. The solar load data has been taken from the published records of the Central Building Research Institute at Roorkee.

Analysis of Rainfed Areas

The flood peaks have been computed using the series system of linear channels. The flow times from different parts of the catchment to the outlet were calculated and isochrones were drawn. Readjustment of areas was carried out in accordance with Eq. (4). Since the rainstorms recorded on June 24-27,

1968 and July 14-17, 1969 appeared to be reasonably uniform, the same were used for deriving the instantaneous unit hydrograph and one-day unit hydrograph. The drainage basin area was divided into subareas as proposed in Eqs. (3) or (4). The distributed input from each of these subareas enters at the upper bound of each linear channel of the system which is shown in Fig. 1(c).

Thiessen's polygons are drawn for the basin area. The average effective depth for each polygon area is computed using Eq. (5). The time distribution of rainfall excess is obtained by projecting the same on to observed hyetograph assuming a constant rate of abstraction. The precipitation input and computed runoffs are given in Table 1 and Table 2 respectively.

Table 1. Weighted Gross and Effective Rainfall Computations (Storm 7/15-16/69)

S. No.	Rain Gauge Station	Kulu		Banjar		Keylong		Total weighted Gross Rainfall (mm)
	Station Weight	0.5424		0.3885		0.0691		
	Date	Obs. RF	Wt. RF	Obs. RF	Wt. RF	Obs. RF	Wt RF	
1	15.7.69	29.0	15.73	36.4	14.14	12.10	0.84	30.71
2	16.7.69	11.0	5.96	15.4	5.98	-		11.94
3	Total	40.0		51.8		12.1		42.65
4	P_{ave}			11.45				
5	P_i	10.72		13.90		3.24		

Table 2. Runoff Computations for Peak Flood for the Storm 7/15-16/69

Input Rain fall (mm) Distribution:

S. No.	Date	Kulu	Banjar	Keylong
1	15.7.69	10.72	13.90	3.24
2	16.7.69	-	-	-
3	Total	10.72	13.90	3.24

Direct Run off (cumecs):

S. No.	Date	Ordinate	Kulu	Banjar	Keylong	Total computed	Observed Run off.
1.	14.7.69	Q_o				0	0
2	15.7.69	Q_1	88.00	192.50		280.50	243.7
3	16.7.69	Q_2	119.00	-	1.32	120.32	164.7
4	17.7.69	Q_3	000.77		5.92	6.69	0
5	18.7.69	Q_4			0.76	0.76	
6	19.7.69	Q_5			0		

Three storms recorded on dates 8/10/68, 8/6/69 and 8/18/69 have been analysed along similar lines and a comparison of observed runoff hydrographs and the computed from model is given in Figs. 2 and 3.

Analysis of Area under Snowcover

As discussed earlier, the contribution of snowmelt to the river runoff has been worked out using the meteorologic parameters. The area under permanent snowcover (above 14,000 ft) is divided into elevation subzones at intervals of 2000 ft and are detailed in Table 3.

Snowmelt from subzones are worked out using meteorologic parameters extrapolating their values for the mid-elevation of each zone and substituting them in Eqs. (7) or (8) whichever

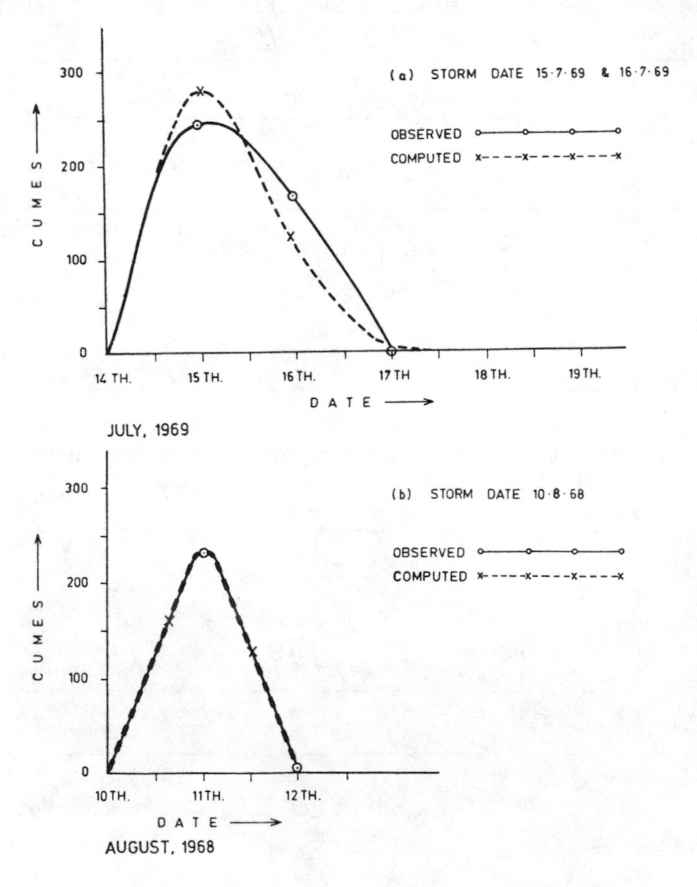

Fig. 2. Comparison of computed and observed hydrographs (storms July 14, 1969 and August 10, 1968)

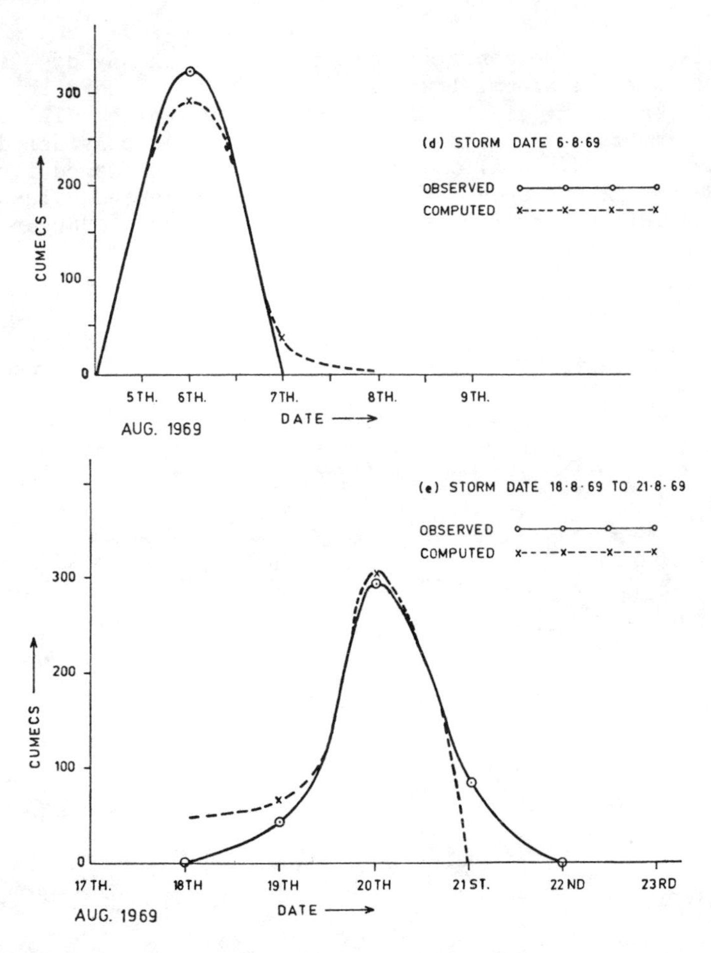

Fig. 3. Comparison of computed and observed hydrographs (storms Aug. 6, 1969 and Aug. 18, 1969)

Table 3. Elevation Subzones of Areas above Snowline

S. No.	Subzone altitude (in 1000 ft)	Mid. Elevation (in 1000 ft)	Area of subzone (Sq.km.)
1	14 -16	15	652
2	16 - 18	17	378
3	18 - 20	19	300
4	20 and above	20.5	72
		Total	1402

is applicable. The contribution of snowmelt on any day is considered to be binomially distributed over the days that follow. The coefficients are computed by using Eq. (11). Travel time from the subzone 1 to the outlet is considered as one day. An additional one-day time lag is considered for the subsequent subzones. The observed and computed values of the total discharges for the period April to June 1969 are compared in Fig. 4.

Conclusions

In Figs. 3 and 4, a close agreement of computed and observed hydrographs implies that a series system of linear

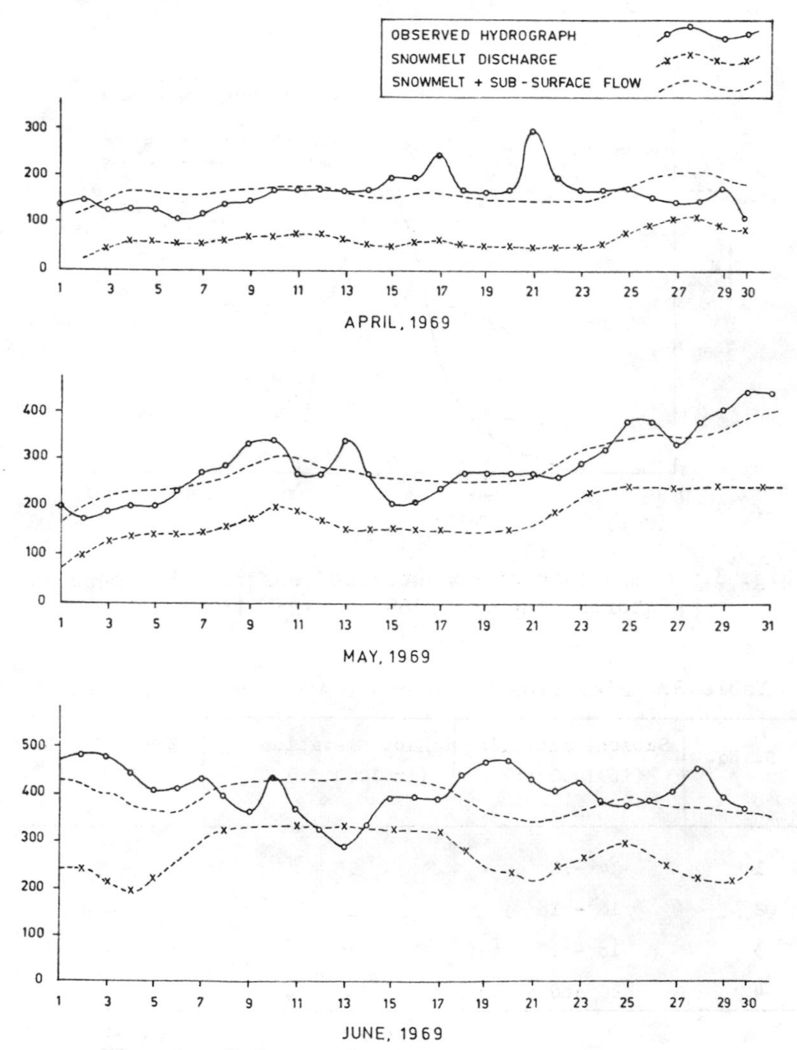

Fig. 4. Hydrographs for snowmelt season

channels is capable of representing the catchment action of a mountainous terrain. It is mainly due to the fact that input is considered as distributed. Also, the snowmelt estimation using meteorologic parameters is fairly suitable, particularly when details of glaciers and snowpacks in high Himalayan reaches are not available for hydrologic and design purposes. The Beas River is an important river of Himachal Pradesh and Punjab states in India. These investigations reveal that in 1969 during the lean flow season (i.e., from April to June) the snowmelt constituted 40 to 67% of the total Beas River discharge.

References

[1] Chow, V. T., 1964. "Handbook of Applied Hydrology," McGraw-Hill Book Company, New York.

[2] Central Building Research Institute Publication, 1959. "Climatological and Solar Data for India," Roorkee, U.P., India.

[3] India Meteorological Department Publication, 1953. "Climatological Tables of Observatories in India."

[4] Dooge, J.C.I., 1959. "A General Theory of Unit Hydrographs," Journal of Geophysical Research, 64(2) 241-256.

[5] Hill, S. A. "The Meteorology of the North-West Himalaya," India Meteorological Memoirs, Vol. 1, Chap. XIII, pp. 377-426.

[6] Corps of Engineers, U. S., 1973. "Hydrograph Analysis Vol. 4," Davis, California.

[7] Mathur, B. S., 1972. "Runoff Hydrographs for Uneven Spatial Distribution of Rainfall," Ph.D. dissertation, Indian Institute of Technology Delhi, New Delhi.

[8] Mokliak, V. I. and T. A. Tchekushkina, 1967. "Calculation of Flood Discharges by Means of Unit Hydrograph," I.A.S.H. Symposia on Floods and Their Computations, Leningrad.

[9] Viessman, W., J. W. Knapp, G. L. Lewis and T. E. Harbaugh, 1977. "Introduction to Hydrology," Dun-Donnelley, 2nd edition, 704 pages.

A MONTHLY WATER BALANCE MODEL APPLIED TO TWO DIFFERENT WATERSHEDS

By

A. Van der Beken
Vrije Universiteit Brussel
Pleinlaan 2
B-1050 Brussels, Belgium

Abstract

A water balance model on a monthly basis uses precipitation and Penman evaporation rates as input. It has six parameters whose initial values can be easily evaluated. The pattern search technique has been used for automatic optimization.

The model has been applied successfully to two different watersheds. The watershed of the Grote Nete has an area of 553 km^2, has a relatively low relief (ca. 55 m) and its soil is sandy. Moreover, deep percolation into an important aquifer is likely to occur and seepage from crossing navigation channels is possible. The watershed of the Zwalm has an area of 115 km^2, a relatively moderate relief (140 m) and a loamy soil. Deep percolation is necessarily restricted.

A comparison of the parameter values optimized for both watersheds indicates the consistency and reliability of the model. For instance, average actual evapotranspiration in the sandy region is calculated for a period of six years as 62% of the evaporation, while in the loamy region the ratio is 86%. In both cases, total stream flow volumes are simulated with a relative error less than 3%.

This model illustrates four guidelines for building a hydrological model. The author's conclusion is that no single model can be fully representative in any detail for a wide variety of streams.

Model Philosophy

Guidelines for building conceptual models in hydrology according to Crawford and Linsley (1966) can be listed as follows:

(1) the model should be representative for the regimes of a wide variety of streams;

(2) it should be easily applied to different watersheds with existing hydrological data;

(3) it should be physically sound so that, in addition to streamflow, estimates such as surface runoff or evapotranspiration, can be made.

The purposes of these requirements are obvious. The first requirement is a goal, supported only to the degree that the model is successful in different areas. The second guideline asks for a minimum of parameters. This allows for a rapid calibration and possibly for selecting parameters when the model is used in areas where data for calibration are lacking. Finally, the third requirement is the most imperative if we want to use the model for evaluation of the effect of watershed changes on different outputs.

We suggest an additional guideline:

(4) the calibration procedure should be an automatic one, or at least, fully reproductive without ambiguity.

This is a classical requirement of scientific techniques. No one should propagate a model when the underlying calibration technique is not clear and accessible.

These guidelines have been kept in mind by constructing a monthly water balance model which will be described in the next sections.

The Model Concept (Fig. 1)

A single reservoir with storage S represents the total water of the basin. All water balance terms have a monthly basis and are expressed in mm. The storage is recharged by different inputs: precipitation P, surface water inflow, and groundwater inflow. It is depleted by evapotranspiration E, surface outflow Q, groundwater outflow G, and by pumping volumes T not returned to the water system within the limits of the basin.

Evapotranspiration. Potential evapotranspiration values EP are available in Belgium for different stations and watersheds (Bultot and Dupriez, 1974). They were calculated by the Penman formula, and adjusted for variable albedo and heat transfer. However, these corrections are based on a limited number of observations in only two hydrometeorological stations. On the other hand, evaporation values E_o for the free water surface are also known but do not need adjustments.

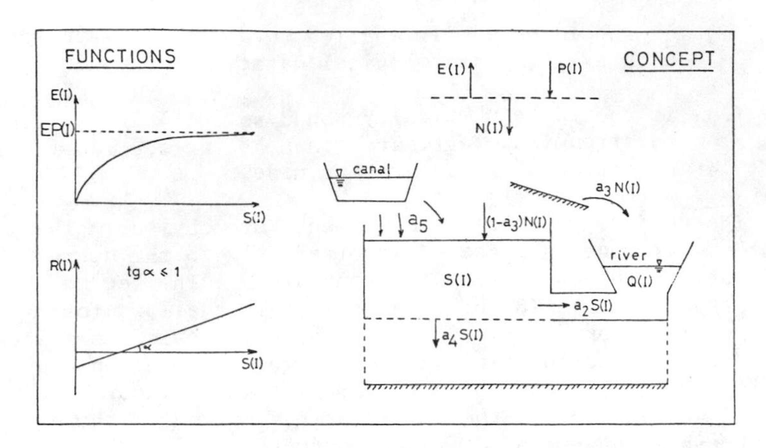

Fig. 1. Model concept and model functions

Hence, we end up with two possibilities: We can estimate the actual evapotranspiration rate E from EP or from E_o. In either case, we must introduce a second variable, which will be obviously the storage S. The proposed equations are:

$$E = E_o \left(1 - e^{-a_1 S}\right) \tag{1}$$

or

$$E = EP \left(1 - e^{-a_1' S}\right) \qquad a_1,\ a_1' > 0 \tag{1'}$$

This equation is physically sound and takes into account the common assumption that E is almost at the potential rate until moisture content is very low. Parameter a_1 or a_1' incorporates the influence of soil texture and vegetation pattern. The albedo and heat transfer corrections, introduced by Bultot and Dupriez (1974), are equally influenced by the soil texture and vegetation pattern. Hence, it seems logical to use Eq. (1) (E_o and a_1) instead of Eq. (1') (EP and a_1'). Values of E_o are unique and only based on meteorological observations, without local influences of soil or vegetation. Parameter a_1 then reflects all local influences on the evapotranspiration process. The value of a_1 will decrease when the soil texture is more sandy.

Infiltration and surface runoff. If $P - E < 0$, neither infiltration nor surface runoff occurs and the storage S is depleted by an amount $N = P - E$.

If $P - E > 0$, the net precipitation N is divided into two parts: a fraction $a_3 N$ is directly diverted to the river and flows out, while $(1-a_3)N$ infiltrates and fills the reservoir. The surface outflow Q has therefore two components when $N > 0$:

$$Q = a_2 S + a_3 N \qquad 0 \le a_2, a_3 \le 1 \tag{2}$$

Parameter a_2 will increase when soil texture is more sandy. Parameter a_3 will increase with urbanization degree and average basin slope. When $N < 0$, surface outflow is:

$$Q = a_2 S \tag{2'}$$

<u>Surface and groundwater inflow or outflow</u>. In a general case many possibilities may exist for inflow or outflow of water which is unknown or cannot be measured directly:

- ill-defined basin limits
- seeping navigation channels crossing the basin
- groundwater inflow and outflow
- unknown pumping activities rejecting water outside the basin
- introduction of water into the basin by water distribution networks.

As long as very little is known about these terms, a very simple equation might be useful:

$$R = a_4 S - a_5 \qquad 0 \le a_4 \le 1 \qquad a_5 > 0 \tag{3}$$

If pumping rates and water distribution rates are known, then a variable T (negative or positive) may be introduced. T is the imported or exported volume of water.

<u>The balance equation</u>. The final balance equation can be written as follows:

$$S(t+1) = S(t) + P(t) - E(t) - Q(t) - R(t) - T(t) \tag{4}$$

Obviously, $S(1)$ must be known initially. There is only one known input variable: $P(t)$. $E(t)$ is calculated from the known variable $E_0(t)$ but is related to $S(t)$. The only known output variable is $Q(t)$. Hence, $Q(t)$ will be used as a calibration measure.

<u>Relationship between infiltration and surface runoff</u>. There is an interesting relationship between a_2 and a_3, which can help us in choosing initial values of a_2 and a_3.

When neglecting $R(t)$ and $T(t)$, Eqs. (2) and (4) yield

$$Q(t) = a_2(1-a_3) \sum_{j=1}^{\infty} (1-a_2)^{j-1} N(t-j) + a_3 N(t) \tag{5}$$

Let us consider a unit net-precipitation at $t=0$:

$$N(0) = 1$$

$$N(j) = 0 \qquad \forall j \neq 0$$

We find then:

$$Q(0) = a_3$$

$$Q(j) = a_2(1-a_3)(1-a_2)^{j-1} \qquad \forall j > 0 \tag{6}$$

Hence:

$$\sum_{t=0}^{\infty} Q(t) = a_3 + a_2(1 - a_3) \sum_{t=0}^{\infty} (1 - a_2)^t = 1$$

There are two possibilities:

a. $Q(1) > Q(0)$ or $a_2 > \frac{a_3}{1-a_3}$ (7)

b. $Q(1) < Q(0)$ or $a_2 < \frac{a_3}{1-a_3}$ (7')

Figure 2 clearly shows how the values of a_2 and a_3 will influence the delay of the net-precipitation.

Calibration Procedure

An automatic procedure for calibration or optimization of the parameter values has the advantage that no subjectivity is introduced when comparing calculated to measured outflows. When the procedure is fully described, anyone may reproduce the results using the same data. There is also strong evidence that all parameters should be optimized simultaneously since interdependency is likely in hydrological models. For instance, in the model described, we will treat $S(1)$ as a parameter a_6 to optimize much in the same way as a_i $(i=1,5)$.

For models such as this with relatively few parameters, the "Pattern Search" technique of Hookes and Jeeves may be well adapted (Wilde, 1964, p. 145). It has been successfully applied to three types of hydrological models by our research group.

The criterion generally used in automatic optimization is the sum of squares of the deviations between calculated and measured observations, i.e., outflows Q:

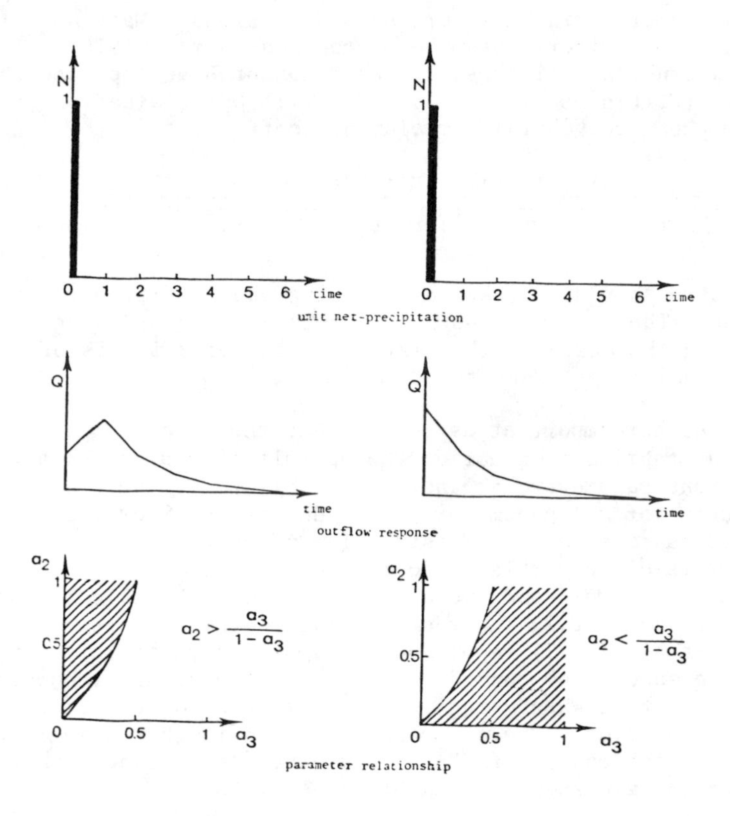

Fig. 2. Effect of a unit net-precipitation input on the outflow Q and its relationship with parameters a_2 and a_3

$$F^2 = \sum_{t=1}^{N} [Q(t) - M(t)]^2 \tag{8}$$

where N = number of observations
M = measured outflow.

A common derived criterion is the standard deviation

$$SD = \sqrt{\frac{F^2}{N-i}} \tag{9}$$

where i = number of parameters. This criterion allows us to compare either different calibration periods with the same model, or the same calibration period with different models (different number of parameters).

Both criteria, however, have dimensions. Many non-dimensional criteria have been proposed (Sarma, 1970). It is well-known that all these criteria do not have the same sensitivity (Fortin et al., 1974). We limit our evaluation by using the more general correlation coefficient:

$$r = \frac{N\Sigma M(t)Q(t) - \Sigma M(t)\Sigma Q(t)}{\sqrt{[N\Sigma M(t)^2 - (\Sigma M(t))^2][N\Sigma Q(t)^2 - (\Sigma Q(t))^2]}} \qquad (10)$$

Care should be taken by choosing the initial parameter values. These values must be calculated on a hydrological basis in the best possible way. The optimization is often very significantly shortened.

Another important aspect is that the F^2-criterion may show a stabilization during the optimization process, although different parameters are not yet stabilized. Therefore, the evolution of all parameter values during the iteration process should be considered. This problem of unstable parameter values is closely related to the sensitivity problem of the parameters. Mein and Brown (1976) propose the calculation of a covariance matrix in order to measure this sensitivity. Bladt et al. (1977) report on the application of this method for the same model considered here. The procedure, however, is prohibitive when using a large number of observations N for calibration. The alternative method in that case is to consider the change of F^2 for a given percentage change in parameter values (Dawdy and O'Donnel, 1965).

Watershed Selection

The model has been applied to two different watersheds with characteristics summarized in Table 1. Both regions have a well-developed agricultural pattern with a little more pasture in watershed I. Forest cover is relatively low and is less than 10%. The urbanization degree can be estimated at 15-20% for both watersheds.

Table 1 Some Characteristics of the Selected Watersheds

Name	I Grote Nete	II Zwalm
area (km^2)	553	115
relief (m)	55	140
soil	sandy	loamy

Watershed I has a thick neogene sandy deposit but its river system is only draining the upper pleistocene and holocene cover of loamy sand. Groundwater inflow and outflow in the underlayer is likely to occur both in the northern and southern part of the watershed. Moreover, an important navigation channel system of about 40 km length crosses the basin. The water level in these channels is everywhere above the groundwater level. Therefore, seepage from these channels is possible. Pumping activities by a water distribution agency and by industry export a small volume of water.

Watershed II has a loamy cover of variable thickness, resting on several clay-sand layers. There have been more than 250 springs registered in the region. A small volume of water is probably imported by the drinking water distribution system.

For both watersheds, the calibration period extended from January 1967 till December 1972.

Results and Discussion

Using the balance equation (4) without term $T(t)$, initial and final parameter values after 100 iterations, are listed in Table 2. Parameter a_1 is less for watershed I than for watershed II which confirms our concept on the soil texture influence. Important also is the comparison of the values of EP, E_0 and E for both watersheds (Table 3). As may be expected, E is much larger in watershed II than in watershed I.

Parameters a_2 and a_3 for both watersheds have values which corroborate our assumptions. The constraint $a_2 < \frac{a_3}{1-a_3}$ is fulfilled for watershed II but not for watershed I, although the difference $a_2 = 0.192/\frac{a_3}{1-a_3} = 0.187$ is small. This could be attributed to the larger area and smaller slope of watershed I.

Parameter a_4 suggests a positive groundwater outflow in watershed I, but almost none in watershed II. This also confirms our knowledge of the geological structure of both watersheds.

Parameter a_5 is almost zero in watershed I which may be realistic since bottom and side wall obstruction of the channels can reduce the seepage rate. In watershed II, however, it takes a rather surprisingly large value, which means that a volume of water is introduced into the basin at a constant rate of 3.95 mm per month. This volume is certainly unrealistic when compared to the drinking water supply. A more logical explanation is the influence of the springs. Therefore,

Table 2 Initial and Final Parameter Values after 100 Iterations

Parameter	Initial Value	Final Values Watershed I	Watershed II
a_1	0.01000	0.01009	0.02860
a_2	0.27000	0.19187	0.15000
a_3	0.20000	0.15725	0.21200
a_4	0.03000	0.03572	0.00060
a_5	1.00000	0.00156	3.95000
a_6	300.00	290.25	357.00
F^2		1606	5897
SD		4.93	9.45
r		0.92	0.85

Table 3 Average Values in mm/year(*)

	Watershed I	Watershed II
P	725.9	709.7
E_o	617.9	603.5
EP	492.4	478.0
E	383.7	521.8
E/E_o	0.621	0.865
M	321.4	264.3
Q	322.5	271.1
Q/M	1.0003	1.026

*Values of EP are those for grass. For watershed I values of station Mol were used. For watershed II values of stations Melle and Ukkel were used to calculate arithmetical averages of EP and E_o.

the model for watershed II should possibly introduce a second reservoir, also discharging into the river system.

Parameter value a_6, which is the initial storage value $S(1)$ is also very well explained in both cases.

Confidence in the model, as applied to both watersheds, is further supported by the results after introduction of a constant rate $T = 0.8$ mm in the balance equation. For watershed I the following parameter values were then obtained:

$a_1 = 0.01009$; $a_2 = 0.19336$; $a_3 = 0.15837$

$a_4 = 0.03083$; $a_5 = 0$; $a_6 = 287.27930$

The important parameters a_1, a_2 and a_3 which should theoretically not be influenced by introducing T remain stable. Only parameters a_4 and mainly a_5 change, which can normally be expected.

Another support for the model is the simple analysis of the F^2-changes when changing one parameter, all others being constant at their optimal value. Results are shown in Fig. 3a and 3b where F^2 is shown versus a percentage change of parameter value. The three most important and well defined parameters a_1, a_2 and a_3 show greater sensitivity than the less defined parameters a_4 and a_5 which play a minor role in the balance.

Conclusion

The results shown in this paper are only a small part of the data available on both watersheds and the model tested (Bladt et al., 1977; Van der Beken and Byloos, 1977). The aim was to discuss the guidelines for building a hydrological model and to show that they have been fairly satisfied by the model presented. That does not mean that the model is perfectly adapted to any watershed. On the contrary, as suggested for watershed II, a better result will be obtained by slightly rebuilding the model according to our knowledge of the physical reality of the system. The author feels that this example may well suggest a general rule: No single model can be built with the goal of encompassing all possible situations in any detail. In this view, guideline (1) in the first section should be modified as follows: "the model _in its main features_, should be representative for the regimes of a wide variety of streams". Any model should be applied with caution and its results carefully interpreted. Minor changes will often be necessary.

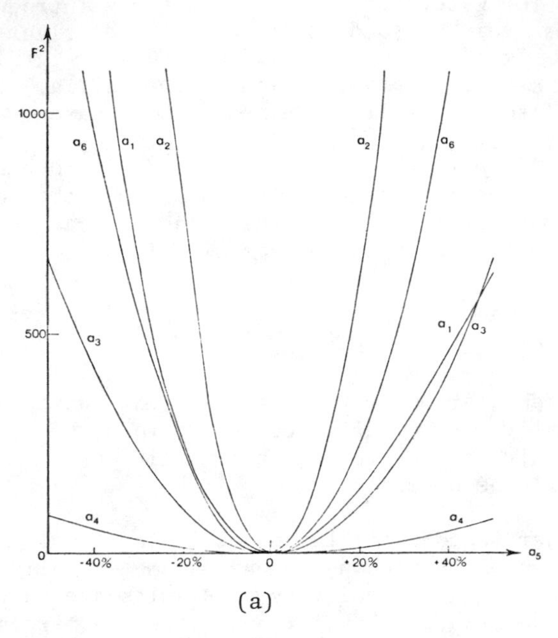

(a)

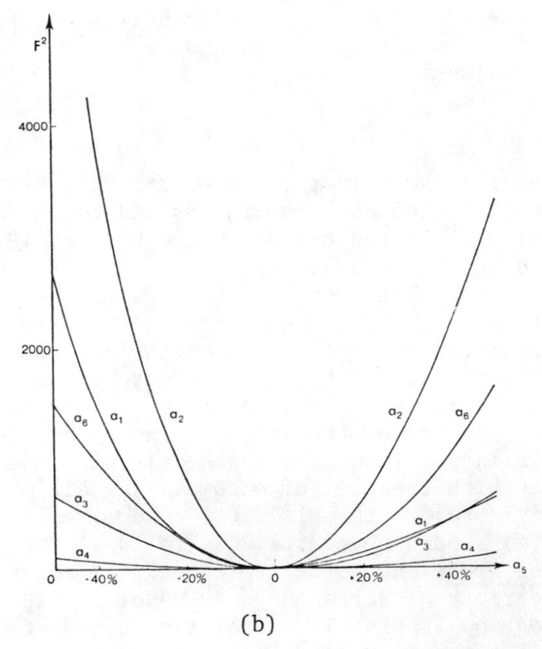

(b)

Fig. 3. Change in F^2-values for a percentage change of the optimized value for all 6 parameters.
a. Watershed I b. Watershed II

Acknowledgment

The data used in this paper were prepared by Mr. A. Bladt. His help is greatly appreciated. Acknowledgment is also due to the Steering Group for Water Management in the Province of Antwerp for support of the research project.

References

[1] Bladt, A., Demaree, G., and Van der Beken, A., 1977. "Analysis of a Monthly Water Balance Model Applied to Two Different Watersheds," IFIP Working Conference on Modelling and Simulation of Land, Air and Water Resources Systems, Ghent, Belgium, August 29-September 2, 1977.

[2] Bultot, F. and Dupriez, G., 1974. "L'évapotranspiration Potentielle des Bassins Hydrographiques en Belgigue," Institut Royal Météorologique, Bruxelles, Publ. Série A n° 85, 61 p.

[3] Crawford, N. H. and Linsley, R. K., 1966. "Digital Simulation in Hydrology: Stanford Watershed Model IV," Dept. of Civil Engineering, Stanford University, Technical Report No. 39.

[4] Dawdy, D. R. and O'Donnell, T., 1965. "Mathematical Models of Catchment Behaviour," Proc. ASCE, 91, HY4, pp. 123-137.

[5] Fortin, J. P. et al., 1974. "Proposition et Analyse de Quelques Criteres Adimensionnels D'Optimisation," Proc. Warsaw Symposium "Mathematical Models in Hydrology", IAHS, Publ. N° 101:548-557.

[6] Mein, R. G. and Brown, B. M., 1976. "Statistical Sensitivity of Parameters in a Rainfall-Runoff Model," Technical Conference Instit. Engrs., Australia, pp. 83-87.

[7] Sarma, P., 1970. "Effects of Urbanization on Runoff from Small Watersheds," Ph.D. thesis, Purdue University, LaFayette, 282 p.

[8] Van der Beken, A. and Byloos, J., 1977. "A Monthly Water Balance Model Including Deep Infiltration and Canal Losses," Hydrological Sciences Bulletin 23, 3-9.

[9] Wilde, D. J., 1964. "Optimum Seeking Methods," Prentice Hall, Inc., N. J., 202 p.

A PHYSICALLY REALISTIC FORMULATION FOR CONCEPTUAL MODEL DEVELOPMENT

By

H. S. Wheater
Department of Civil Engineering
University of Bristol
Queens Building, University Walk
Bristol, 8, England

Abstract

This paper describes the formulation and performance of a conceptual model of flood runoff in small, predominantly rural catchments. The conceptual structure is based on the particular mechanisms of catchment response appropriate to a low-lying area with subsoils of reduced permeability. The dominant soil response is simulated by a linear reservoir, and it is shown that this representation has physical significance. Model structure is simplified by using isolated event simulation for which a soil water budgeting model is used to calculate Soil Moisture Deficit, a physically realistic index of catchment response. The performance of this conceptual model is compared with that of a simpler black box model.

Introduction

Physical realism is a concept which has had wide interpretation by mathematical modellers. It can be used simply to guide the formulation of a non-linear model or to enable model application to ungauged catchments or changing land use by the physical explanation of parametric variation. Few, if any, successful applications of the latter have been recorded.

This paper discusses the extent to which physical realism is applicable to catchment simulation and a physically realistic conceptual model is presented for a particular application to small catchments in which typical soil response is shown to be analogous to drainflow. This leads to the development of a linear reservoir approximation of the active soil phase. The ability of the model to identify the variable response of a small catchment is shown to be comparable with that of a simpler "black box" model, but the physically based model has the potential for wider application.

Physical Realism

A physically realistic model can be one with a conceptual structure based on a physical interpretation of catchment response or a deterministic model which attempts to represent the full complexity of flow processes in terms of fundamental physical properties. The latter is the goal of scientific investigation, but the feasibility of attaining that goal for catchment simulation is questionable.

Woolhiser (1971) examines the prospects of fully-deterministic simulation which he suggests cannot in general be achieved for subsurface flows. Stephenson and Freeze (1974) comment that massive data requirements as well as theoretical and computational limitations preclude the application of fully-deterministic models on other than a small plot scale.

Snyder (1971) considers that parametric hydrology is applicable to the general order of hydrological problems in which some understanding of causal relationships is available but insufficient information exists for a fully deterministic approach. The goal of parametric hydrology is progress towards determinism, but this is impeded by lack of homogeneity in natural systems, which prevents their precise definition. Snyder suggests that the micro-scale differential calculus which provides the mathematical basis of the deterministic approach must be replaced by "macro-scale" relationships. If models based on micro-scale relationships are applied on a small catchment scale, the model elements will still be orders of magnitude greater than the laboratory scale on which such relationships were generally derived. The nonuniformities of natural systems are likely to have a major effect on runoff, influencing the precision of such models accordingly. This will be most pronounced for small catchments, in which the natural damping of channel networks and the effect of spatial averaging is minimised. The difficulty of simulating the flow mechanisms of upland catchments was discussed as a result of a numerical analysis by Freeze (1972). Hewlett (1974) commented on the concentration effect of three-dimensional form, the high permeability of stream banks and the variable nature of channel networks. Other workers have commented on the high effective permeabilities in areas of forest litter and the effect of natural "pipes" in generating rapid downslope runoff in other upland areas. Such effects are not easily included even if they are expected in a micro-scale approach.

If a micro-scale approach is inappropriate, moving to a larger scale of response necessitates averaging flow processes without explaining the total variability. Highly complex relationships are therefore generally not justified, as approximation is inherent in the parametric approach and necessary to reduce the problem of data input. A procedure whereby model complexity must be justified by performance is outlined by Nash and Sutcliffe (1970), and has been adopted in this study.

The application of the parametric approach to small lowland catchments is presented. Particular attention has been given to the physical identification of the mechanisms of response. A primary mechanism has been identified and a parametric representation developed based on the physical processes.

The Gloucester Surface Water Study

The Gloucester Surface Water Study was initiated to consider the problems of flooding with particular reference to an area of North-West Gloucestershire, England. The wider aims of the study include the effect of urban development on drainage and flooding in the area. This paper is confined to the simulation of predominantly rural catchments.

The study area of 400 km^2 is shown in Fig. 1 (Shaw and Wheater, 1977). The instrumentation was installed from 1970. Most streamflow recorders are pressure bulbs calibrated in situ and recording on weekly pen-charts. There are six flumes. The majority of rain gauges are daily autographic tilting-syphon recorders. The catchments are small (< 50 km^2) and predominantly flat and low lying. Landslopes are generally

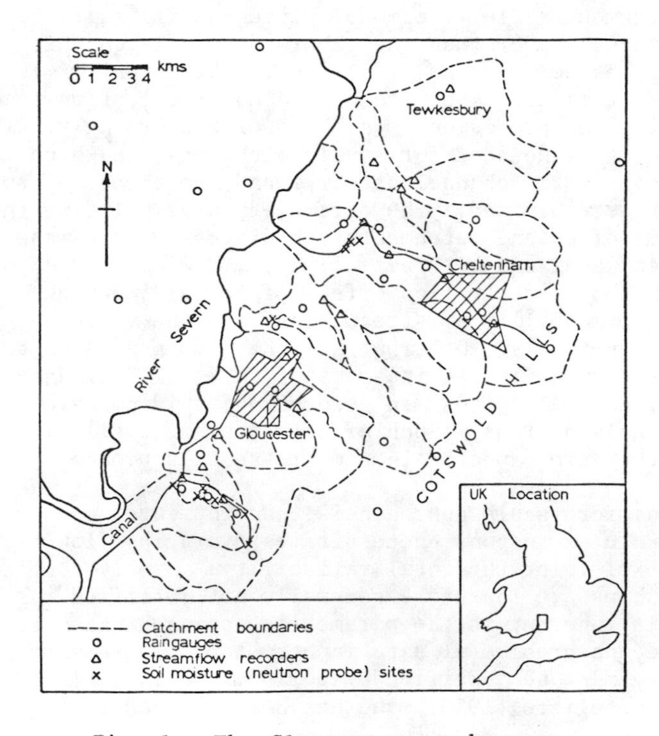

Fig. 1. The Gloucester study area

less than 3%. The soil water regime is Class 3a (Soil Survey, 1975), i.e., soils with impermeable substrata, in drier lowlands with significant moisture deficit in most years. This regime accounts for 32.7% of England and Wales. With the exception of the major urban areas shown on Fig. 1, the land use is predominantly agricultural. Extensive artificial drainage has been installed for agricultural purposes, almost exclusively tile drainage. The streams in the area are small (typically 3 to 4 m wide and 2 m deep) and may be intermittent due to highly seasonal catchment response.

Experimental work in low-lying areas is limited and has mainly come from an interest in tile-drainage design. Site investigation in the area (Wheater, 1977) has shown relatively permeable topsoils (1.5mm/min) overlying relatively impermeable subsoils (1.0×10^{-2}mm/min) with commonly a break in permeability at the ploughing depth of 0.3 m. A similar soil configuration was investigated by Trafford (1973), who concluded that the predominant flow occurred laterally at the permeability interface, so that tile-drains merely interrupted this flow (Fig. 2). The presence of tile drains did not significantly modify the natural flow mechanisms, although they did affect the threshold at which overland flow could occur (due to surface saturation). Soil moisture studies in the Gloucester area (Wheater, 1977) confirmed that response occurred primarily in the topsoil and changes in moisture below this level followed a seasonal pattern, with moisture levels below 1 m unchanging.

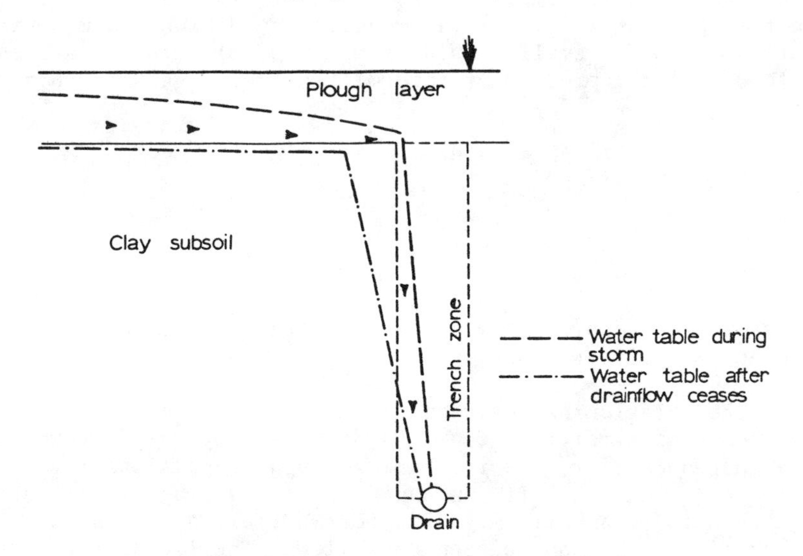

Fig. 2. Flow path to a tile drain in a clay subsoil (after Trafford, 1973)

The physically realistic simulation of the runoff generation process thus rests on the modelling of the surface layer response. The soil configuration described is analogous to the classical field drainage problem of field drains located at an impermeable boundary. In the undrained condition, the drainage outlet is a field ditch; in the drained condition, the outlet is a backfilled trench with the tile-drain actually located at the bottom. In the former case, distance between "drains", i.e., ditches may be relatively large, but field experience indicated that the mechanism would not be substantially different. With this analogy in mind, analytical approximations of drainflow theory were examined.

Drainflow

A number of analytical solutions exist to the non-steady drainflow problem. These are reviewed by Van Schilfgaarde (1970). A development of potential theory leads to a relationship for water-table recession of the form

$$\frac{m}{m_o} = \exp(-Kt/CLfF)$$

where m is the mid-plane water table height at time t after rain ceases, m_o at time zero, K is the hydraulic conductivity, L the drain spacing, f the drainable porosity or specific yield and F represents the initial boundary condition. C is a coefficient which will be $\gg 1$ shortly after initial ponding, $= 1$ for intermediate conditions, and approximately $= 0.8$ for small m ($0.02 < m/L < 0.08$). This has been extended to the problem of intermittent recharge. If $B = FfCL$,

$$\frac{m}{m_o} = \exp(-Kt/B)$$

and

$$\frac{dm}{dt} + \frac{m}{B} = 0$$

Hence for C approximately constant, the system can be represented by a linear reservoir.

The relationship between m and q, the discharge expressed as a vertical flux, is discussed by Childs (1960), who quotes earlier work indicating rough proportionality between q and m. If, in addition, the average yield rate is taken to be a uniformly distributed flux, an exponential recession of m can be seen to follow. The linear relationship between m and q has been approximated by subsequent laboratory experiments and supported by field observation including those of Trafford and Rycroft (1973). This implies that tile-drain discharge may also be represented by a linear reservoir. Confirmation is also obtained from a solution including unsaturated flow, assuming constant diffusivity to apply in the vicinity of the water table, which indicates exponential drainflow recession.

A simple linear reservoir can therefore be seen to be an approximation to the response of the field configuration, and the reservoir coefficient can be defined in terms of diffusivity or permeability and specific yield.

Conceptual Model Structure

A physically realistic conceptual model framework was defined which incorporated the representation of soil response discussed above. Within this framework an optimal development was sought to give the best performance while maintaining as simple a structure (i.e., fewest parameters) as possible, following the procedure of Nash and Sutcliffe (1970).

An important simplification was introduced by using an isolated event formulation. In addition to removing evaporation from the runoff model and reducing the effect of baseflow, this permits the maximum use of discontinuous records. An input of an index of catchment response is required.

The soil moisture budgeting model developed by the British Meteorological Office (Grindley, 1970) provides such an input. This model is well established in the U.K. as a routine agricultural service and has also been applied as an index of catchment response (N.E.R.C., 1975), (Shaw and Wheater, 1977). The model uses a "root constant" as the single parameter for each category of land use. Following the original concept of Penman (1949) a potential soil moisture deficit is calculated by balancing rainfall and potential evaporation. This potential deficit is transformed to give an actual deficit using a postulated relationship (Fig. 3). Few field data are available to validate the model. Wheater (1977) shows that the model is highly sensitive to catchment land use description, but that this simple model gives a realistic representation of soil moisture variation in the Gloucester area for a particular category of land use. The model can therefore provide a physically realistic input of catchment soil moisture to a model using explicit soil moisture accounting.

A lumped representation was adopted for initial model development, which was assessed using 15-minute rainfall and runoff data for a data set of 15 major storms on one of the study catchments, the Normans Brook (area 24.4 km), over the period 1970-1974. These storms included a wide range of catchment conditions, so that considerable variation in the

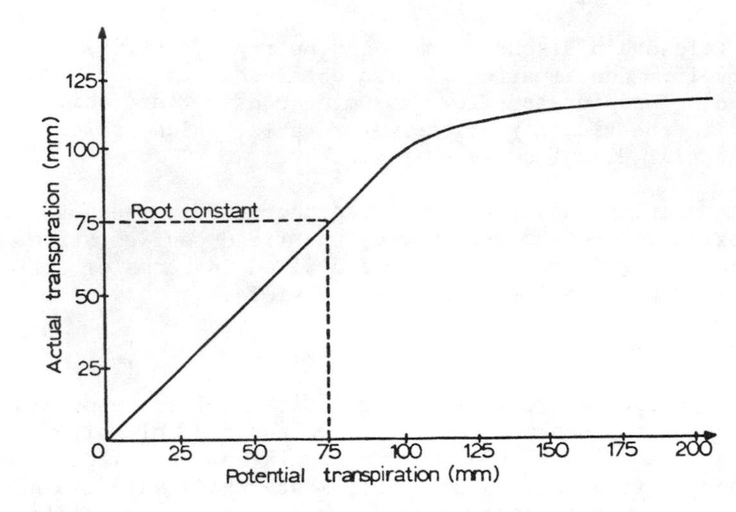

Fig. 3. Drying curve for 75 mm root constant (after Penman, 1949)

magnitude of peak storm flow rates was included. A multivariate stochastic search technique was used during model development as this was independent of irregularities of model response surface anticipated during development. The performance criterion adopted for optimization was the R^2 efficiency defined by Nash and Sutcliffe (1970). This is based on an objective function of the sum of the squares of the difference between observed and simulated flow rates, which accentuates peak fitting - a desirable characteristic for a flood event simulation. A volume criterion was also evaluated.

Model development produced an 11 parameter model, shown in Fig. 4. Rainfall is added to an overland flow reservoir from which infiltration into the surface soil zone is calculated using the Holtan equation (Holtan, 1961). Excess rainfall is routed through a linear reservoir as overland flow. Any initial soil moisture deficit is apportioned to the topsoil as a negative storage subject to a maximum allowable deficit. If a greater deficit exists, it is taken up by the subsoil. Runoff is only permitted from positive storage. The residual infiltration term in the Holtan equation passes as seepage to the subsoil. Lateral flow from the topsoil is calculated using the linear reservoir, as is also the subsoil response generating baseflow. The combined outflows of throughflow, overland flow and baseflow are routed through the channel representation of a further linear reservoir and linear channel (constant time delay). To account for the rapid contribution to runoff from urban and riparian areas, a parameter representing this proportion of the catchment area apportions rainfall directly to channel routing.

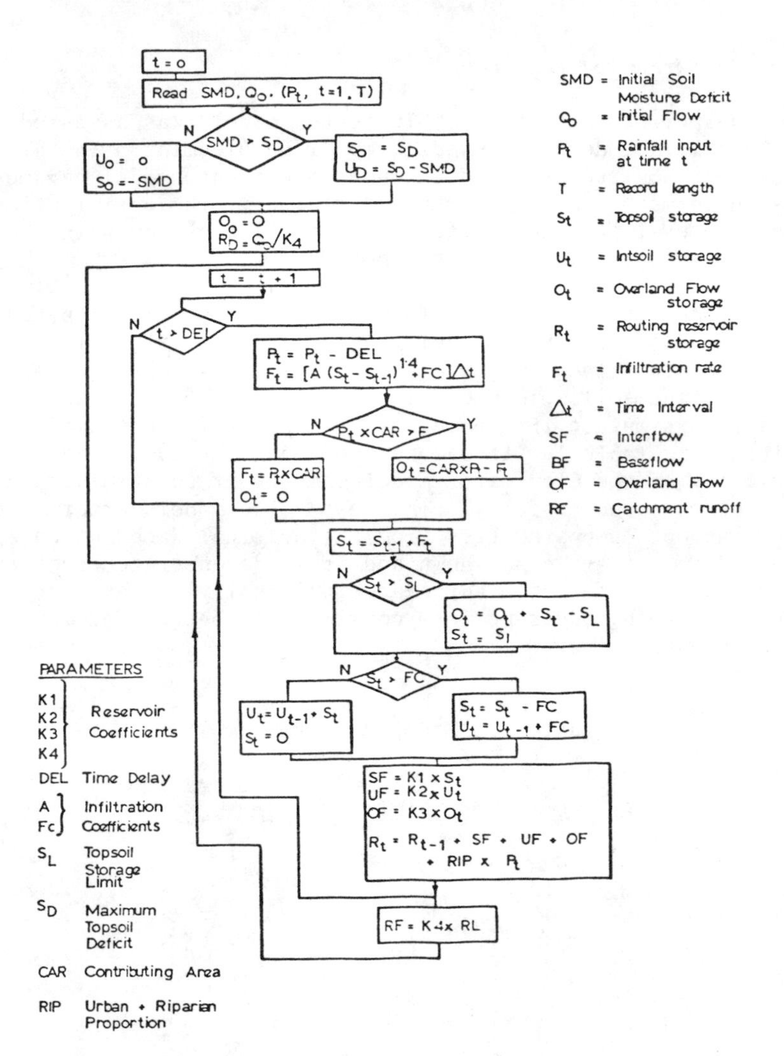

Fig. 4. Flow diagram of Gloucester catchment model

The low-lying character of the area and the general poor drainage results in local areas with no direct drainage outlet. Such areas are not able to contribute to storm response and it is necessary to account for these. In consequence an additional parameter was introduced to represent the actively contributing area of the catchment, CAR, and this was used to proportion the rainfall passing into the active soil components.

The development procedure showed that performance was significantly improved by the use of CAR, the inclusion of an active subsoil reservoir and the introduction of a time delay on the routing procedure. However, various configurations of distributed soil storages proved unsuccessful.

Model Performance

The performance of the "Gloucester model" was assessed on the original data set used in model development. In this assessment no attempt has been made to estimate model parameters in terms of catchment characteristics. The optimisation procedure evaluates the performance of all the storms in the data set on a single set of parameters.

A comparative estimate of model performance is obtained using the Isolated Event Model (N.E.R.C. 1975). This is a four-parameter black box model using the same isolated event structure and soil moisture deficit input. The model was adapted to simulate 15 minute data. Results for both models are shown in Table 1. It can be seen that the additional complexity of the Gloucester model enables it to simulate the data set more successfully. The respective model performance using the optimum parameters from the original data set on a second data set are also shown and it is demonstrated that the I.E.M. has a performance advantage in identifying catchment response. Both models perform considerably better for low S.M.D. events.

Table 1. Model Performance

Storm Date	S.M.D.	Gloucester Model				Isolated Event Model		
		Observed Peak Flow Rate	Simulated Peak Flow Rate	Objective Function	R^2	Simulated Peak Flow Rate	Objective Function	R^2
Year Day No.	mm	m^3/s	m^3/s			m^3/s		
a) Optimised Results								
70218	82.5	0.249	0.533	3.76	0.624	0.162	0.37	0.410
70255	61.6	0.869	0.602	1.81	0.748	0.347	4.95	0.311
70315	22.1	0.762	0.689	6.20	-0.089	0.834	1.57	0.724
70333	3.9	5.163	4.485	18.95	0.938	5.988	23.81	0.922
70339	0.5	1.635	1.521	5.61	0.698	1.153	3.55	0.809
70363	0.0	0.707	0.757	1.61	0.682	0.389	6.67	- 0.317
71022	0.0	2.281	2.095	5.13	0.887	2.005	6.61	0.854
71030	0.0	3.726	3.798	14.45	0.936	3.999	14.28	0.937
71113	31.7	1.214	0.730	16.57	0.293	1.444	31.27	- 0.334
71185	47.0	2.326	2.150	13.83	0.787	1.226	36.32	0.440
71221	96.0	1.319	1.166	1.75	0.895	0.365	24.51	- 0.471
72163	62.9	0.727	0.251	2.73	0.224	0.100	5.11	- 0.454
73043	0.9	0.847	0.952	2.09	0.601	0.577	2.11	0.598
73113	27.0	2.081	1.567	15.59	0.673	1.717	4.00	0.916
74166	101.5	0.722	0.346	1.06	0.527	0.070	4.25	- 0.896
All storms				111.2	0.923		169.38	0.883
b) Reconstruction on Split-Sample Data								
70251	69.1	0.234	0.525	5.12	-16.262	0.211	0.63	-1.132
70310	37.4	1.164	0.646	13.01	0.501	0.876	3.06	0.882
70321	10.1	2.202	1.402	56.70	0.184	1.580	26.23	0.622
71020	0.0	2.618	2.129	4.98	0.889	1.763	10.73	0.756
71076	0.0	2.117	1.987	7.00	0.783	1.618	8.10	0.748
72251	92.0	0.579	0.942	15.11	-2.162	0.200	4.02	0.159
73020	1.3	1.570	2.126	17.92	0.204	1.547	1.57	0.930
73124	27.0	0.633	0.389	1.07	0.493	0.357	1.61	0.240
All storms				120.77	0.698		55.97	0.860

The problem of system identification increases significantly with the number of parameters. The majority of conceptual models have in excess of 10 parameters merely to represent the physical processes. The black box I.E.M. with only 4 parameters can therefore identify those parameters more precisely and this gives the model relatively good performance as shown by split sample testing.

For application to ungauged catchments, model parameters must be derived in terms of catchment characteristics. The object of a physically realistic model is that these parameters are related to physical properties. For a model with few parameters, regional analysis is possible using regression techniques. The performance of the I.E.M. confirms the feasibility of this approach. However, for model application outside the available data base, for ungauged catchments or for changing catchment land use a physically realistic model is necessary. The Gloucester model with a physically based structure has this potential while achieving a performance comparable to that of the I.E.M.

Summary and Conclusions

Fully deterministic simulation is not generally feasible on a catchment scale. A parametric approach is required where the parameters can be explained in terms of physical processes. A physically realistic conceptual rainfall-runoff model for small catchment is developed on which primary soil response is simulated by such a parametric relation. The optimised performance of the model is good, but split sample testing also shows the merits of a simpler black box model. Such a black box model can be applied to ungauged catchments primarily through regional analysis. However, a physically-based model has wider application and in particular can provide a vehicle for the representation of the runoff effects of land use change.

References

[1] Childs, E.C., 1960. "The Nonsteady State of the Water Table in Drained Land," Journal of Geophysical Research, 65, 2:780.

[2] Freeze, R. A., 1972. "Role of Subsurface Flows in Generating Surface Runoff 2, Upstream Source Areas," Water Resources Research 8, 5:1272-1283.

[3] Grindley, J., 1970. "The Estimation and Mapping of Evaporation," Proceedings of the Reading Symposium "World Water Balance," I.A.S.H.

[4] Hewlett, J. D., 1974. "Comments on Letters Relating to 'Role of Subsurface Flows in Generating Surface Runoff. 2, Upstream Source Areas'" by R. Allen Freeze, Water Resources Research, 10, 3: 605-607.

[5] Holtan, H. N., 1961. "A Concept for Infiltration Estimates in Watershed Engineering," U.S. Dept. of Agri., Agri. Res. Serv. 41-51, p. 8.

[6] Nash, J. E., and Sutcliffe, J. V., 1970. "River Flow Forecasting Through Conceptual Models, Part 1 - A Discussion of Principles," Journal of Hydrology 10: 281-290.

[7] Natural Environment Research Council, 1975. "Flood Studies Report," N.E.R.C., London.

[8] Penman, H. L., 1949. "The Dependence of Transpiration on Weather and Soil Conditions," Journal of Soil Science, 1: 74-89.

[9] Shaw, T. L., and Wheater, H. S., 1977. "The Relationship of Unit Hydrograph Parameters to Soil Moisture Conditions, Determined from 8 Small Lowland Catchments in Gloucestershire," University of Bristol Report to Water Research Centre, Jan. 1977, 193 p.

[10] Snyder, W. M., 1971. "The Parametric Approach to Watershed Modelling," Nordic Hydrology 11: 146-166.

[11] Soil Survey of England and Wales, 1975. "Soils and Field Drainage," Technical Monograph No. 7, Soil Survey, Rothamstead, England, 80 p.

[12] Stephenson, G. R., and Freeze, R. A., 1976. "Mathematical Simulation of Subsurface Flow Contributions to Snowmelt Runoff, Reynolds Creek Watershed, Idaho," Water Resources Research 10, 2: 284-294.

[13] Trafford, B. D., 1973. "The Theoretical Aspects of the Relationship Between Field Drainage and Arterial Drainage,"Annual Conference of River Authority Engineers, Cranfield, Ministry of Agriculture, Fisheries and Food.

[14] Trafford, B. D., and Rycroft, D. W., 1973. "Soil Water Regimes in a Drained Clay Soil," Journal of Soil Science, 24, 3: 386-391.

[15] Van Schilfgaarde, J., 1970. "Theory of Flow to Drains," Advances in Hydroscience 6: 42-106.

[16] Woolhiser, D. A., 1971. "Deterministic Approach to Watershed Modelling," Nordic Hydrology 11: 146-166.

[17] Wheater, H. S., 1977. "Flood Runoff from Small Rural Catchemnts," Unpublished Ph.D. Thesis, University of Bristol.

TIME OF PEAK SURFACE RUNOFF FROM RAINSTORMS

By

Ben Chie Yen
Professor of Civil Engineering
Univ. of Illinois at Urbana-Champaign, Ill.

Yung-Yuan Shen
Lead Hydraulic Engineer, Hydraulics Division
Stone and Webster Eng. Corp., Denver, Colo.

Ven Te Chow
Professor of Civil Engineering
Univ. of Illinois at Urbana-Champaign, Ill.

Abstract

Time of occurrence of peak discharge of surface runoff from a watershed under rainfall is a reflection of the response behavior of the watershed to its input. Therefore, it depends on the hydrometeorologic characteristics of the rainstorm and physiographic characteristics of the watershed. Existing empirical formulas diverge considerably on the effects of the different influential parameters. Results of controlled laboratory experiments are utilized to provide basic information which is useful in engineering applications. It has been found that for engineering purposes as an approximation the time to peak $t_p = CA^{\alpha}L^{\beta}e^{bT}/S^{\gamma}i^{\lambda}w^{m}$ in which A is the watershed area; L is a characteristic length, S is the watershed slope, T and i are the duration and intensity of the rainfall excess, and w the speed of rainstorm movement. The coefficient C and exponents α, β, b, γ, λ and m are all positive values depending on the watershed and rainfall characteristics.

Introduction

Numerous previous investigations have been conducted on the magnitude of peak rainstorm runoff rate from a watershed. Such past efforts are clearly commendable and justifiable because in the planning and design of many environmental and water resources projects the peak runoff rate is one of the most important hydrological factors to be considered. However, in many problems such as those for flood mitigation and pollution control, the time of occurrence of the peak discharge is also important. In fact, any analysis involving hydrographs would consider the peak discharge occurrence time. Yet, only limited information exists in the literature concerning the peak occurrence time. This is due partly to the inconsistency in its definition and partly to the difficulties in accurately determining this time from recorded data. It is not difficult to determine from measured data the clock time that the peak discharge of the total runoff occurs, including the case that the data is digitized over finite time intervals. However, considerable uncertainties exist on the time of occurrence and magnitude of the peak discharge of either the direct runoff or the surface runoff (Chow, 1964, p. 14.3) because of the assumption involved in making the base flow deduction. The reference time from which the peak discharge time is measured; be it the centroid of the hyetograph, the incipient of the surface or direct runoff, or the beginning of the rainfall excess of a storm (i.e., the initial time that rainfall exceeds abstractions), or the commencement of the actual rainfall; is difficult to determine accurately in addition to measurement reliability and other reasons, because of the areal and temporal variations of the rainfall or rainfall excess.

One possible course for a more fruitful investigation is to study the rainstorm runoff from a watershed with impervious surfaces so that the effect of infiltration and base flow can be avoided. Such a watershed does not exist in nature. Thus, a controlled laboratory watershed is a possible alternative. Although one should be very careful in interpreting the results because of the scale effect, laboratory experiments do offer the advantage of controlled conditions under which the effect of each of the influential factors can be investigated individually. One of these laboratory watersheds existed in the University of Illinois (Chow and Yen, 1974) on which a large number of systematic experiments were conducted. The objective of this paper is to report the effects of the influential factors on the time of occurrence of peak discharge due to rainstorms on the basis of experimental results.

Measures of Time to Peak Flow

The time of occurrence of peak discharge due to a rainstorm, no matter how it is defined, is affected by two major groups of factors; namely, the watershed physiographic factors and the rainstorm hydrometeorological factors. For a watershed with an impervious surface, the watershed parameters may be represented symbolically by the watershed area A, a representative length L, a shape factor ζ, a representative basin slope S, a surface drainage pattern and storage factor η, and a representative runoff resistance R_f. The rainstorm parameters may be represented by the basin averaged duration and intensity of rainfall excess, T and i, respectively, and symbolic parameters R_a and R_t to represent respectively the areal and temporal distribution of the rainfall excess. Thus, the time of occurrence of peak discharge, t_p, can be expressed schematically as

$$t_p = F_1(i,T,R_a,R_t,A,L,S,\zeta,\eta,R_f) \quad (1)$$

in which F represents a function.

Different measures of the time of occurrence of peak discharge have been given by hydrologists (Fig. 1); namely, the time measured from the commencement of rainfall excess to the occurrence of peak flow, t_p, (McCarthy, 1938, Shen et al., 1974); the time measured from the beginning of direct runoff to the peak flow time, t_5, (SCS, 1973, p. 16-3; Yen and Chow, 1969); the time between the peaks of rainfall excess and runoff, t_3; or the time measured from the centroid of the hyetograph to the occurrence of the peak discharge, t_2, (Snyder, 1938; Hickok et al., 1959; Chow, 1962; Linsley et al., 1975). The latter two measures, t_2 and t_3, have often been referred to as the lag time.

Most of the previously proposed peak discharge lag time formulas can be considered as special forms of the following equation:

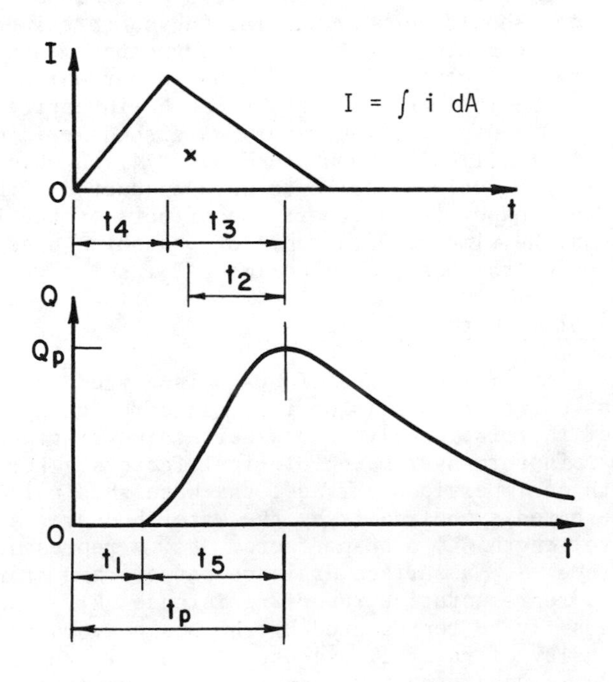

Fig. 1. Definition sketch for time of occurrence of peak discharge

$$t_2 = CA^{\alpha}(LL_c)^{\beta}/S^{\gamma} \tag{2}$$

in which C is a coefficient, α, β, and γ are constant exponents, L is the basin length, and L_c is the distance from the centroid of the watershed to the outlet, presumably as a replacement of ζ in Eq. 1. For the sake of uniformity, in Eq. 2 the time t is measured in hours, A in acres, L and L_c in miles, and S being dimensionless (i.e., ft/ft). Based on data from the Appalachian Mountains region, Snyder (1938) proposed that

$$t_2 = C(LL_c)^{0.3} \tag{3}$$

The coefficient C varies from 1.8 to 2.2. Linsley et al. (1975) suggested that

$$t_2 = C(LL_c/S^{0.5})^{0.38} \tag{4}$$

where C varies from 0.35 to 1.2. Chow (1962) analyzed 20 watersheds in the North-Central States, with size ranging from 2.8 to 4,580 acres and proposed that

$$t_2 = 0.000541(L/\sqrt{S})^{0.64} \tag{5}$$

Taylor and Schwarz (1952) studied 20 rural watersheds, ranging from 20 to 1600 sq mi in size, in the North Atlantic States and suggested that

$$t_2 = \frac{0.6}{\sqrt{S}} \exp\left[\frac{0.212}{(LL_c)^{0.36}} T\right] \tag{6}$$

in which T is the rainfall excess duration in hours. Hickok et al. (1959) analyzed 14 arid watersheds up to 1000 acres in size in Arizona, New Mexico and Colorado and proposed that

$$t_2 = \frac{0.00772A^{0.488}}{[S(\Sigma L)^{0.5}]^{0.61}} \tag{7}$$

in which ΣL is the total length of visible channels in miles.

Edson (1951) used a parabolic time-area curve and a unit hydrograph of "zero" duration to obtain a two-parameter formula for discharge and a time response which is characteristic of the watershed. His relation for time to peak for any rainfall excess duration greater than zero is

$$t_5 = \frac{T}{1-e^{T/2}} \qquad (8)$$

The Soil Conservation Service (1973) SCS method uses the time t_5 which can be approximated by

$$t_5 = 0.67\ t_c \qquad (9)$$

in which t_c is the time of concentration. If t_c is estimated by using Kirpich's (1940) formula, one obtains

$$t_5 = 0.064\ L^{0.77}/S^{0.385} \qquad (10)$$

Referring to Eq. 1, none of these formulas accounts for the effects of the rainfall excess intensity, i, flow resistance parameter, R_f, the surface drainage pattern and storage parameter, η, and the temporal and areal distribution of the rainfall excess, R_t and R_a. Most of these formulas do not consider the duration of rainfall excess, T, or the watershed area, A, or both. Admittedly under many circumstances the effects of i, R_f, η, and perhaps sometimes of R_t, R_a, and A, are secondary. But unless the duration is very long, the effect of T on the time to peak is obviously important.

Experimental Data Utilized

The data utilized in this paper are selected from the controlled experiments performed in a 40 ft by 40 ft square basin of the Watershed Experimentation System (Chow and Yen, 1974). Four types of watershed surfaces were tested on the 40-ft square basin. Only one type will be discussed here, called Type 1. The basin surface was 6061 aluminum plate, and the shape of the watershed from which the data used in this paper were collected was two 20' x 40' overland planes of aluminum plates, meeting symmetrically at the longitudinal x-axis. The overland planes are inclined in both x- (longitudinal) and y- (lateral) directions. Watersheds of different sizes and shapes can be tested within the maximum size of 40' x 40' of the basin. The basin outlet width is 4 feet.

More than two thousand experiments were performed on the Watershed Experimentation System. In reference to Eq. 1, these experiments include three different shape factors ζ, four different types of surface drainage pattern factor η, ten different basin areas, A, three different types of rainfall areal distribution, R_a, and various values of the parameters i, T, R_t, L, and S. The rainfall intensity ranged from 2 to 12 in./hr; duration from 30 to 400 sec. The lateral slope was kept constant at $S_y = 1\%$ whereas the longitudinal slope for different experiments ranged from 0.5% to 3%. From Eq. 1, assuming that the runoff resistance factor R_f can be represented by the kinematic viscosity ν of water, the shape factor by L/W where W is the width of the watershed, and the rainfall time distribution by the velocity of the rainstorm movement, w, then through dimensional analysis, the peak discharge time for an impervious watershed can be expressed as

$$\frac{t_p \nu}{A} \text{ or } \frac{t_p i}{L} = F_{2,3}\left(\frac{iL}{\nu}, \frac{T\nu}{A}, \frac{L^2}{A}, \frac{wT}{L}, S, R_a, \eta\right) \qquad (11)$$

For a given type, η, of impervious watershed of rectangular shape, and with a given stationary rainfall areal distribution, R_a, Eq. 11 can be simplified as

$$\frac{t_p \nu}{A} \text{ or } \frac{t_p i}{L} = F_{4,5}\left(\frac{iL}{\nu}, \frac{T\nu}{A}, \frac{L}{W}, S\right) \qquad (12)$$

The time from the commencement of rainfall excess to the occurrence of peak discharge, t_p (Fig. 1), is defined here as the time to peak. The time between the centroid of the hyetograph and the peak discharge time, t_2, is defined as the peak lag time. Clearly, Eqs. 11 and 12 apply to t_2 as well, simply by replacing t_p with t_2. In this paper the discussions emphasize the variation of t_p with the different influential parameters. For the laboratory experiments, the time from the beginning of direct runoff to the time of peak discharge, t_5 (Fig. 1), is the same as t_p except for the cases of downstream or laterally moving rainstorms, for which $t_p > t_5$. For all the stationary rainstorm cases tested, t_3 cannot be defined because the rainstorms are all with uniform time distribution. For all the moving rainstorm experiments utilized in this paper, $t_3 = t_2$. However, t_2 as well as two other time measures that are often used in hydrology, namely the time of concentration and equilibrium time, however they are defined, will be the subject of other papers and will not be discussed here.

Details of the laboratory watershed equipment, experimental procedures, measurement accuracy and recorded data can be found elsewhere (Chow and Yen, 1974; Shen et al., 1974; Ben-Zvi, 1970; Yen and Chow, 1968; Harbaugh, 1966) and hence are not repeated here. The experimental data were analyzed according to Eqs. 11 and 12. It is beyond the space limitation to present all the analyzed results in this paper and hence only a representative set of the results are shown here. Typical nondimensional experimental hydrographs are shown in Fig. 2 from which the values of t_p are read and t_2 determined. These values are then plotted as functions of influential parameters as suggested in Eqs. 11 and 12 and shown in Figs. 3 to 6.

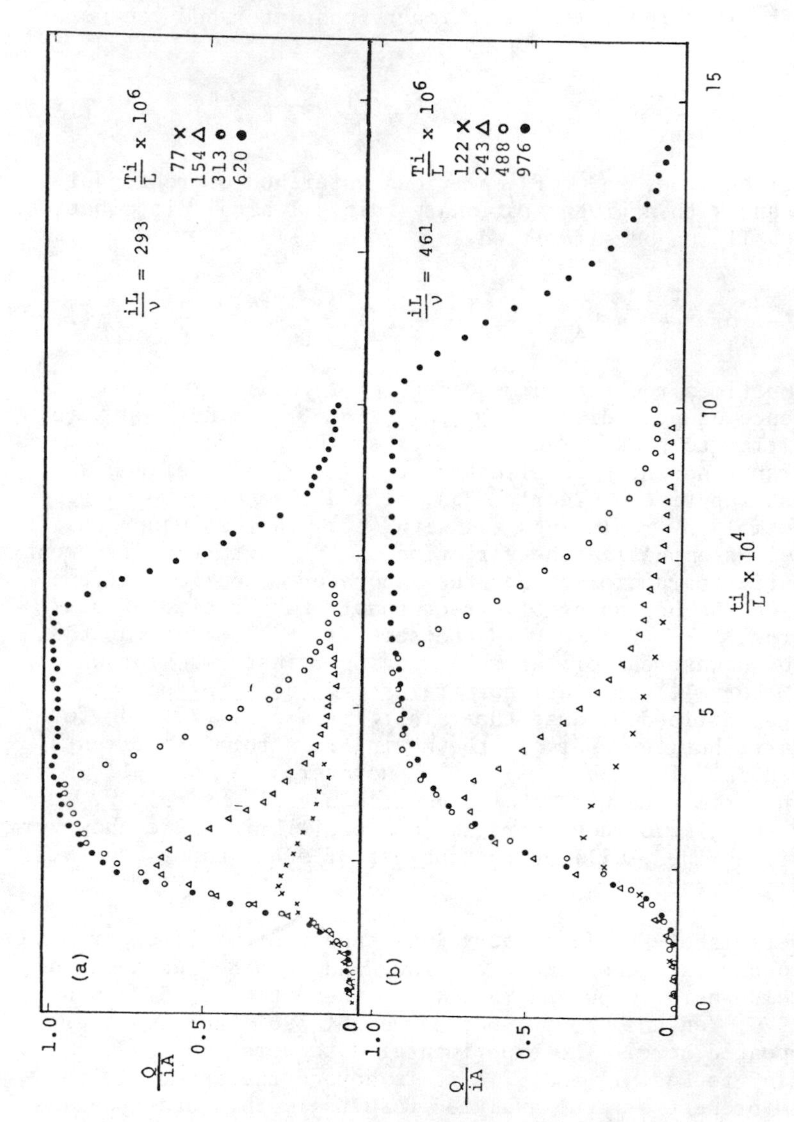

Fig. 2. Typical experimental hydrographs

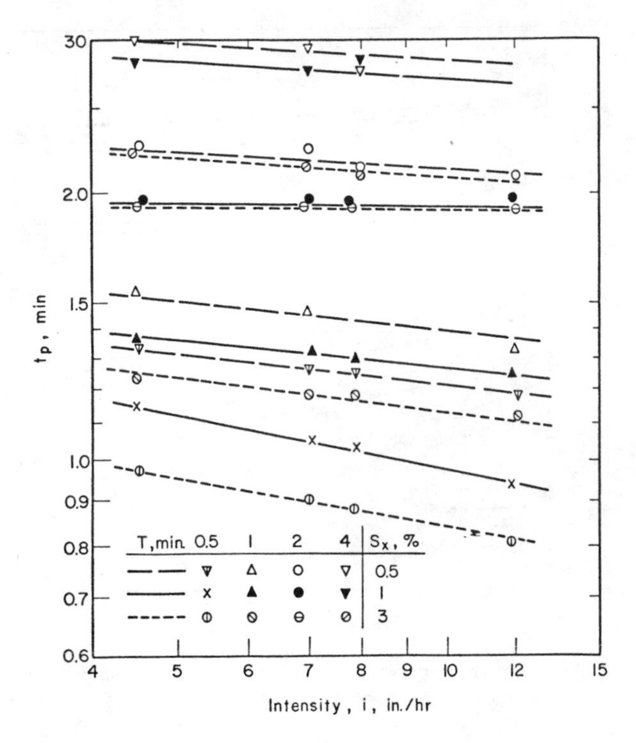

Fig. 3. Variation of time to peak with intensity of rainfall excess for stationary rainstorms on Type 1 40' x 40' watersheds

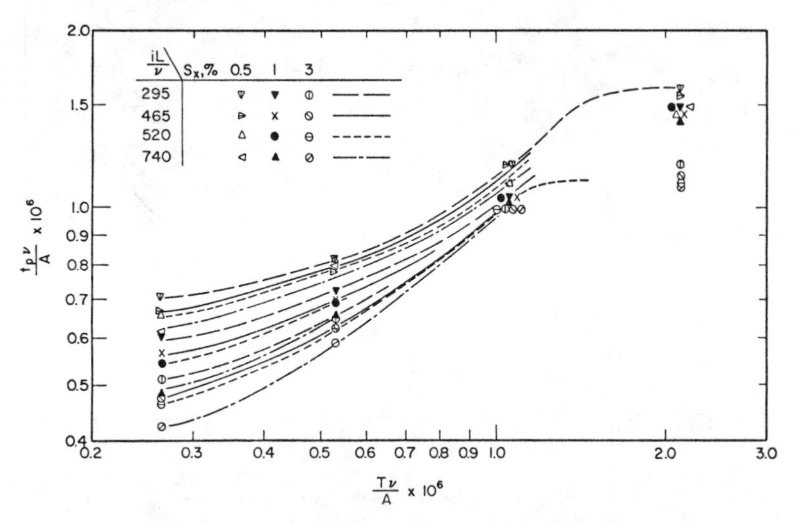

Fig. 4. Variation of time to peak with duration of rainfall excess for stationary rainstorms on Type 1 40' x 40' watersheds

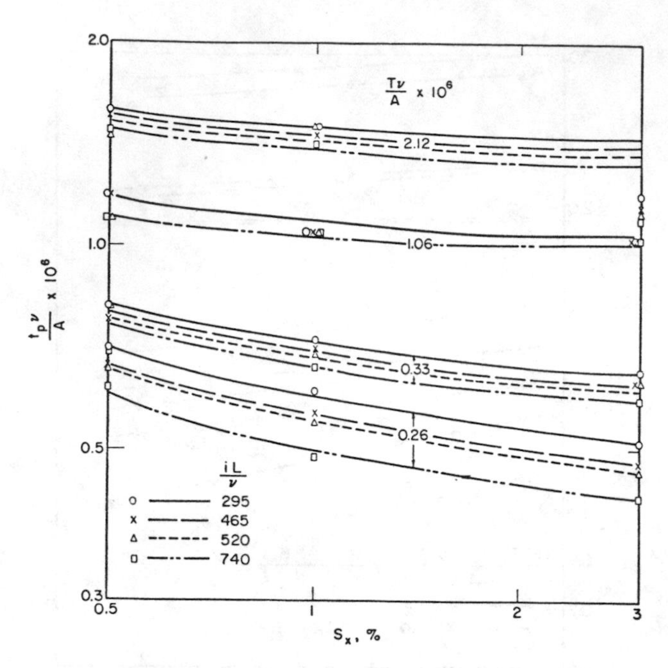

Fig. 5. Variation of $t_p\nu/A$ with watershed slope for stationary rainstorms on Type 1 40' x 40' watersheds

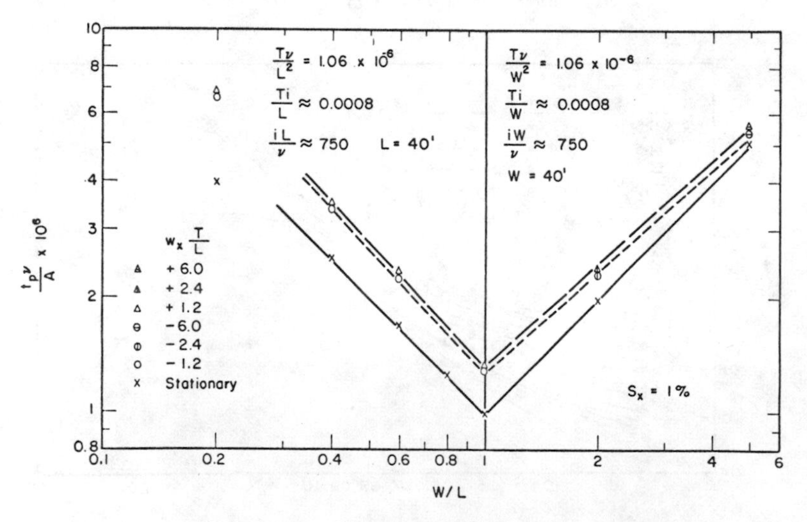

Fig. 6. Variation of $tp\nu/A$ for rainstorms over Type 1 watersheds of various shapes

Time to Peak Discharge

The factors that affect the time to peak discharge, t_p, for the surface runoff from an impervious watershed can be expressed nondimensionally as in Eqs. 11 and 12. The effects of rainfall intensity, duration, and watershed slope are first illustrated in this section by considering the relatively simple case of stationary rainstorm of uniform areal and time distributions over a square Type 1 watershed, i.e., Eq. 12 with $L/W = 1$. The effects of other influential factors are subsequently discussed in order.

Effect of rainfall intensity. The experimental result of t_p as a function of the rainfall excess intensity, i, is plotted in Fig. 3 for different rainfall durations and longitudinal basin slopes for the Type 1 watersheds of 40' x 40' as an example. This plot is given in a dimensional form as an illustration of the original data. It can easily be nondimensionalized by multiplying the values of t_p and i by the corresponding values of ν/A and L/ν, respectively.

As it can be seen from Fig. 3, over the range of experimental conditions, for a given duration T and slope S, the relationship between t_p and i can be represented by a straight line with a negative slope on a logarithmic plot. In other words,

$$t_p \sim i^{-\lambda} \tag{13}$$

where λ is a positive number. Similar plots for moving rainstorms, for watersheds with other shapes and sizes, all indicate similar linear relationship between t_p and i on logarithmic plots. However, the slope and location of these straight lines vary with the values of the influential parameters. The experimental results indicate that the value of the exponent λ in Eq. 13 decreases with increasing watershed slope S_x, rainfall duration T, watershed size A, and increasing speed of rainstorm movement. Moreover, it appears that the value of λ decreases with increasing complexity of the drainage pattern.

The decrease of t_p with i and the variation of λ with the different parameters can be explained by considering the hydraulics of the unsteady flow within the watershed. For instance, for two rainfalls of different i but with otherwise identical conditions on a watershed, the one with a low intensity of rainfall excess may just have the surface barely wetted. The values of the Reynolds and Weber numbers for the flow are very small and the raindrop effect on flow resistance is dominating. Consequently, the flow resistance is high and it takes a long time for the water from an upstream area to arrive at the watershed outlet. Contrarily, the one with a higher intensity will have a greater flow depth and the runoff changed from laminar to turbulent flow at an earlier time, and consequently a relative reduction of the flow resistance (easier to observe through consideration of Weisbach's resistance coefficient and the Moody diagram). Thus, the water reaches the outlet faster and an earlier t_p is expected.

Effect of rainfall duration. The effect of the duration of rainfall excess, T, on the time to peak is more complicated than that by intensity. The selected experimental results of Type 1 watersheds are shown nondimensionally in Fig. 4 as $t_p\nu/A$ vs. $T\nu/A$ as an illustration. The logarithmic plot shows that when the duration of rainfall excess is short, for a given intensity uniformly distributed over the watershed, the time to peak approaches to a constant as the duration decreases. In other words, in the limiting case of "splashing" a small volume of water onto the watershed, if the duration of water supply is considerably shorter than the time of concentration of the watershed, the peak flow time is essentially the same so long as the supply rate is the same, and it is insensitive to the small changes in duration. Obviously, this constant minimum response time for peak flow is smaller and approached at a shorter duration if the intensity is heavier because of the higher Reynolds number of the corresponding flow.

As the duration increases, the time to peak also increases at an increasing rate, asymptotic to a 45° line on the logarithmic plot, i.e., approaching a linear increase of t_p with T. However, because the relative peak discharge Q_p/iA approaches unity asymptotically, in practice it is very difficult to determine reliably the time of occurrence of peak flow if the duration is longer than the time of concentration. Therefore, the experimental results show that the time to peak again approaches a constant for durations much longer than time of concentration. The magnitude of this upper constant time to peak depends on the accuracy of the experimental equipment and measurements.

The curves in Fig. 4 can be replotted on logarithmic paper as $(t_p-t_o)\nu/A$ vs. $T\nu/A$ where t_o is the lower asymptotic constant time for short durations. The result is approximately a straight line indicating t_p-t_o proportional to T^δ where δ is about unity. Alternatively, if $t_p\nu/A$ is plotted on a log scale while $T\nu/A$ on a linear scale for the range $0.1\ t_c < T \leq t_c$, where t_c is the time of concentration, the result is a set of straight lines. In other words, within this range of T, approximately,

$$t_p = C_1 e^{bT} \tag{14}$$

in which C_1 is a coefficient, and b a constant exponent.

Effect of basin slope. It is obvious that as the watershed slope, S, becomes steeper, there is a larger driving force due to gravity to move the water downstream. Consequently, with otherwise identical conditions, a watershed with a steeper slope produces a surface runoff of relatively higher peak in a shorter time than that from a watershed with a milder slope. Experimental results like those shown in Fig. 5 indeed confirm such a general trend of dependence of t_p on the basin slope S.

As shown in Fig. 5, the effect of increasing slope on decreasing t_p diminishes as the slope becomes larger and/or the rainfall duration becomes longer. Many early studies suggested

$$t_p \sim S^{-\gamma} \tag{15}$$

where γ has a value around 0.5. This hypothesis apparently is based on the reasoning that commonly used velocity formulas such as Manning's, Chezy's or Darcy-Weisbach's, the flow velocity is proportional to $S^{0.5}$. However, strictly speaking these formulas are applicable only to steady uniform flow whereas the watershed surface runoff is highly nonuniform and unsteady. Also, even for steady uniform flow the velocity is proportional to $S^{0.5}$ only if it is fully developed turbulent flow. Therefore, as can be observed from Fig. 5, rigorously speaking Eq. 15 is invalid.

However, it should be mentioned here that the nonlinear relationship between t_p and S_x is perhaps exaggerated in the laboratory experiments because of the physical size of the watersheds. As it is commonly known in fluid mechanics, a scaled model satisfying simultaneously the Reynolds, Froude, and Weber criteria cannot be achieved unless the model is of full scale size. Since the surface runoff starts with laminar flow and soon changes to turbulent flow, the effects of the laminar flow and raindrop impact are probably over-emphasized in the laboratory results than in natural watersheds. For natural watersheds the relationship between t_p and S is probably closer to linear on a logarithmic plot than that shown in Fig. 5, and within engineering accuracy Eq. 15 may be considered as an acceptable approximation. However, in practice it may be necessary to use different values of γ for different ranges of watershed and rainfall conditions.

Effects of watershed drainage pattern and surface and storage conditions. Actually, four types of basins of different drainage patterns and surface and storage conditions have been tested. Two of them, (Types 2 and 3) were with clearly defined channels and two (Types 1 and 4) without. Two (Types 3 and 4) were with considerable local surface storage. Nevertheless, the experimental results indicate that the previous discussions on the effects of rainfall excess intensity, duration, and watershed slope on the time to peak prevail. However, the exponents or constants λ, b, and γ in Eqs. 13 to 15 appear to vary with the watershed drainage pattern and surface conditions. It is obvious that t_p increases as the drainage pattern becomes complicated and the surface storage increases. For example, for otherwise corresponding conditions t_p increases successively from Type 1 to Type 4 watersheds as the surface drainage pattern becomes more complicated and the surface roughness increases.

Effects of shape and size of watershed. For the laboratory experiments utilized in this paper, the watershed shape is expressed in accordance with Eq. 12 in terms of L/W. A typical result is shown in Fig. 6 for the Type 1 watersheds with $S_x = 0.01$.

A plot of t_p with watershed size A for Type 1 watershed of 16, 24, and 40 ft squares indicates that approximately

$$t_p \sim A^{\alpha} \tag{16}$$

where α is about 0.05 for the square watersheds tested. Assuming that Eq. 16 with different values of α is applicable to watersheds of other shapes as well, including rectangular watersheds, and by noting that $A = WL$ for rectangular watersheds, from Fig. 6 one obtains

$$t_p \sim L^{\beta} \tag{17}$$

where β has a value less than unity.

Approximate formula for time of peak flow. From an engineering viewpoint and based on the experimental results and previous investigations appearing in the literature, the time of occurrence of peak discharge of surface runoff due to rainfall excess on a small watershed can be approximated by

$$t_p = C \frac{A^{\alpha} L^{\beta}}{S^{\gamma} i^{\lambda} w_x^{m}} e^{bT} \tag{18}$$

in which the coefficient C and the exponents are all positive numbers. Truly speaking they are not constants but vary slightly with the hydrometeorological and physiographic conditions. A multi-regression analysis using a large amount of data is necessary to determine these values for natural watersheds. From existing evidence it appears that Eq. 18 applies for the range of duration $0.1\ t_c < T \leq t_c$ with b much less than unity. The value of m is near but usually less than unity and should be zero for stationary rainstorms. The value of λ is usually less than 0.4 and is smaller for larger watershed area and slope. The value of γ ranges from 0 to 0.6, higher when low Reynolds number flow dominates and lower when the drainage pattern is complex. The exponent α depends heavily on the shape of the watershed and usually is a small value much less than 0.5. The value of β is usually between 0.3 and 0.7.

Conclusions

The time of occurrence of peak discharge of a watershed is a reflection of the response of the watershed to the input-rainfall. Consequently, the time of peak flow depends on all the hydrometeorological and physiographic factors that affect the temporally and spatially varying runoff within the watershed. Through controlled laboratory experiments of relatively simple and fundamental conditions, basic information on rainfall-runoff relationship can be obtained which provides useful implication to the more complex and larger natural watersheds. Through dimensional analysis the influential factors can be grouped as shown in Eqs. 11 and 12 as a guide for data analysis to investigate the effects of these factors.

Based on existing information an approximate general formula for the time of peak surface flow from a small watershed, Eq. 18, is proposed for the range of rainfall excess duration, T, between $0.1\ t_c$ to t_c where t_c is the time of concentration. Briefly, the time to peak flow increases with increasing duration of rainfall excess, size and length of the watershed, and with decreasing intensity of rainfall excess, and watershed slope. It also depends on the shape, drainage pattern, and surface and storage conditions of the watershed. Most of the existing empirical formulas are special cases of the approximate general formula. The most likely ranges of the values of the coefficient and exponents in the formula have been suggested. However, a multiple-regression analysis of a large amount of field and laboratory data, when available, to establish the values of these coefficients and exponents for different ranges of rainfall and watershed conditions would be most useful for engineering purposes.

Acknowledgments

In this paper selected data collected at the University of Illinois at Urbana-Champaign over a ten-year span (1964-1973) are utilized. The experimental program, at its different phases, was partially or totally supported by National Science Foundation grants GP-1464, GK-1155, GK-11292. This paper is partially supported by NSF grant GK-40867.

Notations

A = watershed area;

b = exponent;

C = coefficient or constant;

F = function;

i = intensity of rainfall excess;

i_e = areal averaged intensity of rainfall excess;

L = length;

L_c = length from watershed centroid to outlet;

m = exponent;

n = Manning's roughness factor;

R_a = representative parameter for areal distribution of rainfall excess;

R_f = representative parameter for watershed flow resistance;

R_t = representative parameter for temporal distribution of rainfall excess;

S = slope;

S_x = longitudinal slope;

S_y = lateral slope;

T = duration of rainfall excess;

t = time;

t_c = time of concentration;

t_p = time of peak flow; time to peak measured from commencement of rainfall excess to peak discharge;

t_2 = peak lag time, measured from centroid of hyetograph to peak discharge;

t_3 = time between peaks of hyetograph and hydrograph;

t_5 = peak discharge time measured from beginning of surface runoff to peak discharge;

W = watershed characteristic width;

x = longitudinal direction;

y = lateral direction;

α = exponent;

β = exponent;

γ = exponent;

ζ = watershed shape factor;

η = watershed drainage pattern and surface condition factor;

λ = exponent; and

ν = kinematic viscosity of water.

References

[1] Ben-Zvi, A., 1970. "On the Relationship between Rainfall and Surface Runoff on a Laboratory Watershed," Ph.D. Thesis, Department of Civil Engineering, University of Illinois at Urbana-Champaign, Illinois, 128 p.

[2] Chow, V. T., 1962. "Hydrologic Determination of Waterway Areas for the Design of Drainage Structures in Small Drainage Basins," Engineering Experiment Station Bulletin No. 462, University of Illinois at Urbana-Champaign, Illinois, 104 p.

[3] Chow, V. T., ed., 1964. "Handbook of Applied Hydrology," McGraw-Hill Book Co., New York.

[4] Chow, V. T. and Yen, B. C., 1974. "A Laboratory Watershed Experimentation System," Civil Engineering Studies Hydraulic Engineering Series No. 27, University of Illinois at Urbana-Champaign, Illinois, 196 p.

[5] Edson, C. G., 1951. "Parameters for Relating Unit Hydrographs to Watershed Characteristics," Transactions, American Geophysical Union, 32:591-597.

[6] Harbaugh, T. E., 1966. "Time Distribution of Runoff from Watershed," Ph.D. Thesis, Department of Civil Engineering, University of Illinois at Urbana-Champaign, Illinois, 130 p.

[7] Hickok, R. B., Kepple, R. V., and Rafferty, B. R., 1959. "Hydrograph Synthesis for Small Arid-Land Watersheds," Agricultural Engineering, 40(10):608-611, 615.

[8] Kirpich, Z. P., 1940. "Time of Concentration of Small Agricultural Watersheds," Civil Engineering, 10(6):362.

[9] Linsley, R. K. Jr., Kohler, M. A., and Paulhus, J. L. H., 1975. "Hydrology for Engineers," McGraw-Hill Book Co., New York.

[10] McCarty, G. T., 1938. "The Unit Hydrograph and Flood Routing," presented at the Conference of the North Atlantic Division, U.S. Army Corps of Engineers; revised, U.S. Army Engineer's Office, Providence, R.I., 1939.

[11] Shen, Y. Y., Yen, B. C., and Chow, V. T., 1974. "Experimental Investigation of Watershed Surface Runoff," Civil Engineering Studies Hydraulic Engineering Series No. 29, University of Illinois at Urbana-Champaign, Illinois, 197 p.

[12] Snyder, F. F., 1938. "Synthetic Unit-Graphs," Transactions, American Geophysical Union, 19:447-454.

[13] Soil Conservation Service, U.S. Department of Agriculture, 1972. "SCS National Engineering Handbook," Section 4, Hydrology.

[14] Taylor, A. B. and Schwarz, H. E., 1952. "Unit Hydrograph Lag and Peak Flow Related to Basin Characteristics," Transactions, American Geophysical Union, 33:235-246.

[15] Yen, B. C. and Chow, V. T., 1968. "A Study of Surface Runoff Due to Moving Rainstorms," Civil Engineering Studies Hydraulic Engineering Series No. 17, University of Illinois at Urbana-Champaign, Illinois, 112 p.

[16] Yen, B. C. and Chow, V. T., 1969. "A Laboratory Study of Surface Runoff Due to Moving Rainstorms," Water Resources Research, 5(5):989-1006.

DISCUSSIONS - SECTION I
Catchment Hydrology

Michael C. Quick*

A divergence of opinion on fitting procedures has been expressed and I would like to comment from our experience. In our modelling work (Quick and Pipes, 1976) we use both an objective function and graphical inspection, so that we sit midway between the two camps. We do this because if you use an objective function blindly you will run into problems. It is necessary to examine errors carefully and review the meteorological and flow data for periods of high error. Otherwise you may be biasing your parameters to fit nonexistent or highly erroneous data.

It is useful to contrast snowmelt and rain catchment response. In calculating the response of mountain catchments we have to deal with both snowmelt and rain. Snowmelt modelling has some considerable advantages because snowmelt tends to occur at different elevation ranges as the snowmelt season advances. This allows us to study how mountain catchment parameters vary with elevation. This information might be of interest to rainfall runoff modellers.

To comment on event models compared with continuous simulation models, I feel there is much useful information to be gained from continuous simulation. It is during low flow periods that the slower components of runoff and soil moisture deficit build-up can be adequately studied. These slower components of runoff from a large part of the total runoff and occur when the event modeller may not even be looking.

Reference: Quick, M. C. and Pipes, A., 1976. "A Combined Snowmelt to Rainfall Runoff Model," Canadian Journal of Civil Engineering, Vol. 3, No., pp. 449-460, Sept.

Howard S. Wheater**

I would like to thank the General Reporter for his comments on my paper, and to make the point that the model discussed in the paper is an individual event model. This has particular advantages for flood estimation. A discontinuous

*Address unknown.

**Univ. of Bristol, Queens Bldg, University Walk, Bristol, England.

data record may be used for calibration and some simplification of model structure is possible. The key to the individual event simulation is the measure of antecedent catchment moisture status. A soil water budgeting model has been verified from observed soil moisture profiles and provides an input of soil moisture deficit (S.M.D.) which is used to define the status of the model soil moisture stores.

The use of S.M.D. as an index of catchment moisture in a second study of flood response of the Gloucester catchments, sponsored by the Water Research Centre, England, has led to results which support the work presented by Lynch et al. and in part answer the comments of the General Reporter on that paper. A unit hydrograph analysis was carried out on 152 events over eight catchments. The catchments varied in area from 20-50 km^2 and in urban proportion from 10-35%. Flow separation was carried out using an exponential "master recession" and rainfall losses distributed using a one-parameter proportional loss. The volume of stormflow total rainfall was observed to be a function of S.M.D. for positive S.M.D. and also of rainfall duration for zero S.M.D. The dimensions of the U.H. were also dependent on S.M.D., high S.M.D. giving a high peak, short baselength U.H., low S.M.D. giving low peak, longer duration. Time to peak of the U.H. did not vary. Analysis over the eight catchments showed that the effect of increasing urban proportion on a dry catchment led to a marked increase in peakedness of the U.H. However, for a wet catchment, no such effect could be detected.

A. Van Der Beken*

I would like to add that, in the meanwhile since the paper was written, we extended the sensitivity analysis of the parametric model by calculating the covariance-matrix according to a method proposed by Mein and Brown from Australia. This powerful method gives us the standard deviation of each parameter-value and the correlation coefficients between pairs of parameters. Results will be presented at the forthcoming IFIP Working Conference, Ghent, Belgium. I would suggest that this method should be considered in any sensitivity analysis.

Stephen J. Burges**

I take exception with Dr. Overton's written remarks: "Urban planning .... Any schemes to improve water quality or

*Vrije Universiteit Brussel, Pleinlaan 2, B-1050 Brussels, Belgium.

**Civil Engineering Department, University of Washington, Seattle, Washington.

reduce flooding beyond that of a natural stream can hardly be economically or environmentally justified." This statement seems to indicate that man should not try to improve upon nature; even some animals improve upon nature! Man does not live in a natural environment, it is engineered. Our efforts might well be directed towards improving the engineering we do. How about some imagination in planning rather than supporting the purely pedestrian? Perhaps the reporter, and others, would find valuable perspective in the writings of René Dubos, e.g., René Dubos, "Symbiosis Between the Earth and Humankind," Science, Vol. 193, pp. 459-462, August 6, 1976.

Keith Beven*

Response to General Reporter's comments on papers by Beven and Beven and Kirkby.

The two papers presented differ in approach because they are dealing with fundamentally different objectives. The first aim to further understanding of a well defined system model of hillslope hydrology, whereas the second is directly concerned with approaches to the fuzziness of the real world prototype system. However, common to all our work is the aim of developing models that are physically based in the sense that the parameters may be measured in the field, that reproduce some of the spatially variable process interactions in a realistic way, and that do not require any form of optimisation for calibration. We are therefore directly concerned with the ungauged catchment forecasting problem and are seeking to reduce the dependence on regional generalisation of parameter values (including correlation with catchment characteristics), and make maximum use of information, such as topographic structure, readily available in the catchment itself.

I should like to add to Dr. Overton's comments on the use of optimisation, from experience with our variable contributing area model. Looking at sensitivity analyses for the model parameters we found that our measured values did not correspond to the optimum values, as might be expected when you are reducing a small number of measurements from a distribution to a single value for use in the model. However the optimum values would in fact have changed the intended mode of operation of the model entirely, using the nonlinear subsurface response element of the model as a quick response flow outing element and eliminating the operation of the contributing are components of the model. In a sense it is good to see that

*Institute of Hydrology, Maclean Bldg., Crowmarsh Gifford, Wallingford, Oxon, England.

optimisation of the model would reflect the physical processes occurring in the real catchment which is typically flashy in response but using the measured parameter values the model concepts are used as designed and appear to reproduce the spread of a saturated contributing area reasonably satisfactorily. The point I am trying to make is that optimisation may work to accommodate reality, often in a subtle way, to the detriment of the physical basis of theory on which a model is based.

Response to Dawdy's comments that bias in input and parameter measurements can be accommodated to some extent by optimisations. This was specifically one reason why we have tried to develop models that avoid the use of optimisation and are objective in the sense that the parameters are measurable. However, it is then important to use measurement methods appropriate to the scale of system being simulated. The parameters of our model are derived from recession curve analysis, sprinkler infiltrometer, soil moisture deficit and flow tracing data. I feel the next stage of development will be to incorporate variation in measured parameter values into the model formulation.

V. R. Krishna Murthy*

Comments on paper: "A Monthly Water Balance Model Applied to Two Different Watersheds," by Van Der Beken.

Since the full text of the paper was not available to me, I would like to ask the author as to the number of years of data used in developing the parameters of the model. The total streamflow volumes were simulated with a relative error of less than 3%. It would be very interesting to know the average and maximum error in the predicted monthly streamflow volumes. If the parameters were developed from the same period of data used in testing, then the accuracy of the model should be checked by predicting the streamflows for a period not used in developing the parameters. If the model could predict monthly streamflows for such a period, with reasonable accuracy, the usefulness of the model developed will be enhanced.

J. Martinec**

Author's discussion (comment) to the General Report of Professor Overton. The crucial point of the snowmelt-runoff model is the variable recession coefficient determined by

*Texas Water Rights Commission, Austin, Texas

**Federal Institute for Snow and Avalanche Research, 7260 Weissfluhjoch/Davos, Switzerland.

Eq. (6) of the paper from the current runoff. This enables the model to handle the rainfall-runoff (if it accompanies the snowmelt) as well: If the discharge rises, the recession coefficient decreases and the quicker watershed response is thus taken into account. Rainfall during snowmelt is included in the error assessment and a good timing of peaks is required from the model. The computing procedure is sensitive to the temporal variability of the energy input and needs areal monitoring of the snow cover by aeroplanes or satellites. On the other hand, differences in the time-lag are not critical. The explanation might be a quick propagation of the effect of the massive meltwater infiltration via the subsurface storage.

A. Van Der Beken*

Comments on General Discussion. I would like to comment on the classification of models. According to my opinion, there may be a difference in model structure between off-line models using past-time data and on-line models using real-time data for forecasting goals. In the latter case, real-time data may be used on-line for adjusting model-parameters continuously. This has an influence on model structure and possibly also on optimization procedure.

*Vrije Universiteit Brussel, Pleinlaan 2, B-1050 Brussels, Belgium.

[illegible] differs from the current model. This model [illegible] is used to handle the variable [illegible] in [illegible] [illegible] is the discharge through the [illegible] coefficient [illegible] and the [illegible] is expressed [illegible] [illegible] and the errors [illegible] [illegible] [illegible] The [illegible] procedure [illegible] [illegible] [illegible] [illegible] [illegible] [illegible] the [illegible] [illegible] [illegible] are not [illegible] [illegible] [illegible] [illegible] [illegible] [illegible] the [illegible]

[illegible]

[illegible] [illegible] [illegible] [illegible] [illegible] [illegible] [illegible] [illegible] [illegible] [illegible] [illegible] differences [illegible] model [illegible] [illegible] [illegible] [illegible] [illegible] [illegible] [illegible] [illegible] [illegible] [illegible] [illegible] [illegible] [illegible] [illegible] [illegible] [illegible] [illegible] procedure [illegible]

[illegible]

SECTION II

GROUNDWATER HYDROLOGY

Section Chairman, Richard Cooley
Research Hydrologist
U.S. Geological Survey
Denver, Colorado

Rapporteur, Thomas Maddock III
Research Hydrologist
U.S. Geological Survey
Reston, Virginia

SECTION II

GROUNDWATER HYDROLOGY

SECTION II

GROUNDWATER HYDROLOGY

GENERAL REPORT

By

Thomas Maddock III
USGS, Systems Group, Water Resources Division
Stop 410, Reston, Virginia

Introduction

It has been my charge to review and synthesize a set of six papers into a state of the art report on groundwater hydrology. I must admit that I approach this task with considerable misgivings and with a slight juandiced eye; in that six papers, each confined to 15 pages or less, hardly seems sufficient to even abstract the voluminous reams of papers produced in groundwater hydrology. And to some extent these misgivings have been realized, yet, I am happy to report, not to the appraciable extent I expected. Although, not all inclusive, the six papers I review here cover a wide range of groundwater, hydrologic problems, and the technique used to analyze these problems. Unfortunately, these papers only provide a small keyhole within which to peek at any kind of "state of the art." This is not the fault of the authors. Nor will I be presumptuous, and assume that I can fill in the "gaps." This report is thus a review only of the six papers sent to me.

As their titles indicate, these papers do cover a partial spectrum of subject material in groundwater hydrology.

Report of Papers

I begin my discussion of papers with that of Dr. M. Mirabzadeh, entitled, "The Use of Block Technique in Mathematical Modeling of Ground Water." Mirabzadeh presents a numerical method for solving the flow equation for groundwater. Although the author professes that his technique is based on dynamic programming, I do not believe that it is essential to the development nor the understanding of the recursive relationship developed in the paper. The author has developed an extremely rapid computational technique, but a technique that may be extremely limited in its application, because I believe that it is limited to:

1. A finite difference mesh only.
2. Full rectangular region (or subregion).
3. Dirchlet boundary conditions on the full region.

The latter limitation is critical because it would restrict the technique to only those regions that are enclosed by constant head boundaries. Perhaps the author has extended his technique to other type boundary conditions, but it is my feeling that such an extension would reduce the technique's computational efficiency and may even be impossible.

The technique is based on matrix partitioning using row blocking, and under the assumptions and limitation listed above, its computational efficiency and operations and storage requirements would be excellent.

The paper by Dr. I. Krauss, entitled "Determination of Transmissivity from the Free Water Level Oscillation in Well-Aquifer Systems," is an expansion of work by Cooper et al. (1965) and Bredehoeft, Cooper, and Papadopulos (1966). Dr. Krauss develops a set of differential equations describing the oscillation of water levels in a well due to the rapid release of compressed air that has been pumped into a well. She then relates the damping effects on oscillations and the frequency of oscillation of the water levels in the well produced by the reduction of pressing to the transmissive quality of the aquifer near the well.

Although this relation varies with the storage coefficient, which is also unknown, her analysis indicates that this variation may not be too critical (Fig. 2 of her report). The author has applied her technique to a series of wells and reports excellent agreement between transmissivities calculated by her method and transmissivities calculated by the more traditional methods. I would like very much to know if the author did any cost analysis for her technique, and if it compared favorably with the pumping test. Given that her method and the pumping test give comparable results, the economics plays a crucial role in the application of her technique.

The paper entitled, "Impact of Long-Termed Discharge on an Aquifer with Small Recharge" by Dr. M. Klenke is a true "state of the art" paper in the application of a groundwater model to a flow system. The author describes the application of a hybrid (digital analog) model to a region of potential groundwater development in North Africa. The exploitation of water from an aquifer underlying the region is projected to be greater than the natural recharge and capture potential to the aquifer, hence sustained mining is to be expected. Furthermore there is little, if any, data in the region to provide "accurate" determination of the aquifer parameters. Thus, the author tests his model by simulating the response of the aquifer to several potential well fields and to variety of possible parameter values. The author concentrates his report on the results of his study, which is of obvious interest to the North Africans. My only criticism, which is somewhat selfish, is that I wish he would have concentrated his paper on the hybrid aspects of the model and relate his experiences and tribulation in that aspect. Perhaps he has done that in some of his papers referenced in the report.

Peter Rogers of Harvard University has argued that hydrologists have lost touch with engineering bases of many of the disciplinary concerns. He points out that our elaborate methodology does not produce "rules-of-thumb" that would be useful to practicing and practical engineers, and that it is unreasonable to expect investment of enormous sums of money in complicated and complex algorithms to produce simple results. The papers entitled "The Hydrogeological Assessment of Fresh Water Lenses in Oceanic Islands," and "Analysis of Transient Ground-Water Flow from Seepage Ponds" by Chidley and Lloyd and Glass, Christensen and Rubin, respectively are somewhat successful attempts to provide "rules-of-thumb" for hydrologic problems.

The paper by Chidley and Lloyd presents a simple assessment of fresh-water lenses in the Grand Cayman Islands of the Carribean. The paper presents a groundwater balance for a fresh-water balance for a fresh-water lens resting on saline water. The water balance consists of estimating potential recharge to fresh-water lenses and potential discharge to evapotranspiration. The potential recharge is calculated using existing meteorological data and estimates of cloud cover and incoming radiation. Losses to soil moisture and evapotranspiration are calculated and the differences between precipitation and these values is assumed to recharge the fresh-water lenses. The hydrogeology is described by a simple groundwater model. It is assumed that the lens behavior can be modeled as a series of steady-state systems with a Ghyben-Hertzberg relation holding. The authors report the ability to determine the developable capabilities of the fresh-water lenses with their approach. Although the authors concentrate on presentation of the technique and indicate indirectly that it was successfully applied to the Grand Cayman Island, the authors may wish to comment on some of the implementation problems that occurred and the problem they found with estimating parameters for their model.

The paper by Glass, Christensen, and Rubin presents a method for estimating the size of seepage ponds and the quantity of outflow due to seepage from these ponds. The methodology is composed of two parts. First, a method of storage routing is developed by which the seepage pond size can be calculated. The routing procedure is reported to be simple enough to be done on the programmable calculator. The outflow from the seepage pond is calculated by a modified Green and Ampt equation. This seepage is assumed to form a transient groundwater model. The fluid flow from the mound is analyzed using digital groundwater models. A set of curves are developed relating pond depth to volume and inflow with time. The groundwater model is used to calculate a set of dimensional curves that can be used to determine the mound height with time. As with the Chidley and Lloyd paper, the authors concentrate basically on presenting their technique and because of the restriction in page numbers, did not report an application. The authors may wish to comment on any areas of application that have been completed or that may be under way.

Conclusions

The collection of six papers that I have reported on indicate and illustrate a number of problem areas to be attacked in our field. They are implementation, uncertainty, rules-of-thumb and non-optimization.

1. Implementation problem--In many of our modeling activities we fail to incorporate many of the restrictions that the decision makers may be faced with. This is understandable in that as hydrologists we tend to emphasize physical-factors to the neglect of economic, political and social factors. Chidley and Lloyd point out in the discussion of their paper that even after substantiating that fresh-water lenses were a viable source of water, the lenses were not economically exploitable. Thus it behooves us as hydrologists to interact as freely as possible with the decision makers to try and gain understanding of nonhydrologic constraints.

2. Uncertainty problem--The groundwater hydrologist lives in a world of uncertainty, and unfortunately, many times, fails to recognize this. Even when he does recognize its existence, he may not have the mathematical tools to include it formally in his analysis. He must rely on some surrogate measure of its affects. Klenke's paper presents a good example of using one such surrogate procedure to determine well field design in the face of uncertainty of flow characteristics.

3. Rule-of-thumb problems--There is a great need to develop simple, easily applicable techniques in all phases of groundwater hydrology. I am appalled to say that in many areas of our field we have become numerical analysts and foregone being hydrologists. We cannot seem to answer simple questions without complex and highly sophisticated models which rarely match the understanding level of the decision makers.

4. Non-optimization problems--With the rise in popularity of systems analysis and the "system's approach" has been the rise in the use of the term optimal design or optimal system. Implied in the term optimization is a maximum or minimum. Unfortunately because of uncertainty, what we may find as an optimal may be far from the true set of optimal conditions or designs. Thus we need some other design criteria than that which depends on some maximum or minimum value. Such a criteria might be resilence, the ability of a design to survive or flourish in face of large degrees of uncertainty.

These are just closing thoughts and I thank you for listening and the opportunity of being a reporter.

THE HYDROGEOLOGICAL ASSESSMENT OF FRESH WATER LENSES IN OCEANIC ISLANDS

By

Thomas R. E. Chidley
Senior Lecturer
University of Aston, U.K.

and

John W. Lloyd
Senior Lecturer
University of Birmingham, U.K.

Abstract

A description is given of the quantitative assessment of some thin fresh water lenses resting upon saline water on Grand Cayman Island in the Caribbean Sea. Grand Cayman is a small limestone island, typical of the oceanic island hydrogeological environment. Because they are small, 'big technology' is not appropriate. Maximum use must be made of existing data and methods. Even so, some fairly advanced techniques of investigation were used, and the reasons why they were necessary are shown clearly. The description of the investigation gives a good guide to a possibly general methodology which is appropriate to this kind of situation.

Certain hydrological studies were carried out and the way in which the results of these studies were used in the final assessment is described. Recharge to the lens areas was estimated by using a very simple watershed model approach. Due to limited radiation data, a radiation-cloud cover correlation was established.

The geology of the island is described briefly, especially in relation to the general theoretical background in which decisions were made about the behaviour of the lenses. A practical method for identifying the plan shape of the lenses is described. The vertical dimensions were measured by determining the lens base configuration and relating it to the surface configuration. The lens base configuration was determined by an electric resistivity method.

A simplified differential equation for flow in the fresh water lens was used to relate the recharge, lens shape and permeability to one another. This model was used to calibrate the regional permeabilities and to give guidance on the abstraction patterns that could be sustained with little elaboration.

Introduction

This paper describes some investigations carried out on some thin fresh water lenses resting upon saline water. The lenses are situated in the Caribbean Sea on Grand Cayman Island. In a case where fresh water is abstracted from a lens the problem of maintaining adequate water quality is of paramount importance. An abstraction scheme can be designed only if all elements of the groundwater equation are balanced. The paper describes how each element of this equation was estimated at the investigation stage.

A very wide range of theories from different disciplines are brought together in the investigation. This kind of study will almost always be a small investigation because of the environment and appropriate technology must be used. The application of theory in this situation with a limited budget and limited time is brought out in the discussion. The way in which 'applicable theory' is used to provide a framework for, and offer guidance in, making relevant decisions in a background of uncertainty, and with incomplete information, should be invaluable for someone faced with a similar problem.

Some useful climatological relationships for the islands are presented and the format used in the investigation is a good guide for similar investigations in the future. The problems of using uncontrolled estimates of groundwater recharge, based upon hydrological data are discussed. Unfortunately, in this case they were not capable of exact solution but a basis for decision is given.

It is essential for understanding of the discussion to realize that, where an island of permeable rock exists, rainfall on the island will frequently create a fresh water lens in the porous rock which floats on the saline water that otherwise pervades the rock. Within the lens is a zone of fresh water and a transition zone of varying thickness in which exists a salt concentration gradient. The thickness and nature of this transition zone depends upon a very large number of factors. The relevance of factors is made clear in the text.

Hydrology

The purpose of the hydrological investigations was to provide an estimate of the 'potential recharge' to the fresh water lens aquifers. The phrase 'potential recharge' has been used because there is some discrepancy between what could be estimated with reasonable reliability from hydrological data, and what could be estimated from groundwater measurements. The estimate of 'potential recharge' was made on the basis that vegetation on the recharge areas was mainly fed by water from soil moisture, rather than from the water table. It was not possible in the time available to de-lineate exactly the areas where plant roots could reach to the water table and where they could not. An approximate resolution of this problem had to be made.

Because of the long period of relatively dry weather from December to June the climate of the Cayman Islands is considered to be 'semi-arid tropical'. Convectional storms occur during May to October. Seasonal temperature variation ranges from 25°C to 30°C with relative humidity in the range 70-80%. Annual rainfall varies from between 1000 mm to 2300 mm, with an average of just below 1500 mm.

Figure 1 shows the general layout of Grand Cayman Island together with the location of the various data sources and fresh water lenses. The Penman method was used to estimate potential evaporation rates but a full set of data for this purpose was not available. A meteorological station at the Airport provided a nine-year record for most of the required information but duration of bright sunshine and radiation data were lacking. However, cloud cover measurements were available for the period, and with an overlapping 27-month record of incoming radiation data a method was devised to synthesize the missing radiation data.

A relationship between cloud cover and incoming short wave radiation was established. The method used to form the relationship was to plot the ratio of the ten-day average measured radiation to daily incoming radiation at the top of the atmosphere against average cloud cover for the same period. The daily cloud cover was taken as the average of 15 readings. A regression analysis of the two variables gave the following relationship:

$$R_m/R_t = 0.77 - 0.046\ C \qquad (1)$$

with a coefficient of correlation of 0.75, where R_m is measured incoming radiation, R_t is incoming radiation at the top of the atmosphere, and C is the ten daily average cloud cover (number of tenths of sky covered by cloud). Equation 1 may be compared with the relationships between incoming radiation and duration of sunshine given by Chidley and Pike, 1971. If one assumes that an estimate of the ratio of measured sunshine (n) to maximum possible sunshine (N) is given by $n/N = (10-C)/10$, equation 1 can be written as

$$R_m/R_t = 0.31 + 0.46\ n/N \qquad (2)$$

Equation 2 is similar to what one would expect at these latitudes. Using this relationship an estimate of potential

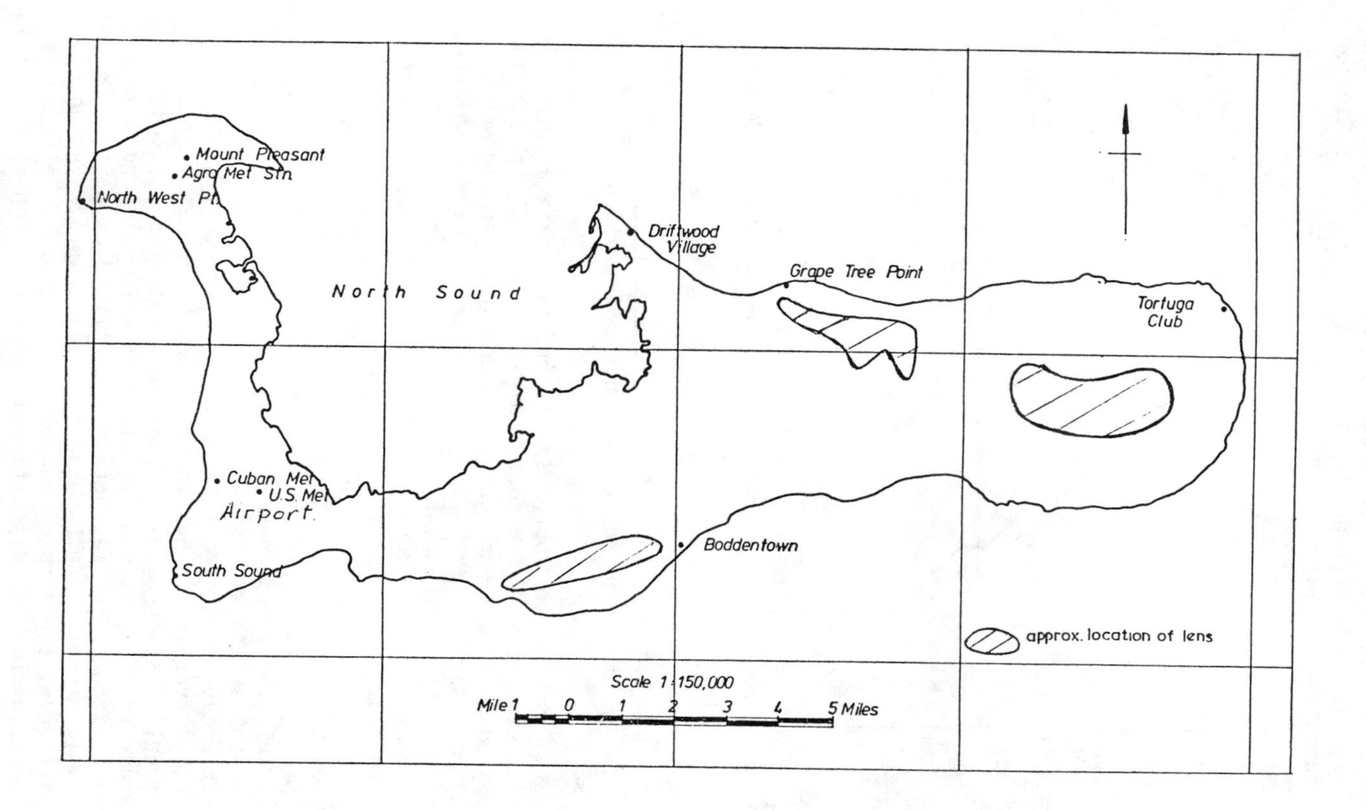

Fig. 1. Location of rainfall gauges

evaporation was prepared for a nine-year period, for which rainfall data was also available at various locations. Fig. 2 compares the computed evaporation using actual and radiation measurements and estimates from cloud cover.

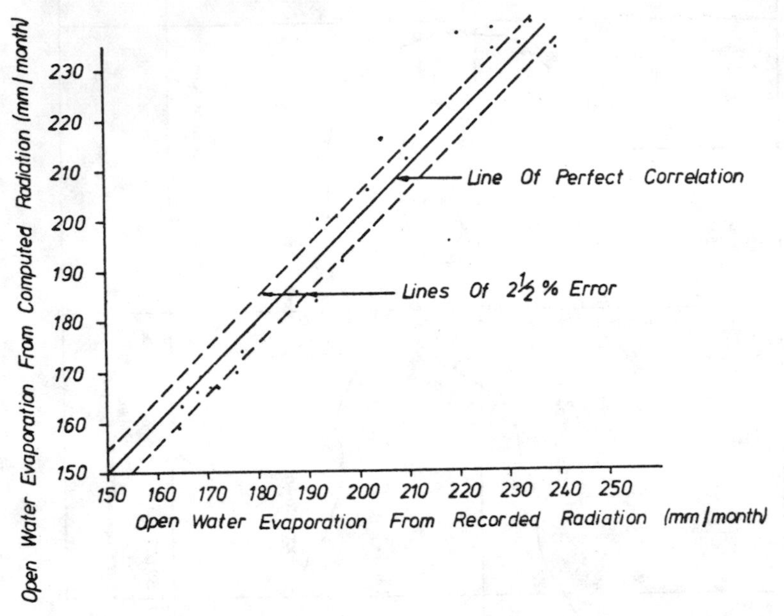

Fig. 2. Evaporation from computed radiation compared to measured evaporation

The rainfall and potential evaporation data were used to form estimates of the potential recharge. A very simple watershed model approach was used, knowing that surface runoff could only form a very small part of the hydrologic balance, because of the micro-Karst terrain, and the fact that there is virtually no well developed drainage. The model was constructed using the concept illustrated by Lloyd, Drennan and Bennell, 1967. The basis for the model is that a proportion (x) of the rainfall is always lost to runoff, and the residual precipitation falls into an interception store of capacity (Z). This interception store represents ponding and storage on vegetal surfaces. Water is permitted to evaporate from the interception store at the potential rate. If after full potential evaporation has taken place the capacity of the interception store is exceeded, the surplus is allowed to increase the soil moisture (SM). The soil moisture zone is characterized by two parameters, the 'field capacity' (C) and the 'wilting point' (D). When the soil moisture is above the wilting point the surplus evaporation potential, after meeting the requirements of evaporating the interception store, is allowed to operate on the soil at the full potential rate.

When the soil moisture falls below the wilting point, residual potential evaporation is cut back to a fraction (F) of it, to give the estimated actual evaporation. If, after allowing for evaporation, the soil moisture content is above the field capacity, then the surplus above field capacity is recharge to the water table.

Thus, there are five variables controlling the process. Because potential evaporation was supplied in three parts a further parameter, albedo, was introduced to allow for the different reflectivity of different surfaces. Daily accounting was used. The main objections to this model are that; a different ordering of the process priorities would produce different results; daily accounting implies that surplus water can drain from the soil moisture reservoir within a daily period; and, little account was taken of the runoff process within the micro-Karst. This latter defect is somewhat compensated for by the rapid drainage permitted from the soil moisture reservoir.

The original concept for use of this model was to operate it on a grid basis as had been done in Jordan by Chidley (1968). In this case, the whole country was divided into a series of squares and different soils and vegetation were assigned to each grid square. A ten-day accounting unit was used and rainfall was interpolated for each grid square for each ten-day period. Normally, it is essential to do this if there is any appreciable spatial variability in the parameters or in the rainfall, because the processes used are nonlinear. Interpolation of the results from spot calculations will produce 'erroneous' results. However, because of the low variability of rainfall and the fact that there were only two types of soil and vegetation cover to consider, the recharge for each lens was worked out on the basis of assuming that parameters were constant for each lens but differing between lenses. Rainfall data for the most appropriate adjacent station was used. Table 1 shows the sets of parameters

Table 1. Catchment Model Parameters

	Ironshore Soils scrub and grass	Bluff Limestone Soils jungle, bush
Interception Store (mm)	2.5	5.0
Field Capacity (mm)	50.0	50.0
Wilting Point (mm)	20.0	20.0
Albedo	0.1	0.1
Evaporation Reduction	0.1	0.1
Runoff Factor	0.0	0.0

actually used. In most system simulation studies where process parameters have to be estimated it is necessary to have a historic record against which the parameter selection can be validated. In this case there is no independent means of measuring several of the hydrologic balance components. This does not invalidate the approach for two reasons; even with no validation a sensitivity analysis can be carried out and in this case even if the absolute value of the recharge could not be determined, the temporal and spatial variability would be given and these would not be very sensitive to parameter choice.

A parameter sensitivity analysis was carried out and for reasonable variations in the values of the parameters, the variation in average annual estimate of recharge of 400 mm was of the order of ±12%. Largely because of the variability of rainfall the average annual 'potential recharge' rate varied over the island from between 620 mm and 400 mm. Within time, the recharge rate varied from about 25% of rainfall for low rainfall years, to 50% in wet years. The nature of the intensity of the rainfall further affects the picture. These estimates of potential recharge appear consistent with estimates made by the authors in Jamaica.

Geology

The Cayman Islands lie in a structurally complex area which has not yet been defined in detail. The islands form prominences on a submarine ridge which is bordered to the north by the Yucatan Basin and to the south by the Cayman Trench which reaches a maximum depth of about 6,500 m. The narrowness of the ridge and the isolated geomorphological location of the Cayman Islands is emphasized by the fact that the 100 fathom isobath seldom occurs outside a distance of one mile from the shore. The immense thickness of rock and proximity of the sea is paramount to the validity of the assumptions made in the modelling of the lenses referred to later.

Grand Cayman is topographically low and flat, with a maximum height above sea level of just over 10 m. For the most part elevations are of the order of 3 m above sea level.

As a consequence of its environment, the geology appears to be relatively simple. The prominences on the submarine ridge have existed in a shallow warm sea eminently suitable for coral formation. The rock types deposited consist, therefore, exclusively of fossiliferous limestones and their degraded derivations.

Hydrogeology

The purpose of the hydrogeological investigations was to delineate the extent of the fresh water lenses and measure or infer some of the hydro-geological properties of the aquifers. A general distribution of the lenses was available from previous work which enabled a rapid start to be made on identifying the plan of the lenses at the water table by using electrical conductivity measurements of groundwater in bore holes, natural sink holes and dug wells. The water levels in these locations were also used to determine the configuration of the water table.

The shape of the saline interface and the thickness of the brackish water transition zones between fresh water and sea water was studied using surface resistivity measurements in conjunction with borehole conductivity control profiles. Estimates of aquifer characteristics, permeability and storage were assessed using traditional pumping test techniques and groundwater level responses to tidal fluctuations.

A significant factor in the study of the fresh water lenses of the Cayman Islands is the extensive dispersive effect caused by variation in recharge and tidal fluctuation in a situation where the aquifer is relatively very thick and the fresh water lens very thin. Apart from minor instances the lenses are entirely land-locked. These factors had an important bearing upon proposals for development.

In order to delineate the lens configuration in the three space dimensions a basis for drawing a line between fresh water and salt water had to be fixed. The two main factors in assessing this value were the potability of the water and the effect on the groundwater hydraulics. From the point of view of potability a chloride level of 500 ppm was used, which is close to the 600 ppm U.N. recommended limit. The 500 was chosen because its presence was easily detected in the electric resistivity studies. This assumption gives a reduced lens size and thickness which would be on the safe side with regard to development of water resources.

Although water levels were measured in boreholes, fluctuations due to tidal and barometric pressure variations and differences in benchmarks made it difficult to maintain the spatial integrity of the measurements. It was decided to determine the free water surface location of the lenses by assuming that it was the mirror image of the configuration of the salt-fresh water interface, with a magnification based upon the Ghyben-Hertzberg ratio.

Correlations were established between the measured water table elevation above sea level and the depth to the 500 ppm salt concentration and also the depth of occurrence of undiluted sea water. This is shown in Fig. 3. It would seem that a 1:20 ratio corresponds to the 500 ppm ratio, with there being little significant difference between the Ghyben-Hertzberg ratio of 1:40 and an observed ratio of 1:50 for the actual base of the transition zone of fresh to salt water. The effect of the choice of this ratio on the resulting analysis is discussed later.

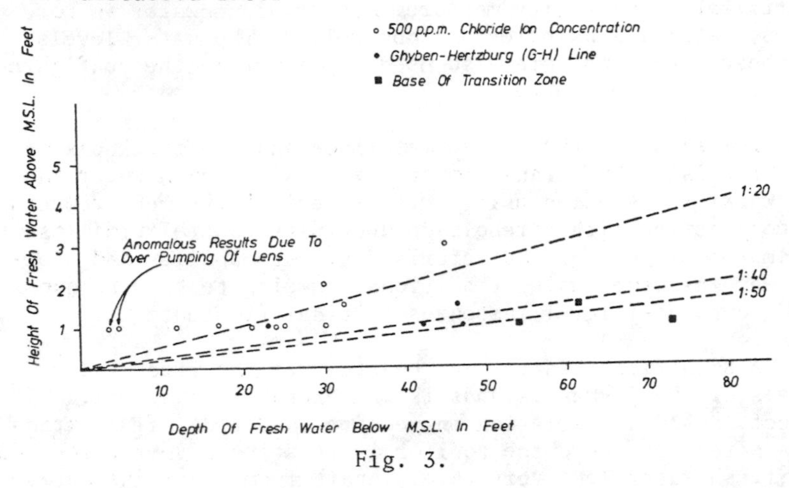

Fig. 3.

The electric resistivity measurements were taken with an A.B.E.M. Terrameter using a Wenner configuration of electrodes. The analysis was performed on a digital computer. Full details of this exercise are given by Lloyd and Bugg (1976).

In a small island of this nature there is unlikely to be very elaborate drilling and pumping equipment available and because of the highly irregular distribution and large range of values of permeability encountered, conventional pump tests can only be expected to give a broad indication of the range and distribution of permeabilities to be expected. Because the data obtained in such a way could not be expected to be definitive and because it had been observed that an apparently steady state situation developed fairly soon after pumping commenced, all pump tests were analysed using the Thiem steady state formula. A borehole conductivity and temperature probe was placed below the point of pump suction.

Other factors to bear in mind in general are, the effect of the movement of the saline interface during the pump test and the fact that the free water surface, water level fluctuations and higher dissolved CO_2 levels would give rise to a more extensive development of solution cavities giving local increase in permeabilities. It was assumed that over the very short, a few hours or days, durations of the pump tests, the aquifer would behave as though the base of the aquifer was at the bottom of the wells. Thus the permeability could be estimated by dividing the transmissibility by the depth of penetration of the well. It is admitted that this is a crude assumption but it does yield information on how a typical partly penetrating well will operate and gives an index of permeability variations.

Groundwater Modelling

A simple form of digital groundwater model was used to investigate regional permeability variations, to form estimates of recharge from groundwater data and to investigate the effect of abstraction schemes. If one assumes uniform recharge and aquifer characteristics and that the permeability is isotropic with a hydrostatic pressure distribution within the fresh water; and also that the Ghyben-Hertzberg relationship (Edelman, 1972) holds for the location of the salt-fresh water interface with all fresh water flow through the edge of the lens; a very simple steady state theory can be developed for groundwater flow in a fresh water lens.

It can be shown that with these simple assumptions the maximum safe rate of abstraction from the lens is equal to the recharge to the lens if the abstraction is made at the very edge of the lens. If the abstraction is at the centre of the lens the safe abstraction rate is a maximum of one-half of the recharge. In practice, if one removes the recharge at the edge of the lens the salt water will eat into the lens edge unless steps are taken to prevent it. Also, if one removes water from the very centre of the lens, which affords maximum protection to the lens edge, a certain thickness of fresh water has still to be maintained, otherwise the wells will go saline; this reduces yield. Various measures can be taken to alleviate the increase of salinity, e.g., concentric wells, but there will always be a limit to the rate of abstraction which is less than one-half of the recharge rate. This suggests that the practical optimum location of wells will be somewhere between the centre and the edge of the lenses, the exact location depending upon the lens thickness which in turn is controlled by the net recharge rate and the permeability. In all other cases of abstraction, more elaborate measures will be required to protect the collapse of the lens. In Grand Cayman the lenses are very susceptible to erosion by salt water at their edges, as evidenced by their being land locked already.

By making a few modifications to the assumptions above, a simple dynamic model of lens groundwater flow can be devised. These assumptions are that the aquifer properties can be allowed to be nonhomogeneous and that recharge rates can vary with time and space. Further assumptions, which are believed to be valid for Grand Cayman, are that the lens is very shallow, compared to the total aquifer thickness and that the permeability is fairly high over this total thickness. In this case, and for time intervals of a few months or years it is assumed that the pressure distribution in the salt water is uniform and invariant. The implication of this is that the lens behaviour can be modelled as a series of steady states with a Ghyben-Hertzberg type relationship holding for all time. The resultant differential equation for two-dimensional flow becomes:

$$\frac{\partial}{\partial x} T_x \frac{\partial h}{\partial x} + \frac{\partial}{\partial y} T_y \frac{\partial h}{\partial y} + \frac{R}{1+a} = S \frac{\partial h}{\partial t} \qquad (3)$$

where x, y, t are independent space and time variables, T_x and T_y are 'directional' transmissibilities, R is the recharge rate, S is the storage coefficient, 'a' is the Ghyben-Hertzberg ratio and h is the elevation of the water surface above datum.

It was shown that it was possible to calibrate regional permeabilities using steady state estimates of recharge, provided the lens was not significantly developed by pumping. Cyclic variation of recharge and long term variation in recharge did not affect the calibrations obtained. The range of permeabilities obtained and their spatial distribution was similar to the range and distributions obtained from the pump tests.

There was a difficulty in relating the computed 'potential recharge' to the recharge that was required to give a sensible calibration. The basis for the calibration was to match the computed saline-freshwater interface with the measured one using steady or cyclic recharge. If the interface defined by the 500 ppm line was used it was apparent that in order to obtain a calibration the potential recharge would have to be reduced in some way, or the lens would be too deep. Alternatively, the permeabilities would have to be increased to permit the extra water to escape. On the other hand, if the very bottom of the transition layer was used as the bottom of the lens, something like the computed 'potential recharge' would give a sensible calibration with reasonable estimates of permeability. The relationship between permeability, Ghyben-Hertzberg ratio and recharge is clearly shown in Eq. 3.

Logically, one should choose a true 1:40 Ghyben-Hertzberg ratio and use the full recharge. This, however, does not allow for the fact that it is highly probable that because of the close proximity of the water table to the land surface that actual evaporation was underestimated. In the event it was decided to use the 1:20 G.H. ratio, corresponding to the 500 ppm chlorides because it would not have mattered if the recharge were much higher, if most of it was to be lost to the transition zone. The effective recharge from the abstraction point of view was the one which gave a calibration at the 500 ppm level. Fig. 4 gives an idea of the degree of matching that was possible between computed and measured salt-freshwater interfaces.

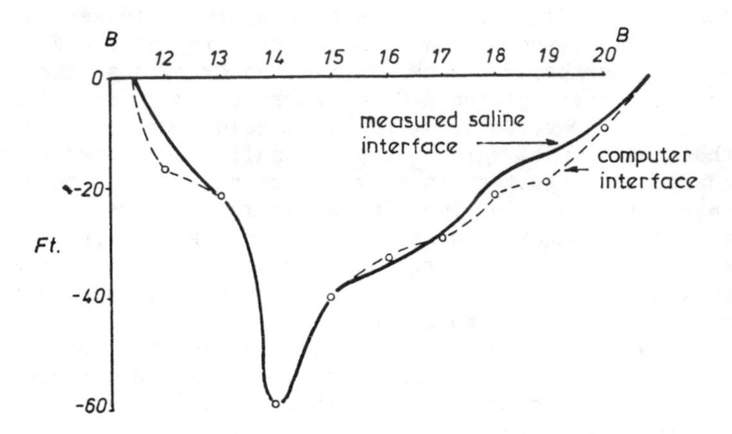

Fig. 4. Typical calibration

Development

During the modelling it was obvious that there was some thinning of the lenses near their edges which was more than could be expected if the aquifer were of uniform properties. A calibration could only be obtained by increasing the permeabilities and/or reducing the recharge in those areas. Both methods were used in calibrating the lenses. Where it was known that the water table was very close to the land surface an increase in evaporation was permitted. In these cases zero net recharge was assumed. It was argued that higher permeability would be found near to the lens edges due to increased tidal action there and due to the fact that more water was passing through an ever thinner section, giving more opportunity for solution of the limestone rocks.

Several methods of development were considered but the main one recommended was that of open ditches. Even with open ditches, sufficient depth of lens to install them with safety from saline encroachment only existed near the centre of the lenses. Different abstraction rates for well fields comprised of ditches were simulated. From the model simulations it was found that about one-quarter to one-third of the average annual recharge could be abstracted without reducing lens thickness to an extent that would permit abstraction to become saline or weaken too much the outflow at the edge of the lens. Greater abstraction rates than this would require positive actions to control saline encroachment.

Conclusions

The paper has shown how a range of theoretical knowledge was necessarily brought together within a single investigation. Approximate theories were developed from more elaborate ones, either because data was missing or because the nature of the investigation did not warrant a more detailed study. Within the validity of the assumptions made, it was shown that the spatial variability of aquifer characteristics and recharge are important factors affecting the development of fresh water lenses in limestone aquifers. However, it was also shown that temporal variations were not so important because of the relatively large storage in such systems compared to long term yield. A steady state analysis will suffice in most cases. The modelling of the dynamic state was simplified to that of modelling a modified single phase system. It was shown that this simplification would reduce to the steady state solution from arbitrary initial conditions. A dynamic analysis, even the simplified one described here would be required if consumption of the storage of water in the lens was intended as a short term measure.

References

[1] Chidley, T.R.E., 1968. "Report on Computer Procedures and Programs for the Analysis of Project Data," UNFAO, Sandstone Aquifers of Eastern Jordan, Project No. 212, Nov, 1968, 60 p.

[2] Lloyd, J. W. and Bugg, S. F., 1976. "A Study of Fresh Water Lens Configuration in The Cayman Islands Using Resistivity Methods," Qu. J. of Eng. Geol., Vol. 9, No. 4, pp. 291-303.

[3] Lloyd, J. W., Drennan, Bennell, B. M. U., 1967. "Groundwater Recharge Study in North Eastern Jordan," Paper No. 6962, Proc. I.C.E., Vol. 37.

[4] Edelman, J. H., 1972. "Groundwater Hydraulics of Extensive Aquifers," Inth. Inst. for Land Reclamation, Wageningen, The Netherlands, p. 215.

[5] Chidley, T.R.E. and Pike, J. G., 1970. "A Generalized Computer Programme for the Solution of the Penman Equation for Evapotranspiration," J. of Hyd. 10, pp. 75-84.

ANALYSIS OF TRANSIENT GROUNDWATER FLOW FROM SEEPAGE PONDS

By

John P. Glass

B. A. Christensen

Hillel Rubin*

Department of Civil Engineering
University of Florida
Gainesville, Florida 32611

*On leave from Technion, Haifa, Israel

Abstract

The increasing use of seepage ponds for disposal of storm water has brought about the need for an analytical method that can be used in pond design. Because of the small size of the typical seepage pond project the analysis should be simple and inexpensive to perform. At the same time, it should be based on the physical principles of groundwater flow and should provide a rational basis for comparison of one site against another.

The analysis presented here is composed of two parts. First, a method of storage routing is developed by which the seepage pond size can be determined. The equation of Green and Ampt, as modified by Bouwer, is used to account for the outflow due to seepage. The routing procedure is simple enough to be done by hand or on a programmable calculator.

The transient groundwater mound that appears in response to infiltration from the pond is analyzed using the Dupuit assumptions. The effect of variable moisture content above the water table is included. The nonlinear partial differential equation which describes axisymmetric mounds in cylindrical coordinates is solved by a finite difference method. Although the solution requires considerable computer time, the results can be extrapolated to any similar pond configuration through the use of dimensionless graphs. Practical design of seepage ponds using these graphs is simple and accurate. They also provide a useful tool for comparing the relative merits of one site against another.

Introduction

In locations having favorable soil and groundwater conditions, seepage ponds can provide a means of storm water disposal that is both economically and environmentally attractive. Most storm water seepage ponds serve a small drainage area and depend on the transient nature of the runoff hydrograph for successful performance. A storm on the catchment area results in pond inflow which is of short duration relative to the time scale of groundwater flow. As a consequence, seepage from the pond, which is often its only provision for outflow, is transient also. The capacity of such a pond to divert storm water runoff into a phreatic aquifer depends on two processes: (a) the rate of seepage, and (b) the reaction of the water table to this seepage.

These facts are well known by scientists but are usually not taken into consideration in pond design. The objectives of this paper are to develop methods by which the designer can estimate seepage pond effectiveness and determine the storage capacity required.

Unsteady Seepage and Pond Capacity

The adequacy of a seepage pond is evaluated by a storage routing procedure, which is basically an account of the inflow, and change of stored volume over successive discrete time increments. The inflow is represented by the runoff hydrograph for the design storm on the area to be drained. For the purposes of the present exposition it will be considered a given function of time. The outflow, on the other hand, is dependent on time, depth of ponding, and the properties of the soil.

Consider an initially dry seepage pond constructed in unsaturated soil. As it fills, the water begins to escape from it by vertical unsatured flow, or infiltration. The most convenient way to describe this flow is by the one-dimensional infiltration formula of Green and Ampt (1911) as modified by Bouwer (1969). It has been shown that this approach, which was long thought to be purely empirical, is soundly based on physical principles (Morel-Seytoux and Khanji, 1974) and can give very good answers (Whistler and Bouwer, 1970).

Green and Ampt based their derivation on a simplified model of infiltration which treats the soil as a bundle of vertical capillary tubes. The vertical hydraulic conductivity and moisture content of the unsaturated flow are considered constant, as is the capillary suction potential of the advancing wetting front. Applying Darcy's law to this idealized flow, the infiltration rate is described as

$$w = K_t(H + L + \psi)/L \tag{1}$$

where w = infiltration rate,
K_t = hydraulic conductivity of the transmission zone,

H = depth of ponded water,
ψ = capillary suction potential, and
L = depth of penetration of the wetting front.

The rate of advance of the wetting front is

$$\frac{dL}{dt} = w/f \qquad (2)$$

where t = time, and
f = the volumetric fraction of fillable pore space.

The equation of continuity applied to the pond yields

$$\frac{dV}{dt} = A\frac{dH}{dt} = I - fA_f\frac{dL}{dt} \qquad (3)$$

where V = volume of stored water in pond,
I = inflow,
A = area of water surface, and
A_f = area through which infiltration takes place.

Since the Green and Ampt infiltration model assumes one-dimensional vertical flow, the area term, A_f, must be treated as a constant in the equation even though it is generally not constant physically. Introducing Eq. 3 into 1 and 2 we obtain, after differentiation, the following nonlinear differential equation

$$f/K_t \frac{d^2}{dt^2}(L^2/2) + (f\Delta_f/A-1)\frac{dL}{dt} - I/A = 0 \qquad (4)$$

which can be solved numerically.

However, in most practical cases H changes more slowly than L so that over a small time step Eq. 2 can be integrated with H held constant. This leads to the well known infiltration equation

$$(K_f/f)t = L + \Gamma \ln\left(\frac{L+\Gamma}{\Gamma}\right) \qquad (5)$$

where $\Gamma = H + \psi$ is considered constant but is varied in discrete steps as the calculation proceeds. This provides a link between outflow and time which, together with Eq. 3, forms the basis for a storage routing procedure that can be carried out by hand.

An adjustment must be made during the process for the fact that Γ is not really constant as was assumed in the derivation of Eq. 5. This is done through the use of a fictitious value of time which must be calculated from Eq. 5 at each discrete time step. The method is similar to that used by Bouwer (1969) in the case where Γ varies due to nonuniform soil. It requires the solution of Eq. 5 for both L and t. At the end of a time step, when the cumulative infiltration is known, Eq. 5 is solved for the fictitious value of time using the newly calculated value of Γ. At the beginning of a new time step it is solved for L by Newton's method. This value of L is then applied to Eq. 3 to find pond depth. The infiltration area, A_f, to be used here is the time average of the pond area over the time of infiltration.

This method of storage routing was checked against experimental data reported by Weaver and Kuthy (1975). In their experiment, a seepage pond was constructed and filled with water at a controlled rate. Figure 1 shows the curves describing the area and volume vs. depth relationships for this pond. Soil test data was reported which led to the following values of soil parameters $K_t = 1.2$ ft/hr (0.366 mm), $f = 0.2$, and $\psi = 0.5$ ft (0.152 m). Figure 2 shows the comparison between experimental results and the prediction based on the Green and Ampt equation.

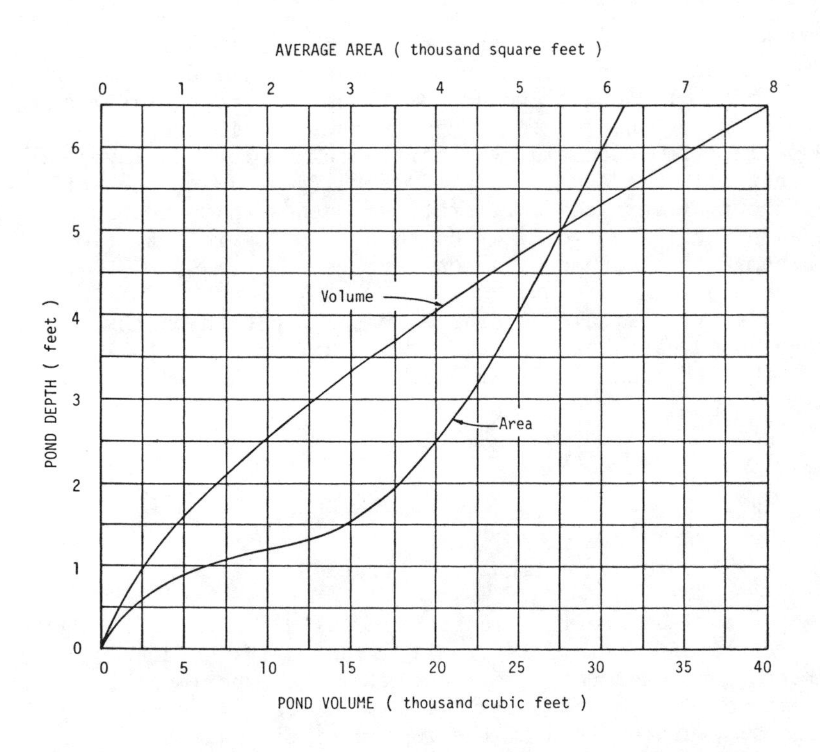

Fig. 1. Depth vs. area and volume of test pond

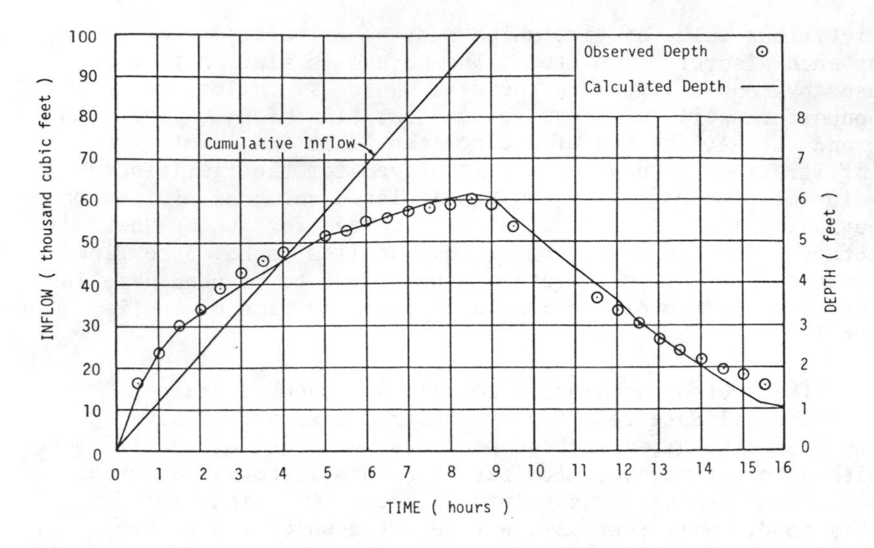

Fig. 2. Comparison of experimental and calculated storage routing

Water Table Response to Seepage

The reaction of the water table aquifer to vertical infiltration from a seepage pond is characterized by the growth of a groundwater mound. For analysis we consider a circular pond constructed in homogeneous and isotropic soil. The aquifer is underlaid by a horizontal imprevious layer and initially has a horizontal free surface. When the wetting front of vertical infiltration reaches this free surface the mound begins to form. From this time the infiltration is assumed to continue at a constant rate.

In a cylindrical coordinate system centered on the seepage pond, as shown in Fig. 3, the axisymmetric flow in the saturated mound is described by

$$\frac{\partial h}{\partial t} = \frac{K_s}{n} \frac{\partial}{\partial r} (h \frac{\partial h}{\partial r}) + \frac{K_s h}{nr} \frac{\partial h}{\partial t} + \frac{w}{n} \tag{6}$$

where K_s = saturated hydraulic conductivity,
n = specific yield, and
w = rate of vertical infiltration.

Both w and n are functions of radial distance, r. The value of specific yield is greatly reduced in the zone of infiltration under the pond. Based on the experimental evidence of Bodman and Colman (1943) it seems reasonable to assign a specific yield to the zone of infiltration with a value of 20% of the specific yield elsewhere. The rate of

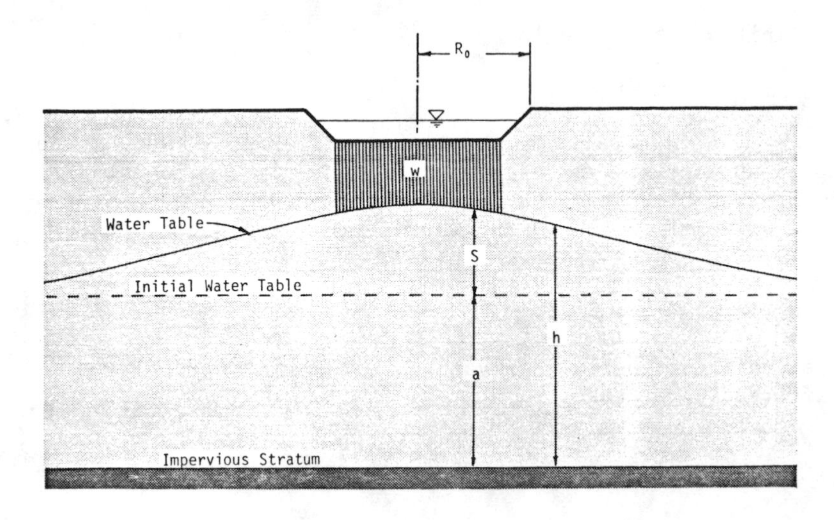

Fig. 3. Configuration of pond and aquifer

infiltration, w, is zero when r is greater than the pond radius, R_o.

Noting from Fig. 3 that $h = S + a$, Eq. 6 can be rearranged:

$$\frac{\partial S}{\partial t} = \frac{K_s}{n} (S + a) \frac{\partial^2 S}{\partial r^2} + \frac{K_s}{n} \left(\frac{\partial S}{\partial r}\right)^2 + \frac{K_s}{rn} (S + a) \frac{\partial S}{\partial r} + \frac{w}{n} \tag{7}$$

To make the equation dimensionless, introduce the following dimensionless variables:

$$t' = fK_s a/R_o^2, \quad r' = r/R_o, \quad S' = S/a$$

Substitution of these variables along with some manipulation and omitting the primes results in:

$$n \frac{\partial S}{\partial t} = \frac{\partial^2}{\partial r^2} (S^2/2) + \frac{\partial^2 S}{\partial r^2} + \frac{1}{r} \frac{\partial}{\partial r} (S^2/2) + \frac{1}{r} \frac{\partial S}{\partial r} + WR_o^2/(K_s a^2) \tag{8}$$

In this form, the nonlinear terms are distinctly separated from the linear terms. They are treated differently in the numerical analysis.

$$\frac{\partial S}{\partial r} = \frac{\partial}{\partial r} \left(\frac{S^2}{2}\right) = 0 \quad \text{at} \quad r = 0 \tag{9}$$

and

$$S = 0 \quad \text{at} \quad r = \infty$$

The initial condition is:

$$S = 0 \quad \text{at} \quad t = 0 \tag{10}$$

To solve the problem by finite differences, we must set up an appropriate space-time grid. The time increments are designated by i's and the space increments by j's. The time derivative of S is approximated by a forward difference.

$$\frac{\partial S}{\partial t} \simeq (S_{i+1,j} - S_{i,j})/\Delta t \tag{11}$$

The nonlinear spatial derivatives are represented by central difference operators:

$$\frac{\partial}{\partial r}(S^2/2) \simeq (S^2_{i,j+1} - S^2_{i,j-1})/(4\Delta r) \tag{12}$$

$$\frac{\partial^2}{\partial r^2}(S^2/2) \simeq (S^2_{i,j+1} - 2S^2_{i,j} + S^2_{i,j+1})/2\Delta r^2 \tag{13}$$

The linear space derivatives are the most influential terms in the equation. If the solution were to become numerically unstable, it would probably be because of these terms. Therefore, they are approximated by the mean of the finite difference representations on the $(i+1)$th and the (i)th time rows. This is the implicit method of solution developed by Crank and Nicolson (1947) which has good stability and convergence characteristics:

$$\frac{\partial S}{\partial r} \simeq \frac{1}{2}\left\{(S_{i+1,j+1} - S_{i+1,j-1})/(2\Delta r) + (S_{i,j+1} - S_{i,j-1})/(2\Delta r)\right\} \tag{14}$$

$$\frac{\partial^2 S}{\partial r^2} \simeq \frac{1}{2}\left\{(S_{i+1,j+1} - 2S_{i+1,j} + S_{i+1,j-1})/\Delta r^2 + (S_{i,j+1} - 2S_{i,j} + S_{i,j-1})/\Delta r^2\right\} \tag{15}$$

When these finite difference approximations are substituted into Eq. 8, a tridiagonal system of equations is generated which can be solved by Gaussian elimination.

The computer program developed from these finite difference operators was tested with several combinations of grid mesh ratio and grid size. It was found to give stable results with a ratio of $\Delta t/\Delta r^2 \leq 0.11$. The mesh size chosen was $\Delta r = 0.03$. To simulate the boundary condition at infinity, the calculations were carried out to $r = 25$ at which point S was required to equal zero always.

For comparison with the linearized analytical solution of Hantush (1967), the finite difference program was run with constant specific yield, n. The results of this run are shown by the dashed lines in Fig. 4. They are practically identical to the results obtained by the method of Hantush. When n was allowed to vary as a function of r, the results were very much different, as shown by the solid lines in Fig. 4. Figure 5 shows a more complete set of such curves which could be used in seepage pond design.

Discussion and Conclusions

An analysis of the hydraulic operation of a seepage pond should be conducted in two phases. The first phase is concerned with the rate at which water seeps out of the pond by vertical infiltration through its bottom. The equation for vertical unsaturated flow developed by Green and Ampt (1911) can be used to describe the outflow in a storage routing process. The storage routing is simple enough to be done by hand when the soil is assumed to be homogeneous. The Green and Ampt equation can also be adapted for use in soil where the hydraulic conductivity varies monotonically with depth (Bouwer, 1969) which is a very common occurrence. This calculation is about as simple as the storage routing procedure but the combination of variable ponded depth and variable hydraulic conductivity would be cumbersome enough to make the use of a computer or a programmable calculator desirable.

The second phase of the analysis deals with the response of the water table to recharge from the seepage pond. The determination of adequate capacity of the groundwater aquifer requires use of the design charts for each specific case. The design chart itself is based on the numerical solution of the nonlinear differential equation.

The importance of the finite difference solution of the groundwater mound is that it makes possible the consideration of an axisymmetric variation of the specific yield. It would also be possible to include the effects of axisymmetric variation in other soil properties.

It should be noted that the saturated hydraulic conductivities referred to in the two phases of the analysis are, very often, not the same. Infiltration is concerned with vertical flow while the groundwater mound is mainly influenced by horizontal flow. The difference between the saturated hydraulic conductivities in the two directions usually stems from horizontal layering of the soil.

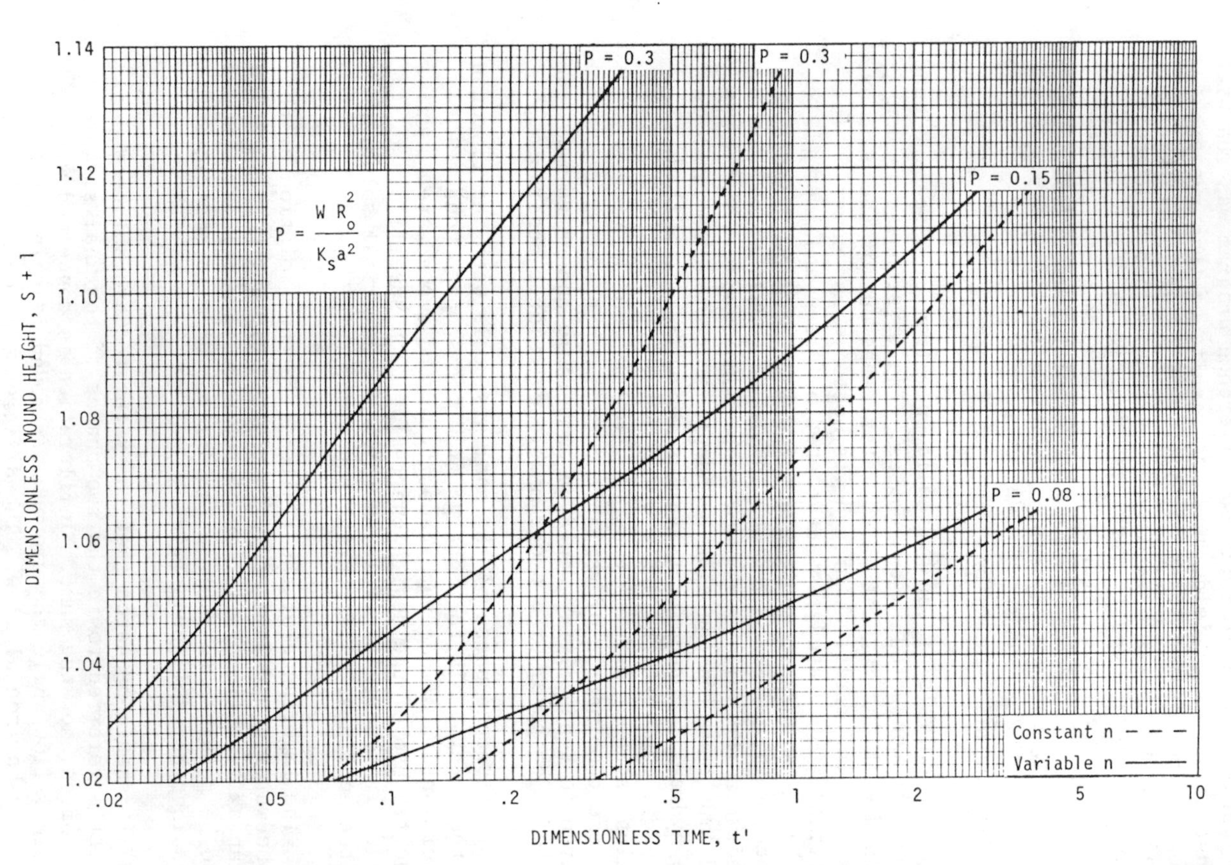

Fig. 4. Mound height at r'=0, comparison of constant and variable n

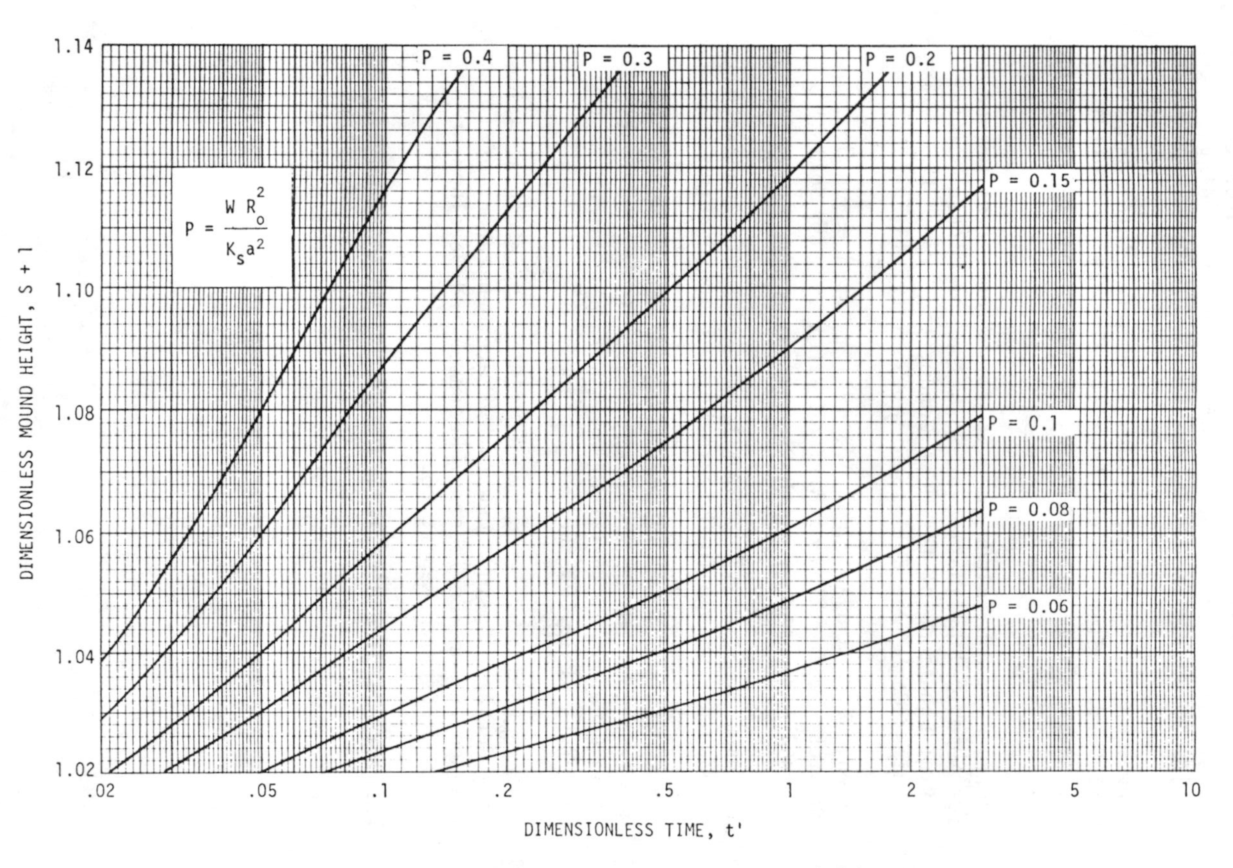

Fig. 5. Mound height at r'=0, variable n

References

[1] Bodman, G. B. and Colman, E. A., 1943. "Moisture and Energy Conditions During Downward Entry of Water into Soils," Soil Science Society of America Proceedings, 8:116-122.

[2] Bouwer, H., 1969. "Infiltration of Water into Nonuniform Soil," Proceedings of the American Society of Civil Engineers, Irrigation and Drainage Division, IR4, 95:451-462.

[3] Crank, J. and Nicolson, P., 1947. "A Practical Method for Numerical Evaluation of Solutions of Partial Differential Equations of the Heat Conduction Type," Proceedings Cambridge Philosophical Society, 43:50-67.

[4] Green, W. H. and Ampt, G. A., 1911. "Studies on Soil Physics, I, The Flow of Air and Water Through Soils," Journal Agricultural Science, 4:1-24.

[5] Hantush, M. S., 1967. "Growth and Decay of Groundwater-Mounds in Response to Uniform Percolation," Water Resources Research, 1, 3:227-234.

[6] Morel-Seytoux, H. J. and Khanji, J., 1974. "Derivation of an Equation of Infiltration," Water Resources Research, 4, 10:795-800.

[7] Weaver, R. J. and Kuthy, R. A., 1975. "Field Evaluation of a Recharge Basin," New York State Department of Transportation, Engineering Research and Development Bureau, Research Report 26.

[8] Whisler, F. D. and Bouwer, H., 1970. "Comparison of Methods for Calculating Vertical Drainage and Infiltration for Soils," Journal of Hydrology, 1, 10:1-19.

IMPACT OF LONG-TERMED DISCHARGE ON AN AQUIFER WITH SMALL RECHARGE

By

Manfred Klenke
Institut fur Wasserwirtschaft
Hydrologie und landwirtschaftlichen Wasserbau
Technische Universitat Hannover
Callinstr. 15 c, 3000 Hannover, Germany

Abstract

The management of groundwater reservoirs for irrigation projects includes investigation of the behaviour of groundwater systems in case of a long-term discharge. One important problem is the determination of the change of groundwater potential. Because of a low rate of recharge in arid regions, the pumping of groundwater leads to a permanent deficit between discharge and recharge. For this reason, a steady state piezometric head will not be reached.

To predict the time dependent change of head, simulation models can be used. They permit calculation of the effects of external influences on the natural system under hydrologic and technical restrictions. Difficulties result from the fact that in most cases the parameters of the groundwater system, transmissivity and storage coefficient, are not exactly known within the investigated area and therefore an accurate calibration of the model is impossible.

An irrigation project in a North African region is taken as an example to demonstrate the development and the use of a simulation model. The effects of discharge and drawdown near the pumping wells for different well arrangements were to be examined.

The computations were made by using several different values of the distribution of transmissivity and storage coefficient, to determine the effect of these parameters on the drawdown. This is of special interest, because all aquifer parameters are not obtained as precise values. Comparison of the results illustrates the sensitivity of the groundwater system to aquifer parameter variation. Variation of the local distribution of the pumping wells made it possible to find satisfactory arrangements of the well fields.

Introduction

A long-term supply of water requires the management of groundwater reservoirs under hydrological and technical restrictions. This problem is of special importance in arid regions, where a large demand conflicts with a limited recharge. It is necessary to predict the effects of external influences on the natural system. For this problem, simulation models can be used. The present investigation of the change of groundwater potential in an extensive aquifer under the special boundary conditions of arid regions, however, shows that the development and use of simulation models include many difficulties. Therefore a critical estimation of the reliability of the results is necessary.

Outline of Hydrologic Situation

Large groundwater rates are needed for irrigation projects in a North African region. To calculate the effects caused by discharge, the time dependent piezometric head inside the investigated region need to be computed. Also, the drawdown near the pumping wells for different well arrangements was to be examined. Figure 1 provides a survey of the investigated area. As the pumping rates are greater than the total subsurface recharge, a permanent sinking of the groundwater level was expected. For the decision, how long a recharge is possible under present conditions and technical restrictions, simple water balances for the well fields are not sufficient. Instead of these, the change of piezometric head with respect to time had to be computed inside the complete area.

The investigated region has a dimension of 130,000 km^2. The groundwater reservoir consists of several layers. The aquifer in which pumping wells are located is a confined aquifer with a thickness of approximately a hundred meters. It is separated from other aquifers by impermeable layers.

The knowledge about the hydrogeologic parameters results from the information about the geological structure and the hydrogeological situation. Several pumping tests give additional information about transmissivity and storage coefficient in some sections of the area.

Field measurements of the actual groundwater potential made it possible to fix the initial conditions.

The boundary conditions were determined approximately. Flow rates across the boundaries had to be specified instead of a constant head since influence of the boundaries caused by discharges was expected. The rates were calculated by

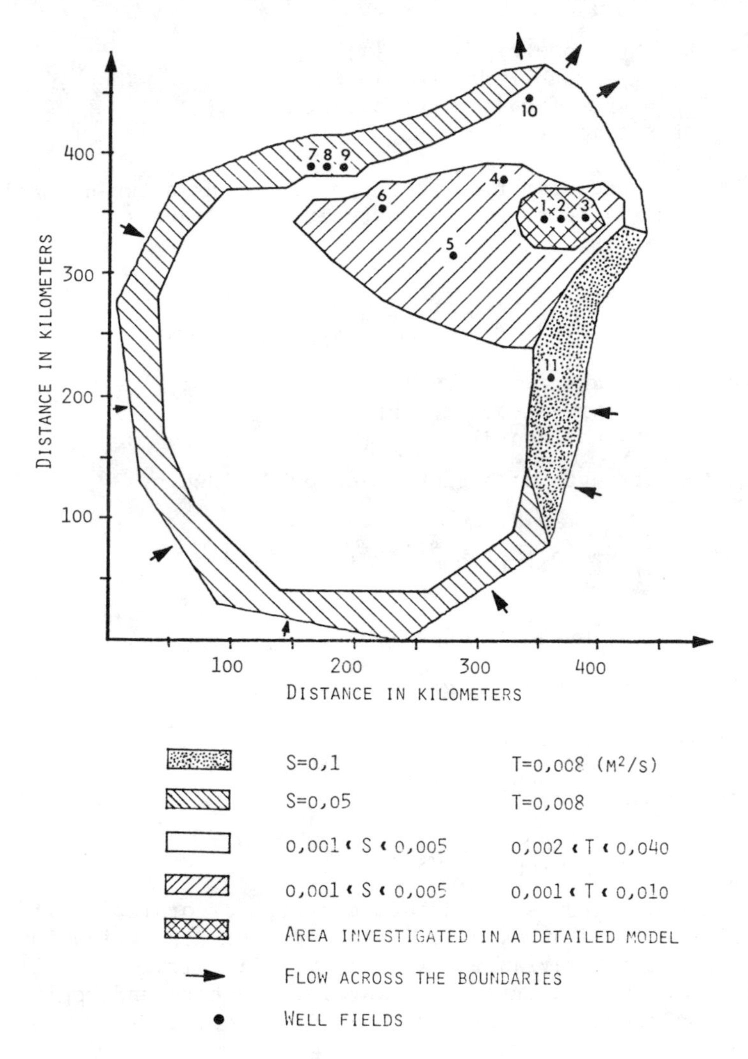

Fig. 1. Investigated region with areas of different hydraulic parameters and boundary for detailed model

evaluating the slope of groundwater potential near the boundaries and using average values of transmissivity.

Development of a Simulation Model

Mathematical description. Groundwater flow in porous media can be described by a partial differential equation of the diffusion type, which is derived from the Darcy law and the condition of continuity (Davis, De Wiest, 1966).

$$\text{div}\ (k_f \cdot \text{grad}\ h) = S_{sp} \cdot \frac{\partial h}{\partial t} \qquad (1)$$

where h = piezometric head (m)
k_f = permeability (m/s)
S_{sp} = specific storage coefficient (1/m)
t = time (s)

For groundwater flow in extensive areas a two-dimensional model is sufficient. Using cartesian coordinates, Eq. (1) can be expressed as

$$\frac{\partial}{\partial x}\left(T \cdot \frac{\partial h}{\partial x}\right) + \frac{\partial}{\partial y}\left(T \cdot \frac{\partial h}{\partial y}\right) = S\frac{\partial h}{\partial t} + q \qquad (2)$$

where T = transmissivity (m^2/s)
S = storage coefficient (-)
x,y = local coordinates (m)

q is the strength of a sink function defined by

$$q = \sum_{i=1}^{n} q_i\,(x_i, y_i) \cdot \delta[(x-x_i),(y-y_i)] \qquad (3)$$

δ = Dirac delta function.

The initial condition is given by

$$h(x,y,t=0) = A(x,y)$$

and the boundary condition is

$$f\,[h(t, x=x_b, y=y_b)] = B(t)$$

Eq. (2) can be transformed into a system of ordinary differential equations by using a finite element discretization of the independent variables x and y (Zienkiewicz, 1971). Approximation by a grid of triangular elements and application of the Galerkin procedure (Pinder and Frind, 1972) gives finally an ODE for each node of the grid

$$0 = \sum_{i=1}^{m} a_i\,h_i + b_i\,\frac{dh_i}{dt} + Q_i + Q_R \qquad (4)$$

Coefficients a and b are dependent of the size and shape of the elements and the hydraulic parameters T, respectively S.

Q_i represents the discharge of water at node i. Q_R represents the flux of water across the boundary. This last term is formed only at nodes, where a boundary condition is given by $Q_R = f(t)$.

Verification of the computer model. Since in the present case the unsteady phase, i.e., the change of piezometric head with respect to time, is of special interest, it is advantageous to compute the relationship between head and time in a continuous form. For this purpose hybrid computation has been shown as an efficient method (Klenke, 1975).

Because of the limited capacity of analog hardware a completely analog programming of Eq. (4) is impossible for a grid with an adequate number of nodes. For this reason a component-sharing method (Bekey and Karplus, 1971; Klenke, 1975) is used and the complete solution is computed in an iteration process.

A solution at the nodes of a subdomain is obtained by analog integration. The functions at nodes on the boundary of such a subdomain are taken from the iteration step before. For the first step these functions are arbitrarily chosen.

The analog part of the program is realized on a patching panel. The mathematical relations between the nodes of a subdomain according to Eq. (4) are formed by analog elements. The digital part of the program performs the following tasks:

- control of the analog computer run
- evaluation of coefficients and other numerical calculation
- data transfer between the digital and analog computer.

Figure 2 shows the principal steps of computing in a simplified flow chart.

Aquifer Analysis

One of the aims of the investigation is to decide if irrigation can be guaranteed over a period of 50 years. To answer this question, long-range effects of discharge and maximum local drawdown had to be simulated.

Large-scale change of groundwater potential. For the simulation the investigated region was approximated by a grid of 196 nodes defining 359 elements (see Fig. 3). It is obvious that it was impossible to represent single wells in a well field with a reasonable number of nodes in a model, enclosing the complete area. Therefore the pumping rates of the wells in each well field were summarized to a single value. The average of annual discharge of the well fields is given in Table 1.

Analysis of field measurements of the hydraulic parameters shows that it is possible to define four sections in which transmissivity and storage coefficient are assumed to

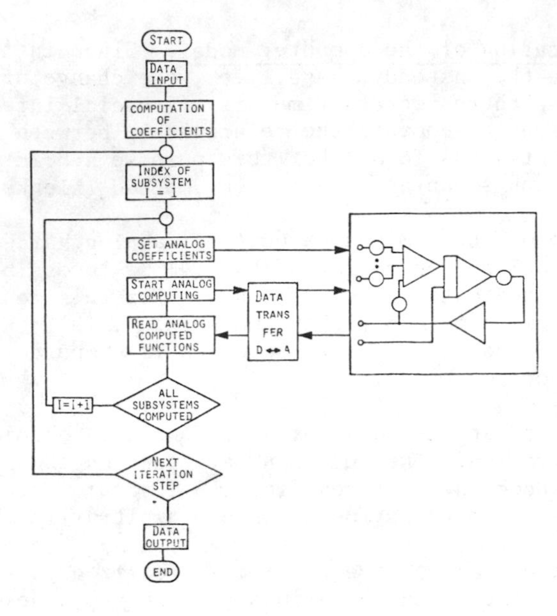

Fig. 2. Simplified flow chart of the simulation model

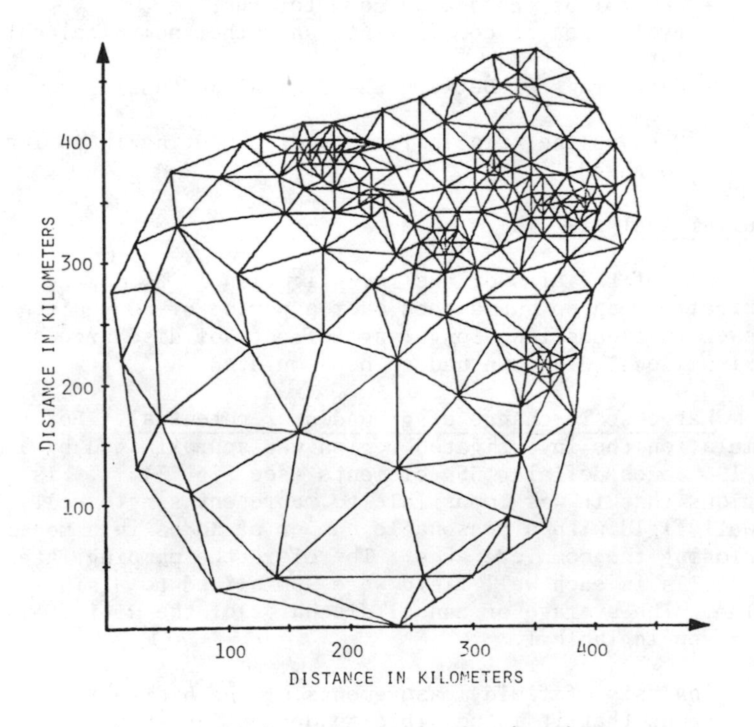

Fig. 3. Finite element approximation

Table 1 Pumping Rates of the Well Fields

Well field (numbers correspond to fig. 1)	Annual discharge (Average in 10^6 m^3)
1	26.58
2	6.25
3	15.64
4	7.82
5	2.21
6	10.20
7	12.65
8	26.74
9	2.30
10	5.99
11	8.55

be constant. This is of course not the actual parameter distribution. The simplification may lead to local differences between simulation results and natural groundwater level but for a large-scale simulation it is deemed a sufficient approximation.

Mean values of measured data were taken as basic parameters and in a first computation the reaction of the aquifer was simulated over a period of 50 years. The groundwater potential after a 50 year discharge is represented in Fig. 4, which shows equipotential lines. The numbers indicate the drawdown in meters. It is recognizable that the area of drawdown has been extended almost all over the investigated region. Additionally, parts of the boundary are influenced by the decreasing groundwater potential. Further investigations show that the flow out of the area in the northern part of the boundary decreases and reverses after approximately 10 years.

Because transmissivity and storage coefficient cannot be seen as precise values it was necessary to determine the sensitivity of model response to parameter variation. Maximum and minimum parameter ratings were fixed for two sections (see Fig. 1) and a series of simulations was performed with transmissivity and storage coefficient values within these limits. At each node of the grid the model response was evaluated. For a selected node within the region of deepest drawdown some functions of piezometric head with respect to time are shown in Fig. 5. The results make it possible to define a reliability interval for the groundwater potential. Additionally the functions show that with the assumed boundary conditions a steady state is not reached. It indicates that a permanent emptying of the reservoir results from discharge.

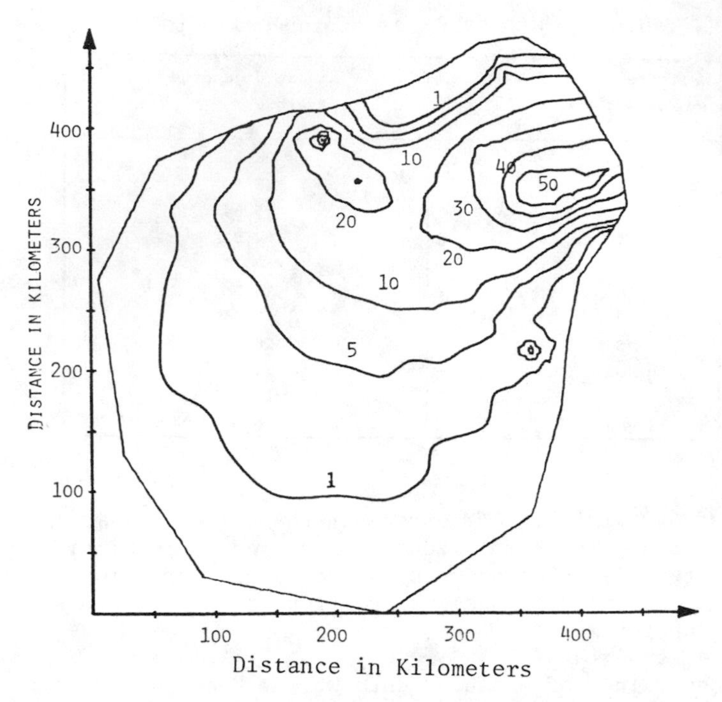

Fig. 4. Equipotential lines of piezometric head after a period of 50 years

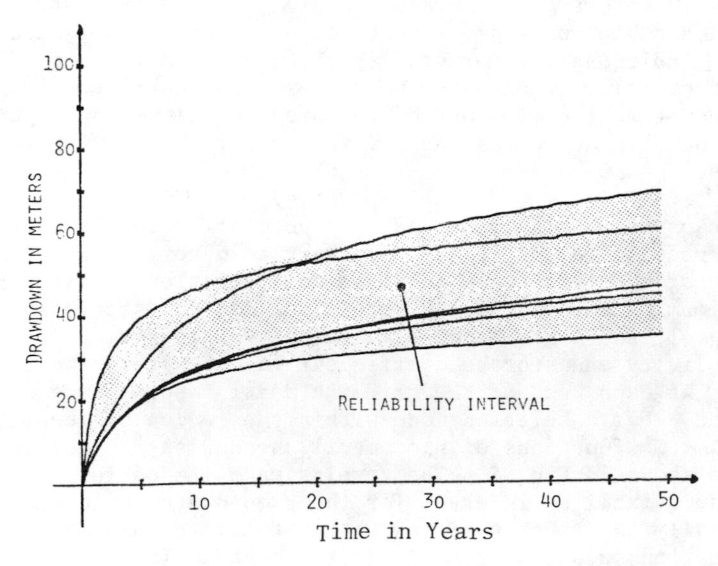

Fig. 5. Computed drawdown vs. time at a selected node for different hydraulic parameters

Groundwater level inside the well fields. For investigation of the aquifer behaviour in the neighbourhood of wells, the areas of irrigation were discretized with a finer grid, so that each well of a well field could be taken into account. For one possible arrangement of wells the distribution of pumping rates for the fields 1, 2 and 3 are given in Table 2.

Table 2 Distribution of the Pumping Rates at Well Fields 1, 2 and 3 to Single Wells

Well field		Annual discharge (10^6 m³)
1	1.1	2.74
	1.2	4.69
	1.3	5.08
	1.4	3.13
	1.5	6.25
	1.6	4.69
2	2.1	2.35
	2.2	1.56
	2.3	1.56
	2.4	0.78
3	3.1	7.04
	3.2	8.60

The result of a simulation with pumping rates according to Table 2 is given in Fig. 6. It shows a decrease of the maximum drawdown inside the well field of about 15 percent,

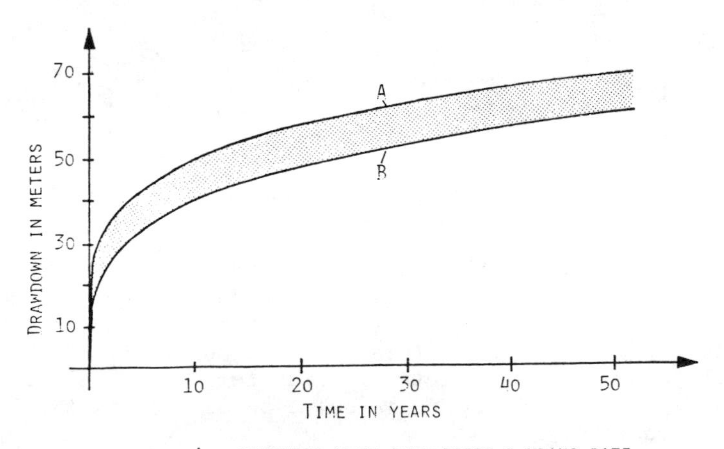

Fig. 6. Maximum computed drawdown in a well field

compared with the result of the large-scale model. Corresponding results were obtained for other sections of irrigation. The solution with lumped pumping rates appears to represent an upper bound to the expected drawdown. Simulations with variation of position and pumping rate of the wells make it possible to find a satisfactory well arrangement.

Finally, Fig. 7 illustrates the effect of some variations of hydraulic parameters in the domain of irrigation project nos. 1, 2 and 3. With constant storage coefficient a change of transmissivity results in a change of drawdown, which is

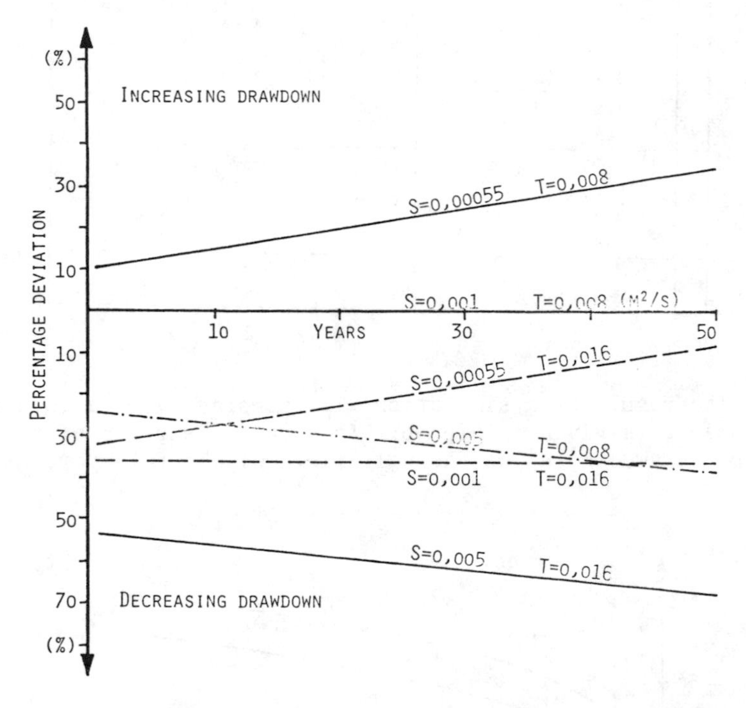

Fig. 7. Effect of hydraulic parameter variation

independent of time. Against that, a modification of the storage coefficient causes a time dependent change of drawdown functions, compared with the solution for a basic distribution of aquifer parameters. Furthermore, Fig. 7 shows that for a probable bandwidth of parameter distribution, the solutions may vary from a basic solution as much as 35% in either direction.

Conclusions

Critical review of the results of the investigation points to two aspects which have to be considered.

In a region with a permanent deficit between recharge and discharge the pumping of groundwater is only possible for a limited period. The length of time depends on the reaction of the natural system and hydrological and technical restrictions. The behaviour of the groundwater potential can be predicted and the prognosis gives helpful data for the management of the reservoir.

On the other hand, the given example of aquifer analysis illustrates the difficulties in obtaining accurate results, because of variability of the assumed hydraulic parameters. In these cases results can be given only within a certain bandwidth, depending on the estimation of extreme parameter values and a sensitivity analysis of the natural system.

References

[1] Bekey, C. A. and Karplus, W. J., 1971. "Hybrid Computation," John Wiley and Sons, Inc., New York.

[2] Davis, S. N. and De Wiest, R.J.M., 1966. "Hydrology," John Wiley and Sons, Inc., New York.

[3] Klenke, M., 1975. "Einsatz eines hybriden Rechnersystems bei Grundwasserhaushaltsuntersuchungen," Mitteilungen aus dem Institut für Wasserwirtschaft, Hydrologie und landwirtschaftlichen Wasserbau der TU Hannover, 32: 255-378.

[4] Pinder, G. F. and Frind, E. O., 1972. "Application of Galerkin's Procedure to Aquifer Analyses," Water Resources Research, 8:108-120.

[5] Zienkiewicz, O. C., 1971. "The Finite Element Method in Engineering Science," McGraw Hill, New York.

DETERMINATION OF THE TRANSMISSIBILITY FROM THE FREE WATER LEVEL OSCILLATION IN WELL-AQUIFER SYSTEMS

By

I. Krauss
Landesamt fuer Gewaesserkunde
Rheinland-Pfalz, Am Zollhafen 9
D-6500 MAINZ/Fed. Rep. of Germany

Abstract

The confined well-aquifer system is able to exhibit a free water level oscillation. The parameters of the free oscillation, damping coefficient and natural frequency are determined by the parameters of the aquifer particularly the transmissibility and the well geometry. The relations between these parameters have been established, which makes it possible to calculate the transmissibility from the measured free oscillation of a well-aquifer system. Transmissibilities determined by this method were found in reasonable agreement with those evaluated from pumping test data. The free oscillations and the evaluation of transmissibilities for different wells are shown as examples.

Introduction

The usual methods to determine the transmissibility of aquifers are based on the evaluation of pumping tests. Pumping tests are of considerable expense and they are time consuming. This paper will explain the possibility of determining the transmissibility of confined aquifers without pumping test. The method is based on the measurement and analysis of free water level oscillation in the well-aquifer system. The advantage of this method consists in low costs and easy execution.

The analysis of seismic wave reaction in confined well-aquifer systems has shown (Krauss, 1974) that the water level behaves similar to a damped oscillator. Such an oscillator can perform a free oscillation. The free oscillation is completely characterized by the damping coefficient and the natural frequency of the system. In a well-aquifer system, on the other hand, these parameters are determined by the parameters of the aquifer, particularly transmissibility and the geometry of the well. When the exact relations between the measured parameters of the free oscillation are defined the parameters of the aquifer can be calculated.

Model

An uncapped, nonflowing artesian well cased to the top of a homogeneous isotropic aquifer and screened or open throughout the thickness d of the aquifer is the example of Fig. 1.

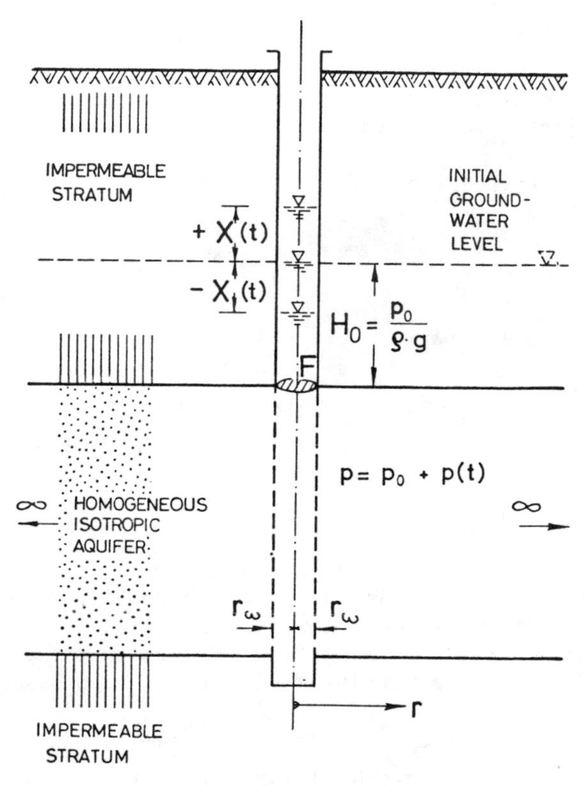

Fig. 1. Model of a well in an ideal artesian aquifer

Initiated by slug or other method, the water level in the well will move and perform the free oscillation of the well-aquifer system.

The movement of the water level $x(t)$ is described by a differential equation of second order:

$$\frac{d^2x(t)}{dt^2} + 2\beta\,\omega_w \frac{dx(t)}{dt} + \omega_w^2\, x(t) = 0 \qquad (1)$$

β = damping coefficient, [-], and
ω_w = natural frequency of the well-aquifer system, [s^{-1}].

Particular solutions of Eq. (1) are

$$x(t) = x_o \cdot e^{-\beta\omega_w \cdot t + i\omega_w \sqrt{1-\beta^2} \cdot t} \quad ; \quad \beta < 1 \tag{2.1}$$

and

$$x(t) = x_o \cdot e^{-\beta\omega_w \cdot t + \omega_w \sqrt{\beta^2-1} \cdot t} \quad ; \quad \beta > 1 \tag{2.2}$$

which describe the free oscillation of the water level in case of undercritical and overcritical damping.

For simplification Eqs. (2.1) and (2.2) can be written as

$$x(t) = x_o \cdot e^{\gamma \cdot t}; \quad \gamma \text{ complex} \tag{3.1}$$

and

$$x(t) = x_o \cdot e^{\tilde{\gamma} \cdot t}; \quad \tilde{\gamma} \text{ real} \tag{3.2}$$

We wish to express the system parameters β and ω_W in Eqs. (1), (2.1) and (2.2) by the parameters of the aquifer and the well geometry.

Equilibrium Relation

Let us suppose that the well-aquifer system (Fig. 1) is performing a free oscillation. Then there is dynamic equilibrium between the forces in well and aquifer. When the acting forces on the cross-section F are considered a dynamic equation of equilibrium (Cooper et al., 1965) is obtained.

$$\rho \cdot F \left[\frac{d}{dt} \left((H_o + x) \frac{dx}{dt} \right) + g(H_o + x) \right] = F \cdot (p_o - p(r,t)_{r_w}) \tag{4}$$

p_o = hydrostatic pressure, $[N/m^2]$;
p = pressure in aquifer caused by the variable water level, $[N/m^2]$;
H_o = height of water column in well casing, [m];
F = cross section of the well, $[m^2]$;
g = gravity acceleration, $[m/s^2]$;
ρ = density of water, $[kg/m^3]$;
r_w = radius of the well, [m];
r = distance from center of well, [m]; and
t = time, [s].

The first term of the left side is the force of inertia and the second term is the weight of the water column. The right side expresses the pressure in the aquifer. The latter consists of the hydrostatic pressure p_o and a variable pressure p, expressing the feedback of variable water level in the well on the aquifer. It depends on time and distance from the well and is to be taken at the radius of the well r_w.

After differentiation and linearization with $H_o = p_o/\rho \cdot g$ Eq. (4) is simplified to

$$\frac{d^2x}{dt^2} + \frac{g}{H_o} \cdot x = - \frac{g}{H_o} \cdot h\,(r,t)_{r_w} \tag{5}$$

$h(r,t) = p(r,t)/\rho \cdot g$ = pressure head [m].

In Eq. (5) $h(r,t)$ must be calculated.

<u>Equation of Groundwater Flow</u>

The differential equation for the change in pressure head h of a compressible viscous liquid in an elastic artesian aquifer is given by De Wiest (1966) as:

$$\Delta h(x,y,z,t) = \frac{S}{T} \cdot \frac{\partial h(x,y,z,t)}{\partial t} \tag{6}$$

T = transmissibility [m^2/s]
S = storage coefficient [-].

Under the assumption that the pressure is cylinder-symmetrical around the well Eq. (6) is reduced to

$$\frac{\partial^2 h(r,t)}{\partial r^2} + \frac{1}{r}\frac{\partial h(r,t)}{\partial r} = \frac{S}{T}\frac{\partial h(r,t)}{\partial t} \tag{7}$$

The boundary conditions are:

1. $h(\infty,t) = 0$
2. $\lim_{r \to 0} \left(-T \cdot r \frac{\partial h(r,t)}{\partial r}\right) = \frac{Q}{2\pi} = \frac{F}{2\pi} \cdot \frac{dx}{dt}$

   Q = discharge from aquifer into the well [m^3/s].

The 2nd boundary condition is a continuity equation which expresses that the discharge from the aquifer and the flow into the well are equal.

With these boundary conditions Eq. (7) can be solved. Separation of the variables and introducing

$$h(r,t) = R(r) \cdot e^{\gamma t} \tag{8}$$

leads to a modified Bessel differential equation for $R(r)$:

$$\frac{d^2R(r)}{dr^2} + \frac{1}{r}\,\frac{dR(r)}{dr} - x^2 \cdot R(r) = 0;\; x^2 = \frac{\gamma \cdot S}{T} \tag{9}$$

The solution of these equations can be combined as a linear combination of the modified cylinderfunction, first kind, of order zero I_o and the modified cylinderfunction, second kind, of order zero K_o (Mc Lachlan, 1955, p. 116).

$$R(r) = A\, I_o\,(x \cdot r) + B\, K_o\,(x \cdot r) \tag{10}$$

For the evaluation of the coefficients A and B the boundary conditions are used: $I_o(x \cdot r)$ increases monotonically for $r \to \infty$, while $K_o(x \cdot r)$ decreases monotonically. Therefore it follows from the first boundary condition, that A=0 and

$$R(r) = B \cdot K_o(x \cdot r) \tag{11}$$

The coefficient B can be evaluated from the second boundary condition. Inserting R(r) in Eq. (8) yields

$$h(r,t) = B \cdot K_o(x \cdot r) \cdot e^{\gamma t} \tag{12}$$

For $r \to 0$, $K_o(x \cdot r)$ is equal to $-\ln\,(x \cdot r)$.
Hence, the second boundary condition leads to

$$\lim_{r \to 0} \left(- T \cdot r\, \frac{d(B \cdot K_o(x \cdot r))}{dr}\right) e^{\gamma t} = \lim_{r \to 0} \left(T \cdot r\, \frac{d(B \cdot \ln(x \cdot r))}{dr}\right) e^{\gamma t} =$$

$$= T \cdot B \cdot e^{\gamma t} = \frac{Q}{2\pi} = \frac{F \cdot dx}{2 \cdot \pi\, dt} \tag{13}$$

For B it follows that;

$$B = \frac{x_o \cdot \gamma \cdot \pi \cdot r_w^2}{2\pi \cdot T} \cdot \frac{e^{\gamma t}}{e^{\gamma t}} = \frac{x_o \cdot \gamma \cdot r_w^2}{2T} \tag{14}$$

with

$$\frac{dx}{dt} = x_o \cdot \gamma \cdot e^{\gamma t} \quad \text{and} \quad F = \pi \cdot r_w^2 .$$

Hence the complete solution of Eq. (7) is

$$h(r,t) = \frac{x_o \cdot \gamma \cdot r_w^2}{2T} \cdot K_o\left[\left(\frac{\gamma \cdot \rho \cdot r^2}{T}\right)^{\frac{1}{2}}\right] \cdot e^{\gamma t} \tag{15}$$

The solution is separated into real and imaginary parts. The real part of Eq. (15) for $\beta<1$ can be written:

$$h(r,t) = \frac{r_w^2}{2T} \left[(KR(\alpha) - \frac{\beta}{\sqrt{1-\beta^2}} \cdot KI(\alpha)) \frac{dx}{dt} - \frac{\omega_w}{\sqrt{1-\beta^2}} KI(\alpha) \cdot x \right] \quad (16)$$

with

$$x = x_o \, e^{-\beta\omega_w t} (\cos(\omega_w \sqrt{1-\beta^2} \cdot t) + \frac{\beta}{\sqrt{1-\beta^2}} \sin (\omega_w \sqrt{1-\beta^2} \cdot t))$$

and

$KR(\alpha)$ real part and

$KI(\alpha)$ imaginary part of $K_o\left[\left((-\beta + i\sqrt{1-\beta^2}) \frac{\omega_w \cdot r^2 \cdot S}{T}\right)^{\frac{1}{2}}\right]$

The real part of Eq. (15) for $\beta > 1$ can be written

$$h(r,t) = \frac{r_w^2}{2T} \cdot KR(\tilde{\alpha}) \cdot \frac{dx}{dt} \quad (17)$$

with

$$x = x_o \, e^{(-\beta\omega_w + \omega_w\sqrt{\beta^2-1}) \cdot t}$$

and

$$KR(\alpha) = \text{real part of } K_o\left[\left((-\beta + \sqrt{\beta^2-1}) \cdot \frac{\omega_w \cdot r^2 \cdot S}{T}\right)^{\frac{1}{2}}\right].$$

Equation of Water Movement in the Well

When $h(r,t)$ for $\beta < 1$ and $\beta > 1$ at $r = r_w$ (immediately outside the screen) are introduced in Eq. (5) the equations for the motion of water in the well are obtained

$$\frac{d^2x}{dt^2} + \frac{g \cdot r_w^2}{H_o \cdot 2T} (KR(\alpha) - \frac{\beta}{\sqrt{1-\beta^2}} KI(\alpha)) \frac{dx}{dt} + \frac{g}{H_o}(1 - \frac{r_w^2 \cdot \omega_w}{\sqrt{1-\beta^2} \cdot 2T} KI(\alpha))$$

$$\cdot x = 0; \quad \beta < 1 \quad (18)$$

$$\frac{d^2x}{dt^2} + \frac{g}{H_o} \cdot \frac{r_w^2}{2 \cdot T} KR(\tilde{\alpha}) \frac{dx}{dt} + \frac{g}{H_o} x = 0; \quad \beta > 1 \quad (19)$$

In both cases they are found to be equations of damped oscillators. The comparison with Eq. (1) yields for the coefficients:

$$2\beta\omega_w = \frac{g}{H_o} \frac{r_w^2}{2T} (KR(\alpha) - \frac{\beta}{\sqrt{1-\beta^2}} KI(\alpha)) \quad (20)$$

$$\beta < 1$$

$$\omega_w^2 = \frac{g}{H_o}\left(1 - \frac{r_w^2 \cdot \omega_w}{2T \cdot \sqrt{1-\beta^2}} KI(\alpha)\right) \tag{21}$$

and

$$2\,\beta\omega_w = \frac{g}{H_o} \cdot \frac{r_w^2}{2T} KR(\tilde{\alpha}) \tag{22}$$

$$\beta > 1$$

$$\omega_w^2 = \frac{g}{H_o} \tag{23}$$

In principle these equations constitute the relation between the system parameters β and ω_W and the aquifer characteristics and the well geometry respectively.

It is not possible to solve the equations explicitly for β and ω_W because of their correlation in the argument of the Kelvin-function. Therefore another solution has been chosen. Introducing

$$C = \frac{r_w^2 \cdot \omega_w}{T} = \text{damping factor } [-] \tag{24}$$

and combining Eq. (20) with Eq. (21) yields:

$$\beta = \frac{1}{4} C \left(KR(\alpha) - \frac{\beta}{\sqrt{1-\beta^2}} KI(\alpha)\right) \cdot \left(1 - \frac{C}{2\sqrt{1-\beta^2}} KI(\alpha)\right)^{-1}; \quad \beta<1 \tag{25}$$

$$\alpha = [(-\beta + i\sqrt{1-\beta^2}) \cdot C \cdot S]^{\frac{1}{2}}$$

By combination of Eqs. (22) and (23) the result is:

$$\beta = \frac{1}{4} C \cdot KR(\tilde{\alpha}); \quad \beta > 1 \tag{26}$$

$$\tilde{\alpha} = [(-\beta + \sqrt{\beta^2-1}) \cdot C \cdot S]^{\frac{1}{2}}$$

The damping factor C has been varied in a computer program until Eqs. (25) and (26) were fulfilled for each particular β. The solution is unequivocal. Each β corresponds with only one C and therefore also only one transmissibility T.

In Fig. 2 the calculated values C versus β for storage coefficients $S = 10^{-3}$, 10^{-4} and 10^{-5} have been plotted. β is almost proportional to 1/T. The dependence on the storage coefficient is small, because it is only found in the argument of the Kelvin-function.

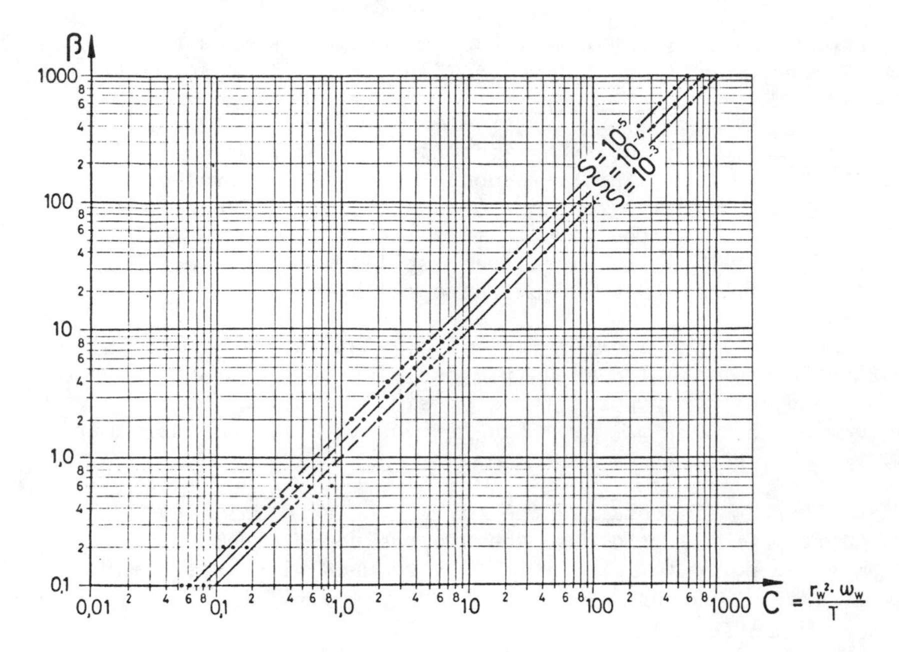

Fig. 2. Damping coefficient β in relation to $C = r_W^2 \cdot \omega_W/T$ for different storage coefficients

The values of C for $\beta < 1$ and $\beta > 1$ do not fit together. The reason is neglect of the nonlinear terms in Eq. (4). The linear approximation is unsatisfactory in the range $\beta > 0.5$ to $\beta < 1.2$. This range was interpolated in the graph of β and C (Fig. 2). The relation between the natural frequency ω_W and the height of the water column H_o can be calculated from Eqs. (21) and (23). It is approximately:

$$\boxed{\omega_W = 3.3/\sqrt{H_o}} \tag{27}$$

<u>So the desired correlation between β and ω_W is given from Fig. 2 and Eq. (27).</u>

<u>Determination of Transmissibility</u>

<u>Stimulation and measurement of the free oscillation.</u> The free oscillation must be stimulated for practical determination of the transmissibility. This can be done by compressed air to lower the water level some centimeters below the initial stage. When the pressure is suddenly released, the water will approach the initial level either oscillating (undercritical damped, $\beta < 1$) or exponentially (overcritical

damped, $\beta > 1$). The movement can be recorded with a pressure transducer. The time of free oscillation may last several minutes.

Evaluation of the transmissibility. The damping coefficient β and the natural frequency ω_W must be evaluated from the recorded movement of the water level. The parameter β is proportional the logarithm of the amplitude ratio in case of under-artical damped free oscillation. ω_W can be calculated from the period of the oscillation.

Several methods exist for the evaluation of β and ω_W for the overcritical damped free oscillation which are not described in this paper. It is possible in all cases to determine β and ω_W from the free oscillation of the water level.

After β is obtained the corresponding C-value can be taken from Fig. 2. In case the storage coefficient is unknown, the curve for $S = 10^{-4}$ may be used with sufficient accuracy. Equation (24) may be transformed to:

$$\boxed{T = r_w^2 \cdot \omega_w / C(\beta)} \qquad (28)$$

The transmissibility T can be evaluated when the radius of the well r_W is known and the natural frequency ω_W is determined from the free oscillation.

Practical experiences. The transmissibility from free oscillation of the well-aquifer system was determined at wells in areas with different hydrogeological properties. The transmissibilities calculated by this method were found in reasonable agreement with those evaluated from pumping test data.

Four wells of different damping characteristics are chosen as examples of practical measurements. Their free oscillations are shown in Fig. 3. The curves have been normalized for better comparison, i.e., the initial amplitude has been set equal -1 and the following amplitudes are reduced to the same scale.

The free oscillations of the four shown wells cover the range from small damping with damping coefficient $\beta << 1$ to overcritical damping with $\beta > 1$. The corresponding transmissibilities range from 10^{-2} m^2/s to 10^{-4} m^2/s. Table 1 summarizes the measured and the calculated characteristics of the Fig. 3 wells. For comparison transmissibilities and permeabilities are added which were determined from pumping tests.

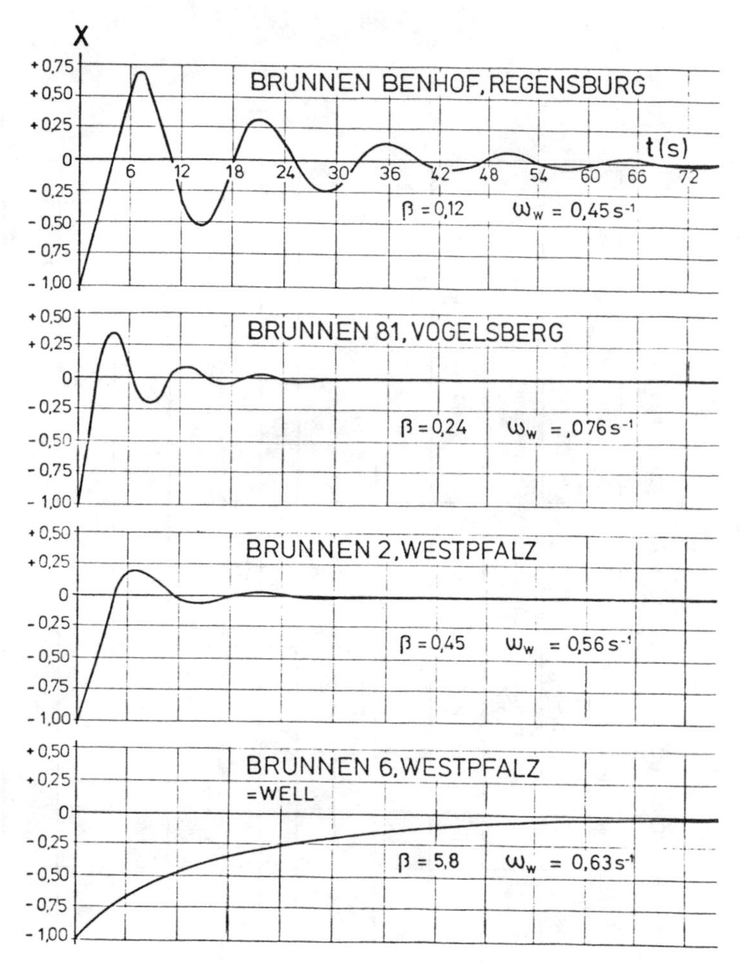

Fig. 3. Free water level oscillation in four different wells (normalized)

The transmissibilities calculated from free oscillation compare reasonably well with those obtained from pumping tests.

Table 1 Characteristics of the Analyzed Wells

	Symbol	Unit	Well Benhof	Well 81	Well 2 deep	Well 6 shallow
Region			North of Regensburg	Vogelsberg	Westpfalz	Westpfalz
Rock of the aquifer			cavernous limestone	cavernous basalt	cavernous sandstone	cavernous sandstone
Thickness of the aquifer	d	[m]	about 23	about 12	about 60	about 60
Height of water column	H_o	[m]	30	18	40	28
Radius	r_w	[m]	$15x10^{-2}$	$7.3x10^{-2}$	$5x10^{-2}$	$5x10^{-2}$
Natural frequency	ω_w	[s^{-1}]	0.45	0.76	0.56	0.63
Damping coefficient	β	[-]	0.12	0.24	0.45	5.8
Damping factor	C	[-]	0.08	0.17	0.35	4.6
Transmissibility from free oscillation	T	[m^2/s]	$1.3x10^{-1}$	$2.4x10^{-2}$	$4.0x10^{-3}$	$3.4x10^{-4}$
Transmissibility from pumping test	T_p	[m^2/s]	$2.0x10^{-1}$	$1.8x10^{-2}$	$1.1x10^{-3}$	$3.1x10^{-4}$
Permeability from free oscillation	$k = T/d$	[m/s]	$5.5x10^{-3}$	$2.0x10^{-3}$	$6.7x10^{-5}$	$5.7x10^{-6}$
Permeability from pumping test	$k_p = T_p/d$	[m/s]	$8.6x10^{-3}$	$1.5x10^{-3}$	$1.8x10^{-5}$	$5.2x10^{-6}$

References

[1] Bodvarsson, G., 1970. "Confined Fluids as Strain Meters," J. Geophys. Res., 75, 2711-2718.

[2] Bredehoeft, J. D., Cooper, H. H. Jr. and Papadopulos, I. S., 1966. "Inertial and Storage Effects in Well-Aquifer Systems: An Analog Investigation," Water Resources Res., 2(4), 697-707.

[3] Cooper, H. H. Jr., Bredehoeft, J. D., Papadopulos, I.S., and Bennet, R. R., 1965. "The Response of Well-Aquifer Systems to Seismic Waves," J. Geophys. Res. 70, 3915-3926.

[4] Krauss, I., 2. 1974. "Die Bestimmung der Transmissivität von Grundwasserleitern aus dem Einschwingverhalten des Brunnen-Grundwasserleitersystems," J. Geophys.-Zeitschrift für Geophysik 40, 381-400.

[5] Krauss, I., 6. 1974. "Brunnen als Seismische Übertragungsysteme," Berichte des Instituts für Meteorologie und Geophysik der Universität Frankfurt/Main. Eigenverlag des Institutes, Frankfurt/Main, 120 p.

[6] McLachlan, N. W., 1955. "Bessel Functions for Engineers," 2nd ed., at the Clarendon Press, Oxford, 239 p.

[7] Tychonoff, A. N. and Samarski, A. A., 1959. "Differential-gleichungen der Mathematischen Physik," VEB-Verlag, Berlin, 660 p.

[8] De Wiest, R. J. M., 1966. "On the Storage Coefficient and the Equation of Groundwater Flow," J. Geophys. Res. 71, 1117-1122.

[9] Wunsch, G., 1962. "Moderne Systemtheorie," Akadem. Verlagsges, Leipzig, 262 p.

THE USE OF BLOCK TECHNIQUE IN MATHEMATICAL MODELLING OF GROUND WATER

By

M. Mirabzadeh
University of Tehran
Department of Irrigation Engineering
College of Agriculture
Karadj, Iran

Abstract

Using dynamic programming, it is shown how the solution of partial differential equations for groundwater flow over subregions can be combined to obtain the solution of a region. The partial differential equation is approximated by a difference equation by discretizing the two independent variables into $N+1$ and L subintervals. In contrast to the standard approaches, the application of dynamic programming converts the problem into that of solving a difference equation or order $N \times N$ over $L-1$ steps and involving $L-1$ matrix inversions of order $N \times N$. These matrices are independent of boundary conditions, hence can be used for any different boundary condition matrices $A^{I}_{(i)}$ and $A^{II}_{(i)}$ from subregions I and II can be used to obtain a combined solution for region III (=I+II).

First a combined solution is obtained for a region consisting of two rectangular subregions and then it is shown how the solutions over subregions, with any shape, can be combined to obtain the solution over whole regions.

Introduction

Under certain problem conditions, it is essential to solve the groundwater flow equation by numerical methods. Presently, the most common technique used if the finite difference method (Todd, 1962; Forsythe and Wasow, 1967). The flow equation is substituted by a system of linear algebraic equations whose solution give the numerical value of hydrodynamic potential in the flow region.

Many efficient iterative techniques are already available to solve the system of linear equations (Young, 1954; Peaceman and Rachford, 1955; Forsythe and Wasow, 1960; Fayer and Sheldon, 1962), but there are difficulties associated with the iterative techniques, which limits their application in certain cases. The principle limitations are:

- Rate of convergence of the iterative techniques is not only dependent upon the geometrical configuration of the problem area but also depends on the distribution of hydrodynamic coefficients, and the initial and boundary conditions. The iterative technique may fail or converge slowly, especially in the irregular region cases.

- The rate of convergence of the more efficient iterative methods such as successive over relaxation (Young, 1954) and alternating direction implicit methods (Peaceman and Rachford, 1955) depend critically upon one or more parameters. There are difficulties associated with the determination of the optimal or near optimal parameters.

- To study the groundwater movement, it becomes essential to solve the same problem with many different boundary conditions. A major limitation of iterative techniques is that a given solution provides no information about other solutions. Hence, the iterative techniques are very inefficient, especially for solving transient flow problem over long periods of time.

The application of dynamic programming is not only free from these limitations, but it is a direct (non-iterative) technique which opens new avenues for solving the groundwater problems efficiently.

From the work of Bellman, 1960; Angel, 1968; Collins and Angel, 1971; dynamic programming was recently used to solve the flow equation in two dimensions and three dimensions (Mirabzadeh, 1973, 1975) in heterogeneous and anisotropic media. The advantage of dynamic programming over iterative techniques was demonstrated in our work.

In this paper it is demonstrated how the solution over two subregions, I and II, are used to obtain the solution for entire region, I + II, by dynamic programming.

Flow Equation

The equation of flow through porous media is obtained by substituting Darcy's Law into the continuity equation. For flow in two dimensions (x,y) the flow equation is as follows:

$$\text{div}\,(\bar{\bar{T}}\ \text{grad}\ h) = S\,\frac{\partial h}{\partial t} + q \tag{1}$$

where h is hydrodynamic head [L],
S is storage coefficient,
q is flux, $[L\text{-}T^{-1}]$,
t is time, [T],

$\bar{\bar{T}}$ is transmissibility tensor, $[L^2 - T^{-1}]$

$$\bar{\bar{T}} = \begin{bmatrix} T_x & T_{xy} \\ T_{yx} & T_y \end{bmatrix}$$

Equation (1) for steady state flow yields:

$$\text{div } (\bar{\bar{T}} \text{ grad } h) = q \qquad (2)$$

For isotropic porous mediums or anisotropic mediums where the principle axes of transmissibility tensor coincides with the coordinate axes, Eq. (2) becomes:

$$\frac{\partial}{\partial x} [T_x(x,y) \frac{\partial h}{\partial x}] + \frac{\partial}{\partial y} [T_y(x,y) \frac{\partial h}{\partial x}] = q(x,y) \ . \qquad (3)$$

<u>Discretization</u>. The numerical solution of Eq. (3), under Dirichleh conditions, over region R is considered, where region R is defined by:

$$(x.y) \ \varepsilon \ R \qquad 0 \leqq x \leqq a, \qquad 0 \leqq y \leqq b$$

The discretization is as in Fig. 1. The mesh spacing is Δx in the x-direction and is Δy in the y-direction. The nodes are designated by i and j as follows:

$$0 \leqq x \leqq L_x \qquad 0 \leqq i \leqq L$$

$$0 \leqq y \leqq (N+1) \ \Delta y \qquad 0 \leqq j \leqq N+1$$

Using the notations in Fig. 2 and the usual finite difference approximations and assuming $\Delta x = \Delta y = \Delta$, the Eq. (3) at a point (i,j) in the region, R, is approximated by:

$$T_x(i-1,j) \ h \ (i-1,j) + T_x \ (i,j) \ h \ (i+1,j) + T_y(i,j-1) \ h \ (i,j-1)$$

$$+ T_y(i,j) \ h \ (i,j+1) - (\Sigma T_{x,y})_{i,j} \ h(i,j) = Q(i,j) \qquad (4)$$

where $(\Sigma T_{x,y})i,j = T_x(i-1,j) + T_x(i,j) + T_y(i,j-1) + T_y(i,j)$

$$Q(i,j) = \Delta^2 q(i,j)$$

Now we define an N-dimensional vector $H(i)$ whose components are hydrodynamic potential, $h(i,j)$, on the interior points located on the ith line.

$$H(i) = [h(i,j)] \qquad J = 1,2,3,...N$$

$$i = 1,2,...,(L-1).$$

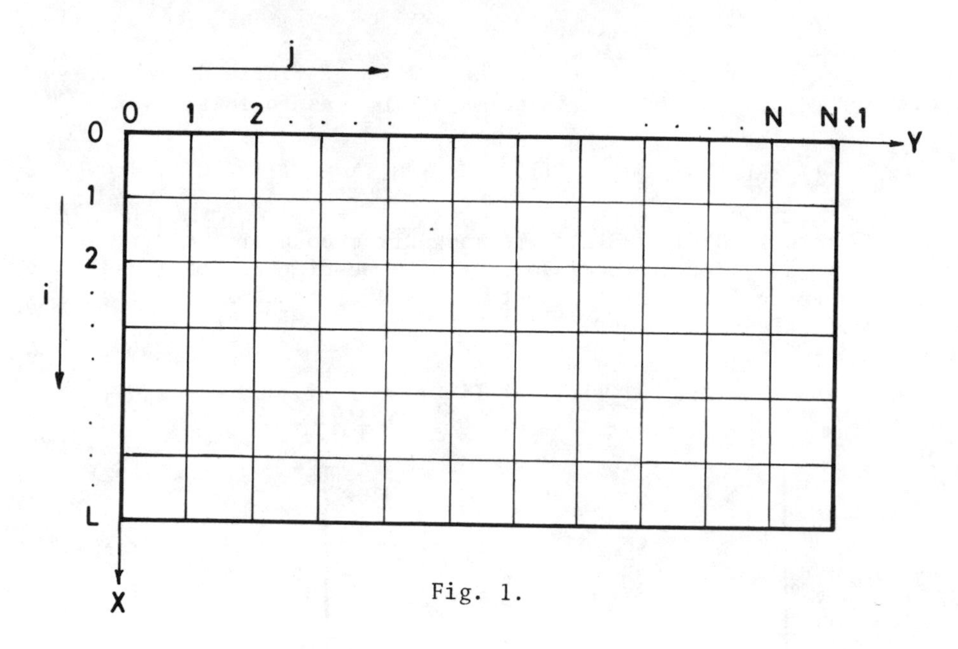
j
0
1
2
N
N+1
Y
0
1
2
i
L
X
Fig. 1.

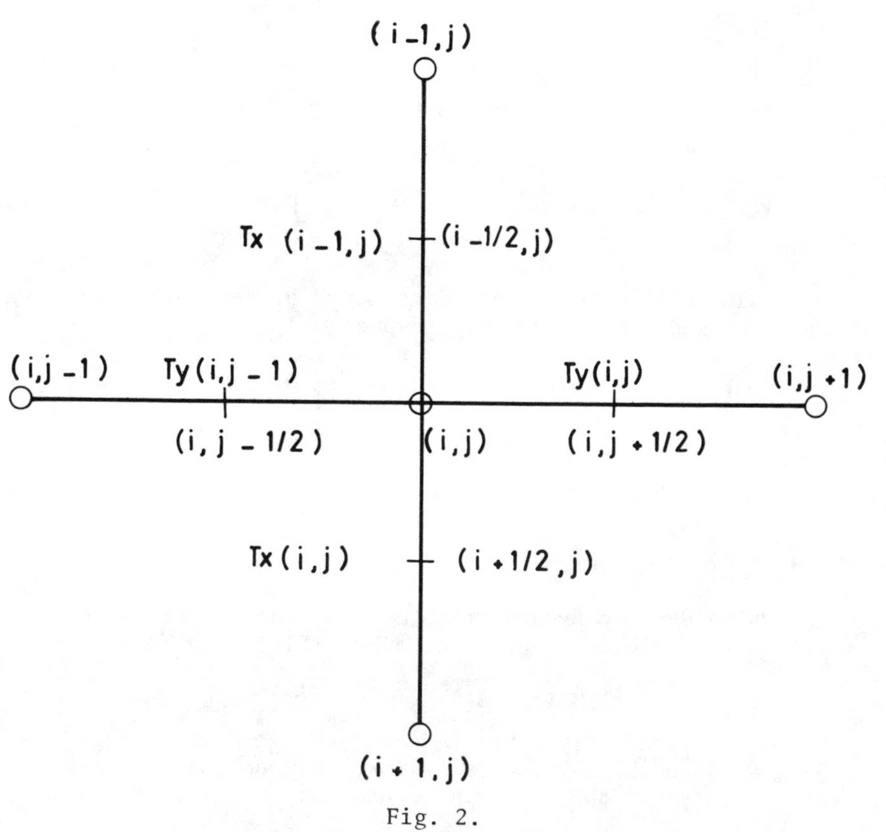
(i-1,j)
Tx (i-1,j)
(i-1/2,j)
(i,j-1)
Ty(i,j-1)
Ty(i,j)
(i,j+1)
(i, j-1/2)
(i,j)
(i,j+1/2)
Tx(i,j)
(i+1/2,j)
(i+1,j)
Fig. 2.

Using the above definition, the system of finite difference Eq. (4) can be written in terms of H_i as follows:

$$F(i-1)\ H(i-1) + F(i)\ H(i+1) - G(i)\ H(i) + C(i) - Q(i) = 0 \quad (5)$$

Vectors $H(0)$ and $H(L)$ are known from boundary conditions $F(i)$ and $G(i)$ are diagonal and three-diagonal matrices, respectively, with components defined by transmissibility coefficients T_x and T_y

$$F(i) = \text{digo}\ [T_x(i,1),\ T_x(i,2) \ldots . T_x(i,j) \ldots . T_x(i,N)]$$

$$G(i) = \begin{bmatrix} \alpha_1^i & \beta_1^i & & & & & \\ \beta_1^i & \cdot & \cdot & & & & \\ & \cdot & \cdot & \cdot & & 0 & \\ & & \beta_{J-1}^i & \alpha_j^i & \beta_J^i & & \\ & & & \cdot & \cdot & \cdot & \\ & 0 & & & & & \\ & & & & \cdot & \cdot & \beta_{N-1}^i \\ & & & & & \beta_{N-1}^i & \alpha_N^i \end{bmatrix} \quad \text{where:} \begin{cases} \alpha_J^i = (\Sigma T_{x,y})_{i,j} \\ \cdot \\ J = 1,N \\ \beta_J^i = T_y(i,j) \\ J = 1,\ N-1 \end{cases}$$

$C(i)$ and $Q(i)$ are vectors determined by boundary conditions and flow conditions, $Q(i,j)$, respectively.

$$C(i) = \begin{bmatrix} T_x(i,0)\ H(i,0) \\ 0 \\ 0 \\ 0 \\ T_x(i,N)\ h(i,N+1) \end{bmatrix} \qquad Q(i) = \begin{bmatrix} Q(i,1) \\ \vdots \\ Q(i,j) \\ \vdots \\ Q(i,N) \end{bmatrix}$$

Dynamic Programming

Using dynamic programming, Eq. (3) can be solved numerically. This yields a solution to Eq. (5) in the form of

$$H(i+1) = A(i)\ H(i) + B(i) \quad (6)$$

where $A(i)$ and $B(i)$ are independent of $H(i)$ (Angel, 1968; Mirabzadeh, 1975). Using Eqs. (6) and (5) and solving for $H(i)$ gives the following equation:

$$H(i) = [F^{-1}(i-1) \left(G(i) - F(i)\ A(i)\right)]^{-1} [H(i-1) + F^{-1}(i-1) \left(C(i) - Q(i) + F(i)\ B(i)\right)] \quad (7)$$

Comparing Eqs. (6) and (7) yields:

$$A(i-1) = [F^{-1}(i-1) \left(G(i) - F(i)\ A(i)\right)]^{-1} \quad (8)$$

with the initial conditions:

$$A(L-1) = 0 \quad (8a)$$

and

$$B(i-1) = A(i-1)[F^{-1}(i-1) \left(C(i) - Q(i) + F(i)\ B(i)\right)] \quad (9)$$

With the initial conditions:

$$B(L-1) = H(L) \quad (9a)$$

The initial conditions (8a) and (9a) are derived from Eq. (6) setting $i = (L-1)$.

The calculation procedure is to solve Eq. (8) using condition (8a) and storing all the matrices $A(i)$ in the memory, then Eq. (9) will be solved to obtain vectors $B(i)$, using condition (9a) and the matrices $A(i)$. Now Eq. (6) can be solved for $H(i)$ utilizing $A(i)$ and $B(i)$.

Combining Solutions

Considering the region R (Fig. 3) it will be shown that solutions over subregions I and II can be combined to obtain the solution over the region R(= I+II).

It is assumed that the computational procedure to obtain the solutions over subregions I and II have been carried out in downward and upward directions, respectively. These solutions can be written as follows:

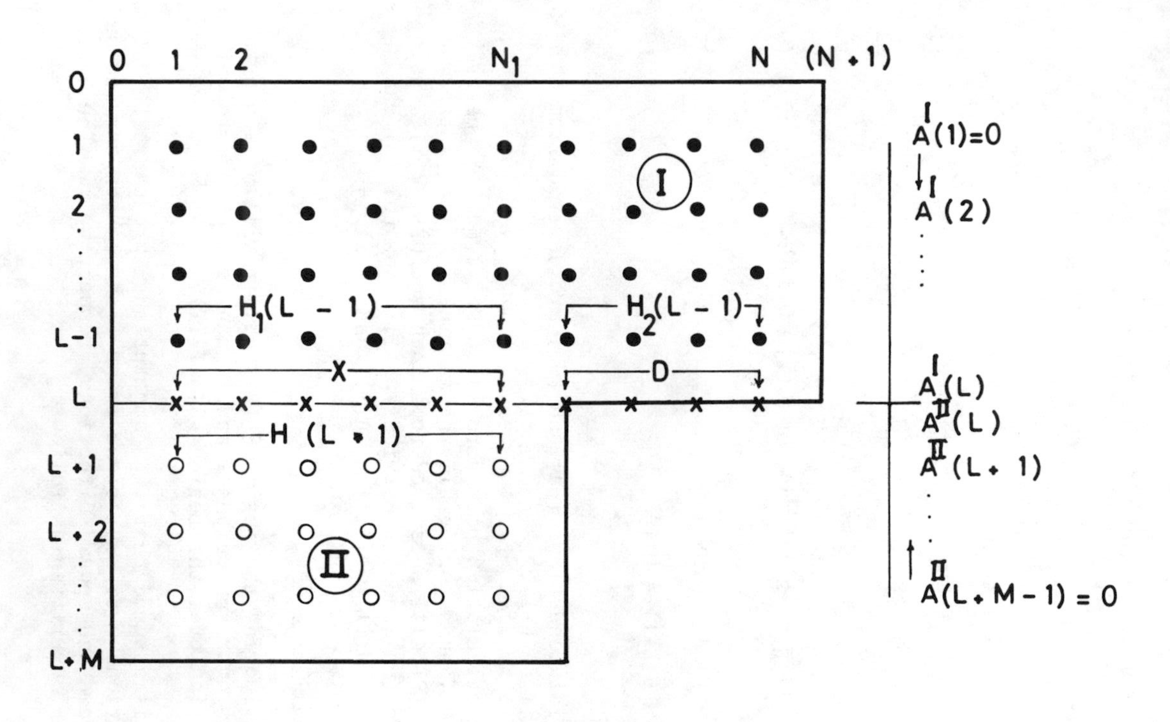
0
1
2
N1
N
(N + 1)
0
1
2
L-1
L
L + 1
L + 2
L + M
I
II
H1(L - 1)
H2(L - 1)
X
D
H (L + 1)
A(1)=0
A (2)
A(L)
A(L)
A (L + 1)
A(L + M - 1) = 0

Fig. 3.

Region I

$$\begin{cases} A^{I}(i+1) = [F^{-1}(i-1)\,\big(G(i) + F(i)\,A^{I}(i)\big)]^{-1} \\ \quad i = 1,2\ldots L-1 \\ \text{with initial conditions:} \\ A^{I}(1) = 0 \end{cases}$$

$$\begin{cases} B^{I}(i+1) = A^{I}(i+1)\,[F(i+1)\,\big(C(i) - Q(i) + F(i)B^{I}(i)\big)] \\ \quad i = 1,2\ldots L-1 \\ \text{with initial conditions:} \\ B^{I}(1) = H(0) \end{cases}$$

$$\begin{cases} H^{I}(i) = A^{I}(i+1)\,H^{I}(i+1) + B^{I}(i+1) \\ \quad i = L-1, L-2, \ldots\, 1. \end{cases}$$

Region II

$$\begin{cases} A^{II}(i) = [F^{-1}(i)\,\big(G(i+1) - F(i+1)\,A^{II}(i+1)\big)]^{-1} \\ \quad i = L+M-2,\ L+M-3\ldots L \\ \text{with initial conditions:} \\ A(L+M-1) = 0 \end{cases}$$

$$\begin{cases} B^{II}(i) = A^{II}(i)\,[F^{-1}(i)\big(C(i+1) - Q(i+1) + F(i+1)B^{II}(i+1)\big)] \\ \quad i = L+M-2,\ L+M-3\ldots L \\ \text{with initial conditions:} \\ B^{II}(L+M-1) = H(L+M) \end{cases}$$

$$\begin{cases} H^{II}(i+1) = A^{II}(i)\,H^{II}(i) + B^{II}(i) \\ \quad i = L,\ L+1, \ldots\, L+M-2 \end{cases}$$

Vector X whose components are hydrodynamic head on the boundary interfacing the two subregions is defined as:

$$X = [h(L,j)] \qquad j = 1,2...N_1$$

To obtain the combined solution components of the vector X must be known. To calculate X, the hydrodynamic head relations on the lines (L-1), (L), and (L+1) are used. The relationship between $H(L)$ and $H(L-1)$ can be written as:

$$H^I(L-1) = A^I(L)\, H^I(L) + B^I(L)$$

It may also be written as:

$$\begin{bmatrix} H_1^I(L-1) \\ \\ H_2^I(L-1) \end{bmatrix} = \begin{bmatrix} A_{11} & A_{12} \\ \\ A_{21} & A_{22} \end{bmatrix} \begin{bmatrix} X \\ \\ D \end{bmatrix} + \begin{bmatrix} B_1^I(L) \\ \\ B_2^I(L) \end{bmatrix} \qquad (10)$$

Where D is known from the boundary condition and is defined by:

$$D = [h(L,j)], \qquad J = N_1 + 1, \ldots, N$$

From Eq. (10) we can write

$$H_1^I(L-1) = A_{11}\, X + A_{12}\, D + B_1^I(L) \qquad (11)$$

The relationship between X and $H^{II}(L+1)$ is

$$H^{II}(L+1) = A^{II}(L)\, X + B^{II}(L) \qquad (12)$$

There is also a relation among $H^I(L-1)$, X, and $H^{II}(L+1)$ in the form of

$$F_1(L-1)H_1^I(L-1) + F(L)\, H^{II}(L+1) - G(L)X - Q(L) + C(L) = 0 \qquad (13)$$

solving Eqs. (11), (12), and (13) the vector X is defined by

$$X = A^* \cdot B^* \qquad (14)$$

where:

$$A^* = [G(L) - F_1(L-1)A_{11} - F(L)\, A^{II}(L)]^{-1}$$

$$B^* = [F_1^I(L-1)(A_{12}D + B_1^I(L)) + F(L)B^{II}(L) - Q(L) + C(L)]$$

Knowing the vector X on the boundary between subregions I and II, the solution over the whole region can easily be obtained. Dynamic programming was also used to obtain the solution for an irregular shaped region under various boundary conditions (Mirabzadeh, 1973). Hence, for an irregular shaped region consisting of two subregions with a straight interface boundary, the solutions over the two subregions has been derived.

For most practical problems the boundary between two subregions is not a straight line (Fig. 4). Similar to the previous case, the value of the vector X on the interface boundary must be known. Linear relations can be written between X and Y_1 (Y_1 is the hydrodynamic head on the points next to interface in subregion I).

$$Y_1 = A_1 X + B_1 \tag{15}$$

Similarly,

$$Y_2 = A_2 X + B_2 \tag{16}$$

where Y_2 is the hydrodynamic head on the points next to the interface in subregion II.

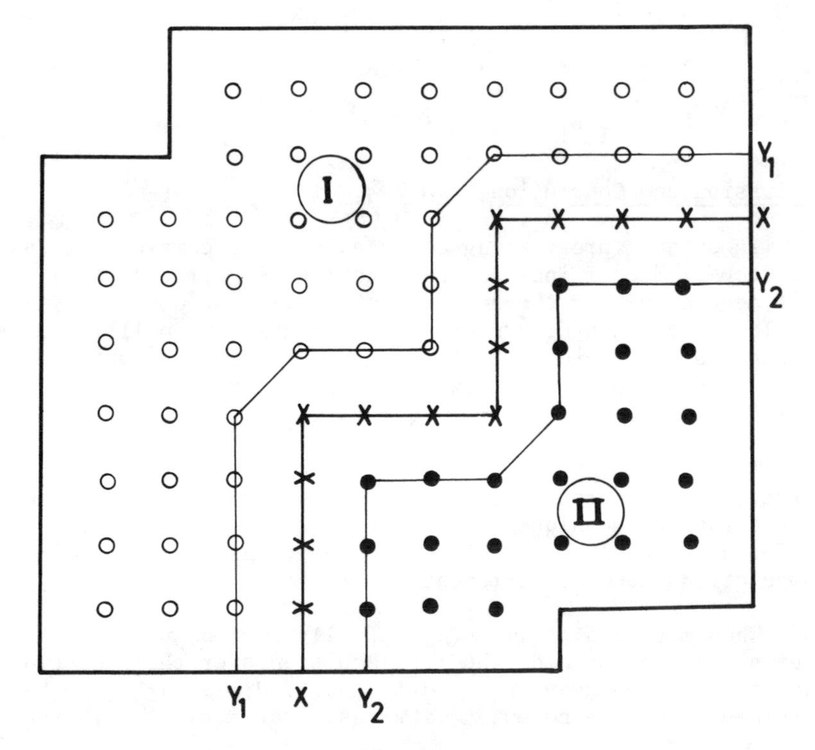

Fig. 4.

Vectors B_1 and B_2 are obtained by setting $X=0$ and solving the flow equation over the two subregions.

Matrices A_1 and A_2 are obtained by solving the flow equation K times over the two subregions (K is total number of points on the interface). The procedure is to solve the flow equation K times, setting discharge equal to zero and the hydrodynamic head equal to zero for all the points on the boundaries except one point on the interface which is equal to one.

Having the finite difference equation on the interface points, the following relationship may be written

$$SX + D_1 Y_1 + D_2 Y_2 + C = 0 \tag{17}$$

Matrices S, D_1, and D_2 are determined by the transmissibility coefficients and vector C is determined by the known hydrodynamic heads at two ends of the interface boundary.

Now, combining Eqs. (15), (16), and (17), the vector X is obtained.

$$X = A^* \cdot B^*$$

$$A^* = -(S + D_1 A_1 + D_2 A_2)^{-1}$$

$$B^* = D_1 B_1 + D_2 B_2 + C$$

Discussion and Conclusion

In dynamic programming computations are carried out in two steps: Step I includes computation of matrices A(i) that depend only on the geometrical configuration of the problem area and hydrodynamic coefficients. Step II includes computation of vectors B(i) and H(i) utilizing matrices A(i), the boundary and flow conditions (for unsteady-state flow initial conditions are required).

This method has the following advantages: The method is direct, hence there is no need to choose parameters as with the iterative techniques. To solve the flow equation for different initial and/or boundary conditions, only step II computations need to be repeated.

When matrices A(1), A(2)..A(L-1) are computed for a region, the solution of the flow equation over the subregions that are defined by vectors H(1), H(2)..H(J): $J \leq L$ can be obtained using the computed matrices. Matrices $\overline{\overline{A}}^I(i)$ and

$A^{II}(i)$ of subregions I and II can be used to solve the flow equation over region III (= I + II).

Combining the solutions over the two subregions facilitates the solution of certain groundwater flow problems and it also eliminates one of the limitations of dynamic programming, the size of the matrices A(i). In this approach, it is essential to invert (L-1) matrices of order N^2 where L and N+1 are the numbers of discretizations in the x and y directions respectively. By increasing N, not only is the computation time and the memory storage requirement increased appreciably, but the computation error also increases. The size of errors can be so high that the method gives incorrect results. Combining the solutions makes it possible to reduce the size of the matrices by increasing their numbers.

References

[1] Angel, E., 1968. "Discrete Invariant Imbedding and Elliptic Boundary-Value Problems Over Irregular Regions," J. Math. Anal. Appl., 23:471-484.

[2] Bellman, R., 1960. "Introduction to Matrix Analysis," McGraw-Hill, New York.

[3] Collins, D. and Angel, E. S., 1971. "The Diagonal Decomposition Technique Applied to the Dynamic Programming Solution of Elliptic Partial Differential Equations," J. Math. Anal. Appl., 33: 467-481.

[4] Fayers, F. J. and Sheldon, J. W., 1962. "The Use of a High-Speed Digital Computer in the Study of the Hydrodynamics of Geologic Basins," J. Geophys. Res., 67: 2421-2431.

[5] Forsythe, G. E. and Wasow, W. R., 1960. "Finite-Difference Methods for Partial Differential Equations," John Wiley and Sons, 443 p.

[6] Mirabzadeh, M., 1973. "Application de la Programmation Dynamique aux Ecoulements Permanents et Transitoires en Milieux Poreux," First World Congress on Water Resources, Chicago, Sept. 1973. (Bulletin des Sciences Hydrologiues, XX, 26(1975), 261-274.

[7] Mirabzadeh, M., 1975. "Application de la Programmation Dynamique dux Modeles Tridimensionnels de Calcul des Ecoulements dans les Milieux Poreux," IAHS Publ. No. 115: 168-175.

[8] Peaceman, D. W. and Rachford, H. H., 1955. "The Numerical Solution of Parabolic and Elliptic Differential Equation," J. Soc. Indus. Appl. Math. 3: 28-41.

[9] Todd, J., 1962. "Survey of Numerical Analysis," McGraw-Hill, New York.

[10] Young, D., 1954. "Iterative Methods for Solving Partial Differential Equations of the Elliptic Type," Trans. Amer. Math. Soc., 76: 92-111.

D. Oakes*

Question to Chidley and Lloyd. Your model uses a Ghyken-Hertzberg relationship to estimate the thickness of the fresh water lens. This relationship is only valid for static conditions and in particular will give erroneous results under pumping conditions. Did you look into the applicability of using this relationship in this case.

*Water Research Centre, Medmenham Laboratory, P.O. Box 16, Medmenham, Marlow, Bucks, SL7 2HD, England.

SECTION III

SOIL WATER PROCESSES

Section Chairman, Roger E. Smith
Research Hydraulic Engineer
Agricultural Research Service
U.S. Dept. of Agriculture
Fort Collins, Colorado

Rapporteurs, Y. Mualem
Senior Research Associate
Engineering Research Center
Colorado State University
Fort Collins, Colorado

and

Eshel Bresler
Head, Soil Physics Division
Institute of Soils and Water
Volcani Center
Bet Dagan, Israel

SECTION III

SOIL WATER PROCESSES

SECTION III

SOIL WATER PROCESSES

GENERAL REPORT

By

Y. Mualem
Senior Research Associate
Engineering Research Center
Colorado State University
Fort Collins, Colorado

and

Eshel Bresler
Head, Soil Physics Division
Institute of Soil and Water
Volcani Center, Bet Dagan, Israel

Introduction

Wetting and drying processes of soils play an important role in the hydrological cycle. The soil is wetted by infiltration of water applied to its surface either by rainfall or irrigation. A drying process follows when the rain has ceased or its rate decreases below the ability of the soil to transfer the water down to a dryer zone. This process is generally referred to as a redistribution of the soil water. Evaporation and water extraction by roots intensify the drying process of the wetted soil.

This report reviews three papers which deal with wetting processes by infiltration, one paper on soil drying due to water uptake by plant roots, and four papers studying the hydraulic properties of soils, required as data for any theoretical analysis of the soil water processes.

A. Wetting Processes

A nonconventional paper on the water infiltration process is contributed by Dixon and Simanton. They present a non-Darcy type theory on "Water Infiltration Processes and Air--Earth Interface Conditions." Their study includes:

(a) Experimental verification,

(b) Compatability with infiltration theories based on Darcy's law,

(c) Applicability to practical control problems,

(d) Justification for prediction purposes, and

(e) Possibility of refining it.

We do not believe that this approach can replace the Darcy based theories. However, this is an extensive study which may contribute to the insight of some aspects of the physical processes of infiltration under field conditions. Thus, such an analysis may be useful as a complementary one to the basic quantitative study based on the Darcy flow equations. Because of its dimension and the features of this study we cannot give a more detailed review. We hope they can advocate their approach during the author's discussion or the general discussion period.

Whisler, Curtis, Niknam and Romkens studied the effect of the crusting phenomenon on infiltration. Soil crusting is caused by the destruction of soil structure at the surface as a result of raindrop action and physio-chemical disperison. Crusting of a cultivated soil is an important factor in agriculture as well as in hydrology of tilled basin. Soil crusting can be observed even in some desert areas following an intensive rainfall. The soil crust has a sealing effect. Due to its low permeability, it impedes both water infiltration and air exchange with the atmosphere.

In this study, the authors incorporate the crusting phenomena into numerical models of infiltration. The one dimensional vertical Richards' equation

$$C \frac{\partial \psi}{\partial t} = \frac{\partial}{\partial z} [K(\psi) \frac{\partial \psi}{\partial z} + K(\psi)] \tag{1}$$

is solved numerically for infiltration with (a) fixed vertical grid system $\Delta z = 0.1$ cm and (b) refined grid system. Two approaches have been used to simulate the crusting phenomena. The first one is to simulate the crust effect on infiltration by adjusting the boundary conditions at the soil surface. A linearly decreasing flux (R) at the surface,

$$R = R_1 - B_5(t - t_{c1}) \ ; \quad t_{c1} \leq t \leq t_{c2} \tag{2}$$

is imposed as an upper boundary condition where B_5 is the prescribed rate of the R-decrease after the crust starts to form at time $t = t_{c1}$ until it is fully structured at time t_{c2}. The second approach is to simulate the soil crusting by modifying the Richards' equation to account for the changes of hydraulic properties of soils as depth-dependent:

$$C(\psi,z) \frac{\partial \psi}{\partial t} = \frac{\partial}{\partial z} [K(\psi,z) \frac{\partial \psi}{\partial z} + \frac{K(\psi,z)}{\partial z}] \ , \tag{3}$$

where $C = \partial\Theta/\partial\psi$, ψ is the capillary head and K is the unsaturated hydraulic conductivity.

In this case, the soil crusting is simulated by linear reduction of the original $K(\psi)$ and $\Theta(\psi)$ as a function of time

$$\left.\begin{aligned} K(\psi,z) &= K(\psi)[1 - B_1(t-t_{c2})] \quad (4) \\ \Theta(\psi,z) &= \Psi_s \; [1 - B_3(t-t_{c2})] \quad (5) \end{aligned}\right\} t_{c1} < t \leq t_{c2}$$

Considering the numerical procedure, the authors found that it is obviously more efficient and accurate to use a refined grid in the crust-region while maintaining a coarse grid for the rest of the soil profile.

The computed results are shown in Whisler et al., Figs. 2-6, assuming some arbitrary values for the unknown B = parameters characterizing the crust formation. Three degrees of crust effect are investigated: (1) full crust effect when both hydraulic conductivity and water content are reduced; (2) partial crust effect, only the hydraulic conductivity is reduced while $\Theta(\psi)$ is maintained unchanged; and (3) no crust is formed and the soil surface properties are not affected. The figures show the profiles of the capillary head and the water content distributions for the three cases of the crust effect. We tend to agree with the authors that the differences between curves (1) and (2) in their Figs. 2 and 3 are attributed mainly to the time differences rather than to the change in the soil characteristic $\Theta(\psi)$. The reduction of Θ values affect the water content distribution only in the zone of the crust (Fig. 3). There are, however, very significant differences between the curves of type (3) where no crust is considered and the other two cases. The capillary head and, consequently, the water content are considerably larger for the noncrusted soil. Similar results are shown in their Figs. 4-6. The crust reduces the water intake a short time after the rain starts (Fig. 4). Hence, the cumulative infiltration is significantly diminished (Fig. 5) followed by substantial increase of the accumulative runoff (Fig. 6).

The authors indicated that the computed results based on simulating the crust phenomenon by adjustment of the flux condition, are not given in their Fig. 2 because this approach is not "physically realistic." The argument given by the authors to support this conclusion is that such an approach "allows the soil to drain when the rainfall rate falls below the saturated conductivity," which is physically impossible since one assumes that when a crust has formed, "air could not enter the soil surface" to allow drainage.

We cannot follow this reasoning because it is physically inaccurate. Even after the rainfall rate falls below the saturated conductivity, infiltration continues. There is an excess of air which has to be driven out to permit the water entry. In this sense, there is no change as long as the rainfall continues. Hence, there is no reason not to apply the "surface flux adjusting" approach. Furthermore, if one assumes that when the crust is formed it does not allow the air to penetrate through, the conclusion should be that infiltration can be significantly impeded by the compressed air (especially when a shallow groundwater table is present). Upon such an assumption, two phase flow problems of both water and air (Morel-Seytoux, 1973) should be considered with a boundary condition of zero air flux at the soil surface.

The flux distribution based on the reduced rainfall approach (the 0 curve) and the reduced hydraulic properties approach are compared in the 1 and 2 curves in Fig. 4 of Whisler et al. This comparison is not informative because in the first case it represents just the prescribed boundary condition and can be assumed, therefore, to take any shape including that of curves (1) and (2). Hence, from a hydrological point of view, which is not concerned about the water content distribution in the thin layer of the crust, the adjusting flux approach may be preferred as a short cut method since the second approach based on reduced hydraulic properties involves more unknown parameters and is considerably more complicated.

Upon these observations, the authors may like to clarify their point of view in the revised version of the paper. A comment on the parameters identification and the accuracy of the computed results based on the two approaches would also be appreciated. The hydraulic properties at the boundary $z = z_c$ between the crust and the soil bulk, should also be discussed.

As the crusting of the soil surface impedes the interchange of air between the atmosphere and the soil bulk, the air effect may be an important factor (Morel-Seytoux, 1973). Parlange and Hill discuss the solutions found in literature of the two-phase flow in a horizontal soil column. Two cases are considered: (i) when the soil column is sealed so that air must escape through the wetting soil surface and (ii) when the air can escape ahead of the wetting front.

The common basic flow equation, when both fluids are incompressible and the gravity flow is neglected,

$$\frac{\partial\Theta}{\partial t} + f'V \frac{\partial\Theta}{\partial x} = \frac{\partial}{\partial x} (D \frac{\partial\Theta}{\partial x}) \tag{6}$$

where V is the total Darcy's flux of the water and the air;

$$V = q_a + q_w \tag{7}$$

D is the two phase (air and water) diffusivity term equivalent to E of Morel-Seytoux (1973, 1975),

$$D = K_a f \frac{d(P_w - P_a)}{d\Theta} = D_w(1-f) \tag{8}$$

and

$$f = K_w/(K_w+K_a) \quad \text{and} \quad D_w = K_w \frac{d(P_w - P_a)}{d\Theta} \tag{9}$$

are the (little) fractional flow function, and the one phase (water) diffusivity respectively. Note that K_a and K_w are the air and the water permeabilities, respectively, and $f' = df/d\Theta$. It is also interesting to note that $D_w \to \infty$ as $\Theta \to \Theta_s$ and that D (or E) has a finite (zero) value at soil saturation (i.e., $\Theta = \Theta_s$).

Parlange and Hill are discussing two cases. In the first case the air and the water flow in opposite directions with equal rate ($q_w = -q_a$) leading to

$$V = 0 \; ; \quad \frac{\partial\Theta}{\partial t} = \frac{\partial}{\partial x} (D \frac{\partial\Theta}{\partial x}) \tag{10}$$

Parlange (1975a,b) and Brutsaert (1976) derived the sorptivity (S)

$$S^2 = 2(\Theta_s - \Theta_i)^{\frac{1}{2}} \int_{\Theta}^{\Theta_s} D(\Theta-\Theta_i)^{\frac{1}{2}} \, d\Theta \tag{11}$$

by solving Eq. (10) for uniform initial condition and an imposed water content at the wetting surface

$$\begin{array}{lll} t = 0 & x \geq 0 & \Theta = \Theta_i \\ t > 0 & x = 0 & \Theta - \Theta_s \end{array} \tag{12}$$

For $x \neq 0$ Eq. (6) can be set in the form

$$\frac{1}{2} Sf' + \frac{d}{d\Theta} (D_w f \frac{d\Theta}{d\phi}) = \frac{1}{2} + \frac{d}{d\Theta} (D \frac{d\Theta}{d\phi}) \tag{13}$$

where $\phi = xt^{-\frac{1}{2}}$ and $S = 2Vt^{\frac{1}{2}}$. The authors discuss the approximate solution derived by Parlange (1971, 1975c).

$$\phi = (2/S) \int_{\Theta}^{\Theta_s} D_w(a)\, da \qquad (14)$$

with

$$S^2 = 2 \int_{\Theta_i}^{\Theta_s} (\Theta - \Theta_i)\, D_w(\Theta)\, d\Theta \qquad (15)$$

They conclude that Eq. (14) "yields an approximation which is reasonably accurate when air effects are negligible and becomes even more accurate when air effects are important" because Eq. (14) strictly annuls the left-hand side of Eq. (13) but only approximately the right-hand side.

As Eq. (14) strictly annuls the left-hand side of Eq. (13) it means that the air flow has no effect. Consequently, such a solution cannot account physically for the effect of the air flow. It should be interpreted, therefore, as an approximate solution for the case of the water flow only. The authors should clarify this point in their revised version.

Two sets of experiments which correspond to the two flow problems discussed above are carried out in a horizontal plexiglass chamber. The water content is computed when the wetting front reaches $x = 45$ cm in two cases: when the holes at the upper side of the horizontal column are sealed and when they are opened, allowing the air to escape. Results are given in Fig. 2 of Parlange and Hill. Both experiments are carried out with water level equal to about 2 cm in the supplying reservoir located at the front of the soil column. The observed cumulative amounts of water, Q, are found to be $Q = 2.15t^{\frac{1}{2}}$ when the holes are sealed and considerably larger $Q = 3.78t^{\frac{1}{2}}$ when they are opened. These results indicate that the air effect may reduce the water intake if the air is not free to escape ahead of the wetting front (Morel-Seytoux, 1973). A second effect which is observed in the authors' Fig. 2 is that the decrease of Θ is more abrupt when the holes are opened.

The dashed profiles plotted in Fig. 2 of Parlange and Hill represent mathematical functions fitted to the experimental data by least square deviation technique.

Applying the equation of Bruce and Klute (1956),

$$2D = -\,(d\phi/d\Theta) \int_{\Theta_i}^{\Theta} \phi\, d\Theta \qquad (16)$$

D is computed using the analytical profile for the sealed holes case. The $\Theta(x)$ profile given in the figure for the open holes case and Eq. (14) are used to derive D_W. With these values of D and D_W, the function f is determined by Eq. (8). The computed $f(\Theta)$ which characterize the air effects is drawn in the authors' Fig. 3.

One may wonder, however, to what extent the so computed f-function complies with the basic definition of f given by Eq. (9). In addition, from the experimental procedure, it seems that the actual boundary condition at $x = 0$ is $P_W =$ constant rather than $\Theta_W =$ constant for which the theoretical solutions are derived. Clarification of these two points in the revised paper would be appreciated.

B. Drying Processes

The only paper on the drying process of soil is that by Molz who discusses theories of the soil-root system with application for water consumption by plants. This is a very important aspect of soil water drying since large percentages of the water added to soil by infiltration is transferred back to the atmosphere by transpiration after crossing the soil-root interface.

In the first part of the paper the author discusses various models of pathways through which water may flow in plant tissues. For each possible pathway (i.e., Vacuole, cell wall, Symplasm pathway) the author presents an expression for the tissue-water diffusivity (D_W). As it is impossible to measure all the quasi-microscopic parameters which appear in the macroscopic property D_W, their use for prediction of transpiration is rather limited.

In this study, the author examines the mechanism of the soil water extraction by a single root system. Investigation is focused on the hydraulic head distribution under which the water flows through the unsaturated soil and the root cortex. A vertical root is considered with a horizontal symmetric radial flow. Flow in the root domain is assumed to be dominated by

$$\frac{\partial p}{\partial t} = D_w \frac{\partial^2 p}{\partial r^2} + \frac{D_w}{r} \frac{\partial p}{\partial r} \tag{17}$$

where p is the water pressure head, r is the radial distance from the root center, and D_W is the cortex (tissue)-water diffusivity. In the soil domain the water flow is governed by

$$C_s \frac{\partial \psi}{\partial t} = K_s \frac{\partial^2 \psi}{\partial r^2} + \frac{K_s}{r} \frac{\partial \psi}{\partial r} + \frac{dK_s}{d\psi} \left(\frac{\partial \psi}{\partial r}\right)^2 \tag{18}$$

where ψ is the soil water capillary head, C_s is the differential soil water capacity, and K_s is the unsaturated hydraulic conductivity of the soil.

Using a numerical technique (Molz, 1976), Eqs. (17) and (18) were solved simultaneously for the prescribed boundary conditions

$$r = r_e \qquad q_r = q_e = \text{constant} \tag{19}$$

$$r = r_s \qquad q_r = 0 \tag{20}$$

where the subscripts e, s, and r denote endodermis, soil and root surface, respectively. The pressure head on the soil-root interface at $r = r_r$, is assumed to be continuous.

The results shown in Molz's Fig. 4 are computed data using the soil-water characteristics (ψ-Θ and K-Θ), and the tissue-water diffusivity value, D_w taken from the literature. The results for the first 33 hours indicate that the pressure gradient in the root cortex is very steep while $\partial\psi/\partial r$ in the soil domain is very small.

Based on these results, the author concludes that the generally accepted idea that the "soil hydraulic properties rather than the plant hydraulic properties, dominate the water uptake process," is inaccurate. Coming to a rather opposite conclusion, he suggests that the extraction term

$$Q(z,t) = L(z)\,[\psi(\Theta/\Theta_s)-p]/(R_c + R_e) \tag{21a}$$

Added to the Richards' equation to account for the water extraction by roots should be differently interpreted. As in previous studies $L(z)$ and Θ/Θ_s are the root density distribution function and the saturation degree, respectively, but R_c and R_e should represent the cortex resistance and the endodermis resistance, respectively, rather than the cortex and the soil resistances.

It seems to us that the authors' conclusions overdraw the physical meaning of the computed results. Considering the flow problem under investigation, there is no reason to solve simultaneously Eqs. (17) and (18) by numerical procedure. As the water flow in the root tissue is tension saturated, the boundary condition (17) applies also for $r = r_r$, namely at the surface of the root. Equation (18) can be solved, therefore, separately to derive the accurate $p(r)$ profiles shown in Molz's Fig. 4. In this sense, the flow in the root, no matter what model is assumed for the flow mechanism in the plant tissue, has no effect whatsoever on the water flow in the soil domain which is influenced by the value of $q_e = q_r$. By solving Eq. (18) alone, $p(r)$ or $\psi(r_r)$ can be computed. Afterward, the $p(r)$ profile in the cortex can be easily determined by downward parallel displacement of the $q_r = q_e$ profile to match the values of $p = \psi$ at the soil-root interface. Hence, it is the soil water flow which dominates the $p(r)$ distribution in the plant tissue.

It is very well accepted that the root can produce relatively high pressure gradients. The problem arises when the soil fails to meet the water requirements dominated by potential evapotranspiration or the climatic conditions in general. The purpose of using the soil-root extraction term is to represent the pattern by which water is extracted from the soil profile to produce the transpiration rate. In this context the models of flow in plant tissue may indicate the way the plant adjusts itself to the environment conditions, namely to the climatic conditions and the soil-water conditions. It cannot affect them. Therefore, the real outcome of this study supports the common agreement that besides the root distribution factor $L(z)$ the most important factor is the soil water availability dominated by the hydraulic properties of soil. The extraction terms of Nimah and Hanks (1973a, 1973b) modified by Feddes et al. (1974) and Neuman et al. (1975) show a proper way to implement the presently known physical knowledge about the soil-water-root system. In these models, the soil water parameters $K(\Theta)$ and $\psi(\Theta)$ in addition to the plants parameters $b(z)$ or $RDF(z)$ are included in the root extraction term as

$$Q(z,t) = K(\Theta)[\psi(z) - p(r_r)]/b(z) \tag{21b}$$

in which $1/b(z) = RDF(z)/(\Delta x \cdot \Delta z)$ is the macroscopic sink term due to plant uptake.

As it is the hydraulic properties of soils that dominate the flow pattern of soil water, Eq. (21b) is much more applicable than Eq. (21a). In addition Eq. (21b) contains physical parameters of soil, $K(\Theta)$, and plant, $b(z)$, which are easier to determine than the plant parameters R_c and R_e of Eq. (21a). Several studies of these hydraulic parameters are reviewed in the following.

C. Soil Water Parameters

Determination of the hydraulic properties of soils constitute prerequisite information for mathematical simulation of infiltration, water uptake by plants and any other process of the soil water flow. Sun-fu Shih studies the variability of the hydraulic conductivity (K) and the uncertainty associated with the prediction of the soil-water potential. This paper describes a procedure of selecting the simulated sample size according to the distribution of (K) and the desired standard deviation (S_ϕ) of the hydraulic head (ϕ).

If the probability density function of K in a given field is log-normal, then a new variable $y = \log K$ is defined and the synthetic K based on this distribution has the form

$$K = \exp\,[c(\mu_y + \sigma_y t)] \tag{22}$$

where $c = \ln 10$ and t is a generator of random number with values between 0 and 1. Obviously K in Eq. (22) is equal to or smaller than the measured K because $\exp(c\mu_y) \leq \mu_K$. The relationship between σ_K and μ_K depending on σ_y is given by

$$\begin{aligned} \sigma_K &> \mu_K \quad \text{if} \quad \sigma_y > 0.3616 \\ \sigma_K &= \mu_K \quad \text{if} \quad \sigma_y = 0.3616 \\ \sigma_K &> \mu_K \quad \text{if} \quad \sigma_y < 0.3616 \end{aligned} \tag{23}$$

To find the sample size which yields the desired S_ϕ of the predicted hydraulic head in ground water flow Freeze's (1975) approach is adapted. The sample size n is defined as the number of the layers (m) of a nonuniform homogeneous media, multiplied by the number M of the Monte Carlo runs, i.e., $n = Mm$. Freeze (1975) has used these parameters for the analysis of the effect of stochastic parameter distribution on the uncertainty in the prediction of the hydraulic head ϕ.

An exponential function in which the power is a second order polynomial of S_ϕ and σ_y is arbitrarily defined to simulate n as

$$n = \exp(A_o + A_1 S_\phi^2 + A_2 S_\phi^2 + A_3 \sigma_y + A_4 \sigma_y^2) \tag{24}$$

where A_o, A_1, A_2, A_3, A_4 are constants obtained from the runs. Thus n as a function of S_ϕ and σ_y can be obtained if A_o, A_1, A_2, A_3, and A_4 are known.

In his Fig. 3, Shih presents the sample size results calculated using the method of Freeze (1975) and Eq. (24), with $A_o = 29.17$, $A_1 = 0.436$, $A_2 = 0.00508$, $A_3 = 8.12$ and $A_4 = -2.162$. It demonstrates that Eq. (24) can be used to select n according to σ_y and S_ϕ. We feel however that the universality of the form of Eq. (24), its parameters, and the optimal method of their determination should be further discussed in the revised version of the paper.

Ganoulis and Thirriot studied the distribution of the nonwetting fluid in a vertical soil column (their Fig. 1) using a stochastic model rather than relating directly to the $\Theta(\psi)$ relationship. The soil is modeled by a capillary network of a varying radius (r) interconnected in series and in parallel (Fig. 3, Ganoulis and Thirriot). The capillary pressure (P_c) at the interface between the two fluids is defined by

$$P_c = P^{(1)} - P^{(2)} = 2T_s \cos \alpha / r_c \tag{25}$$

where (1) and (2) indicate the nonwetting and the wetting fluids, respectively, and T_s, α, and r_c are the interfacial tension, the contact angle and mean radius, respectively. The distribution of $P^{(1)}$ and $P^{(2)}$ is taken to be linear with z representing static equilibrium.

$$P_c(z) = P_o + (\rho^{(1)} - \rho^{(2)})\, gz \tag{26}$$

where $P_o = \rho^{(1)} gh$ is the pressure at the soil surface $z=0$. Two conditions are required for a pore M to be filled by the nonwetting fluid: (a) that the radius of the pore is larger than r_c and (b) that a continuous track exists between this pore and the soil surface through pores of radius $r > r_c$.

The model considers three types of pores within the interval Δz (shown in the author's Fig. 4) (A) Pores of radius $r > r_c$ blocked by pores of radius $r < r_c$, (B) pores which divide within the Δz interval, and (C) unblocked pores. The probability of the three occasions are estimated by Ganoulis et al. (1976) to be a function of the elevation

$$P[A] = \beta_1(z)\ \Delta z = -\frac{\Delta z}{\ell} \log\left(\int_{\sigma_c(z)}^{\sigma} g(r)dr\right) \tag{27}$$

$$P[B] = \beta_2(z)\Delta z = -\ 2\Delta z\ kP \int_{r_c(z)}^{\infty} rf(\sigma)dr / \int_0^{\infty} r^2 f(r)dr \tag{28}$$

$$P[C] = 1 - P[A] - P[B] \tag{29}$$

where ℓ is the average value of the pore length, k and P are the tortuosity and the porosity parameters and $f(r)$ is the density distribution of (r). Denoting the probability that the nonwetting fluid fills the pores at level z by $w(z)$ and elaborating Eqs. (26) to (29), the authors derive the formal solution

$$w(z) = \frac{\exp\{-\int_o^z C(s)ds\}}{1 + \int_o^z B(s)\exp\{-\int_o^s C(t)dt\}ds} \quad (30)$$

for the boundary condition $w(o) = 1$.

To illustrate this solution, the authors assume that

$$f(r) = \frac{r}{r^2} \exp\left\{-\frac{r^2}{2r_o^2}\right\} \quad (31)$$

and hypothesize some arbitrary values for the other parameters: $r_o = 1$ mm, $k = 1.5$, $P = 0.32$, $T \cos\sigma = 0.023$ N/m and $\ell = 2r_o$. Equation (30) is solved for two cases: (i) the fluid on the soil surface is much heavier than the fluid in the column $\rho^{(1)} = 900$ Kg/m^3 and $\rho^{(1)} - \rho^{(2)} = 900$ Kg/m^3 and (ii) the fluid on the surface is the lighter one with $\rho^{(1)} = 900$ Kg/m^3 and $\rho^{(2)} - \rho^{(1)} = 100$ Kg/m^3. Though not definitely stated, the first case may represent oil and air while the second may represent oil and water. In each case, two solutions are derived for $h_o/r_o = 4$ and $h_o/r_o = 8$.

The computed $w(z)$ profiles are described by the authors in their Fig. 6. They show that in both cases the w distribution is very sensitive to the boundary condition h. The second important result is that w has a minimum value at $z/r_o = 4.5$ which appears to be independent of h. It seems unlikely to observe such a phenomenon in soil and the authors may like to ease the concern of the readers by comment on the physical process by which the wetting fluid is retarded from the soil column to permit the computed final distribution of w.

The authors may also like to clarify the necessity of the statistical approach in analyzing the nonwetting fluid infiltration, indicate how the model can be calibrated for a given soil and fluids and comment on the applicability of the results.

Most natural processes of unsaturated flow include cycles of wetting and drying. Simulating such processes has to account for the hysteretical nature of the hydraulic properties of soils. Mualem presents a theoretical model of hysteresis which requires only one measured branch of the hysteretical loop for calibration.

The author makes use of a previous independent domain theory (Mualem, 1974) according to which a porous medium is represented by a bivariant distribution function $f(\overline{r},\overline{\rho})$.

The variables $\bar{r}$ and $\bar{\rho}$ are normalized length scales characterizing the openings and the volume of the pore domains, respectively. Assuming complete independence between the $\bar{r}$ and $\bar{\rho}$, $f(\bar{r},\bar{\rho})$ can be described as a multiple of two functions

$$f(\bar{r},\bar{\rho}) = h(\bar{r})\ \ell(\bar{\rho}) \tag{32}$$

This is the similarity hypothesis, $h(\bar{r})$ and $\ell(\bar{r})$ can be interpreted as the density distribution functions of the openings and the pore-volumes, respectively. In a homogeneous porous medium the areal porosity equals the volumetric porosity and so does their corresponding distribution functions. Consequently, the author assumes

$$h(\bar{r}) = \ell(\bar{\rho}) \quad \text{for} \quad \bar{r} = \bar{\rho} \tag{33}$$

The one unknown function, $h(\bar{r})$, can be now determined by using either the main wetting curve or the main drying curve. Using the capillary law, r and ψ are uniquely related by $\psi = c/r$, where c is a constant.

Taking the main wetting curve, $\theta_w(\psi)$, an experimentally measured soil water retention function, the main drying curve, $\theta_d(\psi)$, can be predicted by

$$\theta_d(\psi) = [2\theta_u - \theta_w(\psi)]\ \frac{\theta_w(\psi)}{\theta_u} \tag{34}$$

where θ is the "effective water content" and θ_u is the resaturated (ultimate) value of θ. Equation (34) can be used to derive the $\theta_w(\psi)$ when $\theta_d(\psi)$ is measured. Similarly, drying and wetting scanning curves are formulated by

$$S_e({}^{\psi_{max}}{}_{\psi_1}{}^{\psi}) = S_e(\psi) + [1 - S_{ew}(\psi)]\ S_{ew}(\psi_1) \tag{35}$$

and the drying scanning curve by

$$S_e({}_{\psi_{min}}{}^{\psi_1}{}_{\psi}) = S_{ew}(\psi) + [S_{ew}(\psi_1) - S_{ew}(\psi)]S_{ew}(\psi) \tag{36}$$

where S_e is the effective saturation and ψ_{min} and ψ_{max} are the minimum and the maximum measured values of ψ, respectively.

Mualem's Fig. 4 shows the comparison between the predicted and the measured drying curves of hysteresis for Molonglo sand

(Talsma, 1970). His Fig. 5 shows this comparison between the wetting curves. In both cases a relatively good agreement is found.

The author mentions that for soils with definite air entry value a considerable discrepancy between the measured and the predicted curves is expected as the independent domain theory cannot account for the air blockage against the water entry. For this reason it is recommended to use the main wetting curve for calibration.

The second author of this report has independently checked this model against several soils. While good agreement is found for some, very poor agreement is found for others. We believe that the model is sensitive to the minimum value of Θ. As Θ_{min} becomes larger than the residual water content, the discrepancies increase. This is in addition to the effect of the air entry value.

The last paper of this report by Gaudu, Hoa, Thirriot and Thore studies the effect of hysteresis on the evolution of water content distribution in soil profiles. The authors devised two procedures for simulating the ψ-Θ hysteresis. The first method is to represent the main hysteretical loop by two polynomial functions of the fourth order. Each scanning curve is determined by the ordinates of the point (A) at which this curve is supposed to merge with the corresponding curve of lower order (secondary drying scanning curve with primary drying scanning curve and so on); by the point from which the scanning curve originates (B) and by requiring that at point (A) the two curves match by the first, the second, and the third derivative. To comply with the domains theory of hysteresis, point A is chosen to be the former reversal point. The use of this method is restricted as a polynomial function cannot allow either ψ or $d\psi/d\Theta$ receive infinite values.

To overcome this problem the authors investigated the function

$$s_e - s_{eo} = \frac{c}{1+\alpha[|\psi| - |\psi^*|]^\beta}; \quad |\psi| > |\psi^*|; \quad s_e = \frac{\Theta-\Theta_r}{\Theta_{sat}-\Theta_r} \qquad (37)$$

In this case there are five parameters to be determined: s_{eo}, c, ψ^*, α and β. Note that for $\beta < 1$, $ds/d\psi \to \infty$ at $\psi = \psi^*$. Therefore, $|\psi^*|$ is taken equal to the value of ψ at which saturation is achieved by infiltration. In principle, for any ψ-Θ curve the points of its origin and its end are known and can be used to eliminate s_{eo} and c. α and β, however, are computed by the least square deviation technique for each curve using the experimental data for this purpose.

The effect of the simulated hysteresis is shown in Figs. 10 and 11 for soil drainage under daily cycles of evaporation and for infiltration through a soil dam under fluctuating water table in the reservoir, respectively. The effect of the initial condition is apparent in Fig. 10 of Gauda et al. In this case, the polynomial method is used to simulate hysteresis. Significant differences are apparent for the use of three initial nonrealistic initial conditions of uniform $\Theta = 0.25$ achieved by: (1) wetting process; (2) draining, and (3) the median $\psi(\Theta)$ curve.

Experimental results for the variation of water content in an earth-dam are compared with the numerical simulation with and without hysteresis. The authors' Fig. 11 shows a good similarity between the simulated and the observed processes when hysteresis is taken into account but manifest considerable deviation when hysteresis is neglected. These results are consistently observed for the three depths investigated.

The reasons for not using the known physical models of hysteresis based either on the independent or the dependent domain theories are not given in the original paper. A comment by the authors in the revised version will be appreciated. It may be also worthwhile to indicate where the measurements were taken in the dam and explain the reasons for the differences in the initial conditions shown in their Fig. 10.

References

[1] Bruce, R. R. and Klute, A., 1956. "The Measurement of Soil-Water Diffusivity," Soil Sci. Soc. Am. Proc., 20, pp. 456-462.

[2] Brutsaert, W., 1976. "The Concise Formation of Diffusive Sorption of Water in Dry Soil," Water Resour. Res., 12, pp. 1118-1124.

[3] Dixon, R. M. and Simanton, J. R., 1977. "Water Infiltration Processes and Air-Earth Interface Conditions," The Third Fort Collins International Hydrology Symposium.

[4] Feddes, R. A., Bresler, E. and Neuman, S. P., 1974. "Field Test of a Modified Numerical Model for Water Uptake by Root System," Water Resour. Res., 10, pp. 1199-1206.

[5] Freeze, R. A., 1975. "A Stochastic Concept Analysis of One-Dimensional Ground Water Flow in Nonuniform Homogeneous Media," Water Resour. Res., 11(5), pp. 725-741.

[6] Ganoulis, J. and Thirriot, C., 1976. "Stochastic Study of the Entrance of a Nonwetting Fluid in a Porous Medium, under the Influence of the Capillarity," Proc. 11th International AIRH Symposium in Stochastic Hydraulics, Lund, Sweden, Aug. 2-4, 1976.

[7] Ganoulis, J. and Thirriot, C., 1977. "Etude theorique Probabiliste de la Dispersion Capillaire en Milieu Poreux," The Third Fort Collins International Symposium.

[8] Gaudu, R., Hoa, N. T., Thirriot, C. and Thore, Ph., 1977. "Comparison de Procedes de Simulation Numerique du Phenomene d'Hysteresis de la Relation Suction-Teneur en eau et Etude Critique de Effets sur les Ecoulements en Milieux non Saturex," The Third Fort Collins International Hydrology Symposium.

[9] Molz, F. J. and Peterson, C. M., 1976. "Water Transport from Roots to Soil," Agronomy Journal, 68, pp. 901-904.

[10] Molz, F. J., 1977. "Theoretical Hydrology of the Soil-Root System with Practical Implications for Water Extraction," The Third Fort Collins International Symposium.

[11] Morel-Seytoux, H. J., 1973. "Systematic Treatment of Infiltration with Application," Completion Report Series No. 50, Colorado State University, Fort Collins.

[12] Morel-Seytoux, H. J., 1975. Reply to a comment on his paper, "Derivation of an Equation of Infiltration," Water Resour. Res. 11: 763-765.

[13] Mualem, Y., 1974. "A Conceptual Model of Hysteresis," Water Resour. Res., 10(3), pp. 514-520.

[14] Mualem, Y., 1977. "Theory of Universal Hysteretical Properties of Unsaturated Porous Media," The Third Fort Collins International Hydrology Symposium.

[15] Neuman, S. P., Feddes, R. A. and Bresler, E., 1975. "Finite Element Analysis of Two-Dimensional Flow in Soils Considering Water Uptake by Roots: I. Theory," Soil Sci. Soc. Am. Proc., 39, pp. 224-230.

[16] Nimah, M. N. and Hanks, R. J., 1973a. "Model for Estimating Soil Water, Plant and Atmospheric Interrelations, 1, Description and Sensitivity," Soil Sci. Soc. Am. Proc., 37, pp. 522-527.

[17] Nimah, M. N. and Hanks, R. J., 1973b. "Model for Estimating Soil Water, Plant and Atmospheric Interrelations, 2, Field Test of Model," Soil Sci. Soc. Am. Proc., 37, pp. 528-532.

[18] Parlange, J. Y., 1971. "Theory of Water Movement in Soils: I. One-Dimensional Absorption," Soil Sci., 111, pp. 134-137.

[19] Parlange, J. Y., 1975a. "On Solving the Flow Equation by Optimization: Horizontal Infiltration," Soil Sci. Soc. Am. Proc., 39, pp. 415-418.

[20] Parlange, J. Y., 1975b. "Determination of Soil-Water Diffusivity by Sorptivity Measurements," Soil Sci. Soc. Am. Proc., 39, 1011.

[21] Parlange, J. Y., 1975c. "Theory of Water Movement in Soils: II. Conclusion and Discussion of Some Recent Developments," Soil Sci., 119, pp. 158-161.

[22] Parlange, J. Y. and Hill, D. E., 1977. "Air and Water Flow in a Horizontal Column - Influence of the Air Boundary Conditions," The Third Fort Collins International Hydrology Symposium.

[23] Shih, S. F., 1977. "Synthetic Hydraulic Conductivity and Its Application," The Third Fort Collins International Hydrology Symposium.

[24] Whisler, F. D., Curtis, A. A., Niknam, A. and Romkens, M.J.M., 1977. "Modeling Infiltration as Affected by Soil Crusting," The Third International Hydrology Symposium.

Acknowledgment

This work is partly supported by funds provided by the National Science Foundation under grant ENG-7800845, "The Role of Hysteresis of the Hydraulic Properties of Soils in Unsaturated Flow Processes." The NSF support is gratefully acknowledged.

WATER INFILTRATION PROCESSES AND AIR-EARTH INTERFACE CONDITIONS

By

Robert M. Dixon and John R. Simanton
Southwest Watershed Research Center
Agricultural Research Service
United States Department of Agriculture
442 E. 7th St., Tucson, Arizona

Abstract

The air-earth interface theory holds that *interfacial roughness and openness control the rates and routes of water infiltration by governing the flow of air and water in underlying macropore and micropore systems. Roughness* refers to the microrelief that produces depression storage, whereas *openness* refers to the macroporosity that is visible at the soil surface. Soil air and free surface water exchange freely across a *rough open* surface with consequent rapid water penetration via the relatively short broad straight paths of the macropore system. In contrast, surface exchange of air and water is greatly impeded by a *smooth closed* surface with consequent slow water penetration via the relatively long narrow tortuous paths of the micropore system. These relative differences in water penetration rates and routes are attributed to corresponding differences in phase continuity within the macropore system. Both air and water phases are maintained continuous by a rough open surface and discontinuous by a smooth closed surface. Discontinuity in the phases causes relatively high soil air back pressures and low soil water pressures, whereas phase continuity produces low air pressures and high water pressures.

The air-earth interface theory that surface roughness and openness control infiltration and the Darcy concept that hydraulic conductivity and gradient control infiltration are reconciled by introducing and defining a new hydraulic parameter, referred to as the *effective surface head*, which controls both the hydraulic conductivity and hydraulic gradient at the soil surface. Transmission characteristics of the soil profile are reflected in the magnitude of the effective surface head.

The air-earth interface concept appears to have considerable potential in the solution of land management problems wherein uncontrolled point infiltration, surface runoff and erosion are contributing factors. Such problems would be alleviated by designing land management systems to achieve a given level of surface roughness and openness or effective surface head.

The air-earth interface theory may be quantified by relating surface roughness and openness or effective surface head to the two parameters in Kostiakov's equation. These relationships were found to take the form of a power function for the coefficient of time in Kostiakov's equation and a linear function for the time exponent.

Introduction

Uncontrolled infiltration often causes the inefficient use and irreversible loss of our vital soil and water resources. For instance, excessive tillage or overgrazing diminishes the soil's ability to absorb water, thereby increasing soil and water losses from the soil surface through the processes of evaporation, runoff, and erosion.

Many other problems are either directly or indirectly related to man's inability to control infiltration at appropriate levels. These include flash flooding of upland watersheds, excessive erosion of upland stream banks, sedimentation of waterways and reservoirs, pollution of surface and groundwaters, excessive evaporation from soil surfaces, inefficient leaching of soluble salts and excessive leaching of plant nutrients, inefficient on-site use of precipitation for vegetal production, inefficient water harvesting for off-site precipitation uses, slow recharge of groundwater and declining water tables, and inefficient irrigation of various land areas. Desertification of most semiarid and arid regions of the world is accelerated by excessive surface runoff and evaporation resulting from uncontrolled infiltration.

According to a new infiltration theory referred to as the *air-earth interface* (AEI) theory, interacting soil surface and water source conditions control water infiltration rates and water penetration routes (Dixon, 1972). In this paper the AEI theory is briefly reviewed and an approach to theory quantification for absolute infiltration control is presented.

AEI Theory

The spatial domain of the AEI theory and its physical models is the *micro-interface* and its physical properties, *microroughness* and *macroporosity*. Micro-interfaces are defined as square or circular surfaces less than 1 m^2 in size; microroughnesses are soil surface irregularities having horizontal periodicities ranging from 1 to 100 cm; and macropores are soil voids assumed to be cylindrical tubes and plane cracks having diameters and widths ranging from 1 to 10 mm at the air-earth interface.

The AEI theory makes the general argument that *soil surface roughness and openness control infiltration of free surface water by governing the flow of air and water in underlying macropore and micropore systems,* wherein *roughness* refers to the microrelief that produces depression storage, and *openness* refers to the macroporosity that is visible at the soil surface. The macropore system includes the space immediately above the AEI and that space within macropores which fills and drains largely by gravity during and after soil surface exposure to free or ponded water (Fig. 1). Macropores include those voids produced by clay shrinkage, tillage,

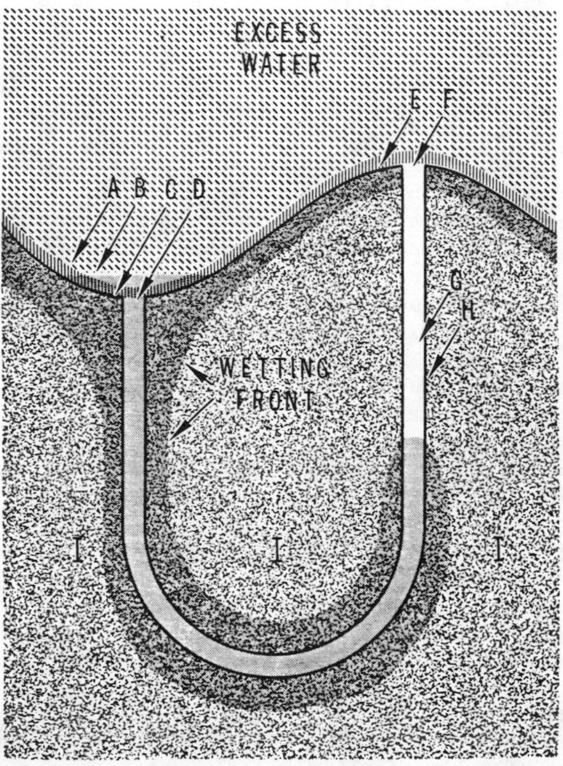

Fig. 1. Soil model containing a micropore system and a macropore system. The macropore system includes the space immediately above the air-earth interface and that within macropores, whereas the micropore system includes the space within and between individual soil individual soil aggregates. Symbol definitions are: A = plant residue cover on air-earth interface; B = free water surface; C = microdepression in air-earth interface; D = water intake port of macropore; E = micro-elevation in air-earth interface; F = soil air exhause port of macropore space; H = macropore wall; and I = micropore space. From Dixon and Peterson (1971).

earthworms, roots, internal erosion, ice lenses, pebble dissolution, and entrapped gas. In contrast, the micropore system includes the spaces within and between individual soil aggregates (textural and structural pores or simple and compound packing voids) that fill and drain largely by capillarity. Thus, during rapid wetting of an initially dry soil, the macropore and micropore systems contain water at pressures of near atmospheric and below atmospheric, respectively. The two systems of pores share common porous borders at the AEI and along macropore walls which allow intersystem flow of water and displaced soil air.

The AEI theory embodies six physical interfacial models (Fig. 2) representing two degrees of surface roughness and

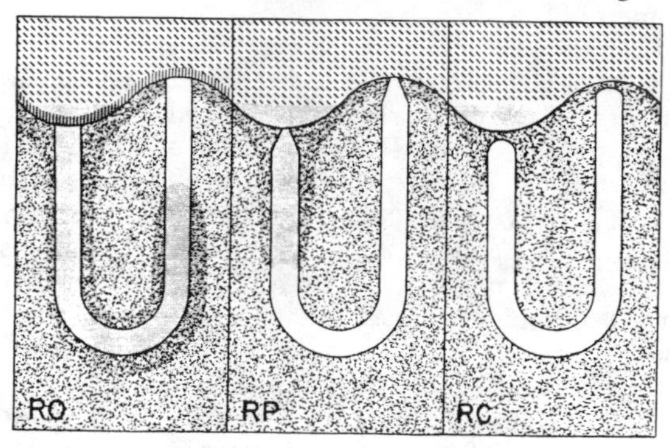

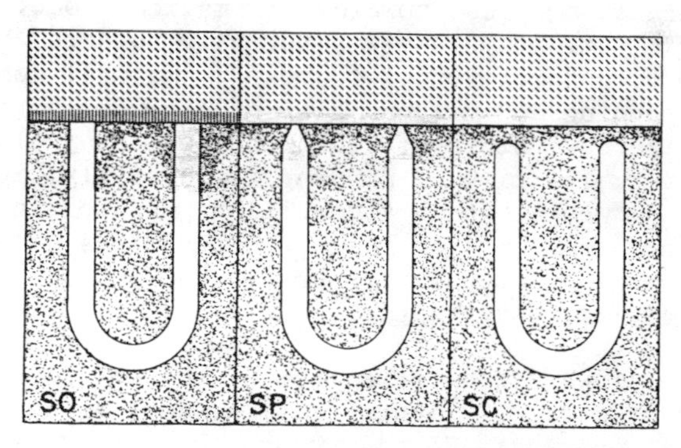

Fig. 2. Air-earth interface models and associated u-shaped macropore for water infiltration into soils. Models RO, RP and RC represent rough interfaces containing open, partly open (unstable) and closed macroports, respectively; whereas models SO, SP and SC represent smooth interfaces containing open, partly open (unstable) and closed macroports. From Dixon and Peterson (1971).

three degrees of surface openness (Dixon and Peterson, 1971). The subterranean part of the macropore system is depicted as a single U-shaped tube to graphically reflect its infiltration role as a water-intake air-exhaust circuit. Models RO, RP, and RC represent rough interfaces with open, partly open (unstable), and closed macropore interfacial openings or macroports, respectively. Models SO, SP, and SC represent plane (smooth) interfaces with open, partly open, and closed macroports, respectively. These models, which have been studied experimentally (Dixon, 1975b), are intended to guide relative infiltration control by serving as a reference framework within which needed modifications in existing surface conditions may be considered.

AEI Theory Quantification

Although the AEI physical models help to explain the wide range in infiltration rates produced by varying surface conditions and provide physical principles upon which to base the design of surface management practices for relative infiltration control, they do not facilitate quantitative prediction for absolute control. Unfortunately, the physical system assumed in the development of the AEI theory is far too complex for detailed mathematical modeling. Even if the simultaneous flow rates and routes of two fluids in two interacting pore systems as affected by two dynamic AEI conditions could be successfully modeled, the large number of parameters required to do so would make the resulting mathematical model too cumbersome for practical infiltration control. Perhaps the most notable progress toward mathematical modeling of complex infiltration systems was recently reported by Morel-Seytoux (1976, and references therein). Natural complexities of delayed ponding and viscous flow of air were both considered in the derivation of equations for rainfall infiltration. However, the time of ponding is spatially highly variable under the upper boundary conditions assumed in the AEI theory. Micropores located in microdepressions saturate quickly under high intensity rainfall, but macropores located on microknolls may never saturate.

Progress toward quantification of the AEI theory involved three major steps: (1) identification, definition, and interpretation of important AEI theory parameters; (2) selection and interpretation of an appropriate two-parameter equation for modeling the AEI theory; and (3) determination of functional relationships between theory and equation parameters.

Theory parameter identification. Surface microroughness and surface macroporosity are the two principal physical parameters of the AEI theory. These two interrelated and interacting properties of the soil surface have yet to be characterized directly in a way that accurately reflects their infiltration roles. Such characterization presents a formidable task because of the great rapidity and intensity of physical and biotic structure-forming processes at the soil surface. A single hydraulic parameter has been chosen that integrates the effects of microroughness and macroporosity on the performance of the U-shaped water-intake air-exhaust circuits or the macropore systems (Dixon, 1975b). This parameter, referred to as the *effective surface head* h_s, is defined as the difference between surface water head h_w and soil air pressure head h_a, or $h_s = h_w - h_a$. It usually has a narrow range of only a few centimeters of water surrounding the reference zero taken as ambient atmospheric pressure. The effective surface head is commonly less than zero where a large surface area becomes saturated, such as during intense rainfall and basin and border irrigation.

Studies of air pressure buildup under border-irrigated alfalfa (Dixon and Linden, 1972) led to the definition of h_s by showing that soil air pressure affects infiltration by opposing the downward force of surface water within the macropore system. Whenever soil air pressure exceeded the sum of the hydrostatic pressure due to surface head and the soil bubbling pressure, macropores would exhaust soil air rather than infiltrate surface water, as evidenced by streams of bubbles emanating from surface openings of macropores. Thus the surface head, effective in driving water into open macropores, was the actual surface head minus the soil air pressure head.

Because of the limited area wetted, conventional infiltrometers and rainfall simulators cannot ordinarily produce measurable soil air back pressures and the resulting negative effective surface heads that are common during natural infiltration. Consequently, the actual surface head and effective surface head associated with these devices are essentially identical and always greater than zero. Several unique new infiltrometers, referred to as *closed-top infiltrometers* (Dixon, 1975a), were developed to simulate negative as well as positive h_s in a narrow range surrounding zero. The design of these infiltrometers was based on the principle that a positive soil air pressure can be simulated by imposing an equivalent negative air pressure above the ponded-water surface.

Data from the closed-top infiltrometers indicated that infiltration is highly dependent on h_s in a narrow range surrounding zero (Dixon, 1975a). Cumulative 30-minute infiltration increased 19% per cm of h_s for one soil and 33% for another within an h_s range of -3 to +1 cm. Such large effects are not consistent with some theoretical studies and some field studies that have been reported. For instance, Philip (1958) suggested about a 2% theoretical infiltration increase per cm of surface head at small times. In field studies, Horton (1940) and Lewis and Powers (1939) found no clear effect of ponded-water depth on infiltration. The observed large infiltration response to h_s is attributed to the control that h_s exerts over fluid flux in soil macropores; i.e., the rate and ultimate degree of macropore water saturation depends on h_s. Thus, h_s determines not only the hydraulic gradients in the macropore system, but also the hydraulic conductivities.

Algebraic equation selection. The next step in quantifying the AEI theory was to select a suitable infiltration equation from those reported in the literature and then mathematically and physically interpret it relative to the AEI theory. The two-parameter time functions that were considered included:

$$I_v = AT^B \qquad \text{Kostiakov (1932)}$$

$$I_v = AT^{\frac{1}{2}} + BT \qquad \text{Philip (1957)}$$

$$I_v = AT^{\frac{1}{2}} + (B) \qquad \text{Ostashev (1936)}$$

$$I_v = AT + (B) \qquad \text{Darcy (1856)}$$

The equations of Ostashev and Darcy were modified slightly by adding a constant as shown in parentheses to improve their fitting ability and to make them more comparable with the other two equations.

The four equations were least-square fitted to data from (1) AEI, effective surface head and soil air pressure experiments; (2) border irrigation infiltrometers; (3) wet and dry infiltrometer runs; (4) sprinkled-water infiltrometers; and (5) ponded-water infiltrometers with both open and closed tops. Thus, a wide diversity of water source and infiltration system conditions were represented in this equation-fitting study, the results of which will be detailed in a subsequent paper. The conclusion was, however, that only Kostiakov's equation gives a consistently accurate fit regardless of the data source. Furthermore, it ranked equal to or better than each of the other equations for several other evaluation criteria. Consequently, Kostiakov's equation was selected for modeling the AEI theory of infiltration.

Infiltration rate I_R and the rate of deceleration I_D are given by the first and second derivative forms of Kostiakov's equation which are:

$$I_R = ABT^{B-1} \qquad I_D = AB(1-B)T^{B-2}$$

The integral and derivative forms of Kostiakov's equation indicate that where $0 < B < 1$:

(1) $I_V = 0$ and I_R and I_D are undefined for $T=0$;

(2) $I_V \to 0$, $I_R \to \infty$ and $I_D \to \infty$ as $T \to 0$; and

(3) $I_V \to \infty$, $I_R \to 0$ and $I_D \to 0$ as $T \to \infty$.

Thus, the infiltration volume increases at a decreasing rate monotonically with increasing time; and the infiltration rate and its deceleration decrease at a decreasing rate approaching zero asymptotically at large times.

The condition $0 < B < 1$ holds for most data sets from natural infiltration systems; however, infrequently the condition $B > 1$ prevails, indicating that the infiltration rate is increasing with time.

The mathematical interpretation of the parameters in the integral and derivative forms of Kostiakov's equation is readily apparent. If the unit for time is hours, then parameter A may be interpreted as either the first-hour infiltration volume I_V or the mean first-hour infiltration rate $\bar{I}_R$; the parameter product AB is the instantaneous infiltration rate I_R at the end of the first hour or at $T=1$, parameter B is first-hour end rate divided by the mean rate or $B = I_R/\bar{I}_R$ for $T=1$, and the time coefficient $[AB(1-B)]$ is the deceleration (negative acceleration) of the infiltration rate at $T=1$. Thus sets of infiltration data may be conveniently and meaningfully summarized in terms of the A and B parameters and the time period upon which they are based. Such summarizations give the first-hour infiltration and its abatement ratio and permit calculation of infiltration volume, rate, and deceleration for any selected time. Parameter A usually ranges from 0 to 20 and gives the integral curve its magnitude, whereas parameter B usually ranges from 0 to 1 and gives the integral curve its shape.

The A and B parameters may be quickly estimated from infiltration data since $A = I_V$ and $AB = I_R$ at $T=1$; however, better estimates are usually obtained by transforming the integral form to obtain the linear equation:

$$\ln I = \ln A + B \ln T ,$$

which can be least-square fitted to infiltration data. Such fits are easily performed with hand calculators programmed for simple linear regression analyses.

A physical interpretation of the Kostiakov equation and its parameters relative to the AEI theory is possible, although not as readily apparent as the preceding mathematical interpretation. The AEI theory assumes that all infiltrating surface water is subsequently stored in the soil profile. Thus, I_V becomes the storage volume of infiltrated water, I_R is the storage rate, I_D is the deceleration in the storage rate, T is the elapsed time after incipient ponding during which storage has been occurring, parameter A is the storage during the first hour, AB is the storage rate at the end of the first hour, and B is a dimensionless ratio of AB and A which reflects the degree of storage rate abatement during the first hour.

Infiltration has long been recognized as a process reflecting the net effect of numerous concurrent decay or abatement factors (Horton, 1940) which cause the decreasing infiltration rates with increasing elapsed time after the onset of the process. In natural soils, under complex initial and boundary conditions, the abatement of capillary pressure gradient (the justification for the $T^{\frac{1}{2}}$ dependency) is often relatively unimportant compared with other infiltration abatement factors, some of which are infiltration-related abatement processes (Dixon, 1975b). These factors include (1) capillary pressure head reduction at the wetting front resulting from increasing moisture content with depth, (2) surface crusting or sealing, (3) soil subsidence or settling, (4) soil air pressure buildup and air entrapment, (5) clay mineral hydration, (6) eluviation and illuviation, (7) surface water head dissipation, (8) decreasing water phase continuity in the macropore system through air entrainment and entrapment, (9) macroporosity extent and continuity reduction with depth in the profile, and (10) anaerobic slime formation. Some other soil conditions, which will be referred to here as infiltration augmentation factors, tend to offset (and infrequently reverse) the normal abatement in infiltration rates. Such conditions include (1) increasing flow dimensionality with time, (2) increasing wettability with depth, (3) decreasing moisture content (or increasing air porosity) with depth, (4) decreasing water repellency with depth, (5) eluviation (micropiping) that increases surface macroporosity and subsurface macropore continuity, (6) increasing ponded water depth, (7) increasing surface area ponded, and (8) increasing water phase continuity in the macropore system through air displacement and absorption.

The magnitude of parameter B in Kostiakov's equation thus reflects the net interacting effect of the preceding abatement and augmentation factors on the time course of infiltration, with the magnitude being inversely related to the number and intensity of infiltration abatement factors and directly related to the number and intensity of augmentation factors that are active in a given infiltration system.

Values for B near zero, near one, and above one, indicate the dominance of abatement factors, little dominance of either abatement or augmentation factors, and dominance of augmentation factors, respectively. Since most of the abatement and augmentation factors are greatly affected by AEI conditions, parameter B may be regarded as a function of such conditions, especially where unfilled storage space is large enough to not dominate infiltration abatement. Parameter B is expected to be relatively large where effective surface head and surface microroughness and macroporosity are relatively large, and relatively small where these AEI conditions are relatively small.

Darcy-based flow theory for simple infiltration systems can also be useful in physical interpretation of parameter A in Kostiakov's equation. The coefficient in Darcy's equation is given by the product of the hydraulic conductivity and hydraulic gradient for a near-saturated stabile porous soil. For such soils, both the conductivity and gradient are relatively constant. However, for unsaturated soils, the conductivity and gradient are not constant, but are interdependent variables with the gradient decreasing and the conductivity increasing as the soil wets by infiltration. Thus, in accordance with Darcy's equation and the view of surface infiltration presented by Childs (1969), parameter A may be regarded as the product of the first-hour time-weighted means for hydraulic conductivity and hydraulic gradient at the soil surface. The surface hydraulic gradient and conductivity are greatly affected by surface microroughness and macroporosity and their hydraulic counterpart, effective surface head. Consequently, parameter A is also a function of these AEI conditions. Parameter A is expected to be relatively large where effective surface head and surface microroughness and macroporosity are relatively large and relatively small where these AEI conditions are relatively small.

In conclusion, the preceding mathematical and physical interpretations are in agreement that parameters A and B are interrelated. The physical interpretation indicates that both parameters depend on AEI conditions.

Theory versus equation parameters. The last step in quantifying the AEI theory was to relate its two parameters to the two parameters in Kostiakov's equation. The families of I_v curves, shown in Figs. 3 and 4, were used for this purpose. Parameters A and B were determined by least-square fitting of Kostiakov's equation to the family of curves generated by varying surface roughness and openness at the AEI. Parameter means and the coefficients and exponents of the first and second derivative forms were then plotted as functions of the AEI condition (Fig. 5). The four AEI

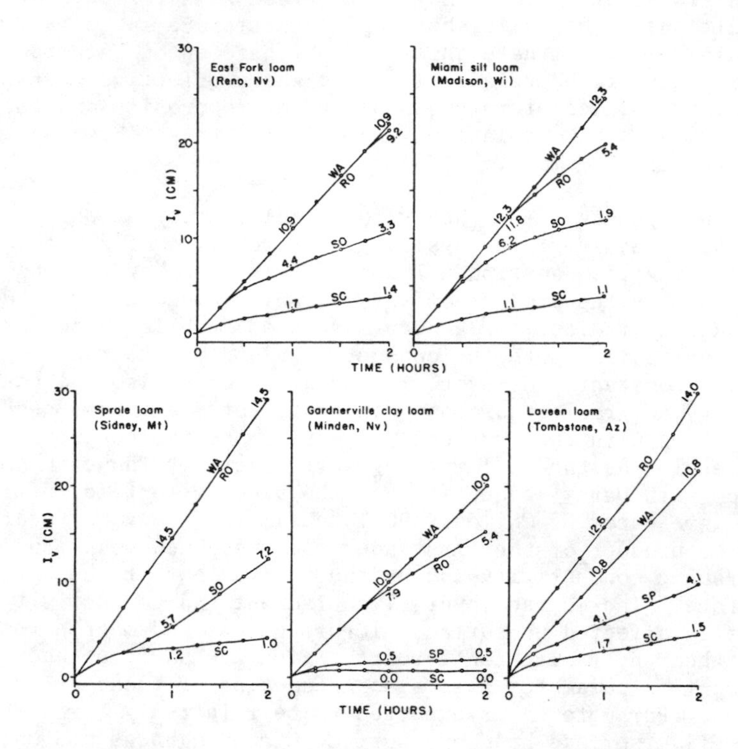

Fig. 3. Sprinkled-water infiltration under imposed air-earth interfaces RO and SC and naturally occurring interface either SO or SP. The curve labeled WA gives the total water applied by the infiltrometer spray nozzle. Numbers near curves at 1- and 2-hour times denote infiltration rates in cm hr^{-1} for these times. From Dixon (1975).

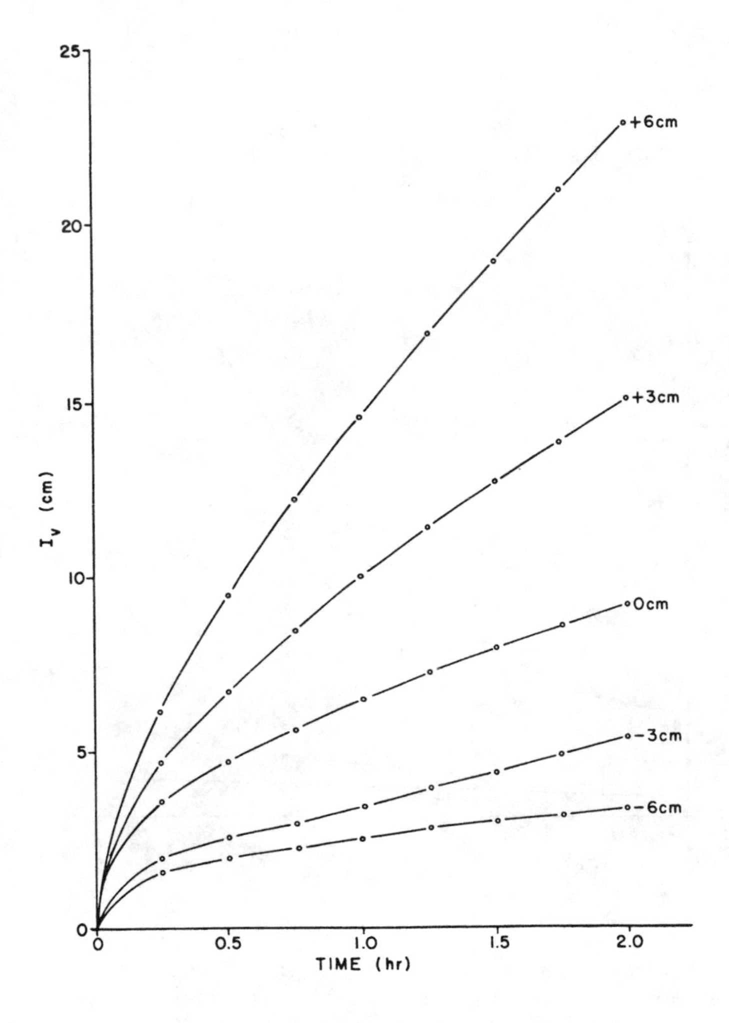

Fig. 4. Ponded-water infiltration I_V as a function of time and effective surface heads ranging from a minus 6 to a plus 6 cm of water as produced by a closed-top infiltrometer.

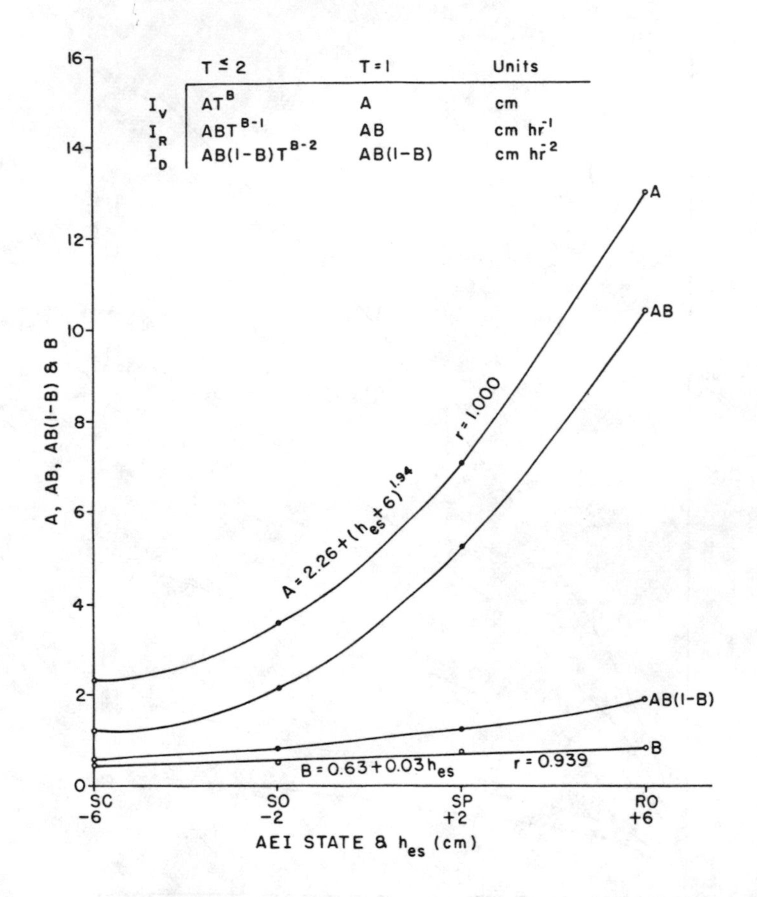

Fig. 5. Parameters for the integral and first and second derivative forms of Kostiakov's equation as functions of the AEI physical state and the estimated equivalent effective surface head h_{es}.

conditions, representing a broad range in surface roughness and openness, were assigned the effective surface head values that would be expected under intense rainfall over a large area. This assignment of approximate numerical values expedited subsequent linear regression analyses and facilitated comparison with the curves presented in Fig. 6.

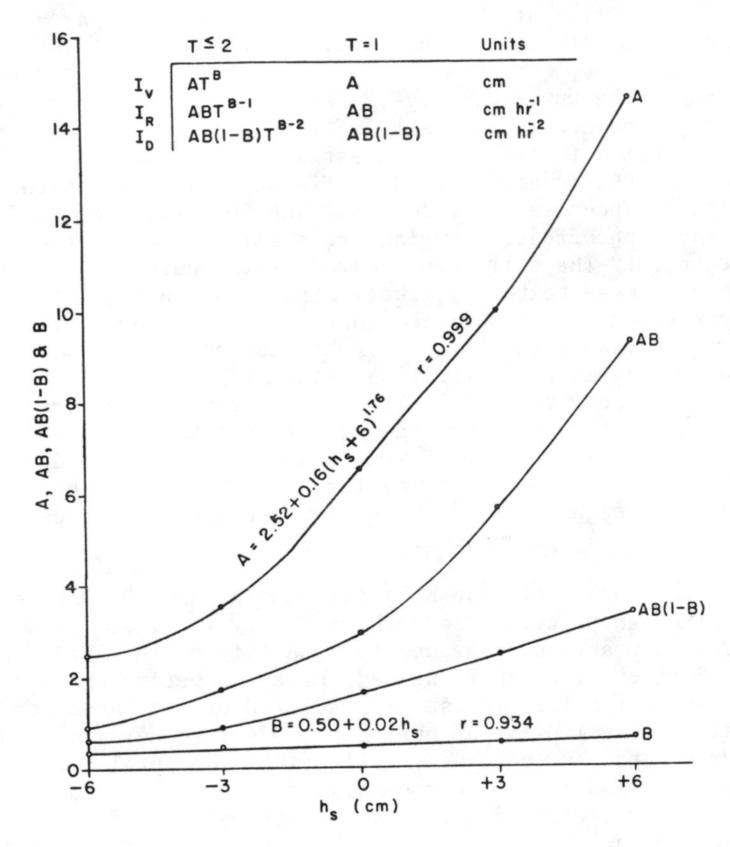

Fig. 6. Parameters for the integral and first and second derivative forms of Kostiakov's equation as functions of effective surface head h_s.

The family of curves generated by varying the effective surface head (Fig. 4) were analyzed similarly to produce Fig. 6. The close correspondence of the shape and magnitude of the curves in Figs. 5 and 6 is consistent with the hypotheses that h_s is the hydraulic manifestation of surface roughness and openness, and that the closed-top infiltrometer may be used on an uncrusted soil to determine the infiltration effects of these two interacting and interrelated physical conditions. Linear regression analyses indicated that parameter A is accurately described as a power function of the numerical estimates of the AEI condition and the effective surface head, whereas parameter B is linearly related to the AEI condition and h_s. Parameter A increases at an increasing rate with increasing surface roughness and openness and with increasing h_s as indicated by power function exponents of 1.94 and 1.76. The coefficients AB and AB(1-B) corresponding to the instantaneous infiltration rate and its rate of deceleration at T=1, respectively, increase at an increasing rate with increasing time.

Although the curves in Figs. 5 and 6 exhibit surprisingly close correspondences, the small differences that do exist may be attributed to (1) error in estimating the numerical range for the AEI conditions, (2) differences in soil texture, and (3) differences in water source. The RO interface would probably have an effective surface head slightly below the estimated 6 cm. The soils represented by the curves shown in Fig. 5 have a mean texture slightly finer than that of the soil represented in Fig. 6. The curves of Fig. 5 are derived from sprinkled-water infiltration, whereas those of Fig. 6 are from ponded-water infiltration. The effect of soil texture would probably be relatively small compared to water source. Inherent to the sprayed-water source is the infiltration augmentation factor of increasing ponded area and depth with time. This factor may largely account for the differences in magnitude and shape of corresponding curves for the A and B parameters.

The functional relationships for the A and B parameters as given graphically and mathematically in Figs. 5 and 6 can provide a practical approach for quantifying the AEI theory. Further research is needed, however, before absolute infiltration for all soils can be predicted by this approach. This includes development of better methods for characterizing surface roughness and openness, evaluation of natural effective surface heads under diverse AEI and water-source conditions, and correlation of the measured effective surface head and corresponding surface roughness and openness. The curves in Figs. 5 and 6 are appropriate for medium-textured soils that are initially dry and well-structured. With the aid of closed-top infiltrometers, similar sets of curves need to be developed for coarse- and fine-textured soils. Methods that facilitate correcting for the infiltration effect of antecedent moisture and single-grain soil structure need to be developed.

Summation. The AEI theory provides a conceptual basis for relative infiltration control at the air-earth interface. Kostiakov's equation can be used in absolute infiltration control by interpreting the coefficient A as a function of effective surface head, with large A values being associated with rough open surfaces and positive effective surface heads and small A values with smooth closed surfaces and negative effective surface heads. Exponent B may be viewed as a function of infiltration abatement-augmentation factors with values near zero, near one, and above one, indicating the dominance of abatement factors, little dominance of either the abatement or augmentation factors, and dominance of augmentation factors, respectively. Since many of the abatement and augmentation factors affect the effective surface head and vice versa, parameters A and B are interdependent. Further theoretical and experimental research is needed to determine the independent effect of various infiltration abatement and augmentation processes on the parameters of Kostiakov's equation. The study of water infiltration as affected by dynamic surface boundary conditions is a fertile field for major experimental and theoretical advances.

References

[1] Childs, E. C., 1969. "Surface Infiltration. An Introduction to the Physical Basis of Soil Water Phenomena," John Wiley and Sons, Ltd., New York, pp. 274-294.

[2] Darcy, H.P.G., 1856. "Les Fontaines Publiques de la Ville de Dijon," Dalmont, Paris.

[3] Dixon, R. M., 1972. "Controlling Infiltration in Bimodal Porous Soils: Air-Earth Interface Concept," Proc. 2nd Symp. Fundamentals of Transport in Porous Media, IAHR, ISSS, University of Guelph.

[4] Dixon, R. M., 1975a. "Design and Use of Closed-Top Infiltrometers," Soil Sci. Soc. Amer. Proc., 39: 755-763.

[5] Dixon, R. M., 1975b. "Infiltration Control through Soil Surface Management," Proc. Symp. on Watershed Management, Irrigation and Drainage Division, ASCE and Utah State University, pp. 543-567.

[6] Dixon, R. M. and Linden, D. R., 1972. "Soil Air Pressure and Water Infiltration under Border Irrigation," Soil Sci. Soc. Amer. Proc., 36:948-953.

[7] Dixon, R. M. and Peterson, A. E., 1971. "Water Infiltration Control: A Channel System Concept," Soil Sci. Soc. Amer. Proc., 35:968-973.

[8] Horton, R. E., 1940. "An Approach Toward a Physical Interpretation of Infiltration Capacity," Soil Sci. Soc. Amer. Proc., 5:399-417.

[9] Kostiakov, A. N., 1932. "On the Dynamics of the Coefficient of Water Percolation in Soils and the Necessity of Studying it from the Dynamic Point of View for the Purposes of Amelioration," Trans. Sixth Commission ISSS, Russian Pt. A., pp. 17-21.

[10] Lewis, M. R. and Powers, W. L., 1939. "A Study of Factors Affecting Infiltration," Soil Sci. Soc. Amer. Proc., 3:334-339.

[11] Morel-Seytoux, H. J., 1976. "Derivation of Equations for Rainfall Infiltration," Journal of Hydrology, 31: 203-219.

[12] Ostashev, N. A., 1936. "The Law of Distribution of Moisture in Soils and Methods for the Study of the Same," International Conference Soil Mechanics and Foundations Engineering Proceedings, Vol. 1 (Sect. K): 227-229.

[13] Philip, J. R., 1957. "The Theory of Infiltration: 4. Sorptivity and Algebraic Infiltration Equations," Soil Science, 84:257-264.

[14] Philip, J. R., 1958. "The Theory of Infiltration: 6. Effect of Water Depth Over Soil," Soil Science, 85: 278-286.

ETUDE THEORIQUE PROBABILISTE
DE LA DISPERSION CAPILLAIRE EN MILIEU POREUX

Par

Jacques Ganoulis
School of Technology
Aristotelian University
Thessaloniki, Greece

Claude Thirriot
Institut de Mechanique des Fluides
2, rue Camichel
31071-Toulouse, France

Abstract

The percolation of a non wetting fluid applied to the surface of a soil is considered. In equilibrium state the microscopic geometry of the pores in connection with the capillary forces, determine a non uniform distribution of the injected fluid into the soil, called capillary dispersion.

The problem is formulated in stochastic terms. The modelling of the pores by a network of capillaries having a variable radius in series and in parallel, leads to a differential equation. Numerical solutions of this equation permit to evaluate the probability that the injected fluid penetrates into a given pore, when this fluid is lighter or heavier than the original one.

Resume

On considère l'infiltration d'un fluide non mouillant répandu à la surface d'un sol.En état d'équilibre, la géometrie microscopique des pores en liaison avec les forces capillaires, déterminent une distribution non uniforme de la teneur en fluide injecté a l'intérieur du sol, que nous appelons dispersion capillaire.

Le problème est formulé en termes stochastiques. La schématisation des pores par un réseau de capillaires à rayon variable en série et en parallèle, conduit à une équation differentielle.La résolution numérique de cette équation permet d'évaluer la probabilité pour que le fluide injecté pénètre dans un pore donné, lorsqu'il est plus ou moins léger que le fluide en place.

Introduction

L'infiltration d'une couche fluide répandu en surface d'un sol,dépend bien sûr des caractéristiques physiques du fluide et de la structure du milieu poreux,mais aussi des forces en présence,telles que les forces de la capillarité et de la pesanteur. Lorsque le fluide entrant est non mouillant, la pression capillaire s'oppose à sa pénétration et le fluide peut être arrêté à la rencontre des pores à faible dimension.

En fait,même si l'un des deux fluides est l'air,dont on néglige souvent la masse spécifique,il s'agira toujours d'un couple de deux fluides.En hydrologie du sol,ce sera l'entrée de l'air dans les pores après abaissement du niveau de la nappe phréatique.Mais on peut aussi examiner la pollution d'un sol, à la suite de l'intrusion accidentelle d'un hydrocarbure répandu en surface ou en profondeur du sol saturé en eau.Lorsque,dans le cas limite,onélimine l'air en faisant le vide,on aura à faire àl'intrusion du mercure dans des échantillons de milieu poreux,au cours des essais de porométrie.

Entre le fluide pénétrant et le fluide en place,les forces capillaires en liaison avec la diversité des tailles des pores créent une zone de transition et nous appelons ce phénomène dispersion capillaire.L'étude du problème est abordée par voie stochastique,permettant de tenir compte du blocage de gros pores, qui ne sont accessibles que par des passages retrécis (Fig. 1). Plus précisement,la dimension des rayons de ces passages doit être inférieure ou égale au rayon critique r_c, défini par l'équation de Laplace:

$$r_c = \frac{2T_s \cos\theta}{p_c} \tag{1}$$

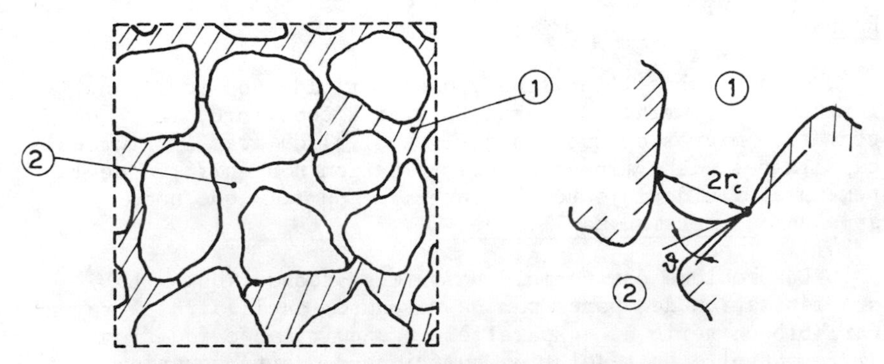

Fig. 1. Blocage des gros pores délimités par des cols et definition du rayon critique. Blocking of wide pores borded by necks and definition of the critical radius.

où T_s est la tension interfaciale, θ l'angle de mouillage, mesuré dans le fluide mouillant 2 , et $p_c = p^{(1)} - p^{(2)}$ est la pression capillaire (Fig. 1). Il est évident que ce phénomène de blocage ne peut pas être décrit par les modèles capillaires uniformes (Scheidegger, 1963, p. 62).

Le modèle stochastique que nous décrivons par la suite, est un modèle en série et en parallèle. Il tient compte de la variation des rayons des pores en série mais aussi des connexions des pores en parallèle. L'analyse stochastique des événements, qui peuvent avoir lieu dans une tranche infiniment petite du milieu poreux, permet d'établir l'équation différentielle pour la probabilité du remplissage d'un pore. Enfin, la résolution numérique de cette équation conduit à calculer cette probabilité, lorsque le fluide pénétré est plus ou moins lourd que le fluide en place.

Le Problème Physique

Dans un milieu poreux homogène et isotrope saturé par le fluide mouillant 2 , on considère l'infiltration d'un fluide non mouillant 1 , répandu en surface sous forme d'une couche d'épaisseur h (Fig. 2). Sous l'influence des forces de gravité et de capillarité supposons que le fluide 1 arive au pore M, situé à la profondeur z. Pour ceci, deux conditions sont nécèssaires:

(a) que le pore M ait un rayon $r > r_c$ (pore surcritique), où r_c est defini selon l'équation (1).

(b) qu'il existe au moins un système des pores surcritiques unis, reliant M à la surface.

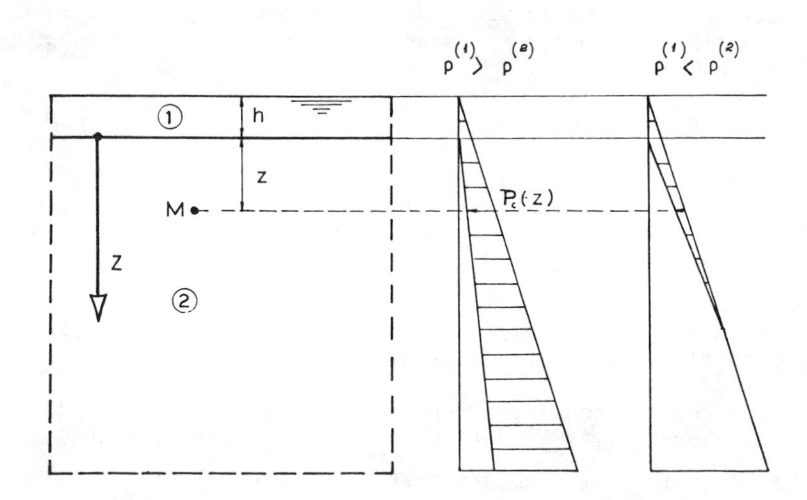

Fig. 2. Schema du problème physique. Sketch of the physical problem.

En comptant les z positifs vers le bas,comme le montre la Fig. 2, la pression capillaire p_c est une fonction linéaire de z, de la forme:

$$p_c(z) = p_c(o) + (\rho^{(1)} - \rho^{(2)}) gz \tag{2}$$

où $p_c(o) = \rho^{(1)} gh$ est la pression en surface.Lorsque $\rho^{(1)} > \rho^{(2)}$ la fonction $p_c(z)$ est croissante, tandis que pour $\rho^{(1)} < \rho^{(2)}$ elle est décroissante.

Pour que les conditions (a) et (b) soient valables, il faut que l'écoulement des deux fluides soit extrêmement lent, de sorte que les forces de viscosité et d'inertie soient négligeables. On peut aussi imaginer, que la couche en surface atteigne son épaisseur h progressivement, selon une succession d'états de quasi-équilibre, au cours desquels l'équation (1) reste valable.

Lorsque la dimension des rayons des pores suit une loi de distribution de densité $f(r)$, il est évident qu'à chaque niveau de pression capillaire p_c il existe des pores ayant un rayon $r \leq r_c$ (pores souscritiques), qui constituent des barrières au fluide non mouillant. Etant donné que ces barrières sont réparties au hazard dans le milieu poreux,on concoit la difficulté du calcul de la probabilité d'un réseau de pores surcritiques $(r > r_c)$ et unis. En tout cas,tous les pores surcritiques situés à la profondeur z ne seront pas remplis et la probabilité de leur remplissage sera calculé d'après une modelisation stochastique.

Le Modele Stochastique

Chaque pore est caracterisé ponctuellement par son rayon r dont la taille suit une loi de distribution de densité $f(r)$.Les rayons des pores peuvent varier en série (Fig. 3) de sorte que des cellules de rayon r_i et de longueur X_i variable peuvent apparaître en succession. Mais chaque pore peut aussi communiquer avec les pores voisins. Nous fairons l'hypothèse que les liaisons sont uniquement binaires.

Selon les considerations précédentes, l'ensemble des pores surcritiques $(r>r_c)$ contenus à la profondeur z peut être divisé en trois événements distincts (Fig. 4):

-l'évènement A, désignant le pore surcritique $(r>r_c)$ qui devient souscritique $(r \leq r_c)$ dans l'intervalle Δz,

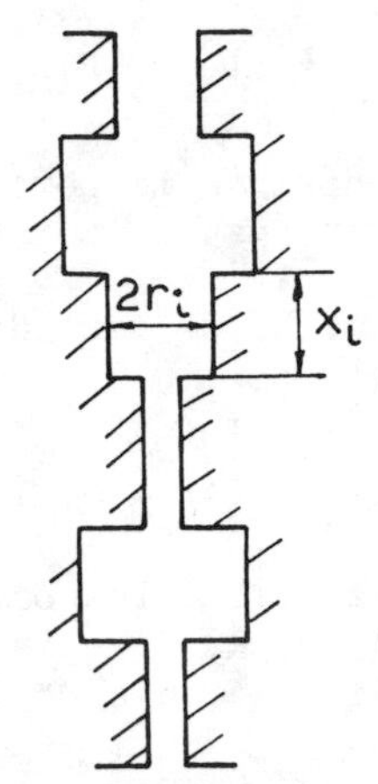

Fig. 3. Variation en série des rayons et des longueurs des pores. Variation in series of pore radius and lengths.

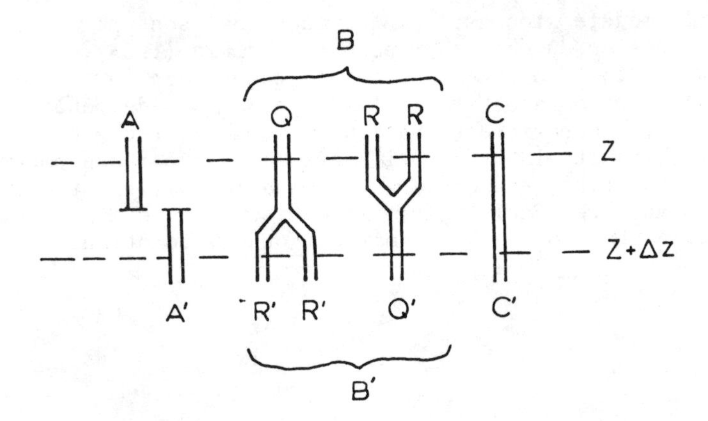

Fig. 4. Les diverses sortes des pores surcritiques à la profondeur z. The different kind of supercritical pores at the depth z.

-l'évènement B, désignant le pore surcritique ramifié. B est la somme des évènements Q et R, où Q est le pore surcritique à z qui se divise en deux à $z + \Delta z$ et R est l'évenement de deux pores surcritiques à z qui se reunissent à un seul à $z + \Delta z$. Il est évident que si $P[\cdot]$ désigne la probabilité, on aura:

$$P[R] = 2\ P[Q]$$

et par conséquent:

$$P[Q = \frac{1}{3}\ P[B] \qquad (3)$$

$$P[R] = \frac{2}{3} P[B] \tag{4}$$

-l'évènement C désigne le pore surcritique à z qui ne se ferme pas et ne se divise pas à $z + \Delta z$.

Etant donné que A, B, C sont disjoints et forment un ensemble complet, on a:

$$P[A] + P[B] + P[C] = 1 \tag{5}$$

Comme le montre la Fig. 4 les pores A', Q', R', C' sont les équivalents des A, Q, R, C mais qui font leur apparition à la profondeur $z + \Delta z$. A cause de l'homogéneité du milieu, on aura les égalités:

$$P[A'] = P[A],\ P[Q'] = P[Q],\ P[R'] = P[R],\ P[C'] = P[C] \tag{6}$$

Le modèle stochastique que nous analysons par la suite est une extension du modèle proposé par Markin (1963). L'extension est du au fait qu'ici, nous tenons compte des forces de pesanteur et on aboutit ainsi à un processus de pénètration non uniforme, à coefficients variables.Comme en recherche opérationnelle (Girault, 1959) on utilisera l'analyse des éléments infiniment petits afin d'établir l'équation differentielle de la probabilité. Ceci étant, introduisons les coefficients <u>d'engorgement</u> β_1 et de <u>ramification</u> β_2, fonctions de z et tels que:

$$P[A] = \beta_2(z)\ \Delta z \tag{7}$$

$$P[B] = \beta_2(z)\ \Delta z \tag{8}$$

Les coefficients β_1 et β_2 caracterisent respectivement l'intensité du blocage et de la ramification d'un pore surcritique dans l'intervalle Δz. Des arguments théoriques et physiques (Ganoulis et al., 1976) permettent l'estimation de ces paramètres selon les expressions suivantes:

$$\beta_1(z) = - \frac{1}{\ell} \text{Log} \left(\int_{r_c(a)}^{\infty} f(r)dr \right) \tag{9}$$

$$\beta_2(z) = 2kP \int_{r_c(a)}^{\infty} rf(r)dr / \int_{o}^{\infty} r^2 f(r)dr \tag{10}$$

où ℓ est la longueur moyenne des pores, K et P sont respectivement les coefficients de rugosité et de porosité du milieu. Compte tenu des relations (5), (7) et (8) on a aussi:

$$P[C] = 1 - \{\beta_1(z) + \beta_2(z)\}\Delta z \qquad (11)$$

Soit maintenant E et E'le remplissage d'un pore surcritique à z et $z + \Delta z$ respectivement, et posons $W(z)$ et $W(z+\Delta z)$ la probabilité de ces évènements. On aura:

$$W(z) = P[E], \; W(z+\Delta z) = P[E']$$

Il est évident que E'est indépendant de la forme des pores surcritiques, donc des A', B', C'. Etant donné la relation (5), on peut écrire:

$$W(z+\Delta z) = P[E'] = P[E' \cap (A' \cup B' \cup C')] = P[E' \cap A'] + P[E' \cap B'] + P[E' \cap C'] \qquad (12)$$

L'examen de la Fig. (4), l'indépendance des évènements et la relation (11) permettent d'écrire:

$$P[E' \cap A'] = 0 \qquad (13)$$

$$P[E' \cap C'] = P[E \cap C] = P[E]P[C] = (1 - \{\beta_1(z) + \beta_2(z)\})W(z)\Delta z \qquad (14)$$

La Fig. (5) montre toutes les situations possibles du remplissage d'un pore ramifié B. On a:

$$P[E' \cap B'] = P[E' \cap R'] + P[E' \cap Q'] \qquad (15)$$

avec:

$$P[E' \cap R'] = P[E \cap Q] = P[Q]P[E] = \frac{1}{3}\beta_2(z)W(z)\Delta z \qquad (16)$$

Par ailleurs, en posant E'' l'arrivée du noeud M et $\overline{E}$ l'évènement contraire à E, comme le montre la Fig. 5, on peut écrire:

$$P[E' \cap Q'] = P[E'' \cap M] = P[E''|M]P[M] = (2P[E] \cdot P[\overline{E}] - P[E]^2)P[R] = \{2W(z)\,(1 - W(z)) - W(z)^2\} \cdot \frac{2}{3}\beta_2(z) \cdot \Delta z \qquad (17)$$

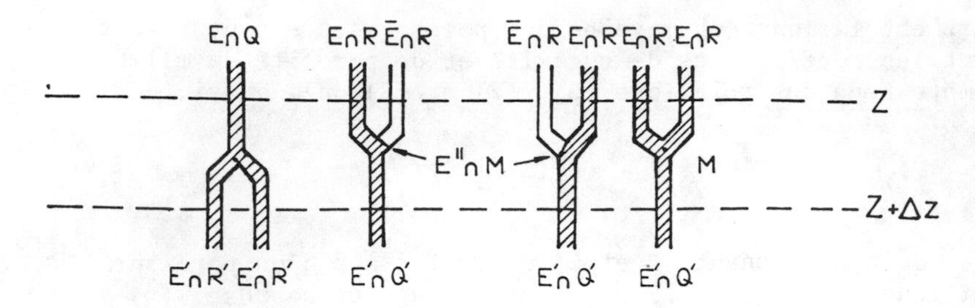

Fig. 5. Remplissage des pores ramifies. Filling of ramificated pores.

En introduisant les resultats partiels (13), (14), (15), (16), et (17) dans l'équation (12), lorsque $\Delta z \to 0$ on abtient l'équation différentielle suivante:

$$\frac{dW(z)}{dz} + C(z)W(z) + B(z)W^2(z) = 0 \tag{18}$$

où $C(z) = \beta_1(z) - \beta_2(z)/3, \quad B(z) = \beta_2(z)/3$ (19)

L'équation (18) avec la condition limite $W(0) = 1$ décrit la distribution de la probabilité du remplissage des pores surcritiques à la profondeur z.

Resolution Numerique Et Applications

En appliquant la transformation:

$$W(z) = \frac{1}{U(z)}$$

l'équation (18) prend la forme linéaire:

$$\frac{dU}{dz} = C(z)U + B(z) \tag{20}$$

La solution générale de (20) est connue. Revenant à la fonction $W(z)$, avec la condition $W(o) = 1$, on trouve:

$$W(z) = \frac{\exp\{-\int_o^z C(s)ds\}}{1 + \int_o^z B(s)\exp\{-\int_o^s C(t)dt\}ds} \tag{21}$$

Les fonctions $B(s)$ et $C(s)$ sont données d'après les expressions (19).

L'équation (21) a été intégrée numériquement sur l'ordinateur. Le calcul des intégrales a été effectué selon la méthode classique de Simpson. Le critère de précision fixé d'avance, a été respecté avec un découpage automatique plus ou moins fin du domaine d'intégration.

Pour les applications, nous avons supposé que la densité de la distribution des rayons des pores suit approximativement une loi de Rayleigh, de la forme:

$$f(r) = \frac{r}{r_o^2} \exp\left\{-\frac{r^2}{2r_o^2}\right\} \qquad (22)$$

Pour l'exemple de calcul, soit un milieu poreux ayant $r_o = 1$ mm, $k = 1.5$ et une porosité $P = 0.32$. Le rayon critique r_c est calculé d'après les expressions (1) et (2), avec $T_s \cos\theta = 0.023$ N/m. Ensuite, les paramètres $\beta_1(z)$ et $\beta_2(z)$ sont calculés en utilisant les formules (9) et (10), avec $\ell = 2r_o$. Au cours de l'intégration numérique de l'équation (21),nous avons distingué deux cas:

(a) Le fluide appliqué en surface est plus lourd que le fluide en place. Lorsque $\rho^{(1)} = 900$ Kg/m^3 et $\rho^{(1)} - \rho^{(2)} =$ 900 Kg/m^3 également,la Fig. 6 montre les resultats numériques obténus pour deux épaisseurs de la couche fluide $h/r_o = 4$ et $h/r_o = 8$. La probabilité de remplissage des pores surcritiques $W(z)$ diminue dans les deux cas jusqu'à la profondeur $z \simeq 4.5\ r_o$, puis elle augmente et tend asymptotiquement vers 1. L'intéprétation de ces résultats peut être possible partant de la définition des paramètres β_1 et β_2. En effet, le paramètre d'engorgement β_1 caractérise le blocage des pores et aux faibles pressions capillaires, près de la surface, il emporte sur le parametre de ramification β_2 qui favorise la circulation en parallèle du fluide pénétrant. Mais au fur et à mesure que p_c augmente en fonction de z, le phénomène inverse provoque de plus en plus le remplissage des pores surcritiques.

(b) Le fluide pénètrant est plus léger que le fluide en place. Par exemple soit un hydrocarbure non mouillant de masse spécifique $\rho^{(1)} = 900$ Kg/m^3 répandu à la surface d'un sol

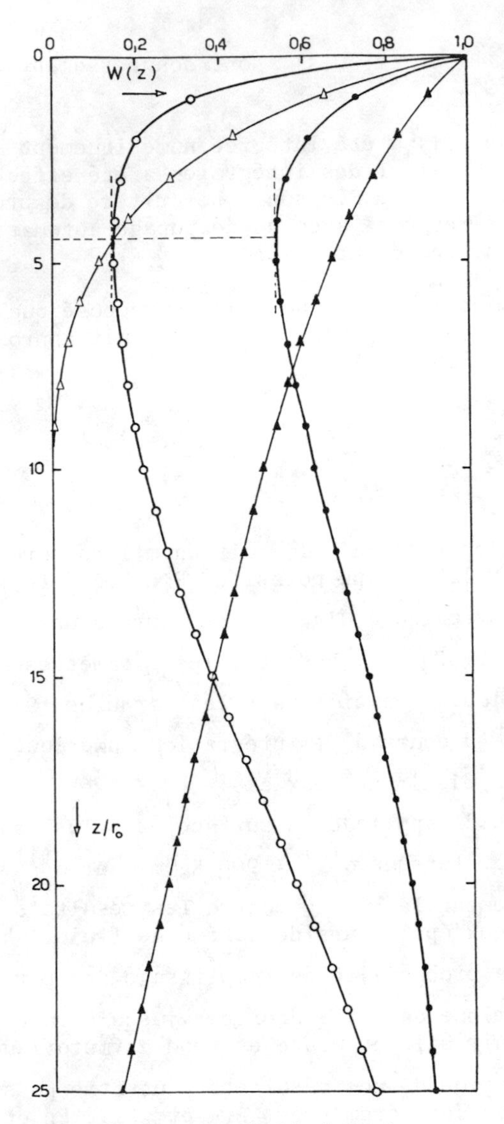

Fig. 6. Variation de la probabilité W(z):

—○— $h/r_o = 4$, $\rho^{(1)} > \rho^{(2)}$;

—●— $h/r_o = 8$, $\rho^{(1)} > \rho^{(2)}$; —△— $h/r_o = 4$, $\rho^{(1)} < \rho^{(2)}$;

—▲— $h/r_o = 8$, $\rho^{(1)} < \rho^{(2)}$.

Variation of the probability W(z) : —○— $h/r_o = 4$, $\rho^{(1)} > \rho^{(2)}$;

—●— $h/r_o = 8$, $\rho^{(1)} > \rho^{(2)}$; —△— $h/r_o = 4$, $\rho^{(1)} < \rho^{(2)}$;

—▲— $h/r_o = 8$, $\rho^{(1)} < \rho^{(2)}$.

saturé par l'eau. Alors, $\rho^{(2)} - \rho^{(1)} = 100$ Kg/m^3 et la Fig. 6 montre les résultats du calcul pour $h/r_o = 4$ et $h/r_o = 8$. Nous pouvons constater que dans ce cas la diminution de la pression capillaire en fonction de la profondeur provoque une augmentation du parametre d'engorgement β_1 et une diminution du paramètre de ramification β_2. Alors le blocage des pores devient total et $W(z) \to 0$ après une certaine profondeur.

Conclusions

Le modèle stochastique proposé contient trois paramètres essentiels: le paramètre en série β_1 le paramètre en parallèle β_2 et la pesanteur. Tandis que les deux premiers paramètres entrent toujours en jeu dans le sens opposé, la pesanteur peut favoriser le blocage ou la pénétration en fonction du contraste de densité des fluides superposés. Lorsque le fluide en surface est plus léger,il y a blocage complet, tandis que dans le cas contraire,la pénètration devient excessive, après l'apparition d'une zone engorgée, près de la surface.

References

[1] Ganoulis, J. and Thirriot, C., 1976. "Stochastic Study of the Entrance of a Nonwetting Fluid in a Porous Medium, Under the Influence of the Capillarity." Proc. IIth International AIRH Symposium in Stochastic Hydraulics, Lund, Sweden, Aug. 2-4 (To be published by Water Resources Publications, Fort Collins, Colorado).

[2] Girault, M., 1959. "Initiation aux processus aleatoires," Dunod, Paris, 107 pp.

[3] Markin, V. S., 1963. "Investija Akad. Nauk. SSSR, Ser. Him.," 151:620-623 (in Russian).

[4] Scheidegger, A., 1963. "The Physics of Flow Through Porous Media." University of Toronto Press, 313 pp.

COMPARAISON DE PROCEDES DE SIMULATION NUMERIQUE DU PHENOMENE D'HYSTERESIS DE LA RELATION SUCCION-TENEUR EN EAU ET ETUDE CRITIQUE DES EFFETS SUR LES ECOULEMENTS EN MILIEUX NON SATURES

By

R. Gaudu, Nguyen Tan Hoa, C. Thirriot
Institut de Mécanique des Fluides
2, rue Charles Camichel
31071 Toulouse Cedex (France)

et

Ph. Thore
Faculté des Sciences, B.P. 812
Yaounde (Cameroun)

Summary

After giving a description, based on experimental data, of the evolution of the hysteresis, the authors propose some mathematical representations of the scanning curves.

The tried and compared algorithms, concern either a differential or an explicit derivation.

In the differential method, two ways are explored:

- the application of the theory of singular points of differential equations,

- interpolation of the values of the pair of distinct tangents according to the sense of evolution of water-content.

The explicit derivation is essentially based on the interpolation with two different classes of functions: polynomials and rational fractions.

At last, examples of the influence of the hysteresis are shown and compared to experimental data.

Resume

Après une description de l'évolution de l'hystérésis à partir des résultats expérimentaux, on propose quelques modes de représentation mathématiques des graphes de passage entre les courbes enveloppes de drainage et d'humidification.

Les algorithmes essayés et comparés concernent soit une description différentielle, soit une description explicite.

Dans la description différentielle, deux voies sont suivies:

- l'application de la théorie des points singuliers des équations differentielles,

- l'interpolation des valeurs en un point du couple de tangentes distinctes suivant le sens d'évolution de la teneur en eau.

La description explicite est essentiellement fondée sur l'interpolation avec comme classes de fonction, soit des polynomes, soit des fractions rationnelles.

Enfin, des exemples sont donnés d'influence de l'hystérêsis avec comparaison éventuelle avec des résultats expérimentaux.

Introduction

Le phénomène d'hystérésis en écoulement diphasique dans les milieux poreux est actuellement un fait d'évidence. Il etait déja pris en considération par Haines en 1930 d'un point de vue agronomique.

Et en Génie Pétrolier, ingénieurs et chercheurs ont très tôt aussi pris en compte la multiplicité de la relation entre "pression capillaire" et teneurs en fluide.

Un essai d'explication peut être donné, de manière illustrative, en considerant des déplacements de ménisque dans des canaux de forme non cylindrique (effet dit de la bouteille d'encre) ou mieux encore dans des assemblages de tels canaux. Hillel cite encore de nombreuses autres causes possibles, telles que la variation de l'angle de contact des ménisques, le passage ensolution ou le dégazage de l'air dissous, les variations de volume du squelette solide du sol.

L'analogie avec des phénomènes similaires apparaissent dans d'autres domaines tels que l'adsorption, la magnétisation, la polarisation, la rhéologie, peut aider à comprendre, au moins de manière conceptuelle, l'hystérésis en milieu poreux.

Un demi-siecle après Duhem et Brillouin, Everett developpa une théorie générale de l'hystérésis fondée sur la théorie des domaines indépendants, théorie que Poulovassilis (1961) appliquera aux milieux poreux non saturés.

Du point de vue expérimental, en France, G. Vachaud et son équipe de Grenoble, cernèrent de plus près la réalité en examinant de manière fine les courbes de passage entre les branches extrêmes du cycle d'hystérésis et en éprouvant la sensibilité à l'effet dynamique dû à des fluctuations rapides.

Dans la simulation mathématique, les idées de Ibrahim et Brutsaert (1968) d'une représentation discrète par balayage de grilles à mémoire, furent reprises par Colonna, Brissaud et Millet (1971) sur des applications pétrolières et par Thoré pour l'étude d'infiltration et d'évaporation sur des sols à caractère argileux.

En nous inspirant des idées précédentes, nous avons envisagé la représentation par des graphes continus qui, à notre avis, doivent donner plus de souplesse à la simulation de l'hystérésis. Dans ce qui suit, nous faisons part de nos tentatives (parfois infructueuses) et de quelques exemples d'application.

Caracterisation de L'Hysteresis en Milieu Poreux

Description de l'évolution. Pour rendre plus concrète la présentation, considérons la planche 1 qui est un exemple de résultats obtenus dans notre Laboratoire sur une colonne verticale. Cette installation a pour objectif essentiel la détermination de la perméabilité relative à l'eau. La méthode d'essai consiste à assurer à la base de la colonne un débit constant qui peut changer de sens quasi instantanément.

L'alternance du sens du débit provoque en tout point de la colonne le changement de sens de variation de l'humidité. Pour une même histoire de débit à la base de la colonne, on peut obtenir en différentes sections à altitude croissante des courbes de passage alternées correspondant à des teneurs moyennes en eau et des amplitudes de fluctuations d'humidité de plus en plus petites.

Bien sûr, la justesse des conclusions dépend de la qualité et donc de la précision des mesures. Mais nos expériences permettent de proposer quelques hypothèses. La principale est le principe de l'indébilité créée par l'état extrême atteint (Ceci rejoint d'une certaine manière les observations de Brissaud Colonna et Millet et la théorie des domaines indépendants).

Pour illustrer concrètement ce principe, considérons sur la planche 2, une évolution de premier drainage jusqu'à une première valeur minimale θ_{1min} de la teneur en eau (où $\psi = \psi_1$) suivie d'évolutions alternées d'augmentation ou de diminution de θ. Toutes les évolutions ultérieures de drainage seront telles qu'elles conduiront les graphes $\psi(\theta)$ à repasser par le point extrême (θ_{1min}, ψ_1) avant de provoquer une baisse de teneur en eau en deçà de θ_{1min}. Aussitôt que le point extrême est dépassé, le milieu poreux perd la mémoire de cet état extrême. Au risque de naiveté, décrivons de manière détaillée l'évolution représentée sur la Fig. 2.

Nous appellerons courbe de passage les graphes d'évolution (θ) intérieurs au cycle principal.

Le drainage démarre depuis l'état de saturation jusqu'au point 1 en suivant la courbe principale. Le changement de sens d'évolution de l'humidité fait apparaître un brusque changement de pente du graphe ( ) au point de rebroussement 1, sur lequel ont aussi insisté Vachaud, Thony et Vauclin. La courbe de passage en infiltration tend vers l'état extrême initial 0 et se rapproche de la courbe principale d'infiltration. Nouvelle alternance du sens de variation de l'humidité en 2. Le nouvel arc du graphe (θ) se dirige vers l'état extrême atteint en drainage, c'est à dire le point 1. Stoppé en 3, nouveau départ en infiltration vers le point extrême le plus proche, c'est à dire 2 (repéré aussi 4 dans la chonologie. Alors, retour vers 3 par le chemin déjà suivi (on pourrait parler de cycle intérieur atteint par une accomodation immédiate). Mais dans ce nouveau drainage, la teneur en eau descend au-dessous de θ_3, le graphe de passage poursuit alors son chemin vers l'état extrême antérieur immédiatement voisin de 3, c'est à dire l'état 1. L'évolution est stoppée en 5 avant d'atteindre 3. Nouvelle infiltration suivant un chemin qui se dirige vers 2, mais est stoppée en 6. Le graphe repart en principe vers 5 et non vers 3. Si on poursuit le drainage, on dépasse 5 et on continue vers 1 qu'on suppose dépasser pour finir de décrire la courbe principale de drainage jusqu'à la teneur en eau résiduelle θ_r.

La planche 1, quant à elle, présente une évolution similaire mais la condition initiale à toute l'histoire est la teneur résiduelle et l'évolution démarre par l'infiltration jusqu'au point A. Bien sûr, le processus est alors moins clair mais parce qu'il correspond à une évolution réele qu'il est tout de même difficile de commander (en fait on effectue comme de la télé-commande puisque l'évolution dans la section d'observation dépend de la loi de débit imposée à une plus grande profondeur).

Principes de la simulation. Dans les deux cas de figure, le principe de description conduit à l'emboitement de graphes et à l'effacement d'état extrême dès qu'on a dépassé un point de rebroussement atteint antérieurement. Ceci sera pris en compte dans plusieurs algorithmes de simulation.

Cependant, l'observation de la Fig. 1 appelle d'autres observations. Le graphe d'infiltration issu du point D ne

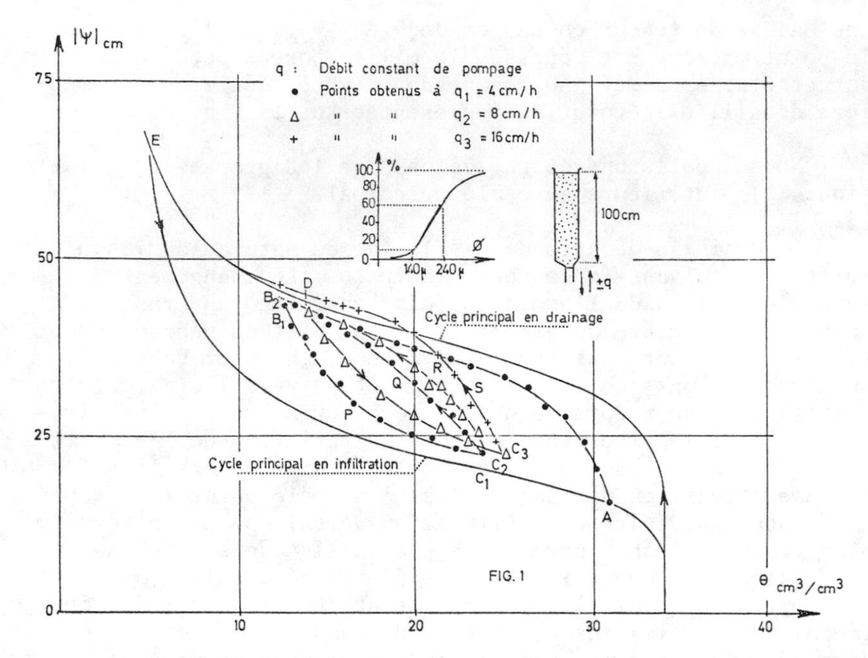

Fig. 1. Cycles $\psi(\theta)$ expérimentaux (sable de Fontainebleau)

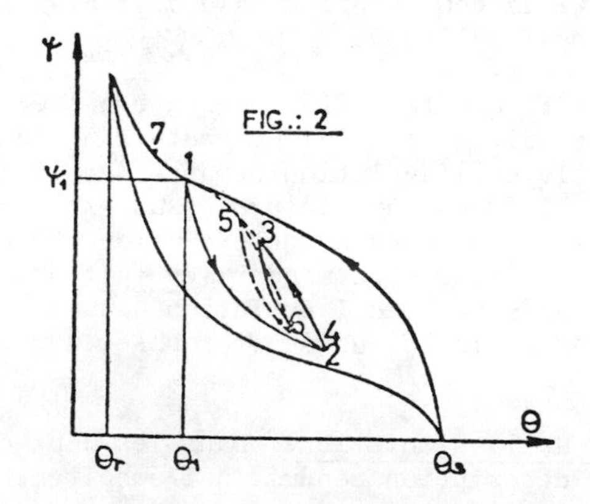

Fig. 2. Exemple d'évolution de l'hystérésis

ne passe pas exactement par l'état extrême immédiatement antérieur, c'est à dire le point C_2. Les courbes de drainage ont une courbure qui augmente avec le débit de pompage. Les courbes issues de A et C_3 se coupent. Ces derniers faits peuvent être vraisemblablement attribués à l'effet dynamique mis en évidence par G. Vachaud et M. Vauclin.

Il faut donc être prudent dans la recherche d'une simulation des phénomènes d'hystérésis et l'incertitude qui demeure quant au déroulement du phénomène physique et à la précision de la mesure afférente, peut autoriser la recherche d'autres voies. Nous avons ainsi envisagé l'hypothèse suivante: l'hystérésis est essentiellement caractérisée par l'existence en un point du plan (θ,ψ) de deux tangentes aux graphes de passage: l'une correspondant au drainage, l'autre a l'infiltration. La simulation consistera alors a se donner les deux champs de dérivées $d/d\theta$ dans le domaine entre les courbes principales qui constituent l'enveloppe. Bien sûr, ce faisant, on nie l'existence d'une fonctionnelle prenant en compte toute l'histoire hydrique du milieu non saturé. Par un point, il ne peut passer qu'une courbe de sens donné qui ne passe pas forcément par l'état extrême antérieur.

Utilisation de grandeurs réduites. Pour alléger la présentation ultérieure et aussi dans un souci de généralisation, nous utiliserons les grandeurs réduites:

$$x = \frac{\theta-\theta_{min}}{\theta_{max} - \theta_{min}} \quad \text{avec} \quad \theta_{min} = \theta_r, \quad \theta_{max} = \theta_s$$

$$\text{et} \quad y = \frac{-\psi}{|\psi_m|}$$

$$\text{avec} \quad |\psi_m| = \frac{-1}{2(\theta_{max}-\theta_{min})} \int_{\theta_{min}}^{\theta_{max}} (\psi_{pi} + \psi_{pd})d\theta$$

Le cycle d'hystérésis en grandeurs réduites peut être caractérisé de différentes manières: par une description variationnelle en utilisant les moments successifs, par l'affichage de quelques paramètres locaux significatifs. Nous avons retenu cette dernière manière en choisissant la flèche maximale et la moyenne des pentes des tangentes d'inflexion. Bien sûr, nous ne nous cachons pas que cette description pourrait être défectueuse dans certaines circonstances, par exemple si les courbes principales ne semblent pas se rejoindre à distance finie (mais pour la simulation pratique, on peut envisager une nouvelle délimitation du domaine utile par une courbe de passage au lieu de la courbe principale comme ce serait le cas sur la Fig. 4 qui reproduit les résultats dus à G. Vachaud, J. L. Thony, M. Vaclin).

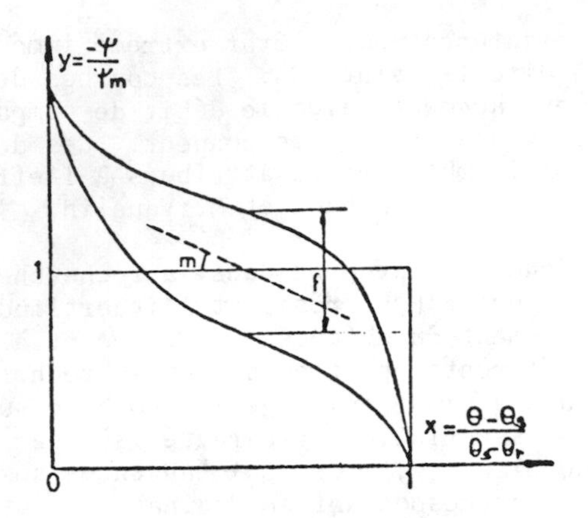

Fig. 3. Cycle d'hystérésis en variables réduites (sable de l'Isere)

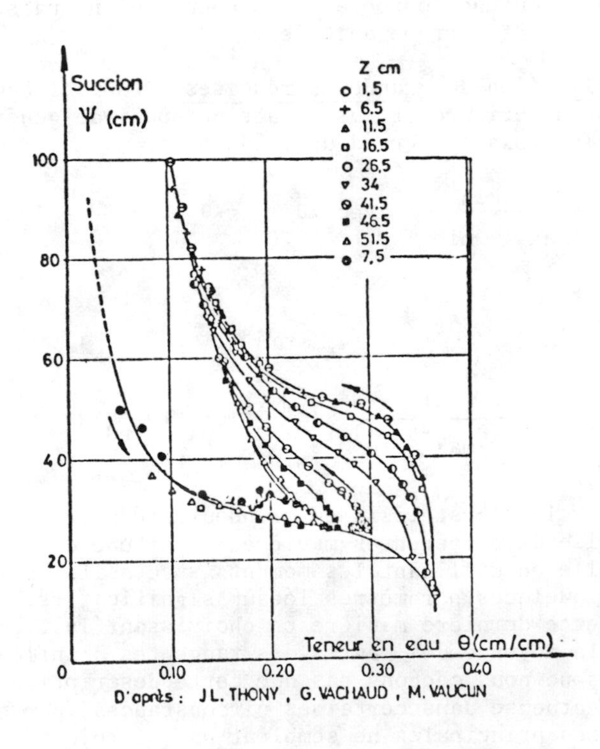

Fig. 4. Résultats expérimentaux d'après Vachaud G., Thony, J. L. et Vauclin, M.

La Definition Differentielle Sans Memoire

Les graphes de passage sont définis par l'équation:

$$\frac{dy}{dx} = m\ (x,y,\varepsilon)$$

ε = 1 à l'infiltration,
ε = - 1 au drainage.

Reste à définir les champs de dérivées. Pour en donner une idée, nous avons représenté sur la planche 5, les dérivées déduites des expériences de l'équipe de Grenoble (Fig. 4). Si on n'est pas trop exigeant sur la précision, on peut admettre une variation grossièrement linéaire de la dérivée, entre les courbes enveloppes, à humidité constante. (En fait, l'effet de proximité de la saturation est très sensible). On peut donc proposer l'idée d'interpolation suivante: pour le drainage, à x fixé, la pente dy/dx part d'une valeur $m_{pi} = m(x, y_{pi}(x), \varepsilon{=}-1)$ sur la courbe principale d'infiltration et atteint la valeur m_{pd-} de la pente de la courbe principale de drainage (le raccordement sur l'enveloppe est donc asymptotique). Une formulation simple de l'interpolation est:

$$m = m_{pi-} + (m_{pd-} - m_{pi-})\quad f(u)$$

avec $$u = \frac{y(x) - y_{pi}(x)}{y_{pd}(x) - y_{pi}(x)}$$

Pour f(u) nous avons essayé quelques formules monômes. La Fig. 5 présente des exemples de résultats obtenus. La faillite relative du procédé apparaît lorsque le rebroussement a lieu trop près du point de départ.

L'Utilisation de la Théorie des Points Singuliers

Dans l'ensemble des types de points singuliers, les noeuds correspondent au passage de toutes les courbes intégrales par le point singulier.

Nous avons donc considéré le tracé d'une courbe de passage comme un problème de tir: du point de rebroussement le graphe doit atteindre le point extrême antérieur le plus proche (qui n'ait pas été effacé). Ce point va jouer dans la formulation le rôle d'un noeud de coordonnées (x_e, y_e).

Avec la translation $x' = x - x_e$, $y' = y - y_e$, l'équation différentielle du graphe de passage est:

$$\frac{dy'}{dx'} = \frac{f(x',y')}{g(x',y')}$$

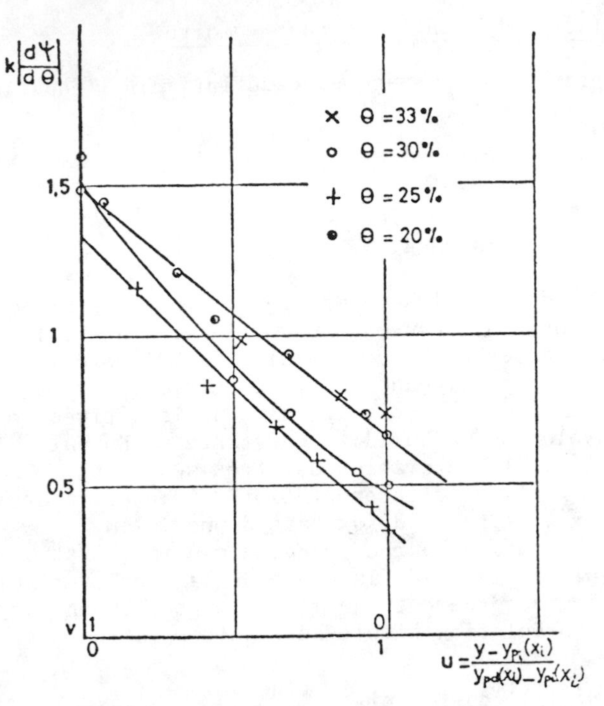

Fig. 5. Variation de dψ/dθ = f(u) à θ constante - Réseau de courbes déduites des résultats expérimentaux de la Fig. 4

dont la forme la mieux étudiée (et la plus simple) est:

$$\frac{dy'}{dx'} = \frac{ax' + by'}{cx' + dy'}$$ avec trois paramètres indépendants.

Cette forme homographique ne permet pas l'apparition de points d'inflexion. C'est une limitation qui disparaîtrait si on prenait des expressions $f(x',y')$ et $g(x',y')$ non linéaires plus riches.

La Fig. 7 donne une idée des graphes de passage que l'on peut ainsi obtenir.

La Simulation Explicite: L'Interpolation

Bien sûr, la solution d'un problème d'interpolation dépend étroitement de la classe des fonctions choisies pour constituer l'espace de projection de l'approximation. Mais l'identification de l'approximation se fait à l'aide d'informations spécifiques du problème étudié. Quelles sont ici ces informations? On sait qu'un graphe de passage part d'un point de rebroussement et atteint un autre point de rebroussement

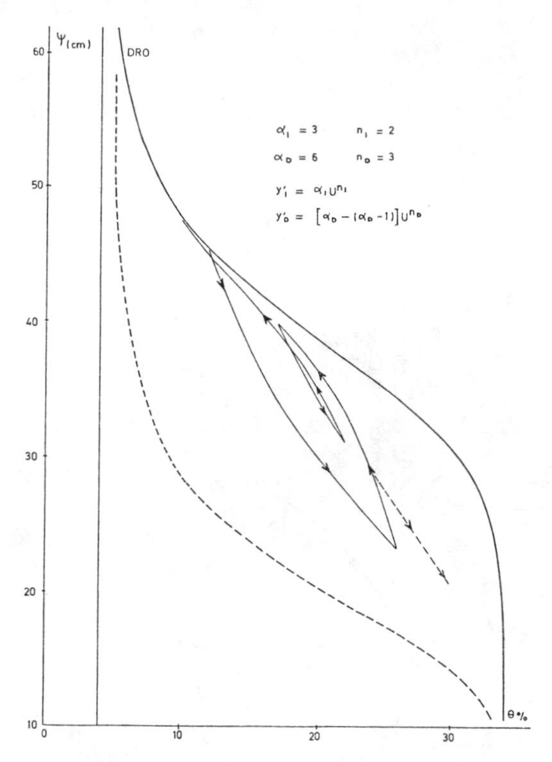

Fig. 6. Représentation des courbes de passage par description différentielle

état extrême immediatement antérieur. Et rien de plus a priori. En particulier, on ne dispose d'aucune information sur les points intermédiaires entre les deux points extrêmes de rebroussement. Tel quel, le problème posé d'interpolation n'a que des solutions équivoques. Nour allons presenter maintenant deux solutions essayées, procédant de deux idées directrices un peu différentes.

<u>L'interpolation polynomiale à arguments répétés</u>. Le polynôme est la fonction à tout faire, facile à manipuler mais limité dans son essence même: il ne peut faire apparaître de valeurs infinies ni pour la fonction, ni pour ses derivées pour une valeurs finie de l'argument. C'est une restriction gênante dans le cas de cycles d'hystérésis mais non redhibitoire. Pour des sols naturels argileux, comme l'indique la Fig. 8, les courbes principales de drainage et d'infiltration présentent une forme suffisamment douce pour être bien représentées par un polynôme de degré peu élevé (nous avons choisi le quatrième degré qui permet un ajustement prenant en compte le point d'inflexion avec la valeur de sa tangente).

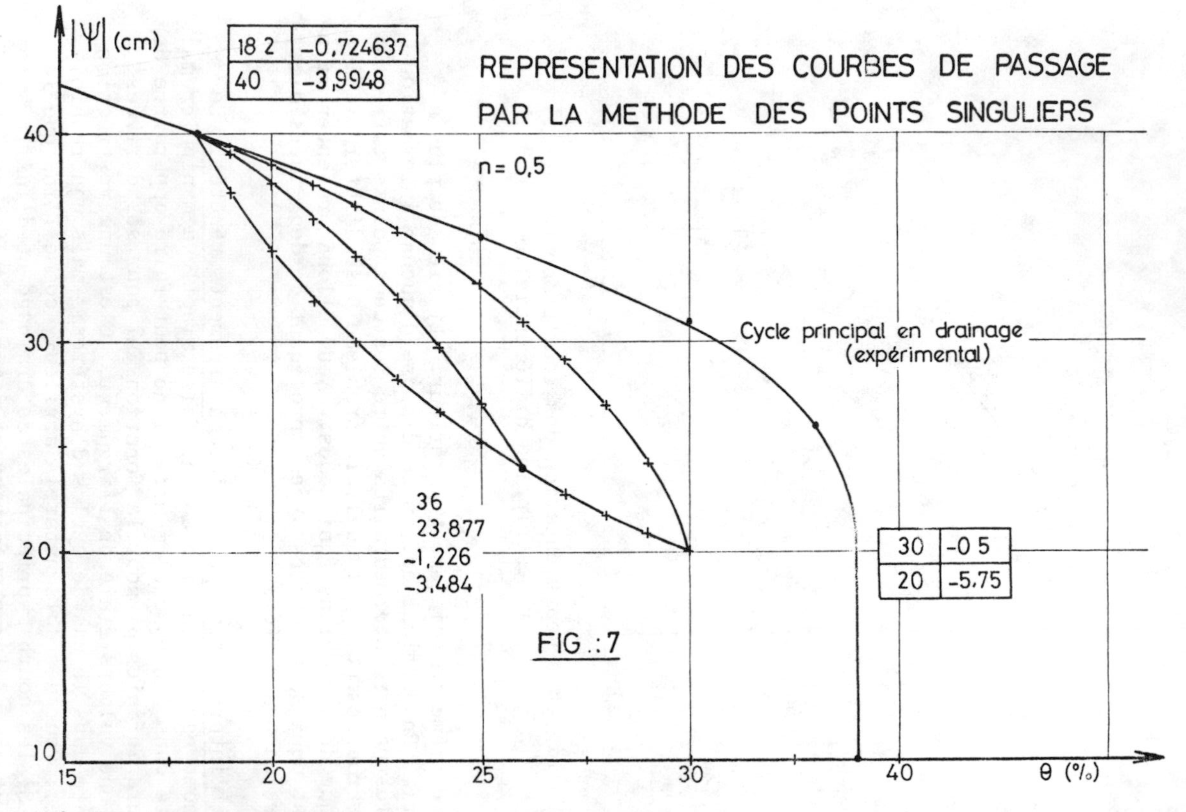

Fig. 7. Representation des courbes de passage par la methode des points singuliers.

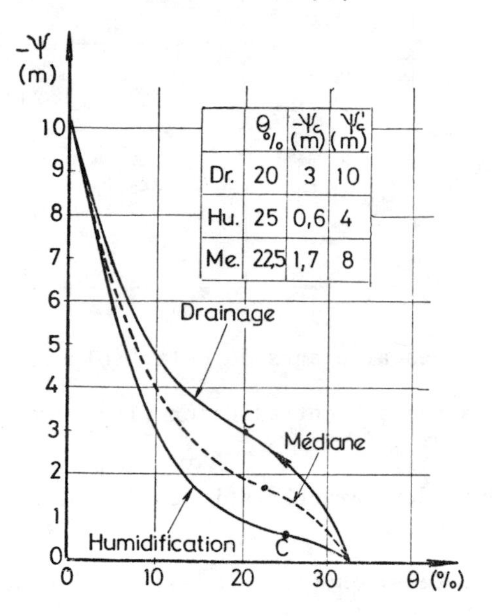

Fig. 8. Cycles $\psi(\theta)$ pour sol de Yaoundé

Pour le tracé des courbes intérieures, nous le considérerons encore comme un problème de tir. L'idée essentielle est d'imposer un raccordement plus qu'osculateur au point extrême d'arrivée A, c'est à dire égalité en ce point de l'ordonnée, de la pente, de la courbure et de la dérivée troisième entre la courbe qui avait permis d'atteindre pour la première fois le point extrême et les courbes qui aboutiront par la suite.

Pratiquement, du point de vue numérique, nous utiliserons un algorithme d'interpolation de Newton aux différences divisées, à arguments répétés trois fois au point d'arrivée A. (Pour chaque point extrême conservé, on met aussi en mémoire dérivées première et seconde).

Le graphe de passage a pour expression:

$$y(x) = y_A + y'_A(x - x_A) + \frac{y''_A}{2}(x - x_A)^2 + \frac{y'''}{6}(x-x_A)^3 + \delta_A^{(4)}(x-x_A)^4$$

La différence divisée d'ordre 4 étant donnée par le tableau

ω	ψ	$\delta(1)$	$\delta(2)$	$\delta(3)$	$\delta(4)$
ω_A	ψ_A				
		ψ'_A			
ω_A	ψ_A		$\psi''_A/2$		
		ψ'_A		$\psi'''_A/6$	
ω_A	ψ_A		$\psi''_A/2$		$\frac{\delta_{AAAB} - \psi'''_A/6}{\omega_B - \omega_A} = \delta$
		ψ'_A		$\frac{\delta_{AAB} - \psi''_A/2}{\omega_B - \omega_A} = \delta_{AAAB}$	
ω_A	ψ_A		$\frac{\delta_{AB} - \psi'_A}{\omega_B - \omega_A} = \delta_{AAB}$		
		$\frac{\psi_B - \psi_A}{\omega_B - \omega_A} = \delta_{AB}$			
ω_B	ψ_B				

Quels sont les avantages de cette solution?

- Il n'y a pas de contradiction flagrante avec les résultats expérimentaux

- La donnée des caractéristiques du sol se limite à celle des deux courbe enveloppes,

- Si les effets d'hystérésis sont négligeables, la continuité est assurée.

L'interpolation à base de fraction. Cette fois-ci, l'idée de base est de considérer une fonction dont le graphe ait l'allure des branches d'hystérésis. Nous avons porté notre choix sur la fonction:

$$x - x_o = \frac{C}{1 + (y - y^*)\beta} \qquad \text{si} \quad y > y^*$$

Contrairement aux polynômes, cette fonction présente une dérivée dy/dx infinie pour $y = y^*$ (si $\beta > 1$) et une asymptote $x = x_o$. L'interpolation non linéaire va consister à ajuster les cinq paramètres x_o, C, α, β et y^*.

Signification physique des paramètres. x_o représente une teneur en eau résiduelle fictive. Pour une bonne représentation des courbes principales, il faut évidemment $x_o = 0$. $C = x_s^*$ fait intervenir une teneur en eau maximale fictive x_s^* (égale à 1 en principe pour les courbes princpales). C est un coefficient d'affinité qui permet de passer d'un arc de la courbe principale à un arc de graphe de passage de même sens.

y^* correspond à une hauteur capillaire fictive.

Les parametres α et β sans signification physique apparente ont été déterminés par la méthode des moindres carres dans l'ajustement des courbes principales sur les résultats expérimentaux.

Détermination des graphes de transition. Comme précédement, on impose que le graphe de passage joigne le point de rebroussement D au point extrême A antérieur. Restent encore à fixer trois conditions. Nous avons admis que y^* avait la même valeur que pour la courbe principale d'infiltration. (x_o a été pris égal à 0).

Pour les graphes de drainage, C a été apprécié à partir de la valeur x_D point de rebroussement de départ à partir d'observations expérimentales.

$C = a\, x_D + 1 - a$ (avec a # 0,94 dans nos expériences numériques).

Pour ce qui concerne les graphes intermédiaires d'infiltration, une légère entorse a été faite au principe du passage par l'état extrême antérieur, car nous avions remarqué sur les résultats expérimentaux que tous les graphes d'infiltration passent au voisinage immediat d'un point M_{RO}.

Sur la Fig. 9, sont représentés les résultats de lissage correspondant à des résultats expérimentaux obtenus au laboratoire. L'accord semble très convenable.

Exemples D'Application

Etude de l'évaporation sur sol sec en zone aride. Après un épisode pluvieux, s'installe une longue période sèche. L'eau qui s'infiltrait, est reprise par l'evaporation superficielle soumise au cycle thermique diurne qui induit à proximité de la surface des changements quotidiens de sens d'écoulement d'où l'intervention impérative des courbes intérieures d'hystérésis.

La simulation a été faite à l'aide de l'interpolation polynomiale. Par rapport à une évolution moyenne sans hystérésis, la Fig. 10 montre l'influence decisive de l'histoire par l'intermédiaire de l'hystérésis.

Echange entre réservoir et flanc amont d'un barrage en terre. Il s'agit d'un problème bidimensionnel d'infiltration dans une digue sous l'effet d'une variation du niveau de l'eau dans le reservoir amont. L'hystérésis ici a été simulée à l'aide de fraction. Sur la Fig. 11, on note à la fois l'excellent accord entre expérience et théorie tenant compte de l'hystérésis et le décalage très important d'avec les résultats numériques sans hystérésis.

Conclusion

Le nombre de nos tentatives est implicitement l'aveu qu'aucune des procédures essayées n'est complètement

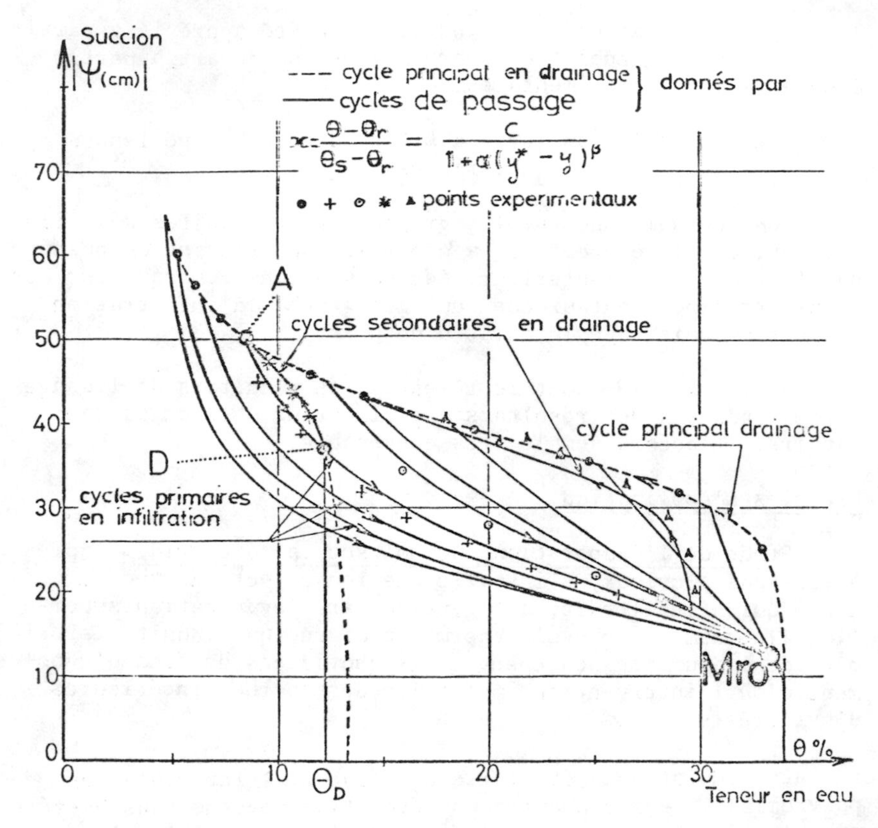

Fig. 9. Représentation des cycles par fraction rationnelle - Comparaison avec résultats expérimentaux obtenus sur du sable de Fontainebleau

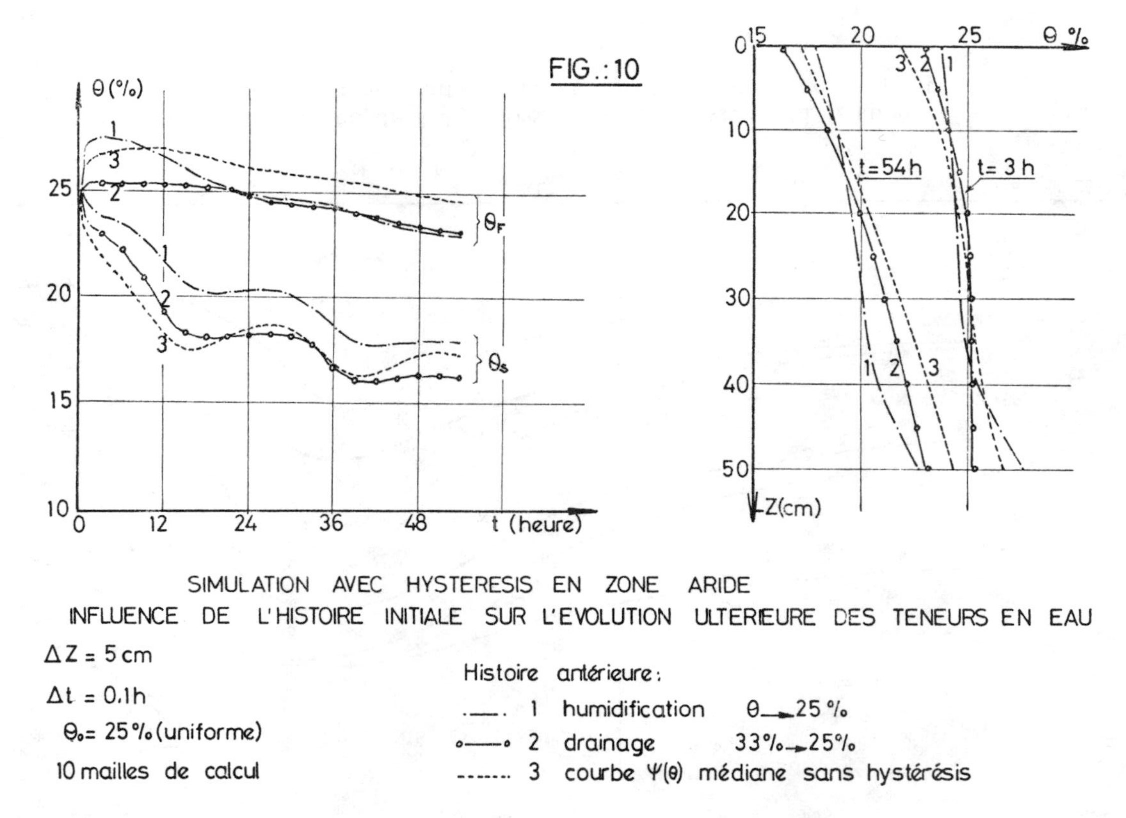

Fig. 10. Simulation numérique avec hystérésis de l'évaporation en zone aride

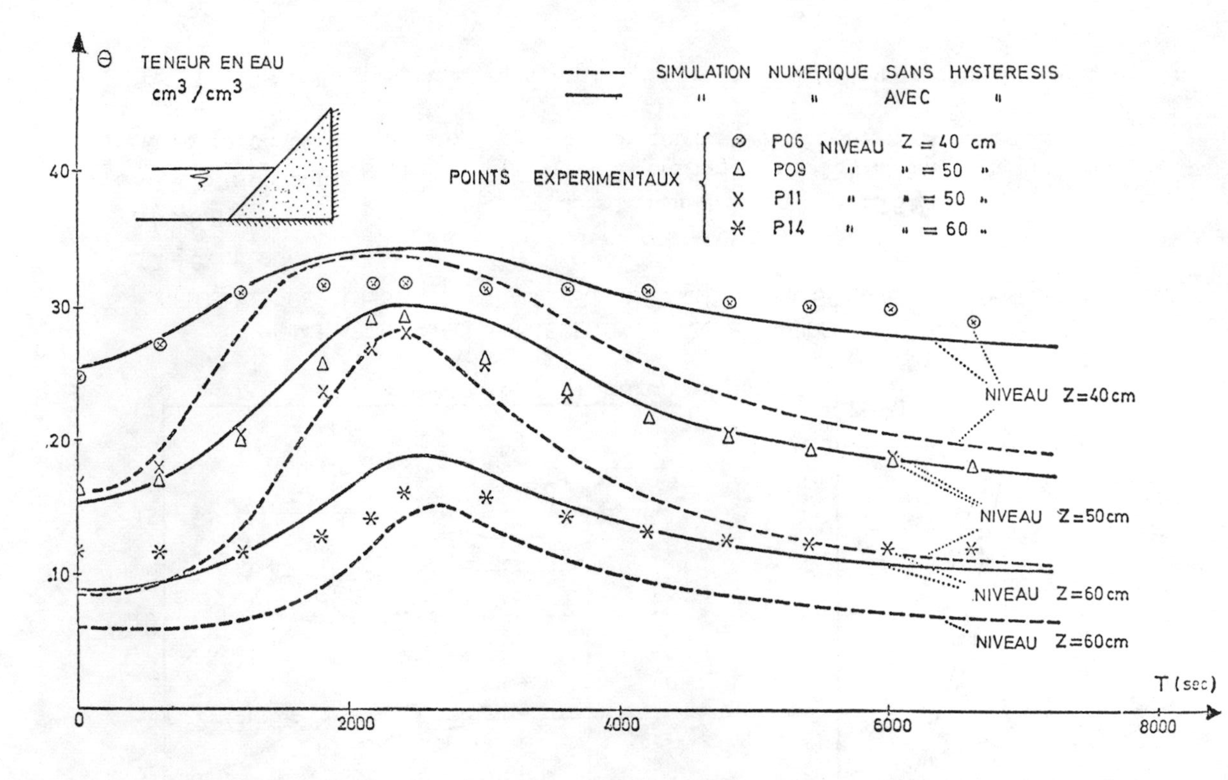

Fig. 11. Influence de l'hystérésis dans la simulation numérique des échanges entre une digue et son réservoir amont

satisfaisante pour la simulation de l'hystérésis. Mais la comparaison des calculs et des observations expérimentales est tout de même franchement encourageante: nous estimons pouvoir rendre compte de l'influence de l'hystérésis dans la reconstitution d'évènements naturels ou la prévision des phénomènes spontanés ou stimulés. Et ceci sans prétention outrancière à l'exactitude mais avec optimisme mesuré et toujours critique.

A l'examen attentif des applications, nous nous rendons compte que ce qui importe c'est la pondération globale: de petits aléas qui se compensent sur la forme des arcs d'hystérésis n'auront pas de répercussion très sensible sur les phénomènes en milieu poreux fortement tempérés par leur caractère diffusif. D'ailleurs, l'hystérésis accentue ce caractère car, comme son homologue mécanique le frottement solide, elle marque la réticence d'un système au changement même sous l'effet d'une forte variation de contrainte provoquée par le milieu extérieur.

Si nous revenons à la philosophie de la simulation, nous pourrons dire que peu ou prou, tous les procédés sont fondés sur le comportement sur les courbes principales enveloppes. D'une manière générale, ces courbes enveloppes jouent explicitement (interpolation) pour les courbes intérieures qu'elles canalisent. Dans le cas de l'interpolation polynomiale, ces courbes enveloppes contiennent toute l'information nécessaire à la figuration de l'hystérésis. Ceci est dû à la très grande force d'attraction et de direction donnée aux états extrêmes aboutissements des courbes intérieures; celles-ci ne disposent que du seul degré de liberté, la position du point de départ qui pèsera d'autant moins sur la destinée de la courbe que plus de conditions seront imposées a l'arrivée.

Bibliographie

[1] Haines, W. B., 1930. "Study in the Physical Properties of Soils. The Hysteresis Effect in Capillary Properties and the Mode of Moisture Distribution Associated Therewith," J. Agri. Sci. 20, 97-116.

[2] Hillel, D., 1974. "L'eau et le sol, principes et processus physiques," Vander Editeur.

[3] Duhem, 1896. "Mém. I, II, III, 54. Mém. IV, V, 56," Mémoires Acad. Belg.

[4] Brillouin, 1888. "Comptes-rendus," 106, 416, 482, 537, 589, Journal Physique, 2e Série, 1889, 8, 169.

[5] Everett, D. H., 1952. "A General Approach to Hysteresis," Trans. Faraday Soc. Part I, 1952, 48, 749; Part II, 1954, 50, 187; Part III, 1954, 50, 1077; Part IV, 1955, 51, 1551.

[6] Poulovassilis, A., 1962. "Hysteresis of Pore Water, an Application of the Concept of Independant Domains," Soil Science T. 93, pp. 405-412.

[7] Vachaud, G. et équipe, 1968. "Contributions à l'Étude des Problèmes D'Écoulement en Milieux Poreus non Saturés," Thèse, Grenoble.

[8] Vachaud, G. et Thony, J. L., 1971. "Hysteresis During Infiltration and Redistribution in a Soil Column at Different Initial Water Contents," Water Resources, 7.

[9] Vauclin, M., 1971. "Effets Dynamiques sur la Relation Succion-Teneur en eau lors D'Écoulements en Milieu non Saturé," Thèse, Grenoble.

[10] Ibrahim, H. A. et Brutsaert, W., 1968. "Intermittent Infiltration into Soils with Hysteresis," Journal of Hydraulics Division, Proc. of ASCE, Jan., pp. 113-137.

[11] Colonna, J. Brissaud, F. et Millet, J. L., 1971. "Evolution of Capillarity and Relative Permeability Hysteresis," Society of Petroleum Engineers, Paper SPE 2941, Houston, Oct. 1971, 45e Congrès Annuel.

[12] Colonna, J. et Millet, J. L., 1971. "Effets de Déplacements Diphasiques Alternés sur les Propriétes Hydrodynamiques des Roches," Association de Recherche sur les Techniques d'Exploitation du Pétrole (ARTEP), 4e Colloque-Rueil Malmaison, Juin 1971.

[13] Thore, Ph., 1976. "Elaboration et Critique de Modeles Mathématiques et Numériques de Tarissement par Évaporation - Application aux sols des Zones Arides," Thèse, Toulouse.

THEORETICAL HYDROLOGY OF THE SOIL-ROOT SYSTEM WITH PRACTICAL IMPLICATIONS FOR WATER EXTRACTION

By

Fred J. Molz
Alumni Associate Professor
Civil Engineering Department
Auburn University
Auburn, Alabama

Abstract

An interface of major significance in the earth's hydrologic system exists at the boundary between soil and the root systems of vegetation. Well over 35 percent of the precipitation that falls on land surfaces is absorbed by plant roots and returned to the atmosphere as transpiration. At the soil-root interface, water transport changes from a normally unsaturated condition to a saturated condition. In the root tissue, there are two parallel pathways available for flow. Water can pass from cell to cell through both the symplasm and cell vacuoles along gradients of water potential, and it can move through the pore space of cell walls along gradients of hydrostatic pressure or matric potential. Exchange of water between these two pathways can occur through the cell membranes.

Over the past eighteen years, several differential equations have been developed in the attempt to describe water transport in plants. These equations are all diffusion equations and differ mainly in the theoretical expressions for the diffusivity. Plant tissue parameters which enter the expression for the diffusivity include the hydraulic conductivity of the cell wall material, storage coefficient for the cell walls, cellular volume, elastic modulus of the cell wall, and permeability of the cell membranes.

The theory of Molz and Ikenberry is used to predict water potential in a cylindrical soil-root system as a function of time and radial distance. Results indicate that there will be small water potential gradients in the soil relative to those in the root, except under relatively dry conditions after approximately 90 percent of the plant-available water has been extracted. A pseudo-steady state water potential distribution in the root cortex is predicted also. These results are used to derive a sink term which can be added to the continuity equation to represent water extraction by entire plant root systems. Parameters entering this extraction function are length of root per unit soil volume, the water potential difference between the root xylem and the root surface, the degree of saturation of the soil at the root surface, and the hydraulic resistance of the root cortex and endodermis.

As far as engineering hydrologists are concerned, the most important problems at the soil-root interface are those that relate to the prediction or simulation of transpiration in some useful way. In the near future, this may involve the development and application of simulation models of the sub-surface hydrologic system utilizing root water extraction functions similar to those discussed in this paper. However, additional experimental studies of water transport in the soil-plant system are needed in order to select the best extraction function.

Introduction

An interface of major significance in the earth's hydrologic system exists at the boundary between soil and the root systems of vegetation. The geometry of the boundary itself is exceedingly intricate, time dependent, and essentially impossible to describe in detail using presently available techniques. However, its complexity is matched by its importance. On a global scale, something in the vicinity of 65 percent of the annual land precipitation is lost to the atmosphere through evapotranspiration, and well over 50 percent of this volume crosses the soil-root interface. It has been estimated that to produce 20 fresh-weight tons of a crop, approximately 2000 tons of water must be drawn from the soil (Penman, 1970). Clearly, one of the long-term goals of both engineering and agricultural hydrologists must be to develop an improved understanding of the hydrologic processes involved in the transport of water from soil to vegetation.

The Water Pathways

At the soil-root interface, water transport abruptly changes from a normally unsaturated to a saturated condition, possibly tension saturated. It is now generally accepted that there are several pathways which water can follow while traversing plant root tissue (Fig. 1). In the early part of this century, pathway A was favored (the vacuolar pathway). This viewpoint held that water movement occurred from vacuole to vacuole, traversing the cytoplasm and membranes of each cell (Newman, 1974). At a later date (Scott and Priestley, 1928) pathway B was suggested. In this one (the cell wall pathway) water moves in the porous cell walls, bypassing the cell membranes. The discovery of plasmodesmata led to still another possible pathway, the symplasm pathway which is labeled as pathway C. Thus, the major water transport routes in Fig. 1 are labeled with the capitol letters A, B, and C.

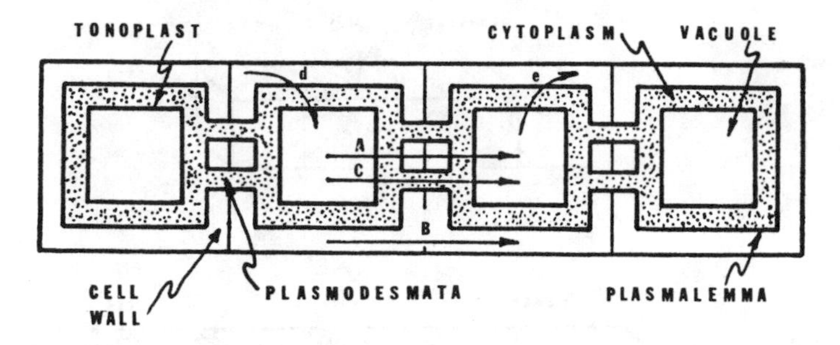

Fig. 1. Idealized diagram of a linear aggregation of plant cells. Each cell consists of a central vacuole surrounded by cytoplasm and a rigid cell wall. The cytoplasm is bounded on the outside and inside by semipermeable membranes and connected to the cytoplasm of contiguous cells through pores calles plasmodesmata. Various water pathways are indicated by letters and arrows. For purposes of illustration, the thickness of the cell walls and the diameter of the plasmodesmata are exaggerated.

It is also possible for water to be exchanged between the vacuole and cell walls, and these sub-pathways are labeled with the small letters d and e.

Quantitative Theories of Water Transport

To the writer's knowledge, Philip (1958a, 1958b, 1958c) was the first investigator to present a quantitative theory of water transport through plant tissue. Philip's development did not consider the effects of diffusible solutes and was based on the assumption that a negligible amount of water flowed through the cell wall pathway (Fig. 2A). Nevertheless, it was applied in a useful manner to certain aspects of water transport in the soil-plant-atmosphere system (Philip, 1966) and to the radial flow of water in cotton stems (Molz and Kelpper, 1972; Molz et al., 1973) and leaf disks (Molz et al., 1973). Using a nonequilibrium thermodynamic approach (Kedem and Katchalsky, 1958; Dainty, 1963), Molz and Hornberger (1973) extended Philip's theory to include the effects of a diffusible solute.

All of the theories discussed so far are based on the assumption that the water flux is primarily from cell to cell, traversing the cell membranes, with little independent flux through the cell walls. Because a considerable body of experimental evidence indicates that this is not the case, Molz and Ikenberry (1974) developed a theory which allowed for a water flux in the cell wall pathway as well as the classical vacuolar pathway (Fig. 2B). However, no consideration was given

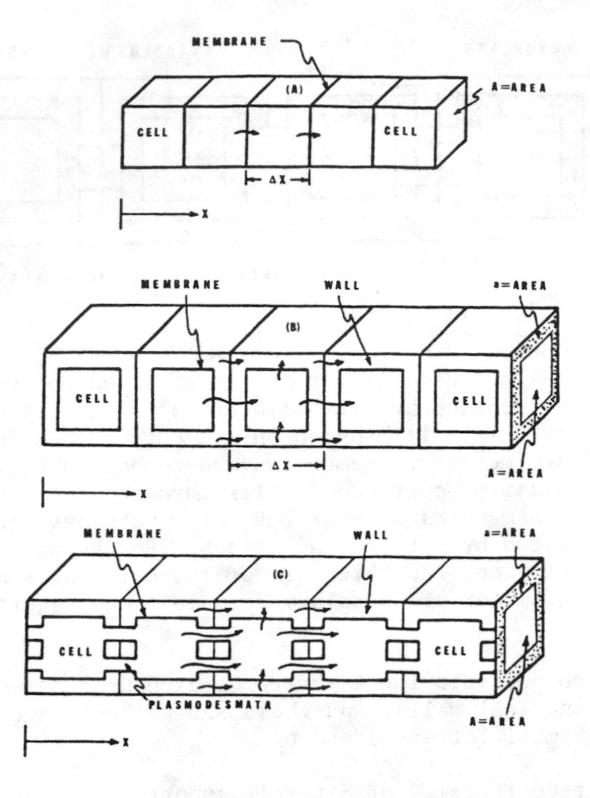

Fig. 2. Three different models of plant tissue that have been used to develop quantitative theories of water transport. Model A was used by Philip (1958a), model B by Molz and Ikenberry (1974) and model C by Molz (1967a). In model C, the apoplasm is exterior to the membrane, and the symplasm is interior. Model A does not allow for cell wall water transport while models B and C do. The symbol X denotes tissue thickness, and the wavy arrows indicate water flow.

explicitly to the symplasm, which could constitute an important pathway, especially in roots (Newman, 1974).

In a quantitative sense, the concept of the vacuolar pathway as distinct from the cell wall pathway is somewhat artificial. This is because water must enter the cell walls in order to get from vacuole to vacuole. If one allowed the resistance of the cell wall pathway to approach infinity, water transport would be halted also in the vacuolar pathway. Thus, the two pathways are not truly distinct. The problem resolves itself if one introduces the concept of the symplasm.

Two fairly distinct water pathways then emerge, the symplasm pathway which is interior to the plasmalemma and the apoplasm pathway which is exterior (Fig. 2C). (For practical purposes, it is assumed that the apoplasm pathway and cell-wall pathway are essentially identical.) This viewpoint is the basis of a third quantitative treatment of water transport in plant tissue (Molz, 1976a). However, the hydraulic properties of the symplasm are relatively unknown, and this limits application of the theory at the present time.

The differential equations which constitute the various theories that have been developed over the past eighteen years to describe water transport through plant tissue in the absence of solute effects are quite similar in that they are all diffusion equations. In the one-dimensional Cartesian case, they can be written in the form

$$\frac{\partial \phi}{\partial t} = D \frac{\partial^2 \phi}{\partial x^2} \tag{1}$$

where ϕ = tissue water potential (bar), t = time (sec), D = tissue diffusivity ($cm^2\ sec^{-1}$), and x = distance (cm). The several equations do differ, however, in the expressions that are obtained for the diffusivity. (Details can be obtained from the appropriate references.) Based on the idealized tissue model shown in Fig. 2A, Philip (1958b) obtained the expression

$$D = \frac{AK(\Delta x)^2(\varepsilon + \pi_o)}{2V_o} \tag{2}$$

where A = cross-sectional area of a cell (cm^2), K = permeability of "membrane" separating cells ($cm\ sec^{-1}\ bar^{-1}$), Δx = length of cell (cm), ε = elastic modulus of cell wall (bar), π_o = osmotic pressure of cell contents at zero turgor pressure (bar), and V_o = cell volume at zero turgor pressure. Obviously, Eq. (2) contains only parameters relating to the so-called vacuolar pathway.

Using the concept of a cell wall pathway in parallel with the vacuolar pathway (Fig. 2B) along with the assumption of negligible water potential differences between the interior of a cell and its outer wall, Molz and Ikenberry (1974) obtained a diffusivity given by

$$D = \frac{\Delta x(Pa + K\Delta xA/2)}{1.5a\Delta xS + V_o/(\varepsilon + \pi_o)} \tag{3}$$

where P = hydraulic conductivity of the cell wall material ($cm^2\ sec^{-1}\ bar^{-1}$), a = cross-sectional area of the cell wall pathway (cm^2), A = cross-sectional area of the vacuolar pathway, and S = storage coefficient of the cell wall material (bar^{-1}). If $a = 0$ (i.e., no cell wall pathway), then Eq. (3) reduces to Eq. (2). Still another expression for tissue diffusivity can be obtained based on the tissue model shown in Fig. 2C. Starting from this viewpoint, Molz (1976a) obtained

$$D = \frac{\Delta x(pa + K_p A \Delta x)}{1.5a\Delta xS + V_o/(\varepsilon + \pi_o)} \qquad (4)$$

where K_p = permeability of plasmodesma membrane (cm sec^{-1} bar^{-1}). Although Eqs. (3) and (4) are superficially quite similar, the possibility of a plasmodesma membrane of high permeability allowed in Eq. (4) could lead to considerable differences in an estimate for D (Tyree, 1969, 1970).

Application of Theory

The theories discussed briefly above have been applied to various transport processes in plant stems and leaves. However, in this paper the writer will be concerned mainly with applications to water transport in the soil-root system. The system of interest, shown in Fig. 3, consists of a cylindrical root of radius r_r surrounded by a cylinder of homogeneous soil of radius r_s. A constant flux boundary condition will be applied at the endodermis of the root located at

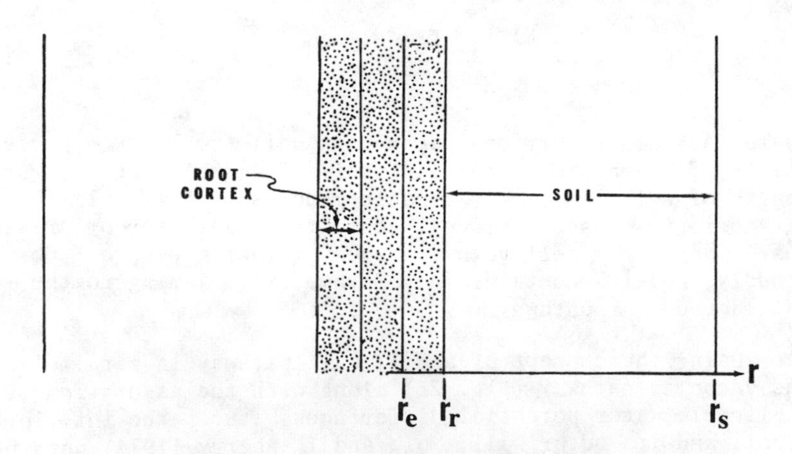

Fig. 3. Diagram of a root-soil system with cylindrical symmetry. r_e is the radius of the endodermis, r_r is the root radius, and r_s is the radius of the surrounding soil cylinder. Thus r_e r r_r defines the tissue domain where Eq. (5) applies, and r_r r r_s defines the soil domain where Eq. (6) applies.

$r = r_e$, and a zero flux condition at $r = r_s$. These conditions are similar to those utilized in previous studies (Gardner, 1960, 1965; Cowan, 1965).

The equation governing water transport in the root tissue domain is given by the radial form of Eq. (1), i.e.,

$$\frac{\partial \phi}{\partial t} = D \frac{\partial^2 \phi}{\partial r^2} + \frac{D}{r} \frac{\partial \phi}{\partial r} \tag{5}$$

where r = radial distance (cm). In the soil domain, water flow should be approximately radial in the vicinity of a root, and the radial form of the Darcy-Richards equation is given by

$$C_s \frac{\partial \tau}{\partial t} = K_s \frac{\partial^2 \tau}{\partial r^2} + \frac{K_s}{r} \frac{\partial \tau}{\partial r} + \frac{dK_s}{d\tau} \left(\frac{\partial \tau}{\partial r}\right)^2 \tag{6}$$

where τ = soil water potential (bar), C_s = specific soil water capacity (bar^{-1}), and $K_s = K_s(\tau)$ = hydraulic conductivity ($cm^2\ sec^{-1}\ bar^{-1}$).

Equations (5) and (6) were solved simultaneously using a numerical technique (Molz, 1976b) with values or functions for D (Eq. 3), C_s, and K_s obtained from the literature. For a relatively high flow rate into the root (Newman, 1974) and a variety of soil types, water potential distributions similar to those shown in Fig. 4 were obtained. The figure indicates that it takes about a half hour for the root cortex to respond fully to the boundary condition at the endodermis. Thereafter, the potential profiles in the root are essentially steady-state distributions, i.e., the curve connecting the potentials at r_e and r_r at any time can be obtained from the steady-state solution for flow in a cylinder with constant potential boundary conditions. Until the water potential at the root surface reaches approximately -5.5 bars, there is little water potential gradient in the soil relative to that in the root. Because many soils will have over 90 percent of the plant-available moisture removed at a potential of -5.5 bars, the calculation represented in Fig. 4 predicts small water potential gradients in the soil relative to those in the root except under relatively dry conditions with only 10 percent or so of the available water not extracted.

Some Practical Implications

A detailed quantitative analysis of water flow in the vicinity of a single, idealized root may provide some interesting results without too much computational effort, but such an approach applied to an entire root system with hundreds or

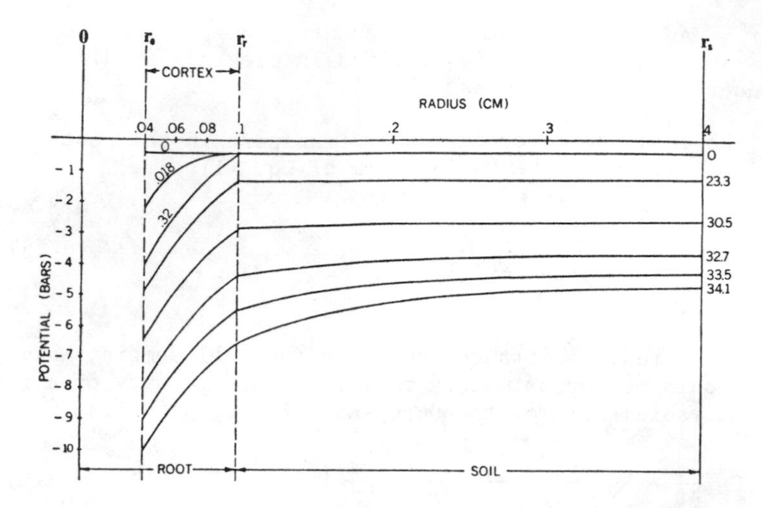

Fig. 4. Root and soil water potentials as a function of radius and time for a root extracting water at a constant rate. The initial water potential in the system was -0.4 bars, and the numbers next to the curves represent the time in hours. In solving Eqs. (5) and (6) to obtain the figure, a flux of 10^{-6} cm sec^{-1} was specified at r_e, continuity of flow and potential was maintained at r_r and a zero flux was specified at r_s.

thousands of individual sub-roots would be overwhelmingly complex. To get around this problem, investigators have turned increasingly to the use of extraction functions to describe water uptake by plant root systems. In this integrated approach, plant water uptake is represented by a volumetric sink term in the continuity equation. For the one-dimensional vertical case, this may be written as

$$\frac{\partial\theta}{\partial t} = - \frac{\partial v}{\partial z} - Q(z,t) \qquad (7)$$

where θ is volumetric water content (cm^3 cm^{-3}), v is the vertical water flux in the soil (cm sec^{-1}), z is a vertical coordinate (cm) and $Q(z,t)$ is water uptake by plant roots (cm^3 cm^{-3} sec^{-1}). The sink term often contains a parameter called "root density," but individual root geometry is not involved (Gardner, 1964; Molz and Remson, 1970, 1971; Nimah and Hanks, 1973a, 1973b; Feddes et al., 1974).

The results shown in Fig. 4 lend a measure of support to such approaches. Because the water potential distribution in the root cortex corresponds to a steady state distribution at a given time, one can write (Crank, 1956, p. 62)

$$q = 2\pi K_t \ [\tau(t) - \phi(t)] \ \ln(r_r/r_e) \tag{8}$$

where q = water uptake per unit length of root ($cm^3 \ cm^{-1} \ sec^{-1}$) and K_t is the bulk tissue hydraulic conductivity. (In terms of the wall and membrane parameters, K_t is given by $(Pa + K\Delta xa/2)/(a+A)$ with units $cm^2 \ sec^{-1} \ bar^{-1}$ (Molz, 1976b).) In a complete root, the endodermis separates the cortex from the xylem and is thought to represent a significant barrier to water flow. If the endodermis is viewed as a "skin" of resistance at the inside boundary of the cortex, then one can write

$$q = \Delta\phi/(R_c + R_e) \tag{9}$$

where $\Delta\phi$ = the water potential difference between the root surface and the root xylem, $R_c = [2\pi K_t \ln(r_r/r_e)]^{-1}$ = resistance of the cortex (bar sec cm^{-2}) and R_e = resistance of the endodermis (bar sec cm^{-2}). Given that $L(z)$ is the "effective" length of root per unit volume of soil, the water uptake rate per unit soil volume becomes

$$Q(z,t) = L(z)\Delta\phi/(R_c + R_e) \tag{10}$$

For a soil containing roots of different radii, the resistance terms in Eq. (10) would have to be interpreted as some type of average. Also, under drying conditions, the portion of $L(z)$ subjected to a water flux would be expected to decrease with decreasing water content. This is because water contained in the soil pores might not be uniformly available to all areas of root surface. According to Herkelrath (1975), this effect can be taken into account by modifying Eq. (10) to read

$$Q(z,t) = L(z)\Delta\phi(\theta/\theta_s)/(R_c + R_e) \tag{11}$$

where θ is volumetric moisture content and θ_s is the saturation moisture content.

To the writer's knowledge, an extraction function such as Eq. (11) has not been utilized in a data-calibrated simulation model of the subsurface hydrologic system although Hillel et al. (1975) used a related model in a sensitivity study of water and salt movement to plant roots. In an algebraic sense, similar expressions have been studied (Nimah and Hanks, 1973a, 1973b; Feddes et al., 1974), but soil hydraulic conductivity was used instead of tissue hydraulic conductivity. This was due to the fact that for many years it had been tacitly assumed by many (including the present writer) that soil hydraulic properties rather than plant hydraulic properties dominated the water uptake process. Such an assumption is inconsistent with the results shown in Fig. 4, especially in the higher water content range. Recent experimental work by Herkelrath (1975) also supports this viewpoint. It would thus appear that extraction functions similar to Eq. (11) deserve further study, although care must be taken not to oversimplify the extraction phenomenon (Hillel et al., 1975).

Future Research Needs

At the present time, the most pressing research needs appear to be experimental in nature. More information is needed concerning the magnitude of the resistances in the various root water pathways, and careful measurements of water extraction by plant roots and its relationship to water potential, water content, and root geometry are required. Some research efforts in these areas have already been performed (Herkelrath, 1975; Taylor and Klepper, 1975; Molz and Peterson, 1976), but more are needed. Only in this way can the existing theories be tested and realistic values for tissue diffusivity and other tissue parameters obtained (Molz, 1975).

In the area of theoretical advances, the next step will probably involve the application of nonequilibrium thermodynamic concepts to the simultaneous transport of water, solutes, and other materials in the soil-root system. It may also be necessary to consider the inclusion of additional potentials such as the electrical potential (Katchalsky and Curran, 1965). Due to the dominance of water flow, such approaches may not be necessary in problems dealing only with the transpiration stream. However, this is not certain, and single-root, steady-state analyses have already been developed which apparently explain some long-standing conductivity anomalies observed at relatively low rates of water uptake (Dalton et al., 1975; Fiscus, 1975).

As far as engineering hydrologists are concerned, the most important problems at the soil-root interface are those that relate to prediction or simulation of evapotranspiration in some useful way. In the immediate future, the most useful approaches in a practical sense will probably involve the development and application of simulation models of the subsurface hydrologic system utilizing root water extraction functions of some type, possibly related to Eq. (11). Nimah and Hanks (1973a, 1973b) and Feddes et al. (1974) have taken important steps in this direction, and further efforts appear warranted. Perhaps it may be fruitful to combine a subsurface simulator of the Nimah and Hanks (or related) types with one of the several surface hydrologic simulators such as the Stanford Watershed Model (Crawford and Linsley, 1966).

References

[1] Cowan, I. R., 1965. "Transport of Water in the Soil-Plant-Atmosphere System," Journal of Applied Ecology, 2: 221-239.

[2] Crank, J., 1956. "The Mathematics of Diffusion," Oxford University Press, London, 345 p.

[3] Crawford, N. H. and Linsley, R. K., 1966. "Digital Simulation in Hydrology: Stanford Watershed Model IV," Stanford University Department of Civil Engineering Technical Report 39.

[4] Dainty, J., 1963. "Water Relations of Plant Cells," Advances in Botanical Research, 1: 279-326.

[5] Dalton, F. N., Raats, P.A.C. and Gardner, W. R., 1975. "Simultaneous Uptake of Water and Solutes by Plant Roots," Agronomy Journal, 67: 334-339.

[6] Feddes, R. A., Bresler, E. and Neuman, S. P., 1974. "Field Test of a Modified Numerical Model for Water Uptake by Root Systems," Water Resources Research, 10: 1199-1206.

[7] Fiscus, E. L., 1975. "The Interaction between Osmotic and Pressure-Induced Water Flow in Plant Roots," Plant Physiology, 55: 917-922.

[8] Gardner, W. R., 1960. "Dynamic Aspects of Water Availability to Plants," Soil Science, 89: 63-73.

[9] Gardner, W. R., 1964. "Relation of Root Distribution to Water Uptake and Availability," Agronomy Journal, 56: 35-41.

[10] Gardner, W. R., 1965. "Dynamic Aspects of Soil-Water Availability to Plants," Annual Reviews of Plant Physiology, 16: 323-342.

[11] Herkelrath, W. N., 1975. "Water Uptake by Plant Roots," Doctoral Dissertation, University of Wisconsin, Madison.

[12] Hillel, D., van Beek, C.G.E.M. and Talpaz, H., 1975. "A Microscopic-Scale Model of Soil Water Uptake and Salt Movement to Plant Roots," Soil Science, 120: 385-399.

[13] Katchalsky, A. and Curran, P. F., 1965. "Nonequilibrium Thermodynamics in Biophysics," Harvard University Press, Cambridge, 215 p.

[14] Kedem, O. and Katchalsky, A., 1958. "Thermodynamic Analysis of the Permeability of Biological Membranes to Nonelectrolytes," Biochemica et Biophysica Acta, 27: 229-246.

[15] Molz, F. J., 1975. Comments on "Water Transport through Plant Cells and Cell Walls: Theoretical Development," Soil Sci. Soc. of Am. Proc., 39: 597.

[16] Molz, F. J., 1976a. "Water Transport through Plant Tissue: The Apoplasm and Symplasm Pathways," Journal of Theoretical Biology, 59: 277-292.

[17] Molz, F. J., 1976b. "Water Transport in the Soil-Root System: Transient Analysis," Water Resources Research, 12: 805-808.

[18] Molz, F. J. and Remson, I., 1970. "Extraction Term Models of Soil Moisture Use by Transpiring Plants," Water Resources Research, 6: 1346-1356.

[19] Molz, F. J. and Remson, I., 1971. "Application of an Extraction Term Model to the Study of Moisture Flow to Plant Roots," Agronomy Journal, 63: 72-77.

[20] Molz, F. J. and Klepper, B., 1972. "Radial of Water Potential in Stems," Agronomy Journal, pp. 469-473.

[21] Molz, F. J. and Hornberger, G. M., 1973. "Water Transport through Plant Tissue in the Presence of a Diffusable Solute," Soil Science Society of America Proceedings, 37: 833-837.

[22] Molz, F. J. and Ikenberry, E., 1974. "Water Transport through Plant Cells and Cell Walls: Theoretical Development," Soil Science Society of America Proceedings, 38: 699-704.

[23] Molz, F. J. and Peterson, C. M., 1976. "Water Transport from Roots to Soil," Agronomy Journal, 68: 901-904.

[24] Molz, F. J., Klepper, B. and Browning, V. D., 1973. "Radial Diffusion of Free Energy in Stem Pholem: An Experimental Study," Agronomy Journal, 65: 219-222.

[25] Molz, F. J., Klepper, B. and Peterson, C. M., 1973. "Rehydration versus Growth-Induced Water Uptake in Plant Tissues," Plant Physiology, 51: 859-862.

[26] Newman, E. I., 1974. "Root-Soil Water Relations," In: E. W. Carson (ed.), The Plant Root and Its Environment, The University Press of Virginia, pp. 363-440.

[27] Nimah, M. N. and Hanks, R. J., 1973a. "Model for Estimating Soil Water, Plant and Atmospheric Interrelations, 1, Description and Sensitivity," Soil Sci. Soc. Am. Proc., 37: 522-527.

[28] Nimah, M. N. and Hanks, R. J., 1973b. "Model for Estimating Soil Water, Plant and Atmospheric Interrelations, 2, Field Test of Model," Soil Sci. Soc. Am. Proc., 37: 528-532.

[29] Penman, H. L., 1970. "The Water Cycle," Scientific American: 222, 99-108.

[30] Philip, J. R., 1958a. "Propagation of Turgor and Other Properties through Cell Aggregations," Plant Physiology, 33: 271-274.

[31] Philip, J. R., 1958b. "Osmosis and Diffusion in Tissue: Half Times and Internal Gradients," Plant Physiology, 33: 275-278.

[32] Philip, J. R., 1958c. Correction to the paper entitled "Osmosis and Diffusion in Tissue: Half Times and Internal Gradients," Plant Physiology, 33: 443.

[33] Philip, J. R., 1966. "Plant Water Relations: Some Physical Aspects," Annual Reviews of Plant Physiology, 17: 245-268.

[34] Scott, L. I. and Priestley, J. H., 1928. "The Root as an Absorbing Organ. I. A Reconsideration of the Entry of Water and Salts in the Absorbing Region," New Phytologist, 27: 125-140.

[35] Taylor, H. M. and Klepper, B., 1975. "Water Uptake by Cotton Root Systems: An Examination of Assumptions in the Single Root Model," Soil Science, 120: 57-67.

[36] Tyree, M. T., 1969. "The Thermodynamics of Short-Distance Translocation in Plants," Journal of Experimental Botany, 20: 341-349.

[37] Tyree, M. T., 1970. "The Symplast Concept: A General Theory of Symplastic Transport According to the Thermodynamics of Irreversible Processes," Journal of Theoretical Biology, 26: 181-214.

AIR AND WATER FLOW IN A HORIZONTAL COLUMN - INFLUENCE OF THE AIR BOUNDARY CONDITION

By

J.-Y. Parlange and D. E. Hill
Department of Ecology and Climatology
The Connecticut Agricultural Experiment Station
Box 1106, New Haven, Connecticut 06504

Abstract

Horizontal infiltration of water into a column is considered. The water potential at the soil surface is imposed. An analytical solution predicts that the rate of infiltration, when the air can escape ahead of the wetting front, is larger than the rate of infiltration, when the air must escape through the soil surface. The theoretical results are illustrated by comparison with experimental observations, and the influence of air movement is measured directly.

Introduction

It has long been recognized that if there is restriction to air flow during wetting, infiltration is slowed down and air is compressed (Powers, 1934; Baver, 1937; Lewis and Powers, 1939; Free and Palmer, 1940). The effect of compressed air has been discussed theoretically (Elrick, 1961; Adrian and Franzini, 1966; Brustkern and Morel-Seytoux, 1970; McWhorter, 1971) and the validity of the conclusions have been checked by careful observations (McWhorter, 1971; Vachaud et al. 1973, 1974).

It is also well understood that "if the infiltrated waters leave pores that are in contact with the outside air, the soil air pressure remains unchanged" (Baver, 1937). In that case the traditional flow equation can be used (Vachaud et al., 1973, 1974) leading to the general belief that if air remains at the atmospheric pressure ahead of the wetting front then air flow is entirely negligible. It has been the fundamental contribution of McWhorter (1971, 1975) and Morel-Seytoux and Khanji (1974, 1975) to show that even if air is free to escape and can be considered incompressible, air flow is still important. That is, the apparent soil-water properties entering the diffusion equation depend on air properties as well, e.g., air conductivity.

In the following we shall consider the case when any increase in air pressure is much smaller than one atmosphere.

We shall compare the effect of air flow when the air escapes through the surface and when it can escape ahead of the wetting front.

Theory

For simplicity the effect of gravity is ignored and we assume that Darcy's law applies to both phases, or, (Elrick, 1961),

$$q_a = -k_a \nabla p_a, \quad q_w = -k_w \nabla p_w \tag{1}$$

where q, k and p are the volume flux, the conductivity, and the pressure for the air, subscript a, and water, subscript w. Since air pressure remains close to one atmosphere, both air and water are taken as incompressible, hence

$$q_a + q_w = V \tag{2}$$

with

$$\nabla \cdot V = 0 \tag{3}$$

The water potential Ψ is defined as usual by

$$\Psi \equiv p_w - p_a \tag{4}$$

Conservation of mass for water can be written as,

$$\partial\theta/\partial t + \nabla \cdot q_w = 0 \tag{5}$$

where θ is the moisture content per unit volume. Eqs. (1), (2) and (4) yield

$$q_w = Vf - k_a f \nabla\Psi \tag{6}$$

where

$$f = 1/(1 + k_a/k_w) \tag{7}$$

is a function of θ. Replacing q_w in Eq. (5) by its value from Eq. (6) we finally obtain, using Eq. (3),

$$\partial\theta/\partial t + f'V \cdot \nabla\theta = \nabla \cdot [D\nabla\theta] \tag{8}$$

where $f' \equiv df/d\theta$ and

$$D = k_a f \, d\Psi/d\theta \tag{9}$$

Eq. (8) has the form of the usual diffusion equation with the added term $f'V \cdot \nabla\theta$. It is interesting that the latter is very similar to the usual gravity term, e.g., see McWhorter (1975), except that in the case of gravity $\nabla\theta$ operates in the vertical direction, while it acts in the V direction in the present case. We now consider the properties of the one-dimensional solutions of Eq. (8), i.e., when $\nabla \equiv \partial/\partial z$. Eq. (8) then reduces to,

$$\partial\theta/\partial t + f'V\partial\theta/\partial z = \partial[D\partial\theta/\partial z]/\partial z \tag{10}$$

as first obtained by McWhorter (1971).

Case V = 0

This occurs with a finite column, closed at the bottom, with enough empty large pores at the soil surface to permit air escape with some air compression ahead of the front. Eq. (3) shows that V is independent of z and since the bottom of the column is closed $V = 0$ everywhere, and Eq. (10) reduces to

$$\partial\theta/\partial t = \partial[D\partial\theta/\partial z]/\partial z \tag{11}$$

The only difference between this equation and the diffusion equation usually solved is that, D, instead of increasing with θ up to saturation, decreases to zero just before saturation is reached (McWhorter, 1971). However, as long as D has a sharp maximum integral techniques developed earlier when D increases up to saturation, remain valid (Parlange, 1971, 1972, 1975a). In particular if the initial moisture content is θ_i, and the moisture content at the surface is imposed and is equal to θ_s, the sorptivity S obtained by optimization (Parlange, 1975a) can be written as (Parlange, 1975b; Brutsaert, 1976)

$$S^2 = 2(\theta_s - \theta_i)^{\frac{1}{2}} \int_{\theta_i}^{\theta_s} D(\theta-\theta_i)^{\frac{1}{2}}\, d\theta \tag{12}$$

Eq. (11) also permits the direct measurement of D by the standard Bruce and Klute method (1956). Note that in their original study Bruce and Klute did find a sudden decrease of D near saturation suggesting that their measurements might have been affected by air movement. Note also that, as pointed out by Morel-Seytoux and Khanji (1975) the very large values of the diffusivity near saturation are difficult to measure accurately, whether D in Eq. (11) is affected by air or not. Although we shall discuss Morel-Seytoux's alternate approach in the next section, let us point out here that the simple approach to this problem suggested by Talsma et al. (1972) can always be used: Recognizing that D above a certain value of θ is essentially indeterminate, profiles are calculated only when D is reasonably accurate. For θ near saturation, D is treated essentially as a delta function in such a way that the integral of D in Eq. (12) keeps its appropriate value. Finally if D is measured by a two-dimensional method it is possible to obtain accurate values of D closer to saturation (Turner and Parlange, 1975). In conclusion if air escapes through the surface and air compression is negligible the air effect is real but affects the value of the diffusivity rather than the form of the equation. We are now going to discuss the case $V \neq 0$, when the form of the equation is also affected by air flow.

Case $V \neq 0$

Morel-Seytoux and co-workers have investigated extensively this case when air escapes freely ahead of the front, rather than through the surface. This must be the case if ponding occurs at the surface for instance and there is no appreciable air pressure build up. Since only water moves through the soil surface, V in Eq. (10) represents the flux of water entering the soil, since V can only be a function of t, from Eq. (3).

The fundamental approach followed by Morel-Seytoux and co-workers, e.g., see Morel-Seytoux and Khanji (1974, 1975) consists in neglecting the diffusion term in Eq. (10), at least as a first approximation. This first approximation then obeys the equation

$$z = f' \int_0^t V dt \tag{13}$$

near the soil surface, with a discontinuous wetting front. It is clear (Morel-Seytoux and Khanji, 1975), that the diffusivity must smooth out the discontinuous front. Their method requires a knowledge of f, f' and also f''. However, f, and *a fortiori* its derivatives, are difficult to measure accurately. For instance, Morel-Seytoux and Khanji (1975) deduced f from air permeability measurements for values of θ where the air permeability could not be distinguished significantly from zero. In that case there would not seem to be any reason not to replace the "real" sigmoid f-function by one that increases rapidly to one for an effective saturation. (Note that at the effective saturation some residual air is entrapped in large pores, so that θ_s is less than the water content of the truly saturated soil.) With this simpler f-function, f' is essentially a delta-function and using the construction of Morel-Seytoux and Khanji (1975) the profile is saturated, i.e., $\theta = \theta_s$, up to the front where θ drops discontinuously to zero. Any improvement on the shape of the profile must then come entirely from the diffusivity. In the following we take diffusivity into account for our first approximation. It is then possible to obtain f for any soil by comparing observations for the two cases $V=0$ and $V \neq 0$, rather than from measurements of air permeability.

We are trying to solve Eq. (10) with the boundary conditions

$$\theta = \theta_i \quad \text{for} \quad t < 0$$

$$\theta = \theta_s \quad \text{for} \quad x = 0, \quad t > 0 \tag{14}$$

For these conditions Eq. (10) has a similarity solution (McWhorter, 1971) with the variable

$$\phi = zt^{-\frac{1}{2}} \tag{15}$$

Then Eq. (10) becomes

$$\frac{1}{2}(Sf' - \phi) = d[Dd\theta/d\phi]/d\theta \tag{16}$$

where $2V = St^{-\frac{1}{2}}$. Let us first define D_w as the diffusivity that would arise if air was replaced by a gas with zero viscosity, or,

$$D_w \equiv k_w d\Psi/d\theta \tag{17}$$

Hence, from Eq. (9)

$$D = D_w(1 - f) \tag{18}$$

We now rewrite Eq. (16) as

$$\frac{1}{2} Sf' + d[D_w fd\theta/d\phi]/d\theta = \frac{1}{2}\phi + d[D_w d\theta/d\phi]/d\theta \tag{19}$$

Let us assume for an instant that the left-hand side of this equation, which represents the effect of air flow, being the terms dependent on f, is strictly zero. Then, Eq. (19) would give

$$\frac{1}{2}\phi + d[D_w d\theta/d\phi]/d\theta = 0 \tag{20}$$

We recognize in this equation the standard diffusion equation ignoring air effects. Since D_w increases very rapidly with θ, a first approximation $\phi_o(\theta)$ is known to be an adequate solution of Eq. (20) with (Parlange, 1971, 1975c)

$$\phi_o = (2/S) \int_{\theta}^{\theta_s} D_w(a)da \tag{21}$$

with

$$S^2 = 2 \int_{\theta_i}^{\theta_s} (\theta-\theta_i)\, D_w(\theta)d\theta \tag{22}$$

Although adequate, the expression in Eq. (21) satisfied Eq. (20) approximately, i.e., it annuls the right-hand side of Eq. (19) approximately only. However with this approximation the left-hand side of Eq. (19) is *strictly* zero. Consequently ϕ_o is just as good an approximation to Eq. (19) as it is to Eq. (20). Better approximations than ϕ_o to Eq. (20) were found (Parlange, 1975a,c). However the latter do not annul the left-hand side of Eq. (19) exactly, hence they may not be an improvement over Eq. (21) for the present case when air effects are significant. If f is essentially zero up to θ_s, i.e., if air flow has no retarding effect, then higher approximations to Eq. (20) would also improve the solution of Eq. (19). However, if the effect of f, near θ_s, is significant then it becomes crucial to use approximations to Eq. (20) that keep the left-hand side of Eq. (19) as close to zero as possible. Hence Eq. (21) yields an approximation which is reasonably accurate when air effects are negligible, and becomes even more accurate when air effects are important.

We first note, by comparison of Eqs. (12) and (22) that, since $D=D_w(1-f)$, the sorptivity and hence the water intake, when air must escape through the surface, is less than when air is free to move ahead of the front. The comparison must naturally be made for the same value of θ_s. Only when f is essentially zero, up to θ_s, i.e., when air has no effect anyway, do the two results become indistinguishable. Looking at Eqs. (21) and (22) we also see the effect of air flow in the case $V\neq0$. As already mentioned above the larger the air effect the more accurate Eqs. (21) and (22) become. However when air flow has no effect, i.e., $f=0$ up to θ_s, the best estimate of the sorptivity is given by Eq. (12) (Parlange, 1975b; Brutsaert, 1976), but with D replaced by D_w. This shows that the air effect reduces the water intake but only by a few percent when air is free to escape ahead of the front. A second effect is that the shape of the profile predicted by Eq. (21) is less dispersed, i.e., the decrease of θ is more abrupt, than when air has no effect (Parlange, 1975a). Hence the assumption of Morel-Seytoux and Khanji (1975) about a discontinuous front is indeed a better assumption when air flow has a significant effect. The present results will now be illustrated experimentally, and the f-function measured.

Materials and Methods

A horizontal Plexiglas chamber (65 x 4.5 x 4.5 cm), shown in Fig. 1, was fitted with a very coarse sintered glass porous plate to provide a reservoir (4.5 x 4.5 x 4.5) at one end. Seven holes (1 cm diameter) were spaced 10 cm apart along the upper surface of the chamber to provide avenues of air escape and for sampling ports for gravimetric moisture measurements upon completion of each experiment. The port farthest from the water reservoir was connected to a water manometer to record air pressures that develop ahead of the wetting front. The main body of the chamber was filled with a known weight of air-dry very fine sand (100-74 μ) packed to a known volume (B.D. 1.45). Water placed in a hole in the reservoir passes through the very coarse porous plate and proceeds as a wetting front along the sand-filled column.

Water was added with a pipette at varying time intervals throughout the experiment to maintain a constant depth during each experiment. During one set of experiments, all 6 sampling holes were left open to allow free escape of air ahead of the wetting front. In another set of experiments 5 or 6 sampling holes were plugged with rubber corks and sealed with silicone rubber. The sampling port next to the water reservoir area was left unsealed to allow air to escape behind the wetting front.

Results and Discussion

The results of six experiments, three with the sampling holes sealed and three with the holes open are presented.

Fixed amounts of water were added at varying time intervals. For example, with the holes sealed 1 cc was added at intervals such that the cumulative amount of water, Q, obeyed $Q = 2.15\ t^{\frac{1}{2}}$. With the holes opened, 2 cc was added such that $Q = 3.78\ t^{\frac{1}{2}}$. These values of Q were chosen by trial and error so that the water level in the reservoir should be fairly constant in the two sets of experiments. Indeed the level was about 1.9 ± 0.3 cm in all cases. Minor variations in the level were probably caused by slight differences in packing the soil before each experiment.

Due to the method of introduction of water, the movement is not truly one-dimensional until about 400 sec have elapsed.

When the holes were opened the water manometer showed no increase in air pressure. However, when the holes were sealed the pressure rose rapidly to about 65 cm ± 5 and then varied with slow oscillations between 70 and 60 cm.

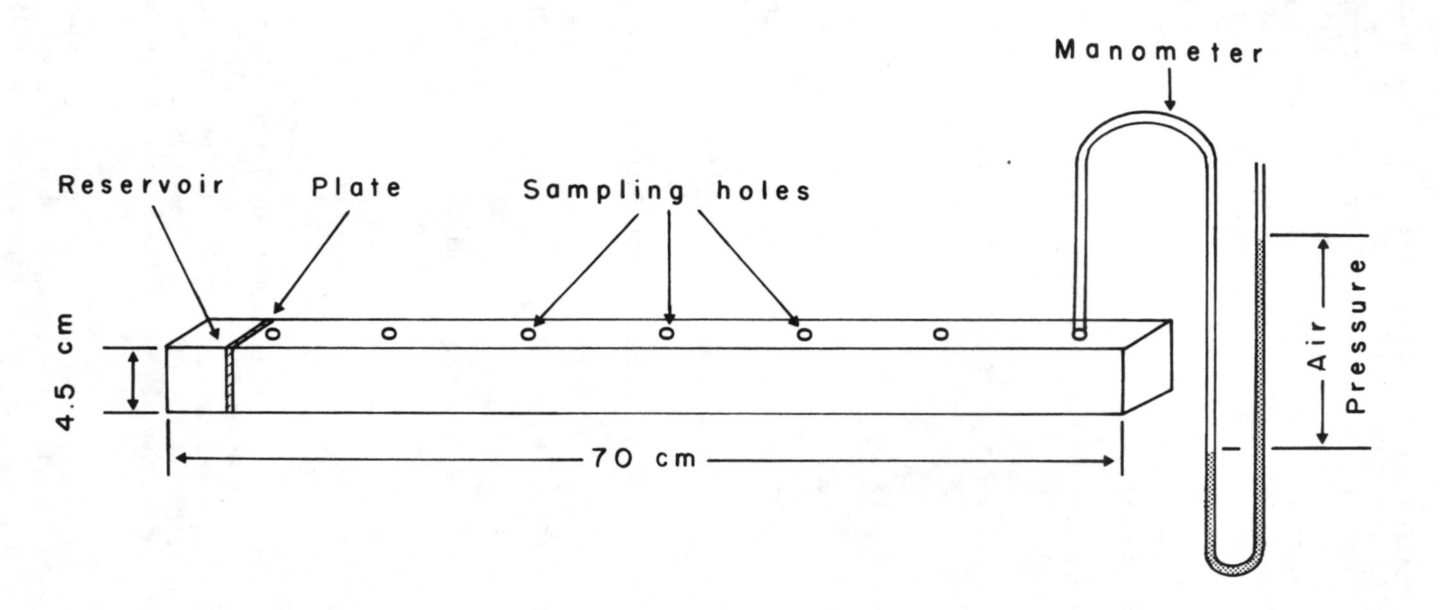

Fig. 1. Sketch of chamber showing the reservoir, sampling holes and manometer.

Other experiments, not reported here, were carried out at various Q's. For example, if $Q = 2.15\ t^{\frac{1}{2}}$ with the holes open the water level in the reservoir dropped to essentially zero. On the other hand, if $Q = 3.78\ t^{\frac{1}{2}}$ with the holes sealed, water practically filled the reservoir. All experiments were stopped when the wetting front had reached a distance of 45 cm from the porous plate. This distance was reached at $t = 2 \times 10^4$ sec with the holes sealed and $t = .87 \times 10^4$ sec with the holes open. Soil samples were then taken and the moisture content determined. The moisture profiles are shown in Fig. 2.

Moisture profiles in Fig. 2 were represented analytically with a least square fit by

$$z = 45 - 1.094 \times 10^3\ \theta^{3.72} \tag{23}$$

and

$$z = 45 - 0.014218 \exp 18.821\ \theta \tag{24}$$

In both cases the correlation coefficient was higher than 0.99. Note that the extrapolated values of θ at $z=0$ are close in both cases, i.e., 0.424 and 0.428 respectively. The decrease of θ is more gradual when the holes are sealed than when the holes are opened. This is in qualitative agreement with the theoretical analysis.

From the profile in Eq. (23) and the Bruce and Klute equation

$$2D = -\ (d\phi/d\theta) \int_{\theta_i}^{\theta} \phi d\theta \tag{25}$$

we can derive D or $D_w(1 - f)$. Calculating d /d from Eq. (24) and comparing with Eq. (21) we can also calculate D_w. By taking the ratio of $D_w(1 - f)$ and D_w we calculate $(1 - f)$ and hence f. Figure 3 shows the value of f thus measured. Again the qualitative behavior of f vs. θ is as expected. Quantitatively we expect the values of f to be most reliable for $0.3 < \theta < 0.4$ where the profiles are measured accurately. Values for higher and lower values of θ are only extrapolations like the profiles themselves. Note that the maximum value of f is 0.83 and is less than one. This may be due either to the error in the estimation of f for large θ's or because $\theta \simeq 0.43$ is significantly less than saturation.

In conclusion we have developed two analytical solutions for the horizontal infiltration of water when air movement is considered. In one case air must escape through wet soil

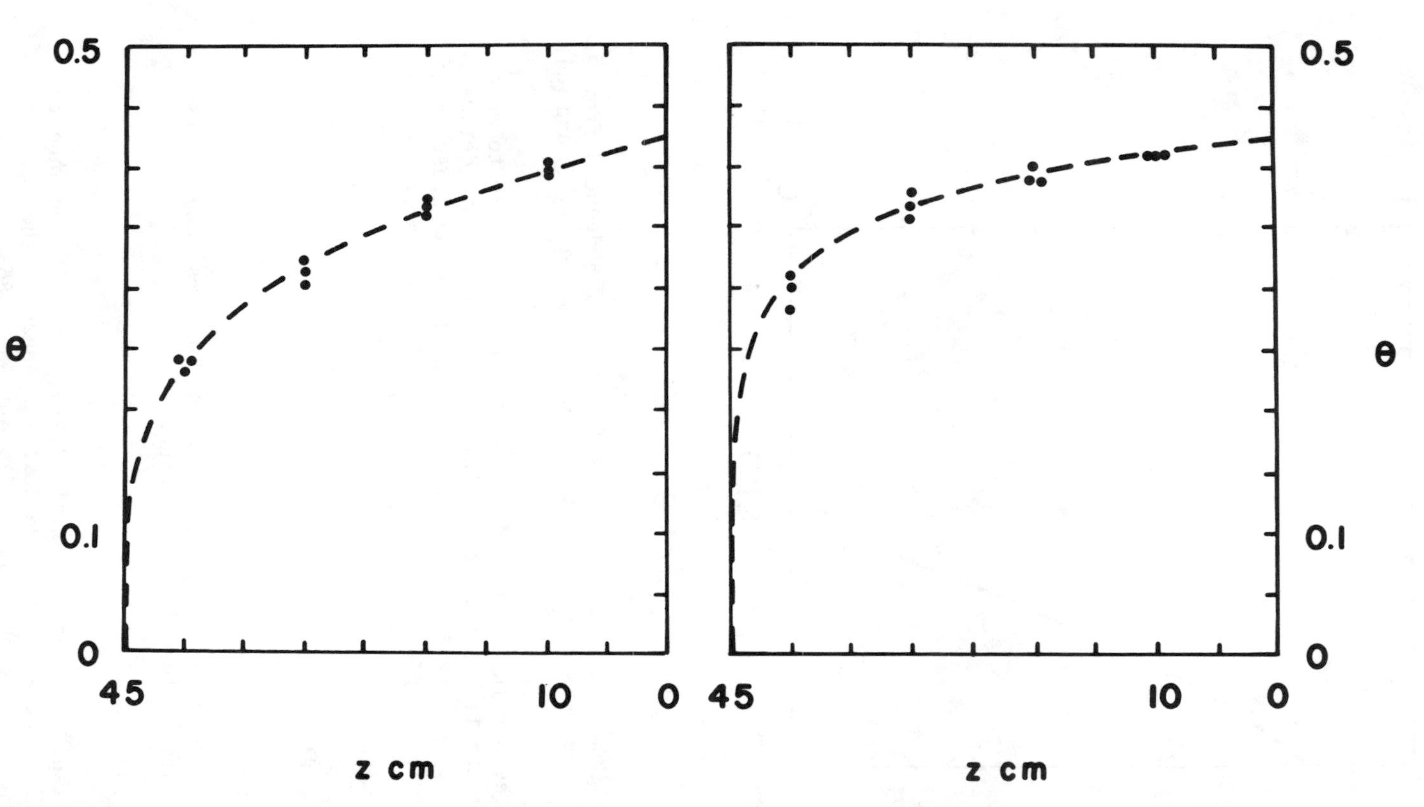

Fig. 2. Moisture profiles after the wetting front has reached 45 cm from the source. Data for six experiments are shown by dots. The dashed curves correspond to Eq. (23) when the holes were open (left) and to Eq. (24) when the holes were sealed (right).

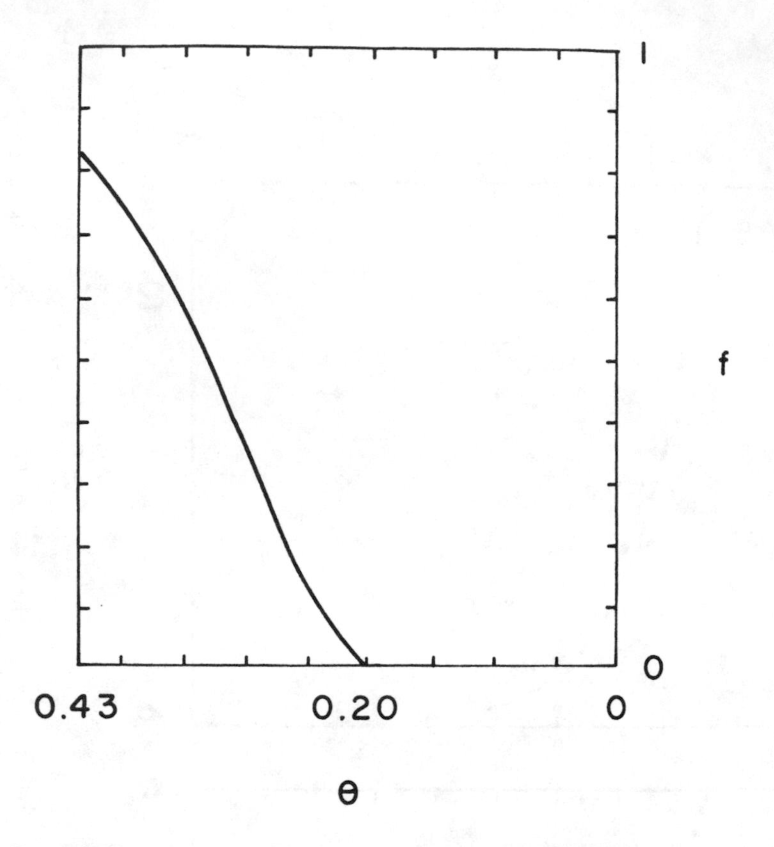

Fig. 3. Measured f-function as a function of θ deduced from the moisture profiles represented by Eqs. (23) and (24).

behind the wetting front and in the other air is free to move ahead of the front. The two solutions can be used to determine the effective soil-water diffusivity from the measurement of the moisture profiles in each case. Comparison of the two results permits the determination of the f-function which characterizes air effects. Application of the method was illustrated experimentally by considering water movement into a horizontal chamber filled with very fine sand.

Acknowledgment

The authors are gratefully indebted to Mrs. Joan Bernstein for her careful technical assistance.

References

[1] Adrian, D. D. and Franzini, J. B., 1966. "Impedance to Infiltration by Pressure Build-up Ahead of the Wetting Front," J. Geophys. Res., 71, pp. 5857-5863.

[2] Baver, L. D., 1937. "Soil Characteristics Influencing the Movement and Balance of Soil Moisture," Soil Sci. Soc. Am. Proc., 1, pp. 431-437.

[3] Bruce, R. R. and Klute, A., 1956. "The Measurement of Soil-Water Diffusivity," Soil Sci. Soc. Am. Proc., 20, pp. 456-462.

[4] Brustkern, R. L. and Morel-Seytoux, H. J., 1970. "Analytical Treatment of Two-Phase Infiltration," Proc. ASCE J. Hydro. Division, 96, pp. 2535-2548.

[5] Brutsaert, W., 1976. "The Concise Formulation of Diffusive Sorption of Water in Dry Soil," Water Resour. Res., 12, pp. 1118-1124.

[6] Elrick, D. E., 1961. "Transient Two-Phase Capillary Flow in Porous Media," The Physics of Fluids, 4, pp. 572-575.

[7] Free, G. R. and Palmer, V. J., 1940. "Interrelationship of Infiltration, Air Movement, and Pore Size in Graded Silica Sand," Soil Sci. Soc. Am. Proc., 5, pp. 390-398.

[8] Lewis, M. R. and Powers, W. L., 1939. "A Study of Factors Affecting Infiltration," Soil Sci. Soc. Am. Proc., 3, pp. 334-339.

[9] McWhorter, D. B., 1971. "Infiltration Affected by Flow of Air," Hydrology Papers, Colorado State University, Fort Collins, Colorado, No. 49, 43 p.

[10] McWhorter, D. B., 1975. "Vertical Flow of Air and Water with a Flux Boundary Condition," Annual Meeting ASAE, Davis, California, Paper No. 75-2012.

[11] Morel-Seytoux, H. J. and Khanji, J., 1974. "Derivation of an Equation of Infiltration," Water Resour. Res., 10, pp. 795-800.

[12] Morel-Seytoux, H. J. and Khanji, J., 1975. "Prediction of Imbibition in a Horizontal Column," Soil Sci. Soc. Am. J., 39, pp. 613-617.

[13] Parlange, J.-Y., 1971. "Theory of Water Movement in Soils: I. One-Dimensional Absorption," Soil Sci., 111, pp. 134-137.

[14] Parlange, J.-Y., 1972. "Analytical Theory of Water Movement in Soils," Proc. 2nd Symp. IAHR-ISSS, 1, pp. 222-236.

[15] Parlange, J.-Y., 1975a. "On Solving the Flow Equation by Optimization: Horizontal Infiltration," Soil Sci. Soc. Am. Proc., 39, pp. 415-418.

[16] Parlange, J.-Y., 1975b. "Determination of Soil-Water Diffusivity by Sorptivity Measurements," Soil Sci. Soc. Am. Proc., 39, p. 1011.

[17] Parlange, J.-Y., 1975c. "Theory of Water Movement in Soils: II. Conclusion and Discussion of Some Recent Developments," Soil Sci., 119, pp. 158-161.

[18] Powers, W. L., 1934. "Soil-Water Movement as Affected by Confined Air," J. Agr. Res., 49, pp. 1125-1133.

[19] Talsma, R., Parlange, J.-Y. and Groenevelt, P. H., 1972. "Soil Moisture Distribution During Two-Dimensional Absorption from a Cylindrical Source," Aust. J. Soil Res., 10, pp. 209-214.

[20] Turner, N. C. and Parlange, J.-Y., 1975. "Two-Dimensional Similarity Solution. Theory and Application to the Determination of Soil-Water Diffusivity," Soil Sci. Soc. Am. Proc., 39, pp. 387-390.

[21] Vachaud, G., Vauclin, M., Wakil, M. and Khanji, M., 1973. "Effects of Air Pressure on Water Flow in an Unsaturated Stratified, Vertical Column of Sand," Water Resour. Res., 9, pp. 160-173.

[22] Vachaud, G., Gaudet, J. P. and Kuraz, V., 1974. "Air and Water Flow During Ponded Infiltration in a Vertical Bounded Column of Soil," J. Hydro., 22, pp. 89-108.

THEORY OF UNIVERSAL HYSTERETICAL PROPERTIES OF UNSATURATED POROUS MEDIA

By

Yechezkel Mualem
Senior Research Associate
Engineering Research Center
Colorado State University
Fort Collins, Colorado 80523

Abstract

Theory of capillary hysteresis is presented which yields a unique, universal, nondimensional hysteresis, independent of the soil type. The theoretical model permits prediction of the hysteretical domain and the water content--capillary head relationship for any process, which includes successive wetting and drainage, from one measured retention curve only. The theoretical model is checked against experimental data of sand soil for which no significant effect of water blockage against air entry was observed. A reasonable agreement is found between the measured main drying curve, the measured primary scanning curve and the corresponding predicted curves.

The model is very convenient for calibration and operation as a practical tool for representing capillary hysteresis.

Introduction

Most processes in nature and applications which are related to unsaturated flow include successive cycles of wetting and drainage. Any theoretical study, aimed at simulating such cyclic unsaturated flow problems, has to take into account the hysteretical nature of the hydraulic properties of the soils. Hysteresis of the water content (Θ) as a function of the capillary head (ψ) is particularly important due to its considerable magnitude. However, for two main reasons hysteresis is rarely considered in hydrological studies. First, the complexity of the numerical solution of unsaturated flow problem, in the aeration zone, is greatly increased with alternating boundary conditions, and even more so by attempts to account for the hysteretic phenomena. Second, before 1973 models of hysteresis required large amounts of measured data for calibration and were therefore inconvenient for practical use. Hence most of the works which analyze cyclic flow problems of the aeration zone are based on some heuristic simplifications.

Whisler and Klute (1965) are apparently the first investigators who took into consideration the effect of hysteresis on infiltration into vertical soil column at equilibrium under gravity. A thin layer of ponded water was assumed to be applied at the upper end of the soil column. Hysteresis was introduced because each point in the column was wetted up along a different wetting scanning curve. This study, however, used arbitratily drawn hysteresis loops as well as the inside $\Theta(\psi)$ scanning curves in computation. Nevertheless, the artificial data permitted a detailed analysis of the physical phenomenon. It showed that ignoring hysteresis affects the computed results. A similar problem but with constant flux at the soil surface was thoroughly investigated by Whisler and Klute (1967).

Staple (1966, 1969) studied one cycle of infiltration and redistribution in vertical columns. During infiltration the soil surface was assumed to be saturated. Measured main loops and scanning curves of the capillary hysteresis were used in computation. Intermediate values were computed by linear interpolation. He found that the hysteretic property of the soil influences considerably the water distribution in the soil profile. Rubin (1967) also investigated the effect of hysteresis on soil moisture redistribution following infiltration. Mathematical formulae were employed for the soil moisture characteristics to avoid two-way linear interpolation. He found, that due to hysteresis, a higher water content is ascertained in the upper part of the soil and the wetted zone is, therefore, shallower than the one computed with a unique soil characteristic.

Bresler et al. (1969) and Hanks et al. (1969) studied evaporation and soil water redistribution following three different infiltration rates. The scanning curves within the hysteresis were taken to be straight lines, each line corresponding to both wetting and drying. The slopes of the scanning curves were chosen arbitrarily with the restriction that they must be less than the slopes of the main curves at the point of intersection. They confirmed Rubin's finding and inferred that the influence of hysteresis becomes stronger as the rate of infiltration is larger and that evaporation is enhanced by hysteresis. This computed result was similar to experimental observations.

Ibrahim and Brutsaert (1968) and Staple (1970) used the Neel-Everett model of hysteresis in the computational procedure of a flow problem which included successive processes of wetting and drainage. The first authors, however, used hypothetical hydraulic properties of soil for the analysis. This way one can insure applicability of the theoretical model of hysteresis. But for a practical case, when the dependence of the water content (Θ) on the capillary head (ψ) is measured (as in the second case) one has to check first the adaptability of the Néel-Everett model to the examined soil. The reason is that this model often yields results which are in contradiction with the observed phenomena of the capillary hysteresis. This test was not conducted, however, before Staple (1970) employed the model.

Dane and Wierenga (1975) studied the effect of hysteresis on infiltration and redistribution in a layered soil. Arbitrary mathematical formulae were used to derive the hysteretical scanning curve inside the main loop. The authors concluded that redistribution and drainage were better predicted when hysteresis was taken into account.

Reeves and Miller (1975) studied infiltration in the case of erratic rainfall. They employed an arbitrary model that has no theoretical basis. As a result, disturbing anomalies in the computed results were found which suggests the need for another model of hysteresis.

It is clear from this review that hysteresis of the hydraulic properties of soil plays a significant role in hydrology of the aeration zone. However, it also indicates some of the problems of handling hysteresis in such theoretical research. The objective of this study is to present a theoretical model of hysteresis which requires only one measured branch of the hysteretical loop for calibration and to check this model against measured data of a sand soil. Detailed reviews of other models of hysteresis and theories of capillary hysteresis and hysteresis of the hydraulic conductivity were presented in a series of previous publications (Mualem, 1973, 1974a, 1974b, 1976a, 1976b; Maulem and Dagan, 1972, 1975, 1976; Mualem and Morel-Seytoux, 1976, 1977).

Theory

The basic hypothesis employed in this study has been presented by Mualem (1974a). A porous medium is viewed as an assembly of groups of pores. Each group is characterized by having such apertures on its boundary surface so that the water menisci when located there have the same mean radius of curvature r. $h(r)$ represents the distribution function of these subunits of groups of pores. Within each group the volumetric distribution function of the elementary pore domains is identical to that of the whole porous medium $\ell(\rho)$. This is the similarity principle. It is further assumed that r and ρ have the same range of distributions $R_{min} < \rho$, $r < R_{max}$. Using normalized variables

$$\bar{r} = \frac{r - R_{min}}{R_{max} - R_{min}} ; \qquad \bar{\rho} = \frac{\rho - R_{min}}{R_{max} - R_{min}} \tag{1}$$

the distribution function (f) of the elementary pore domain of radius $\bar{\rho}$ which has an access to the surrounding through opening of radius $\bar{r}$ is

$$f(\bar{r}, \bar{\rho}) = h(\bar{r})\ell(\bar{\rho}) \tag{2}$$

Function f is defined in the square region $0 \leq \bar{r} \leq 1$; $0 \leq \bar{\rho} \leq 1$. Using the capillary law, $\psi = \alpha/R$, the variables ρ and r can be related to the capillary head (ψ) at which the pore domain is filled and ψ at which air can penetrate through the counter surface and permit drainage of the pore, respectively. α is a coefficient of proportion. Accordingly, the functions $h(\bar{r})$ and $\ell(\bar{\rho})$ represent distribution functions of areal and volumetric properties (apparent porosities) respectively. As the areal porosity equals the volumetric porosity in a homogeneous porous media, Mualem (1976b) deduced that the two distribution functions should be identical, namely,

$$h(\bar{r}) = \ell(\bar{r}) \tag{3}$$

This latter assumption leads to a new version of Model II (Mualem, 1974). For convenient identification the new version is labeled Model V.

The water content as a function of the capillary head, for a given process, can be determined by integration over the corresponding diagram of the filled pore domains. The region of the filled pore domains for the main wetting process is described in Fig. 1b. By wetting from ψ_{min} to ψ all the pore domains of radius $0 \leq \bar{\rho} \leq \bar{R}(\psi)$ are filled. Accordingly, the water content along the main wetting curve Θ_w is

$$\Theta_w(\bar{R}) = \Theta_{min} + \int_0^{R} h(\bar{\rho})\, d\bar{\rho} \int_0^{1} h(\bar{r})\, d\bar{r} \tag{4}$$

where Θ_{min} is the lowest measured water content. As a matter of convenience three new variables are defined as follows: the "effective" water content θ, the saturation S_e and the integral distribution function H, where

$$\theta(\psi) = \Theta(\psi) - \Theta_{min}; \qquad S_e(\psi) = \frac{\Theta(\psi)-\Theta_{min}}{\Theta_u-\Theta_{min}} \tag{5}$$

$$H(\psi) = \int_{\psi_{min}}^{\psi} h(\psi)\, d\psi \tag{6}$$

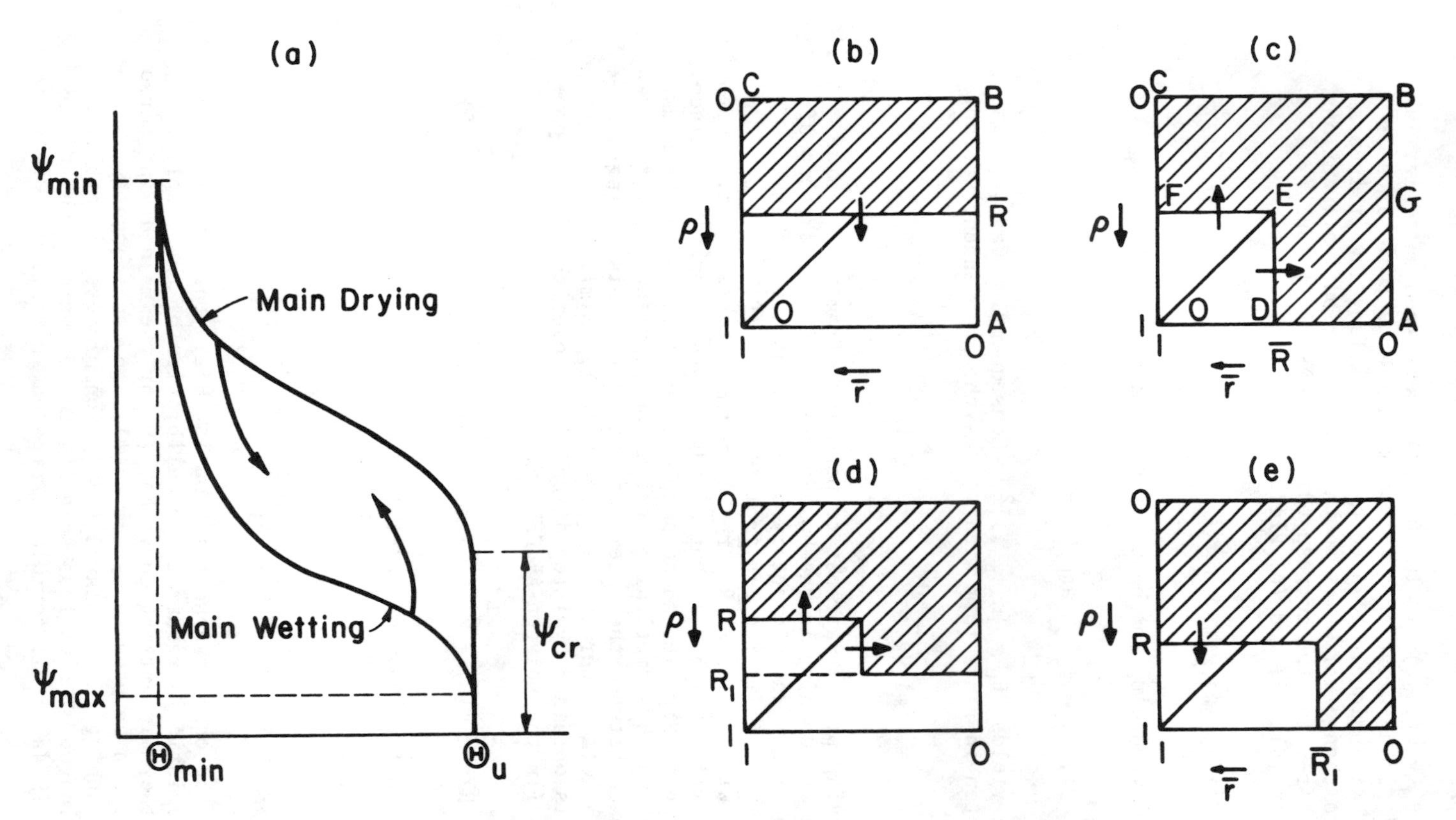

Fig. 1. Nomenclature of capillary hysteresis (a) and Mualem's (1974) diagram of the filled pore domains shadowed region for: the main wetting process (b); the main drying process (c); the primary drying process (d); and the primary wetting process (e).

where $\theta_u = \theta(\psi_{max})$ is the ultimate water content (Fig. 1a) and ψ' is a variable of integration. Note that since r and ψ are uniquely related by the capillary law it is possible to change variables so that the distribution functions $h(\psi) = h(r)$ for the corresponding values of ψ and r. Equations (4), (5) and (6) yield

$$\theta_w(\bar{R}) = H(\bar{R})\ H(1) \tag{7}$$

and by introducing the boundary condition

$$\bar{R} = 1, \quad \psi = \psi_{max} \quad \text{and} \quad \theta = \theta_u \tag{8}$$

Eq. (8) yields $\theta_u = [H(\psi_{max})]^2$. Consequently, Eq. (4) can be written in final form with ψ as the independent variable

$$\theta_w(\psi) = H(\psi)\ \sqrt{\theta_u} \tag{9}$$

In a drying process air has to penetrate from the outside through the opening in the boundary surface of the group of pores to reach the elementary pore domains located inside. Hence, when ψ is reduced from $\psi(1)$ to $\psi(\bar{R})$ only subregions with aperture $\bar{R} \leq \bar{r} \leq 1$ permit access to air. Pore domains with $\bar{R} \leq \rho \leq 1$ but having $r \leq \bar{R}$ cannot be drained. Referring to Fig. 1c, these pore domains located in the square DEGA are the hysteretical contribution represented by the difference between the water content on the main drying and the main wetting curves $(\theta_d(\psi) - \theta_w(\psi))$. Similar to Eq. (4), $\theta_d(\psi)$ is obtained by integration of f over the filled pore domains (in the polygon ABCFED, Fig. 1c).

$$\theta_d(\bar{R}) = \theta_w(\bar{R}) + \int_{\bar{R}}^{1} h(\bar{\rho})\ d\bar{\rho} \int_{0}^{\bar{R}} h(\bar{r})\ d\bar{r} \tag{10}$$

By introducing Eqs. (6) and (9) in Eq. (10) and applying the capillary law, θ_d obtains its final form as a direct function of (ψ).

$$\theta_d(\psi) = [2\ \theta_u^{\frac{1}{2}} - H(\psi)]\ H(\psi) \tag{11}$$

As the only unknown function is $H(\psi)$ only one soil characteristic is required for calibration of the model. Either the main wetting curve Eq. (9) or the main drying curve Eq. (11) can be used to derive $H(\psi)$, and consequently, any other scanning curve related to a given process of wetting and drainage. The various scanning curves were formulated by Mualem (1974a). All formulae expressed in terms of $H(\psi)$ and

$L(\psi)$ are valid for this version as well, but with $L(\psi)=H(\psi)$. Moreover, all scanning can be expressed as a direct function of either $\theta_w(\psi)$ or $\theta_d(\psi)$. Taking the main wetting curve as the given function Eqs. (11) and (9) yield

$$\theta_d(\psi) = [2\ \theta_u - \theta_w(\psi)]\ \frac{\theta_w(\psi)}{\theta_u} \tag{12}$$

By introducing Eq. (5) into Eq. (12) the main drying curve, expressed in terms of effective saturation, obtains the form

$$S_{ed}(\psi) = [2 - S_{ew}(\psi)]\ S_{ew}(\psi) \tag{13}$$

Enderby's (1955) notation of hysteretical process is used to designate the various scanning curves. Accordingly,

$$\theta(\psi) = \theta(\psi_{min} \overset{\psi_1}{} \psi_2 \overset{\psi}{})$$

indicates a process that follows wetting from ψ_{min} up to ψ_1, then drainage and a decrease of ψ to ψ_2 and rewetting again to a final value of ψ. By integration of f over the filled pore domain described in Fig. 1d and using a technique similar to that used to derive Eq. (13), the primary drying scanning curve is obtained.

$$S_e(\psi_{min} \overset{\psi_1}{} \psi) = S_{ew}(\psi) + [S_{ew}(\psi_1)-S_{ew}(\psi)]\ S_{ew}(\psi) \tag{14}$$

Similarly, integration of the filled pore domains diagram in Fig. 1e yields the primary wetting scanning curve

$$S_e(\overset{\psi_{max}}{} \psi_1 \overset{\psi}{}) = S_{ew}(\psi) + [1 - S_{ew}(\psi)]\ S_{ew}(\psi_1) \tag{15}$$

Discussion

The most important feature of results is that a unique relationship is obtained between the various hysteretical curves. In the S_{ew} - S_{ed} plane a universal hysteresis is displayed by Eqs. (13), (14) and (15) which is independent of the soil type. Figures 2 and 3 show the drying and the wetting curves, respectively, as a function of S_{ew}.

As discussed in several previous publications (see recent reviews by Mualem and Morel-Seytoux (1976, 1977) theories based on independent domains theory are not efficient in cases where a considerable portion of the hysteretical region is observed within the range of the air entry value. At high saturation there is a pronounced effect of the pore-water

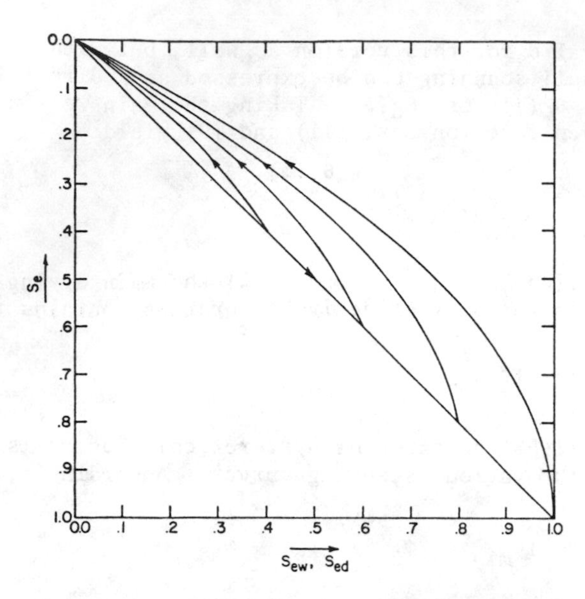

Fig. 2. Theoretical nondimensional main hysteretical loop and primary drying scanning curves for unsaturated porous media.

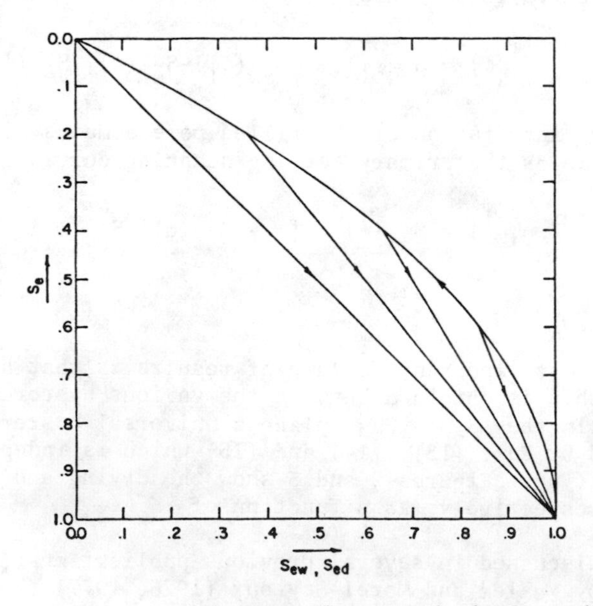

Fig. 3. Theoretical nondimensional main hysteretical loop and primary wetting scanning curves for unsaturated porous media.

blockage against air entry. The behavior of the elementary pore domains becomes dependent on the state of the surrounding pores. This phenomena is greatly responsible for the observed shape of drying curves with $d\theta/d\psi \simeq 0$ as θ tends to θ_u. As the independent domains theory disregards this phenomenon, the agreement between the predicted and the measured drying curves is relatively poor in these cases (Mualem, 1973, 1974a, 1976b). As far as the present model is concerned, it implies that in such cases the drying curve cannot appropriately represent the pore distribution function of the soil. Therefore, one may expect better results when the main wetting curve is used rather than when the main drying branch of the hysteretical loop is used for calibration.

The soil used in the present study to examine the theoretical model is Molonglo Sand investigated by Talsma (1970). A stable, time independent hysteresis is assumed. Figures 4 and 5 show comparison between the experimental hysteretical curves and theoretical ones, derived from the measured main wetting curve solely. Figure 4 shows the predicted main drying curve and three drying scanning curves, using Eq. (11). A reasonable agreement is found between the measured and the computed main drying curves. A greater deviation is observed, however, between the experimental and the theoretical drying scanning curve. It should be noted that the form of the measured drying curves indicates two particularly important properties. First, the measured drying scanning curve, branching from the wetting boundary at $\psi = -5$ cm, intersect the main drying curve. This phenomenon does not comply with the domains theory of hysteresis and contributes to the discrepancies between the measured and predicted curve. One of the basic requirements of the theoretical model is that the main drying curve (branching from $\psi = 0$) should be also the boundary drying curve. Second, in other experimental studies drying scanning curves starting at high saturation attain smaller slopes ($d\theta/d\psi$) at the reversal point than drying curves starting at lower saturation (see for example the detailed experiment of Bomba, 1968). This phenomenon is attributed to the effect of the water blockage against air entry. In the present case, however, an opposite phenomenon is manifested for which there is no theoretical explanation. Figure 5 shows comparisons between the measured and predicted primary wetting scanning curves, using Eq. (15). In this case the maximum deviation in water content for any ψ is equal to the difference between the measured and the predicted main drying. Consequently, the accuracy of the model in this case is much better than that found for the computed drying scanning curves.

Concluding, if one takes into account that the whole domain of the capillary hysteresis is predicted from only one measured curve of $\theta(\psi)$, the accuracy of the theoretical

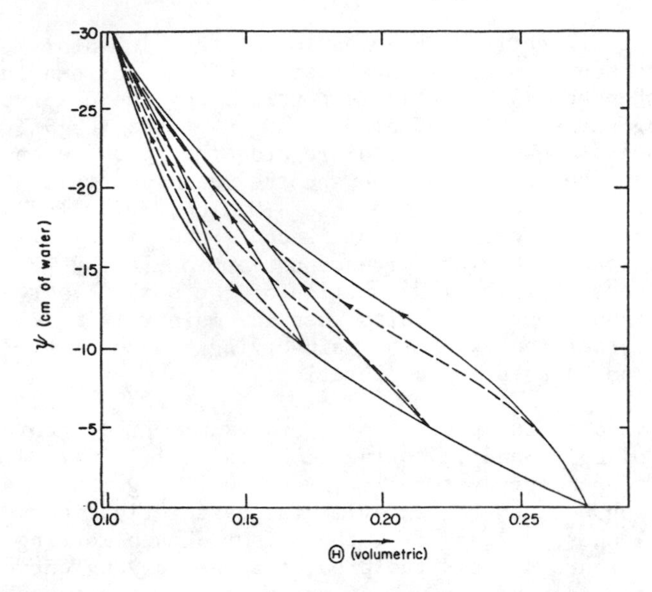

Fig. 4. Measured drying curves for Molonglo sand (solid lines) and the corresponding predicted ones (dashed lines) derived from the measures main wetting curve only by using Eq. (14)

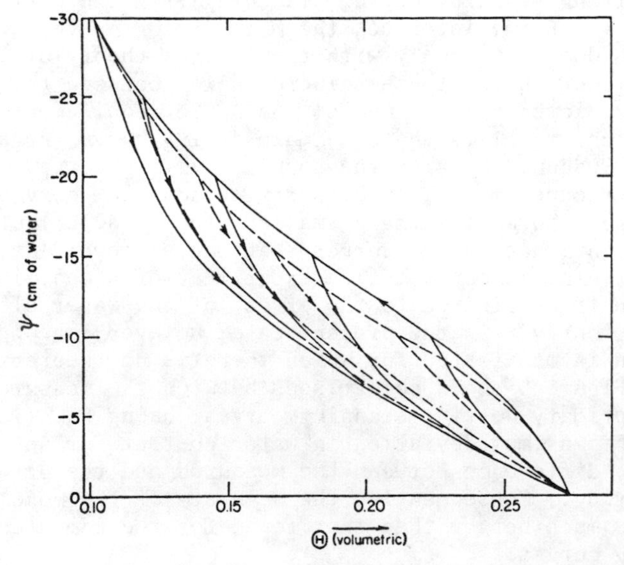

Fig. 5. Measured primary wetting scanning curves for Molonglo sand (solid lines) and the corresponding predicted ones (dashed lines) derived from the measured main wetting curve only by using Eq. (15).

model can be considered a good one. Due to the reduced amount of data required to calibrate the theoretical model and the extremely simple analytical relationship between the various hysteretical curves, one may find this model a very efficient tool for practical use in the theoretical solution of unsaturated flow problems in the aeration zone.

Acknowledgments

The work upon which this paper is based was supported by funds provided by the National Science Foundation under Grant ENG-7611542, "Hysteresis of the Hydraulic Properties of Soils." The NSF support is gratefully acknowledged.

References

[1] Bomba, S. J., 1968. "Hysteresis and Time Scale Invariance in Glass-Bead Medium," Ph.D. Thesis, University of Wisconsin.

[2] Bresler, E., Kemper, W. D., and Hanks, R. J., 1969. "Infiltration, Redistribution, and Subsequent Evaporation of Water from Soil as Affected by Wetting Rate and Hysteresis," Soil Sci. Soc. Am. Proc., 33, pp. 832-840.

[3] Dane, J. H. and Wierenga, P. J., 1975. "Effect of Hysteresis on the Prediction of Infiltration Redistribution and Drainage of Water in a Layered Soil," J. of Hydrology, 25, pp. 229-242.

[4] Enderby, J. A., 1955. "The Domain Model of Hysteresis, 1, Independent Domains," Trans. Faraday Soc., 51, pp. 835-848.

[5] Hanks, R. J., Klute, A., and Bresler, E., 1969. "A Numerical Method for Estimating Infiltration, Redistribution, Drainage, and Evaporation of Water from Soil," Water Resour. Res., 5, pp. 1064-1069.

[6] Ibrahim, H. A., and Brutsaert, W., 1968. "Intermittent Infiltration into Soils with Hysteresis," J. Hydraulics Division, ASCE, 94, pp. 113-137.

[7] Mualem, Y. and Dagan, G., 1972. "Hysteresis in Unsaturated Porous Media: A Critical Review and New Simplified Approach," Rep. 4, Project A10-SEC-77, Technion-Israel Inst. of Technol., Haifa, Israel.

[8] Mualem, Y., 1973. "Modified Approach to Capillary Hysteresis Based on a Similarity Hypotehsis," Water Resour. Res., 9(5), pp. 1324-1331.

[9] Mualem, Y., 1974a. "A Conceptual Model of Hysteresis," Water Resour. Res., 10(3), pp. 514-520.

[10] Mualem, Y., 1974b. "Hydraulic Properties of Unsaturated Porous Media: A Critical Review and New Models of Hysteresis and Prediction of the Hydraulic Conductivity," (Hebrew), Project No. 38/77, Technion-Israel Inst. of Technol., 235 p., Haifa, Israel.

[11] Mualem, Y. and Dagan, G., 1975. "A Dependent Domain Model of Capillary Hysteresis," Water Resour. Rese., 11(3), pp. 452-460.

[12] Mualem, Y., 1976a. "Hysteretical Models for Prediction of the Hydraulic Conductivity of Unsaturated Porous Media," Water Resour. Res., Vol. 12, No. 6, pp. 1248-1254.

[13] Mualem, Y., 1976b. "Extension of the Similarity Hypothesis Used for Modeling Soils Characteristics," (submitted for publication in the Water Resources Research).

[14] Mualem, Y. and Dagan, G., 1976. "Methods of Predicting the Hydraulic Conductivity of Unsaturated Soils," Technion, Israel Institute of Technology, Research Project 442, Haifa, Israel.

[15] Mualem, Y. and Morel-Seytoux, H. J., 1976. "Capillary Pressure," to appear in the Encyclopedia of Soil Science, CEP76-77YM-HJM18, Engineering Research Center, Colorado State University, Fort Collins, Colorado, 44 p.

[16] Mualem, Y. and Morel-Seytoux, H. J., 1977. "A Critical Analysis of a Capillary Hysteresis Model Based on a One Variable Distribution Function," CEP76-77YM-HJM20, Engineering Research Center, Colorado State University, Fort Collins, Colorado, 35 p.

[17] Reeves, M. and Miller, E. E., 1975. "Estimating Infiltration for Erratic Rainfall," Water Resour. Res., 11(1), pp. 102-110.

[18] Rubin, J., 1967. "Numerical Method for Analyzing Hysteresis Affected, Post-Infiltration Redistribution of Soil Moisture," Soil Sci. Soc. Am. Proc., pp. 13-20.

[19] Staple, W. J., 1966. "Infiltration and Redistribution of Water in Vertical Columns of Loam Soil," Soil Sci. Soc. Am. Proc., 30, pp. 553-558.

[20] Staple, W. J., 1969. "Comparison of Computed and Measured Moisture Redistribution Following Infiltration," Soil Sci. Soc. Am. Proc. 33, pp. 840-847.

[21] Staple, W. J., 1972. "Predicting Moisture Distribution in Rewetted Soils," Soil Sci. Soc. Am. Proc., 34, pp. 387-392.

[22] Talsma, T., 1970. "Hysteresis in Two Sands and the Independent Domain Model," Water Resour. Res., 6(3), pp. 964-970.

[23] Whisler, F. D. and Klute, A., 1965. "The Numerical Analysis of Infiltration, Considering Hysteresis, into a Vertical Soil Column at Equilibrium under Gravity," Soil Sci. Soc. Am. Proc., 29, pp. 489-494.

[24] Whisler, F. D. and Klute, A., 1962. "Rainfall Infiltration into a Vertical Soil Column," Trans. ASCE, 10, pp. 391-395.

MODELING INFILTRATION AS AFFECTED BY SOIL CRUSTING

By

F. D. Whisler, A. A. Curtis, A. Niknam, and M.J.M. Römkens
Department of Agronomy
Mississippi Agricultural and Forestry Experiment Station
Mississippi State University
and U.S.D.A. Sedimentation Laboratory
Mississippi State, MS 39762

Abstract

Soil crusting reduces the infiltration rate of water on many tilled and cropped soils. This soil surface phenomena is a dynamic one following tillage operations and subsequently influences the total hydrology of the watershed.

There are at least two approaches to incorporating the transient soil crusting phenomena into numerical models of infiltration. One approach is to impose a boundary condition of a decreasing flux as a function of time. Another approach is to let the soil water characteristic functions, such as the hydraulic conductivity-pressure head and the water content-pressure head relationships, and other soil properties, such as porosity, etc., change with time. Both approaches have been compared for the same soil and base boundary conditions.

The decreasing flux approach allows the soil to drain when the rainfall rate falls below the saturated conductivity. Such results do not appear physically realistic unless air can enter the soil profile at the surface from the sides or around the crust. In the second approach the calculated surface flux is reduced by decreasing the hydraulic conductivity and the total porosity of the crust. This appears to be a much more realistic approach when considering experimental data.

Numerical models are compared where the numerical grid spacing is held fixed throughout a short profile (10-20 cm) and where variable grid spacings are used. Runoff initiation and accumulated runoff are also calculated during infiltration, both with and without crust formation.

These techniques will be useful in building larger watershed models where tilled crops are grown.

Introduction

The problem of soil crusting is worldwide. Soils that are normally tilled for row crop production are exposed to the unhindered kinetic energy of raindrops for a considerable period of time unless minimum tillage or surface residue management practices are followed. In areas of the world where storms have a higher than average intensity and the soils are low in organic matter and thus structurally weak, soil crusts quickly develop and influence water movement.

All phases of soil water movement as well as plant growth are influenced by soil crusts. Water infiltration rates are decreased 10 to 20 fold by crust formation (Edwards and Larson, 1969; Hillel and Gardner, 1970; Farrell and Larson, 1972; Ahuja, 1973; Ahuja and Römkens, 1974). As a consequence of lower infiltration, there is more effective drought and more of the rainfall leaves the field as runoff, carrying with it soil, nutrients, and pesticides which contribute to non-point pollution of streams (Epstein and Grant, 1967; Moldenhauer and Koswara, 1968; Falayi and Bouma, 1975).

There are some reports where the physical characteristics of the soil crusts have been included in predictions of water infiltration rates and amounts. The amount of water infiltrated into a crusted soil is treated analytically by using simplifying assumptions to the Richards' equation by Hillel and Gardner (1970); Farrell and Larson (1972); and Ahuja and Römkens (1974).

Ahuja (1973) developed a numerical analysis model which predicts the soil water content profiles with time but only for fixed soil crust values. In nature, however, the crust forms with time shortly after the start of a storm until it reaches constant bulk density, hydraulic conductivity, and strength values. Thus, the Ahuja model is insufficient. The Edwards and Larson (1969) model develops the soil water content profiles and crust characteristics with time but in essence does so by controlling the water flux across the top interval in the numerical finite differencing scheme. It also limits the thickness of the crust to the numerical model interval, but crust thicknesses may vary from 0.5 to 2 cm or more depending upon the soil (Bresler and Kemper, 1970). Experimentally, Epstein and Grant (1967) have shown that certain soil physical properties, such as porosity, bulk density and modulus of rupture, change rapidly at the beginning of a rain storm. Thus the decreasing flux of water through the soil surface sould also be represented as a consequence of changes of soil properties rather than by being imposed as a boundary condition.

The purpose of this paper is to present two methods of solving the soil water flow equation, i.e., using a fixed grid system and a refined grid system; and using two representations of the top boundary condition, i.e., a time variant flux condition (Edwards and Larson, 1969) or time dependent soil properties for a surface layer.

Theory

The equation for isothermal, constant concentration vertical flow in a noncompressible porous material (often referred to as the Richards' equation) is:

$$c(h) \frac{\delta h}{\delta t} = \frac{\delta}{\delta z} (K(h) \frac{\delta h}{\delta z}) + \frac{\delta K(h)}{\delta z} \tag{1}$$

in which $c(h) = d\Theta/dh$ = volumetric water capacity, Θ = volumetric water content, h = pressure head (negative for unsaturated conditions), t = time, z = vertical dimension defined as positive in the upward direction and $z = 0$ at the surface, and K = hydraulic conductivity, considered to be a function of the pressure head. In following the logic of Bruce and Whisler (1973) and references therein, Eq. (1) can be rewritten to include soil depth dependent changes in physical properties.

$$c(h,z) \frac{\delta h}{\delta t} = \frac{\delta}{\delta z} (K(h,z) \frac{\delta h}{\delta z}) + \frac{\delta K(h,z)}{\delta z} \tag{2}$$

Fixed intervals. Equation (2) can be solved using the Crank-Nicholson method of finite differencing with iterations at each time step (Whisler and Klute, 1965). Using this scheme, the vertical direction is composed of N segments each Δz in size. In the cases reported herein, $\Delta z = 0.1$ cm and $N = 200$. The interrelationships between $K(h)$ and $\Theta(h)$ were taken from Brooks and Corey (1964). They are:

$$K(h) = K_s \ (h_B/h)^{\eta} \qquad h < h_B \tag{3a}$$

$$K(h) = K_s \qquad h \geq h_B \tag{3b}$$

and

$$(\Theta(h)-\Theta_r)/(\Theta_s-\Theta_r) = (h_B/h)^{(\eta-2)/3} \qquad h < h_B \tag{4a}$$

$$\Theta(h) = \Theta_s \qquad h \geq h_B \tag{4b}$$

where h_B = water entry value = -75 cm of water; K_s = saturated hydraulic conductivity = 1.5×10^{-2} cm min^{-1}; η = a soil characteristic parameter = 6.94; Θ_s = saturated water content (and also the total porosity) = 0.423 cc/cc; and

Θ_r = the residual water content = 0.136 cc/cc. The values of the parameters are similar to those reported for the Touchet silt loam by Laliberte et al. (1966); and hysteresis is ignored in this model except for using the water entry value instead of the air entry value.

The initial condition of the soil profile was assumed to be a uniform water content or mathematically:

$$\Theta(z,0) = \Theta(z) \quad \text{at} \quad t=0 \quad \text{and} \quad -L \leq z \leq 0 \tag{5}$$

where L is the length of the flow system under consideration.

The lower boundary condition was assumed to be a constant water content or pressure head (a semi-infinite flow system) or mathematically:

$$h(-L,t) = h(-L,0) \tag{6}$$

The top boundary conditions considered were:

For no surface crust (Bruce and Whisler, 1973)

$$R = At \qquad 0 < t \leq t_1 \tag{7a}$$

$$R = R_1 \qquad t_1 < t \leq t_2 \tag{7b}$$

$$h(o,t) = 0 \qquad t_2 < t \leq t_R \tag{7c}$$

where A = the rate of rainfall, R, increase (or rainfall simulator), cm min^{-2}, from zero up to time t_1 = 0.25 min; R_1 = a constant rate of rainfall, cm min^{-1}; t_2 = time at which surface ponding starts if $R_1 > K_s$; t_R = time at which the rainfall stops (or rainfall simulator is shut off). It is assumed that for Eqs. (7a-b) that R = flux through the soil surface.

For surface crusting;

adjusting the hydraulic conductivity

$$K(h,z) = K_i \qquad 0 < t \leq t_{c_1} \tag{8a}$$

$$K(h,z) = K_i - B_1(t-t_{c_1})K_i \qquad t_{c_1} < t \leq t_{c_2} \tag{8b}$$

$$K(h,z) = B_2 K_i \qquad t_{c_2} < t \leq t_R \tag{8c}$$

where B_1 governs the rate of hydraulic conductivity decrease after the crust starts to form at time, t_{c_1}; B_2 is the fractional amount of hydraulic conductivity decrease in the crust after it is fully formed at time, t_{c_2}; and K_i is the unadjusted value of the conductivity. Equations (8a-c) are applied to each successive depth interval at a time interval Δt, starting at the surface and working down to the depth, z_c, at the bottom of the crust. For the studies reported herein, $t_{c_1} = 2$ min, $B_1 = 0.48$ min^{-1}, $B_2 = 0.01$, $\Delta t =$ 0.5 min and $z_c = 0.5$ cm. Thus Eq. (8b) was applied to the top node at 2 min, to the second node at 2.5 min, etc. The value of t_{c_2} is determined in the program by comparing (8b) to (8c) and selecting which ever value of $K(h,z)$ was greater.

Adjusting the porosity:

$$\Theta(h,z) = \Theta_s \qquad 0 < t \leq t_{c_1} \tag{9a}$$

$$\Theta(h,z) = \Theta_s - B_3(t - t_{c_1})\Theta_s \qquad t_{c_1} < t \leq t_{c_2} \tag{9b}$$

$$\Theta(h,z) = B_4\ \Theta_s \qquad t_{c_2} < t \leq t_R \tag{9c}$$

where B_3 governs the rate of porosity decrease after crusting starts; and B_4 is the fractional amount of porosity remaining in the crust after it is fully formed. For the studies reported herein $B_3 = 0.083$ min^{-1}, and $B_4 = 0.75$. The t values were the same or determined similarly as for the hydraulic conductivity.

Equations (8a-c) and (9a-c) give a linear change in soil properties with time for the crust development and then a leveling off or constant value for these properties. This roughly fits the results of Epstein and Grant (1967). Other relationships could be used however, as better experimental results become available.

To simulate a crust by adjusting the surface flux, R, Eq. (7b) is modified as:

$$R = R_1 \qquad t_1 < t \leq t_{c_1}$$

$$R = R_1 - B_5(t_{c_1} - t) \qquad t_{c_1} < t \leq t_{c_2} \tag{10b}$$

$$R = B_6\ R_1 \qquad t_{c_2} < t \leq t_2 \tag{10c}$$

where $B_5 = 0.006$ cm min^{-2}, $B_6 = 0.1$, $t_{c_1} = 2$ min, $t_{c_2} =$ 20 min. This value of R was then applied across the top interval of the profile (Edwards and Larson, 1969).

The accumulated infiltration can be calculated by integrating the water content profile at each time step and subtracting the initial, integral profile amount. When the top boundary condition changes from (7b) to (7c), time t_2 is the start of runoff. The amount of runoff at each time step can be calculated as the difference between the rainfall rate, (7b), and the actual flux entering the soil profile. Accumulated runoff is calculated by summation.

Grid refining. Grid refining was used to analyze the crust formation in detail. It is advantageous both for computational efficiency and accuracy to refine the grid size in the region of the seal while maintaining a coarse grid for the remainder of the profile. Increased efficiency is a result of having less nodes than for a uniform grid of the same Δz. Increased accuracy is due to a better finite difference approximation of the flow equations. This improvement applies not only to the dependent variable h, but also to the terms containing $K(h)$ and $c(h)$.

The implementation of grid refining involved the replacement of the uniform grid spacing, Δz, with spacings annotated by either P_i or Q_i about each node. At the depth of the change of grid size, P_i is the large grid spacing and Q_i is the small grid spacing. The finite difference equations were then developed in terms of P_i and Q_i and thus the interface between two different size grids could be accommodated. For example, for the detailed analysis of the crust, a uniform grid size of 0.25 cm was refined to 0.025 cm for the top 1.0 cm of the profile.

Results and Discussion

The results in terms of pressure head profiles versus depth as a function of time are shown in Fig. 1 for the development of a complete crust. As infiltration started, the pressure head at the soil surface increased (became less negative). After the crust started to form (2 min) the pressure head within the crust became almost linear. In fact between 4 and 10 min, the pressure head just below the crust became more negative and then increased again with increasing time. Such an observation would be difficult, if not impossible, to measure and verify experimentally. This is both one of the strengths and weaknesses of numerical modeling, in that one can observe some interesting phenomena that may be impossible to physically measure.

A comparison of pressure head profiles at essentially the same time are shown in Fig. 2 for the cases of: curve (1), a full crust, i.e., both the hydraulic conductivity and porosity were changed, curve (2) where only the hydraulic conductivity was changed, and curve (3) where no crust was formed. From other evidence the differences between curves (1) and (2) are due more to time differences than physical properties. However for the case of no crust it can be observed that the wetting front went deeper, faster than with a crust. This would be expected. Also with no crust the pressure head at

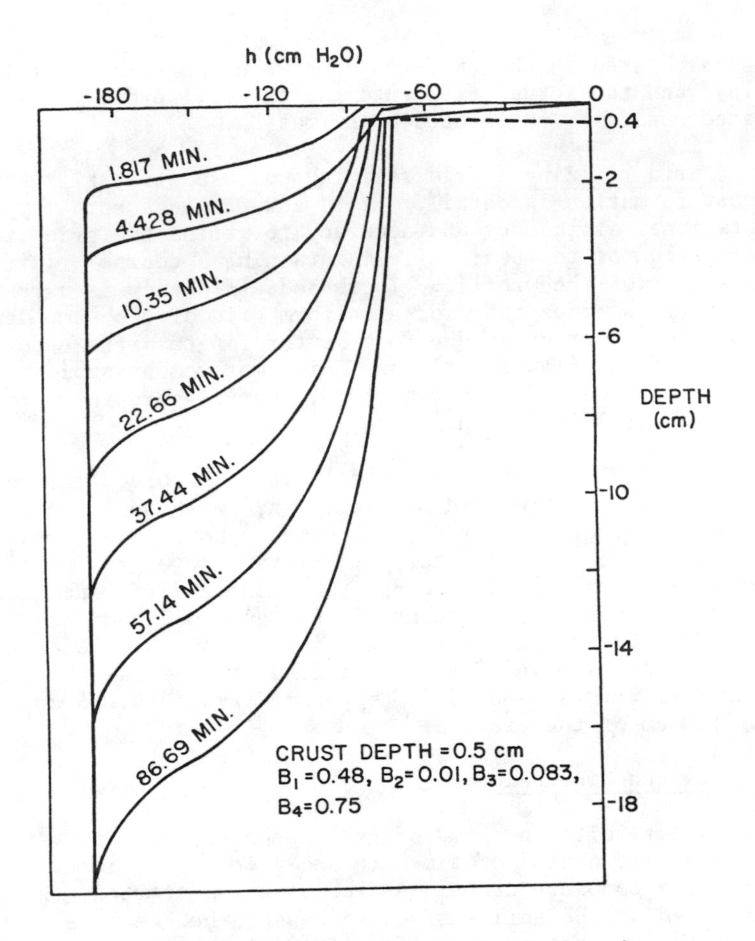

Fig. 1. Pressure head versus depth at the times given for a crust of reduced hydraulic conductivity and porosity.

the soil surface was not as great as with a crust. Thus when a crust develops, ponding would be expected to occur sooner, less water would enter the soil, and more runoff would occur. The case in which the crust was modeled by a change in the surface flux is not shown on this graph for the following reasons: (1) when the flux dropped below the saturated conductivity of the soil, the surface soil began to drain, i.e., the surface pressure head became less than the water entry value, (2) such could only happen physically if air could enter the crust, and (3) since one assumes a crust has formed, air could not enter the soil surface.

A comparison of water content profiles which correspond to the pressure head profiles in Fig. 2 is shown in Fig. 3. Here one sees an unexpected behavior within the crust for the

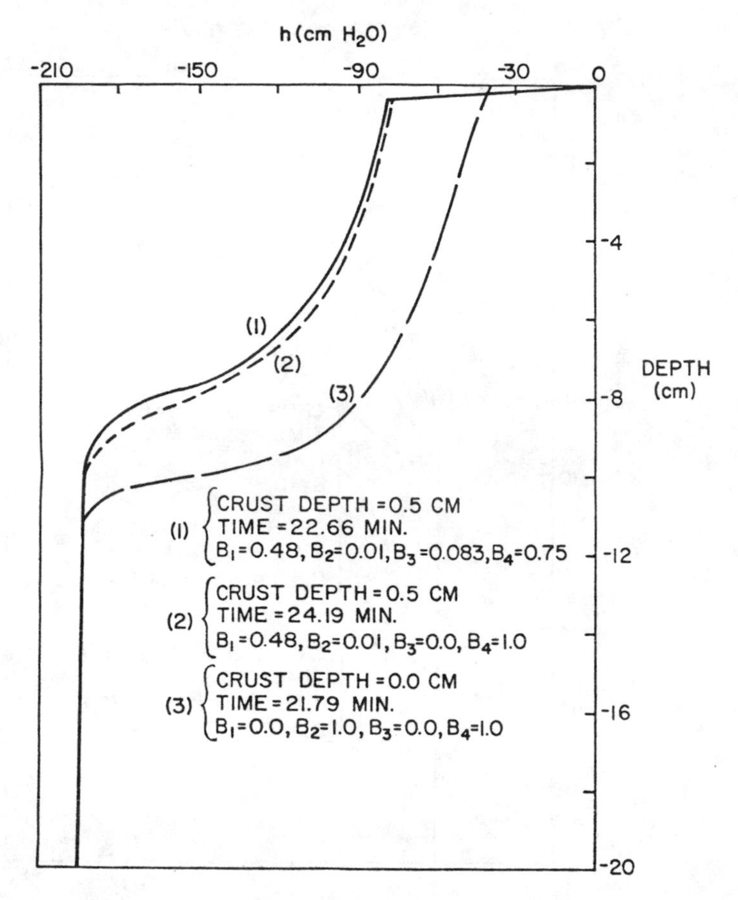

Fig. 2. Pressure head versus depth for two different crusts and no crust. The constants for the seal coefficients are given.

case of a fully developed crust, i.e., the bottom of the crust has a lower water content than the bulk soil immediately below it. This is explained by the fact that the porosity of the crust is lower and thus its volumetric water content would be lower at the same pressure head. Thus there are differences between curves (1) and (2) within the crust in this figure when no difference was observed in Fig. 2. This deviation of water content profiles within the crust is probably due to the fact that the water entry value was not changed. This problem is under further investigation. Curve (3), no crust, showed the steeper, and deeper wetting front as expected.

The infiltration fluxes are compared in Fig. 4. All of the curves are identical up to 2 min. Curve (0) is for the model of a controlled flux, i.e., using strictly Eqs. (10a-c).

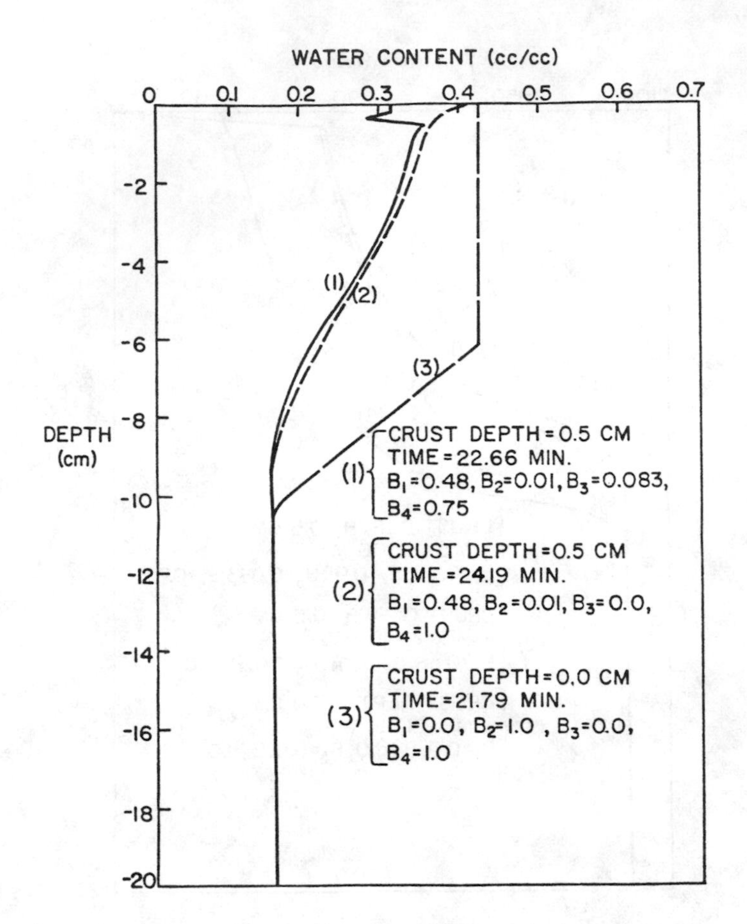

Fig. 3. Water content versus depth for two different crusts and no crust. These correspond to the pressure head profiles in Fig. 2.

Curves (1) and (2) could not be distinguished from one another for this crust. It can be seen that they drop off more steeply than curve (0) but to not as high a value (less negative). Even the no crust curve (3) showed some increase (decrease in magnitude) after 40 minutes of infiltration. This results from the unit hydraulic gradient at the soil surface decreasing as the wetting front progresses deeper into the soil profile.

The accumulated infiltration curves as a function of time for the cases shown in Figs. 2 and 3 are compared in Fig. 5. The amount of accumulated water is greater for the no crust case (3) as would be expected from the earlier discussion of Figs. 2 and 3. The linear relationship with time is also expected since this flow system is a semi-infinite system with constant saturated hydraulic properties (Klute et al., 1965).

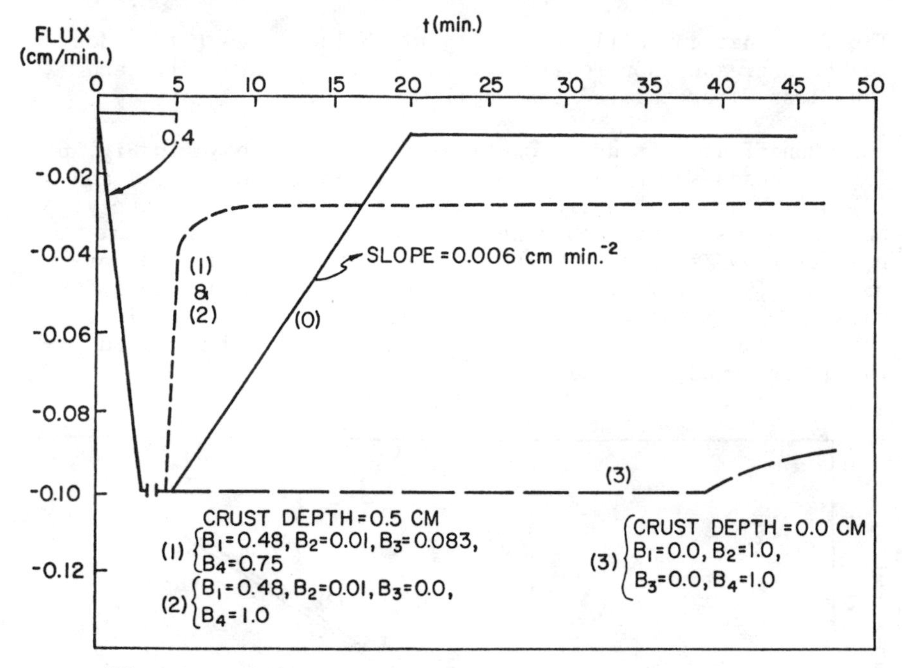

Fig. 4. Infiltration flux versus time for three different crusts and no crust.

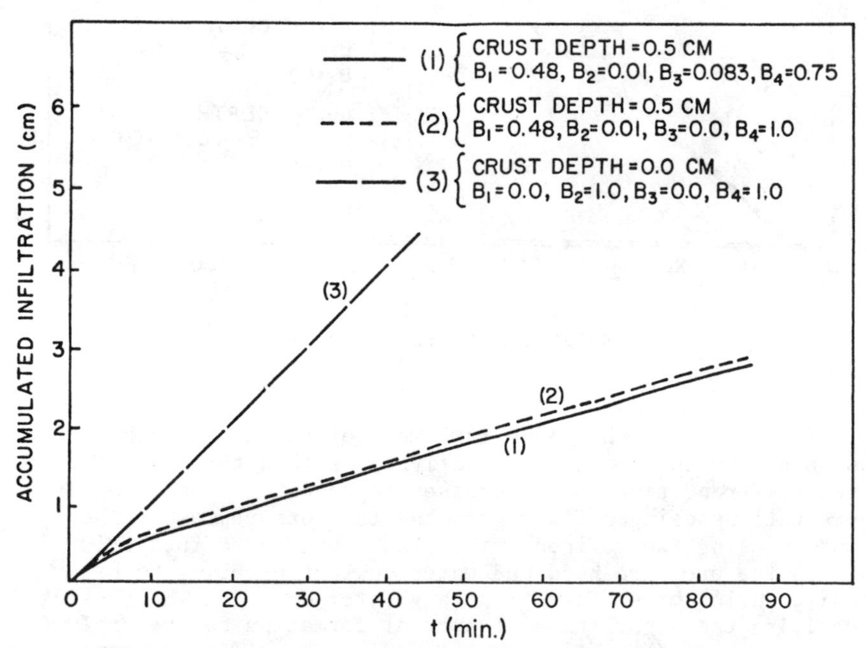

Fig. 5. Accumulated infiltration versus time for two different crusts and no crust.

The fact that curve (1) lies slightly below curve (2) is due to the decrease in porosity within the seal, curve (1), and thus less water entered that soil.

Runoff amounts as a function of time are shown in Fig. 6 for the cases shown in Fig. 5. The curves for cases (1) and (2) are hardly distinguishable and started at approximately the same time. This would be expected based upon the flux results shown in Fig. 4. The runoff amounts calculated for no crust (3) barely show on this scale. The computations for this case terminated after 45 minutes of infiltration since the bottom of the profile was starting to wet and violated the lower boundary condition of Eq. (6).

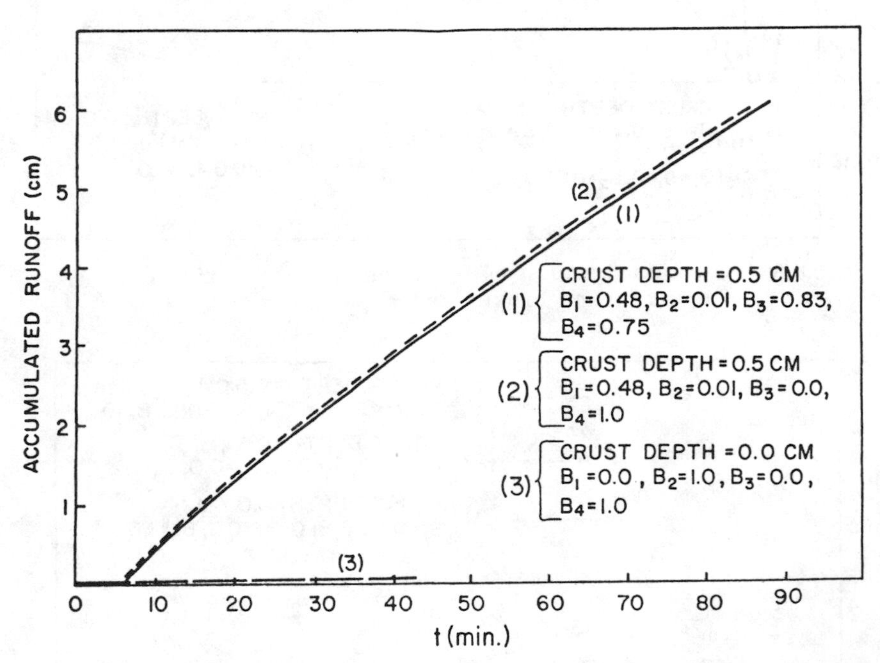

Fig. 6. Accumulated runoff versus time for two different crusts and no crust.

Figs. 7 and 8 show an enlargement of the pressure head and water content profile respectively, within the crust for three different times. The crosses represent the solution of the equations using a fixed grid and the dots represent the solution using the refined grid. The use of a refined grid enables the pressure head and water content profiles to be studied during crust formation in greater detail. The physics underlying the algorithm of the crust formation can be explored also in more detail. There are few distinct differences between the two solutions at any time, but the refined grid

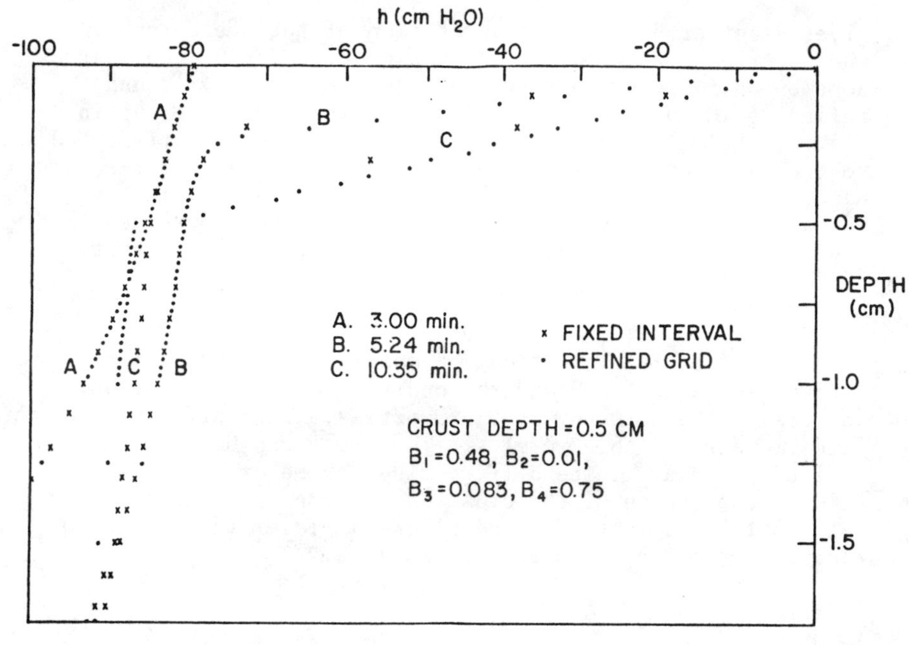

Fig. 7. Pressure head versus depth for a fixed grid and refined grid at three times.

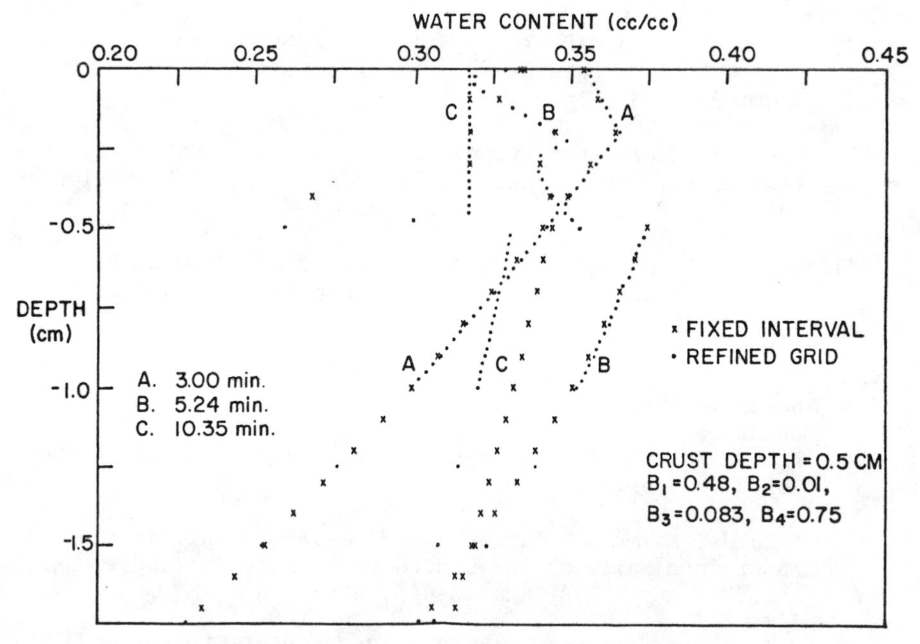

Fig. 8. Water content profiles corresponding to the pressure head profiles of Fig. 7.

gives a better spatial representation of the pressure head and water content gradients. In Fig. 8, Curve C, there are apparent differences observed between the fixed grid and refined grid curves immediately below the crust whereas in Fig. 7, the differences were slight. This may be attributed to the fact that there is a more expanded scale in terms of water content in Fig. 8 than for the pressure head in Fig. 7. The actual maximum magnitude of the differences were about 2.4% for the pressure head and 3.6% for the water content.

Conclusions

A soil surface crust can be modeled as either a time variant surface flux boundary condition or as time variant changes in the soil physical properties. The latter method gives more reasonable results. The crust can be modeled with a fixed interval finite differencing scheme or as a refined grid in the region of the crust and coarser grid in the lower bulk soil. The latter method allows a closer examination of the physics of the simulated flow system.

References

[1] Ahuja, L. R., 1973. "A Numerical and Similarity Analysis of Infiltration into Crusted Soils," Water Resour. Res. 9, pp. 987-994.

[2] Ahuja, L. R. and Römkens, M.J.M., 1974. "A Similarity during Early Stages of Rain Infiltration," Soil Sci. Soc. Amer. Proc. 38, pp. 541-544.

[3] Bresler, E. and Kemper, W. D., 1970. "Soil Water Evaporation as Affected by Wetting Methods and Crust Formation," Soil Sci. Soc. Amer. Proc. 34, pp. 3-8.

[4] Brooks, R. H. and Corey, A. J., 1964. "Hydraulic Properties of Porous Media," Hydrology Papers, Colorado State University, Fort Collins, 3, pp. 1-27.

[5] Bruce, R. R. and Whisler, F. D., 1973. "Infiltration of Water into Layered Field Soils," (In: A. Hadas, D. Swartzendruber, P. J. Ritjtema, M. Fuchs, and B. Yaron.) Ecological Studies 4, Physical Aspects of Soil Water and Salts in Ecosystems, pp. 77-89.

[6] Edwards, W. M. and Larson, W. E., 1969. "Infiltration of Water into Soils as Influenced by Surface Seal Development," Trans., ASAE 12:463-465, 470.

[7] Epstein, E. and Grant, W. J., 1967. "Soil Losses and Crust Formation as Related to Some Soil Physical Proper-Ties, Soil Sci. Soc. Amer. Proc. 31, pp. 547-550.

[8] Falayi, O. and Bouma, J., 1975. "Relationships between the Hydraulic Conductance of Surface Crusts and Soil Arrangement in a Typic Hapludalf," Soil Sci. Amer. Proc. 39, pp. 957-963.

[9] Farrell, D. A. and Larson, W. E., 1972. "Dynamics of the Soil-Water System during a Rainstorm," Soil Sci. 113, pp. 88-95.

[10] Hillel, D. and Gardner, W. R., 1970. "Transient Infiltration into Crust-Topped Profiles," Soil Sci. 109, pp. 69-76.

[11] Klute, A., Whisler, F. D. and Scott, E. J., 1965. "Numerical Solution of the Nonlinear Diffusion Equation for Water Flow in a Horizontal Soil Column of Finite Length," Soil Sci. Soc. Amer. Proc. 29, pp. 353-358.

[12] Laliberte, G. E., Corey, A. T., and Brooks, R. H., 1966. "Properties of Unsaturated Porous Media," Hydrology Papers, Colorado State University, Fort Collins, 17: pp. 1-40.

[13] Moldenhauer, W. C. and Koswara, J., 1968. "Effect of Initial Clod Size on Characteristics of Splash and Wash Erosion," Soil Sci. Soc. Amer. Proc. 32: pp. 875-879.

[14] Whisler, F. D. and Klute, A., 1965. "The Numerical Analysis of Infiltration, Considering Hysteresis, into a Vertical Soil Column at Equilibrium under Gravity," Soil Sci. Soc. Amer. Proc. 29: pp. 489-494.

DISCUSSIONS - SECTION III
Soil Water Processes

Claude Thirriot*

As answer to the questions raised by the General Reporter concerning the paper presented by J. Ganoulis: The physical process is shown by a short motion picture exhibiting the intrusion of mercury in a schematic plane artificial model with a fixed equivalent porometry. Mercury content measured is in good agreement enough with the theoretical prediction allowed by the stochastic model.

Concerning influence of gravity, when a heavier unwetting fluid is flowing through the porous media, capillary pressure in increasing with depth. If a little volume succeeds to go down through the big pores in the upper region, then in the lower zone it invades pores which are smaller and smaller when depth increases because of the increase in capillary pressure between the two fluids. Thus the saturation in the entering fluid is larger in the lower zone and goes on increasing.

For summarizing, in the upper zone, near the entrance the blockade is preponderant and provides a progressive reduction of saturation in entering unwetting fluid (this phenomenon is to compare at the complete blockade that can appear in the horizontal case) and in the lower zone gravity effect leads to a saturation graph similar to that one of capillary fringe of wetting fluid (it is only a resemblance of curve due to a common effect of gravity upon capillary pressure change).

One of the applications of our model is the determination of the true porometric graph from the capillary pressure curve obtained with the apparatus of Purcell and forced introduction of mercury in a porous media sample.

Concerning the paper presented by R. Gaudu, Hoa, Thirriot and Thore, we use interpolation methods to represent hysteresis and scanning curves because their simplicity and rapidity in computation and their good agreement with observed cycles. But we also used (especially Mr. Thore) independent or dependent domain method.

Concerning the dam problem, experiments were led on a sand model one meter high. Measurements of water content were realized with photoelectrical gauges made in our laboratory. These gauges are very cheap, with reproducible results and electrical instantaneous response easy to record but with a nonlinear calibration curve. Unit price is about twenty dollars.

---

*Institut de Mecanique des Fluides de Toulouse, France.

Concerning the evaporation phenomenon, we examine different initial conditions to test the sensibility of the nearly stationary phenomena against variation of initial state. In this topic we chose, of course, extreme situations that are evidently perhaps unrealistic.

Robert M. Dixon*

Dr. Bresler, I agree with your brief general comments. I did not come prepared to advocate rigorously the paper authored by Mr. Simanton and myself; however, I will atempt to introduce a bit of our philosophy on the nature of infiltration systems.

About 10 years ago we came to the shocking conclusion that the earth is not a thin flat slab of uniform microporous soil material with a thin uniform layer of water ponded on top and open to the atmosphere at the bottom as is commonly assumed by Darcy-based infiltration theory. Instead the natural infiltration system has a highly dynamic and nonuniform upper boundary and the soil profile, by no stretch of the imagination, is uniform. In fact, it exhibits a highly complex geometry of interconnected micropores and macropores. Furthermore, the lower boundary of the infiltration system is highly irregular and may be regarded as more or less closed to the atmosphere.

Nevertheless, Darcy's Law need not be abandoned since the product of the hydraulic conductivity and gradient still gives the water infiltration rate at the soil surface. The nonconventional aspect of the air-earth interface theory is that these hydraulic characteristics are controlled by the dynamic air-earth interface rather than by the soil profile.

Thus the air-earth interface theory does not contradict Darcy-based theory; instead it complements and extends such theory to natural soils having variable surface conditions.

F. D. Whisler**

Especially when one uses the Brooks and Corey functional relationships between K-h-θ when the surface flux decreases in magnitude below the saturated hydraulic conductivity, the soil surface begins to drain. Thus a "flux only" representation or model of a surface seal is not physically realistic. It is observed that a real seal does have considerable changes

*Southwest Watershed Research Center, USDA-ARS, 442 E. 7th St., Tucson, Arizona.

**Dept. of Agronomy, Miss. State University, Mississippi State, Mississippi.

in its hydraulic properties and thus as long as it is raining at a rate greater than the seal saturated hydraulic conductivity, it will not drain.

It is true that the shape of the flux is purely arbitrary in the "flux only" representation of the surface seal. It is shown here only to compare with the other fluxes where the soil parameters were changed since all had the same "applied rainfall" rate.

The runoff calculations are probably the most interesting results from a hydrologists point of view (Fig. 2c of the review). These results show a distinct difference between sealed and unsealed surfaces while the "flux only" model results are not shown (we will show them in the final version), they were closer to the unsealed soil than with the surface seal.

<u>Yechezkel Mualem*</u>

Comment about the work of Parlange. I would like to comment on Dr. Parlange's solution of two phase flow described by our Eq. (13). The author indicates that the left hand side of this equation represent the air effect, namely, when the air effect is zero and air does not imped the water flow the left hand side equal to zero. I cannot understand, therefore, how the approximate solution given by Eqs. (14) and (15) which annuls the left hand side of Eq. (13) indicating no air effect can be considered at the same time a better solution when the air effect becomes important.

I have a short complementary comment to the review given by the reporter, Dr. Bresler, regarding my paper, "Theory of Universal Hysteretical Properties of Unsaturated Porous Media." The main contribution of my new model of hysteresis is that it requires only one retention curve, either in the main drying process or the main wetting process, for calibration of the model. Any other scanning curve can be then derived on the basis of this one measured curve. This is the minimum experimental data one may think of. However, this save in the experimental data is earned (for some soils) at the expense of accuracy. The model cannot account for the effect of the air blockage against water entry as this is an independent domain model. The poor accuracy of the model is found mainly in the range of the air entry value while for lower ψ its accuracy is improved considerably. The use of this model should be limited, therefore, for generally heavy soil which does not indicate a relatively high air entry value.

---

*Engineering Research Center, Colorado State University, Fort Collins, Colorado.

Sun-Fu Shih*

Further explanation of "Synthetic Hydraulic Conductivity and Its Application" by S. F. Shih.

The comments given by the general reporter are appreciated. But some parts are needed to explain further. Equation (6)

$$k = e^{c\mu_y + c\sigma_y} \qquad (6)$$

had a typing error which must be corrected as

$$k = e^{c\mu_y + c\sigma_y t} \qquad (6)$$

If the $k_1 = k_2$ is existing, then Eq. (7) is correct. However, as indicated in the original paper, the k_1 is not equal to k_2 in most practical sample data, then a new relationship as given in Eq. (10) was developed.

The statement given by the reporter, the mean of the logarithms is less than the logarithm of the mean does not necessarily imply that an experimental effective permeability based on a heterogeneous matrix will be more than the mean permeability, is true in certain case. But, the synthetic hydraulic conductivity is generated based on the values of μ_y and σ_y (as shown in Eq. 6) which were computed based on the data with log transformation. Therefore, as the conclusion given in Eq. (10), the mean of synthetic hydraulic conductivity based on the log normal distribution has resulted in lower value than experimental data. This fact also implies that the mean of synthetic data based on the normal distribution has a similar result as experimental data because the data used in normal distribution are without transformation.

Based on Eqs. (1), (2), (3), (4), and (5), the μ_k and σ_k are correlated each other. The example A also showed that both synthetic and experimental data give a same conclusion as shown in Eq. (5), i.e., the σ_k values were greater than μ_k when the values of σ_k were greater than 0.3616.

As Fig. 3 shows, the computed data are based on the Monte Carlo simulation which has involved a major statistical error. Therefore, based on the multiple regression technique, a best fit relationship between sample size n and uncertainties (S_ϕ σ_y) from the computed data was shown in Eq. (15).

*Agricultural Research and Education Center, IFAS, University of Florida, Belle Glade, Florida

SECTION IV

SUBSURFACE WATER QUALITY

Section Chairman, Leonard Konikow
Research Hydrologist
U.S. Geological Survey
Denver, Colorado

Rapporteur, Lynn Gelhar
Professor of Hydrology
New Mexico Institute of Mining and Technology
Socorro, New Mexico

SECTION IV

SUBSURFACE WATER QUALITY

SECTION IV

SUBSURFACE WATER QUALITY

GENERAL REPORT

By

Lynn W. Gelhar
New Mexico Institute of Mining and Technology
Socorro, New Mexico

In this session on subsurface water quality there are six papers which deal with a broad range of topics relating to solute transport; included are investigations of unsaturated flow, thermal convection, adsorption, numerical methods and salinity modeling. The papers in this session could be categorized in several ways; I have chosen to consider the following three classifications: i) investigations of basic physical phenomena, ii) development of methods of analysis and simulation and iii) applications to prediction under field conditions. Beginning with papers which emphasize basic investigations and ending with field application, this report provides the summary of each paper along with my comments, questions and suggestions relating to each.

The paper by Starr, Parlange, Sawhney and Frink, "Diffusivity of Solutes in Adsorbing Porous Media," deals with the basic question of how adsorption will effect dispersive mixing in a porous medium. The theory of longitudinal dispersion for laminar flow in a straight capillary tube is developed following the approach of G. I. Taylor; the effect of a nonlinear adsorption isotherm is included. They find that the resulting one-dimensional transport equation has an additional term which is proportional to the square of the longitudinal concentration gradient (see eq. 20) and arises because of the nonlinearity of the adsorption relationship. The dependence of the dispersion coefficient on adsorption is the same as found by Golay (1958) assuming a linear isotherm (k = const.). The results of experiments on dye adsorption in a Teflon tube show that the effect of the nonlinearity of the isotherm is minor but that the effect of adsorption on the dispersion coefficient is important. The experiments confirm the dependence of the dispersion coefficient on adsorption as predicted by the theory (see Fig. 1). Experiments with a calcium saturated soil indicate only a very weak dependence of the dispersion coefficient on adsorption. The authors attribute this observation to large scale heterogeneity; the medium's pore size distribution and the interconnected pore structure.

The analysis of the capillary tube problem provides some insight on the effect of nonlinear adsorption on dispersion in porous media. Many readers will find it difficult to follow the overall rationale of the iterative expansion method use in the theoretical analysis. The single capillary tube model does not simulate the behavior of a real soil; a multiple random tube model (see Haring and Greenkorn, 1970) would provide a more realistic representation of natural media.

The paper by Oakes, "The Movement of Water and Solutes through the Unsaturated Zone of the Chalk in the United Kingdom," introduces some important basic questions about solute movement in the unsaturated zone. The paper first describes observations of tritium in the unsaturated chalk that imply a downward rate of movement of aroung 1 m/yr. Infiltration and groundwater level data are then analyzed assuming a linear response relationship; the results imply that the effect of surface infiltration moves down to the water table at a rate on the order of 1 m/day. The author explains this two order of magnitude difference in rate of movement in terms of a fissure system which allows rapid movement and a porous matrix containing immobile water into which solutes diffuse very rapidly. Laboratory experiments are reported showing that the time scale of diffusion into a typical porous matrix block is on the order of a day or less. A model simulating the movement of nitrates and tritium is described briefly; results show favorable agreement with field observations (see Fig. 7).

The description of the model is generally inadequate. Was steady flow assumed? How was the interchange between the two phases handled? What numerical technique was used to solve the transport equation? Such ambiguities would be eliminated if the equations forming the basis for the model were given. It would also be helpful if the method of determining the monthly infiltration was described.

The observed difference in the rate of downward movement might also be explained from classical concepts of unsaturated flow as follows (see Wilson and Gelhar, 1974). The vertical distribution of moisture content $\theta(z,t)$ is described by

$$\frac{\partial \theta}{\partial t} = \frac{\partial}{\partial z} \left[D(\theta) \frac{\partial \theta}{\partial z} \right] - \frac{dK(\theta)}{d\theta} \frac{\partial \theta}{\partial z} \qquad (1)$$

where z is measured downward, $D(\theta)$ is the soil moisture diffusivity and $K(\theta)$ is the hydraulic conductivity. If a small moisture pulse is introduced as a perturbation on a uniform moisture content θ_0,

$$\theta = \theta_0 + \theta' \quad , \quad \theta' << \theta_0$$

Eq. (1) can be approximated by

$$\frac{\partial\theta'}{\partial t} + K' \frac{\partial\theta'}{\partial z} = D_0 \frac{\partial^2\theta'}{\partial z^2}, \quad D_0 = D(\theta_0), \quad K' = \left.\frac{dK}{d\theta}\right|_{\theta_0} \tag{2}$$

A small pulse of moisture will thus propagate downward with a velocity K' and spread with a diffusivity D_0.

For a conservative solute the equation describing the concentration $C(z,t)$ is

$$\theta \frac{\partial C}{\partial t} + q \frac{\partial C}{\partial z} = \frac{\partial}{\partial z}\left(\theta E \frac{\partial C}{\partial z}\right) \tag{3}$$

where q is the specific discharge and $E(q,\theta)$ is the dispersion coefficient. Again considering small changes in and q around θ_0 and $q_0 = K(\theta_0)$, the steady state moisture flux, the transport equation is approximated by

$$\frac{\partial C}{\partial t} + U_0 \frac{\partial C}{\partial z} = E_0 \frac{\partial^2 C}{\partial z^2}, \quad U_0 = \frac{q_0}{\theta_0}, \quad E_0 = E(q_0,\theta_0) \tag{4}$$

A tracer introduced with the pulse of moisture will therefore move downward with a velocity U_0 and disperse according to E_0. If an exponential conductivity relationship is assumed

$$K(\theta) = K_1 e^{a\theta}, \quad K_1, \quad a = \text{consts.}$$

then the ratio of the speed of the moisture pulse to that of the solute is

$$\frac{\theta_0 K'}{K(\theta_0)} = a\theta_0 \tag{5}$$

For soils the parameter a might typically be in the range 20 to 100 depending on the moisture content (Warrick et al., 1971; DeSmedt and Wierianga, 1977), so that at higher moisture contents the moisture pulse would propagate at least 10 times as fast as the solute.

Thus classical unsaturated flow characteristics can explain at least some of the difference in rates of downward movement reported in this paper. If the bulk unsaturated conductivity relationship of the chalk are determined it may be that the entire difference can be explained by the elementary concept. A large increase in hydraulic conductivity would be expected as the fissures become saturated. The phenomenon is one of wave propagation wherein the effect of a moisture disturbance in unsaturated flow can travel downward much more rapidly than the actual water displacement. The effect of infiltration at the surface will be felt at the water table long before that infiltrating water actually reaches the water table. Some of the retardation of the tritium noted in this paper may also be due to adsorption, Wierenga, et al., 1975, have observed significant retardation of tritium in laboratory tests with unsaturated sandstone cores.

I consider the unsaturated flow phenomenon reported in this paper to be extremely important. The propagation phenomenon is of fundamental and applied significance and deserves more attention in future research.

The paper by Souers, Rubin and Christensen, "On the Migration of Contaminants in an Aquifer Subject to Geothermal Activity," outlines analytical and numerical solutions for thermal convection and mass transport in an idealized system. The paper gives the formulation of the problem of thermal convection in a horizontal layer of porous medium heated from below; included are the effects of nonlinear resistance and velocity dependent dispersion coefficients in three dimensions. Some results of previous linear stability analyses of this convection problem by Rubin are then summarized. The theory predicts that the instability will be in the form of longitudinal cells with helical streamlines. An analytical perturbation approach using Fourier expansion is used to describe the supercritical convective motion. The paper then outlines the numerical approach to contaminant motion in the analytically determined convection field using the method of characteristics and presents some numerical results for transport of an ideal tracer in the convection field.

This paper has several features which detract from its utility. First of all, several important physical assumptions of the formulation are not given. Thermal equilibrium between the fluid and the solid matrix is implicit in Eq. (3); this may not be a realistic assumption under the high Reynolds number conditions which are being treated in terms of flow resistance. The ideal tracer assumption is not presented and the boundary conditions are not stated for the contaminant transport problem. Also it is apparently being assumed that the contaminant is injected without affecting the flow field; however, this is not stated. Secondly, the discussion of application to the Boulder Zone of the Floridian Aquifer is largely irrelevant to the analysis undertaken in the paper. In the introduction it is stated this zone is highly anisotropic; yet the analysis assumes isotropy of hydraulic conductivity. Sea water intrusion is noted in the field but salinity effects are neglected in the analysis. The paper contains no comparison of theory with observations in the Floridian Aquifer.

The paper also has several ambiguous features which leave the reader guessing. Several undefined or inadequately defined symbols appear (ℓ_i and β, p. 5; ε_{ij} and θ, p. 7; a and a_0 in (8)). Is $\bar{b}$ a constant in $\bar{R}$ (p. 5)? Is u_i a perturbation in (5)? The presentation and discussion of results is incomplete and confusing. On p. 9 the influence of d/d_p is mentioned but is not demonstrated or explained. On p. 10 an isothermal state is indicated to occur at large Rayleigh number but is not shown. In that same paragraph a better choice of wave number is mentioned. What is it? Is the wave number arbitrary? Table 1 does not adequately describe the cases presented. What are the values of d/d_p, R and Re in each case? Where are the injection points in Fig. 3; should these streamlines close? What is the location of the concentration in Fig. 4; which streamline is referred to in Fig. 5?

The reference list contains numerous entries which are not cited in the text. How are these related to the work? Missing from the literature review is any discussion of related stability analyses of convection in porous media. How do the results of this paper compare with earlier work; in particular, that of Prats (1966) who also considered the effect of a mean horizontal current? How do the results of the stability analysis compare with laboratory experiments? Are longitudinal cells observed in the lab or in the field?

In summary, the authors appear to have taken on a significant problem with some vigor, but the form of presentation leaves me rather uncertain about what has been shown.

The paper by DeSmedt and Wierenga, "Simulation of Water and Solute Transport in Unsaturated Soils," deals with improved explicit finite difference schemes for solving the moisture and solute transport equations numerically. The stability requirements of the classical explicit finite difference scheme are first presented. Improved finite difference representations of the two equations are developed by including terms of first order in the time increment; the corresponding stability requirements of the improved schemes are also presented. Larger space and time increments are permissible with the improved schemes, thereby decreasing the computer time. Three application examples are presented to illustrate the effectiveness of the improved schemes; 75 to 95 percent reductions in computer time are found with no loss of accuracy.

This paper nicely demonstrated how the problems of numerical dispersion and oscillation can be handled systematically when using an explicit finite difference scheme. However, it is not clear that the improved explicit scheme will be competitive with implicit schemes in terms of computational effort. Direct comparisons of the explicit and implicit schemes are suggested for some examples. The examples of this paper also serve to illustrate the solute lag effect which I referred to on the Oakes paper; in comparing Figs. 7 and 8 we see that the moisture front has moved 3 to 4 times as far as the solute front.

The paper by Schwartz, "Applications of Probabilistic-Deterministic Modeling of Mass Transfer in Groundwater Systems," presents a numerical method of simulating dispersion in aquifers using a random walk model. The model calculates the motion of a large number of particles from the known mean convective motion and superimposed normally distributed random perturbations which simulate dispersion. The concentration is determined by the number of particles in a specified volume. Results of the random walk simulation are compared with the analytical solution for the one-dimensional step input; it is shown that the accuracy of the method increases with number of particles per time step. The method is also applied to a two-dimensional vertical regional circulation system into which a contaminant is introduced locally; this example illustrates the effects of the number of particles and the size of the time step.

The random walk simulation method is attractive because of its conceptual simplicity; it provides direct physical insight on the dispersion process. However, there are some important limitations of the method which are not brought out in the paper. The medium must be isotropic with respect to hydraulic conductivity in order to use the dispersion scheme in the paper; most natural materials are not isotropic especially when a vertical section is considered as in the application example. The method also does not apply to systems in which the mean flow is affected by the contaminant distribution, as would be the case with a change of density. There are also some questions about the accuracy of the computational scheme. Even with the largest number of particles, the one-dimensional results (Fig. 3) show significant errors similar to those sometimes encountered due to numerical oscillation with finite difference and finite element methods. The two-dimensional results should converge to a unique solution as the number of particles is increased and the size of the time step and cells are decreased; this has not been demonstrated in this paper. I would suggest that comparisons with exact analytical solutions in two and three dimensions be used to evaluate the accuracy of the scheme. Computer time and storage requirement should be given for the problems in the paper. The paper by Alhstrom and Foote (1976) gives more information on the use of this type of model on groundwater problems.

This method has potential as an alternative to numerical solution of partial differential equations of transport provided that its limitations are recognized and accuracy is evaluated; the method deserves additional investigation. However, optimism of the paper in referring to the method as "...probably the most versatile and powerful mass transfer model existing in North America." hardly seems justified at this point.

The paper by Riley, Bowles, Chadwick and Grenney, "Preliminary Identification of the Salt Pick-up and Transport Processes in the Price River Basin, Utah," outlines the anticipated approach to modeling the salinity production due to diffuse natural sources in predominantly semi-arid basin of the Western United States. The paper first reviews the salinity problem in the Colorado River Basin and then describes the hydrologic conditions of the Price River Basin. The general conceptual structure of a salt transport model is presented with emphasis on the surface runoff aspects; included are an overland flow model and a channel flow model, both of which include components for suspended and dissolved salt transport. In the ephemeral channels several mechanism of salt interchange are considered including mass wasting of channel bank; interflow, groundwater flow and bank storage; and dissolution or precipitation of salts on the bed. The various aspects of salt pick-up or deposition during the channel flow process are discussed in terms of a generalized functional relationship. The authors conclude that more research is needed on salt pick-up mechanisms in channels and suggest more emphasis on salinity sources associated with subsurface flow.

As the authors indicate, the paper is largely a statement of the intended direction of current research; no specific results are presented. The total mass balance equation, (3), of the channel flow model is valid only for steady flow. The storage term $\partial A/\partial t$ should be included or some justification for its ommission should be given. This paper represents a useful effort to conceptualize salt pick-up phenomena in channels; however, a great deal more research on the salt transfer process in open channels will be needed to implement a quantitative model. One might question why such great emphasis is placed on salinity sources associated with surface water if, as indicated in the section Work Needed, there are "...relatively small salt contributions from surface flows...". The emphasis on a stochastic bank wasting model might also be questioned; all of the processes involved in the salinization could as well be viewed stochastically. The authors might also be more specific about the subsurface salinization processes they plan to emphasize in future work.

I would like to close with some brief comments on the state of the art and future direction of research in the area of subsurface water quality, and the relationship of the papers in this session to that direction. I will structure my comments according to the three categories I referred to initially: basic phenomena, simulation methods, and field applications. In the basic area, the fundamental transport equation is well established through theoretical anaylsis and laboratory testing. A basic area which needs more attention is the chemical and biological interactions which are reflected in the source term of our transport equation; the paper by Starr et al. on adsorption is an effort in that direction. Another important basic problem is the role of natural heterogeneities in the dispersion process; this problem is referred to in the discussion by Starr et al., and is currently being investigated at New Mexico Tech using stochastic methods (Gelhar, 1976). There are several basic questions relating to mass transport in unsaturated flow which remain to be resolved; the solute lag effect as illustrated by Oakes' paper is an important example. Also to be resolved are the dependence of dispersion coefficient on moisture content and velocity (see eg. Wilson and Gelhar, 1974) and the role of immobile moisture (see eg. Gaudet et al., 1976 of Oakes' paper). There is also a need for better understanding of systems in which the flow and mass or heat transport equations are coupled; the paper by Souers et al. is an effort in that direction.

In the area of simulation techniques, the last decade has seen a rapid growth of research on the application of numerical methods in subsurface mass transport. However, the implementation of this methodology in practice has been slow in part because of the usual inertia of the system but also because of the personalized nature of computer programming and the competitive environment of a developing research area. In the area of numerical simulation I see a need for more work which improves the reliability and accessibility of existing methods. The papers by DeSmedt and Wierenga, and Schwartz are useful efforts in that direction.

The area of field applications is clearly the most challenging and important when we recognize that hydrology is basically an applied science. The field situation typically involves several potentially important processes; we are usually forced to deal with limited data and seldom have control over experimental conditions. The paper by Riley et al. nicely brings out the complexities that one faces when attempting to model a large scale natural system. However, none of the papers in this session have dealt with the important problem of model calibration or parameter estimation, i.e., the inverse problem. This is an important area of research which should include the stochastic nature of the processes and ultimately should give the reliability of predictions in a statistical sense.

From this brief review of some research directions in subsurface water quality, I judge that this session involved several forward looking papers. My comments on the papers are intended to sharpen the discussion, not detract from the merit of a generally good group of papers.

References

[1] Ahlstrom, S. W. and Foote, H. P., 1976. "Transport Modeling in the Environment Using the Discrete-Parcel-Random-Walk Approach," Proceedings of the EPA Conference on Environmental Modeling and Simulation, Cincinnati, Ohio, April, (in press).

[2] Gelhar, L. W., 1976. "Stochastic Analysis of Flow in Aquifers," Proceedings AWRA Symposium on Advances in Groundwater Hydrology, Chicago, September, (in press).

[3] Haring, R. E. and Greenborn, R. A., 1970. "A Statistical Model of a Porous Medium with Non-uniform Pores," Amer. Inst. Chem. Eng. J., Vol. 16, No. 3, pp. 477-483.

[4] Prats, M., 1966. "The Effect of Horizontal Fluid Flow on Thermally Induced Convection Currents in Porous Media," J. Geophysical Research, Vol. 71, No. 20, pp. 4835-4838.

[5] Warrick, A. W., Bigger, J. W. and Nielsen, D. R., 1971. "Simultaneous Solute and Water Transfer for an Unsaturated Soil," Water Resources Research, Vol. 7, pp. 1216-1225.

[6] Wierenga, P. J., van Genuchten, M.Th. and Boyles, F. W., 1975. "Transfer of Boron and Tritiated Water through Sandstone," J. Environ. Qual., Vol. 4, No. 1, pp. 83-87.

[7] Wilson, J. L. and Gelhar, L. W., 1974. "Dispersive Mixing in a Partially Saturated Porous Medium," Report No. 191, Ralph M. Parsons Laboratory, M.I.T.

SIMULATION OF WATER AND SOLUTE TRANSPORT IN UNSATURATED SOILS

By

F. De Smedt and P. J. Wierenga
Graduate Student and Associate Professor
Department of Agronomy
New Mexico State University
Las Cruces, New Mexico 88003

Abstract

A simulation model programmed in the CSMP (Computer Simulation Modeling Program) computer language, is presented for water flow in the unsaturated zone. The stability and convergence criteria of the model, which is basically an explicit finite difference approximation of the flow equation, are rather restrictive. These criteria are less restrictive with an improved scheme proposed in the paper. It is shown that the improved scheme is particularly useful for simulating the simultaneous transport of water and solute. Use of the improved scheme considerably reduces the required computer time. The theoretical considerations are illustrated with numerical examples.

Introduction

The flow of water and solutes in unsaturated soil can be described by second order partial differential equations of the one dimensional convection diffusion type. Numerical techniques are used for solving these equations, under transient flow conditions. A special numerical technique is provided by the S/360 CSMP simulation language (Speckhart and Green, 1976). This programming language has been used in several studies of transport processes in soils, e.g., unsaturated water movement (Van Keulen and Van Beek, 1971; Bhuiyan et al., 1971), solute movement (Van Keulen and Van Beek, 1971; van Genuchten and Wierenga, 1974), pesticide movement (van Genuchten et al., 1974), and heat transfer (Wierenga and De Wit, 1970). General discussions of the simulation of transport processes in soils were presented by De Wit and Van Keulen (1972) and by Beek and Frissel (1973).

Simulation of these transport processes involves dividing the soil into a number of layers of a given thickness Δx. For each time step Δt, the flow of matter is calculated over the layer boundaries by means of the mass transfer equations. Net gain of mass in each layer is determined by integration of the net fluxes in each layer. The layer thickness Δx and time step Δt depend upon stability and convergence criteria of the particular problem. Of particular interest is that for solute transport it was found that the dispersion of the solute front could not be accounted for by the dispersion coefficient alone. This phenomenon has been experienced frequently when numerical approximations were made of the convective diffusion equation (Peaceman and Rachford, 1962; Stone and Brian, 1963; Gardner et al., 1969; Pinder and Cooper, 1970; Lai and Jurinak, 1972; Bresler, 1973; and Goudriaan, 1973).

Chaudhari, 1971, presented a finite difference approximation which corrected for numerical dispersion in the case of steady solute flow in a soil of uniform water content. Van Genuchten and Wierenga, 1974, showed that when such corrections were made, stability and convergence criteria changed. They were able to use a layer thickness Δx twice the size of that which could be used in the classical numerical solution without loss of accuracy.

Although the simulation of transport processes with CSMP, which is basically an explicit finite difference approximation, consumes more computing time than the implicit finite difference approximations, it is a very useful method as is indicated by the increasing number of applications of CSMP found in the literature. The main advantage of the method is its programmation. We believe that with the new fast computers, use of CSMP as a simulation method will become very convenient.

The purpose of this paper is to investigate the convergence and stability criteria for the simultaneous flow of water and solutes in unsaturated soils, under transient conditions. It will be shown that improved approximation of the transport equations considerably reduces the required computer time.

The equation for water flow. For unsaturated conditions the water flow equation can be written as:

$$\frac{\partial \theta}{\partial t} = \frac{\partial}{\partial z}\left[D(\theta)\frac{\partial \theta}{\partial z}\right] - \frac{\partial K(\theta)}{\partial z} \qquad (1)$$

where θ is the volumetric water content (cm^3/cm^3), t is the time (days), z is the depth measured positively downwards (cm), $K(\theta)$ is the unsaturated conductivity (cm/day) and $D(\theta)$ is the diffusivity (cm^2/day); both are functions of the water content.

The numerical simulation of Eq. (1) in CSMP is basically an explicit finite difference approximation. The classical approach is as follows: from the Taylor series expansion of $\theta(z,t+\Delta t)$ we obtain:

$$\frac{\theta(z,t+\Delta t) - \theta(z,t)}{\Delta t} = \frac{\partial\theta(z,t)}{\Delta t} + O\ (\Delta t) \qquad (2)$$

where $O(\Delta t)$ means: terms of order Δt. With Eq. (1), this becomes

$$\frac{\theta(z,t+\Delta t) - \theta(z,t)}{\Delta t} = \left[\frac{\partial}{\partial z}\left(D\frac{\partial\theta}{\partial z}\right) - \frac{\partial K}{\partial z}\right]_{z,t} + O(\Delta t) \qquad (3)$$

The right hand side of Eq. (2) is approximated with central finite difference equations with respect to z. This yields

$$\frac{\theta(z,t+\Delta t) - \theta(z,t)}{\Delta t} = \left[\frac{\Delta}{\Delta z}\left(D\frac{\Delta\theta}{\Delta z}\right) - \frac{\Delta K}{\Delta z}\right]_{z,t} + O(\Delta t) + O(\Delta z^2) \qquad (4)$$

Hence, the water content at time $t+\Delta t$ can be calculated explicitly with Eq. (4) from the water content distribution at time t. Convergence and stability of this approximation have been discussed by Fried and Combarnous (1968). In this case, the conditions are:

$$\frac{1}{2}\left(K_\theta - \frac{\partial D}{\partial z}\right)\Delta z \leq D \leq \frac{\Delta z^2}{2\Delta t} \qquad (5)$$

where $K_\theta = \frac{dK}{d\theta}$. The numerical solution technique gives oscillations when conditions (5) are not met. For the right hand side inequality, these oscillations are very severe, while for the left hand side inequality, they are usually rather small.

The numerical solution presented by Eq. (4) is basically a rectangular integration with respect to time. The CSMP simulation language can provide the user with more sophisticated integration methods, which all consume more computing time. It is our experience that conditions (5) also have to be met for trapezoidal and Adams second order integrations, while for higher integration methods only the left hand inequality has to be satisfied.

For the discussion later on, it will be convenient to represent conditions (5) in a $(\Delta t, \Delta z)$ plane (Fig. 1). The left hand side inequality is satisfied for all points laying under the line AB. The right hand inequality is satisfied for all points above the parabola, going through O and B. Hence, all the points $(\Delta t, \Delta z)$ in the domain OAB satisfy both inequalities. The most economical choice, with respect to computing time, is point B, with coordinates:

$$\Delta t = \frac{2D}{(K_\theta - \partial D/\partial z)^2} \qquad (6a)$$

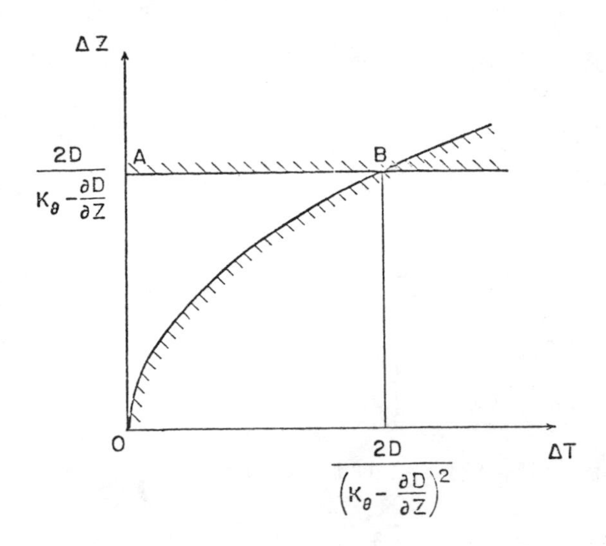

Fig. 1. Schematical representation, in the $(\Delta t, \Delta z)$ plane, of the stability and convergence criteria of the classical explicit finite difference approximation of the water flow equation.

$$\Delta z = \frac{2D}{K_\theta - \partial D/\partial z} \tag{6b}$$

We have treated conditions (5) as being uniquely defined while they depend upon the value of the water content. So, Δt and Δz should be chosen such that (5) is satisfied for all values of θ appearing in the simulation. The maximum value of θ usually yields the most restricted conditions and the corresponding increment Δz and Δt will satisfy condition (5) for all other θ. The conditions also depend upon the term $\partial D/\partial z$. From Eq. (6) it can be seen that taking this term equal to zero will give the smallest increments, Δz and Δt. Improvement of the numerical approximation, Eq. (4), can be obtained in an analogous way as has been presented by Chaudhari (1971) and van Genuchten and Wierenga (1974) for steady state solute flow. Again consider the Taylor series expansion of $\theta(z,t+\Delta t)$:

$$\frac{\theta(z,t+\Delta t) - \theta(z,t)}{\Delta t} = \frac{\partial\theta(z,t)}{\partial t} + \frac{\Delta t}{2}\frac{\partial^2\theta(z,t)}{\partial t^2} + O(\Delta t^2) \tag{7}$$

We will now evaluate the second term of the right hand side of Eq. (7). In this evaluation higher derivatives will be neglected.

$$\frac{\partial^2\theta}{\partial t} = \frac{\partial}{\partial t}\left[\frac{\partial}{\partial z}(D\frac{\partial\theta}{\partial z}) - \frac{\partial K}{\partial z}\right] \tag{8a}$$

$$= \frac{\partial}{\partial z}\left(\frac{\partial D}{\partial\theta}\cdot\frac{\partial\theta}{\partial z}\cdot\frac{\partial\theta}{\partial t} + D\frac{\partial^2\theta}{\partial z\partial t} - K_\theta\frac{\partial\theta}{\partial t}\right) \tag{8b}$$

$$= \frac{\partial}{\partial z}\left[\left(\frac{\partial D}{\partial z} - K_\theta\right)\left(D\frac{\partial^2\theta}{\partial z^2} + \frac{\partial D}{\partial z}\cdot\frac{\partial\theta}{\partial z} - K_\theta\frac{\partial\theta}{\partial z}\right)\right] \tag{8c}$$

$$= \frac{\partial}{\partial z}\left[\left(\frac{\partial D}{\partial z} - K_\theta\right)^2\frac{\partial\theta}{\partial z}\right] \tag{8d}$$

Combining Eqs. (1), (7) and (8d) we get:

$$\frac{\theta(z,t+\Delta t) - \theta(z,t)}{\Delta t} = \left[\frac{\partial}{\partial z}(D^*\frac{\partial\theta}{\partial z}) - \frac{\partial K}{\partial z}\right]_{z,t} + O(\Delta t^2) \tag{9}$$

where $D^* = D + \frac{\Delta t}{2}(\frac{\partial D}{\partial z} - K_\theta)^2$ (10)

And approximation of the derivates with respect to z yields:

$$\frac{\theta(z,t+\Delta t)-\theta(z,t)}{\Delta t} = \left|\frac{\Delta}{\Delta z}(D^*\frac{\Delta\theta}{\Delta z}) - \frac{\Delta K}{\Delta z}\right|_{z,t} + O(\Delta t^2) + O(\Delta z^2) \tag{11}$$

Comparison with Eq. (4) shows that we now have incorporated the terms of order Δt in the numerical approximation.

Analogous to conditions (5), we now have

$$\frac{1}{2}(K_\theta - \frac{\partial D}{\partial z})\Delta z \leq D + \frac{\Delta t}{2}(K_\theta - \frac{\partial D}{\partial z})^2 \leq \frac{\Delta z^2}{2\Delta t} \tag{12}$$

Figure 2 presents these conditions in the $(\Delta t,\Delta z)$ plane. The points in the domain ACO satisfy conditions (12). Theoretically, infinitely large values for Δt and Δz are possible, but we have to consider the fact that the derivation is only approximate and that there are still errors of the order Δt^2 and ΔX^2 in the numerical solution. Van Genuchten and Wierenga (1974) found that in the case of steady state solute flow, the best choice for Δt and Δz is given by point B. This yields:

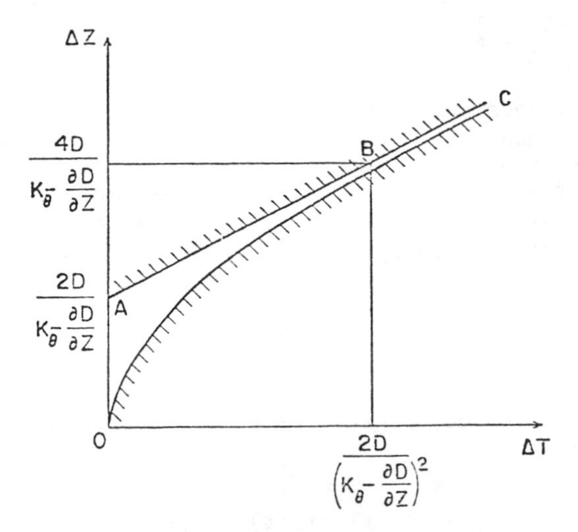

Fig. 2. Schematical representation, in the (t, z) plane, of the stability and convergence criteria of the improved explicit finite difference approximation of the water flow equation.

$$\Delta t = \frac{2D}{(K_\theta - \frac{\partial D}{\partial z})^2} \tag{13a}$$

$$\Delta z = \frac{4D}{K_\theta - \frac{\partial D}{\partial z}} \tag{13b}$$

These equations also have to be evaluated for the maximum water content and for $\partial D/\partial z = 0$, in order to satisfy the stability and convergence criteria for all θ.

Comparison of Eq. (13) with Eq. (6) shows that with the improved numerical approximation, Eq. (11), it is possible to use layers twice as big as for the classical approximation, Eq. (4).

The equation for solute flow. The equation for a non-interacting solute, in unsaturated soil is:

$$\frac{\partial \theta c}{\partial t} = \frac{\partial}{\partial z} (\theta Dap \frac{\partial c}{\partial z}) - \frac{\partial qc}{\partial z} \tag{14}$$

where c is concentration of the solute in the soil solution (meq/liter) and Dap is the apparent diffusion coefficient, (cm^2/day) and q is the Darcian flux of the soil water (cm/day), given by:

$$q = - D \frac{\partial \theta}{\partial z} + K \qquad (15)$$

Equation (14) does not account for diffusion into immobile water (van Genuchten and Wierenga, 1976). The apparent diffusion coefficient Dap, is the sum of the coefficient of molecular diffusion, which is generally assumed constant, and the coefficient of mechanical dispersion, which is generally assumed proportional to the average pore water velocity $v = q/\theta$ (cm/day):

$$Dap = Do + \varepsilon v \qquad (16)$$

The classical explicit finite difference approximation of Eq. (14) is:

$$\frac{\theta(z,t+\Delta t)\ c(z,t+\Delta t) - \theta(z,t)\ c(z,t)}{\Delta t} = \left[\frac{\Delta}{\Delta z}\left(\theta Dap \frac{\Delta c}{\Delta z}\right) - \frac{\Delta qc}{\Delta z}\right]_{z,t} + O(\Delta t, \Delta z^2) \qquad (17)$$

Neglecting the first order terms in Δt in Eq. (17), has a more severe effect, as in the water flow equation (1). Numerous studies report the occurrence of abnormal dispersion of the solute front with numerical approximations. This phenomenon has been called numerical dispersion. Chaudhari (1971) showed that the numerical dispersion was caused by neglecting of first order terms in Δt or Δz. The stability and convergence criteria of this classical approximation is:

$$\frac{1}{2} v \Delta z \leq Dap \leq \frac{\Delta z^2}{2\Delta t} \qquad (18)$$

These conditions have to be satisfied together with conditions (6) when the water and solute flow equation are solved simultaneously. In the $(\Delta t, \Delta z)$ plane we usually obtain the situation presented in Fig. 3. The most optimal choice is given by point A with coordinates

$$\Delta t = \frac{2\ Dap^2}{Dv^2} \qquad (19a)$$

$$\Delta z = \frac{2\ Dap}{v} \qquad (19b)$$

It is clear that the approximation of the solute flow equation determines the maximum value of Δz and the approximation of

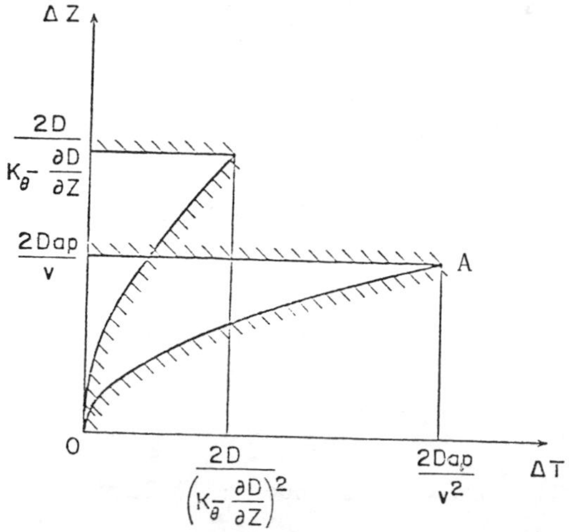

Fig. 3. Schematical representation, in the (t, z) plane, of the stability and convergence criteria of the classical explicit finite difference approximation of the water and solute flow equation.

the water flow equation the maximum value of Δt. It is also obvious that a discretization scheme is obtained which uses much computer time, because of the resultant small value of Δt. The time increment Δt can be so small that it is not necessary to correct for numerical dispersion in the solute flow simulation.

Similar to the water flow equation, the approximation for solute flow can be improved by taking into account the first order terms in Δt:

$$\frac{\theta(z,t+\Delta t)\ c(z,t+\Delta t) - \theta(z,t)\ c(z,t)}{\Delta t} = \frac{\partial \theta c}{\partial t}\bigg|_{z,t} + \frac{\Delta t}{2}\left[\frac{\partial^2 \theta c}{\partial t^2}\right]_{z,t} + O(\Delta t^2) \qquad (20)$$

where $$\frac{\partial^2 \theta c}{\partial t^2} = \frac{\partial}{\partial t}\left[\frac{\partial}{\partial z}\left(Dap\ \theta\ \frac{\partial c}{\partial z}\right) - \frac{\partial qc}{\partial z}\right] \qquad (21a)$$

$$= \frac{\partial}{\partial z}\left[\frac{\partial Dap\theta}{\partial t}\ \frac{\partial c}{\partial z} + Dap\ \theta\ \cancelto{0}{\frac{\partial^2 c}{\partial z \partial t}} - \frac{\partial q}{\partial t}\ c - q\ \frac{\partial c}{\partial t}\right] \qquad (21b)$$

$$= \frac{\partial}{\partial z}\left[\frac{\partial Dap\theta}{\partial t}\ \frac{\partial c}{\partial z} - \frac{\partial q}{\partial t}\ c - \frac{q}{\theta}\ \frac{\partial}{\partial z}\ \cancelto{0}{\left(Dap\ \frac{\theta \partial c}{\partial z}\right)} - \frac{q^2}{\theta}\ \frac{\partial c}{\partial z}\right] \qquad (21c)$$

Combining Eqs. (14), (20), and (21c), and approximation of the derivatives versus z gives:

$$\frac{\theta(z,t+\Delta t)\ c(z,t+\Delta t) - \theta(z,t)\ c(z,t)}{\Delta T} = \left[\frac{\Delta}{\Delta z}\left((\theta\ \text{Dap})^* \frac{\Delta c}{\Delta z}\right)\right]_{z,t} - \frac{\Delta q^* c}{\Delta z} + O(\Delta t^2,\Delta z^2) \qquad (22)$$

where $(\theta\ \text{Dap})^* = [\theta(\text{Dap} + 1/2\ \Delta t\ v^2)]_{z,t\ +\ \Delta t/2}$ (23a)

$$q^* = q\ (z,t + \Delta t/2) \qquad (23b)$$

with stability and convergence criteria:

$$\frac{1}{2}\ v\ \Delta z \leq \text{Dap} + \frac{1}{2}\ \Delta t\ v^2 \leq \frac{\Delta z^2}{2\Delta t} \qquad (24)$$

The situation in the $(\Delta t,\Delta z)$ plane is given in Fig. 4. From this figure it can be seen that for the simultaneous solution of solute and water flow no favorable discretization scheme can be obtained even when we use improved numerical solutions. However, a solution to this problem which can be easily

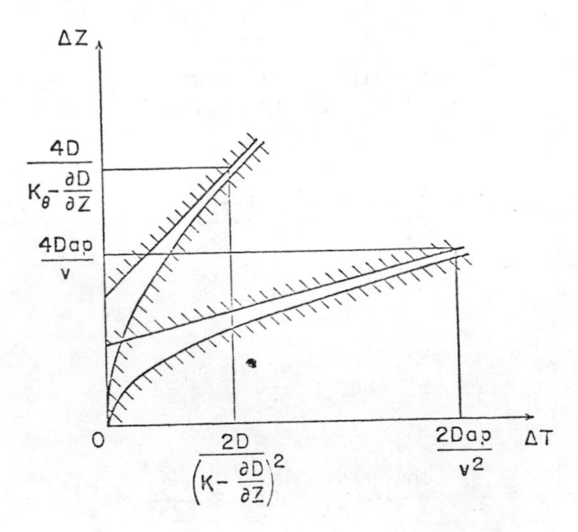

Fig. 4. Schematical representation, in the (t, z) plane, of the stability and convergence criteria of the improved explicit finite difference approximations of the water and solute flow equations.

programmed in CSMP, is possible by allowing different time-steps Δt for the water flow and the solute flow equation. In view of the fact that the restriction on Δz imposed by the approximation of the solute equation is more severe than that of the water flow equation, two situations arise, as is shown in Fig. 5.

For a slowly dispersing solute:

$$\frac{2\ Dap}{v} \leq \frac{D}{K_\theta - \partial D/\partial z} \qquad (25)$$

The best numerical approximation scheme is obtained by using the classical approximation to the water flow equation and the improved approximation for the solute flow equation. The following values for the increments are obtained:

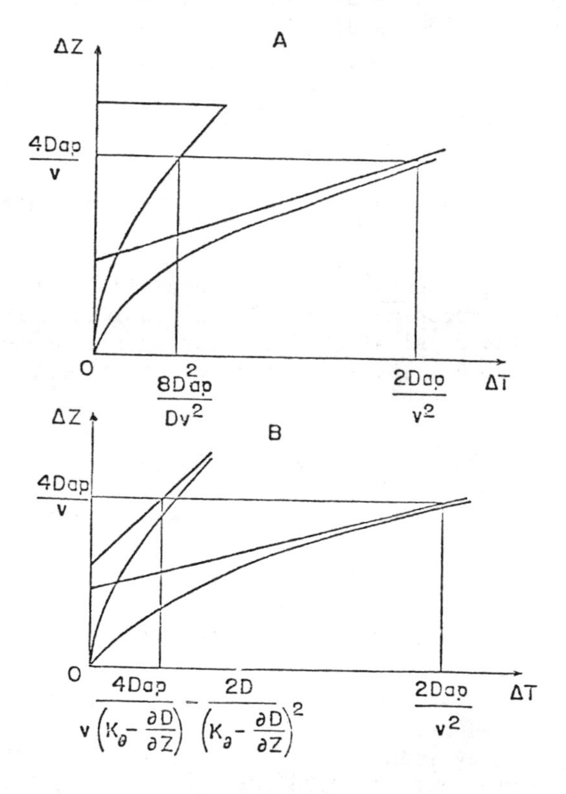

Fig. 5. Schematical representation, in the (t, z) plane, of the discretization scheme for the improved explicit finite difference approach for: a) water and a slowly dispersing solute, and b) water and a fast, dispersing solute.

$$\Delta z = \frac{4\ Dap}{v} \tag{26a}$$

$$\Delta t_\theta = \frac{8\ Dap}{Dv^2} \tag{26b}$$

$$\Delta t_c = \frac{2\ Dap}{v^2} \tag{26c}$$

For a fast dispersing solute:

$$\frac{2\ Dap}{v} > \frac{D}{K_\theta - \partial D/\partial z} \tag{27}$$

The best numerical approximation is now obtained with improved approximations for both equations. The same values for Δz and Δt_c are obtained, and:

$$\Delta t_\theta = \frac{4\ Dap}{v(K_\theta - \partial D/\partial z)} - \frac{2D}{(K_\theta - \partial D/\partial z)^2} \tag{28}$$

In practice, Δt_θ will always be taken as a submultiple of Δt_c.

Examples and discussion. As an example of the use of these numerical approximations, let us consider a 150 cm long soil column. The physical characteristics of the soil are presented in Fig. 6. The following equations were found to describe the relationship:

$$\theta \leq 0.325 \quad K = 10^{-13} \exp\ (92.1\theta)$$

$$\psi = -\ 7.84.10^5 \exp\ (-31.3\theta)$$

$$\theta > 0.325 \quad K = 1.93.10^4 \exp\ (26.3\theta)$$

$$\psi = 4.58.10^3 \exp\ (-15.5\theta)$$

where ψ is the pressure potential from which the diffusivity D can be obtained by: $D(\theta) = K(\theta)\ \partial\psi/\partial\theta$. The soil is a Glendale silty clay loam. The initial water content of the soil column varies between 0.287 cm^3/cm^3 at the surface to 0.318 cm^3/cm^3 at the bottom. As a first example, water is applied at a constant rate of 50 cm/day for 0.2 days and is then allowed to redistribute.

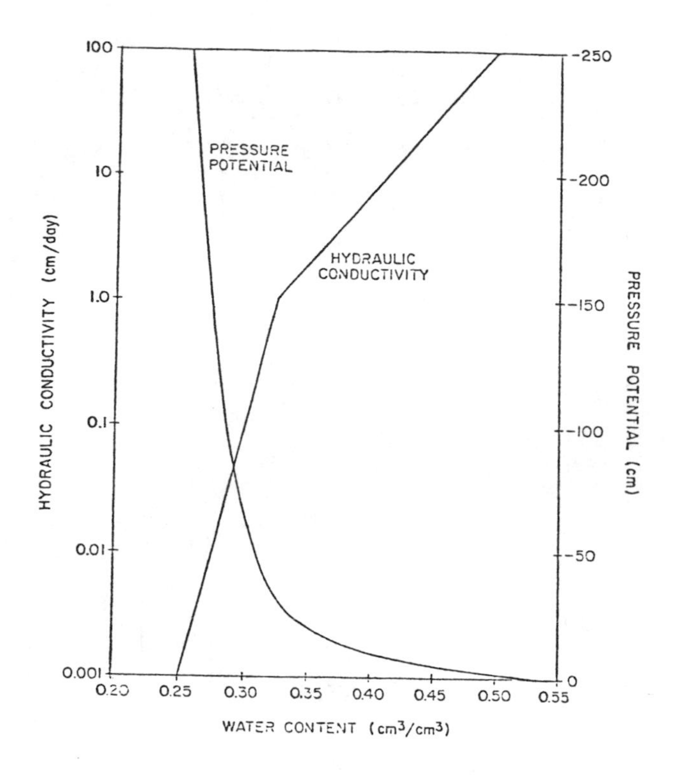

Fig. 6. Hydraulic conductivity and pressure potential versus water content relationships for a Glendale silty clay loam.

The results of the simulation of this problem are shown in Fig. 7. The solid line represents the water content profiles at different times obtained with the classical numerical explicit finite differences approximation, Eq. (4). The values of the increment were $\Delta z = 3$ cm and $\Delta t = 0.002$ days, calculated with Eq. (6). Doubling Δz and still using the classical approach results in oscillation as is shown by the broken lines. The dots present the solution with the improved discretization scheme, Eq. (11), and using a value of $\Delta z = 6$ cm and $\Delta t = 0.002$ days, calculated with Eq. (13). We can see that the oscillations are greatly reduced. The fact that there are still small oscillations is a result of neglecting the higher derivatives in the improved approximation. The computing time with the improved scheme is almost 50 percent less than with the classical scheme.

For the second example, we will assume that initially the soil columns were filled with pure water. The infiltrat-

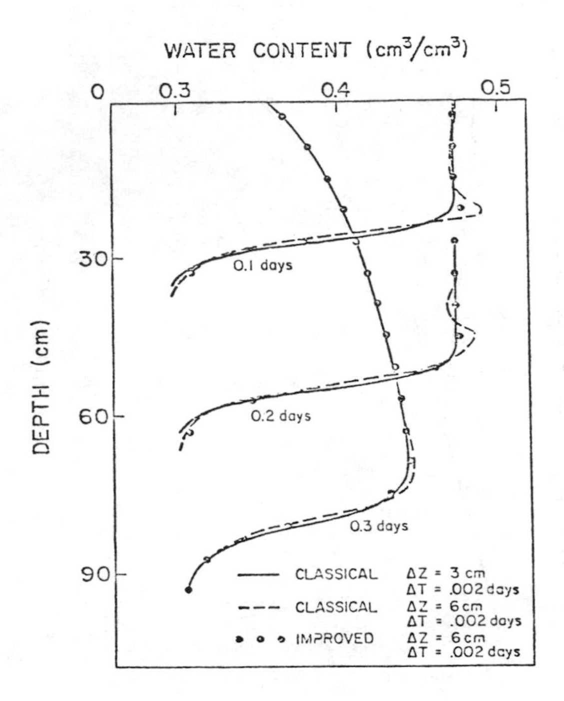

Fig. 7. Example of the explicit finite difference approximation of the water flow equation. Infiltration of 50 cm/day during 0.2 days in a Gelndale silty clay loam.

ing water now contains a dissolved tracer. We will first consider the case where Dap is given by Eq. (16) with Do = 1 cm^2/day and ε = 0.5 cm.

The classical numerical simulation, Eqs. (4) and (17), of this problem of simultaneous water and solute flow uses a $\Delta z = 1$ cm and $\Delta t_\theta = \Delta t_c = 0.0002$ days, calculated with Eq. (19). It is clear that the incorporation of the solute flow in the simulation results in a tremendous increase in computing time because of the small increments which results from stability and convergence criteria. The solid lines in Fig. 8 present the relative concentration profiles at different times. The time increment Δt is so low that the influence of numerical dispersion is negligible. We can calculate that the most economical simulation can be obtained by using a classical approximation, Eq. (4), for the water flow equation and an improved approximation, Eq. (22), for the

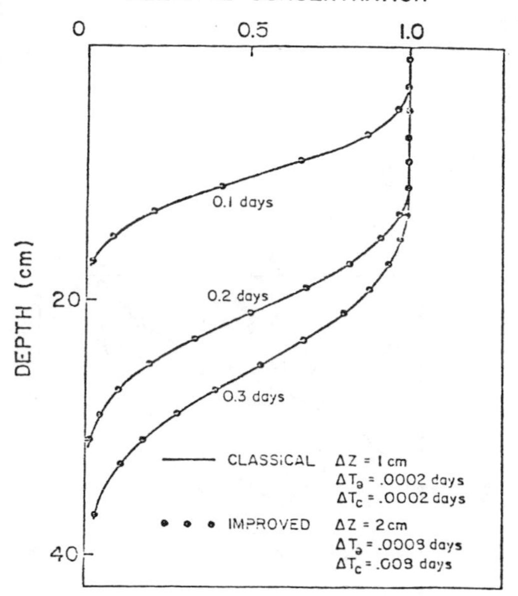

Fig. 8. Example of the explicit finite difference approximation of simultaneous flow of water and a slowly dispersing solute. Infiltration of water and solute of 50 cm/day during 0.2 days in a Glendale silty clay loam.

solute flow equation. The values of the increment are Δz = 2 cm, Δt_θ = 0.0008 days and Δt_c = 0.008 days calculated with Eq. (26). This results in a simulation which uses almost 95 percent less computing time than the classical approach. The dots in Fig. 8 give the results of the improved simulation, which show to be identical with the foregoing results. One problem arises during the simulation with the improved scheme. The time increment Δt_c cannot be greater than Δt_θ when the flux at the surface changes in a discontinuous way (i.e., at t = 0.2 days, when the infiltration stops) or fluctuations in the solute concentration profiles will result. Hence, when the infiltration ends, Δt_c is taken equal to Δt_θ for a time period of 0.01 day, after which Δt_c is increased to its original value.

In the third example, we assume a fast dispersing solute by taking Dap according to Eq. (16) with Do = 1 cm^2/day and

ε = 1.5 cm. The classical numerical simulation equations (9) and (17) of this problem uses Δz = 3 cm and $\Delta t_\theta = \Delta t_c$ = 0.002 days, calculated with Eqs. (19).

The results are presented by the solid lines in Fig. 9. A better simulation is obtained with the improved approximation, Eqs. (11) and (22) for the water and solute flow

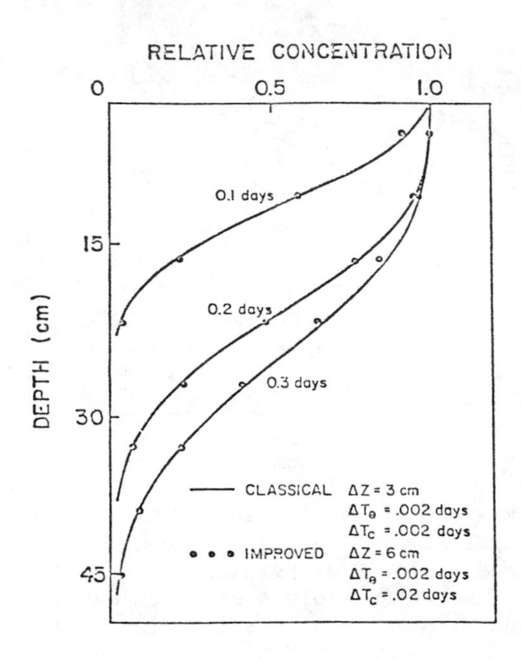

Fig. 9. Example of the explicit finite difference approximation of simultaneous flow of water and a fast dispersing solute. Infiltration of water and solute of 50 cm/day during 0.2 days in a Glendale silty clay loam.

equation, resulting in Δz = 6 cm, Δt_θ = 0.002 days and Δt_c = 0.02 days, calculated with Eqs. (26a), (26b) and (27). Again, Δt_c has to be taken equal to Δt_θ for a small period at t = 0.2 days when the flux at the top of the column changes abruptly. The improved simulation requires almost 75 percent less time than the classical simulation. The results are represented by the dots in Fig. 9. The small differences in both solutions are due to the numerical dispersion in the classical solution and the small oscillations of the water content profiles with the improved finite difference approximation.

Conclusions

An improved explicit finite difference approximation was presented for the water movement and the solute movement in unsaturated soils, under transient conditions. The improvement is obtained by taking into account the first order terms in Δt, in the simulation scheme. The convergence and stability criteria for the improved approximation were shown to be less restrictive than in the classical approximation. For the simulation of water movement alone, a decrease in computing time of almost 50 percent could be obtained in comparison with the classical simulation.

When taking different time steps for the water and solute flow approximation, a reduction of computing time of almost 75 percent for a fast dispersing solute, and almost 95 percent for a slowly dispersing solute, were obtained. Improvement of the approximation of the solute flow equation also corrects for the errors introduced by numerical dispersion.

Acknowledgments

The work upon which this report is based was supported in part by funds obtained from the Commission for Education Exchange between the United States, Belgium and Luxemburg, the Scientific Committee of the N.A.T.O., and U.S. Environmental Protection Agency Grant No. S-80156.

References

[1] Beek, J. and Frissel, M. J., 1973. "Simulation of Nitrogen Behavior in Soils," Centre for Agr. Publ. and Documents Pudoc. Wageningen, 76 p.

[2] Bhiuyan, S. E., Hiler, E. A., van Bavel, C.H.M., and Aston, A. R., 1971. "Dynamic Simulation of Vertical Infiltration into Unsaturated Soils," Water Resour. Res. 7(6):1597-1606.

[3] Bresler, E., 1973. "Simultaneous Transport of Solutes and Water under Transient Unsaturated Flow Conditions," Water Resour. Res. 9(4):975-986.

[4] Chaudhari, N. M., 1971. "An Improved Numerical Technique for Solving Multi-Dimensional Miscible Displacement Equations," Soc. Petrol. Eng. J. 11(3):277-284.

[5] De Wit, C. T. and Van Keulen, H., 1972. "Simulation of Transport Processes in Soils," Centre for Agric. Publ. and Documents. Pudoc. Wageningen, 108 p.

[6] Fried, J. J. and Combarnous, M. A., 1968. "Dispersion in Porous Media," Advan. Hydrosci. 7:170-282.

[7] Gardner, A. O. Jr., Peaceman, D. W., and Pozzi, A. L., 1964. "Numerical Calculation of Multidimensional Miscible Displacement by the Method of Characteristics," Soc. Petrol. Eng. J. 4(1):26-36.

[8] Goudriaan, J., 1973. "Dispersion in Simulation Models of Population Growth and Salt Movement in the Soil," Neth. J. Agric. Sci. 21:269-282.

[9] Lai, S. H. and Jurinak, J. J., 1972. "Cation Adsorption in One Dimensional Flow through Soils: A Numerical Solution," Water Resour. Res. 8(1):99-107.

[10] Peaceman, D. W. and Rachford, H. H. Jr., 1962. "Numerical Calculation of Multidimensional Miscible Displacement," Soc. Petrol. Eng. J. 2(4):327-339.

[11] Pinder, C. F. and Cooper, H. H. Jr., 1970. "A Numerical Technique for Calculating the Transient Position of the Salt Water Front," Water Resour. Res. 6(3):875-882.

[12] Speckhart, F. H. and Green, W. L., 1976. "A Guide to Using CSMP - The Continuous System Modeling Program," Prentice-Hall, Inc., Englewood Cliffs, N.J., 325 p.

[13] Stone, H. L. and Brian, P.L.T., 1963. "Numerical Solutions of Convective Transport Problems," A.I.C.M.E. J. 9(5):681-688.

[14] van Genuchten, M. Th. and Wierenga, P. J., 1974. "Simulation of One-Dimensional Solute Transfer in Porous Media," Agr. Exp. Sta. Bull. 628, New Mexico State University, Las Cruces, N.M.

[15] van Genuchten, M. Th., Davidson, J. M., and Wierenga, P.J., 1974. "An Evaluation of Kinetic and Equilibrium Equations for the Prediction of Pesticide Movement through Porous Media," Soil Sci. Soc. Amer. Proc. 38:29-35.

[16] Van Keulen, H. and van Beek, C., 1971. "Water Movement in Layered Soils - A Simulation Model," Neth. J. Agric. Sci. 19:138-153.

[17] Wierenga, P. J. and de Wit, C. T., 1970. "Simulation of Heat Transfer in Soils," Soil Sci. Soc. Amer. Proc. 34:845-848.

THE MOVEMENT OF WATER AND SOLUTES THROUGH THE UNSATURATED ZONE OF THE CHALK IN THE UNITED KINGDOM

By

D. B. Oakes
Water Research Centre
Medmenham Laboratory
P.O. Box 16, Medmenham
Marlow, Bucks, SL7 2HD, England

Abstract

Groundwater is a major source of public supply in England and Wales where it meets about 30% of the demand. In some areas, particularly southeast England, it is the major source. Much concern has been expressed in recent years about the pollution of groundwater resources by nitrates, and the Water Research Centre has undertaken an extensive programme of drilling and sampling to determine the reason for the rise in nitrate levels and to assess future trends. To assist with the interpretation of field data, mechanisms of water and solute movement through the unsaturated zone have been investigated, and the Chalk, our major aquifer, has received particular attention. The dating of groundwater by tritium measurements has led to considerable speculation regarding flow processes in the Chalk, and the additional information gained from the nitrates investigations has led to the formulation of a flow model in which water and its solutes move at very different velocities. It has been deduced that in the unsaturated zone water moves downwards through the fissure system at about 0.7 m/d, while any solutes carried from the land surface move downwards at about 1 m/yr. Laboratory experiments have yielded evidence in support of this hypothesis though further research is needed before a comprehensive model of the flow mechanism can be established.

A similar flow mechanism is expected to be operative in many other aquifers, and has important implications for groundwater pollution trends.

Introduction

The quality of groundwater pumped from the major aquifers in the United Kingdom, the Chalk and the Bunter Sandstone, is generally very good, requiring only minimal treatment. However, increasing concentrations of nitrate in groundwater have been recorded in the last few years (Greene et al., 1970; Satchell et al., 1972), and the Water Research Centre has embarked on a major field investigation to determine the extent of pollution, and to collect information to permit the estimation of future trends in nitrate levels. As part of this study mathematical models of solute movement through the unsaturated zone have been developed, and considerable attention has been focused on the mechanisms of water flow and solute transport in our major aquifer, the Chalk.

Hydrogeology of the Chalk

The Chalk of southeast England is the major aquifer in the UK, providing over 40% of the total groundwater abstracted. Consisting generally of soft, white limestone, the Chalk extends from the east coast of England to the south coast, either at the surface or concealed beneath younger strata (Fig. 1). Boulder clay and other superficial deposits obscure much of the outcrop. The thickness of chalk varies considerably from 100 m up to 500 m, but satisfactory well

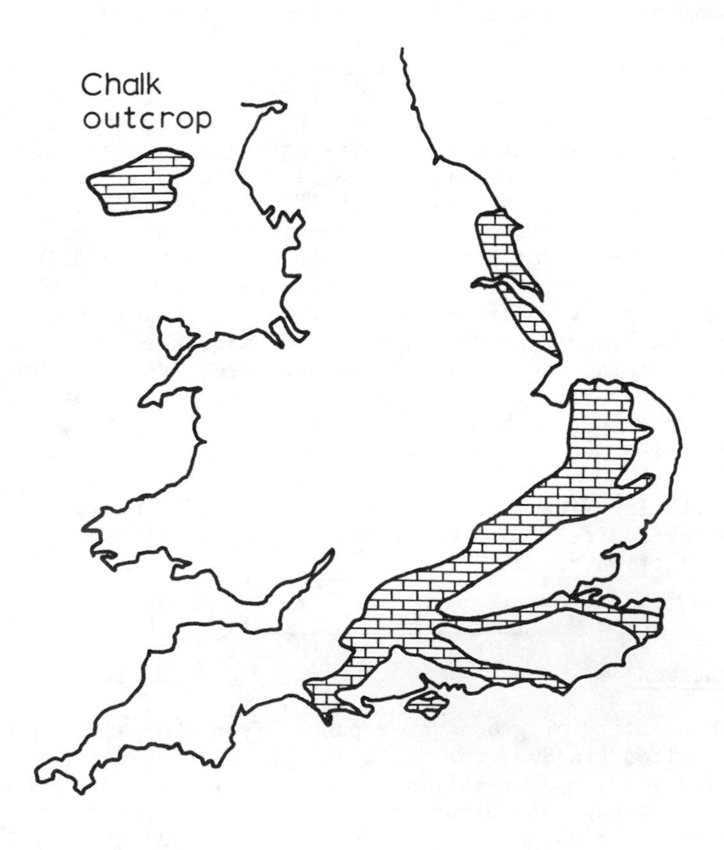

Fig. 1. Outcrop of the Chalk

yields are only obtained at certain horizons where fissuring increases the bulk permeability of the rock. It has been common practice to sink large diameter wells into the Chalk and drive adits horizontally below the water table to intercept the maximum number of water bearing fissures. Yields of up to 15 ml/d have been obtained by this method. Lithological variations of the different layers of the Chalk have considerable influence on its transmissivity and porosity. The stratigraphical levels of hard bands are also responsible for the pronounced escarpments formed at outcrop.

The mass of Chalk as observed in quarry exposures is frequently traversed by horizontal and near-vertical joints and fissures. It is these discontinuities which give rise to the locally high transmissivity values near to the water table, where solution of the calcium carbonate has been most rapid. The rock matrix is composed of fossil debris ranging in diameter from less than 1 μm up to 100 μm (Water Resources Board, 1972), and pore diameters are generally less than 1 μm. This gives rise to a very low intergranular permeability, but a very high porosity. Measurements on core samples of Chalk have yielded permeabilities of the order of 10^{-3} m/d, and porosities ranging from 0.3 to 0.4 (Young et al., 1976). The specific yield, however, is generally less than 0.02 and may be identified with the volume of draining fissures. The jointing pattern observed in quarry exposures is generally very irregular, with block sizes varying from a few centimetres to more than one metre. The size of joint openings is particularly difficult to estimate because of the effects of weathering on quarry faces. Horizontal discontinuities up to 10 mm in width have been reported by Ward et al. (1968), while Foster and Milton (1974) have noted tightly closed joints at relatively shallow depths. Examination of Chalk cores suggests that very small or microscopic cracks are profuse throughout the rock mass so that no part of the pore space is remote from the fissure system.

A consequence of the unusually high specific retention of the Chalk matrix is that even above the water table the pore space is almost fully saturated. Fig. 2 shows the water content profile for a cored, Chalk borehole near Winchester, Hampshire (Young et al., 1976). Water content was estimated by drying core samples in an oven. The profile shows no significant change at the water table; the variations with depth reflect, among other things, the presence of fissuring which will reduce the volumetric water content, and lithological features. The saturated zone is characterised by complete saturation of the fissure system, which will only be partially filled above the water table.

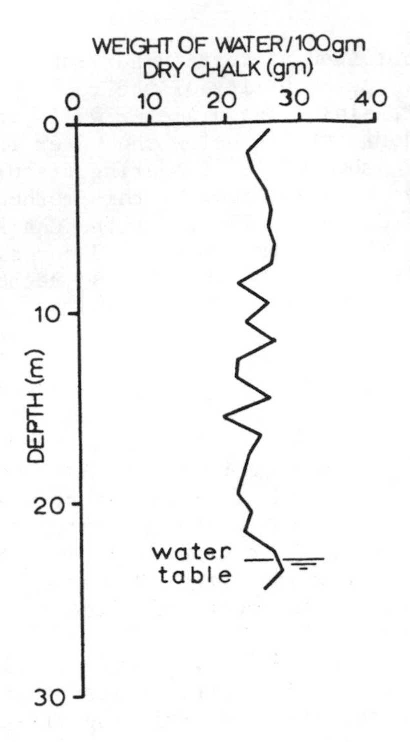

Fig. 2. Water content profile -
Chalk borehole in Hampshire

Mechanisms of Water and Solute Movement

Given these characteristics there has been considerable speculation regarding the mechanisms of rainfall recharge, and of solute movement through the Chalk (Smith et al., 1970; Foster, 1975). In a fully saturated Chalk column with a vertical permeability of 10^{-3} m/d, and a porosity of 0.35, water would migrate downwards through the pore space at about 1 m/yr. This is a similar rate to that reported by Smith et al. (1970) and by Young et al. (1976) for the movement of tritium through the unsaturated zone of the Chalk. Profiles have now been obtained from a number of sites located on the Chalk outcrop and the rates of movement of tritium inferred by correlation to the time variation of tritium in rainfall are all similar at between 0.8 m/yr, and 1.05 m/yr. Tritium is considered to be a particularly suitable tracer for water since it is part of the water molecule. It may therefore be inferred that if recharge is occurring through the pore space the water molecules themselves migrate downwards at about 1 m/yr. This downward movement rate allows the storage in the pore space of the average annual infiltration to the Chalk of 350 mm, but it is not certain whether the maximum daily rates of infiltration could be as readily accommodated. It is also difficult to discount the multitude of fissures that are so evident in Chalk exposures as playing no role in the flow process.

It has been suggested (Foster, 1974; Young et al., 1976) that infiltration through the unsaturated zone may take place along the fissures, and that solutes may diffuse between this moving water and the relatively static water in the matrix. This could perhaps account for the observed slow downward movement of tritium and other solutes. Under natural rainfall conditions the fissures would be partially filled by slowly moving water. High infiltration rates resulting from intense rainfall could be accounted for by rapid flow down the fissure system resulting in some solutes being carried rapidly down to the water table.

The Rate of Rainfall Recharge

An estimate of the rate of movement of water down the fissures to the water table has been obtained by considering the delay in water table response to infiltration. At two sites on the Chalk outcrop well levels were correlated to lagged distributions of calculated monthly infiltrations. The postulated correlation equation was

$$h_i = \sum_{j=1}^{N} a_j \, p_{i-j+1} + b_1 + b_2 \, h_{i-1} \tag{1}$$

where h_i = water level at end of month i,

p_j = infiltration at the surface in month j.

The last two terms on the right-hand side of Eq. (1) describe groundwater level recession, and the coefficients a_j apportion unit surface infiltration in some month to the consequent water level rises in the same month and in succeeding months. Thus, a_1 is the water level rise in the current month, a_2 the rise in the next month, and so on. It will be apparent that the sum of the coefficients a_j must equal the reciprocal of the specific yield. For each of the two sites four years of monthly data was available, and the coefficients were determined by least-squares fitting. The number of lags, N, required in each case was determined as the maximum number which gave all positive coefficients a_j, negative coefficients having no physical meaning. Fig. 3 shows the unit response function for each of the two sites, giving the distribution in time of water table infiltration resulting from unit input in one month at the surface. As validation of the technique Fig. 4 shows a comparison between actual levels at one of the sites, and levels reconstituted

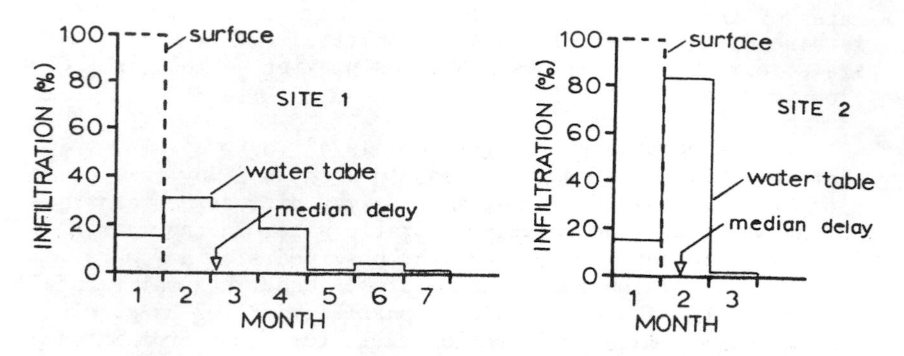

Fig. 3. Unit response functions for infiltration

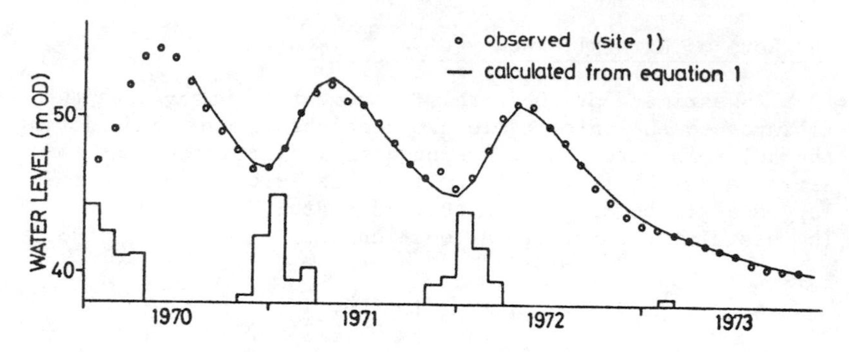

Fig. 4. Well hydrograph calculated from unit response function

using Eq. (1) and the calculated infiltration data. Median delay times between surface and water table infiltrations are shown on Fig. 3, and the corresponding median vertical flow rates were calculated from the thickness of unsaturated chalk to be 0.6 m/d and 0.8 m/d respectively. The spreads of the unit response functions reflect the local heterogeneity of the Chalk and the range of fissure and joint sizes. It is interesting to note that both correlations indicate that 15% of the infiltration reaches the water table within about one month. This may be equated to a rapid fissure flow component arising from recharge through the larger fissures, and high intensity rainfall events.

The Rate of Solute Transport

It is postulated that water and its solutes move slowly through the fissure system, and have a large area of contact with static water held in the Chalk matrix. If the mass transfer of solutes between the mobile and static water is assumed to be instantaneous a simple solution to the equation of solute movement may be obtained. It is found that in the absence of mechanical dispersion solute pulses migrate downwards from the ground surface without change in shape at a velocity v given by

$$v = u\,\theta_m/(\theta_m + \theta_s) \tag{2}$$

where u = velocity of flow in the fissure system,

θ_m = porosity of the mobile phase, and

θ_s = porosity of the static phase.

In reality, there will not be a unique flow velocity in the fissure system, and a range of velocities in inter-linked fissures will result in some dispersion of the solute profile.

If u and v of Eq. (2) are assigned average values of 0.7 m/d and 1.0 m/yr respectively, then we require

$$\theta_m/(\theta_m + \theta_s) \simeq 4.10^{-3}$$

Total porosity averages about 0.35 so that for the assumed migration rates $\theta_m \simeq 1.4\ 10^{-3}$ which indicates that the fissure space cannot be more than about 1/10th full during normal infiltration events.

The assumption of instantaneous equilibrium between the solute concentrations in the pores and in the fissures is, of course, very stringent. A rapid, but not infinite, rate of mass transfer would result in some dispersion of the input solute pulses. The movement of solutes in this case could be assessed with a numerical model and data is currently being collected at the Water Research Centre to facilitate this analysis. Some preliminary results from laboratory experiments are presented in the next section.

Laboratory Experiments

Unfortunately it is not possible to carry out comprehensive laboratory experiments on Chalk columns because of the difficulty in obtaining in a laboratory scale sample a completely representative range of fissure sizes. However, experiments have been undertaken to measure the rate at which solutes will diffuse from fissures into the pore space. In these experiments Chalk cores approximately 100 mm diameter by 75 mm long were saturated with a NaCl solution and sealed in plastercine leaving one face of the core exposed. A small quantity of fresh water was placed in contact with the exposed face, and the rate of increase of NaCl concentration in the water was measured with a specific ion electrode. Within the Chalk block the concentration c of the tracer obeys the diffusion equation:

$$\frac{\partial c}{\partial t} = D \nabla^2 c$$

where D is the diffusion coefficient (L^2T^{-1}) and is a property of the pore structure of the Chalk matrix and of the self diffusion coefficient of NaCl. For the particular boundary conditions applied in this case there is an analytical solution to the change with time of concentration of tracer in the water in contact with the exposed face of the core (Carslaw and Jaeger, 1959, p. 128). Fig. 5 shows the measured concentration variation and the 'best-fit' analytical solution,

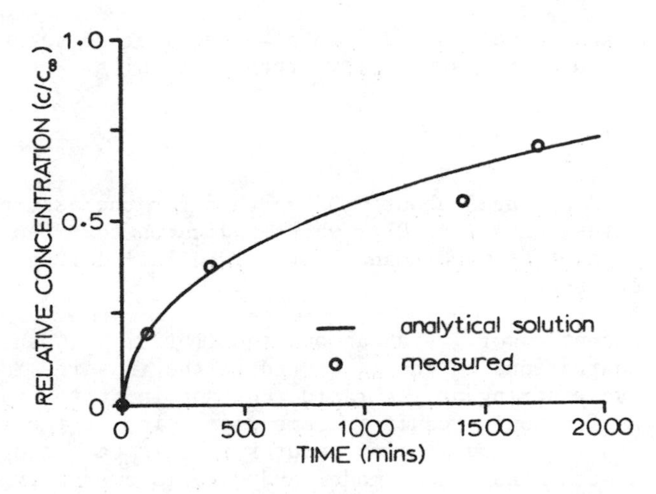

Fig. 5. Diffusion of NaCl into a Chalk block

obtained with a value of the diffusion coefficient $D = 1.3\ 10^{-5}\ cm^2/sec$. This may be compared with a self diffusion coefficient for chloride of $4.7\ 10^{-5}\ cm^2/sec$. Further experiments are being planned using different tracers such as tritium and nitrate. The author is unaware of any experimentation on the relative mobility of various ions in the Chalk. However, in other media similar migration rates for nitrate and chloride have been reported from both field and laboratory studies (Kurtz and Melstead, 1973).

The analytical solution may be used to estimate the effects of diffusion between fissures and blocks of varying sizes. The two curves of Fig. 6 were produced using the diffusion coefficient calculated from the chloride experiment.

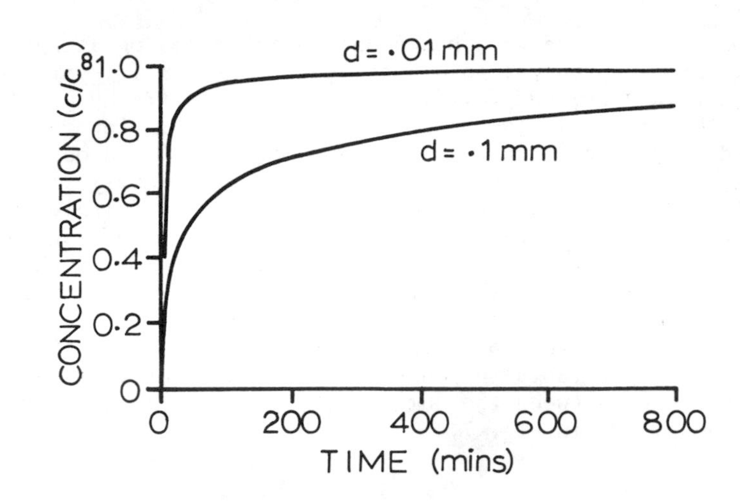

Fig. 6. Predicted diffusion of NaCl from the pores into the fissures

The graphs show the rate at which the concentration of some solute in the fissure water approaches the equilibrium concentration from an initial condition where all of the solute is in the pore water. The dimension of the Chalk block is assumed to be 0.2 m and the depth of water in the fissure in contact with the block is .1 mm in one case and .01 mm in the second case. It is seen that the concentration in the fissure water reaches 90% of its equilibrium value in approximately 0.5 d in the first case and 10 mins in the second case. The concentration in the fissure water reaches 98% of its equilibrium value after about 220 minutes in the second case. The physical dimensions used in these two examples were based on a rather superficial knowledge of the fissure geometry of the Chalk. Further research is required before a detailed analysis of the postulated diffusion process can be undertaken. However, the results illustrate that under certain circumstances the assumption of instantaneous equilibrium between pore and fissure concentrations may not be unrealistic.

Solute Transport Model

It is apparent that in attempting to model the movement of solutes through the Chalk from the surface to the water table, the usual model of unsaturated flow (e.g., Bresler, 1973) is inappropriate. However, both the intergranular flow and fissure flow concepts indicate that a uniform velocity, downward flow model may be used to represent the movement of contaminants input at the ground surface. Such a model has been used to simulate the movement through the Chalk of tritium, and of nitrate resulting from agricultural activities. The surface inputs of tritium are known (Smith et al., 1970), and those for nitrate were based on control rules relating the mass of nitrogen released each year in the soil layers to antecedant land use and fertilizer applications (Young et al., 1976). The vertical flow rate in the model simulations was adjusted in each case to give the best match to field observations, but in all cases a constant value of dispersivity, equal to 0.2 m, was used to account for dispersion within the mobile and static phases and for non-equilibration of concentration between the two phases. It was also assumed in the simulations that 15% of the surface input of solutes travels rapidly to the water table and does not interact with the pore water. This fraction corresponds to that derived by correlating water table response to antecedant infiltration. The comparisons between the numerical model and field observations are shown in Fig. 7.

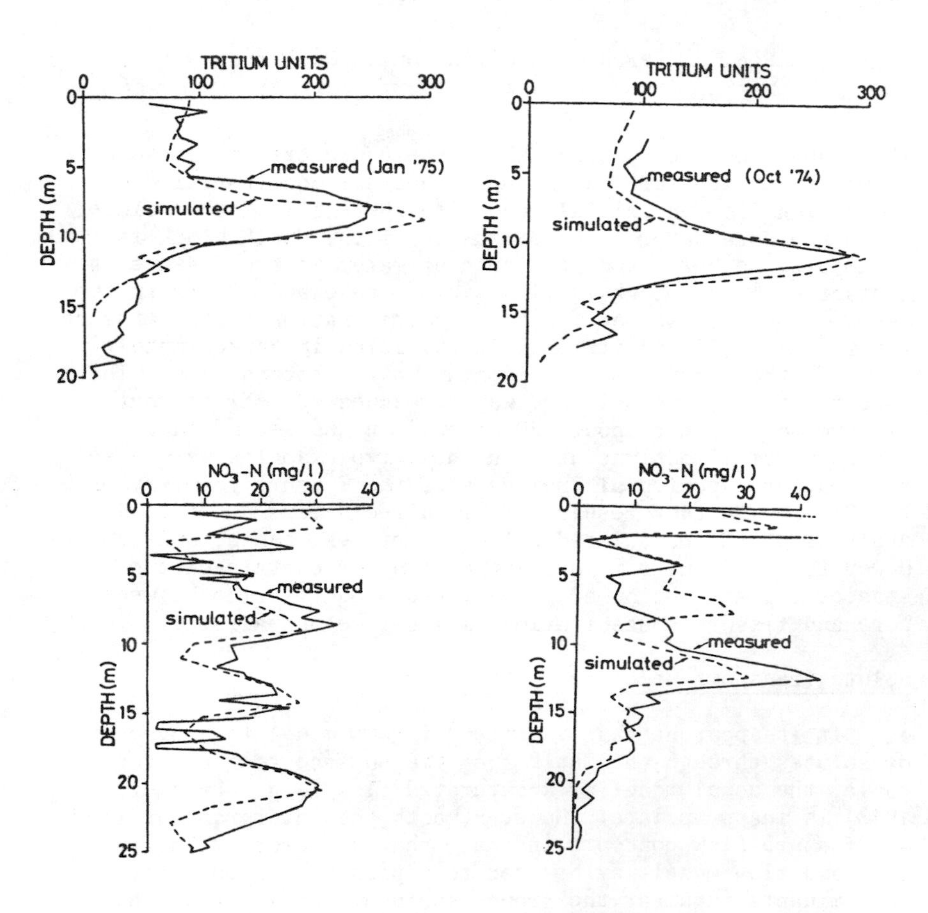

Fig. 7. Measured and simulated solute profiles in unsaturated Chalk

The tritium profiles are particularly important, as the surface inputs are accurately known. An excellent comparison between the computed and measured results for both peak position and peak size is obtained. The computed nitrate profiles do not match the measured profiles so well as those for tritium, but the peak positions are generally well predicted. Uncertainties concerning the quantities input could account for the discrepancies in peak magnitudes. The downward migration rates used in the model simulations all fall in the range 0.8 to 1.05 m/yr and there appear to be no significant differences between the rates for tritium and nitrate.

The overall behaviour of the model in matching the nitrate and tritium profiles is generally good, and supports the hypothesis of a uniform downward migration rate. The results cannot, however, differentiate between the relative merits of the intergranular flow and fissure flow concepts.

Conclusions

Field investigations on the Chalk have indicated that solutes move through the unsaturated zone at a rate of about 1 m/yr. Two distinct mechanisms of water and solute movement have been proposed to explain this phenomenon. One invokes flow of water and solutes through the Chalk matrix and largely ignores the presence of fissuring. The properties of the matrix are such that average annual rates of recharge could be transmitted through the pore space, but it is doubtful whether infiltration from high intensity rainfall events could be as readily accommodated. An alternative mechanism has been proposed which admits recharge via the fissure system and can, therefore, explain the rapid infiltration process which must occur after high intensity rainfall. In this model water and its solutes move down through the unsaturated zone at markedly different rates. Water moves downwards through the fissures at about 0.7 m/d, and above the water table the fissures are, on average, about 1/10th full. Solutes diffuse between the water in the fissures, and the water in the pores under the action of concentration gradients. Instantaneous equilibration of concentrations in the fissures and pores results in surface inputs of solutes moving slowly down through the Chalk in an unaltered profile. A coefficient of diffusion of NaCl through the Chalk matrix, equal to about 1/3rd of the self diffusion coefficient, has been estimated in laboratory experiments. Further research is needed to quantify the physical dimensions of fissures and blocks in the mass of the Chalk. It has been shown that in certain circumstances the mass transfer of solutes between fissure water and pore water may be extremely rapid compared with the rate of flow of water down the fissures.

The retardation of pollutant movement by diffusion between mobile and static water has been investigated by other workers (e.g., Gaudet et al.) particularly in unconsolidated media. The Chalk is probably exceptional, however, in its ability to retard the movement of solutes by a factor of about 250 relative to the rate of transmission of water. Retardation is also expected to occur in the saturated zone where it is generally recognized that the bulk of water movement in the Chalk occurs in the fissures. Flow velocities are much greater than in the unsaturated zone, however, and further research is needed to quantify the transport mechanisms. It is also expected that solute movement will be retarded in other cemented aquifers where both fissure flow and intergranular flow are operative. The implications for groundwater pollution are evidently very important, and model studies are now in progress to provide predictions of the rate at which nitrate now in the unsaturated zone will cause concentrations in pumped groundwater and in springs to rise.

Acknowledgments

The author is grateful to Dr. R. G. Allen, Director of the Water Research Centre, for permission to present this paper.

References

[1] Bresler, E., 1973. "Simultaneous Transport of Solutes and Water under Transient Unsaturated Flow Conditions," Water Resources Research, 9(4), pp. 975-986.

[2] Carslaw, H. S. and Jaeger, J. C., 1959. "Conduction of Heat in Solids," Oxford University Press, London, 510 p.

[3] Foster, S. S. D., 1975. "The Chalk Groundwater Tritium Anomaly - A Possible Explanation," Journal of Hydrology, 25, pp. 159-165.

[4] Foster, S.S.D. and Milton, V. A., 1974. "The Permeability and Storage of an Unconfined Chalk Aquifer," Hydrological Sciences Bulletin, 19(4), pp. 485-500.

[5] Gaudet, J. P., Jegat, H., and Vachaud, G., 1976. "Study of Water and Dissolved Substance Transfers in Non-Saturated Ground, Considering a Partly Stagnant Body of Water," La Houille Blanche, 3/4, pp. 243-252.

[6] Greene, L. A. and Walker, P., 1970. "Nitrate Pollution of Chalk Waters," Water Treatment and Examination, 19 (2), pp. 169-182.

[7] Kurtz, L. T. and Melstead, S. W., 1973. "Movement of Chemicals in Soil by Water," Soil Science, 115(3), pp. 231-239.

[8] Satchell, R.L.H. and Edworthy, K. J., 1972. "Artificial Recharge: Bunter Sandstone," Trent Research Programme, Vol. 7, Water Resources Board, Reading.

[9] Smith, D. B., Wearn, P. L., Richards, H. J. and Rowe, P.C., 1970. "Water movement in the Unsaturated Zone of High and Low Permeability Strata by Measuring Natural Tritium," Proceedings of a Symposium on Isotope Hydrology, IAEA, Vienna, pp. 73-87.

[10] Ward, W. H., Burland, J. B. and Gallois, R. W., 1968. "Geotechnical Assessment of a Site at Mundford, Norfolk, for a Large Proton Accelerator," Geotechnique, 18, pp. 339-431.

[11] Water Resources Board, 1972. "The Hydrogeology of the London Basin," Water Resources Board, Reading.

[12] Young, C. P., Hall, E. S. and Oakes, D. B., 1976. "Nitrate in Groundwater - Studies on the Chalk near Winchester, Hampshire," Water Research Centre, Technical Report TR31, 56 p.

PRELIMINARY IDENTIFICATION OF THE SALT PICK-UP AND TRANSPORT PROCESSES IN THE PRICE RIVER BASIN, UTAH

By

J. Paul Riley, Professor of Civil & Environmental Engineering
David S. Bowles, Research Engineer
D. George Chadwick, Graduate Research Assistant

and

W. J. Grenney, Associate Professor of Civil & Environmental Engineering and Head, Environmental Engineering Division
Utah Water Research Laboratory
Utah State University, Logan, Utah 84322

Abstract

The Price River is a significant contributor of salt to the Colorado River. Relatively pristine waters leaving the upper elevations of the basin degenerate into highly saline waters entering the Green River. The primary reason for this deterioration is the contact of the water with the Mancos shale, a marine shale, which underlies most of the central basin. This paper presents the structure of an evolving model of the salt pick-up and transport processes in the Price River Basin. The initial purpose of the model is to aid in the identification of the natural and man-modified hydro-salinity-sediment system of the basin. This identification procedure will result in both a better qualitative understanding of the important physio-chemical processes, and in a mathematically more precise description of these processes. When the identification stage is complete, the model will be used as a management tool for such purposes as examining various strategies for reducing salt loads in the Price River and in other similar rivers.

Résumé

La rivière Price est un collaborateur de sel significantif de la rivière Colorado. Comparativement, Ces eaux de la rivière Price quittaut les altitudes supérrieures de Bassin, descendent dans Ces hautes eaux salines quientrent dans la rivière green. La raison primaire de cette deterioration est le contact de l'eau auec l'argile schisteuse Mancos, une argile schisteuse marine, qui est en-dessous de la plupant du bassin central. Cet article presente la structure d'un modèle développant le ramassage et le transport du sel du bassin riverain de Price. Le but initial de ce modéle est d'aider á l'identification du systéme de sédiment hydro-salin du bassin, natural et modifié par l'homme. Cette procédure d'identification aura comme résultat á la fois une meilleure compréhension qualitative des importants procédés physio-chemiques et une description précise et mathemétique de ces procédés. Lorsque l'étape d'identification sera terminée, le modèle sera utilisé comme un outil de direction pour des buts tels que, examiner différentes stratégies pour réduire Ces charges de sel dans la rivière Price et dans d'autres rivières similaires.

Introduction

Background of the problem. Control of the rising salinity levels in the waters of the Lower Colorado River is becoming economically and politically advisable. Failure to do so may result in annual damages as high as 1.24 billion dollars (United States Department of the Interior, 1974). The predicted salinity impacts are: 1) agricultural productivity is reduced, 2) the suitability of Colorado River water for municipal and industrial use is reduced, and 3) salinity concentrations in the Colorado River water reaching Mexico exceed established standards. Eliminating increases in salinity will require the formation of new strategies to combat the problem. It is hoped that through research an understanding of the major processes and causes of salinity in the Colorado River will be obtained, thus creating the foundation from which the necessary techniques for salinity reduction might be developed.

One such area of needed research are the natural diffuse sources of the Upper Colorado River Basin. Blackman et al. (1973) estimated that 37 percent of the total salt loading in the Colorado River occurred from the upper basin diffuse sources. Mountainous areas of the upper basin yield most of the total river flow as relatively high quality water. As the streams traverse the immense semi-arid lowlands very little flow is contributed, except during storms, and the water quality deteriorates as the streams interreact with the natural salt bearing geological formations. The Price River Subbasin of central Utah (Fig. 1) exemplifies these diffuse loading conditions. Relatively high quality flows originate in the mountainous areas above the river, averaging a TDS concentration of less than 500 mg/l. As the river and tributaries traverse the low lands, a near continuous decrease in water quality occurs attributed to diffuse sources. Water quality of the river monitored at Woodside near its confluence with the Green River has a weighted average of 4000 mg/l.

Rationale of the paper. In the summer of 1976 research was begun at Utah State University concerning the processes which contribute salinity to the waters of the Price River. To date the work under this research study has emphasized the salt pick-up phenomena associated with surface runoff. However, as an important part of the study it was necessary to

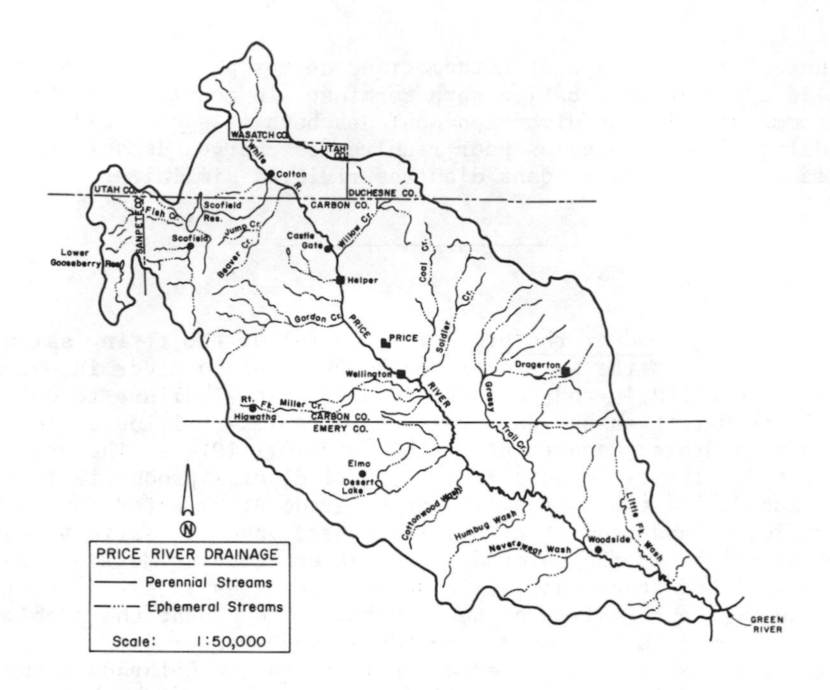

Fig. 1. Map of the Price River Basin, Utah

conduct the functions which form the basic rationale of this paper, namely to: 1) identify the entire system, 2) identify state-of-the-art description of the various processes, and 3) identify future work needed.

Description of the Price River Basin

The Price River, a significant tributary of the Green River, is located mainly in Carbon and Emery counties in East-Central Utah (see Fig. 1). The drainage area of the river is about 1950 square miles (5050 km^2). Elevations range from about 4,200 to 10,450 feet (1280 to 3190 m) above mean sea level.

The climate of the Price River Basin varies considerably with elevation. Annual precipitation varies from less than 8 inches (200 mm) in the lower elevations to over 30 inches (760 mm) in the mountains. Mundorff (1972) estimates that 70 percent of the precipitation falls as snow. In the lower elevations of the basin, approximately 50 percent of the annual precipitation falls in the period of May through September. In contrast with this, only 25 percent of the mountain precipitation occurs during the same period. Summer precipitation is commonly in the form of high intensity thunderstorms which may cause flash floods and considerable erosion.

Average annual yield for the Price River Basin ranges from less than 1 inch (25 mm) in the valley to over 12 inches (300 mm) in the mountains. Although about 50 percent of the total basin is below 6,400 feet (1950 m) only 10 percent of the total yield originates from these elevations which are typical of the central basin. Annual runoff from the central basin is estimated to be 1.08 inches (27.4 mm) which is about 9 percent of the average annual precipitation of 11.7 inches (297 mm). Average yearly outflow from the Price River at Woodside is about 75,000 acre-feet (92 x 10^6 m^3) (Utah Division of Water Resources, 1975). The flow in the Price River is highly regulated by Scofield Reservoir. Figure 2 is a flowchart of the hydrologic system of the Price River Basin and illustrates the relationships between the mountains and the natural and irrigated parts of the valley.

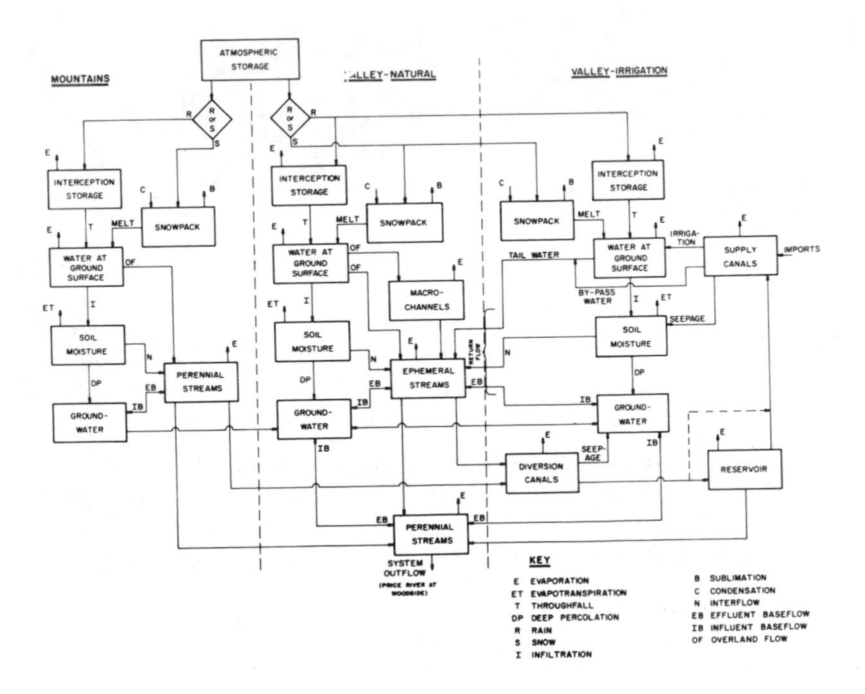

Fig. 2. Flowchart of the hydrological system of the Price River Basin

Outline of the Salt Pick-Up and Transport Model

Introduction. Water is the means of transport of both the dissolved salt and the sediment. Sediment, in turn, carries salt that is attached to it. Therefore, both the water and sediment components may be considered as driving the dissolved salt component. The hydro-salinity-sediment system is divided into three parts: 1) the watershed, 2) the channel banks, and 3) the channels. Each of these parts of the system will be represented by separate, but linked models. Figure 3 illustrates the linkage between these models. Each model is briefly described below, with some detail of explanation given to the important channel flow model. Some parts of the model are currently at the conceptual stage.

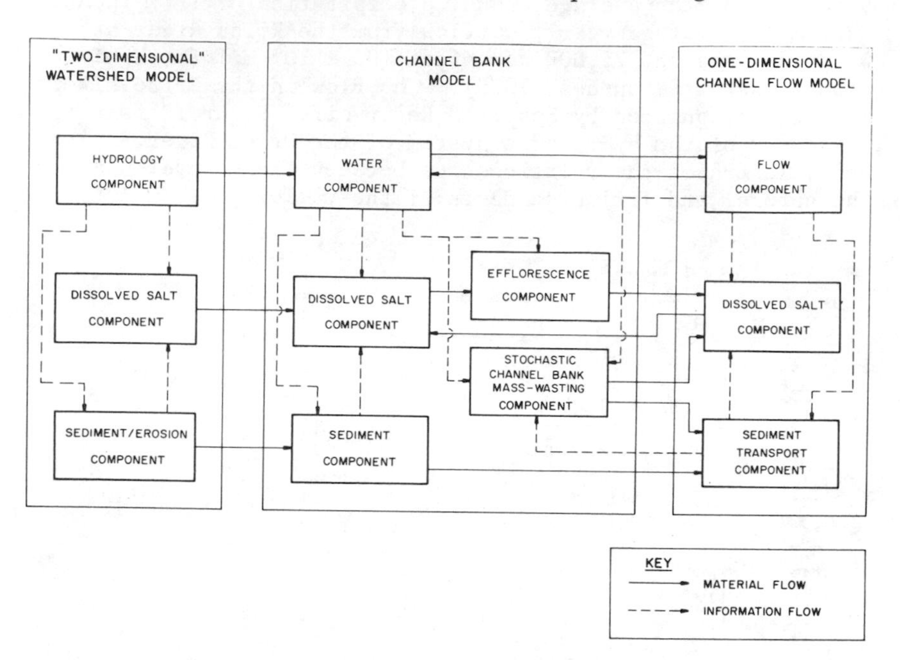

Fig. 3. Linkage of the models representing the hydro-salinity-sediment system of the Price River Basin

Streamflows resulting from short-duration thunderstorms are an important means of transporting salt and sediment in the Price River Basin. To represent the highly transient nature of these events, the proposed model will be a dynamic model. A variable time step feature may be included to improve the computational efficiency by using longer time steps to simulate the periods between hydrologic events.

Watershed model. For the purposes of this study the "two-dimensional" watershed model will simulate the movement of water, salt, and sediment through the surface and subsurface parts of the Price River Valley with the exception of the channel bank zone. Micro-channels are lumped into the overland flow process represented by the watershed model. A micro-channel is here defined as a channel receiving negligible interflow or groundwater flow, and having the highest stream order in the part of the watershed in which it is located.

Therefore, channels in different parts of the watershed, with different stream orders, can be classified as micro-channels provided each channel is of the highest stream order in its part of the watershed. Macro-channels, ephemeral streams, and perennial streams are simulated with the channel flow model. A macro-channel is here defined as a channel receiving negligible interflow or groundwater flow, and with a lower stream order than the micro-channels in the part of the watershed in which it is located. Ephemeral and perennial streams have the usual definitions based on the intermittent or continuous nature of streamflow which can be attributed to contributions from overland flow, interflow, and groundwater flow.

The hydrology component of the watershed model will be based on the compartment type models already developed and successfully applied by Utah State University. Examples of these models include USUWSM (Bowles and Riley, 1976) and BSAM (Huber et al., 1976). Basic inputs to the hydrology component will be precipitation, air temperature, and surface and subsurface inflows to the valley from the hydrologic system of the mountains. Outputs will include overland flow, interflow, and groundwater flow. These outputs will become inputs to the channel bank and channel flow models (see Fig. 3).

Simulated values of overland flow and observed rainfall intensities will be used to drive the erosion component of the watershed model which will calculate the sediment loss from the watershed. Models that will be considered when developing the erosion component will include, the universal soil loss equation (Wischmeier and Smith, 1965), the Meyer model (Meyer and Wischmeier, 1968; Meyer, 1971) and the Hydrocomp-EPA model (Donigan and Crawford, 1976).

The dissolved salt component of the watershed model will be based on previous hydro-salinity simulation models developed at Utah State University. Examples of these include Hyatt et al. (1970), Thomas et al. (1972), and Narasimhan (1975). Further work will be necessary to better define the mechanisms in the subsurface part of these models. In addition, a nonpoint source loading function developed by Ponce (1975) for the Mancos shale wildlands of the Price River Basin may be incorporated.

Channel bank model. The channel banks are an important source of salt and sediment in the stream. They act as an interface through which interflow and groundwater flow originating in the watershed, enter the channels. Erosion of the channel banks by flowing water adds to the sediment load of the stream and the release of salt associated with the sediment increases the salt load.

During the period of the rising limb of a hydrograph water in the channel enters bank storage (Todd, 1955). After the crest of the flood wave has passed, water starts to leave bank storage, thus supplementing flow in the channel. Flow in many of the channels in the Price River Basin is intermittent. After flow in these channels has ceased, water stored in the channel banks pools on the stream bed and is evaporated, or is evaporated at the soil-air interface of the channel bottom and banks. Water that has been stored in the bank zone is usually highly saline and when it evaporates a salt crust called efflorescence remains. Efflorescence also occurs on the berms and bars of perennial streams during periods of low flows, and to a lesser extent on the soil surface at any location in the basin. These salt crusts, which are often conspicuous by their white appearance, determine the salt concentrations at the beginning of a storm, whereas the inherent characteristics of the soil determine the equilibrium salt output of a given hydrologic event (White, 1976). In addition to efflorescence, the initial concentration of salt is greatly influenced by the accumulation of loose weathered material on the watershed and in the dry parts of the channel network. Burge (1974) attributes the rapid weathering of the Mancos shale to cyclic dehydration-hydration of the entrained salts, particularly mirabilite and thenardite.

Another way in which the channel banks are an important source of salt and sediment is by the process of mass wasting. Soil that enters the stream channel by this process becomes part of the sediment load of the stream and also augments the salt load through the release of salts. Although mass wasting is random in its spatial and temporal occurrence, field work on the Price River Basin has indicated that the sloughing of channel banks can be an important factor in determining the amount of salt and sediment yielded in a particular event.

The proposed structure for the channel bank model is contained in Fig. 3. Processes to be represented by this model include:

1. The interface function of the channel banks for interflow and groundwater flow entering the channel.

2. The bank storage function by which water from the channel is stored in the channel banks during the period of the rising hydrograph limb and subsequently release.

3. The transport of dissolved salts between the banks and the channel.

4. The formation of efflorescence at the soil-air interface of the channel.

5. The transport of sediment into the channel from the bank zone.

6. The process of mass wasting of parts of the channel bank into the channel.

Channel flow model. The network of stream channels act as a collector system for the water, salt, and sediment originating on the watershed and in the channel bank zone. The channel network also adds to the salt and sediment loads through channel scour. Streamflow, and the transport of dissolved salt and sediment in macro-channels, ephemeral channels, and perennial channels will be simulated with a dynamic channel flow model. The channel network will be divided into reaches of relatively uniform flow, salt, and sediment characteristics. Reach boundaries will be consistent with the subbasin divisions in the watershed model.

The basis for the channel flow model will be one-dimensional transport equation:

$$\frac{\partial AX}{\partial t} = - \frac{\partial QX}{\partial y} - \frac{\partial (DA \frac{\partial X}{\partial y})}{\partial y} + qu + AJ + wB \quad (1)$$

in which

X = concentration of a water quality parameter $\{ML^{-3}\}$,
A = cross sectional area of the stream $\{L^2\}$,
t = time $\{T\}$,
Q = streamflow rate $\{L^3 T^{-1}\}$,
y = distance along the longitudinal stream axis $\{L\}$,
D = longitudinal dispersion coefficient $\{L^2 T^{-1}\}$,
q = lateral inflow rate $\{L^2 T^{-1}\}$,
u = concentration of the water quality parameter in the lateral inflow $\{ML^{-3}\}$,
J = physio-chemical transformations associated with the flowing volume of water $\{ML^{-3} T^{-1}\}$,
B = physio-chemical transformations associated with the stream bottom and banks $\{ML^{-2} T^{-1}\}$,
w = wetted perimeter of the stream $\{L\}$,
b = width of the stream bottom $\{L\}$, and
h = depth of flow $\{L\}$.

Equation (1) represents the temporal change in concentration of the parameter X within the stream. The first term on the right-hand side is the dispersion term and represents the transport of the water quality parameters due to nonuniform velocity gradients in the stream profile. The second term represents downstream advection of the parameter in the flowing water. Diffuse sources of the parameter are represented by the third term. The fourth term represents the physio-chemical transformation affecting the parameter and associated with the flowing volume of water. Physio-chemical transformations associated with the stream bottom and banks are represented by the fifth term.

Differentiating the advection term in Eq. (1) yields:

$$\frac{\partial AX}{\partial t} = - Q \frac{\partial X}{\partial y} - X \frac{\partial Q}{\partial y} + \frac{\partial (DA \frac{\partial X}{\partial y})}{\partial y} + qu + AJ + wB \qquad (2)$$

Now considering the mass balance for flow and the parameter X we obtain the following equations:

$$\frac{\partial Q}{\partial y} = q_o + q_N + \overset{+}{q_G} - \overset{-}{q_G} + \overset{+}{q_B} - \overset{-}{q_B} - q_E = \sum_i q_i \qquad (3)$$

and

$$qu = q_o u_o + q_N u_N + \overset{+}{q_G} u_G - \overset{-}{q_G} X + \overset{+}{q_B} u_B - \overset{-}{q_B} X = \sum_i q_i u_i \qquad (4)$$

in which the suffixes are defined as follows:

o = overland flow,
N = interflow,
G = groundwater,
B = flow between the channel and bank storage,
E = evaporation,
+ = effluent flow,
- = influent flow.

Rewriting Eq. (2) for dissolved salt and for salt associated with the suspended sediment, we obtain the following equations:

$$\frac{\partial AX_1}{\partial t} + Q \frac{\partial X_1}{\partial y} + X_1 \sum_i q_i - \frac{\partial (DA \frac{\partial X_1}{\partial y})}{\partial y} = \sum_i q_i u_{i1} + AJ_1 + wB_1 \qquad (5)$$

and

$$\frac{\partial AX_2X_5}{\partial t} + Q \frac{\partial X_2X_5}{\partial y} + X_2X_5 \sum_i q_i \frac{\partial (DA \frac{\partial X_2X_5}{\partial y})}{\partial y} =$$

$$= \sum_i q_i u_{i2} + AJ_2 + wB_2 \qquad (6)$$

in which

X_1 = concentration of dissolved salt (TDS) $\{ML^{-3}\}$,

X_2 = ratio of the concentration of salt associated with the suspended sediment to the concentration of suspended sediment,

X_5 = concentration of suspended sediment (from sediment transport component) $\{ML^{-3}\}$.

The subscripts 1 and 2 are also used on u_i, J, and B to distinguish with which water quality parameter these symbols are associated. Figure 4 is a schematic diagram of the state variables, physio-chemical transformations, and non-point loads represented by Eqs. (5) and (6). In this figure the state variable X_3 is the mass of efflorescence per unit area of the stream perimeter and X_4 is the concentration of salt associated with the stream bottom and banks. f_{12} represents the dissolution or precipitation processes by which salt is released from, or deposited on the suspended sediment in the stream. Current research at Utah State University has

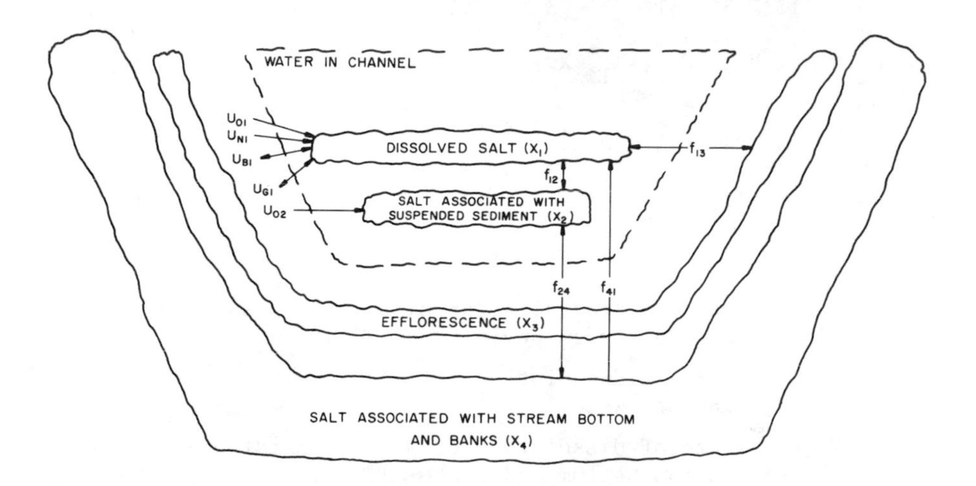

Fig. 4. Schematic diagram of the state variables, physio-chemical transformations, and non-point loads represented by the dissolved salt component of the channel flow model.

indicated that the fluvial sediments are an important source of dissolved salts. Whitmore (1976) has observed that after 0.5 minutes the rate of salt release from saline sediments derived from Mancos shale decreases from a fast initial rate, which accounted for 80 to 90 percent of the total salt released, to a rate three or four orders of magnitude less. Analysis of field data supports the hypothesis that the salt load attributable to the suspended sediments is directly proportional to the amount of suspended sediment (R. H. Hawkins, personal communication, 1977); hence, X_2 is defined as the ratio of the concentration of salt associated with the suspended sediment to the concentration of suspended sediment. f_{13} represents the dissolution or precipitation process by which efflorescence is removed from the wetted channel perimeter, or by which salt is deposited on that perimeter. The process by which salt associated with the stream bottom and banks is dissolved by the stream water is represented by f_{14}. Salt associated with the suspended sediment is resuspended or settled with the suspended sediment and is represented by f_{24}. The sediment transport component of the channel flow model will simulate the resuspension and settling processes for suspended sediment and will be linked to f_{24}.

The transformations f_{12}, f_{13}, and f_{41}, and f_{24} are related to the B and J terms in Eqs. (5) and (6) as follows:

$$B_1 = f_{13}(K_{13}, K_{31}, X_1, X_3) + f_{41}(K_{41}, X_1, K_4) \tag{7}$$

$$J_1 = f_{12}(K_{12}, K_{21}, X_1, X_2) \tag{8}$$

$$B_2 = f_{24}(K_{24}, K_{42}, X_2, X_4, X_5) \tag{9}$$

$$J_2 = - f_{12}(K_{12}, K_{21}, X_1, X_2, X_5) \tag{10}$$

in which

K_{13} = rate of precipitation of dissolved salt on the wetted stream perimeter $\{T^{-1}\}$,

K_{31} = rate of dissolution of efflorescence by the channel water $\{T^{-1}\}$,

K_{41} = rate of dissolution of salt associated with the stream bottom and banks $\{T^{-1}\}$,

K_{12} = rate of precipitation of dissolved salt on the suspended sediment $\{T^{-1}\}$,

K_{21} = rate of dissolution of salt associated with the suspended sediment $\{T^{-1}\}$,

K_{24} = rate of settling of the suspended sediment (from the sediment transport component) $\{T^{-1}\}$,

K_{42} = rate of resuspension of the suspended sediment (from the sediment transport component) $\{T^{-1}\}$,

K_{12} = rate of precipitation of dissolved salt on the suspended sediment $\{T^{-1}\}$, and

K_{21} = rate of dissolution of salt associated with the suspended sediments $\{T^{-1}\}$.

Work Needed

Further work is needed to better define the processes represented by the watershed model, the channel bank model, and the channel flow model. The relatively small salt contributions from surface flows suggest that emphasis should be placed on examining those processes associated with subsurface water movement as represented by the watershed model.

Extensive new research involving field survey work is necessary to provide a basis for the stochastic model of the channel bank mass wasting process. An understanding of factors which affects this process must be gained and frequency distributions for the quantity of soil lost during mass wasting events must be developed for the different soil types in the basin.

In the preceding discussion a framework was structured for identifying and linking the salt producing processes within the channel system of the Price River Basin. These processes were represented by generalized functions, f_{ij}. The formidable task ahead is to discern quantitative relationships which can be used to test management alternatives for salinity control in the basin. Several current research projects are directed toward this objective. For example, laboratory and field studies are underway at Utah State University to estimate the rate of formation of efflorescence in channels and to further understand the causal processes. Preliminary data indicate efflorescence salt masses in stream channels ranging from 1 to over 1100 g/m^2 in the upper 2 cm of soil crust.

Summary and Conclusions

Marine shales are a major contributor of salts to the waters of the Colorado River Basin. From its point of origin on a watershed in the form of precipitation, water follows many different flow paths, and along each of the possible patterns the potential is high for contact between the moving waters and the shales. However, contact with the shales, and therefore, salt pick-up, occurs in two main locations: 1) on the land surface, with the water occurring as overland and channel flows; and 2) beneath the surface of the soil, with the water moving in both the saturated and nonsaturated regimes.

Because the precipitation input characteristics and the watershed conditions are continually changing in both the time and space dimensions, water movement (or drainage) from a catchment area is a highly dynamic phenomenon. There is a continuous interchange between the surface and subsurface flows, so that during its traverse of the watershed a particular volume of water is subject to a broad spectrum of salt pick-up opportunities, including both natural and man-induced conditions. Figure 2 depicts to some extent the highly complex nature of the hydrologic system as it occurs in the real world. On this figure the "boxes" represent storage locations within the system, while the lines and arrows indicate flow processes or paths of water movement between storage locations. Associated with most of the hydrologic flow processes is a salt pick-up mechanism, and this paper is a preliminary attempt to identify a theoretical description of these mechanisms. Much additional research in the library, the laboratory, and the field is needed to adequately identify all the relationships suggested by this paper. Perhaps through an improved understanding of the basic salt pick-up mechanisms associated with both surface and subsurface flows it will be possible ultimately to adopt management procedures so as to minimize the salt loads contributed from rangelands, agricultural lands, coal mines, and other areas of water use within the Colorado River Basin.

Acknowledgments

The work on which this paper is based was funded by the U.S. Bureau of Reclamation, Agreement No. 14-06-D-7961, and the Office of Water Resources and Technology, Project No. A-039-Utah, Agreement No. 14-34-0001-7094.

References

[1] Blackman, W. C., Jr., Rouse, J. V., Schillenger, G. R., and Shafer, W. H., Jr., 1973. "Mineral Pollution in the Colorado River Basin," Journal of Water Pollution Control Federation, 45, (7): 306.

[2] Bowles, D. S. and Riley, J. Paul, 1976. "Low Flow Modeling in Small Steep Watersheds," Journal of the Hydraulics Division, ASCE, 102 (HY9): 1225-1239.

[3] Burge, D. L., 1974. Unpublished Notes, Department of Geology, College of Eastern Utah, Price, Utah.

[4] Donigan, A. S. and Crawford, N. H., 1976. "Modeling Pesticides and Nutrients on Agricultural Lands," Publication No. EPA-600/2-76-043, U.S. Environmental Protection Agency, Washington, D. C.

[5] Huber, A. L., Israelsen, E. K., Hill, R. W. and Riley, J. P., 1976. "BSAM: Basin Simulation Assessment Model Documentation and User Manual," Publication PRWG202-1, Utah Water Research Laboratory, College of Engineering, Utah State University, Logan, Utah, 30 p.

[6] Hyatt, M. L., Riley, J. P., McKee, M. L., and Israelsen, E. K., 1970. "Computer Simulation of the Hydrologic-Salinity Flow System Within the Upper Colorado River Basin," Publication No. PRWG54-1, Utah Water Research Laboratory, College of Engineering, Utah State University, Logan, Utah, 255 p.

[7] Meyer, L. D., 1971. "Soil Erosion by Water on Upland Areas," In: H. W. Shen (ed.), River Mechanics, H. W. Shen, Engineering Research Center, Colorado State Univ., Fort Collins, Colorado.

[8] Meyer, L. D. and Wischmeier, W. M., 1969. "Mathematical Simulation of the Process of Soil Erosion by Water," Paper No. 68-732, presented at the winter meeting of the American Society of Agricultural Engineers, Chicago, Illinois.

[9] Mundorff, J. C., 1972. "Reconnaissance of Chemical Quality of Surface Water and Fluvial Sediment in the Price River Basin, Utah," Dept. of Natural Resources, State of Utah, Tech. Publ. No. 39, 55 p.

[10] Narasimhan, V. A., 1975. "A Hydro-Quality Model to Predict the Effects of Biological Transformations on the Chemical Quality of Return Flow," Ph.D. dissertation, Utah State University, Logan, Utah, 189 p.

[11] Ponce, S. L., II, 1975. "Examination of a Non-Point Source Loading Function for the Mancos Shale Wildlands of the Price River Basin, Utah," Ph.D. dissertation, Utah State University, Logan, Utah, 177 p.

[12] Thomas, J. L., Riley, J. P., and Israelsen, E. K., 1971. "A Computer Model of the Quality and Chemical Quality of Return Flow," Publ. No. PRWG77-1, Utah Water Research Laboratory, College of Engineering, Utah State University, Logan, 94 p.

[13] Todd, D. K., 1955. "Groundwater Flow in Relation to a Flooding Stream," Proceedings of the American Society of Civil Engineeirng, Vol. 81, separate issue 628.

[14] U.S. Department of the Interior, 1974. "River Water Quality Improvement Program," Bureau of Reclamation report, 125 p.

[15] Utah Division of Water Resources, 1975. "Hydrologic Inventory of the Price River Basin," 63 p.

[16] White, R. B., 1976. "Salt Production from Micro-Channels in the Price River Basin: A Progress Report," submitted to the U.S. Bureau of Land Management by Utah Water Research Laboratory, College of Engineering, Utah State University, Logan, 52 p.

[17] Whitmore, J. C., 1976. "Some Aspects of the Salinity of Mancos Shale and Mancos Derived Soils," M.S. thesis, Utah State University, Logan, 69 p.

[18] Wischmeier, W. H. and Smith, D. D., 1965. "Predicting Rainfall Erosion Losses form Cropland East of the Rocky Mountains," Agriculture handbook No. 282, U.S. Department of Agriculture.

APPLICATIONS OF PROBABILISTIC - DETERMINISTIC MODELING TO PROBLEMS OF MASS TRANSFER IN GROUNDWATER SYSTEMS

By

Franklin W. Schwartz
Department of Geology
University of Alberta
Edmonton, Alberta, Canada T6G 2E3

Abstract

It is possible to simulate multi-dimensional mass transfer using a hybrid probabilistic-deterministic technique. This approach differs from conventional numerical methods in that the model has a stochastic component and does not operate directly on a partial differential equation. The groundwater seepage velocity is calculated at node points arranged within a groundwater region. Moving reference particles with an associated quantity of mass are introduced to the region. These particles are moved with a deterministic component of motion which is controlled by the groundwater velocity field and a random motion which is controlled in part by the dispersive character of the porous medium. The simulation proceeds by allowing the particles to progress through the region subject to conditions along the boundaries. At any time, the mass density within the region can be related to concentration.

The accuracy of such a transport model has been verified by accurately reproducing the analytical solution to a simple one-dimensional, dispersion-convection problem. It can be shown that the accuracy of the procedure decreases as the number of moving particles decreases and as the size of the time steps become exceptionally large. Extension of this technique to multi-component and multi-dimensional problems is straightforward. The simplicity, versatility and inherent numerical stability of the approach make it ideal for application to a variety of hydrogeological problems.

Introduction

Many human activities have a potential to contaminate subsurface waters. While generalizations are difficult, contamination results most often from the release of mass either by design or accident into active groundwater flow systems. In assessing the significance of groundwater contamination, one of the most difficult problems is predicting the degree and extent of contamination. This problem arises because of the complex array of factors influencing the dispersal of the contaminant in the subsurface.

Bear (1972) identifies the most important of these factors and reviews the developmental histories of transport concepts. The body of work that he discusses in practice provides the theoretical basis for the quantitative analysis of mass transport. In spite of considerable progress, much of our present understanding is rudimentary and empirical. The scale of investigation most often has been limited unfortunately to laboratory-scale experiments. However, the success in recent years of sophistocated techniques for the mathematical description and predictive analysis of groundwater systems has provided the impetus to develop and to apply regional scale mass transfer models.

Two processes, convection and dispersion, are responsible for the physical transport of mass from one point to another in a groundwater system. Convective transport occurs when mass is transferred by the movement of the groundwater in which the mass is dissolved. This is a primary transport mechanism that determines the extent of spreading from the site of mass entry to the system. The direction and velocity of transport generally is assumed to be identical to that of the groundwater. Accordingly, those features of the hydrogeological setting such as water table configuration or hydraulic conductivity development which control the pattern of groundwater flow also control the pattern of convective transport.

Dispersion refers to phenomena that act primarily to produce solute mixing within a porous medium. On a regional hydrogeological scale, dispersion is produced by gross changes in the character of groundwater flow patterns which result from porous medium non-idealities. The most important non-ideality in this respect is porous medium heterogeneity with respect to hydraulic conductivity. Generally, the magnitude of dispersion increases as the contrast in hydraulic conductivity among the various elements comprising a geological unit becomes more marked. The tendency for dispersion to spread a tracer or contaminant within a flow region results in a dilution of the tracer mass. This process thus provides an important mechanism for concentration attenuation.

Chemical and biological processes may act in addition to the physical transport processes to retard mass transfer. Cation exchange and radioactive decay are two examples of chemical processes. The efficiency of these processes in retarding the spread of mass depends primarily on the character of the mass being transported and, for cation exchange, the character of the porous medium.

Mathematically, one can state this array of physicochemical processes in terms of the following equation (1), known as the dispersion convection equation:

$$\delta/\delta x_{\alpha\tau}\ (D_{\alpha\tau}\ \delta c_i/\delta x_\tau) - \delta/\delta x_\alpha\ (c_i\ v_\alpha) + \sum_{j=1}^{n} \varepsilon R_{ij} = \delta(\varepsilon c_i)/\delta t \quad \alpha,\tau = 1,2 \tag{1}$$

where: $D_{\alpha\tau}$ = dispersion coefficient,

c_i = mass concentration ith constituent,

v_α = Darcy velocity in α direction,

ε = porosity,

R_{ij} = rate of production of constituent i in reaction j from n different processes,

x_α = cartesian coordinate system,
$\alpha = 1,2$
$\tau = 1,2$

All important transport processes are incorporated in Eq. (1). The first term on the left-hand side of Eq. (1) accounts for dispersion and the second accounts for convective transport. The last term of the left-hand side is a source term that accounts for all the significant chemical and biological processes.

Equation (1) is solvable if estimates of the parameters that control the transport processes can be made and boundary and initial conditions are established. The solution to Eq. (1) provides descriptions of concentration distributions in both time and space for the region of interest. These results can be interpreted to yield, for example, the residence time of mass in the system or a prediction of the extent of possible contamination.

Most methods of solving an equation such as (1) involve a direct numerical solution. The method of characteristics and the Galerkin finite-element method are probably the most successful of the techniques. The purpose of this paper is to describe and demonstrate a powerful hybrid probabilistic-deterministic method for modeling mass transfer in ground-water systems. Instead of solving the differential equation directly, the hybrid method addresses the equivalent but more fundamental problem of describing the spread of a large number of moving reference particles within a region. In practice, the moving reference particles, each with a given quantity of associated mass, are introduced to a region in zones or along boundaries where mass inflow occurs. The reference particles are transported within the region. It is impossible to produce an exact mathematical description of individual particle motion because of the large number of unknown factors contributing to the motion. Statistical features of the motion of the particle assemblage as a whole however do provide a basis for representing an idealized pattern of motion for individual reference particles. Because the position of individual particles is known within the region, one can easily determine the distribution of associated mass. This distribution can be simply interpreted in terms of concentrations.

The hybrid method, while not new, has seen only very limited development and application in hydrogeology (Ahlstrom, 1975; Schwartz, 1975). Models of this type that have been used in practice (Ahlstrom, 1975) probably represent the most versatile and powerful mass transfer models existing in North America.

Formulation and Verification

To develop a stochastic technique that in essence depends on the history of motion of a large number of reference particles requires that the particle motions be related to the physical transport processes, convection and dispersion. The convective and dispersive components of the particle motion in fact can be considered separately. In this respect, the approach does not differ appreciably from the method of characteristics (Gardner et al., 1964; Reddell and Sunada, 1970).

The convective portion of the transport in practice can be represented by a deterministic particle motion. The moving reference particles follow groundwater streamlines with a magnitude of velocity that is the same as the seepage velocity. By itself, this type of transport produces what is commonly referred to as 'plug flow' of a tracer.

Convective transport in shallow groundwater systems results from groundwater motion. For example, this motion can be related to regional groundwater flow from recharge areas to discharge areas or to the withdrawal or injection of groundwater by well systems. The actual pattern of groundwater flow is described by the groundwater velocity field. This field, for the general case, is determined by substituting distributed values of hydraulic head, which are calculated by solving the appropriate groundwater flow equation, and hydraulic conductivity values into the Darcy equation. It is assumed that the groundwater flow pattern is independent of mass distributions within the system. This assumption is made for nearly all mass transfer models to eliminate the necessity of iterations during any one time step.

Dispersion or the interaction of the tracer with the porous medium is accounted for in the particle motion by adding a random velocity component to the deterministic motion. The magnitude and direction of this random motion is related to the dispersive character of the porous medium.

Josselin de Jong (1958) investigated tracer migration from a point source. He determined that a swarm of tracer particles will be normally distributed in three dimensions around a center travelling with the mean groundwater seepage velocity. The variance σ_L^2 and σ_T^2 of this distribution in the longitudinal and transverse directions and a time variable is related to the longitudinal and transverse dispersion coefficients as follows (Bear, 1972):

$$D_L' = \sigma_L^2/2t$$

$$D_T' = \sigma_T^2/2t \qquad (2)$$

where: D_L' = dispersion coefficient in the longitudinal direction,

D_T' = dispersion coefficient in the transverse direction,

σ_L^2 = longitudinal variance of the distribution, and

σ_T^2 = transverse variance of the distribution.

The longitudinal and transverse dispersion coefficients in practice can be related empirically to a characteristic parameter of the porous medium termed the dispersivity ε and the magnitude of the seepage velocity:

$$D_L' = \varepsilon_L\ v$$

$$D_T' = \varepsilon_T\ v \qquad (3)$$

where: ε_L = longitudinal dispersivity,

ε_T = transverse dispersivity (often assumed to be .2 or .3 $_L$), and

v = magnitude of the seepage velocity.

It will become clear in the following discussion how the relationships in Eqs. (2) and (3) provide a basis for generating the dispersion-affected motion for each of the reference particles.

In order to more clearly understand how the model functions a flow chart of the computational steps has been included (Fig. 1). A specified number of reference particles with their associated quantities of mass are positioned randomly along the mass inflow boundaries of the region at the beginning of each time step. A velocity is calculated for each reference particle in the region by interpolating values from surrounding nodes. The user generally supplies the nodal

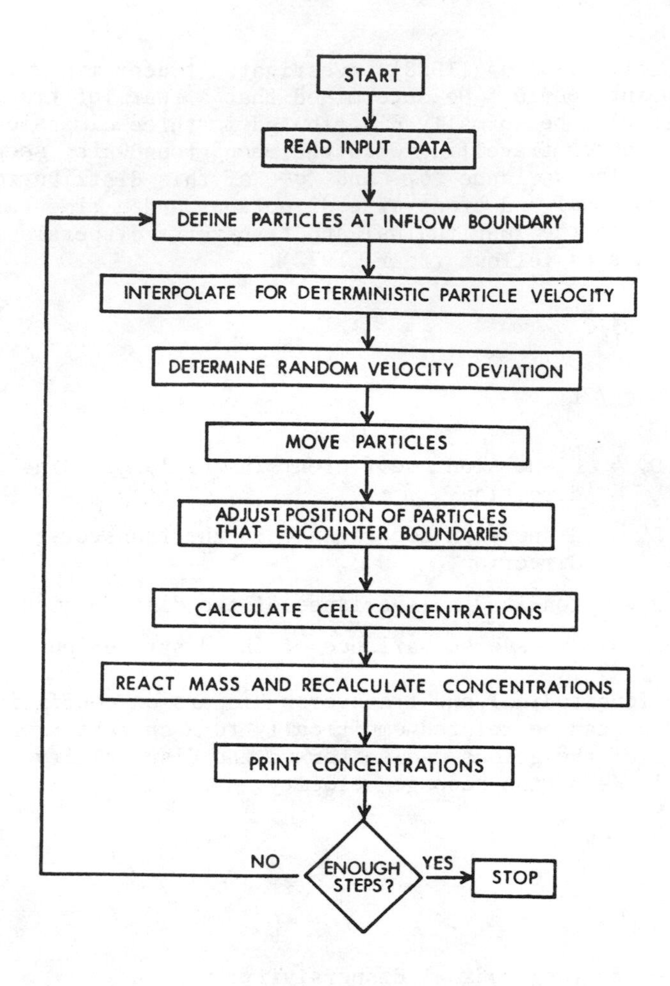

Fig. 1. Flow chart outlining computational steps in the model procedure

velocity values as input data. The reference particles move along their respective vectors a distance that is fixed by the magnitude of the time step and the velocity. This displacement represents the deterministic component of the motion. The new particle position is in effect only temporary. The random component of the particle motion is applied to relocate it in the vicinity. The displacements involved with this random adjustment reflect the magnitude of dispersion effects. Relocation is accomplished by moving the particle first in a direction that coincides with the groundwater flow vector and second in a direction that is normal to the flow vector. The magnitude of the random displacements is determined by generating a normally distributed random number with a mean of 0 and a standard deviation that is related to the porous medium dispersivity by the following equations:

$$\sigma_L = (\varepsilon_L \, v \cdot 2\Delta t)^{\frac{1}{2}}$$

$$\sigma_T = (.2\varepsilon_L \, v \cdot 2\Delta t)^{\frac{1}{2}}$$

where Δt = magnitude of the time step. The following pair of equations describe how the moving particle position at the n+1 time step is calculated from parameters at the nth step.

$$x^{n+1} = x^n + \Delta t \cdot v_x^n + v_x^n \cdot r_1 \cdot \sigma_L / v^n + v_z^n \cdot r_2 \cdot \sigma_T / v^n$$

$$z^{n+1} = z^n + \Delta t \cdot v_z^n + v_z^n \cdot r_1 \cdot \sigma_L / v^n + v_x^n \cdot r_2 \cdot \sigma_T / v^n$$

where x, z = cartesian spatial coordinates,

v = magnitude of seepage velocity,

v_x, v_z = x and z seepage velocity components, and

r_1, r_2 = normally distributed random numbers with a mean of 0.

At the beginning of each new time step, another set of particles are defined along the zone of mass inflow and these and existing particles are all moved in that time step. Concentrations of individual components can be calculated from the total number of reference particles in each cell, their associated quantities of mass and porosity. The grid cells in practice are defined by the nodes at which the discretized values of velocity are defined. In the technique described by Ahlstrom (1975), the various quantities of mass which are associated with the moving particles may undergo a variety of instantaneous chemical or biological reactions.

To gain confidence in the accuracy of the method, the analytical solution to a simple mass flow problem is compared to the model result. Ogata and Banks (1961) give the following approximate solution to the problem of one-dimensional dispersion in a semi-infinite, homogeneous and isotropic porous medium with a step input of a single component:

$$c_i / c_o = \frac{1}{2} \left[\text{erfc } (x - v_x t) / 2 (D_L t)^{\frac{1}{2}} \right]$$

where c_i = concentration at $x > 0$,

c_o = concentration of the tracer at $t = 0$, and

v_x = seepage velocity in the x direction.

The solid lines in the relative concentration versus distance plot (Fig. 2) are the resulting analytical solutions for four different times. The closed circles represent solutions from the hybrid model for the same four time steps.

The parameters for both the analytical and hybrid problems are included in Fig. 2. When you compare the two different solutions, it is apparent that there is reasonable agreement between the two.

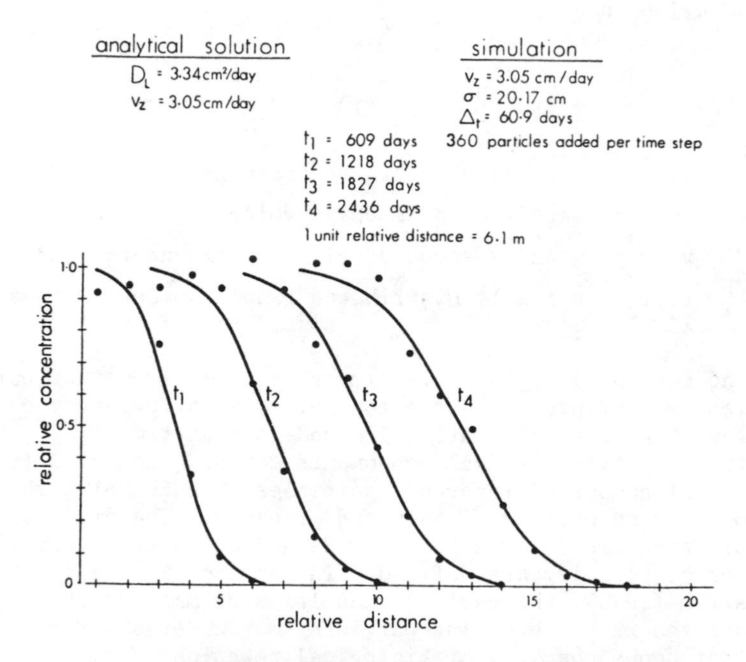

Fig. 2. Comparison of model trial results to the analytical solution of a simple, one-dimensional dispersion-convection problem

Unique Features and Application

This section will examine unique features of the method and briefly demonstrate some of the possibilities for practical application. The basic simplicity of the underlying concepts and ease in programming are the greatest advantage. The algorithm is more or less a simple accounting procedure. Only very simple difference expressions are necessary to describe mathematically movement of a reference particle within the region. Moderate coding skill is all that is necessary to develop the program because the necessity of solving a large set of simultaneous equations, as in most of the other methods, is eliminated.

The accuracy of the solution depends upon two factors, the relative number of reference particles added and the size of the time steps. Problems of numerical dispersion which sometimes characterize the finite difference and finite element methods are eliminated from the hybrid method. Generally, the accuracy of the hybrid method increases as the mass added to the system is distributed among a larger number of moving reference particles. Figure 3 compares the analytical solution for the one-dimensional case to the model results. The mass in the hybrid technique is associated with 36, 90, 180, and 360 reference particles per time step while all other conditions remain the same (Fig. 3). The best correspondence

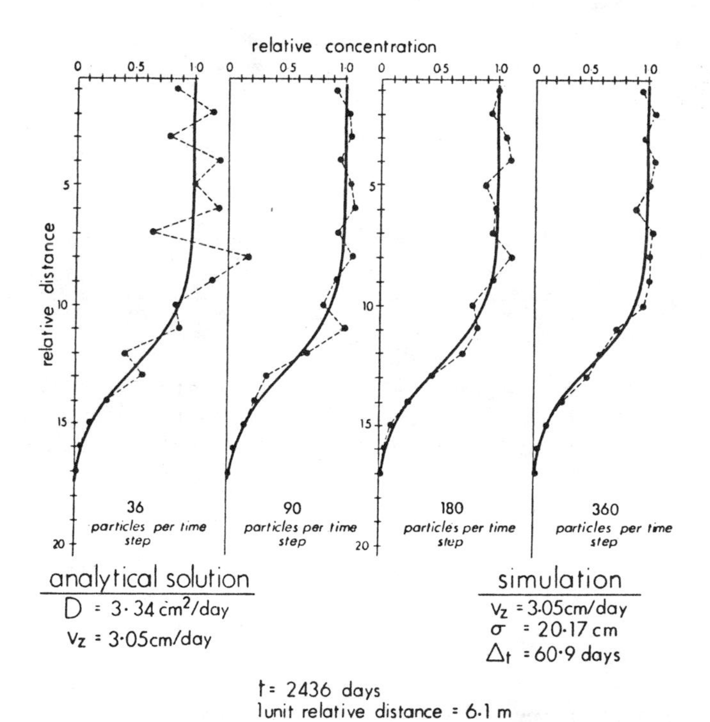

Fig. 3. Trial results that show the deterioration in the accuracy of solutions that result from decreasing the number of moving reference particles

between the model and analytical solutions is achieved with the largest number of moving reference particles added per time step. When very few particles are added (Fig. 3), the model solution tends to oscillate markedly around the analytical solution.

A similar type of analysis is carried out with the two-dimensional region depicted in Fig. 4A. A steady-state equipotential pattern (Fig. 4B) is obtained from a solution of Richard's equation. The velocity field as previously mentioned is obtained from hydraulic head and hydraulic conductivity distributions using the Darcy equation. The boundary conditions for the groundwater flow problem are straightforward. The lateral and bottom boundaries are no flow boundaries. Along the upper boundary, hydraulic head varies linearly as a function of space and is constant in time. This flow region and the assumptions implicit in its definition are discussed in detail by Freeze and Witherspoon (1966, 1967).

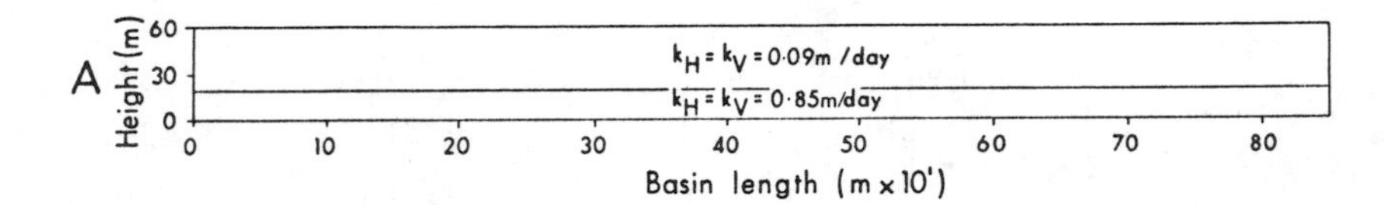

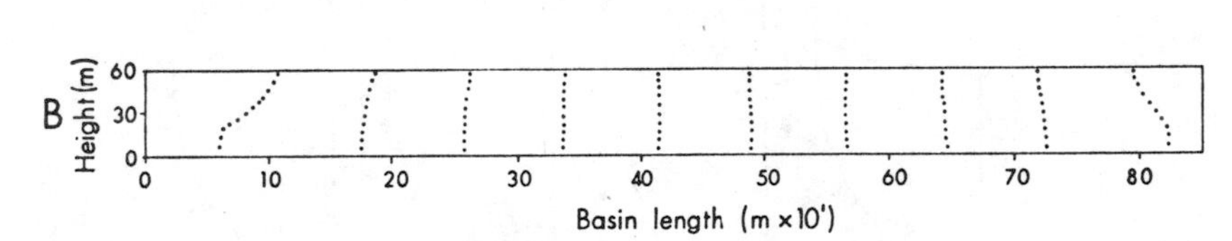

Fig. 4. A. The region that forms the basis for a two-dimensional analysis of mass transfer. A lower conductivity unit overlies a higher conductivity unit. The water table configuration in two-dimensions is a straight line sloping from left to right with a gradient of 0.01

B. The pattern of hydraulic head distributions obtained by solving Richard's equation for the region above. The contour interval is .76 m.

The boundary conditions for the mass transfer problem are somewhat more complex. The lateral and bottom boundaries are assumed to be no-flow boundaries. Moving reference particles are reflected by these boundaries. Along the upper boundary, no-flow conditions coincide with those parts of the boundary where recharge conditions exist. The reasoning here is that in recharging parts of the flow system there are no mechanisms by which mass can leave the system. In discharge portions of the system, direct discharge to the ground surface is able to carry mass out of the system. Accordingly, moving reference particles are allowed to move across portions of the upper boundary where discharge conditions occur. The initial concentration of the single constituent in this system is assumed to be zero at the start of the simulation.

The zone of mass inflow to the system is located below the water table in the recharge portion of the flow field. The results in Figs. 5A, 5B and 5C depict the concentrations of the component in the system after 100 years. Relative concentration distributions are represented by thick wavy bands rather than lines because the bands in each case, for example 5A, enclose the range of results from three simulations. The three runs are different because a different set of random numbers is used to control the dispersive component of the particle motion. The rationale for this approach is that because no analytical solutions are available for the problem, one way to consider the accuracy of the solution is to investigate the consistency of the results. As the number of reference particles added to the inflow grid cell per time step (Γ) is reduced in Figs. 5A, 5B and 5C from 50 to 25 to 10, respectively, the consistency of results is reduced. This conclusion is evident especially by comparing Figs. 5A and 5C. ξ is

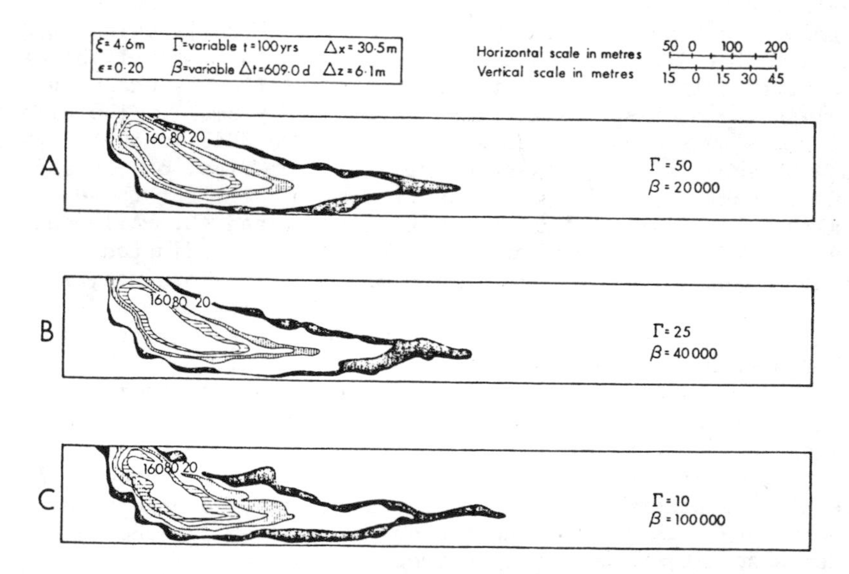

Fig. 5. Increasing relative thickness and overlap of concentration bands illustrate the reduction in consistency of three solutions in each case as the number of moving reference particles (Γ) added to one inflow cell per time step are reduced

the longitudinal dispersivity, ε is the porosity, β is the mass associated with each reference particle, Δx and Δz are the distances between nodes and Δt is the time step size. The range of results for the three trials in Fig. 5A can be depicted by rather narrow bands. However, in Fig. 5C the concentration bands are wider and tend to overlap.

While it is apparent that some deterioration of the accuracy of the solution is the result of decreasing the number of moving reference particles, the deterioration is gradual. Approximate solutions are possible with relatively few reference particles. Criteria for selecting the actual number of moving particles are therefore not straightforward. Usually the number of moving reference particles can generally be considered adequate if the solution appears smooth and continuous.

The hybrid method, unlike many numerical approaches, is inherently stable with respect to time step size. Some deterioration in the accuracy of the result will occur, however, if the time step size is excessively large. The most common indications of an excessive step size are discontinuities in the solution. This problem arises because the direction of convective transport may deviate from the streamlines whenever reference particles move a large distance before the next velocity vectors are calculated. In other words, the deterministic component of the particle motion which is calculated as a finite difference approximation may not accurately be determined in zones where the direction of flow is changing, unless the time step size remains relatively small. The effect of gross increases in step size from 609 to 3045 to 6090 days may be seen in a comparison of results from Figs. 6A, 6B and 6C, respectively. The groundwater flow problem for these cases is similar to that shown in Figs. 4A and 4B. All model parameters with the exception of β, which is adjusted so that a constant mass inflow per time step is maintained, remain constant in Figs. 6A, 6B and 6C. The progressive development of discontinuities in the solution is apparent as the time step size is increased from 609 days (Fig. 6A). This problem is most noticeable at the extremities of the tracer plume (Figs. 6B and 6C) where the relative concentrations of less than 5 are approximately 2% or less of the maximum concentrations found in the plumes. The correspondence between results is otherwise reasonably good. Again, deterioration in the accuracy is gradual and the solution, even for the largest step size, is approximately correct.

One of the most significant features of the modeling approach is its mathematical generality. This feature facilitates its application to a variety of theoretical and practical mass transfer problems. The application that is considered

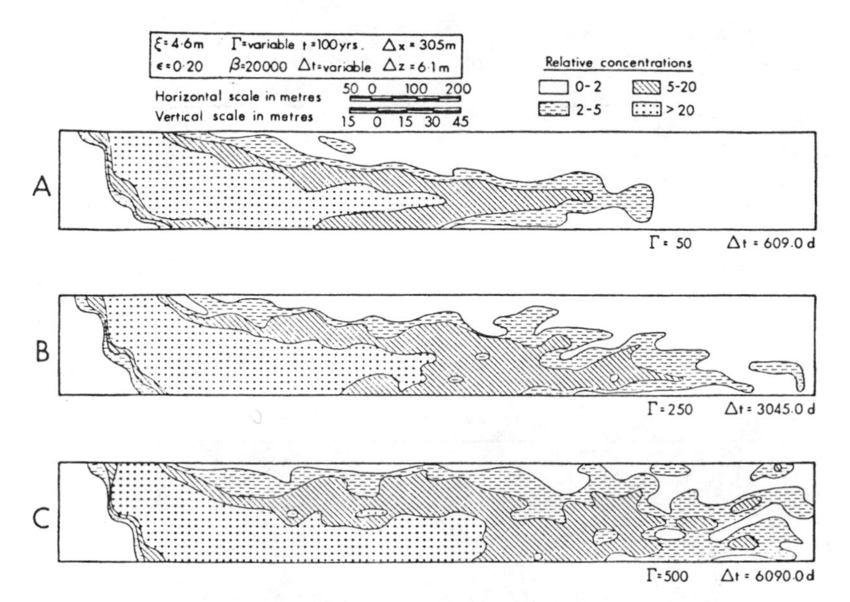

Fig. 6. The discontinuities in the solution become apparent as the time step size Δt is progressively increased to reach a maximum in trial C.

here to demonstrate its versatility is the case of contaminant spread in a shallow groundwater system. A region identical to Fig. 4A forms the basis for the analysis. A water soluble contaminant is assumed to enter the system from the leaching of solid radioactive waste and to move through the system. The three sets of simulation results (Figs. 7, 8 and 9) show the pattern of concentration distribution for three different sets of conditions. In the first case (Fig. 7), mass inflow is maintained during the entire simulation. The second case (Fig. 8) is similar except the mass inflow stops after 38.3 yrs. Comparison of the simulation results in Fig. 7 and Fig. 8 indicates that the effect of eliminating the source of contamination is to cause a pulse of contaminant to move through the system rather than a contaminant plume.

Elimination of the contaminant source during a simulation trial is accomplished in one of two ways. The most preferable method is to assign a zero mass to each moving reference particle added to the system after the source is eliminated. A second method is simply to stop adding the moving reference particles after the required time. The first method is preferable because the larger number of moving reference particles present in the system will increase the accuracy of the solution.

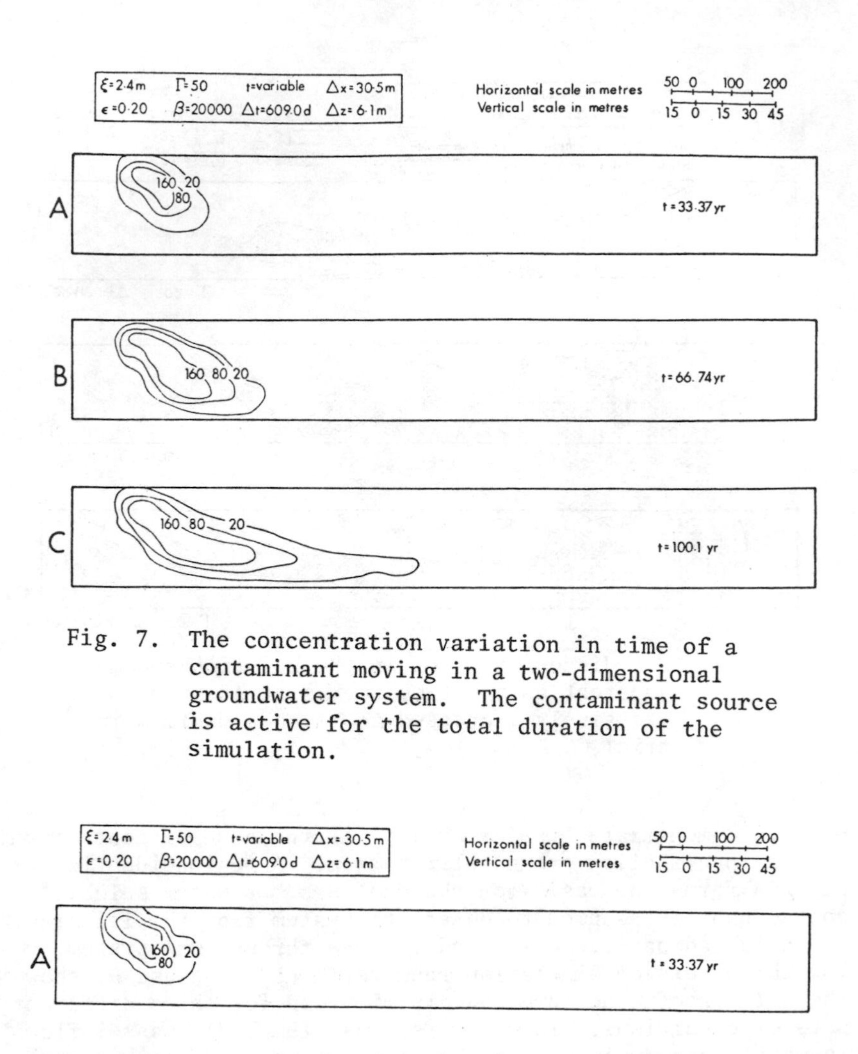

Fig. 7. The concentration variation in time of a contaminant moving in a two-dimensional groundwater system. The contaminant source is active for the total duration of the simulation.

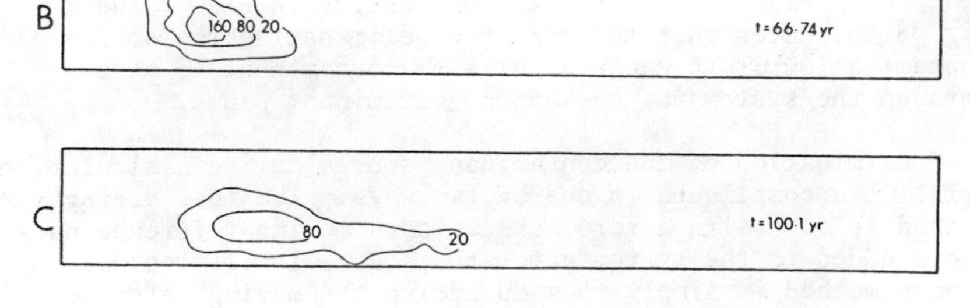

Fig. 8. The concentration variation in time of a contaminant moving in a two-dimensional groundwater system. The contaminant source is only active for 38.3 yrs.

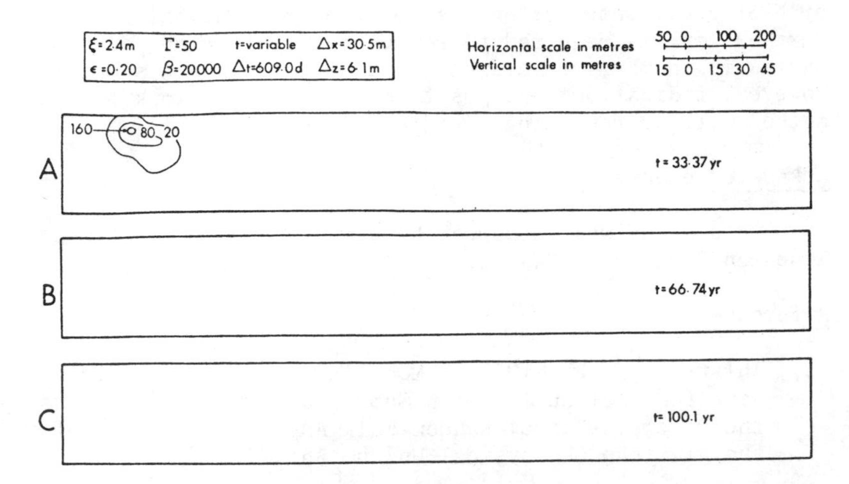

Fig. 9. The concentration variation in time of a contaminant moving in a two-dimensional groundwater system. The contaminant source is active for 38.3 yrs. and the contaminant decays radioactively with a half-life of 10 yrs.

The third case (Fig. 9) is similar to the second (Fig. 8) except that the contaminant is assumed to undergo radioactive decay with a half-life of 10 yrs. This radioactive decay reaction causes the concentration of contaminants to fall below contourable levels in Figs. 9B and 9C. The decay process is accomplished in the model by removing some of the mass associated with each moving reference particle after each time step. The rate of mass removal obeys the common first-order rate expression for radioactive decay.

One of the most important present limitations of the procedure is the increasing computer time required to complete each new time step in the simulation. This situation arises because the total number of moving particles to be moved increases with each time step in the simulation. Although my work has not proceeded beyond this point, indications are that this limitation can be removed. The technique would involve moving a constant number of reference particles and adding successive particle distributions together.

Conclusions

The hybrid probabilistic-deterministic method is a powerful tool for modeling mass transfer in groundwater systems. It combines simplicity and versatility to the extent it can be applied to a variety of practical and theoretical problems. Extension of the technique to three-dimensional problems or multi-component transport problems involving complex chemical or biological interactions is relatively straightforward. Further work is required to remove the limitation posed by the increasing number of reference particles during a simulation. However, indications are positive that further work on the method will be rewarding.

Acknowledgements

This study was supported by a grant from the National Research Council of Canada.

References

[1] Ahlstrom, S. W., 1975. "Modeling the transport of Selected Radionuclides in Subsurface Water Systems Using the Discrete-Parcel-Random-Walk Approach," Trans. of the American Geophysical Union, 56:979.

[2] Bear, J., 1972. "Dynamics of Fluids in Porous Media," Elsevier, New York, N.Y., 764 p.

[3] Freeze, R. A. and Witherspoon, P., 1966. "Theoretical Analysis of Regional Groundwater Flow, 1, Analytical and Numerical Solutions to the Mathematical Model," Water Resources Research, 2: 641-656.

[4] Freeze, R. A. and Witherspoon, P., 1967. "Theoretical Analysis of Regional Groundwater Flow, 2, Effect of Water Table Configuration and Subsurface Permeability Variation," Water Resources Research, 3: 623-634.

[5] Gardner, A. O., Peacemen, A. L., and Pozzi, R., 1964. "Numerical Calculation of Multi-Dimensional Miscible Displacement by the Method of Characteristics," Society of Petroleum Engineers Journal, 4(1): 26-38.

[6] de Jong, J., 1958. "Longitudinal and Transverse Diffusion in Granular Deposits," Trans. of the American Geophysical Union, 39: 67-74.

[7] Ogata, A. and Banks, R. B., 1961. "A Solution of the Differential Equation of Longitudinal Dispersion in Porous Media," United States Geological Survey Professional Paper, 411A.

[8] Reddell, D. L. and Sunada, D. K., 1970. "Numerical Simulation of Dispersion in Groundwater Aquifers," Colorado State University Hydrology Paper, 41: 79.

[9] Schwartz, F. W., 1975. "A Probabilistic Mass Transfer Model," Canadian Hydrology Symp. - 75 Proc., Winnipeg.

ON THE MIGRATION OF CONTAMINANTS IN AN AQUIFER SUBJECT TO GEOTHERMAL ACTIVITY

By

Thomas M. Souers, Former Graduate Student
Hillel Rubin, Past Visiting Professor

and

B. A. Christensen, Professor of Civil Engineering
Hydraulic Lab, Department of Civil Engineering
Weil Hall, Room 241, University of Florida
Gainesville, Florida 32611

Abstract

Florida, like many other sections in the United States and the world, is facing a critical environmental situation. The disposal of waste materials into man's usable environment is rapidly causing severe damage to several basic resources. Some of the damage includes lowering the quality of man's available water sources, polluting the air, and irreversibly spoiling the land. At present, there are several studies underway to study the feasibility of injecting much of this waste into deep, underground aquifers. One such aquifer that is particularly suitable for receiving the injected pollutants is the Boulder Zone in Southern Florida. Due to the thickness of this zone, the common geothermal gradient may lead to groundwater circulation and thus all water that is injected into this aquifer will be affected by the associated convective processes. Due to the large pore size and high transmissivity, there is very intensive solute and heat dispersion and the laminar Darcy law may be invalid even for very small flow velocities. Before the waste substances can be injected, a method must be devised to predict flow patterns and movement of the contaminants as they travel through the aquifer. It is the object of this study to develop such a method.

The solution to this problem is shown to involve two steps. The first step is to solve the equations of motion and heat transport so that values for lateral and transverse velocity components can be calculated. The respective equations are to be solved by expanding finite amplitude perturbations through truncated eigenfunctions. Using the velocities that are calculated in this fashion, the movement of the contaminant is then solved numerically using the characteristic approach and finite difference scheme. These methods have been found to be very stable, and can be applied even at high Rayleigh numbers when more conventional numerical approaches are impractical.

Introduction

The deep zone of the Floridan aquifer, referred to as the Boulder Zone, is an Eocene age formation and is characterized by many large caverns and high values of transmissivity. These two characteristics make this zone particularly unique compared to other aquifers which lack one of these two conditions.

The numerous caverns found in the Boulder Zone range in size from a few inches to as much as 50, 70, or 90 feet in height. The cavities occur not only singly but also in multiples forming a "pancake" stack with only inches of dense dolomite separating them. Even though many of the caverns are rather large in height, vertical permeability is practically nonexistant through the dense dolostone. This is attributable to the lack of vertical fractures or fault zones. The Boulder Zone is, therefore, suitable for deep well injection of pollutant material. Due to the lack of vertical connections, the waste cannot leak upward to contaminate freshwater aquifers. Even if there is some vertical leakage, the thick limestone beds having low permeabilities above this zone will act as filters to remove chemical and physical impurities.

Henry and Kohout (1972) report on observations of a negative thermal gradient in the southern Florida area. From their samples they observed a 1° F decrease each 250 feet of descent as opposed to the normal 1° F increase each 300 feet of descent. Based on this data, Kohout has hypothesized that cold dense seawater from the lower depths of the Florida Bay enters the lower portion of the Floridan aquifer through submarine connections and travels northward at a depth of several thousand feet. Geothermal heat, generated from the earth's core, slowly warms the seawater which begins to rise to the top of the Eocene Zone due to the convection cell motion. The upward component of the convection cell will bring the warmed seawater into contact with the freshwater that enters the aquifer through sinkholes that are present in this karst region. The diluted seawater then flows back to the sea by leakage or submarine springs.

The characteristic transmissivity in the Boulder Zone ranges from 3 to 16 x 10^6 GD/ft with a porosity of between 33-35 percent. The combination of the large pore size and high transmissivity leads to intense solute and heat dispersion, in addition to turbulent porous media flow. This condition is true even if the primary flow velocity is extremely small.

The Boulder Zone model, as shown in Fig. 1, is assumed to be homogeneous, isotropic, horizontal, and infinite in areal extent. The upper and lower boundaries of the aquifer are assumed to be impermeable aquicludes. The porous medium is completely saturated with the primary flow velocity being in the longitudinal x direction. The z and y components represent the vertical and transverse directions, respectively.

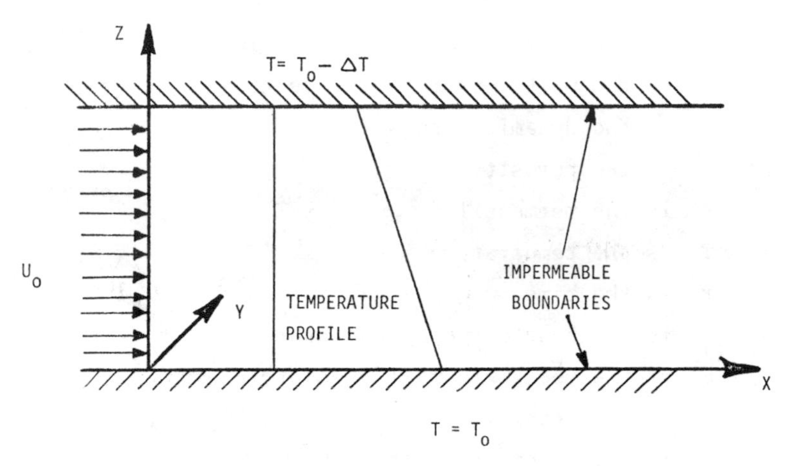

Fig. 1. Flow field model

Basic Equations

In order to simplify the modeling criteria, it is assumed that all density gradients in the Boulder Zone are induced by the temperature component only. Due to the large cavity sizes and high transmissivity, even small velocities in the primary flow direction will cause turbulent motion as well as some degree of mechanical heat dispersion. To calculate this phenomenon, the following necessary equations are approximated using the Boussinesq approach.

$$\frac{\partial u_i}{\partial x_i} = 0 \tag{1}$$

$$\frac{\partial p}{\partial x_i} + \rho g n_i + \frac{\mu \phi}{K} (1 + b)\, u_i = 0 \tag{2}$$

$$\gamma \frac{\partial T}{\partial t} + u_i \frac{\partial T}{\partial x_i} = \frac{\partial}{\partial x_i} \left(E_{ij} \frac{\partial T}{\partial x_j} \right) \tag{3}$$

$$\rho = \rho_o \,[1 - \alpha (T - T_o)]$$

where $\gamma = \dfrac{\rho C_f \phi + \rho_s C_s(1-\phi)}{\rho C_f \phi}$,

u_i is the velocity vector,

x_i is a coordinate,

p is the pressure,

ρ is the fluid density,

g is the gravitational acceleration,

n_i is the unit vector in the vertical direction,

μ is the dynamic viscosity,

ϕ is the porosity,

K is the permeability,

T is the temperature,

t is the time,

E_{ij} is the heat dispersion tensor,

α is the coefficient of volumetric expansion,

ρ_o is the density at $z=0$,

T_o is the temperature at $z=0$,

ρ_s is the solid density,

C_f is the specific heat of fluid, and

C_s is the specific heat of solid.

The coefficient b in Eq. (2) is the turbulent friction function which is directly proportional to the Reynolds number, where the Reynolds number is defined as

$$Re = \frac{\phi U d_p}{\nu}$$

where U is the module of velocity vector, d_p is the characteristic pore size, ν is the kinematic viscosity and the value of b is defined as $b = 0.014$ Re (Rubin, 1976).

If it is assumed that the porous medium is isotropic, as it is in this study, then the dispersion tensor E_{ij} is the numerical sum of the isotropic molecular diffusivity and the second order symmetric mechanical dispersion tensor. The form of the dispersion tensor is

$$E_{ij} = \kappa \delta_{ij} + E^*_{ij}$$

where $E^*_{ij} = E^*_t \delta_{ij} + (E^*_\ell - E^*_t)u_i u_j/U^2 = (E_t - \kappa)\delta_{ij}$

$$+ (E_\ell - E_t)u_i u_i/U^2 \qquad (4)$$

and κ is the thermal diffusivity of a saturated porous medium, δ_{ij} is the Kronecker's delta, E^*_{ij} is the mechanical heat dispersion tensor, E^*_t and E_t are the transverse heat dispersion coefficients, and E^*_ℓ and E_ℓ are the longitudinal heat dispersion coefficients. Since the medium is isotropic, the mechanical dispersion coefficients form a symmetric tensor whose magnitude is dependent on the magnitude of the flow velocity. The dispersion tensor E_{ij} therefore has principal directions parallel and perpendicular to the flow velocity.

Rubin (1976) states that if the Reynolds number is within a moderate range, then the values of E^*_t and E^*_ℓ can be approximated by Saffman's model (1959).

$$E^*_{t/\nu} = \frac{(s+2)Re/\phi}{2(s+1)(s+3)}$$

$$E^*_{\ell/\nu} = \frac{(s+1)^2 Re/\phi}{2(1-s)(s+2)(s+3)}$$

The value of s is a power coefficient which describes the dependence between the flow velocity and the hydraulic pressure drop in the direction of flow. The value of s varies between one for laminar flow and one-half for turbulent flow. The numerical value of s can be approximated by:

$$s = \frac{\ln Re}{\ln(1+b) + \ln Re}$$

as described by Rubin (1976). This value of s can only be used when the Reynolds number is greater than or equal to 2.77, if the value of $b = 0.014\ Re$ has been utilized.

The method employed to solve the equations of motion and heat transport involves a moving coordinate system and uses the following primed values representing the nondimensional forms of their respective quantities.

$$x'_i = (x_i - u_o t \ell_i/\gamma)/d$$

$$u'_i = u_i d/E_t$$

$$t' = tE_t/d^2$$

$$E'_{ij} = E_{ij}/E_t$$

$$p' = \frac{(p-\rho_o gz)K}{\mu\phi E_t(1+b)} + \frac{xu_o}{E_t}$$

$$T' = (T - T_o)/\Delta T$$

$$U' = Ud/E_t$$

where u_o is the unperturbed velocity and d is the porous layer thickness. By substituting these dimensionless variables into Eqs. (1)-(3) and eliminating the primes, the following equations are obtained:

$$\frac{\partial p}{\partial x_i} - RTn_i + (1 + \beta)u_i = 0 \tag{5}$$

$$\gamma \frac{\partial T}{\partial t} + u_i \frac{\partial T}{\partial x_i} = \frac{\partial}{\partial x_i}(E_{ij} \frac{\partial T}{\partial x_j}) \tag{6}$$

where β is the perturbation in the friction function and R is the Rayleigh number, defined as:

$$R = \frac{\alpha g\Delta TKd}{\nu\phi E_t(1+b)}$$

For the situation in which there is no convection taking place within the aquifer, Eqs. (5) and (6) reduce to:

$$u_i = 0$$

$$\beta = 0$$

$$T = -z$$

$$P = P_o - Rz^2/2 \tag{7}$$

$$E_{xx} = E_\ell/E_t$$

$$E_{yy} = E_{zz} = 1$$

$$E_{xz} = E_{xy} = E_{yz} = 0$$

Rubin (1977) utilizes the linear stability analysis to determine the stability criteria for the flow field. This procedure evaluates the reaction of the flow field to small perturbations in all velocity components, temperature, dispersion tensors, friction function, friction power coefficient, and the pressure. The linear stability analysis assumes that second order terms are negligible and the perturbations are very small. By substituting the small disturbances in Eqs. (5) and (6) the point of instability yields

$$\chi = 1 \qquad a_x = 0 \qquad a_y = a$$

$$a_o = \pi \qquad R_o = 4\pi^2 \tag{8}$$

where a_x and a_y are the wave number components. R_o is the minimal value of the Rayleigh number that satisfies the equation at the point of instability, and χ is defined as:

$$\chi = \frac{(E_\ell/E_t)(a_x/a_y)^2 + 1}{(a_x/a_y)^2 + 1}$$

From these results, it can be seen that the convection cells are two-dimensional rolls whose axes are parallel to the unperturbed velocity vector. A helical flow is created when the unperturbed velocity and the convection velocity are superpositioned in the flow field.

The effect of the convection motion on the transport of the contaminant can be predicted by solving the nonlinear equations that describe motion and heat transport. As has been stated, the convection is two-dimensional, so the velocity components can be expressed in terms of the stream function as follows:

$$v_y = \frac{\partial\psi}{\partial z} \qquad v_z = -\frac{\partial\psi}{\partial y}$$

where v_y and v_z are the velocity components in the y and z directions, respectively, and ψ is the stream function.

The form of the equations for the stream function are developed by first substituting the finite amplitude disturbances into Eqs. (5) and (6). By eliminating the pressure perturbation, the equations resolve to

$$R\frac{\partial\theta}{\partial y} + \nabla^2\psi = B(\psi,\beta) \tag{9}$$

$$-\gamma\frac{\partial\theta}{\partial t} + \nabla^2\theta - \frac{\partial\psi}{\partial y} = H(\psi,\theta) + D(\varepsilon_{ij}) + F(\varepsilon_{ij},\theta) \tag{10}$$

The variables B and H are the friction and heat advection spectra, and D and F are two parts of the heat dispersion spectrum. Here ε_{ij} is the dispersion tensor perturbation, and θ is the temperature perturbation.

The general form for all four variables is

$$B(\psi,\beta) = -\frac{\partial}{\partial x_i}\left(\beta\frac{\partial\psi}{\partial x_i}\right) \qquad D(\varepsilon_{ij}) = \frac{\partial}{\partial x_i}(\varepsilon_{iz})$$

$$H(\psi,\theta) = -\frac{\partial(\psi,\phi)}{\partial(y,z)} \qquad F(\varepsilon_{ij},\theta) = -\frac{\partial}{\partial x_i}\left(\varepsilon_{ij}\frac{\partial}{\partial x_j}\right)$$

If the following homogeneous boundary conditions are introduced in Eqs. (9) and (10), then these equations can be solved by a set of truncated eigenfunctions expressing the finite amplitude disturbance. The boundary conditions are

$$\psi,\theta = 0 \quad \text{at} \quad z = 0,1$$

A double Fourier series expansion can be applied for such purposes as follows:

$$\psi = \sum_{p,q=1}^{\infty} \Psi_{p,q} \sin(pay)\sin(q\pi z)$$

$$\theta = \sum_{\substack{p=0\\q=1}}^{\infty} \Theta_{p,q} \cos(pay)\sin(q\pi z)$$

where Ψ is a coefficient in series expansion for ψ and Θ is a coefficient in series expansion for θ.

According to Eq. (8), steady convection follows the point of instability. The first term in Eq. (10) vanishes, therefore, when steady convection is attained. With this condition, the coefficients $\Psi_{p,q}$ and $\Theta_{p,q}$ can be expressed through a power series expansion.

$$(\Psi_{p,q}\Theta_{p,q}) = \sum_{n=1}^{N} (\Psi^n_{p,q},\Theta^n_{p,q})\eta^n$$

where η is defined as $\eta = [(R-R_o)/R]^{0.5}$, and

N is the total number of terms in the series expansion.

By substituting this power series into Eqs. (9) and (10), expressions for solving ψ and θ are obtained. Only coefficients $\Psi^n_{p,q}$ and $\Theta^n_{p,q}$ with even values of $|p| + q$ have nonvanishing values. In addition, the value of $|p|$ and q must be less than or equal to N. The form of the equations for solving ψ and θ are

$$R_o p\pi\Theta^n_{p,q} + R_{os}\, p\pi \sum_{i=1}^{\infty} \Theta^{(n-2i)}_{p,q} + \pi^2(p^2+q^2)\,\Psi^n_{p,q} + B^n_{p,q} = 0$$

$$\pi^2(p^2+q^2)\,\Theta^n_{p,q} + p\pi\,\Psi^n_{p,q} + H^n_{p,q} + D^n_{p,q} + F^n_{p,q} = 0$$

where $R_{os} = R_o/(1 - n^{2S})$ and $X = N/2$.

The application of the power series can also be applied for other field perturbations as well. This includes H, B, D, F, velocity, and dispersion coefficients. If Eq. (4) is subjected to a Taylor's and power series expansion, then the dispersion tensor can be expressed as

$$E_{ij} = (1+\alpha_1 V^2)\delta_{ij} + \alpha_2 u_i u_j + \alpha_3(u_i \ell_j + \ell_i u_j) + (E_\ell/E_t - 1$$

$$+ \alpha_4 V^2)\ell_i \ell_j + \ldots$$

as presented by Rubin et al. (1976). The coefficients α_1, α_2, α_3, and α_4 are defined as

$$\alpha_1 = 1/2u_o\ (\partial E_t/\partial u_o)$$

$$\alpha_2 = (E_\ell - 1)/u_o^2$$

$$\alpha_3 = (E_\ell - 1)/u_o$$

$$\alpha_4 = (1-E_\ell)/u_o^2 + 1/2u_o[\partial(E_\ell - E_t)/\partial u_o]$$

and the value of V^2 is the sum of $v_z^2 + v_y^2$. These approximations are valid for convection velocities slower than the velocity that is induced by the hydraulic gradient.

Discussion of Basic Equations

Convection effects are determined not only by the Rayleigh (R) number but also by Reynolds (Re) and Prandtl (Pr) numbers, too. The ratio between the porous layer thickness and the characteristic pore size has an influence on convection as well. The interdependence of all these variables is applicable only when mechanical dispersion is comparable with molecular diffusion.

If the Prandtl number is greater than one, the dispersion effects are significant even for a Reynolds value of approximately three. If Pr is less than one, then the Re number will determine the dispersion effects. At low Pr and high Re, dispersion effects are significant. As the Re number increases, the effect of the Pr number on dispersion is negligible because mechanical dispersion affects the transport processes more than the molecular diffusion.

Rubin (1977) explains that the coefficients, α_1, α_2, and α_3 are almost inversely proportional to the square of the ratio of the layer thickness to the pore size d/d_p. The variation of the power coefficient s, due to the convection motion, is very small and almost negligible.

If the flow field is subject to a high R number, then a thick zone in the center of the porous layer should achieve an isothermal state. Such conditions will, however, lead to a positive temperature gradient which is counter to the negative temperature gradient observed in the deep zones of the Floridan aquifer. This phenomenon can be eliminated if a better choice of wave numbers is chosen.

The net effect of the mechanical dispersion induced by the convection process reduces the heat transport through the aquifer. The dispersion and turbulence of the flow field that is created by the convection motion acts like a stabilizing mechanism. This apparent phenomenon is associated with increased values of the higher modes of the series expansions. This leads to a reduction in the convergence of the series expansions.

The singly diffusive analysis, as it is presented in this section, is applicable when the density gradients are created by a single component, being temperature in this case. It is also justifiable when boundary conditions and effective dispersion coefficients are identical for all components in a multicomponent system. This is true when mechanical dispersion is more profound than the molecular diffusion in the aquifer.

Transport and Dispersion Contaminants

The formulation of the equations used in the model to calculate the position and dispersion of the contaminant front is a combination of the method of characteristics (MOC) and the explicit central finite difference (FD) schemes in conjection with the equations previously developed. Together, these equations predict the motion of the front as it is affected by geothermal heating. These two methods have been chosen because of their relative simplicity and in order to avoid numerical dispersion of the calculation. The two schemes are also quite stable which is a positive advantage over several other methods.

An area of influence, or a cube, is constructed around a central, stationary node point at an arbitrary time, t. At this same instant, mobile particles enter the aquifer, and the incremental distance that each particle travels during one time step, Δt, is defined as

$$x_{i_{(t+1)}} = x_{i_{(t)}} + [\Delta t_{(t+1)}][u_i(x,z,y)]_{(t)} \tag{11}$$

The value of the time step $\Delta t_{(t+1)}$ is dependent on the maximum velocity U_{max}, in the x, z, or y direction and on an arbitrary spatial distance, delta. The time step is defined as

$$\Delta t_{(t+1)} = \text{delta}/U_{max} \tag{12}$$

The velocity component u_i in Eq. (11) is the velocity corresponding to the velocity in the x_i direction. If $x_i = z = y$, then the velocity components are calculated according to the following:

$$v_z = -\frac{\partial\psi}{\partial y} = -\sum_{\substack{p=1\\q=1}}^{N} \Psi_{p,q}\ p\pi\cos(p\pi y)\ \sin(q\pi z)$$

and

$$v_y = \frac{\partial\psi}{\partial z} = \sum_{\substack{p=1\\q=1}}^{N} \Psi_{p,q}\ q\pi\sin(p\pi y)\ \cos(q\pi z)$$

The summation is carried out with N = up to 16. The value of N is determined by the accuracy that is desired. If the value of the Nusselt number (Nu), (Nusselt, 1915), varied less than 2 percent as N is increased from N to N + 2, then the expansion is terminated. Otherwise, higher values of N are required.

Once the time increment has been delineated, the change in concentration can be calculated using the finite difference approach. The form of the change in concentration is

$$\frac{\partial C}{\partial t} + u_i\frac{\partial C}{\partial x_i} = \frac{\partial}{\partial x_i}\left(D_{ij}\frac{\partial C}{\partial x_j}\right) \tag{13}$$

where $D_{ij} = E^*_{ij}$ which is the solute dispersion tensor and C is the concentration of the contaminant.

In solving the equations of dispersion and transport, the coordinate system is stationary so that the form of the nondimensional numbers is

$$x_i' = x_i/d$$
$$u_i' = u_i d/E_t$$
$$D_{ij}' = D_{ij}/E_t$$
$$C' = (C - C_o)/\Delta C$$
$$t' = tE_t/d^2$$
$$\text{delta}'' = \text{delta}/d$$

where C_o is the concentration of the contaminant at time = 0 and ΔC is the difference between the contaminant and ambient water concentrations.

By substituting these expressions into Eqs. (11), (12), and (13) and eliminating the primes, the following nondimensionless equations are formed:

$$x_{i_{(t+1)}} = x_{i_{(t)}} + [\Delta t_{(t+1)}][u_i(x,z,y)]_{(t)} \tag{14}$$

$$\Delta t_{(t+1)} = \text{delta}/U_{max} \tag{15}$$

$$\frac{\partial C}{\partial t} + u_i \frac{\partial C}{\partial x_i} = \frac{\partial}{\partial x_i} (D_{ij} \frac{\partial C}{\partial x_j}) \tag{16}$$

The method of characteristics is often used for the solution of hyperbolic differential equations. According to this method, the variation of a disturbance of an undulating medium is followed by an observer. The observer's location coincides at all times with the location of the undulation.

The equation of contaminant dispersion due to the convection process cannot be solved adequately according to procedures typical to parabolic equations. This is due to the numerical dispersion that is induced by the convection terms. The conditions for the convergence of a numerical solution to parabolic and hyperbolic equations are entirely different and contradictory. The adaptation of the method of characteristics for the solution of the contaminant dispersion yields the concentration of the injected mobile particle while following its movement through the aquifer. Through this process, the convection terms in Eq. (16) disappear. This leads to

$$\frac{\partial C}{\partial t} = \frac{\partial}{\partial x_i} (D_{ij} \frac{\partial C}{\partial x_j}) \tag{17}$$

In this equation x_i represents the coordinate that is attached to the moving particle. Equation (17) is fully parabolic and is not affected by the numerical dispersion that is usually induced by the convection terms. Since a fixed grid is used in solving for the dispersion, the location of the fluid particles should be traced through the calculations.

To begin the modeling process, a cell with width, length, and height equal to unity is constructed and represents one time phase. This cell is then divided into three subsections, or subcells, in each direction. Within each subcell is a centrally located stationary node representing the concentration of the ambient waters. Each subcell can be divided into even smaller sections to facilitate a more accurate tracing of the pollutant cloud. In this study, each subcell was divided into four separate sections, as shown in Fig. 2. The point for injection is completely arbitrary, depending only upon how far the disposal well penetrates the aquifer.

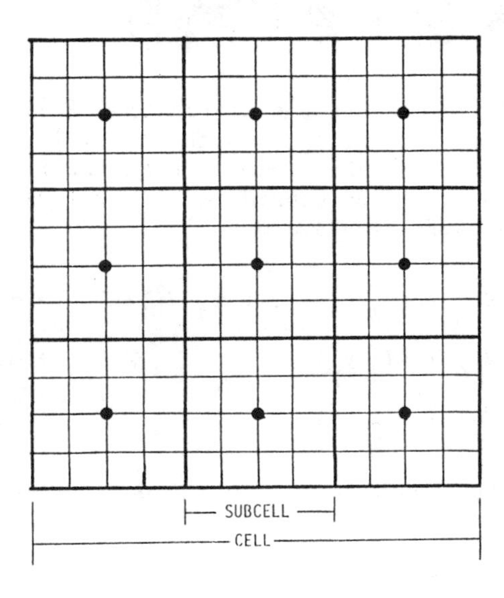

Fig. 2. Aquifer cell grid system

Once the particles have been moved, according to Eqs. (14) and (15), the position of the particle in relation to one of the stationary nodes within a subcell must be defined. A particle with coordinates x, z, y falls within a particular subcell whose node is at x', z', y', if

$$x' - \frac{\Delta x}{2} \leq x < x' + \Delta x/2$$

$$z' - \frac{\Delta z}{2} \leq z < z' + \Delta z/2$$

$$y' - \frac{\Delta y}{2} \le y < y' + \Delta y/2$$

where Δx, Δz, and Δy are the characteristic length, height, and width of each subcell, respectively. Each stationary node is then assigned a new concentration which is the average of the stationary node plus all mobile particles falling within that subcell. The change in concentration of the particles is then calculated according to Eq. (17). In its expanded form it appears as

$$\frac{\partial C}{\partial t} = \{D_{xx} \frac{\partial^2 C}{\partial x^2} + D_{yy} \frac{\partial^2 C}{\partial y^2} + D_{zz} \frac{\partial^2 C}{\partial z^2}\} + [2D_{xy} \frac{\partial^2 C}{\partial x \partial z} + 2D_{xz} \frac{\partial^2 C}{\partial y \partial z}$$

$$+ 2D_{yz} \frac{\partial^2 C}{\partial y \partial z} + (\frac{\partial D_{xx}}{\partial x} + \frac{\partial D_{xy}}{\partial y} + \frac{\partial D_{xz}}{\partial z}) \frac{\partial C}{\partial x} + (\frac{\partial D_{zz}}{\partial z} + \frac{\partial D_{xz}}{\partial x}$$

$$+ \frac{\partial D_{yz}}{\partial y}) \frac{\partial C}{\partial z} + (\frac{\partial D_{yy}}{\partial y} + \frac{\partial D_{xy}}{\partial x} + \frac{\partial D_{zy}}{\partial z}) \frac{\partial C}{\partial y}]$$

If $d/d_p > 10^2$ then the terms within the { } are the dominant expressions, and the terms with the [ ] are neglibible. If $d/d_p < 10^2$, then each expression on the right hand side of the equation should be considered. The form for the various dispersion coefficients are

$$D_{xx} = (1 + \alpha_1 V^2) + (E_\ell/E_t - 1 + \alpha_4 V^2) \qquad D_{yz} = \alpha_2 v_z v_y$$

$$D_{yy} = (1 + \alpha_1 V^2) + (\alpha_2 v_y^2) \qquad D_{xz} = \alpha_3 v_z \tag{18}$$

$$D_{zz} = (1 + \alpha_1 V^2) + (\alpha_2 v_z^2) \qquad D_{xy} = \alpha_3 v_y$$

The form for the change in concentration at some point is

$$\Delta C_{(t+1)} = \Delta t_{(t+1)} (\frac{\partial C}{\partial t})_{(t)}$$

Each moving particle with concentration C'' is then assigned a new concentration according to

$$C''_{(t+1)} = C''_{(t)} + \Delta C_{(t+1)}$$

The stationary node with concentration C' is also assigned a new concentration according to

$$C'_{(t+1)} = C'_{(t)} + \Delta C_{(t+1)}$$

The entire process, which includes the MOC and FD schemes, is repeated for as many phases as is required.

Results and Discussion

The various situations in which the model was tested under are shown in Table 1. The trace of the helical flow that is produced due to the convection motion is shown in Fig. 3. Both traces represent the same time period, but the smaller trace is for injection in the center of the aquifer and the larger trace denotes injection closer to a boundary. It is clearly evident that the larger velocities occur near the boundary conditions. Practically identical traces were obtained in all situations in which the convection process was evident, whether it was negligible or significant. Position, therefore, is the overriding factor in determining the velocity of the contaminant material.

The convection motion created by the geothermal activity had its greatest effect on the dispersion of the pollutant. For the first four tests, in which the convection motion was either negligible or nonexistent, $d/d_p > 10^2$ and the rate of dispersion of the contaminant front is identical in all cases. This is because the value of E_ℓ/E_t was assumed to be equal in all four instances and, according to Eq. (7), all other dispersion parameters are also identical. A summary of the dispersion coefficients are shown in Table 2 and the rate of dispersion is shown in Fig. 4 for the first four cases.

For the last two tests, cases 5 and 6, the convection process is much more significant as it was assumed for these two situations that $d/d_p < 10^2$. The rate of dispersion for

Table 1. Case Study Parameters

Case	d/d_p	Re	Injection Location	Dispersion due to Convection
1	$>10^2$	<1.0	Center	None
2	$>10^2$	<1.0	Boundary	None
3	$>10^2$	<1.0	Center	Negligible
4	$>10^2$	<1.0	Boundary	Negligible
5	$<10^2$	>3.0	Boundary	Significant
6	$<10^2$	>3.0	Center	Significant

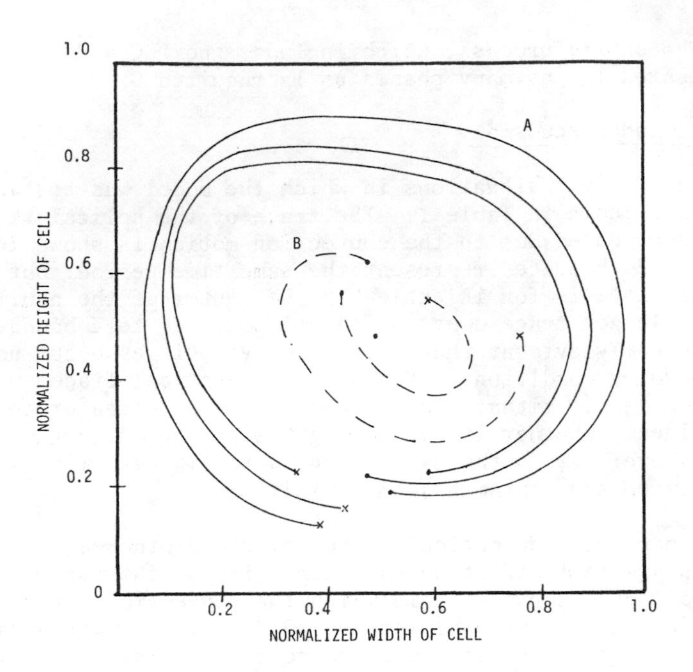

Fig. 3. Helical flow pattern in center and lower portion of aquifer
B - - - - - Center Injection Site
A ________ Lower Injection Site

Table 2. Dispersion Coefficients for Various Test Cases

CASE	DISPERSION DUE TO CONVECTION	Average D_{xx}	Average D_{zz}	Average D_{yy}	Average D_{xy}	Average D_{xz}	Average D_{yz}
1	None	24.0	1.0	1.0	0	0	0
2	None	24.0	1.0	1.0	0	0	0
3	Negligible	24.0	1.0	1.0	0	0	0
4	Negligible	24.0	1.0	1.0	0	0	0
5	Significant	30.87	5.52	5.80	.0088	.0012	.032
6	Significant	23.89	1.01	1.01	.0025	.0021	.0023

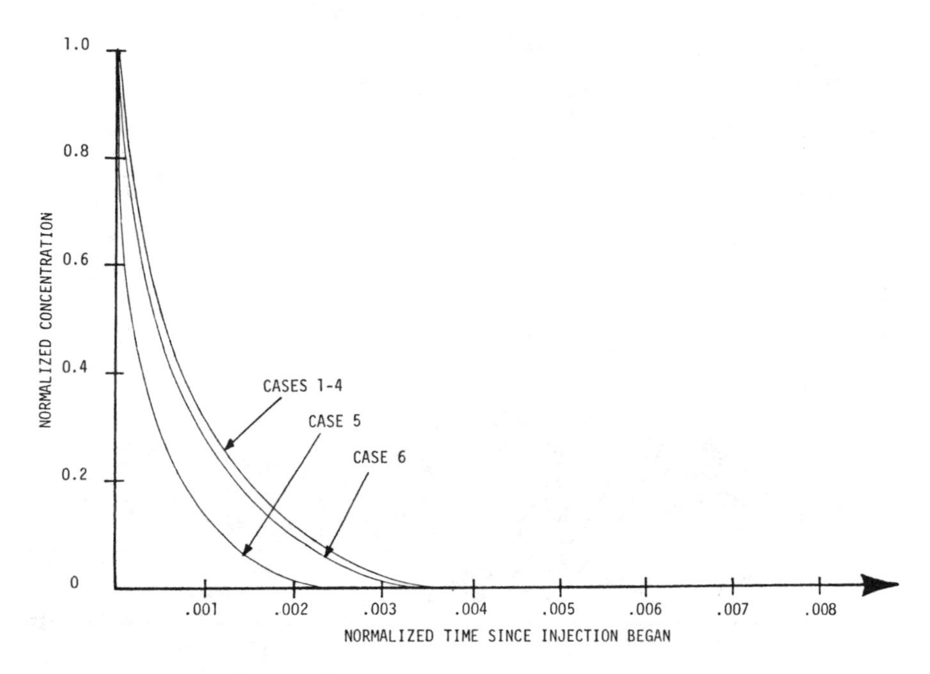

Fig. 4. Concentration curves for various testing situations

case 5 increased over that of the first four tests as shown in Fig. 4, and it is due to the increase in the value of the dispersion coefficients. This increase is caused through the higher degree of dispersion and turbulence in the porous medium caused by the increased pore size. Combined with the high velocities near the boundary walls a higher degree of dispersion is attained. Even though convection motion is significant in case 6, the overall rate of dispersion is only slightly greater than for the first four cases, as shown in Fig. 4. Even though the minor dispersion terms were increased somewhat, the lower velocities occurring near the center of the aquifer are not sufficient to increase the longitudinal dispersion term over that assumed for cases 1-4. The relation between the flow velocities and the dispersion coefficients is clearly evident in Eq. (18).

The values for the dispersion coefficient for cases 5 and 6 are also shown in Table 2. If d/d_p is decreased the values for all the dispersion coefficients increase.

A test was conducted concerning the continuous injection of a pollutant material over a long period of time, which is shown in Fig. 5. The curves show that while the front of the cloud disperses as predicted in the previous instantaneous injection, the particles following the front disperse at a continually slower rate. As the groundwater concentration

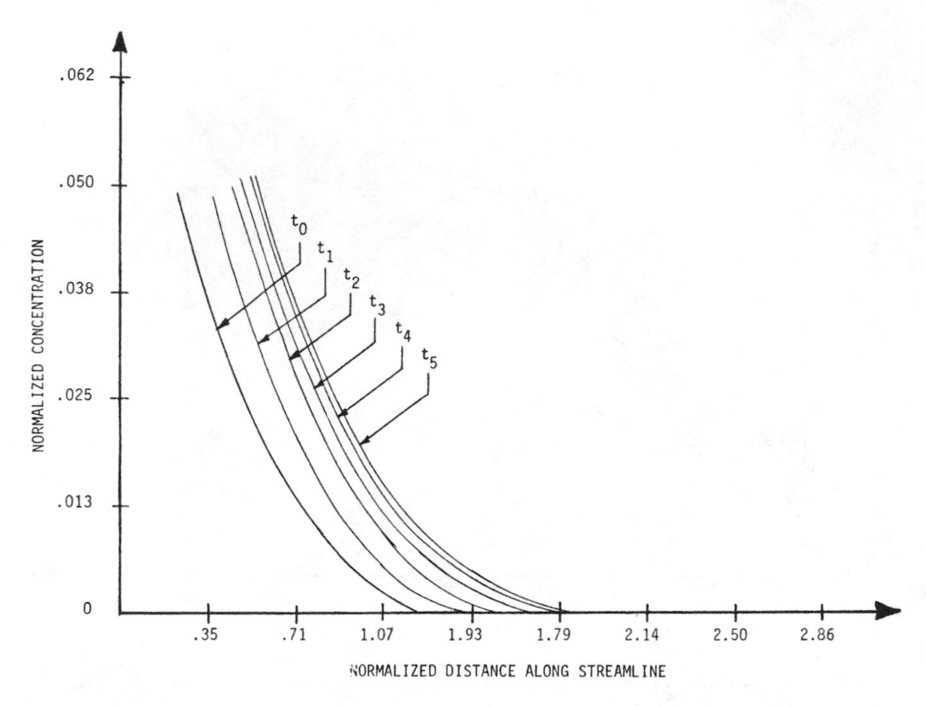

Fig. 5. Concentration curves for continuous injection

increases within the vicinity of the injection well, there is a decrease in the concentration gradient between two points at some time t and the same two points at t+1. This decrease in the gradient causes a continual decrease in the rate of dispersion of the pollutant. Eventually, over a very long time period, steady state conditions will prevail in the well vicinity.

Summary

The spatial coordinates of the contaminant are directly related to its dispersion. In all cases, velocity and spatial orientation remain fairly constant except when the location of the injection is changed. The transverse and lateral velocities, and their respective locations with the aquifer, are primarily a function of the injection well location. The dispersion of the contaminant varies whenever the Rayleigh number and/or d/d_p is varied. The parameter d/d_p has significant importance if its value is less than 10^2. The smaller that the ratio d/d_p is, the larger the values of the dispersion coefficients become, and this will greatly increase the rate of dispersion. For larger values of d/d_p, i.e., $> 10^2$, only the value of the Rayleigh number affects the dispersion phenomenon.

Conclusions

1) In a thick aquifer, such as the Boulder Zone, the effect of the geothermal gradient on flow conditions should be considered.

2) The geothermal gradient may induce convection motions that lead to a helical flow pattern in the aquifer.

3) Spectral expansion of flow field perturbations may supply an efficient method for the simulation of flow conditions in the aquifer subject to density gradients.

4) Spectral methods are applicable even when conventional numerical methods suffer from numerical dispersion or solution instability.

5) The migration of contaminants in an aquifer subject to geothermal activity can be calculated by applying the characteristics approach for the solution of the contaminant dispersion equation.

6) Results which are good in accuracy, convergence, and stability can be obtained by applying a combination of the spectral method for the solution of the dynamic problem and the characteristic approach for the solution of the contaminant dispersion.

7) Due to the geothermal activity, contaminants may migrate upward, downward, or in the lateral direction instead of traveling only in the longitudinal direction.

8) The effect of the convection motion on the contaminant migration depends on the injection location and the flow field parameters.

9) The intensity of the contaminant dispersion depends primarily on the value of the Rayleigh number and the ratio d/d_p. Reynolds and Prandtl numbers have a less significant effect on the dispersion phenomenon.

References

[1] Anonymous, 1971. "Bulletin of Southwest Florida Water Management District," Vol. 2, No. 11, November.

[2] Faulkner, G. L. and Pascale, C., 1975. "Monitoring Regional Effects of High Pressure Injection of Industrial Wastewater in a Limestone Aquifer," Groundwater Journal of the Technical Division, 13:2.

[3] Henry, H. R. and Kohout, 1972. "Circulation Patterns of Saline Groundwater Affected by Geothermal Heating--as Related to Waste Disposal," In: T. D. Cook (ed.), Proceedings of the First Symposium on Underground Waste Management and Environmental Implications, pp. 202-221.

[4] Holley, E. R. and Harleman, D.R.F., 1965. "Dispersion of Pollutants in Estuary Type Flows," Hydraulics Laboratory Report, 74.

[5] Kaufman, M. I., 1973. "Subsurface Wastewater Injection, Florida," Journal of Irrigation and Drainage Division, Proc. ASCE, 99: 53-70.

[6] Kaufman, M. I. and McKenzie, D. J., 1975. "Upward Migration of Deep Well Waste Injection Fluids in Floridan Aquifer, South Florida," Journal Research U.S. Geological Survey, 3:261-271.

[7] Nusselt, W., 1915. "Das Grundsetz des Warmeuberganges," Ges. Ing. 38, 477.

[8] Pinder, G. F. and Cooper, H. H. Jr., 1970. "A Numerical Technique for Calculating the Transient Position of the Saltwater Front," Water Resources Research, 6:875-882.

[9] Puri, H. S. and Winston, G. O., 1974. "Geological Framework of the High Transmissivity Zones in South Florida," Bureau of Geology, Division of Interior Resources in cooperation with the U.S. Geological Survey.

[10] Reddell, D. L. and Sunada, D. K., 1970. "Numerical Simulation of Dispersion in Groundwater Aquifers," Colorado State University Hydrology Paper No. 41.

[11] Rubin, H., 1976. "Onset of Thermohaline Convection in a Cavernous Aquifer," Water Resources Research, 12: 141-147.

[12] Rubin, H., 1977. "Thermal Convection in a Cavernous Aquifer," Water Resources Research, (accepted for publication).

[13] Rubin, H. and Grief, S., 1976. "Numerical Simulation of Singly Dispersive Convection in Groundwaters," Proceedings of the Summer Simulation Conference.

[14] Saffman, P. G., 1959. "A Theory of Dispersion in Porous Medium," Journal of Fluid Mechanics, 10: 321-349.

[15] Schroder, M. C., Klein, H., and Hay, N. D., 1958. "Biscayne Aquifer of Dade-Broward Counties, Florida," Report of Investigations, 17.

[16] Smith, G. D., 1965. "Numerical Solution of Partial Differential Equation," London, England.

DIFFUSIVITY OF SOLUTES IN ADSORBING POROUS MEDIA

By

J. L. Starr, J.-Y. Parlange,
B. L. Sawhney and C. R. Frink
The Connecticut Agricultural Experiment Station
Department of Soil and Water, P.O. Box 1106
New Haven, Connecticut 06504

Abstract

Theoretical and experimental investigations of dispersion in a capillary tube, i.e., a porous medium with only one pore size, show that adsorption has a strong influence on the value of the diffusion coefficient. Similar experiments using a column of non-aggregated soil show the dispersion coefficient to be less dependent on adsorption. The observations in this last case are consistent with the existence of irregularities in the flow which are much wider than the grain size.

Theory

Dispersion of a solute in a porous medium with adsorption is usually described by the diffusion equation

$$\partial c/\partial t + v\partial c/\partial x = D\partial^2 c/\partial x - \partial c_a/\partial t \quad , \qquad (1)$$

where c is the solute concentration in the liquid, c_a is the adsorbed concentration, v the average pore velocity, and D the dispersion coefficient. While D may depend on the soil structure (grain size), on v, and on the direction of flow if instability can develop, it is normally assumed that D does not depend on the exchange capacity of the soil.

This last result cannot be true in general. For instance, if the porous medium is composed of a bundle of capillary tubes, Golay has shown that

$$D = D_M + (1 + 6k + 11k^2)v^2r^2/D_M\ 48(1 + k)^2 \qquad (2)$$

where D_M is the molecular diffusivity, r the radius of the tubes and k is the slope of the adsorption isotherm,

$$k = dc_a/dc \quad , \qquad (3)$$

which is taken as a constant in Eq. (2). Hence, Golay was able to show from first principles that the movement of the solute obeyed Eq. (1), as postulated by soil physicists, but with a D which depends on the exchange capacity.

In the case of a nonlinear adsorption isotherm, i.e., when k is a function of c, D as given in Eq. (2) becomes a function of c. We can ask in that case whether the solute movement will still obey a diffusion equation, now of the form,

$$\partial c/\partial t + v\partial c/\partial x = \partial[D\partial c/\partial x]/\partial x - \partial c_a/\partial t \qquad (4)$$

where D may or may not be given by the expression in Eq. (2). The following solution with k a function of c is a simple generalization of Golay's solution following Taylor's method. We first write the diffusion equation for the concentration $C(x,z,t)$ in a tube, with x along the axis and z normal to it, or,

$$\partial^2C/\partial x^2 + \partial^2C/\partial z^2 + z^{-1}\partial C/\partial z = D_M^{-1}\partial C/\partial t + 2D_M^{-1}v(1-z^2)\partial C/\partial x \;, \qquad (5)$$

where the radius, r, of the tube has been taken equal to one for simplicity. We then express C in terms of its average value c across the tube, or,

$$C = c + f\;\partial c/\partial x_1 + \ldots \qquad (6)$$

where x_1 is a distance measured in a moving framework, such that the term in $\partial c/\partial x$ be removed from Eq. (4) when C in Eq. (6) is used, with

$$(\partial/\partial t)_x = (\partial/\partial t)_{x_1} - 2\alpha v(\partial/\partial x_1) \;, \qquad (7)$$

where α is as yet unknown. Clearly we must have

$$\partial^2f/\partial z^2 + z^{-1}\partial f/\partial z = 2D_M^{-1}v(1 - \alpha - z^2) \qquad (8)$$

with $f(z=0)$ finite and $\int_0^1 fzdz = 0$. Hence,

$$f = (v/2D_M)\;[(1-\alpha)z^2 - z^4/4 - (1-\alpha)/2 + 1/12] \;. \qquad (9)$$

At the wall, conservation of mass requires that

$$-2D_M\partial C/\partial z = k\partial C/\partial t \;. \qquad (10)$$

Using C from Eq. (6) with f from Eq. (9) in Eq. (10), the term in $\partial c/\partial x_1$ gives

$$2\alpha = 1/(k + 1) \quad . \tag{11}$$

We may note that this result is compatible with the convective term of Eq. (4).

We are now looking for the next order term in the expansion of Eq. (6). When $\partial C/\partial x$ in Eq. (5) is replaced using the expression of Eq. (6), terms $\partial^2 c/\partial {x_1}^2$ and $\partial[(1-\alpha)\partial c/\partial x_1]/\partial x_1$ appear in the resulting equation. These terms must be cancelled by the next order terms in the expansion of C. Hence we write

$$C = c + f\partial c/\partial x_1 + g_1\partial^2 c/\partial x_1^2 + g_2\partial[(1-\alpha)\partial c/\partial x_1]/\partial x_1 \ , \tag{12}$$

with

$$D_M^{-2}v^2(1-\alpha-z^2)(z^4/4 - 1/12) + \partial^2 g_1/\partial z^2 + z^{-1}\partial g_1/\partial z = D_M^{-1}K_1 \tag{13}$$

and

$$D_M^{-2}v^2(1-\alpha-z^2)(\tfrac{1}{2}-z^2) + \partial^2 g_2/\partial z^2 + z^{-1}\partial g_2/\partial z = D_M^{-1}K_2 \ , \tag{14}$$

where K_1 and K_2 are independent of z. The average solute concentration is then governed by the equation,

$$\partial c/\partial t + v(1+k)^{-1}\partial c/\partial x = D_M\partial^2 c/\partial x^2 + K_1\partial^2 c/\partial x^2 + K_2\partial[(1-\alpha)\partial c/\partial x]/\partial x \ . \tag{15}$$

To find K_1 and K_2 we first solve Eqs. (13) and (14) and obtain,

$$g_1 = K_1(z^2-\tfrac{1}{2})/4D_M - (v^2/D_M^2)(1-\alpha)\ [(z^6 - \tfrac{1}{4})/4 \times 36 - (z^2- \tfrac{1}{2})/4 \times 12] + (v^2/D_M^2)[(z^8 - 1/5)/4 \times 64 - (z^4 - 1/3)/12 \times 16] \tag{16}$$

and

$$g_2 = K_2(z^2 - \tfrac{1}{2})/4D_M + (v^2/D_M^2)(3/2 - \alpha)(z^4 - 1/3)/16 - (v^2/D_M^2)(1-\alpha)(z^2 - \tfrac{1}{2})/8 - (v^2/D_M^2)(z^6 - \tfrac{1}{4})/36 \ . \tag{17}$$

Using C from Eq. (12), with f, g_1, g_2 from Eqs. (9), (16), (17) in Eq. (10), the terms in $\partial^2 c/\partial x_1^2$ and $\partial[(1-\alpha)\partial c/\partial x_1]/\partial x_1$ give

$$K_1 = -D_M k/(1+k) - (v^2/48D_M)[4k/(1+k) + 1]/(1+k) \tag{18}$$

and

$$K_2 = (v^2/48D_M)[4 + 12k/(1 + k)]/(1 + k) \quad . \tag{19}$$

Using these values of K_1 and K_2 in Eq. (15) we obtain the equation governing the movement of c as,

$$\partial c/\partial t + v\partial c/\partial x = \partial[D\partial c/\partial x]/\partial x - \partial c_a/\partial t - E(\partial c/\partial x)^2 \tag{20}$$

where D is still given by Eq. (2) and

$$E = (r^2 v^2/24D_M)(1 + 4k)(1 + k)^{-1} \, d(1 + k)^{-1}/dc \tag{21}$$

We have shown here that the simple diffusion given in Eq. (4) does not apply for a capillary tube when the adsorption isotherm is not linear. In that case Eq. (20) holds. Whether an equation of the same form holds for a more general porous medium requires further study.

In the experimental investigation which follows we try to measure the dependence of D on k directly. We do so in two cases: the first is a tube so that the dependence should follow Eq. (2). The second is a soil column where the dependence, if it exists, is unknown *a priori*.

Case of a Capillary Tube

We used a teflon tube of length $L = 124$ cm and ID = 0.127 cm to study the adsorption of acid fuchsin. The breakthrough curve was determined by measuring the optical density of the liquid at a wave length of 540 mμ in a Gilford Model 203 flow cell. Care was taken to ensure thorough mixing in the cell. The liquid was buffered at pH 7 since the color is slightly pH dependent. Experiments with two fluxes are reported, a low one of 0.09 ml/min and a high one of 0.19 ml/min. The time necessary to fill the tube was 17.4 ± .1 and 8.2 ± .1 min at the two fluxes respectively. A variety of dye concentrations, c_o, were used starting from a low of 5.7×10^{-6} g/ml to a high of 52×10^{-6} g/ml.

Fig. 1 shows 6 distinct breakthrough curves differing by the flow rate, the concentration c_o, and whether adsorption or desorption occurs. In the latter case $1 - c/c_o$ is plotted rather than c/c_o.

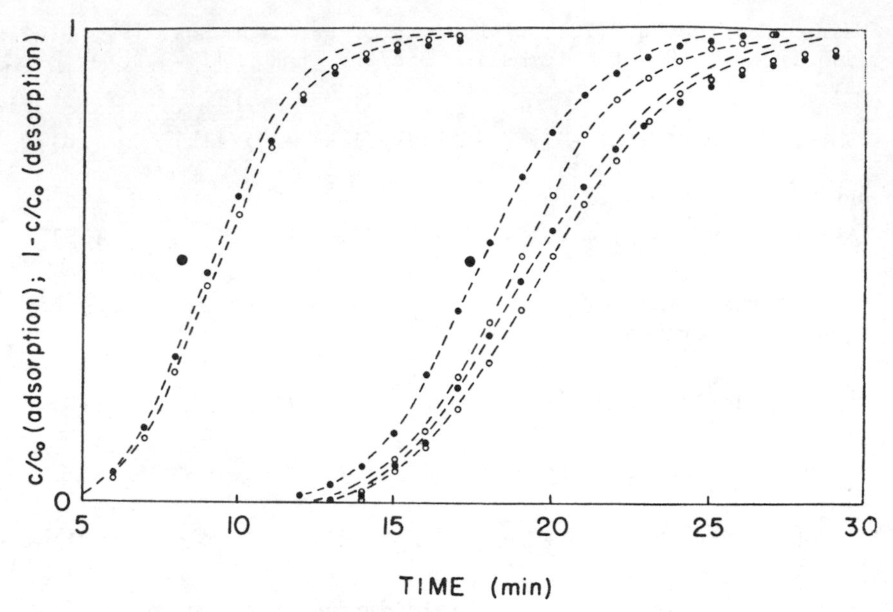

Fig. 1. Breakthrough curves with capillary tube. Dotted lines correspond to desorption (•) and adsorption (o). Dashed lines correspond to Eq. (22). Large dots (●) represent the points for $c/c_o = 0.5$ in the absence of adsorption.

We first note that the mid point in the breakthrough curve is always reached soon after one pore volume has flowed by. This indicates that the amount of adsorbed material is relatively small for our particular example, i.e., k is small compared to one. In addition we shall see later that the dependence of k on c is also small for adsorption. For the present example, the corrections introduced by dk/dc, as given by Eq. (21) can be shown to be negligible. Hence, even though the breakthrough curves depend on c_o we can describe them with Eq. (1) for an appropriate k, taken constant, with the value of the constant depending on the experiment. This great simplification allows us to discuss the observations in terms of the following solution to Eq. (1)

$$\begin{matrix} 2c/c_o \\ \text{(adsorption)} \\ 2(1 - c/c_o) \\ \text{(desorption)} \end{matrix} = \operatorname{erfc}\left\{[L - tv/(1 + k)]/[4Dt/(1 + k)]^{\frac{1}{2}}\right\} \quad (22)$$

The predicted profile is nearly symmetric with respect to the mid point at $L = tv/(1+k)$. Clearly this is not entirely true here; all profiles show some tailing for long times.

This often occurs and is due to a non-equilibrium process with a long relaxation time. For instance, dispersion in aggregated media shows tailing due to slow diffusion in dead-end pores. One could rationalize here that the dye diffuses slowly into the walls of the tube. This long time behavior also explains the observation of hysteresis loops between the adsorption and desorption curves: the adsorption process with a long relaxation time binds material which is not released at once during desorption. As a result the desorption curve is mostly above the adsorption curve. However, since we must have conservation of mass, for very long times the adsorbed material is finally released and the two curves must cross each other. Hence the non-equilibrium process explains both the asymmetry of the breakthrough curves (tailing) and the hysteresis phenomenon. The present theory, even with k a function of c, assumes the system to be in equilibrium and hence cannot describe the tail of the breakthrough curves. However, the main part of the curve can be considered in frozen equilibrium and should be described by Eq. (22), with k obtained from the time where $c/c_o = 1/2$ and D given by Eq. (2).

The first and last two profiles obtained for the low c_o yield essentially the same k: 0.12 ± 0.01 on desorption and 0.15 ± 0.01 on adsorption. We noted that at the low flow rate the walls of the tubes were very slightly pink after 30 min of desorption. The amount of material left in the wall was estimated from the area of the hysteresis loop, and was found equal to only 10^{-9} g per cm of tube.

The adsorption curve at high concentration gave $k \simeq 0.09$, a lower value than at low concentration as expected.

In principle the slope of any of the profiles at $c/c_o = 1/2$ could be used to calculate D_M, from Eqs. (2) and (22). For instance the desorption curve at high concentration, having the lowest $k \simeq 0.02$ is the least influenced by the adsorption-desorption process. Its slope at $c/c_o = 1/2$, $t = t_{1/2}$, is given by

$$\partial(c/c_o/\partial t = [v^2/4\pi D t_{1/2}(1 + k)] \qquad (23)$$

We measure from Fig. 1 $\partial(c/c_o)/\partial t \simeq 6.8^{-1}\ \text{min}^{-1}$, and from Eqs. (23) and (2) we deduce

$$D_M \simeq 4.7\ 10^{-4}\ \text{cm}^2/\text{min} \qquad (24)$$

Using this value we can predict all the profiles from Eq. (22) as shown in the figure. The agreement is obviously quite good except in the tail end as expected. Finally we may note that the effect of adsorption is rather significant even though k is always small. For instance at the low flow rate the slope of the two extreme breakthrough curves would yield a value of D_M differing by a factor two, had we neglected the change of k between those two cases.

In conclusion we have illustrated the value of Golay's equation, even when k is not truly constant. The effect of adsorption on the dispersion coefficient was not negligible for the present example. We are now going to determine whether this is also the case for a soil.

Case of a Soil

Calcium saturated Merrimac fine sandy loam was uniformly packed to a bulk density of 1.72 g cm^{-3} in a 5-cm long column with an ID of 6 cm. The soil, initially saturated with both water and calcium, was leached with .01N, .1N and 1N $CaCl_2$ solutions. The column design allowed for abruptly changing the influent solution. The two solutions were at the same concentration but a tracer, ^{45}Ca, was added to either the displacing or the displaced solution. Samples of the column effluent were obtained with an automated fraction collector and analyzed for ^{45}Ca by liquid scintillation.

This technique was used earlier (Starr et al., 1976) and it was found that the tracer concentration, c, obeyed the diffusion equation. For instance, if the tracer is in the displacing solution, c, obeyed

$$2c/c_o = \text{erfc}\left\{[x - vt/(1 + c_S/c_L)]/[4Dt/(1 + c_S/c_L)]^{\frac{1}{2}}\right\} \quad (25)$$

where c_o is the tracer concentration of the influent solution, c_L is the constant liquid concentration, and c_S is the meq/ml of adsorbed ions. A similar equation can be written if the tracer is present in the displaced solution (Starr et al., 1976). In the present case the slope of the adsorption isotherm k is equal to c_S/c_L. The value of c_S is nearly constant in these experiments, i.e., it is nearly independent of the value of c_L. Hence by changing the value of c_L we effectively change the value of k and we can determine the dependence of D on k. In the previous study we avoided any possible dependence of D on k by using such a low velocity, 1 cm/hour, that the longitudinal diffusive process was essentially molecular, i.e., $D \simeq D_M$. In the present investigation we take $v \simeq 8$ cm/hour and, as we can check later, D is then an order of magnitude larger than D_M.

Fig. 2 shows the breakthrough curves obtained for the three values of c_L when the tracer is in the displacing

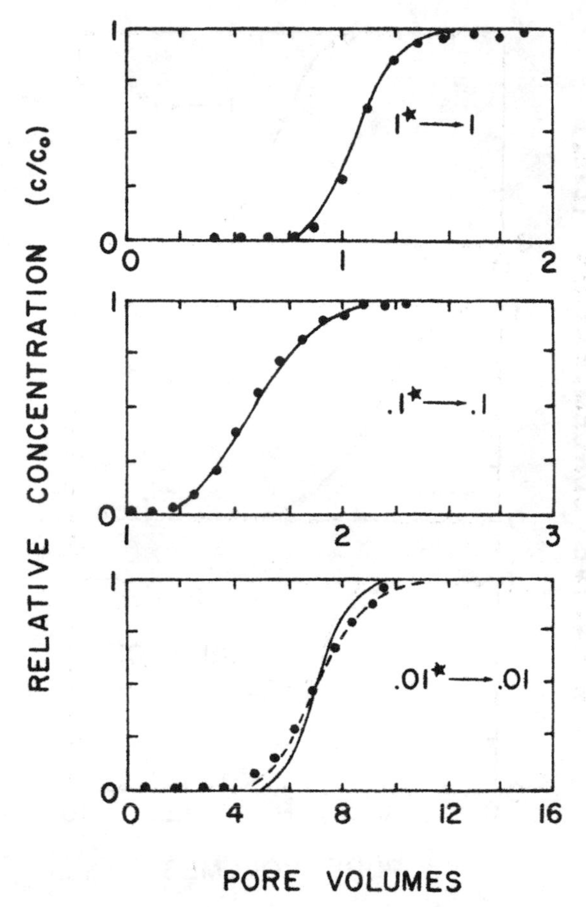

Fig. 2. Breakthrough curves for the indicated concentrations when the tracer is in the displacing solution (★). Solid lines correspond to Eq. (25) and $D/v = 0.055$ cm. Dashed line for $c_o = 0.01$ corresponds to Eq. (25) and $D/v = 0.055 \times 2$.

solution. The solid lines correspond to Eq. (25) with $c_S \simeq 0.06$ meq/ml and

$$D/v \simeq 0.055 \text{ cm} \tag{26}$$

We observe a very good fit for the two largest concentrations. However at the lowest concentration a better fit is obtained if the diffusivity is twice as large. The same property is observed when the tracer is present in the displaced solution, as shown in Fig. 3. These data show clearly that the dependence of D on k is much stronger for a capillary tube than for a soil. For instance for the soil when c_o changes

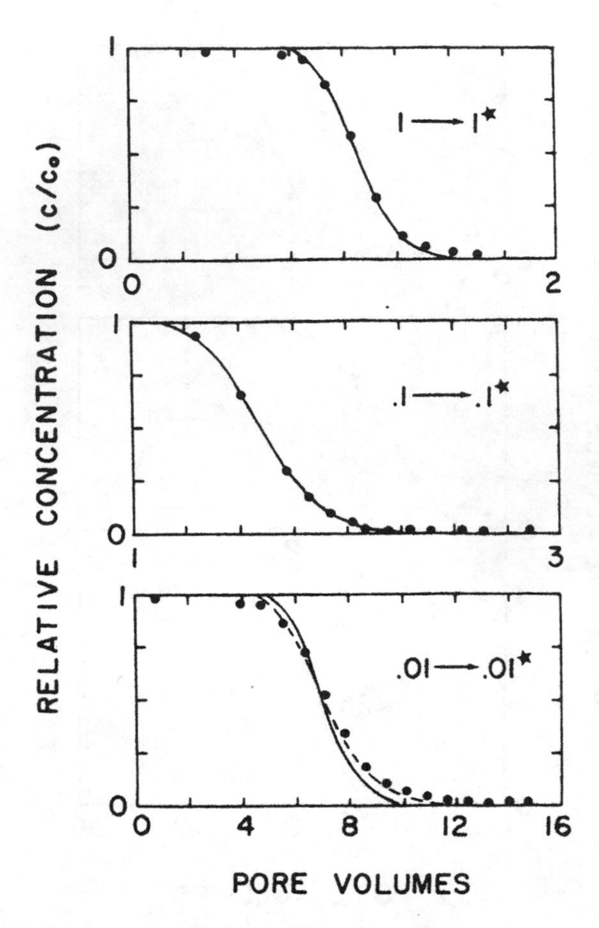

Fig. 3. Breakthrough curves for the indicated concentrations when the tracer is in the displaced solution (★). Solid lines represent the diffusion solution for $D/v = 0.055$ cm. Dashed line for $c_o = 0.01$ represents the diffusion solution for $D/v = 0.055 \times 2$.

from 1 to 0.01, k increases from 0.06 to 6 and D only doubles, while D would have increased by an order of magnitude for a capillary tube according to Eq. (2).

Such different behavior for the soil must be due to the interconnection of pores with different sizes. It is well known that this interconnection leads to D proportional to v (Boussinesq, 1877) as indicated by Eq. (26), while D is proportional to v^2 for a single tube as indicated by Eq. (2). However, the study of dispersion under unstable configurations (Starr and Parlange, 1976) has also shown the importance of pore size distribution. Because of this distribution, irregularities or fingers are created in the flow which have the size of the Darcy scale. If the size of the finger were of one pore size, as for the capillary tube and possibly for aggregated soils with large aggregates, then the base of the finger would be closer to the adsorbing walls than the tip. Consequently more material would be removed at the base of the finger than at the tip, enhancing dispersion. On the contrary if the finger is much larger than one pore size, as for most soils, all parts of the finger are in contact with adsorbing walls. In that case we expect adsorption to increase dispersion only slightly, in agreement with the results shown in Figs. 2 and 3.

In conclusion, on the basis of theoretical and experimental evidence we have shown that for a porous medium with one pore size, i.e., a bundle of capillaries, the diffusion coefficient is strongly affected by adsorption. On the other hand, for a porous medium which is made up of interconnected pores with variable sizes, the dispersion coefficient is less dependent on adsorption.

References

[1] Boussinesq, J., 1877. "Essai sur la theorie des eaux courantes," Mem. Pre. Div. Sav. a l'Acad. des Sci., 23:1-680.

[2] Golay, M., 1958. "Theory of Chromatography in Open and Coated Tubular Columns with Round and Rectangular Cross-Sections," In: Gas Chromatography, D. H. Desty (ed.), Academic Press, New York, pp. 36-53.

[3] Starr, J. L. and Parlange, J.-Y., 1976. "Solute Dispersion in Saturated Soil Columns," Soil Sci., 121: pp. 364-372.

[4] Starr, J. L., Sawhney, B. L. and Parlange, J.-Y., 1976. "Calcium Adsorption-Desorption and Movement During Continuous Leaching of Heavy Soils," In: Proc. Symp. Water in Heavy Soils ICID-ISSS, M. Kutilek and J. Sutor (eds.), Czech. Sci. Tech. Soc., Praha, 1:145-155.

[5] Taylor, G., 1953. "Dispersion of Soluble Matter in Solvent Flowing Slowly through a Tube," Proc. Roy. Soc., 219:186-203.

DISCUSSIONS - SECTION IV
Subsurface Water Quality

David S. Bowles*

Firstly, I would like to thank Dr. Gelhar for his accurate report on our paper. As inferred by the title and as described in the text our paper is an initial attempt at identifying the important salt pick-up and transport processes. The preliminary model identification is based on research performed in the field and the laboratory by Utah State University for the Bureau of Land Management and the Bureau of Reclamation. The main purpose of the paper is to draw together with a systems approach the individual processes studied, and thereby to understand more about their complex interrelationships and the areas in which future research is needed.

The following is in response to questions raised in the general report:

1. Equation 3 should include a storage term so that the equation applies to the dynamic case.

2. Although the channel bank mass-wasting component of the system is the only component which will be represented stochastically (see Fig. 3) this does not imply that mass-wasting is the only stochastic process in the system. On the contrary, in general almost every natural process is a stochastic process. However, to reduce the complexity of the model all processes will be represented deterministically with the exception of mass-wasting which can only be treated stochastically given the time and space resolution possible with a realistic level of data availability.

3. Our paper is not more specific in its discussion of the subsurface salination processes since the research on which the paper is based was directed towards the land surface and channel salination process. Subsurface salination processes are important in the Price River Basin but poorly understood and should be the subject of further research.

V. Yevjevich**

Difference in the celerity of the water wave and the velocity of the tracer in unsaturated zone in the English Chalk may be analogous to the difference encountered in the

*Utah Water Research Laboratory, Utah State University, Logan, Utah.
**Department of Civil Engineering, Colorado State University, Fort Collins, Colorado

saturated zones of highly permeable (karstified) limestone. Dead zones, interaction between the walls or solid particles and the tracer, etc., help to make this phenomenon more evident.

Sun-Fu Shih*

The random walk techniques have been used by Shih to solve the anisotropic porous media and the results showed that the random walk techniques are very powerful and useful tool to solve the groundwater flow problem. But one shortage is taking a longer computing time than finite difference at present technique.

F. D. Whisler**

The concept of water content changes in an unsaturated porous media due to a wave movement and not to mass flow seems inconsistent with the equation of continuity.

D. Oakes***

Before refuting a number of Dr. Gelhar's comments on my paper, I should tell you something about the Chalk which is probably not very familiar to you. It is the most important aquifer in the U.K. and is also very important in NW Europe. The Chalk is a fine grained limestone traversed by many fissures, some of which may be large but the majority of which will be very small and profuse in the Chalk matrix. Pore size is very small, perhaps about 1 μm, and because of this the Chalk matrix is fully saturated above and below the water table. The fissure system on the other hand is full below the water table but mainly dry above. Field evidence supports this picture and the paper shows a profile of water content obtained from core samples. Apart from some noise, possibly due to the presence of fissures and due to sampling problems, the water content is uniform at about 35% by volume.

Measurements of tritium in Chalk cores has led to estimates of the rate of downward movement of tritium of about 1m/year, and correlations of water level fluctuations to antecedent rainfall has indicated water flow rates of up to about 1m/day. Dr. Gelhar has suggested that water and solutes may

---

*Agricultural Research & Education Center, IFAS, University of Florida, Belle Glade, Florida.

**Dept. of Agronomy, Mississippi State University and USDA Sedimentation Lab, Mississippi State, Mississippi.

***Water Research Centre, Medmenham Laboratory, P.O. Box 16, Medmenham, Marlow, Bucks, SL7 2HD, England.

be moving at the same velocity through the matrix, and that percolation at the water table can be due to a pressure effect. This concept of flow in the Chalk has been discussed in the U.K. for many years and is discussed in my paper. I postulate a different mechanism in the paper in which water and solutes move through the fissure system while water in the matrix is static. Solutes can diffuse between the mobile and static water under concentration gradients to account for the slower rate of movement of the tritium. We have also found rates of movement of nitrate and chloride of about 1 m/year, and laboratory experiments on Chalk cores indicate that diffusion of solutes between fissure and pore water can be very rapid.

I think this mechanism is more likely than that postulated by Dr. Gelhar. Vertical permeability of the Chalk matrix is measured to be 10^{-3} - 10^{-4} m/d typically and vertical flow of water is unlikely. Additionally runoff is very rare on the Chalk outcrop even during heavy rain which indicated a relatively rapid downward penetration rate. This can only be via the fissure system.

Although our research cannot yet be considered conclusive the available evidence does suggest that water movement is via the fissure system at a relatively rapid rate. Solute movement can only be retarded by adsorption or by diffusion into static water in the matrix and the evidence does not support an adsorption/desorption mechanism.

The nature of the mechanism is somewhat academic as far as unsaturated flow is concerned but below the water table the fissure system has been extensively developed by solution of the calcium carbonate and the flow is quite definitely in the fissure system. The transport of solutes will then be retarded if there is diffusion into the matrix water.

This mechanism of solute retardation has been considered by other workers in unconsolidated media where its importance has been recognized. I think it is particularly important in the Chalk because of the very large fraction of static water in the unsaturated zone.

A. Nir*

The theoretical indication of the difference in time of arrival of a water pulse and a concentration pulse in unsaturated flow is of great interest. This effect is by now well established experimentally and conceptually in groundwater

---

*Isotope Department, The Weizmann Institute of Science, Rehovot, Israel.

and surface water flow as well as in many other dynamic flow system where mass response is separated from concentration response. However, it seems that it has to be rediscovered from time to time.

As to the specific example of tritium and nitrates, their transit times or rather the transit time distribution may be different in the unsaturated zone for quite different reasons again. Tritium may be lost with the evaporation affecting the unsaturated zone so there may be a counter current of these two components. On the other hand, dispersion of tritium may be assisted by vapor phase transfer assisted by pressure and temperature differentials.

SECTION V

SURFACE WATER QUALITY

Section Chairman, L. Scott Tucker
Executive Director
Urban Drainage and Flood Control District
Denver, Colorado

Rapporteur, Donald E. Evenson
Vice President
Water Resources Engineers
Walnut Creek, California

SECTION V

SURFACE WATER QUALITY

SECTION V

SURFACE WATER QUALITY

GENERAL REPORT

By

Donald E. Evenson, Vice President
Water Resources Engineers, Inc.
710 South Broadway
Walnut Creek, California 94596

Introduction

Purpose. The purpose of this general report is to present the state-of-the-art in surface water quality as derived from a review of 13 papers presented at the symposium. In addition, I have elected to point out what future directions the authors should pursue in additional work or additional research. However, before beginning the start-of-the-art review I would like to add a few comments on the international flavor of the symposium, and the broad range of study objectives that were addressed in the papers submitted for review.

International flavor. In this session, 13 papers were submitted from 8 countries. These countries were: Australia, Canada, France, India, New Zealand, Switzerland, United States, and West Germany. As a consequence of this broad representation, these 13 papers should be typical of the research efforts and studies that are being conducted throughout the world and should represent the international state-of-the-art in surface water quality.

Because so many countries were represented in this session, there were obvious problems related to interpretation into English. In addition, two papers were presented in French and only had English abstracts. There were also problems to mathematical terminology and symbols. These considerations alone presented difficulties in trying to review and critically assess papers. Communications problems were obvious. In addition, the technical content of the papers varied widely; one paper was a broad general discussion on water quality planning whereas another paper had 89 separate equations.

Study objectives addressed. From the review of these papers I found that an extremely broad range of water quality issues were addressed. The technical methods that were presented varied widely. These methods included interpretation of data, simple and advanced statistical methods, and complex mathematical models. And thirdly, the objective or purpose of each of these studies was quite different. Objectives ranged from testing rather straightforward conservative transport mathematical models to developing a broadly based data bank to predict pollution loads. Types of watersheds investigated varied from small urban areas to large, complex river basins. Forested areas were of particular importance. If there was a common thrust throughout these papers it was a concern with pollutant loads carried by storm runoff, especially from nonurban areas.

State-of-the-Art

For purposes of this session I have chosen to define the state-of-the-art as consisting of five separate, but highly related elements. In what I believe to be the increasing order of importance, they are:

1. Technical Analysis
2. Water Quality Measurements
3. Parameter Estimation
4. Wasteloads
5. Water Quality Problems or Issues

The most commonly used measurement of the state-of-the-art is the technical analyses or technical procedures used to analyze problems. The second item, water quality measurements, has to do with how we measure water quality to support technical analysis and how we design water quality sampling programs in a cost effective manner. Nearly all technical procedures require estimates of parameters of some sort. As we have developed more and more complex mathematical and statistical procedures, the number of parameters have increased and the problem of estimating proper model parameters has increased. Therefore it is essential that we address the problem of adequately defining and estimating model parameters. The fourth area is one that is commonly overlooked. How do we estimate wasteloads that are entering receiving water systems? Point sources are the easiest to measure and consequently point source problems have gotten the bulk of the attention in recent years. However, as we become more and more concerned with stormwater runoff, wasteloads become more dispersed and we are now faced with the difficult problem of estimating contributions of wasteloads from nonpoint sources--a problem that is of international concern. The fifth element, and the one that I believe is the most important, is water quality problems or issues. In other words, what are the water quality problems that the United States or the world is now concerned with, i.e., where are we going in the general field of surface water quality? Without problems or issues we have no need for wasteload estimates, parameter identification, water quality measurements or technical analyses. The state-of-the-art in identifying water quality problems or issues tells us where we are going in this broad topic area.

I believe the wide range of issues and approaches taken to water quality problems in the papers presented in this session tells us that there are numerous issues that have to be resolved in surface water quality. This healthy blend of technical procedures, water quality issues, concerns with water quality measurements and sampling programs tells us that the state-of-the-art in surface water quality is indeed a very broad one.

I would now like to present what I believe these papers indicate is the state-of-the-art in each of the five categories mentioned above. As a result, I am not going to review nor critique each of the 13 papers individually. However, I will bring out key points of each of the papers as they affect one of the five areas mentioned above. As a consequence, some papers will be mentioned more than once. All papers will be discussed at least once.

Technical Analysis

Under this broad category, I will present the results of my review in accordance with the following four subcategories: mathematical models, statistical analysis, physical models, and hybrid approaches.

Mathematical models. Four papers dealt specifically with the development and/or application of mathematical models. Brazil, Sanders and Woolhiser (1), in their paper entitled, "Kinematic Parameter Estimation for Transport of Pollutant in Overland Flow," dealt with the problem of defining the two key parameters that are necessary for developing a transport model of overland flow on impervious surfaces. The paper is a logical extension of previous work done at Colorado State University on mathematical modeling of overland flow with a kinematic wave model. Their paper shows that kinematic wave equations can be used to duplicate laboratory conditions for sheet runoff in the transport of soluble pollutants. In this paper, they demonstrate that they can accurately estimate the time of concentration and travel time of overland flow.

The next logical step in their study appears to be to move from the laboratory under controlled conditions to field applications to further demonstrate the validity of the kinematic wave approach in overland flow. This paper represented a sound research approach to document the justification of taking additional steps in kinematic wave modeling. Hopefully it will take the next step and move into field conditions.

Huthmann (3), in a paper entitled "Modeling Water Quality Systems by Multiple Frequency Response Analysis," addressed the problem of whether or not a water quality measurement station could either be replaced or sampled less frequently if a mathematical model could be developed of the river system. In his paper, he showed how an impulsive response function could be used to predict chloride levels at a downstream station on the River Rhine in West Germany. Put quite simply, an impulsive response function shows the response of a system, in this case the Rhine River, to a unit input at some other point in the system.

In this paper, a procedure for calculating an impulsive response function for a river system with two inputs is shown. The impulsive response functions were developed from measured data. On the other hand, Thomann (14) and Orlob and Amarocho (15) have shown how impulsive response functions can be calculated from physical data. I think it would be interesting to compare the two approaches to see if improvements could be made in the impulsive response functions developed by Huthmann (3). In this latter case he was able to demonstrate that he could predict chloride levels at a downstream station within a 10 percent error.

In a more traditional mathematical modeling approach, DeWalle, Taylor and Lynch (7), in their paper entitled, "Predicting the Effect of Shade Removal on Stream Temperature," developed a transient temperature model to predict time varying changes in temperature, including diurnal fluctuations. A detail they added to the traditional stream temperature models that I found of particular interest was the conduction of heat through the stream channel bottom. They looked at both mud and rock channel bottoms and found that taking this factor into account could decrease equilibrium temperatures by as much as 3 or 4 degrees Centigrade. This effect is significant and, if it is not properly taken into account, could have a direct effect on the parameters used in models that do not include heat conduction through the channel bottom. However, the results presented by the DeWalle, Taylor and Lynch paper were hypothetical. I believe their next step should be to move into field conditions to verify the formulation of the mathematical model and to measure the impacts of different bottom materials and of shaded vs. unshaded river reaches. It would also be helpful if they could demonstrate what parameters in their models need to be adjusted to show the effects of shaded vs. unshaded river reaches.

The fourth paper dealing with mathematical models was a paper entitled, "Modeling of River Pollution by Final Effluents from Waste Treatment Plants," by G. Romberg (12). This paper describes a linear transport model that can be used to predict river pollution from individual waste dischargers. In many respects this approach ends up with the same type of predictive tool that Huthmann (3) did with his impulsive response function. However, the method is completely different. The intent of

Mr. Romberg's model is to develop a procedure that could be used in conjunction with a waste discharge pricing policy. In this policy, each discharger must pay in proportion to the amount of water quality degradation that he causes. In other words, it becomes important to relate the volume of discharge to water quality levels at different points in the stream. Mr. Romberg's paper dealt primarily with development of the analytical procedures and his obvious next step is to extend it to field conditions and to parameters other than inorganic constituents.

Statistical Analysis

Statistics was clearly the most commonly and frequently used technical procedure contained in the 13 papers reviewed. The procedures reported varied from simple linear regressions to complex canonical and discriminant analysis. I also believe this is indicative of the state-of-the-art. We are not sure of many of the functional and physical relationships between rainfall, runoff and the complex physical and chemical interactions that take place during the runoff process. Consequently, there are attempts to develop relationships through statistical procedures. The following discussion presents material contained in these papers in accordance with the commonly used types of statistical procedures.

Linear regression. Keller (4) presented a very straightforward approach to the use of statistics in surface water quality studies. His paper entitled, "The Estimate of Ionic Discharge during High Flows in Small Torrent Catchments," describes how he used statistics in combination with the variety of models or formulations of surface runoff to estimate the calcium load that occurs from intense storms in small forested watersheds. In this case, he found that a simple relationship between calcium concentration and flow did not provide an adequate correlation to be reliable for predictive purposes. He also found that intense storms from these small catchments contribute a significant portion of the calcium loads in this watershed. By combining five different methods of calculating calcium loads with some simple linear regression equations, he was able to determine the best procedure for use in estimating calcium loads from small storms. He also concluded that an electrical conductivity recorder with calcium correlated to electrical conductivity provided the best estimate of calcium loads. These comparisons were done under field conditions for five different storms of varying intensities and duration. I believe it would be interesting if, in his next studies, he could compare other parameters or verify the relationships that he has currently developed. In any event, he was able to demonstrate that through some simple relationships he could estimate calcium loads from small forested watersheds.

In a similar approach, Singh (6) describes some relationships between flow and quality that he developed for a forest catchment in Canada. His paper entitled, "Streamflow Quality in Quantity Relationships on a Forest Catchment in Alberta, Canada," involved the use of multiple linear regression techniques to predict concentrations of 8 different water quality parameters from commonly measured hydrologic variables. Water quality parameters that were studied included calcium, magnesium, sodium, potassium, bicarbonate, sulfate, chloride and silicate. In total, he postulated 28 different relationships between these parameters and hydrologic variables and found that multiple linear regression models that incorporated the current flow, the specific conductance and a lag variable of streamflow gave the best fit for most of the constituents mentioned above.

Of the parameters studied, potassium and chloride were the only parameters that did not give statistically reliable results. Singh also notes that additional work in dividing the streamflow into its base flow and storm flow components are presently underway to hopefully correct this problem area.

In a similar approach, A. S. Zakaria (9), in his paper entitled, "Effects of the Hydrometeorological Factors on Suspended Sediment Output Catchment in New Zealand," presented a methodology to predict suspended sediment loads from a small catchment. In this case, he used both multiple linear regression and principle component analysis as statistical procedures. He found that the hydrometeorological factors by themselves could only explain about 40 percent of the total variance observed in the suspended sediment output and that consequently other environmental factors must be equally as important. He also found that because suspended sediment concentrations were low in the watershed that the data were very sensitive to minor erosional events, and, hence, there could be a significant data error. In addition, he found in reviewing the data that most of the sediment derived from the watershed occurs from relatively infrequent high intensity storm events. As a consequence of these two effects, he concluded that it may be very difficult to reliably predict suspended sediment loads.

The multiple regression model that Zakaria (9) found to best explain or predict suspended sediment concentration included the maximum discharge of the storm hydrograph, an antecedent precipitation index and water temperature. The use of the maximum discharge as a measure of suspended sediment concentration is relatively obvious. However, the other two indices, although not obvious, are very instructive. The antecedent precipitation index points out the importance of soil moisture while the temperature parameter indicates that season of the year is important. The variance explained by this regression model is only 36 percent, which is rather low.

To develop an improved relationship between suspended sediment and hydrometeorological factors, he then used principle component analysis to take into account the underlying interrelationships between the parameters. The principle component analysis resulted in two particularly critical components. One was a measure of the magnitude and rate of the preceding flood, and the second was related to energy input from storms. These two components could account for approximately 76 percent of the observed variance in suspended sediment concentrations.

The most complex of the papers that relied on statistical analyses was by Herrmann, Bolz, Symader, and Rump (8). Their paper, entitled "Interpretation and Prediction of Spatial Variation in Trace Metals in Small Rivers by Canonical and Discriminant Analysis," presented an approach to predict trace metal levels as a function of the key economic and natural variables in a complex watershed in Germany. In this case they were trying to predict trace metal concentrations in a plant (moss) which absorbs trace metals directly from the overlying water. Their goal was to relate trace metal concentrations in the moss to the different sources of trace metals throughout the river basin. In all, they investigated 39 different natural and economic variables. Their first step was the use of principle component analysis to find or determine groups of variables which were not related. They then used canonical correlation analysis to determine relationships between different groups of variables within the set of 39. The principal components analysis showed that the water quality variables were highly intercorrelated and that the water quality sampling could be reduced from 14 variables to only 4 with very little loss of information. In other words, the other 10 could be predicted from the 4 being measured. The canonical correlation showed the chemical behavior of the rivers was highly correlated to the trace metal concentrations in the water moss, and that there was little correlation between the trace metal concentrations and variables describing the natural and economic character of the river basin. Finally the use of discriminant analysis showed that the 4 trace metal concentrations could be predicted based upon variables that reflected the economic and natural character of the basin. These characteristics included such things as settlement density, river density, forests, and area with quarternary rocks. Their paper presents a good indication of how complex statistics can be used to develop meaningful relationships to further our understanding of the complex processes that affect water quality in a watershed.

Physical models. Of the 13 papers reviewed, there was only one paper that used a physical or laboratory model for their research or study. This paper was presented by Brazil, Sanders and Woolhiser (1). They used a rainfall-runoff experimental facility at Colorado State University to study overland flow phenomenon on impervious surfaces. I believe that as we learn more through use of statistical analyses and mathematical models that are now being developed, it will be necessary to once again return to the laboratory for more fundamental research to determine more specifically the complex physical and chemical reactions that take place in natural watersheds.

Hybrid approaches. By hybrid I refer to combinations of physical, mathematical or statistical approaches to address a particular problem. Historically there have been very few hybrid approaches and the indications are from the papers reviewed that we are continuing along this same path. I believe there are future possibilities in surface water quality using the hybrid approaches. For example, there are some very valid approaches for the use of advanced statistical procedures to interpret and explain model calibration errors. Typically, we formulate a mathematical model and the only statistics that we use are to quantify the errors resulting from model calibration. Rather than just quantifying errors, I think we should use statistical procedures to try to determine what is causing these errors. Is it related to measurement errors? Or model formulation? Or perhaps there may be some other important variables external to the model that can explain why we have a calibration error. Another possibility is the use of both mathematical and physical models of the same prototype system. Both mathematical and physical models clearly have their advantages and there are certain obvious problems that should not be addressed with either. However, often we can learn a great deal through the joint use of both mathematical and physical models of the same prototype. Although in many cases there have been both mathematical and physical models developed of the same prototype, typically they have not been used conjunctively to help offset the weaknesses in each. I think this would be a very fruitful area of research and one in which we should be directing our future effort.

Water Quality Measurements

Although all the papers relied on some form of water quality measurement, there were two papers in particular that directly addressed the problem of water quality measurement.

Selection of sampling procedure. Although the primary thrust of Keller's paper (4) was the evaluation of different predictive methods, he also indirectly evaluated five different sampling programs necessary to get adequate information to define calcium pollutant loads from small forested watersheds. Based on the results of his analysis he was able to conclude that a sampling program that combined a continuous flow recorder with a continuous electrical conductivity recorder would produce the most accurate estimate of calcium load. In this case the concentration of calcium was obtained from an EC recorder with a simple linear regression. It was a very straightforward analysis and illustrates what can be done in the selection of sampling procedures.

Reduction of sampling frequency. Huthmann (3) used an impulsive response function model to show that in the case of chlorides on the Rhine River, you could greatly reduce the sampling interval without a great loss of information. In his study the loss of information would amount to approximately ±10 percent. In other words, he was able to show that you could use a model to predict chloride levels and could, therefore, significantly reduce the frequency of measurements. However, since the model was based on historical data there is a continuing need to at least collect data at some regular basis to continually improve or verify the model. As pointed out earlier, another improvement would be to go back and develop the impulsive response functions based upon the physical stream system rather than through the use of statistical analyses. This alternative would allow the user to study different types of conditions (e.g., high flow or low flow conditions), gain more confidence in the model, and make even greater reductions in the sampling frequency.

Design of sampling program. Although some papers addressed improvements in sampling programs, there were no papers that specifically dealt with the problem of how to use mathematical or physical models, or statistical analysis to design more efficient and more cost-effective data collection programs. As we will see later and has been mentioned previously, the trend in surface water quality is toward problems related to stormwater runoff. These are highly transient problems and require a considerable amount of data. We are also studying problems with more and more water quality parameters. We are now finding ourselves concerned with heavy metals, pesticides, insecticides, and specific inorganic constituents. All of these require expensive analyses. Therefore, it is becoming more and more important to design data collection programs that are cost effective and to use all the information that is collected in field programs. In the general area of water quality measurements, I believe that a future direction and a great deal of research is required in the design of more effective and efficient sampling programs.

Parameter Estimation

As we develop more complex mathematical models and more involved statistical procedures, more and more parameters must be estimated to develop reliable technical procedures. The recent thrust in developing models that include biological components and their interrelationships with water quality components has added dozens and dozens of additional parameters that must be estimated. As our models cover broader and broader surface water systems, more and more spatially dependent parameters must be estimated. As we get more involved in the highly transient problems of stormwater runoff, there are parameters that must be estimated and frequently these can vary with time or with the intensity of the storm. This has raised the whole problem of how do you estimate multiple parameters in large surface water quality systems. Unfortunately this problem was not directly addressed in any of the papers presented. There was, however, one paper which was written in French that did, at least, identify if not address the problem of estimating reaeration rate constants. This paper was presented by Bernier (11) who discussed the problems of determining appropriate reaeration rate constants when there are errors or uncertainty in the pollution loads that are entering the surface water stream system. Unfortunately, there was only an English abstract available and a complete review of his work was not possible. It does, however, introduce an additional parameter estimation problem that must be addressed in the future. This is the problem of estimating parameters when there are uncertainties in the basic data with which you must work. Here is an obvious need of combining mathematical models with statistical procedures. We not only frequently have uncertainties in the input data, but the measurements that we are trying to calibrate the model with are also subject to error. In addition, we have a third form of error that has to do with the model or the statistical procedure itself. In total, then, there are four sources of errors related to predicting surface water quality. These are:

1. Parameters
2. Input Data
3. Water Quality Measurements
4. Model Formulation

The problem that we must address is how to differentiate between the sources of errors and how to account for them in interpreting results. One error that can be addressed is the error associated with parameter estimation. We have some procedures that can minimize errors between calculated and measured water quality levels. In other words, for a given model formulation, and for the input data that is provided and for the water quality levels that have been measured, we can estimate the best parameters. What we need to do next is determine how to estimate these parameters when we also have estimates of the uncertainties in the other input data. In general, the whole problem area of parameter estimation is one in which a great deal of future direction is needed. The state-of-the-art is young and future research should be directed along these lines.

Wasteload Estimation

Many of the papers directly or indirectly address the problem of determining pollutant loads from watersheds, both under storm and nonstorm conditions.

Nonstorm wasteloads. Perhaps the most comprehensive approach to this problem is that presented by Cluis (10). He described a data system to predict both point and nonpoint source loads of nitrogen and phosphorus. These predictions were based on readily available data and included such data sources as land use, population and employment. Although it may not be as advanced mathematically as some receiving water models, it does provide a vehicle to close the gap between wasteloads entering receiving waters and the data that is readily available to predict them. In this study, he used the Yamaska River Basin in southern Quebec as a pilot basin. This area was under intensive agricultural development and it became important to look at the nitrogen and phosphorus loads that originate from all sources of land use including agriculture. In this example, he was able to demonstrate that the bulk of the nitrogen and phosphorus loads were derived from fertilized areas and livestock. Sewered population represented only approximately 12 percent of the total nitrogen load. He was also able to show that the nonpoint sources contributed in excess of 60 percent of the phosphorus load and over 80 percent of the nitrogen load. This information is extremely important not only for use in mathematical or statistical models, but also in the formulation of water quality control programs and policies. In this case it was clear that attention should be given to the nonpoint sources and, in particular, those related to agriculture.

Stormwater wasteloads. This area received the most attention in the papers reviewed. Several papers showed relationships between different water quality parameters and different hydrologic or hydrometeorological parameters. These results were mixed. In general they were not totally successful and indicated that factors other than hydrologic parameters must be considered to predict water quality loads during storms. For example, Zakaria (9) in his work on suspended sediments found that he could only account for approximately 40 percent of the variance of suspended sediment loads through relationships between the hydrometeorological parameters and suspended sediments. Keller (4) was able to obtain better results for the case of calcium, while Singh (6) looked at eight different cations and anions and found many could be reliably predicted. However, there were some that had very poor predictive capabilities. All of these were based primarily on single or multiple linear regressions and did not directly attempt to define the complex physical and chemical processes that cause the pollutant loads to wash off urban, rural and natural watersheds. This is an area where much additional work is needed and where future directions should be addressed.

In a comparable study, but addressing urban areas, Cordery (2) found that in three separately sewered residential areas in Sydney, Australia, the annual pollutant load from stormwater was about the same as that from treated municipal water flow. He also compared the treatability of stormwater and found that significant removals of suspended sediments and other parameters could be achieved by simple settling. As a consequence, he concluded that it is possible and more cost effective to treat urban stormwater than it would be to provide advanced treatment to municipal effluents. Here we have an excellent case of how a study that is not complex from a mathematical or statistical point of view can be used to focus on the real problems and have the results provide a direct impact on water quality management practices and programs.

Another interesting aspect of Cordery's results shows that many parameters are unrelated to suspended solids levels. A common assumption in urban stormwater models is that they are related. For this reason his results should be looked at carefully. For example, the total phosphorus load appears to be completely independent of flow or suspended solids loads and occurs at a relatively constant rate. He also found that the critical storms from a suspended sediment point of view are those with less than 15 millimeters of rain. And after that it appears that most of the pollution has been washed from the urban watershed and that the level of contribution of pollutants is decreased considerably.

Cordery's paper points out again the need for additional research and direction in the general area of defining the physical and chemical processes that occur during stormwater runoff. The papers presented at this symposium have shown the difficulties that people have encountered in trying to relate stormwater runoff quantity with water quality. It is a very complex process and one that has to be defined in terms of the physical and chemical processes that actually take place.

Water Quality Issues

This is an area where the papers submitted to this symposium really covered a broad range. It is also the area that is most important when dealing with surface water quality. The papers pointed out that the concern that much of the world has at present is in the area of stormwater runoff. They are concerned with a variety of different water quality parameters and the impacts of stormwater runoff. Five different papers address this problem directly. The problem is very complex. It involves detailed hydrologic and hydraulic analyses that must precede any water quality studies. It involves soil and water interactions. It is one that is difficult to measure and requires expensive sampling programs. Another area that was pointed out as a potential problem area was related to agricultural waste. Cluis (10) showed the importance of agriculture to the total nonpoint and point source pollutant load in the Yamaska River Basin in southern Quebec. Agricultural wasteloads are also difficult to measure. In addition, the necessary institutional and legal mechanisms to correct agricultural wasteloads problems does not exist in many countries. It is one problem area which the future direction of surface water quality must address.

The overall importance of surface water quality is amply summarized in two papers that address programs for water quality planning. Chaturvedi (13) outlined the availability and quality of water in the Uttar-Pradesh state of India, and how they plan to use models, planning and data collection programs to resolve some of the water quality issues and problems in the state. James and Steele (5) in their paper described a very comprehensive program designed to evaluate potential effects of coal development in the Yampa River Basin in Colorado and Wyoming. Both of these papers demonstrate the need for adequate water planning. They describe two approaches. The important thing is that we must plan to maintain or enhance surface water quality.

Summary

In summary, I believe these papers have shown that there are technical procedures whether they be mathematical, statistical or physical models that can be used to relate causes and effects in surface water quality problems. They also show that there is a broad range of issues that people throughout the world are concerned with and that the key issue people presently are addressing is related to stormwater problems. These papers have also shown that we are just now beginning to understand the relationships between rainfall and the resulting pollutant loads that enter our surface streams. The future direction of surface water quality is clearly going in this direction.

List of Papers Reviewed

[1] Brazil, L. E., Sanders, T. G., and Woolhiser, D. A., "Kinematic Parameter Estimation for Transport of Pollutants in Overland Flow."

[2] Cordery, Ian, "Urban Stormwater--A Major Contributor to Water Pollution."

[3] Huthmann, Gerhard, "Modeling of Water Quality Systems by Multiple Frequency Response Analysis."

[4] Keller, Hans M., "The Estimate of Ionic Discharge During High Flows in Small Torrent Catchments."

[5] James, I. C. II and Steele, Timothy D., "Assessing the Impacts of Alternative Coal-Development Plans on Regional Water Resources."

[6] Singh, T., "Streamflow Quality and Quantity Relationships on a Forest Catchment in Alberta, Canada."

[7] DeWalle, D. R., Taylor, S. L., and Lynch, J. A., "Predicting the Effect of Shade Removal on Stream Temperature."

[8] Herrman, R., Bolz, U., Symader, W., and Rump, H., "Interpretation and Prediction of Spatial Variation in Trace Metals in Small Rivers by Canonical and Discriminant Analysis."

[9] Zakaria, A. S., "Effects of the Hydrometeorological Factors on Suspended Sediment Output of a Catchment in New Zealand."

[10] Cluis, D. A., "A Nutrient Transport Model Based on Export by Land-Uses."

[11] Bernier, J., "Point De Vue Aleatoire Dans Les Phenomenes D'Autoepuration Des Rivieres."

[12] Romberg, G., "Modeling of River Pollution by Final Effluents from Waste Treatment Plants."

[13] Chaturvedi, A. C., "Water Quality and Availability in Uttar Pradesh (India)."

Additional References

[14] Thomann, R. V., 1963. "Mathematical Model for Dissolved Oxygen," Journal of the Sanitary Engineering Division, ASCE, Vol. 89, No. SA5, October.

[15] Orlob, G. T. and Amorocho, J., 1961. "Nonlinear Analysis of Hydrologic Systems," Water Resources Center Contribution No. 40, University of California, Berkeley, Nov.

STOCHASTIC POINT OF VIEW IN OXYGEN
BALANCE MODELS IN STREAMS

By

J. Bernier
Laboratoire National d'Hydraulique
6, quai Watier 78400 - Chatou, France

Abstract

In a stochastic context, there are difficulties for estimating the parameters of oxygen balance, including reaeration, photosynthesis, respiration... using the only dissolved oxygen measurements; the consideration of the variability of pollution load as error term in the balance equation, introduce a large bias in the estimation of reaeration coefficient namely. This bias is function of the unknown pollution removal coefficient. It is necessary to estimate together the two coefficients using further informations on dissolved oxygen and also further hypothesis. The estimation problem is discussed according several types of random influences on the balance equation.

Résumé en français

Dans un contexte aléatoire, des difficultés apparaissent pour estimer les paramètres du bilan d'oxygène en utilisant les seules mesures d'oxygène dissous. L'introduction des variations de la charge polluante comme termes d'erreur introduit un biais important dans l'estimation du coefficient de réaération, biais fonction du coefficient d'autoépuration inconnu. Ces deux coefficients doivent être estimés ensemble en utilisant des données d'oxygène supplémentaires et des hypothèses supplémentaires. Le problème d'estimation est discuté en considérant plusieurs types d'influences aléatoires sur l'équation du bilan.

Introduction

Depuis Streeter et Phelps les modèles déterministes d'autoépuration des rivières ont été compliqués en introduidant divers termes complémentaires dans les sources et puits du bilan d'oxygène: photosynthèse, respiration de la communauté aquatique, benthique, sédimentation de la charge polluante, etc... Ces efforts n'ont pas encore accru le réalisme pratique de ces modèles compte tenu des difficultés, sinon des impossibilités fréquentes d'estimer les paramètres supplémentaires pris en compte. En fait pour prétendre décrire de façon fiable des observations, les modèles de bilan d'oxygène doivent tenir compte de divers aléas représentant:

- l'erreur dans les équations résultant de la non prise en compte de certains phénomènes et de la schématisation trop grande des phénomènes pris en compte;

- les fluctuations liées aux phénomènes de diffusion-dispersion;

- la nature aléatoire des paramètres du modèle: coefficients de réoxygénation, de dégradation fonction des aléas du débit, de la température de l'eau, etc...;

- les fluctuations des entrées dans le système notamment des rejets polluants.

La prise en compte de ces aléas amène à adopter un point de vue tres différent du point de vue déterministe en ce qui concerne les problèmes de collecte des données, de vérification et d'estimation des modèles. Dans ce dernier cas notamment des précautions doivent être prises au niveau de l'estimation.

Nous avons rencontré ces problèmes à l'occasion du traitement des mesures en continu d'oxygène dissous en diverses stations situées sur la rivière Seine. Compte tenu des difficultés de mesures des variables de l'activitè biochimique telle que la DBO notamment, l'idée de n'utiliser que les seules mesures d'oxygène n'est pas nouvelle. On citera seulement un travail récent très intéressant, celui de Hornberger et Kelly (1976). Le manque de fiabilité des multiples formules de la littérature concernant notamment le coefficient K_2 de réoxygènation accentue l'intérêt de méthodologie d'estimation directe de ce coefficient. Il importe cependant d'étudier avec soin ces méthodologies, de type statistique, qui peuvent dans certains cas, entraîner des erreurs systématiques et aléatoires très importantes.

Le Bilan D'Oxygene

L'équation de bilan sur la concentration moyenne d'oxygène $C(x,t)$, dans la section x d'une rivière et au temps t, peut s'écrire, en ne prenant en compte que le transport par convection:

$$\frac{DC}{dt} = \frac{\partial C}{\partial t} + u \frac{\partial C}{\partial x} = K_2 \, (C_s - C) + P - R \tag{1}$$

où - $\frac{DC}{dt}$ est la derivée totale,

- C_s la concentration de saturation de l'oxygène,
- P le terme de production d'oxygène,
- R le terme de respiration incluant la respiration bactérienne associée à une charge polluante biodégradable,

- u la vitesse moyenne du courant dans la section.

Sous forme intégrale et en considérant deux points (x,t) et (x_o, t_o) liés par la relation:

$$x - x_o = u\,(t-t_o) \qquad (x > x_o,\ t > t_o)$$

On obtient:

$$C(x,t) = C(x_o,t_o)\, e^{-K_2(t-t_o)} + \int_{t_o}^{t} e^{-K_2(t-\tau)} [K_2 C_s + P - R] d\tau \qquad (2)$$

L'idée intéressante de Hornberger et Kelly (1970) est de représenter le terme P, essentiellement la photosynthèse par un terme proportionnel au rayonnement incident I recu par le plan d'eau et de considerér la respiration R comme constante et égale à R_o au niveau d'une journée. En considérant des valeurs moyennes $\overline{C}_s$ et $\overline{I}$ pour tout l'intervalle $[t_o,t]$ pris en compte, il vient alors:

$$C(x,t) = C(x_o,t_o) e^{-K_2(t-t_o)} + [1-e^{-K_2(t-t_o)}]\,[\overline{C}_s + \frac{\alpha}{K_2}\,\overline{I} - \frac{R_o}{K_2}] + \varepsilon \qquad (3)$$

Bien entendu on considère $\overline{I}$ comme nul pour les périodes de nuit et nous avons introduit un terme d'erreur ε représentant les ecarts du modèle à la réalité.

Une telle relation semble suggérer qu'on puisse estimer les coefficients K_2, α, R_o [par l'intermédiaire des transformations:

$$e^{-K_2(t-t_o)},\ \frac{(1-e^{-K_2(t-t_o)})}{K_2}\,\alpha,\ \frac{(1-e^{-K_2(t-t_o)} R_o)}{K_2}]$$

à partir de la méthode des moindres carrés, soit sur deux stations (x_o et x différents), soit sur une seule station (pour autant que les variations entre stations soient faibles).

Une telle procédure utilisée par Hornberger et Kelly peut entraîner des erreurs importantes. Il ne faut pas oublier en effet que le terme d'erreur ε doit nécessairement inclure les variations de la respiration (notamment celle associée à la charge polluante) autour de R_o. Or les variations de cette charge polluant sont nécessairement corrélées avec la variable $C(x_o,t_o)$ d'où le fait que le résidu aléatoire ε est corrélé avec la variable explicative $C(x_o,t_o)$. Il est bien connu que cette correlation peut entraîner un biais très important dans l'estimation par la méthode des moindres carrés qui traite automatiquement le résidu comme indépendant des variables explicatives.

Un modèle d'évolution aléatoire de la charge polluante restante $L(x,t)$ dans la rivière peut permettre de préciser ce biais.

Modele Aleatoire de la Charge Polluante

Supposons que la respiration R contient un terme proportionnel à la charge restante $K_1 L$, en accord avec le modèle classique de Streeter et Phelps, charge polluante qui évolue selon l'équation différentielle stochastique en fonction du temps d'écoulement:

$$D L = [-K_1 L + \mu(t)]dt + \eta\sqrt{dt} \tag{4}$$

$\eta(t)$ est un résidu aléatoire, indépendant des résidus $\eta(t')$ pour $t' \neq t$, de moyenne nulle et de variance σ^2 constante.

L'interprétation concrète de ce modèle peut être mise en évidence avec le concept de masses d'eau. On suppose que tout le long de la rivière, et en particulier à l'amont du point x_o, il existe des injections de charge polluantes distribuées selon une valeur moyenne $\mu(t)\,dt$ fonction du temps d'écoulement. Entre les différentes masses d'eau successives qui passent au niveau de chaque section, il existe des écarts aléatoires à la moyenne $\mu(t)$, représentés par les résidus $\eta\sqrt{dt}$, qui globalement ont une variance constante égale a $\sigma^2\,dt$.

Il résulte de ce modele les propriétés suivantes:

- le coefficient de corrélation partiel (à production P constante) entre la concentration d'oxygène $C(x_o,t_o)$ et la charge polluante restante $L(x_o,t_o)$ au même point et au même instant est égal à:

$$\rho = -\sqrt{\frac{K_2}{K_1 + K_2}} \tag{5}$$

  Dans la mesure où le résidu ε représente les aléas de cette charge polluante, ce coefficient de corrélation représente la liaison entre ce résidu et la variable explicative $C(x_o,t_o)$;

- le coefficient de régression partiel (à production P constante) de $C(x,t)$ en fonction de $C(x_o,t_o)$ est égal à:

$$a' = \frac{K_1\, e^{-K_2(t-t_o)} - K_2\, e^{-K_1(t-t_o)}}{K_2 - K_1} \tag{6}$$

C'est l'estimation de ce coefficient qui fournira l'application de la méthode des moindres carrés à la formule (3) et non pas l'estimation:

$a = e^{-K_2(t-t_o)}$ comme le laisserait penser cette formule, entendue dans le sens déterministe. Le biais, exprimé en %, sur l'estimation du paramètre a peut être calculé par: $100\ (\frac{a'}{a} - 1)$. La Fig. 1 illustre ce biais comme fonction du paramètre $K_2(t-t_o)$ et du paramètre de Fair: $F = K_2/K_1$.

On observe sur cette figure que ce biais est toujours positif et peut être considérable (de 50 à plus de 100%) pour des valeurs usuelles de F.

Dans la mesure où l'erreur ε du modèle (3) n'est pas corrélée avec le rayonnement incident I, l'estimation du coefficient:

$$\frac{\alpha(1 - e^{-K_2(t-t_o)})}{K_2}$$

ne sera pas biaisée.

Erreur Aleatoire D'Echantillonnage

On sait que la précision d'échantillonnage d'un coefficient de régression calculé sur un échantillon limité est d'autant plus petite que la taille de cet échantillon est petite et que la variance de la variable explicative correspondante est plus petite. Dans le cadre de notre modèle, on peut montrer que l'autocorrélation r entre des observations successives $C(x_o,t_{oj})$ successives au point x_o, espacées par des intervalles constants $t_{o,j+1} - t_{o,j} = \delta$ est égale a:

$$r = \frac{K_1\ e^{-K_2\delta} - K_2\ e^{-K_1\delta}}{K_1 - K_2}$$

On constate sur le graphique II que cette autocorrélation peut être très grand pour des pas de temps de l'ordre de grandeur de l'heure. C'est ici encore une autocorrélation partielle (à production P constante) que réglera la variance des $C(x_o,t_{oj})$, variance généralement petite pour des pas de temps courts. Il en résultera donc une dispersion d'échantillonnage importante pour l'estimation de a'.

On observe également que biais et variance d'échantillonnage de l'estimation varient dans des sens opposés. Le biais est d'autant plus petit que l'intervalle $t-t_o$ est petit mais la dispersion d'échantillonnage est d'autant plus grande que l'intervalle δ est plus petit.

Elimination du Biais

Il apparaît que le biais sur l'estimation du coefficient de réoxygènation est trop considerable pour être négligé. Mais si par ailleurs le coefficient K_1 est connu, la formule (6) peut permettre le calcul de K_2 à partir de a'. Mais K_1 est généralement un paramètre aussi mal, sinon plus mal connu que K_2. L'analyse précédente montre qu'il n'est pas si aisé de se débarrasser de la charge polluante et de son taux d'évolution qui règle en fait les propriétés statistiques de variations d'oxygène dissous. Il importe donc d'estimer à la fois K_1 et K_2, conjointement avec les autres paramètres de photosynthèse et de respiration.

Bien entendu, on pourrait atteindre l'ensemble des coefficients en mesurant conuointement aux concentrations d'oxygène $C(x,t_j)$ et $C(x_o,t_{oj})$, les variations de la charge polluante initiales $L(x_o^j,t_{oj})$. Les difficultés pratiques de mesure de cette charge, représentée par la DBO par exemple, sur des pas de temps comparables à ceux qui permettent les mesures d'oxygène, rendent illusoire une telle perspective. Il est cependant possible de remplacer cette information manquante, par une information complémentaire sur les courbes d'oxygène et par des hypothèses complémentaires sur le mécanisme de l'autoépuration. Le modèle aléatoire de charge polluante du paragraphe III peut représenter de telles hypothèses complémentaires. Les résultats du calcul d'estimation serait, bien entendu, conditionnés par la validité d'un tel modèle.

L'idée de principe de la méthode est de relier statistiquement le terme de respiration R de la formule (2) qui est en corrélation avec (x_o,t_o) non pas seulement avec la concentration d'oxygène en x_o mais également avec la concentration d'oxygène en un point supplémentaire amont. En d'autres termes on utilise une méthode à 3 stations situées en x_2, x_1, x_o, basée sur les concentrations observées en t_o, t_1 et t_2 tels que:

$$x_2 - x_1 = u(t_2-t_1)$$

et $$x_1 - x_o = u(t_1-t_o)$$

auquel cas on pourra écrire la formule suivante:

$$C(x_2,t_2) = a_o C(x_o,t_o) + a_1\, C(x_1,t_1)$$

$$+ \left[1-e^{-K_2(t_2-t_o)}\right] \left[\overline{C}_s + \alpha \frac{\overline{I}'}{K_2} - \frac{R_o}{K_2}\right] + \varepsilon \qquad (7)$$

Dans cette formule $\overline{I}'$ est le rayonnement incident moyenné sur l'intervalle de temps total (t_o-t_2).

Si ce terme variait trop dans cet intervalle, il comporterait de décomposer l'intégrale:

$$\alpha \int_{t_o}^{t_2} e^{K_2(t-\tau)}\, I(\tau)d\tau$$

en intervalles plus courts.

Les coefficients a_o et a_1 s'interprêtent comme les coefficients de règressions partiels de $C(x_2,t_2)$ en fonction de $C(x_o,t_o)$ et $C(x_1,t_1)$ (à production $P(t)$ ou $I(t)$ fixée sur l'intervalle $[t_o,t_2]$). Dans ce cas ε est indépendant des variables explicatives et la méthode des moindres carrés peut être utilisée pour estimer la formule (7).

Dans le cadre du modèle de charge polluante aléatoire, les coefficients a_o et a_1 s'expriment en fonction des coefficients K_1 et K_2. Ils sont en effet solution du système suivant:

$$\left.\begin{aligned} &(K_1-K_2)a_o + K_1e^{-K_2(t_1-t_o)} - K_2e^{-K_1(t_1-t_o)} a_1 = K_1e^{-K_2(t_2-t_o)} - K_2e^{-K_1(t_2-t_o)} \\ &[K_1e^{-K_2(t_1-t_o)} - K_2e^{-K_1(t_1-t_o)}]a_o + (K_1-K_2)a_1 = K_1e^{-K_2(t_2-t_1)} - K_2e^{-K_1(t_2-t_1)} \end{aligned}\right\} \quad (8)$$

Ainsi dans le cadre de ce modèle peut-on estimer l'ensemble des paramètres du bilan y compris le coefficient de dégradation K_1 a partir de mesures systématiques d'oxygène à 3 stations ou si les variations spatiales d'oxygène sont petites, à partir de 3 observations successives d'oxygène à intervalles de temps constants sur une seule station.

Cependant si le biais peut être éliminé, les difficultés liées aux erreurs d'échantillonnage sur l'estimation des coefficients a_o et a_1 et par conséquent sur K_1 et K_2, demeurent. Si les intervalles de temps t_2-t_1 et t_1-t_o sont trop petits, les variances d'échantillonnage des estimations de a_o et a, seront généralement importantes. De plus dans ce cas le système (8) devient presque singulier ce qui tend à amplifier les erreurs d'échantillonnage de K_1 et K_2.

Diminution des Erreurs D'Echantillonnage

Pour diminuer les imprécisions d'estimation sur les coefficients, on peut augmenter les intervalles de temps pris en compte. Il y a cependant des limites à ces intervalles imposées par les temps de propagation hydrauliques dans le cas de la methode à 3 stations. Il faut surtout augmenter la taille des échantillons d'observations chronologiques d'oxygène dissous pris en compte, Hornberger et Kelly (1976) ont appliqué la méthode des moindres carrés sur des échantillons journaliers de données horaires en séparant d'ailleurs les valeurs diurnes et nocturnes utilisant ainsi des échantillons qui nous paraissent insuffisants dans le cas de rivières polluées. La prise en compte de données a l'échelle de plusieurs jours ou plusieurs semaines pose cependant une difficulté de fond. La validité d'un modèle d'oxygène ou les coefficients K_1 et K_2 seraient constants à cette échelle peut être mise en doute. En effet ces coefficients peuvent être soumis aux fluctuations de la vitesse moyenne d'écoulement, de la température de l'eau, etc... De plus l'occurrence de masses d'eau incluant certaines pollutions chimiques toxiques peut inhiber les réactions de biodégradation et même de réoxygènation introduisant une variabilité supplémentaire sur K_1 et K_2 par exemple.

L'estimation du modèle sur des périodes de temps plus longues doit laisser la possibilité d'une évolution aléatoire des coefficients K_1, K_2, α, R_o du modèle ou au moins des seuls coefficients K_1 et K_2 et des coefficients de régression a_o et a_1 qui leur sont liés. L'évolution de ces coefficients doit avoir une certaine structure temporelle. Dans la mesure où ils sont fonction de facteurs tels que la vitesse et la température de l'eau, facteurs très autocorrélés, on peut envisager une structure temporelle du type suivant:

$$\left.\begin{aligned} a_o(t_{j+1}) &= \lambda_o\ a_o(t_j) + z_o(t_j) \\ a_1(t_{j+1}) &= \lambda_1\ a_1(t_j) + z_1(t_j) \end{aligned}\right\} \qquad (9)$$

où les z_o et z_1 sont des variables aléatoires successives de variances V_o et V_1, indépendantes entre elles et avec les a_o et a_1 du même pas de temps. On est ramené au problème d'estimation des états de systèmes stochastiques dynamiques (cf. Aoki, 1967) dont la résolution dans le cas linéaire peut être demandée à la technique classique du filtre de Kalman. Dans le domaine des problèmes de qualité des eaux, l'utilisation de cette technique a été proposée récemment par

Koivo et Phillips (1976). Ces auteurs considèrent un système où les états à estimer sont les niveaux d'oxygène, de DBO à plusieurs stations ainsi que l'amplitude de la photodynthèse et un paramètre équivalent à notre R_o; ils supposent cependant connus exactement les paramètres de réoxygènation K_2 et de biodagradation K_1, ce qui peut apparaître irréaliste dans de nombreux contextes. Nous pensons préférable d'inverser le point de vue en supposant connus les observables qui sont les concentrations d'oxygène $C(x_o,t_{oj})$, $C(x_1,t_{1j})$ et $C(x_2,t_{2j})$ et en considérant comme états du système inconnus à estimer, les grandeurs qui règlent les variations du système entre chaque intervalle, soient les coefficients a_o, a_1, α, R_o, grandeurs dont certaines évoluent de facon aléatoire comme nous l'avons vu.

L'application de la technique du filtre de Kalman à ce problème est en cours de développement.

Conclusions

L'exemple du traitement de données d'oxygène dissous nous a permis d'évoquer certains problèmes soulevés par l'introduction d'un point de vue aléatoire dans les phénomènes d'autoépuration. Il n'est pas si aisé d'éliminer certains facteurs et paramètres en les considérant comme termes d'erreurs pour autant que ces erreurs sont corrélées avec des grandeurs prises en compte. Les données absentes ne peuvent être remplacées que par des hypothèses, par un modèle aléatoire dont le réalisme peut d'ailleurs être discuté; la validité des estimations des coefficients K_1 et K_2 notamment est conditionnée par ce réalisme. A cet égard l'utilisation du modèle de charge polluante aléatoire que nous avons donné suppose que le terme d'erreur dans le bilan d'oxygène dépend essentiellement des fluctuations de cette charge polluante. En fait ce terme d' erreur peut dépendre d'autres facteurs, comme par exemple la dispersion-diffusion hydraulique que ne prend pas en compte le modèle de convection pure (ce qui peut être important en étiage pour de faibles vitesses d'écoulement). La prise en compte de cette diffusion, estimée par ailleurs amènerait à modifier le modèle et à diminuer, mais sans l'annuler, la corrélation entre le terme d'erreur et le niveau d'oxygène $C(x_o,t_o)$.

References

[1] Aoki, M., 1967. "Optimisation of Stochastic Systems," Academic Press, New York, London, p. 354.

[2] Bernier, J., 1973. "Données Inadéquates et Modèles Mathématiques de la Pollution en Rivière"- Dans: Comite Expagnol pour la DHI: Colloque sur l'élaboration des projets d'utilisation des ressources en eau sans données suffisantes, Madrid, pp. 51-61

[3] Hornberger, G. M. and Kelly, M. G., 1976. "Atmospheric Reaeration in a River using Productivity Analysis," Journal of the Environmental Engineering Division, ASCE, EE5, pp. 729-739.

[4] Koivo, A. J. and Phillips, G., 1966. "Optimal Estimation of DO, BOD and Stream Parameters Using a Dynamic Discrete Time Model."

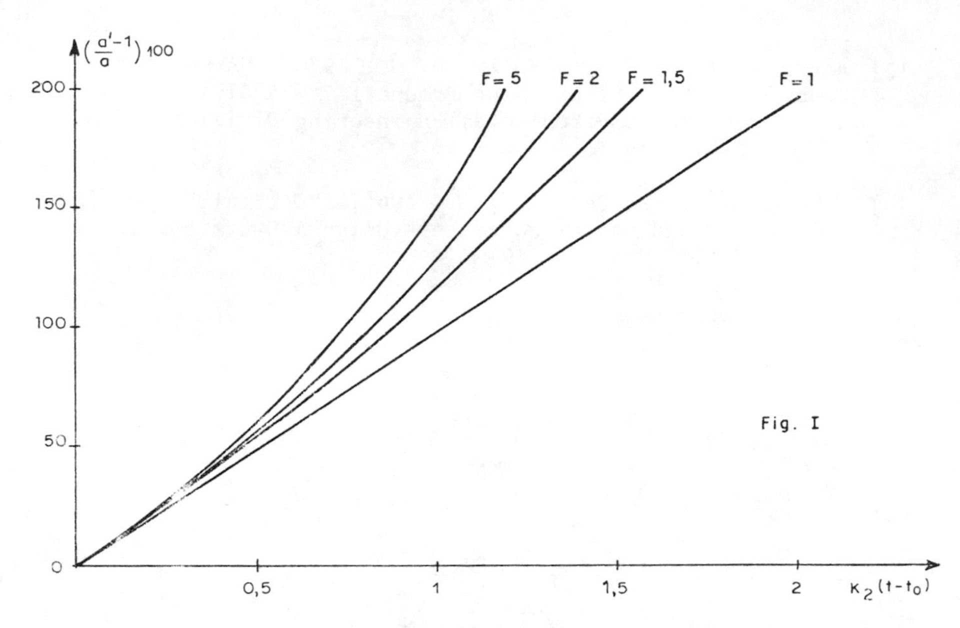

Fig. 1. Biais sur l'estimation de $e^{-K_2(t-t_o)}$
Bias on $e^{-K_2(t-t_o)}$ estimate

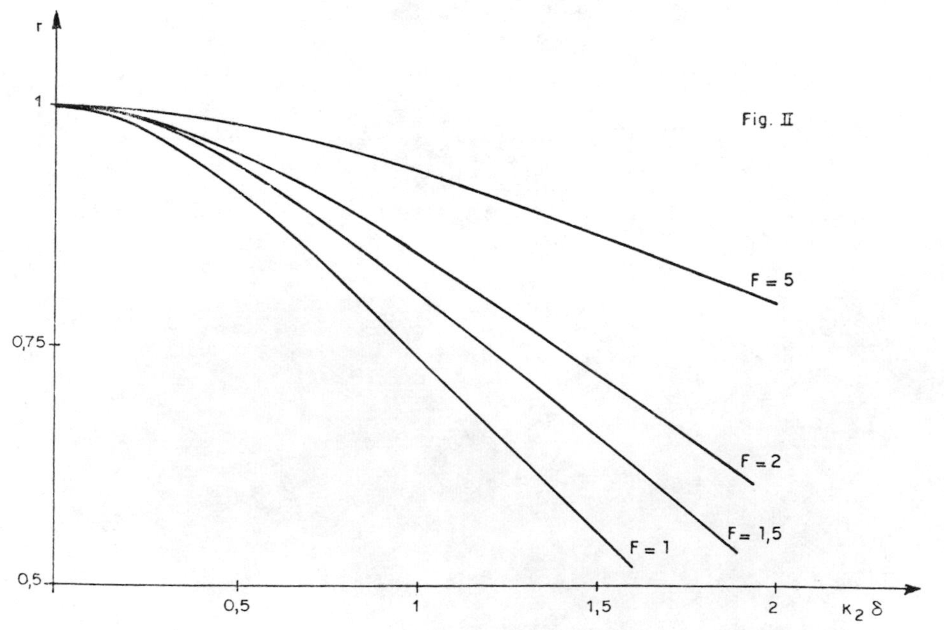

Fig. 2. Autocorrélation de l'oxygène dissous en fonction du pas de temps δ. Dissolved oxygen autocorrelation function of time step δ.

KINEMATIC PARAMETER ESTIMATION FOR TRANSPORT OF POLLUTANTS IN OVERLAND FLOW

By

L. E. Brazil, Water Resources Engineer
Harza Engineering Company
Chicago, Illinois

T. G. Sanders, Asst. Professor of Civil Engineering
Colorado State University, Fort Collins, Colorado

and

D. A. Woolhiser, Hydraulic Engineer,
Agricultural Research Service-USDA
Fort Collins, Colorado

Abstract

A water quality time of travel model was developed to simulate the movement of nonpoint source pollutants in overland flow. The model was derived from basic kinematic wave theory which has been used by recent researchers to describe shallow flow in open channels and flow from surface runoff. The basic kinematic wave theory was adapted to describe the movement of water particles. Because particle motion is analogous to the movement of a conservative soluble pollutant, theory which defines this motion may be an elemental key to understanding the movement of pollutant particles in surface runoff.

Experiments to test the model were conducted on the half-acre Rainfall-Runoff Experimental Facility at Colorado State University. The Facility's size is such that it is an intermediate step between laboratory models and natural watersheds. The experimental watershed is impermeable and can be subjected to a variety of rainfall intensities and duration; also surface roughness and watershed configuration can be controlled. Six hundred experiments were performed to verify the model's predictive capabilities. A rectangular plane 30 meters long and 2 meters wide was selected as the test section in the water quality studies. Rhodamine WT was used to simulate a soluble, conservative pollutant. This fluorescent dye was injected at several points on the watershed surface so that travel times of the dye in overland flow for various conditions could be measured. The experiments were conducted using four surface roughnesses and four different rainfall intensities. The roughnesses ranged from a smooth plane surface to a surface of rills and eroded channels. Other variables investigated included rainfall duration and the quantity and location of the tracer within the watershed. A statistical regression analysis was performed on the data and indicated that the travel times were mainly a function of rainfall intensity and location of tracer input. These two variables are the primary components in the theoretical model. A curve fitting technique was used on the observed data to obtain the best estimate in a least squares sense of the kinematic wave parameters. The theoretical model with these calculated parameters accurately predicted the observed tracer travel times.

Three separate techniques for quickly estimating the parameters of the kinematic wave equation were also developed. Each of these simple estimation techniques was tested using the observed data. The estimation method which assumed Chezy's relationship for turbulent flow proved quite precise and generated estimates of the kinematic parameters which could be used in the model to accurately predict travel times of the soluble tracer.

Results of the study serve as further verification of the validity of kinematic wave theory equations for overland flow and produce a model which could be utilized in the prediction of pollutant movement on watersheds and in the estimation of kinematic parameters. The research indicates that the theoretical kinematic equations with estimated parameters can predict the immediate impact of an instantaneous trace of pollutant carried in overland flow.

Introduction

Pollutants transported by storm water runoff from nonpoint sources are major contributors to the degradation of water quality in the nation's rivers. Studies have revealed that at certain times urban storm water runoff is a more serious threat to water quality than raw sewage discharges. During some storm events, runoff from street surfaces has been found to contribute significantly more BOD, (Biochemical Oxygen Demand) to area waters than the effluent from sewage treatment plants (1). Urban nonpoint pollution results primarily from debris and contaminants on the streets, contaminants from open land areas, publicly use chemicals, and dirt and contaminants washed from vehicles. Water soluble chemicals and pollutants have been found in urban storm water runoff at environmentally significant concentrations (2).

Much attention has been given in recent years to the development of models which can simulate storm water runoff from watershed surfaces. Theoretical equations have been developed to describe shallow flow in open channels and several adequately approximate overland flow. Although these models have been tested for their ability to describe unsteady, free-surface flow, none have been adequately tested as chemical transport models. A number of models have been used to simulate the accumulation and removal of insoluble and partially soluble nonpoint source pollutants (3,4,5). None of these models, however, have components which can simulate the transport of soluble pollutants. The purpose of this research was to develop a transport model for conservative soluble pollutants on an impervious plane and to test it with data from carefully controlled experiments.

Kinematic Wave Theory

One set of equations used to describe overland flow is based on kinematic wave theory. First introduced by Lighthill and Whitham in 1955 (6), the theory has been applied to surface runoff by Henderson and Wooding (7), Wooding (8), and Woolhiser (9).

Derivation of the kinematic equations is based upon the governing equations for unsteady free-surface flow introduced by DeSaint Venant (10). The equation for continuity on a plane surface is:

$$\frac{\partial h}{\partial t} + \frac{\partial (uh)}{\partial x} = q \qquad (1)$$

and the simplified momentum equation is:

$$S_o = S_f \qquad (2)$$

where: h = depth of flow,
q = lateral inflow rate,
x = distance from upstream boundary,
u = local average velocity,
t = time,
S_o = bed slope,
S_f = friction slope.

Substitution of the Darcy-Weisbach equation or the Chezy equation into Eq. (2) will produce the following parametric relationship for depth of flow and velocity:

$$u = \alpha h^{m-1} \qquad (3)$$

where: α = roughness and slope parameter,
m = flow regime parameter.

The parameter m has a value of approximately 1.5 for turbulent flow. Equation (3) has been used by researchers as the basic relationship in numerous watershed models (11,12). Models by Harley et al. (13) and Schaake (14) have demonstrated the applicability of the kinematic wave theory in prediction of runoff hydrographs.

Formulation of Theoretical Model

The water quality model presented herein is derived to predict the convective movement of soluble, conservative pollutants in overland flow. Dispersion is ignored. The basic prediction model is derived by Brazil (15) from Eq. (3). The water particle velocities are represented by:

$$u = \left(\frac{dx}{dt}\right)_p \tag{4}$$

Consider spatially uniform rainfall at a constant rate i beginning at time 0, falling on an impervious surface. Prior to the time of equilibrium (within the shaded zone in Fig. 1 the theoretical depth of the water (h'), is a function of the intensity of the rainfall (i) and the time (t)

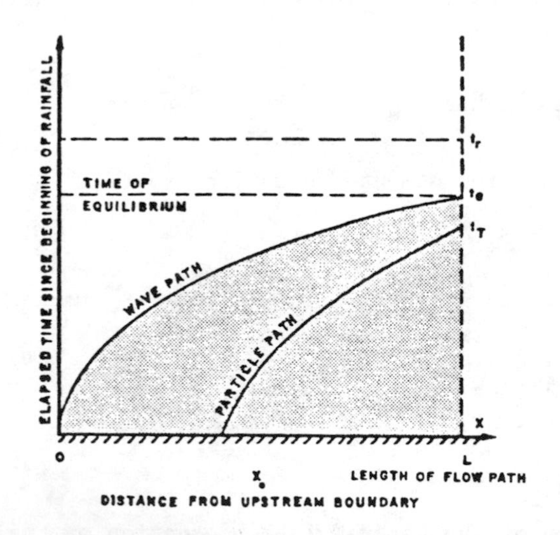

Fig. 1. Characteristic wave and particle paths in the x-t plane

since the start of rainfall input. If the surface is initially dry, water depth at time, t, after rainfall begins is given by:

$$h' = i\,t \tag{5}$$

where: h' = theoretical depth,
i = rainfall intensity.

The theoretical depth (h') will vary from the actual depth (h) by a factor determined from the concentration of objects blocking flow paths on the flow plane. For most practical situations, however, the use of theoretical depths causes no problems in model applications. Substituting Eqs. (4) and (5) into Eq. (3) yields:

$$\left(\frac{dx}{dt}\right)_p = \alpha(it)^{m-1} \tag{6}$$

Integration and rearranging terms gives the time required for a particle starting at (x_o,t) to reach the lower boundary (x = L).

$$t_T = \left(\frac{m(L-x_o)}{\alpha(i)^{m-1}}\right)^{1/m} \tag{7}$$

where: t_T = travel time (seconds),
i = intensity of rainfall (meters/second),
L = length of plane (meters),
x_o = distance from top of plane (meters).

Eq. (7) is applicable only on the rising limb of the hydrograph. After equilibrium is reached on the watershed, the depth no longer increases with time and the flow is considered to be steady and nonuniform. Therefore, it is necessary to compare the observed and predicted t_T with time to equilibrium (t_e) and with the time at which rainfall stops (t_r) for partial equilibrium hydrographs. Time to equilibrium is dependent upon the characteristic wave velocities, which can be determined with an expression derived from the continuity equation and Eq. (6). Substituting Eq. (3) into Eq. (1) yields:

$$\frac{\partial h}{\partial t} + \alpha m h^{m-1} \frac{\partial h}{\partial x} = q \tag{8}$$

Solving Eq. (8) and the total differential of $h(x,t)$ in matrix form yields a solution to the characteristic velocity in the x-t plane:

$$\left(\frac{dx}{dt}\right)_w = \alpha m h^{m-1} \tag{9}$$

Eq. (9) represents the velocity of the characteristic wave down the plane. The notation $(dx/dt)_w$ is used to denote the wave velocity as opposed to the particle velocity $(dx/dt)_p$. Note that the wave velocity is m times greater than the particle velocity. The characteristic paths for both the wave and particle velocities are shown in Fig. 1.

By substituting Eq. (5) into Eq. (9) and integrating we obtain an expression for time to equilibrium, t_e.

$$t_e = \left(\frac{L}{\alpha(i)^{m-1}}\right)^{1/m} \tag{10}$$

It should be noted that for given values of i and L the time to equilibrium is determined by the kinematic parameters α and m. A similar expression can be derived for the time at which recession begins for a partial equilibrium hydrograph.

Verification of Theoretical Model

An experimental procedure was designed to collect data which could test the model's accuracy in predicting time of travel for soluble, conservative pollutants. Because water particle movement was investigated using a fluorescent dye, Rhodamine WT, its travel time across a watershed could be accurately recorded using a fluorometer.

Experimental facility. The Rainfall-Runoff Experimental Facility at Colorado State University was used as the model watershed. An impervious area of approximately 25,000 square feet, the facility simulates a watershed without infiltration, intermediate in size between laboratory models and natural watersheds. The general arrangement of the Rainfall-Runoff Facility is shown in Fig. 2. Characteristics such as surface roughness, imperviousness, and geometry can be changed to represent a wide variety of natural catchments. In addition, simulated rainfall can be generated at rates of 13, 28, 58, and 108 millimeters per hour. Holland (16) gives further details of the facility.

To simplify the conditions on the watershed, one plane section of the surface, two meters wide by thirty meters long, was partitioned from the rest of the facility for use in the experimental runs. Like the plane section of the facility, the experimental section was oriented so as to have a five percent slope. Discharge was measured through a 0.6 foot HS flume and the concentration of the fluorescent dye in the water was analyzed with a continuous flow Turner Model 111 fluorometer.

Experimental procedure. Equilibrium experimental runs were conducted with the dye injected at a point or a transverse line on the plane surface after the system had reached a steady state; for nonequilibrium runs the dye was injected

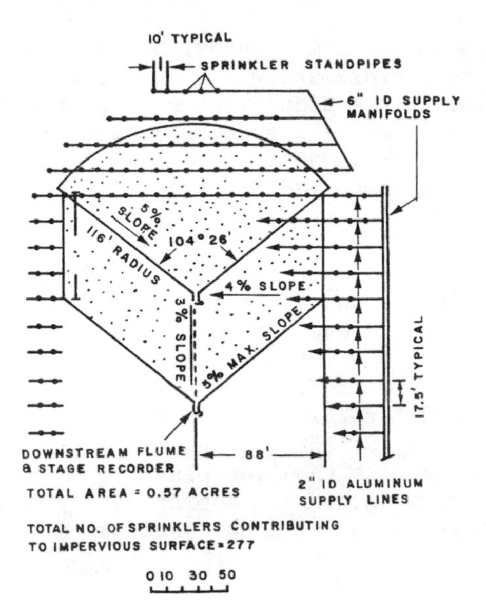

Fig. 2. Experimental Rainfall-Runoff Facility-general arrangement

simultaneous with the start of rain. Thus, for the nonequilibrium runs the dye represented an immediately soluble pollutant already on the surface at the beginning of a rainstorm. During each run, a stopwatch was used to determine the length of time the dye was actually traveling on the exposed plane. This time was equal to the time interval between tracer injection and peak concentration readout minus the time the tracer was pumped and transmitted to the fluorometer.

Variables investigated included surface roughness, rainfall intensity, rainfall duration, and the quantity and location of tracer input. Four surface roughnesses were included in the tests. These were:

Set A - watershed covered with butyl rubber.
Set B - 2 kg/m^2 of 1½-in. dia. gravel spread on the butyl surface.
Set C - 10 kg/m^2 of 1½-in. dia. gravel spread on the butyl surface.
Set D - artificial rilled surface (a smooth surface of a wet sand and cement mixture was eroded by simulated rainfall and allowed to harden for two days).

All four of the available rainfall intensities were utilized for each set of runs. Rainfall was applied for three durations: approximately one-half the time before the watershed runoff reaches equilibrium, a time approximately equal to equilibrium time, and a time approximately 1.3 to 1.5 greater than the equilibrium time. Injections were made at two locations: distances of six meters and fifteen meters from the upper boundary. The four roughnesses, four intensities, three durations, and two locations made a total of ninety-six combinations of nonequilibrium travel time runs.

A few experiments were also conducted to collect data to examine the variables under conditions different from the normal travel time runs. One series included tracing the dye down the plane with the Set D roughness while the intensity of the rainfall was varied. These runs allowed the prediction model to be tested for rainfall patterns more characteristic of nature. Another set of runs was made with the dye injected as a line source across the flow plane to examine the effects of nonuniform cross sections. Approximately six hundred runs were made to test the model under various conditions.

Results

Statistical analysis. A statistical regression analysis was performed on the travel time data to make sure it was predictable and could be correlated with the variables in the model. The variables examined included intensity, initial tracer location, rainfall duration, time of concentration, peak time, amount of dye injected, and dye concentration. The analysis indicated that the most important variables were rainfall intensity and the injection location and that the correlation coefficient was 0.90 or greater for combinations of these two variables alone. The theoretical model, Eq. (7) uses both intensity and injection location as input variables.

Theoretical analysis. The theoretical analysis was divided into three phases. The first phase of the study used equilibrium run data for finding the best estimate of the kinematic parameters α and m. Next, the travel time data for nonequilibrium runs were compared with the predicted travel times using the best estimates of α and m in Eq. (7) and finally, an analysis was conducted to see how well the model could predict travel times when rainfall intensity was varied during a single run.

For equilibrium runs the discharge at any unit width of cross section was only a function of rainfall intensity and distance from the top of the plane. The plane was divided into several reaches and during subsequent runs dye injections were made at a different section to determine the average velocity for each section. This information was used along with equilibrium discharge quantity for each section to calculate average depth of flow, the principle variable in Eq. (3). An expression for discharge per unit width, Q, can be derived by substituting h' into Eq. (3) and by multiplying through by h'. The new equation is:

$$Q = \alpha(h')^m \tag{11}$$

The log form of this equation plots as a straight line on log-log paper with a dependent variable intercept of log α and a slope of m. Steady state data were taken for each surface roughness and showed that Q was highly correlated to h'. In every case the slope of the line, m, was approximately 1.5 indicating that the flow was turbulent. Equation (7) indicates that travel time is inversely proportional to α and that in general a rougher surface will have a smaller value of α. An analysis of the best estimates of α and m from the steady-state runs showed that because of the intercorrelation of α and m, and the fluctuation that can exist in the actual value of m, the roughest surface will not always have the smallest value of α.

Best estimates of α and m obtained from steady-state tracer studies were used in the model, Eq. (7), to predict travel times for the ninety-six combinations of input variables for unsteady flow. The model equations and observed data points were plotted on log-log paper with travel times as a function of rainfall intensity and location of injection. Figure 3 shows a graphical representation of the runs for Set C. A mean error analysis of the predicted travel times was determined using a sum of the errors square criterion for the observed data. The mean error for all four sets was 21 seconds using the best estimate of α and m from the steady state runs. The observed data were generally within 10% of the predicted times and shorter travel times had lower deviations so that the percentage difference between the predicted and observed times was fairly consistent.

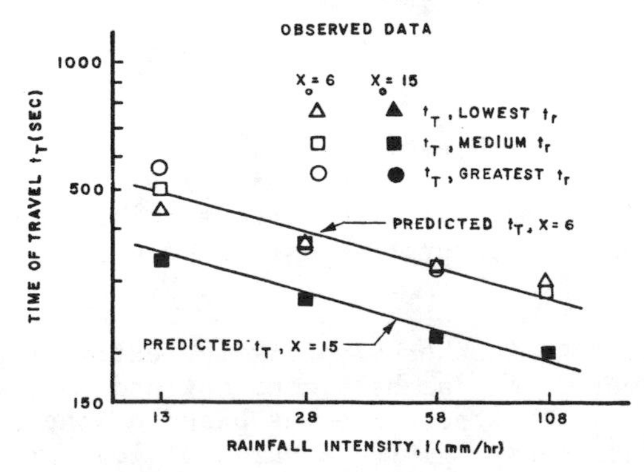

Fig. 3. Predicted and observed travel times versus rainfall intensities for Set C

For variable rainfall intensity the water depth within the shaded area on Fig. 1 is given by

$$R = i_1 t_1 + i_2 t_2 + \ldots \tag{12}$$

where t_1 is the duration of the rainfall of intensity i_1 and t_2 is the duration of the rainfall of intensity i_2, etc. Substitution of this expression into Eqs. (3) and (4) gives an expression which can be used to route the water particles down the plane. The mean error of the predicted travel times using the observed data from the variable intensity runs was 16 seconds. Equation (7) was also used to predict times with the average intensity calculated from total rainfall accumulation used as the input. The mean error for these travel times was 27 seconds, 11 seconds higher than the times using the routing technique.

Equation (7) was also used to predict time for the pollutants injected as a transverse line source. The mean error was 13 seconds, indicating that the model was even more accurate for the line source data examined than for the point source data.

Estimation techniques for kinematic parameters. Several techniques were examined to utilize tracer data to estimate the parameters in the kinematic equation. For injections of dye made at different distances, x_1 and x_2 from the top of the watershed but having equal rainfall intensities the travel times to the lower boundary will be t_1 and t_2 respectively. Substituting these values simultaneously into Eq. (7) yields

$$m = \frac{\ln(L-x_1) - \ln(L-x_2)}{(\ln t_1 - \ln t_2)} \tag{13}$$

$$\alpha = \frac{(L-x_1)m}{i^{m-1}\ t_1^m} \tag{14}$$

The prediction model with estimated values of α and m was tested against data from the ninety-six travel time runs. A mean error of 28 seconds was calculated showing that the pair of parameters could be accurately estimated with a small amount of data.

In an attempt to further simplify the estimating technique, the parameter m was held constant and Eq. (14) was solved to give α. The technique was based on the Chezy turbulent flow relationship, giving a value of 1.5 for m. This method requires only one dye injection and observed travel time to solve for α. The model equation with m equal to 1.5 and α estimated from the simplified technique gave predicted travel times with a mean error of 20 seconds. For the four surface roughnesses, Set A, B, C, and D, and assuming m=1.5; the values of α were 8.62, 4.71, 1.68 and 4.03 respectively. The method assuming a constant m value proved more accurate for the observed data than the technique which allowed both parameters to vary.

Two estimating methods were also tested with m assumed to be 1.67, the value obtained from Manning's equation. The model with m = 1.67, had a greater mean error indicating that Chezy relationship better represented the data. The one injection parameter estimation techniques were also used with variable intensity rainfall. Results were similar to the runs with constant intensity.

A nomograph was prepared to quickly estimate α with data from one observed travel time run and is shown in Fig. 4. The graph displays a solution to Eq. (14) with input variables of travel time, length of flow plane, and rainfall intensity. Because the model is designed to predict only time for flows on the rising limb of the hydrograph, dye injection's must be made at a location within the watershed in order that the kinematic wave from the farthest boundary reaches the bottom

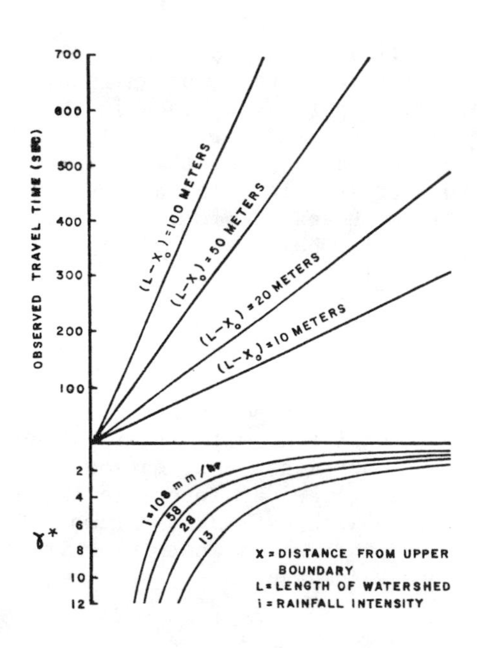

Fig. 4. Nomograph for predicting α, m = 1.5

of the watershed after the dye concentration peak arrives. To insure that this condition is met, the dye must be injected at a distance greater than $L(m-1/m)$ from the top of the plane (15). For $m = 1.5$, the injection must be made within the lower two-thirds of the watershed. The six meter injection point violates this condition but an analysis showed that the expected deviations are small (15).

Applications of the theoretical model. Numerous papers have been written on the value of kinematic wave theory in synthesizing hydrographs and on methods of calculating the kinematic parameters. The parameter estimation techniques described herein can be useful for selecting parameter values for the hydrograph equations as well as for the pollutant travel time model.

The correlation of pollutant strength and sediment combined with the simulation of sediment transport in a watershed has been recently proposed as a methodology for assessing the movement of insoluble and partially soluble nonpoint pollutants (5). Curtis has presented a model that uses the kinematic wave formulation and a set of relationships describing soil detachment and transport processes to simulate the discharge of sediment from an urban area (17). The experimental evidence presented herein suggests that the kinematic model is appropriate for pollutant transport as well as runoff hydrograph prediction.

One specific application of the model would be its use in the prediction of salt transport in urban runoff. Water soluble de-icing salts are applied to roads in the United States during the winter when snow and ice accumulate. The use of road salts has been increasing over the past three decades and has been linked to increases in chloride content of streams (18). The kinematic wave and chemical transport equations in finite difference form may be a useful tool in further understanding the processes involved in salt transport.

Summary

A water quality time of travel model was developed based on the assumption that soluble pollutants in overland flow travel at velocities consistent with the relationships governed by the kinematic wave equations. Estimation techniques were presented for calculating the parameters in the model. The model equation was used with estimated parameters to predict pollutant travel times and was shown to be quite accurate when compared with the observed data.

Conclusions

This study further verifies the validity of the kinematic wave equations for overland flow and also shows that the kinematic wave equations can be used to describe convective transport of pollutants. Data from the analyses indicate that the kinematic form for turbulent flow derived using Chezy's relationship is valid and that the presented model equations can be useful tools in the estimation of kinematic parameters as well as to predict movement of soluble pollutants on planes and, by inference, watersheds represented as networks of planes and channels.

References

[1] URS Research Co., 1974. "Water Quality Management Planning for Urban Runoff," PB-24 689, U.S. Environmental Protection Agency.

[2] American Public Works Association, 1969. "Water Pollution Aspects of Urban Runoff," U.S. Department of the Interior, Federal Water Pollution Control Administration WP-20-15.

[3] Metcalf and Eddy, Inc., University of Florida, and Water Resources Engineers, Inc., 1971. "Storm Water Management Model," Water Pollution Control Research Series, 11024 DOC, 1-4, U.S. Environmental Protection Agency.

[4] Sutherland, R. C. and McCuen, 1976. "A Mathematical Model for Estimating Pollution Loadings in Runoff from Urban Streets," Mathematical Models for Environmental Problems, Pentech Press, London, pp. 283-295.

[5] Donigian, A. S., Jr. and Crawford, N. H., 1976. "Modeling Nonpoint Pollution from the Land Surface," Ecological Research Series, EPA-60013-76-083, U.S. Environmental Protection Agency.

[6] Lighthill, M. J. and Whitham, G. B., 1955. "Kinematic Waves, I," Proc. Royal Society of London, A. (228), 281-316.

[7] Henderson, F. M. and Wooding, R. A., 1964. "Overland Flow and Groundwater Flow from a Steady Rainfall of Finite Duration," Journal of Geophysical Research, 69(8), pp. 1531-1540.

[8] Wooding. R. A., 1965. "A Hydraulic Model for the Catchment-Stream Problem I. Kinematic Wave Theory," J. Hydrol. 3(3/4), pp. 254-267. "II. Numerical Solution," J. Hydrol. 3(3/4), pp. 269-282 (1965). "III. Comparison with Runoff Observations," J. Hydrol. 4(1), pp. 21-37 (1966).

[9] Woolhiser, D. A., 1969. "Overland Flow on a Converging Surface," Trans. of the ASAE, 12(4), pp. 460-462.

[10] Chow, V. T., 1959. "Open-Channel Hydraulics," McGraw-Hill, New York.

[11] Kibler, D. F. and Woolhiser, D. A., 1970. "The Kinematic Cascade as a Hydrologic Model," Hydrology Papers, No. 39, Colorado State University, Fort Collins, Colorado.

[12] Sherman, B. and Singh, V. P., 1976. "A Distributed Converging Overland Flow Model," Water Resources Research, 12(5), pp. 889-908.

[13] Harley, B. M., Perkins, F. E., and Eagleson, P. S., 1970. "A Modular Distributed Model of Catchment Dynamics," Hydrodynamics Lab, Report 133, MIT.

[14] Schaake, J. C., 1970. "Deterministic Urban Runoff Model," Institute on Urban Water Systems, Fort Collins, Colorado.

[15] Brazil, L. E., 1976. "A Water Quality Model of Overland Flow," M.S. Thesis, Colorado State University, Fort Collins, Colorado, 141 p.

[16] Holland, M. E., 1969. "Design and Testing of Rainfall System," Colorado State University, Department of Civil Engineering Report CER69-70MEH71.

[17] Curtis, D. C., 1976. "A Deterministic Urban Stream Water and Sediment Discharge Model," National Symposium on Urban Hydrology, Hydraulics, and Sediment Control, Univ. of Kentucky, Lexington.

[18] Wulkowicz, G. M. and Saleem, Z. A., 1974. "Chloride Balance of an Urban Basin in the Chicago Area," Water Resources Research, 10(5), pp. 974-981.

WATER QUALITY AND AVAILABILITY IN UTTAR PRADESH (INDIA)

By

Abinash Chandra Chaturvedi
Sinchai Buawan, Canal Colony
Lucknow-226001, India

Abstract

The rapidly expanding population of Uttar Pradesh is expected to double by end of the century. The water consumption is expected to go up from 50 million acre feet today to 120 million acre feet by end of the century. Water quality changes from region to region and is definitely related to the long period history and long range hydrological changes. Fresh sands has been deposited by streams in floods. Plain areas only have the potentiality of ground water. There is very little percolation in rocky areas, and it varies from 50% to 100%. Chlorides have been found in the depth zone of 3.75 to 31.14 metres and chloride concentration is between 250 and 850 ppm. Water from zone of saturation have less feri genus material deposited in layers of zeolite/clay. Iron deposits have been located with the zone of aeration and zone of saturation. The coefficient of transmissivity is computed from the measurement of the residual head when the well is instantaneously charged with water. Optimum depth of drilling has been worked out where transmissivity is no longer perceptible. The degree of turbidity and the amount of suspended matter in the spring and wells indicate the precipitation in the intake area.

Introduction

The State of Uttar Pradesh has a population of 925 million which is expected to double by end of the century. At present, the consumption of water is 50 million acre feet and is expected to be 120 million acre feet by the year 2005 A.D. and 150 M A ft by 2025 A.D. There are various changes in water quality as we move from region to region. This has a definite relationship with the availability of water and the long period history. This brings us to the prediction and detection of long range hydrological changes.

Geology

Uttar Pradesh, with the exception of the Hill region and some parts of Bundelkhand, is fortunate in the matter of water resources. The prosperity of this state depends primarily on easy water availability. The economy is still predominantly agricultural. With the introduction of high yielding varieties of seeds, the necessity of irrigation at short intervals of time is felt all the more keenly. Three relief regions stand out clearly from north to south: (1) the snow capped Himalayas, (2) the Gangetic plains, and (3) the Vindhyan hills and plateau. The hills cover an area of 50,000 sq. kms. of which only one-fifth is covered with crops. This mountainous tract is highly dissected and rugged, and is generally overlain by a very thin soil cover. The hills and plateau in the southern part of the state form part of the foreland of the Deccan peninsula. The northern boundary of the foreland is broadly formed by the Rivers Ganga and Yamuna.

It is generally very irregular because of the protrusion of the Ganges alluvion. The low rounded hills extending east-west are remnants of the old table land which have withstood weathering and denudation. The altitude generally does not exceed 300 m and rarely 450 m. The Ganga plains cover almost three-fourths of the entire state. The plains are homogeneous and the general slope of the land is imperceptible. The only distinction in physiographic variation is between the upland older alluvion of the plains and the fingers of newer alluvium of fine silt along the main streams and their tributaries. There are some micro-regional differences of slope and aspect. At places, there is development of sandy soil especially along the east bank in the Bijnore and Moradabad districts. The great detrital piedmont skisting the Siwalik hills where the stream profiles suddenly flatten out and the coarser detrital-boulders and gravel is deposited. In this tract, 30 km wide in the west and narrowing eastwards, the smaller streams are lost in loose talus, except when in spate during the rains. They seep out again where the slope is still flatter and the finer material is deposited in the marshy and jungle lands below.

The Himalayas are young folded mountains, uplifted in the early tertiary period in a series of tectonic movements. The tract of the Ganges plain is the youngest of the three divisions and is characterised by an immense thickness of alluvium derived from the Himalayas. The Vindhyan Hills and plateau division in the south is the oldest and the most stable land mass. It is composed of horizontal rock beds.

Tubewells and Other Wells

The state has 20^{0} thousand tubewells and 90^{0} thousand other wells (masonry and non-masonry) which account for more than half of the net irrigated area in the state. Their distribution in different parts of the state is very uneven. The Ganga valley has far more numerous wells than the hills and the plateau in the South. Even in the Ganga valley, the distribution is far from being uniform. The western and eastern parts have greater concentration than the eastern and central parts. There is great concentration in the middle Ganga-Yamuna plains. The Himalayan rivers not only depend on monsoons but also get plentiful supply of 80 m s ft from the melting of the Himalayan snows. The peninsula rivers entirely depend on monsoon rainfall and almost go dry in the hot weather. These rivers are therefore less useful for irrigation than those of the north.

Soils

The state extends over an area of 294 thousand sq kms. The soils tend to become heavier from the northwest to the southeast. In the districts of Agra, Mathura, Aligarh and Meerut, where the conditions are semi-arid, patches of alkaline soils, as well as alluvium covered by wind borne sand, are quite common. In eastern regions, the areas are low lying and subject to damage by floods. In the high lying parts, the soil conditions of old alluvium, made up of yellowish clay with frequent deposits of hard clay, show marks of continued denudation by streams rushing down from the Himalayas to join the Ganga in the South.

In some places, fresh sand has been deposited by these streams when in flood stages. The soils are classified as clay, loam, and sand. A light loam spreads over large areas whereas the stiffest cultivable clays are suitable for only inferior rice. In the hilly regions, only plain areas have the potentiality of ground water in the districts of Nainital and Dehradun. The slopes are steep in this region and major precipitation goes as runoff with only ten percent of it percolating down to the groundwater. In the southeast Bundelkhand region, the percolation of rain water has been taken only in cultivated area as most of the area is rocky and there is no percolation there. These percentages are 50% for Jhansi, 100% for Jalaun, 60% for Hamirpur and 60% for the Banda district.

Chemical Quality of Water

The quantity of groundwater available and its quality both offer challenge to planning. The central Ground Water Board and the state authorities have tested the quality of groundwater and analysed the results. The chemical quality maps of groundwater have been made which have established an integrated picture of the spatial distribution of chlorides in the groundwater at depths of 3.75 to 31.14 metres tapped by the observation wells. The concentration of chlorides in almost three-fourths of the state does not exceed 250 ppm, which is well within the permissible limits. The remaining area shows chloride concentration between 250 and 850 ppm.

A couple of localized patches in the districts have chloride contents of 1025 ppm. Flourides are also found in scattered pockets of the state. In Unnao, near Lucknow, water content caused several cases of paralysis, several waters having PH less than 4.5.

These waters have high concentration of CO_2. This is expressed as an equivalent amount of calcium carbonate. Alkalinity is due primarily to the presence of hydroxide, bicarbonate and carbonate and is the ability of water to neutralize acid. The presence of borate, phosphate, silicate and other ion constituents imports additional alkalinity to the water and are of significance in certain areas. Alkalinity is expressed as an equivalent amount of calcium carbonate. Excessive alkalinity is undesirable in water for some uses, mainly because of its association with waters having excessive hardness or high concentration of sodium salts.

Groundwater in Mineral Areas

A large number of wells have been bored in these areas. The water has an iron content up to 10 ppm which is beyond the permissible limit of 1-3 ppm. The yield of these tubewells is reduced by 1/2 to 1/3 within a period of 5-7 years. The groundwater in these reaches is a complex chemical solution. It is extremely difficult to determine the state of chemical equilibrium at various stratas. The quality of groundwater is affected by the solution effects, cationic exchange reactions and precipitation of chemicals. In the outcrop zone, the solution effect is very prominent and circulation of water is very quick. When the groundwater moves down below the overlying, relatively impermeable argillaceous beds into those parts of the aquifer which have slower and restricted circulation, the base or cation exchange takes place in mineral soils. These soils absorb ions and exchange them with others in the water. Clay minerals, humic acids and zeolites are the most common natural cation exchangers. These ion exchange reactions involve replacement of calcium and magnesium by sodium when waters move below in certain argillaceous covers into regions of restricted circulation.

Precipitation reactions are associated with changes in temperature and pressure and take place particularly in deep strata. Here velocity changes in water result in deposition or precipitation of nitrous compounds.

Zone of Saturation

Water from the zone of saturation has less ferrugenious material deposited in layers of zeolite and clay. The zone of saturation is below the zone of aeration. Groundwater movement in this zone is governed by hydrostatic pressure.

Its lateral component is more important. The velocity is less and the water is in contact with the soil cations for longer periods. Testing of soil samples at various depths, especially in the interface of the zone of aeration and zone of saturation, helped to locate the iron deposits.

Coefficient of Transmissivity

In aquifers, the coefficient of transmissivity is computed from the measurement of the residual head when the well is instantaneously charged with water. Wells in hard rock areas of Vinhyaachal in South and Southeastern U.P. yield water having poor transmissivity. The coefficient of transmissivity is measured by slug tests in a number of cases and compared with pump test data. We have to arrive at an optimum depth of drilling where an increase in the value of T is no longer perceptible. The thickness of the weathered portion is about 10 m and is mainly composed of clay. The permeability of the weathered portion is low due to the high percentage of clay, although these reaches possess intergranular porosity. The residual head after the injection of the slug is measured by the point source heat flow equation and the recovery equation. In the Feris method of analysis, the residual head is plated against time.

Generally the dispersion of the slug is very fast. In two minutes, 95% dispersion takes place. The rate of increase in transmissivity and of yield with depth is obtained by tests at different depths. The average value for transmissivity obtained by the Feris method is 35 m^3/day for each kilometre length of weathered and fractured granite gniess with a hydraulic gradient of 1m per kilometre. Where the gradients are greater, especially in undulating terrain, greater groundwater flow is indicated. The operational yields of wells are worked out from the graphs. The relationship is not linear for high drawdowns. The theoretical head is computed by the extrapolation of the time versus residual head curves.

Groundwater Potential

Complete evaluation of the groundwater resources requires a comprehensive study of the balance between the total water and gains and losses of the basin for a period of time. For unconfined aquifers, pump test data were analysed by the methods of Theis, Jacob, Chow and recovery. The coefficient of transmissivity, T varied from 182 cm^2/S to 215 cm^2/S and the specific yield, S varies from 132 to 168. To select a value from the above, strata logs of representative wells in the region are examined and fine to medium sand is found between clay and hard clay. As the aquifers are unconfined and interconnected, the clay layers tend to reduce the specific yield value. The lowest value of S should be adopted and T, by the Jacobs method, works out to be 215 cm^2/S.

Groundwater reservoirs have an important role in the hydrological cycle. Excessive withdrawals increase the draft with loss of water from storage. To study the groundwater movement, maximum hydraulic gradient at different periods at definite locations are computed. The water table rises with heavy rainfall and goes down in periods of low rainfall. Drainage has an important effect on water table fluctuations. Where drainage is restricted, the water table rises with rain or excess irrigation. Underground flow also influences the water table. The change in groundwater storage is the precipitation minus the total of stream flow, evapotranspiration, subsurface flow and change in soil moisture. This is also determined by multiplying the change in mean groundwater stage by the gravity yield. The value of groundwater storage multiplied by the area gives the overall change in groundwater storage during each month. There was found wide variation in monthly changes of groundwater storage. It is necessary to monitor hydrological network stations in a state programme and collect seasonal and annual data on water levels. Such data should be published periodically in groundwater bulletins, records, popular and semi-popular journals, and made available to hydrogeology and allied disciplines.

Quality for Agriculture

Water is called C_1, and S_1 if it has low salinity and low alkalinity. When the salinity is below 250 micros per cm at 25° C, it is classified as C_1 or low salinity water. The sodium absorption ratio is used as an index of the alkali hazard of irrigation water. As conductivity goes up, S A R must come down. Water has been classified in 20 groups of C_1 to C_5 and S_1 to S_4. The U.S. diagram is generally used for classifying irrigation waters. Water is suitable in the C_1 S_1 and C_2 S_2 groups.

Run of Models

The time of travel is a probabilistic variable. Isotopic techniques are used to determine the movement of water particles. The catchment is divided into different areas. The hydrologic system or water balance work on the wandering models. State space differentiates the state of various particles, concluding in the transition probability matrix. The interval of travel at the end of the section for a unit amount of water is chosen rather small, so that the time of arrival at the end of a section for a unit of water is exponentially distributed. Knowing the maximum velocity and the length of reach, the minimum time of travel is calculated. The faster water particle reaches the end of the section in a given time. The complete discharge in a section is obtained by summing up the sub-discharges. Time invariant models can be obtained by non-stationary transition probability matrices. There is complete harmony of stochastic and deterministic (physical) models. Space invariance, subsurface runoff and evaporation are considered as well by the model.

Basin Plans

The basin plan or management document identifies the water quality problems of a particular basin and the effective program has been worked out on the model. The broad water land resources plan and the basinwise facilities plan have been tested on the model. Our state water quality management basin plan differs from facilities and area management plan as the latter is limited in geographical area under local jurisdiction. In present practice, 30% to 40% of the water delivered is utilized for irrigation and 50% of it is utilized by plants.

Specific water consumption reduction by 25% to 30% by use of complex technical, agrotechnical and plant breeding measures is tested by system analysis. Return water use methods have been used with advantage especially for rice irrigation. Reduction of percolation losses in rice fields has been considered reduced by model tests by subsoil compaction, bentonite application, clay soil in corporation, intensive puddling and by application of tank silts.

Information Service

We have a water resources information service to define the system development; and fully automated water surveys supported by the most dependable and accurate instrumentation. Measurements are made on a large number of points. The accurate location of these recording stations is equally important and ensures a reliable plot of the parameters. Innumerable difficulties have been experienced in water resources use. Half of the water of the minor and medium rivers could not be harnessed and is inevitably wasted. The amount of water carried in the rivers and the underground is nearly two-thirds of the annual precipitation.

Conclusion

Groundwater resources have been worked out on the basis of the study of the balance between the total water and gains and losses in a particular period. Pump test data is analyzed for unconfined aquifers. In such aquifers, the lowest value of transmissivity adopted is 215 cm^2/S. The quality of water in limestones terrains has been studied and the heavy degree of turbidity has been indicated and an increased rate of discharge. Extended groundwater circulation under artesian conditions has indicated water free of mechanically carried material. Clear water has indicated that increase in recharge has not followed the increased recharge. There is high infiltration in limestone areas and high bacterial contamination arises from soil and surface waste. This contamination is not high in regions of naturally filtering medium in large storages of underground water. Higher mineralized water indicates the great depth of water in Palezoic rocks including water bearing sandstones and limestones of unproductive shales. Much water in springs and wells is found in the tertiary volcanic rocks of the northern Himalayan regions. Salt grass shows the availability of water within 3.5 metres of the surface in the Terai Regions (low lying patches). Luxurious growth of woods has indicated water availability to depths of 6 to 8.5 metres. Structural models of hydrologic processes have indicated the presence of water and its duration in a large number of cases.

References

[1] Chaturvedi, A. C., 1976. "Irrigation Possibilities in Bundelkhand," Daily Navjiwan.

[2] Chaturvedi, A. C., 1975. "Instrumentation in Irrigation," National Symposium on Instrumentation and Automation at Allahabad by Institution of Engineers, India.

[3] Chaturvedi, A. C., 1975. "Irrigation Development in Bundelkhand - A New Approach," Annual Number, Institution of Engineers, U.P. Centre, December.

[4] Chaturvedi, A. C., 1975. "Water Resources Council," talk , Technical Division, Office of the Chief Engineer, I.D.U.P. at Lucknow.

[5] Chaturvedi, A. C., 1975. "Irrigation Development in Hills of U.P.," Navjiwan dated 2.12.75.

[6] Chaturvedi, A. C., 1975. "Irrigation Development in Bundelkhand Division," Swatantra Bharat dated 15.11.75.

[7] Chaturvedi, A. C., 1975. "Resources for Irrigation and Water Rates," Navjiwan dated 22.12.75.

[8] Chaturvedi, A. C., 1974. "Irrigation Development in Robilkhand' Power and River Valley Development," October.

[9] Chaturvedi, A. C., 1974. "Drought Affected Area Policy for Irrigation Development," Annual number, U.P. Centre Institution of Engineers, India, April.

[10] Chaturvedi, A. C., 1974. "Planning and Execution of Irrigation Works," Hindi Section Journal, Institution of Engineers, India, December.

[11] Chaturvedi, A. C., 1975. "Development and Future Prospects of Irrigation in U.P.," Hindi Section Journal, Institution of Engineers, India, August.

[12] Chaturvedi, A. C., 1975. "Flood Problems of U.P.," Journal of the South African Institution of Civil Engineers, October.

[13] Chaturvedi, A. C., 1962. "Irrigation Development in U.P.," Sitapur Samchar, December.

[14] Chaturvedi, A. C., 1974. "Irrigation Development in Awadh," Swatantra Bharat, Lucknow, June.

[15] Chaturvedi, A. C., 1975. "Planning Rural Water Supply in Backward Areas," Institution of Engineers, India, Public Health Journal Special number, National Symposium in Nainital, October.

[16] Chaturvedi, A. C., 1974. "Water Resources in U.P. and Their Utilisation," talk at U.P. Centre, Institution of Engineers, India.

[17] Chaturvedi, A. C., 1974. "Area Policy for Irrigation Development," talk at U.P. Centre, Institution of Engineers, India.

[18] Chaturvedi, A. C., 1974. "Rural Development in U.P.," Seminar at U.P. Centre, Institution of Engineers, India.

[19] Chaturvedi, A. C., 1974. "Water Problems of U.P.," talk at U.P. Centre, Institution of Engineers, India.

[20] Chaturvedi, A. C., 1975. "Economic and Optimum Utilisation of Water," talk at U.P. Centre, Institution of Engineers, India.

[21] Chaturvedi, A. C., 1975. "Flood Estimation, Forecasting and Warning," talk at U.P. Centre, Institution of Engineers, India.

[22] Chaturvedi, A. C., 1967. "Irrigation Water Management," talk at National Symposium in Central Board of Irrigation and Power New Delhi, India.

[23] Daily Navjiwan, 1976. "Water Availability in U.P."

[24] Chaturvedi, A. C., 1976. "Prospects of Irrigation in Bundel Khand," Bhagirathi.

[25] Chaturvedi, A. C., 1976. "Development of Underground Water," Navjiwan.

[26] Chaturvedi, A. C., 1976. "Water Supply Schemes in U.P.," Navjiwan.

[27] Chaturvedi, A. C., 1976. "Twenty-Five Year Plan for Flood Control," Navjiwan.

[28] Chaturvedi, A. C., 1976. "Planning for Water Supply - A New Approach," Annual souvenir number of the Second National Convention on Environmental Engineering, New Delhi, India.

[29] Chaturvedi, A. C. "National Reconstruction and the Engineer," Journal of the Association of Engineers, Vol. 51, No. 1, India.

[30] Chaturvedi, A. C., 1976. "System Analysis in Water Management," papbr for the International Conference at Wroclaw, Poland.

UN MODELE DE TRANSPORT D'ELEMENTS NUTRITIFS BASE SUR LES CONTRIBUTIONS DES UTILISATIONS DU TERRITOIRE

(A NUTRIENT TRANSPORT MODEL BASED ON EXPORTS BY LAND-USES)

By

Daniel A. Cluis
Professor, INRS-Eau
Université du Québec
C.P. 7500, Québec 10
Québec, Canada

Abstract

Following a detailed study of the nutrient loads exported by various land-uses, a simple transport scheme has been devised. Using the resulting model, nutrient budgets (nitrogen and phosphorus) can be predicted on an annual or seasonal basis at any point on a river, and the relative contributions ofthe different nutrients sources can be evaluated.

Application of the model to a given river basin involves the following steps:

(a) drainage - the drainage pattern is established via hydrographic subdivision of the unit areas (10 km x 10 km) defined by the Universal Transverse Mercator Grid;

(b) land-use - municipal statistics concerning the various land uses within the river basin are compiled, and the pattern of land-use in each drainage unit is established. Included in this compilation are both point and non-point nutrient sources: human population (sewered and unsewered); livestock populations; crop fertilization; forested areas; industries;

(c) nutrient export - the unit nutrient contributions of the various land-uses are obtained by reference to published literature values;

(d) nutrient transport - two coefficients are used for the transfer of the nutrient loads of drainage unit:

C_I = internal loss coefficient for the transfer of non-point sources to the river;

C_T = transport coefficient from one drainage unit to another.

The contribution of one drainage unit to a sampling station downstream can then be written:

$$L = C_T^K \cdot \{P + (C_I \times NP)\}$$

where: K = distance in number of drainage units;
P = nutrient contributions from point sources;
NP = nutrient contributions from non-point sources.

Field data obtained during a two-year (1973-1975) study of the Yamaska River (Quebec, Canada) have been used for calibration purposes. Stable sets of coefficients C_I and C_T have been established for nitrogen and phosphorus in each basin.

As anticipated from chemical and biological considerations, different coefficients are obtained for the two nutrients elements, nitrogen being considerably more mobile than phosphorus.

Resume

A la suite d'une étude détaillée des charges en éléments nutritifs exportées par les utilisations du territoire d'un bassin, un schéma simple de transport a été établi. Le modèle en résultant permet de prédire les transports d'éléments nutritifs (azote et phosphore) sur une base annuelle ou saisonnière en tout point d'une rivière et d'évaluer la contribution relative des différentes sources.

L'application du modèle à un bassin donné comprend les étapes suivantes:

(a) drainage - Le sens des écoulements est établi par l'intermédiaire des subdivisions hydrographiques de la grille Universelle Transverse de Mercator de 10 km de côté;

(b) utilisations du territoire - Les statistiques municipales des utilisateurs sont compilées et distribuées sur les unités de drainage. Les sources ponctuelles et diffuses d'éléments nutritifs sont prises en compte dans ce calcul: population humaine (avec et sans égout), population animale, surfaces fertilisées, apports des forêts et rejets des industries;

(c) contributions en éléments nutritifs - L'apport spécifique des differentes sources a été tiré des chiffres publiés dans la littérature;

(d) transport des éléments nutritifs - Le schéma de transport utilise 2 coefficients constants:

C_I = coefficient interne de perte pour le transfert de source diffuse vers la rivière;

C_T = coefficient de transport d'une unité de drainage à la suivante.

La contribution d'une unité de drainage à une station d'échantillonnage située en aval s'écrit alors:

$$L = C_T^K \cdot \{P + C_I \cdot D\}$$

où: K = distance en nombre d'unités de drainage;
P = contribution en éléments nutritifs des sources pontuelles;
D = contribution en éléments nutritifs des sources diffuses.

Des données mesurées, obtenues par une étude de 2 ans (1973-1975), de la rivière Yamaska ont été utilisées pour calibrer le modèle. Des valeurs stables de C_I et C_T ont été obtenues pour l'azote et le phosphore sur chaque bassin et pour chaque période. Comme des considérations chimiques et biologiques le laissaient prévoir, des couples de coefficients differents ont été obtenus pour les deux éléments nutritifs, l'azote étant considérablement plus mobile que le phosphore.

But Du Modele

Par une étude détaillée des charges en substances nutritives rejetées par les différents utilisateurs sur le territoire, le modèle propose un schéma simple de cheminement de ces substances. Il permet d'établir sur une base annuelle ou saisonnière les quantités de substances nutritives transportées en tout point du réseau hydrologique et de mettre en évidence les contributions relatives des différents utilisateurs du territoire à ces apports. Il permet également de prévoir les répercussions d'un aménagement projeté sur la diminution relative des quantités d'éléments nutritifs transitées dans les tronçons de rivière situés en aval.

Travaux récents dans le domaine de l'eutrophisation des rivières et des lacs. L'étude de la littérature montre que la teneur en éléments nutritifs des eaux d'un bassin est reliée étroitement à l'utilisation du territoire drainé.

- Vollenweider (1968) a mis en évidence la possibilité de classer les lacs selon leur fertilité en fonction de la vitesse d'apport en phosphore par unité de surface et de leur profondeur moyenne. Ce modèle a été amélioré par la suite (Dillon et Rigler, 1975) en introduisant la notion de temps de renouvellement de l'eau.

- Dans le but d'évaluer les apports en phosphore à une cinquantaine de lacs ontariens peu développés, Dillon et Kirchner (1975) ont tenté d'améliorer la formule présentée par Patalas et Salki (1973) en précisant, par des mesures effectuées sur le terrain, la gamme des charges en phosphore exportées selon la géologie (roches sédimentaires ou ignées) et l'utilisation du territoire (forêt, forêt + pâturage, agriculture) (Kirchner, 1975).

Deux travaux très récents s'inscrivent encore plus dans la ligne de pensée du modèle d'apports:

- L'Agence américaine de Protection de l'Environnement (EPA, 1974) vient de publier son premier rapport concernant l'échantillonnage national sur l'eutrophisation. Ce rapport porte sur 143 bassins de drainage de la partie du Centre-Nord et du Nord-Est des Etats-Unis. Son but est d'établir des relations générales entre l'utilisation du territoire et les éléments nutritifs (N, P) des rivières, dans la perspective d'établir des coefficients d'exportation par ruissellement de ces corps chimiques, reliés à l'utilisation du territoire et à certaines caractéristiques géographiques. Cette étude insiste surtout sur les sources diffuses et la densité des animaux d'élevage. L'interprétation axée sur la méthode des régressions entre les charges mesurées et les types d'utilisation du sol n'a pas donné les resultats souhaités; l'étude se poursuit avec, pour objectif, d'étudier environ 1000 bassins à la grandeur des Etats-Unis.

- Uttomark et al. (1974) ont publié, dans le cadre de la même étude, une revue très complète de la littérature sur les charges en éléments nutritifs provenant des sources diffuses au sens large (précipitations, zones urbanisées, forêts, territoires agricoles). Cette étude insiste sur la partie exportée de ces charges.

La préoccupation commune des recherches précédentes est de préciser, pour un bassin donné (lac ou rivière), les contributions des différentes utilisations du territoire drainé, aux charges cheminant dans le réseau hydrographique d'un bassin.

Terminologie. *Carreau entier*: unité de découpage cartographique de dimensions 10 km par 10 km, développée par Environment Canada (1973) et tracée sur les cartes à projection Universelle Transverse de Mercator (UTM).

Carreau partiel: unité de drainage, comprise dans un carreau entier. La banque physiographique du Canada fournit la couverture végétale du carreau partiel et les sens des écoulements de carreau partiel en carreau partiel, permettant ainsi de suivre le cheminement des charges dans le réseau hydrographique.

Ce découpage est compatible avec le modèle hydrologique CEQUEAU utilisé sur les mêmes bassins (Girard, 1970; Girard, Charbonneau et Morin, 1972).

Producteur: utilisation du territoire génératrice d'éléments nutritifs (N, P) (ex.: population bovine); par extension, ce terme recouvre aussi les contributions naturelles (ex.: précipitations).

Apports spécifique: quantité d'éléments nutritifs rejetée par jour, en moyenne, par chacun des utilisateurs sur le territoire (en kg/jour/unité de producteur).

Production: quantité journalière d'éléments nutritifs fournie par les producteurs (en kg/jour de N ou P).

Apport réel: production dont la totalité atteint le réseau de drainage (ex.: les rejets d'égouts, la pluie sur un lac, les charges exportées par une forêt).

Apport potentiel: production dont une partie seulement atteint le réseau de drainage (ex.: apport des animaux, apport des cultures engraissées, rejets humains non collectés).

Débit massique: quantité journalière d'éléments nutritifs transitée dans la rivière (en kg/jour de N ou P).

Utilisation du territoire. Les statistiques canadiennes sont disponibles tous les 5 ans sur une base municipale (Statistique Canada, 1971). Elles sont utilisées pour identifier les apports réels (ex.: populations) et potentiels (ex.: ruissellement agricole) en elements nutritifs.

D'autres sources fournissent, par municipalité, le nombre de personnes desservies par un réseau d'égouts (Québec. Bureau de la Statistique, 1971) et la liste des principales industries, leur type et le nombre de leurs employés (Scott's, 1975).

L'approche fondamentale de distribution des apports est la suivante:

> *les apports ponctuels sont affectés directement aux carreaux partiels où ils se jettent, alors que les apports diffus sont distribués sur les carreaux entiers proportionnellement aux surfaces des municipalités qui s'y trouvent.*

Apports spécifiques. Pour chacun des apports (égouts domestiques, rejets industriels, ruissellement agricole, élevage, précipitation, forêts), les productivités spécifiques en azote et en phosphore ont été établies à partir de la littérature, en séparant les apports réels à la rivière des apports potentiels, c'est-à-dire non totalement transférés (Tableau 1). En ce qui concerne les rejets industriels, les données n'étant pas disponibles, il a fallu utiliser des méthodes indirectes d'estimation, comme le type d'industrie et le nombre d'employés.

Tableau 1. Apports Specifiques Moyens Annuels

	PRODUCTEUR	TYPES D'APPORTS	APPORTS DE N	APPORTS DE P
Apports naturels	Lacs - pluie sur le lac	réels	2.1 kg/jour-km^2	0.02 kg/jour-km^2
	forêts - ruissellement	potentiels[1]	5.3 kg/jour-km^2	0.6 kg/jour-km^2
	marécages - ruissellement	réels	1.5 kg/jour-km^2	
	sol nu - pluie sur sol nu	potentiels	2.1 kg/jour-km^2	0.02 kg/jour-km^2
Apports de la population humaine	population avec égout	réels	14 g/jour-habitant	1.8 g/jour-habitant
	population sans égout	potentiels	14 g/jour-habitant	1.8 g/jour-habitant
Apports de la population animale	bovins	potentiels	187 g/jour-animal	33.6 g/jour-animal
	porcs	potentiels	31	9.5
	moutons	potentiels	24	4.4
	poulets	potentiels	2	0.7
	chevaux	potentiels	159	23.1
Apports des engrais chimiques	blé	potentiels	15 kg/jour-km^2 fertilisé	9.4 kg/jour-km^2 fertilisé
	avoine	potentiels	4.6	9.4
	orge	potentiels	12	12
	fruits de vergers	potentiels	3.4	7.1
	petits fruits	potentiels	29	28
	pâturage défriché	potentiels	15.4	9.1
	foin cultivé	potentiels	15.4	9.1
	maïs-grain	potentiels	40	13
	tabac	potentiels	11	18
	patates	potentiels	25	23
	betteraves	potentiels	25	24
	légumes	potentiels	15	12
	autres	potentiels	31	11
Apports des industries	chacune des industries	réels	dépendant du nombre d'employés et du type d'industrie	

[1] La littérature donne des apports spécifiques réels (0.64 kg/jour-km^2 d'azote et 0.032 kg/jour-km^2 de phosphore); cependant, comme la forêt est une source diffuse, son apport est considéré comme potentiel par un artifice de calcul (voir le Chapitre 4).

Hypothèses du modèle. Les hypothèses suivantes ont été adoptées:

a) les différentes sources sont additives;

b) la densité de drainage est suffisamment grande pour qu'une source potentielle atteigne un cours d'eau à l'intérieur de son carreau partiel d'origine;

c) une fois dans un cours d'eau, une charge reste dans le lit et se déplace vers l'aval avec possibilité saisonnière de sédimentation ou de remise en suspension ou solution, d'assimilation ou de relâchement par la matière vivante (organismes planctoniques et benthiques);

d) comme le modèle cherche à etablir les charges transportées sur une base saisonnière et non journalière (épisode hydrologique), nous considérons que le transfert des charges est suffisamment tamponné pour que les apports spécifiques puissent être considérés comme constants durant la période considérée;

e) faute de relevés géologiques à petite échelle, nous considérons que les deux bassins étudiés sont relativement homogènes, de ce point de vue;

f) une autre hypothèse importante est qu'il n'y ait pas eu de variations trop rapides de l'utilisation du territoire dans le temps; en effet, les données utilisées sont celles de Statistique Canada (1971). Ces données pourraient être réévaluées avec le recensement de 1976.

Un organigramme général des différentes opérations nécessaires à l'établissement du modèle d'apports est représenté sur la Fig. 1.

Schema General

Support topographique et drainage. Pour stocker les informations, nous avons utilisé comme support cartographique le quadrillage Universel Transverse de Mercator (grille UTM), disponible au Canada et décrit par Sebert (1972). Les dimensions du carreau de base choisi sont de 10 km par 10 km, ce qui le rend compatible avec des données fournies sur une base journalière, qui sont utilisées dans le modèle de génération des débits.

L'usage de la grille carrée représente une amélioration pour le stockage et le traitement automatique des données par rapport aux découpages en sous-bassins de taille variable. Ses applications en hydrologie, développées récemment grâce aux travaux de Solomon (1972), se sont avérées fructueuses à

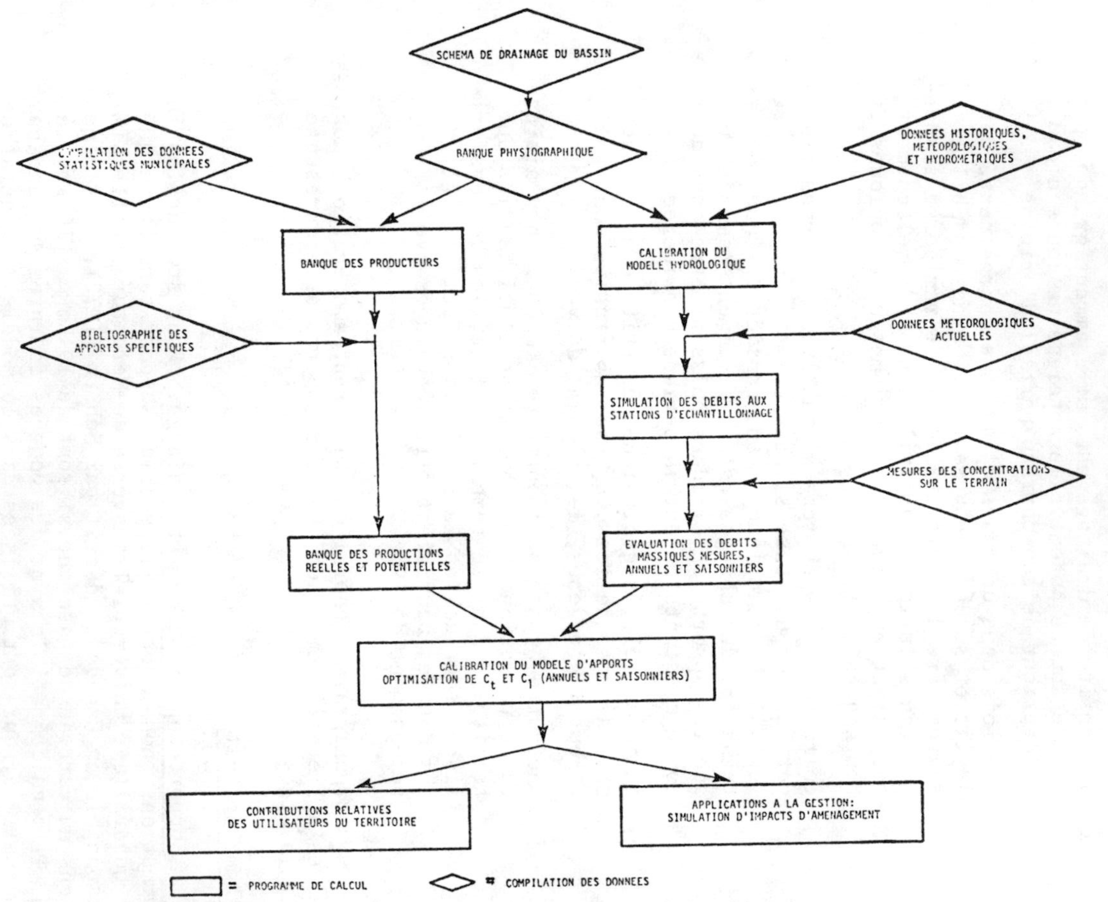
SCHEMA DE DRAINAGE DU BASSIN
BANQUE PHYSIOGRAPHIQUE
COMPILATION DES DONNEES STATISTIQUES MUNICIPALES
DONNEES HISTORIQUES, METEOROLOGIQUES ET HYDROMETRIQUES
BANQUE DES PRODUCTEURS
CALIBRATION DU MODELE HYDROLOGIQUE
BIBLIOGRAPHIE DES APPORTS SPECIFIQUES
DONNEES METEOROLOGIQUES ACTUELLES
SIMULATION DES DEBITS AUX STATIONS D'ECHANTILLONNAGE
MESURES DES CONCENTRATIONS SUR LE TERRAIN
BANQUE DES PRODUCTIONS REELLES ET POTENTIELLES
EVALUATION DES DEBITS MASSIQUES MESURES, ANNUELS ET SAISONNIERS
CALIBRATION DU MODELE D'APPORTS OPTIMISATION DE C_t ET C_l (ANNUELS ET SAISONNIERS)
CONTRIBUTIONS RELATIVES DES UTILISATEURS DU TERRITOIRE
APPLICATIONS A LA GESTION: SIMULATION D'IMPACTS D'AMENAGEMENT
= PROGRAMME DE CALCUL
= COMPILATION DES DONNEES

Fig. 1. Organigramme general du modele d'apports

cause des relations entre la banque physiographique des données primaires (altitude, couverture végétale, pente) et les phénomènes secondaires (évaporation, précipitations, débits).

Chacun des carreaux de base est divisé en carreaux partiels ayant les mêmes caractéristiques physiographiques que leur carreau de base, mais faisant partie de sous-bassins différents. En conservant en mémoire, pour chaque carreau partiel, les références des carreaux partiels situés à l'amont et du carreau partiel situé à l'aval, on peut représenter très simplement la circulation de l'eau sur le bassin. Ce schéma de drainage est identique à celui du Guide d'utilisation des banques de données (Environnement Canada, 1973).

Banque des données des producteurs à l'échelle du carreau partiel. Les données d'utilisation du territoire a l'échelle du carreau partiel constituent la banque des producteurs; si l'on multiplie le nombre de producteurs de chaque classe, par l'apport spécifique journalier correspondant en azote et phosphore, on obtient les banques de production (réelle et potentielle) d'éléments nutritifs.

Transport des charges. A partir des banques de productions (réelles et potentielles) et de drainage établies pour les carreaux partiels, et an tenant compte des hypothèses du modèle énoncées au paragraphe 1.5, nous définissons 2 coefficients de transfert:

C_I = coefficient interne de transfert des apports potentiels au cours d'eau (selon l'hypothèse a);

C_T = coefficient de transport de carreau partiel en carreau partiel (selon l'hypothèse b).

Soit AP et AR les apports potentiels et réels d'un carreau partiel situé à k carreaux partiels en amont d'un point d'échantillonnage. La contribution de ce carreau partiel à la charge transitée pour une période de temps donnée, à la station, s'écrira:

$$Q_M = C_T^k \times (AR_i + C_I \cdot AP_i) \times \Delta t$$

Le transport total en éléments nutritifs à la station d'échantillonnage sera donc la somme des contributions de ce type pour l'ensemble des carreaux partiels situés à l'amont. Ainsi, ce même schéma permet de simuler en tout point du réseau hydrographique les charges transportées.

Sens physique des coefficients de transport et de transfert.

- C_T représente le transport des charges une fois acheminées dans le cours d'eau. Nous nous attendons à ce que

C_T soit proche de 1 sur une base annuelle, inférieur à 1 aux étiages d'hiver et d'été (sédimentation et/ou assimilation), supérieur à 1 à la débâcle de printemps et aux crues d'automne (mise en suspension et/ou relâchement).

- C_I représente la fraction des apports potentiels atteignant la rivière. Nous nous attendons à ce que ce coefficient soit très faible en hiver (sol gelé), fort au printemps (déstockage) et intérmediaire durant la saison végétative. De plus, compte tenu de la chimie des substances nutritives, on s'attend à ce que le coefficient relatif à l'azote soit plus élevé que celui relatif au phosphore.

De plus, de meilleurs résultats seront obtenus pour les stations d'échantillonnage situées suffisamment en aval pour regrouper un nombre adéquat de carreaux partiels. En effet, à cause de l'effet de compensation des erreurs d'affectation des données statistiques, on bénéficie alors d'une meilleure représentation de la réalité. A titre de référence, le bassin complet de la rivière Yamaska comprent 123 carreaux partiels.

Modèle hydrologique utilisé pour la simulation des débits. Un modèle déterministe de simulation de débits utilisant la même schématisation du drainage a été développé par Girard et al. (1970; 1972). A partir des données physiographiques de chaque carreau (altitude moyenne, surface de lacs, de marais, de forêts et de sol nu) et de données météorologiques mesurées, il permet d'évaluer en tout point du bassin, sur une base journalière, les éléments suivants:

a) les données météorologiques probables, compte tenu des données observées aux stations météorologiques situées sur les carreaux voisins et des corrélations existantes entre ces données et les caractéristiques propres de chaque carreau;

b) le transport de l'eau produite par les précipitations liquides ou la fonte de neige compte tenu des caractéristiques topographiques et géologiques;

c) l'écoulement traversant chaque carreau.

Après calage des coefficients du modèle par comparaison avec des débits observés aux stations hydrométriques, le modèle peut simuler, avec une précision suffisante (10-15% durant l'année, 20-25% sous couvert de glace), les débits traversant chaque carreau partiel en tout temps, à partir des données météorologiques.

Ce modèle a été appliqué sur le bassin étudié et a servi à simuler les débits moyens journaliers aux stations d'échantillonnage. Il aurait été souhaitable de travailler avec des débits mesurés; cela n'a pas été possible. Dans le cas de station d'échantillonnage correspondant à une station hydrométrique, nos contraintes de temps nous imposaient de travailler avant que les enregistrements de niveaux soient dépouillés et que les débits soient rendus disponibles; dans le cas des autres stations d'échantillonnage, il était impossible, pendant la durée de cette étude, d'installer des stations temporaires et de les calibrer valablement à cause de leur site souvent inadéquat (absence de section de contrôle hydraulique).

Mesures des concentrations d'azote et de phosphore totaux. Un échantillonnage systématique a été réalisé par le Service de la qualité des eaux du Ministère des Richesses naturelles pendant les deux années de l'étude (MRN-INRS, 1974). Une banque de données a été créée dont nous avons extrait, pour l'usage de ce modèle, les concentrations en azote total et en phosphore total des échantillons naturels (non filtrés). En utilisant des concentrations mesurées ainsi que les débits moyens journaliers simulés par le modèle hydrologique, nous avons évalué sur une base annuelle et saisonnière les charges transportées.

Calibration. La calibration s'effectue en optimisant sur chaque bassin, pour chacune des périodes étudiées, la paire de coefficients C_I et C_T (coefficient interne de transfert et coefficient de transport de carreau partiel à carreau partiel), de facon à ce que les charges calculées se rapprochent des charges mesurées sur le bassin.

Description du Bassin Versant

Le bassin versant de la rivière Yamaska est situé au Québec, au sud du fleuve Saint-Laurent; il est compris entre les latitudes nord 45°05' et 46°05' et les longitudes ouest 72°12' et 73°07'. Les eaux de la rivière Yamaska se jettent dans le lac Saint-Pierre qui constitue un élargissement du fleuve Saint-Laurent entre les villes de Montréal et Québec. Le bassin, couvrant 4,843 km^2, se divise en deux régions distinctes: l'amont, drainé par la rivière Noire et par les branches nord, sud-est et centrale de la rivière Yamaska, appartient aux Appalaches; l'aval, qui comprend le tronçon principal de la rivière Yamaska, se trouve dans les basses-terres du Saint-Laurent.

La Fig. 2 représente la localisation des 19 stations d'échantillonnage, ainsi que la subdivision en 123 carreaux partiels et leurs sens de drainage.

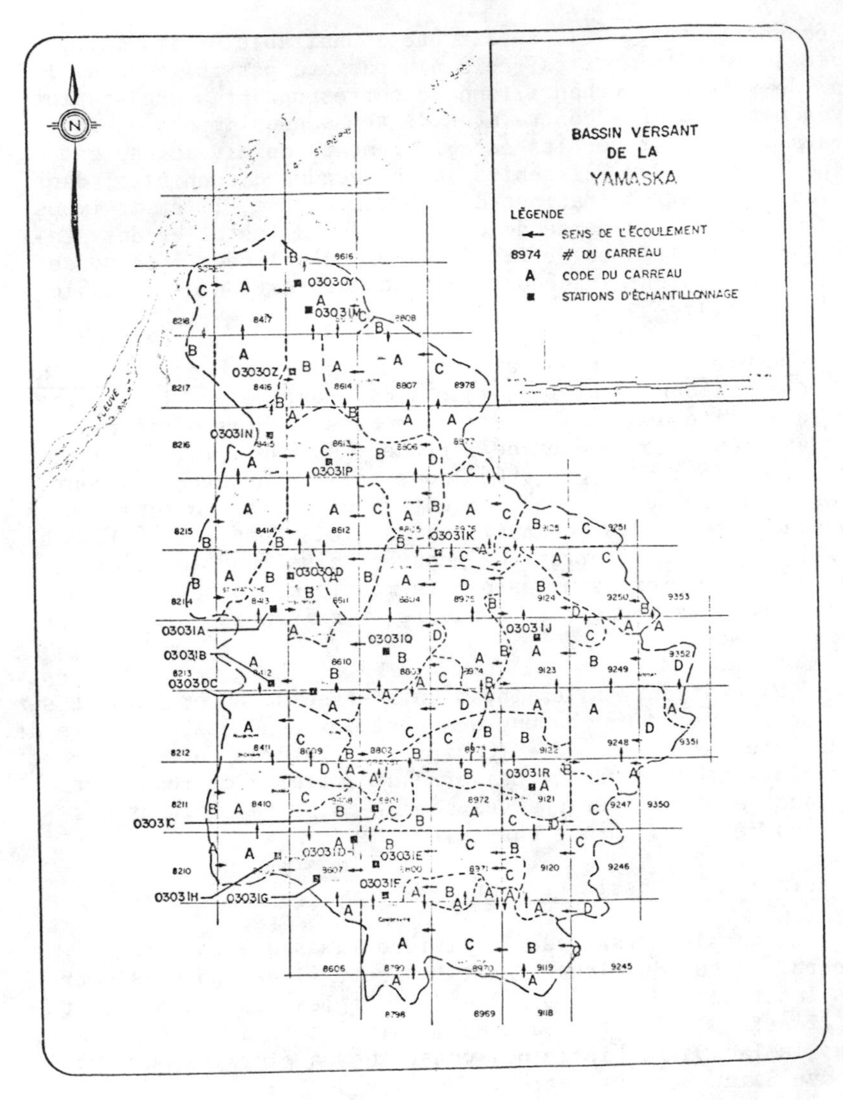

Fig. 2. Schéma de drainage du bassin versant de la rivière Yamaska

Resultats

Les différentes étapes décrites précédemment ont conduit, par simulation des coefficients C_I et C_T, à la calibration du modèle. Ces résultats apparaissent au Tableau 2.

Sur ce tableau des ajustements annuels, l'ordre de grandeur des coefficients de transfert C_I est très comparable à

Tableau 2. Simulations Moyennes et Pourcentage Moyen D'ajustement

SIMULATION	Année			Printemps			Eté			Automne			Hiver			Ajustement annuel pour C_T = 1	
	C_I	C_T	%	C_I	C_T	%	C_I	C_T	%	C_I	C_T	%	C_I	C_T	%	C_I	%
Azote	0.145	1.030	14	0.300	1.055	20	0.095	0.960	24	0.085	1.005	13	0.091	1.04	22	0.155	17
Phosphore	0.063	0.995	29	0.145	1.020	21	0.028	0.975	28	0.017	1.010	33	0.045	0.95	35	0.060	29

ceux qui sont rapportés dans la littérature; notamment, Vollenweider (1968) estime que les coefficients de perte se situent dans une gamme de 10 à 25% pour l'azote et de 1 à 5% pour le phosphore.

Si l'on considère maintenant le coefficient de transport C_T, sa cyclicité saisonnière correspond à des stockages et déstockages liés aux épisodes hydrologiques; on vérifie, en particulier, l'hypothèse que les fortes valeurs de C_T ont lieu au printemps et à l'automne (crues, donc érosion) et que les valeurs plus faibles ont lieu en été et en hiver (étiage, donc sédimentation).

La période de calibration (03-74/02-75) constitue une année hydrologiquement normale pour le Québec. Pour une année plus sèche comme l'été 1975, on peut prévoir que les différences saisonnières de C_I et C_T seraient encore plus marquées.

Quant à la précision d'ajustement du modèle, on peut la considérer commme globalement satisfaisante, compte tenu des erreurs intrinsèques à chacune des opérations qui la constituent, principalement:

- l'erreur d'affectation spatiale des statistiques et leur évolution temporelle;
- l'erreur de génération des débits (surtout en hiver);
- les erreurs liées à l'échantillonnage et à l'analyse (surtout pour le phosphore);
- l'imprécision du mode de calcul du débit massique;
- les incertitudes sur les rejets industriels;
- les erreurs sur les apports spécifiques et leurs variabilités saisonnières.

Utilisations du Modele Calibrè

Le modèle permet de déterminer à chaque station, sur des bases annuelles et saisonnières, les contributions relatives aux apports en azote et phosphore, des différents utilisateurs. A titre d'exemple, le Tableau 3 montre l'origine des charges en azote sur une base annuelle. On peut y noter la grande importance des animaux d'élevage.

Applications du Modèle D'Apports Pour La Gestion

Considérations générales. Une fois les coefficients C_I et C_T établis, on peut utiliser le modèle d'apports pour la gestion des éléments nutritifs au niveau du bassin. Il suffit de modifier les chiffres de production dans les carreaux

Tableau 3. Contribution Relative Annuelle en Azote des Utilisateurs du Territoire (en %)

		Apports réels						Apports potentiels						
		Population avec égout						Population sans égout						
Numéro	Station	Phys.	Déter.	Pluie	Maré-cages	Industrie	Total	Phys.	Déter.	Animaux	Engrais	Pluie	Forêts	Total
1	3031L	8.57	0	0.73	1.21	3.32	13.84	1.16	0	56.08	11.05	7.96	9.90	86.15
2	3031M	1.50	0	0.68	0.00	0.05	2.24	0.97	0	62.75	16.44	9.85	7.76	97.76
3	3030Z	9.12	0	0.66	1.31	3.59	14.68	1.18	0	55.54	10.66	7.77	10.18	85.31
4	3031N	1.67	0	0.55	0.00	0.00	2.22	1.19	0	65.21	15.36	7.81	8.21	97.78
5	3031P	0.48	0	0.00	0.00	0.28	0.75	0.91	0	68.77	16.94	8.33	4.30	99.25
6	3030D	10.27	0	0.60	1.55	4.23	16.65	1.21	0	53.75	9.51	7.59	11.28	83.34
7	3031A	5.70	0	0.64	1.70	3.64	11.68	1.28	0	57.05	9.73	8.00	12.24	88.30
8	3031B	7.34	0	1.01	0.64	6.46	15.46	1.41	0	53.99	7.90	8.09	13.14	84.52
9	3030C	3.41	0	0.29	3.18	0.53	7.40	1.16	0	60.38	10.27	8.06	12.73	92.60
10	3031H	8.60	0	1.32	0.51	8.50	18.93	1.47	0	49.27	5.95	8.42	15.93	81.04
11	3031G	10.85	0	0.16	0.00	19.76	30.78	1.30	0	40.67	5.61	7.18	14.46	69.22
12	3031D	4.22	0	0.73	0.73	1.78	7.45	1.91	0	58.31	5.45	9.14	17.58	92.38
13	3031E	1.99	0	4.19	0.37	0.36	6.92	1.70	0	52.56	4.68	10.47	23.66	93.08
14	3031F	14.91	0	0.23	0.00	27.15	42.29	1.38	0	29.30	2.94	7.02	17.07	57.71
15	3031C	4.22	0	0.73	0.73	1.78	7.45	1.91	0	58.31	5.45	9.14	17.58	92.38
16	3031R	3.69	0	1.47	3.24	1.68	10.09	1.94	0	50.59	3.10	11.97	22.33	89.91
17	3031Q	4.19	0	0.33	4.11	0.56	9.20	1.09	0	57.18	10.34	8.14	14.06	90.80
18	3031K	7.88	0	0.00	6.78	1.53	16.19	0.90	0	56.86	9.82	8.52	7.71	83.81
19	3031J	4.40	0	0.91	1.20	0.16	6.68	1.32	0	50.46	8.82	7.78	24.95	93.32

partiels concernés et d'effectuer le cheminement des charges d'amont vers l'aval; ceci permet d'évaluer les alternatives d'aménagement possibles. Par comparaison avec la situation originale, on est alors en mesure d'évaluer, à priori, les améliorations ou détériorations relatives induites par l'aménagement projeté. Ceci permet, entre autres, d'évaluer l'impact du traitement des eaux usées d'une municipalité (par exemple: Granby), la contribution d'une industrie (par exemple: Domtar), ou la réduction de charge en phosphore induite par la création d'un réservoir artificiel (par exemple: Savage Mills). Ces modifications de charge peuvent être suivies en tout point du bassin versant aval, et s'appliquent directement aux concentrations.

Impact de l'implantation d'usines de traitement à Waterloo et Granby. A ce titre, nous avons évalué l'influence d'un traitement à 95% des rejets d'azote et de phosphore des equx usées des municipalités de Waterloo et Granby (industries incluses). Les résultats apparaissent aux Tableaux 4 et 5.

Ils mettent en évidence, saison par saison, la réduction en pourcentage des charges transportées, donc des *concentrations*, en aval des installations de traitement projetées. La Fig. 3 montre la réduction, en pourcentage par rapport à la situation actuelle, des charges en azote et phosphore transitées, tout le long de la rivière jusqu'à son embouchure. On note que, sur une base annuelle, la réduction de charge est la plus importante directement à l'aval des usines de traite, l' amélioration diminue graduellement, avec des détériorations brutales à la confluence de sous-bassins importants non modifiés (rivière Yamaska-centre et rivière Noire) et, à l'embouchure, la réduction de charge n'est plus que de 7% pour l'azote et de 10% pour le phosphore. Ce schéma annuel se retrouve à chaque saison et, durant l'été qui est la période critique, les réductions des charges, donc des concentrations, sont encore plus marquées.

On remarque aussi que les réductions des charges de phosphore sont plus importantes que celles d'azote, confirmant ainsi l'origine surtout ponctuelle du phosphore (municipalités et industries), par rapport a l'origine essentiellement diffuse (cultures et élevage) de l'azote.

Le cas des élevages industriels. Sur le Tableau 3, on remarque que la contribution relative de l'élevage aux apports est particulièrement importante et dépasse souvent 50% des apports totaux. Reconnaissant la tendance à la concentration d'animaux dans des élevages industriels dont la contribution risque alors de devenir ponctuelle, il serait intéressant, dans l'optique d'une gestion à l'échelle du bassin, de comparer le coût, par unité de charge réduite, des installations de traitement des rejets industriels, municipaux ou d'élevage.

Tableau 4. Reduction, le Long de la Riviere Yamaska, du Debit Massique D'Azote[1], Causee par un Traitement, a 95%, des Rejets des Villes de Waterloo et de Granby

Numéro de Station	année			printemps			été			automne			hiver		
	Avant	Après	%	Avant	Après	%	Avant	Après	%	Avant	Après	%	Avant	Après	%
03031R	203	106	48	305	208	32	171	73	57	164	67	59	168	70	58
03031C	1314	579	56	1948	1201	38	1059	351	67	1068	343	68	1125	386	66
03031H	3883	3079	21	7116	6239	12	2404	1777	26	2627	1892	28	3051	2219	27
03031B	5153	4300	17	10064	9088	9.7	2860	2282	20	3263	2520	23	3973	3073	23
03031A	9734	8856	9.0	20438	19409	5.0	4967	4413	11	5732	4985	13	7173	6237	13
03030D	11197	10266	8.3	23917	22771	4.8	5318	4807	9.6	6539	5784	11	8538	7526	12
03031Z	14756	13708	7.1	34330	32912	4.1	5678	5244	7.6	7874	7104	9.8	11415	10231	10
03031L	16849	15736	6.6	40751	39172	3.9	5933	5532	6.8	8638	7861	9	13126	11845	9.8

[1] Exprimé en kg j^{-1}.

Tableau 5. Reduction, le Long de la Riviere Yamaska, du Debit Massique de Phosphore[1], Causee par un Traitement, a 95%, des Rejets des Villes de Waterloo et de Granby

Numéro de Station	année			printemps			été			automne			hiver		
	Avant	Après	%	Avant	Après	%	Avant	Après	%	Avant	Après	%	Avant	Après	%
03031R	18	4	78	23	8	65	16	2.3	86	16	1.7	89	17	3.2	81
03031C	197	41	79	243	86	65	177	22	88	173	17	90	185	30	84
03031H	383	228	41	654	488	25	265	121	54	262	100	62	287	155	46
03031B	504	349	31	953	780	18	310	173	44	308	143	54	345	225	35
03031A	1043	889	15	2076	1900	8.5	615	482	22	589	423	28	702	588	16
03030D	1210	1055	13	2372	2189	7.8	733	607	17	747	577	23	788	686	13
03031Z	1453	1299	11	3138	2939	6.3	779	665	15	868	692	20	800	717	10
03031L	1589	1435	9.7	3574	3367	5.8	808	700	13	936	757	19	817	742	9.2

[1] Exprimé en kg j^{-1}.

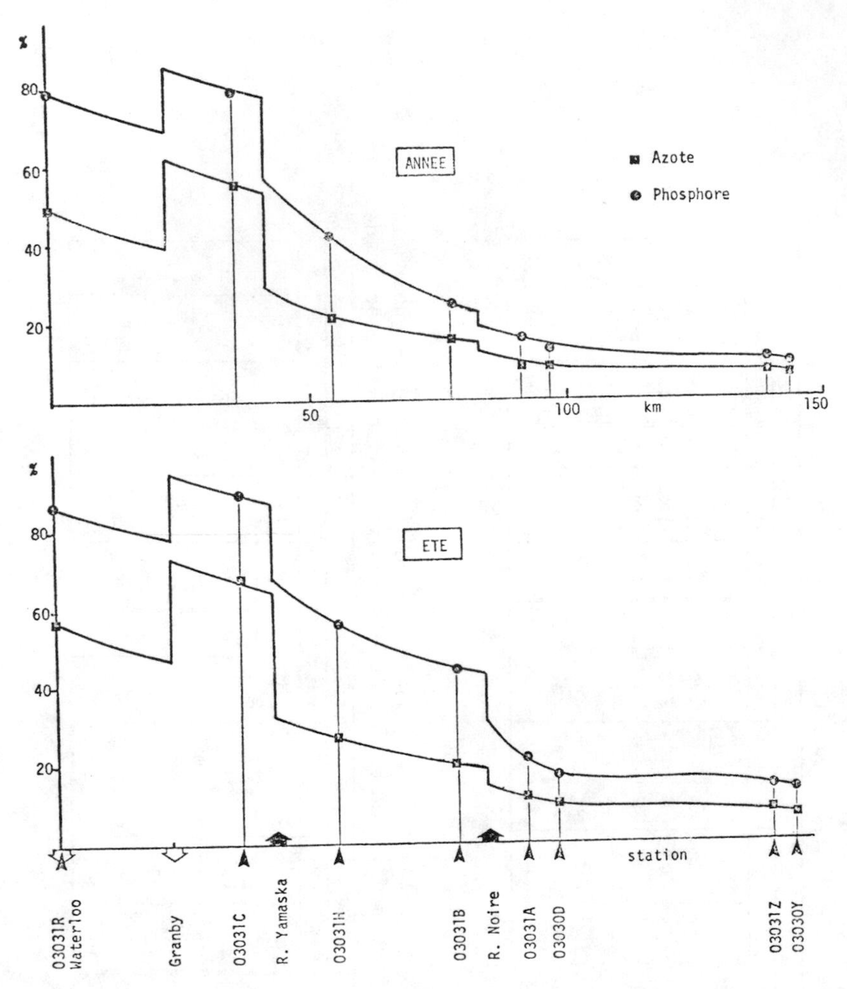

Fig. 3. Profil en long de la réduction des charges transitées

Dans ce contexte, on pourrait être amené à comparer, par exemple, l'efficacité globale d'une usine de traitement d'eaux usées urbaines de $5,000,000 à 1,000 installations de stockage et recyclage de rejets d'élevage de $5,000 chaque. On doit noter que nos résultats s'attaquent à un préjugé très répandu, qui associe l'origine principale de la "*pollution*" par les substances nutritives aux rejets municipaux et industriels.

Limites D'Utilisation du Modele D'Apports

L'application du modèle d'apports au bassin versant de la rivière Yamaska a mis en évidence certaines limitations de la technique dont il faut tenir compte pour obtenir des résultats d'une précision raisonnable. Ces limitations dépendent essentiellement de la relation entre les variabilités spatio-temporelles des débits massiques des éléments nutritifs transportés par la rivière, d'une part, et des variabilités spatio-temporelles des productions rejetées par les utilisateurs d'un territoire donné, d'autre part.

D'une façon générale, la précision du modèle est d'autant meilleure que la simulation se rapporta à l'*état moyen dans le temps et dans l'espace* (permanent en moyenne).

Ainsi, pour un niveau fixé de connaissance de l'utilisation du territoire, la précision des apports journaliers calculés par le modèle diminue, d'une part, par rapport à la réalité, avec la durée de période de temps considérée (année, saison, épisode hydrologique) et, d'autre part, avec la surface du territoire étudié (bassin total, sous-bassin, parcelle).

Pour l'utilisateur, deux cas peuvent se présenter selon qu'il dispose ou non d'un échantillonnage de contrôle avec des mesures de débits.

Cas 1: Sans échantillonnage de contrôle

Malgré l'absence de vérification possible, certaines techniques du modèle sont applicables pour calculer théoriquement les apports en azote et phosphore et pour établir les contributions des utilisateurs à ces apports.

Les limitations sont les suivantes:

sur le plan temporel, compte tenu du fait que les apports spécifiques sont des apports moyens annuels, il est recommandé d'effectuer le calcul *sur une base annuelle* et de poser $C_T = 1$ (année moyenne, transfert interannuel nul);

sur le plan spatial, à cause de l'imprécision dans l' affectation des statistiques d'utilisation du territoire vers les unités de drainage, il est souhaitable que le bassin dont on simule les apports comporte *au moins* 10 de ces unités (dans notre cas, 10 carreaux partiels construits sur la grille UTM de 10 km par 10 km), ce qui permet de bénéficier de la compensation des erreurs d'affectation. Les unités de drainage doivent être du même ordre de grandeur que les unités de recensement; dans notre cas, la surface des municipalités est comparable à celle des carreaux partiels. Celà peut conduite à rechercher des données d'utilisation du territoire à une plus petite échelle. On doit noter ici une application importante de cette idée, utilisée pour l'étude des lacs: quand le bassin drainé est trop petit pour le critère choisi (10 unités de drainage), on précise l'utilisation du territoire localement; ensuite, on peut appliquer le modèle, sur une base annuelle (C_T=1), malgré le changement de dimension de l'unité de drainage.

Case 2: Avec un échantillonnage de contrôle

Quand on dispose de mesures de contrôle (concentrations et débits), l'objectif du modèle n'est plus de prédire les débits massiques transportés par la rivière, mais de préciser les contributions annuelles et *saisonnières* des différentes utilisations en plusieurs points du bassin et de permettre ainsi de proposer des correctifs adéquats.

Là encore, certaines limitations s'appliquent:

sur le plan temporel: la période de temps choisie (année, saison, épisode hydrologique) doit être notablement supérieure à la période entre deux échantillonnages, pour que la calibration ait un sens;

sur le plan spatial: les réserves concernant la définition de l'utilisation du territoire exprimées au cas 1 s'appliquent; on peut noter aussi que si l'on connaît l'utilisation globale du territoire d'un bassin entier, on ne peut, à moins d'homogénéité reconnue, tirer aucune conclusion concernant les contributions relatives sur un de ses sous-bassins.

Enfin, dès que l'on cherche à chiffrer les apports sur une base de temps plus courte que l'année, une saison critique par exemple, on doit absolument utiliser le modèle complet à 2 coefficients et le découpage cartographique qui s'impose.

Conclusion

A partir d'hypothèses simples, concernant l'origine et le transport des éléments nutritifs, le modèle d'apports permet de relier, à l'échelle du bassin versant, les *causes* (l'intensité et la distribution spatiale de l'utilisation du territoire considéré) aux *conséquences* (l'enrichissement des eaux de surface) dans le temps et dans l'espace. Contrairement aux modèles traditionnels de qualité de l'eau, cette approche ne s'intéresse pas a l'évolution de la teneur en oxygène dissous dans l'eau, phénomène très important en climat tempéré, mais qui n'est pas critique au Québec.

- En l'absence de toute mesure sur le terrain et sans définition de grille, la connaissance des statistiques générales d'utilisation du territoire permet une évaluation de la charge annuelle transitée. Si l'on s'intéresse à la variabilité temporelle, il est alors nécessaire de définir un découpage cartographique et de distribuer, sur ce découpage, les statistiques d'utilisation du territoire. Un échantillonnage de contrôle est alors nécessaire pour calibrer le modèle.

- Le modèle d'apports constitue on *outil en voie de développement*; il peut être amélioré, à mesure que des progrès seront réalisés dans l'évaluation des effluents industriels, de l'effet de rétention causé par les lacs, et dans la connaissance de la variabilité temporelle des apports spécifiques des utilisateurs du territoire. Malgré ses faiblesses, il semble bien adapté au problème d'eutrophisation des eaux de surface du Québec, permet d'évaluer l'origine des charges ainsi que la responsabilité des usagers de la ressource et, finalement, rend possible la simulation, à l'échelle du bassin, des conséquences de modifications de l'utilisation du territoire et des mesures de restauration de la qualité de l'eau.

Bibliographie

[1] Dillon, P. J. et Kirchner, N. B., 1975. "The Effects of Geology and Land Use on the Export of Phosphorus from Watersheds," Water Research, 9(2): 135-148.

[2] Dillon, P. J. et Rigler, F. H., 1975. "A Simple Method for Predicting the Capacity of a Lake for Development Based on Lake Trophic Status," Jour. Fish. Res. Board Can., 32(9): 1519-1531.

[3] Environnement Canada, 1973. "Guide to Data Holdings - 3.0 Hydrologic Square Grid System," Electronic Data Processing Committee, Environnement Canada.

[4] EPA, 1974. "Relationships between Drainage Area Characteristics and Non-Point Sources in Streams," National Eutrophication Survey Staff, Report NERC-EPA, 50 p.

[5] Girard, G., 1970. "Essai pour un modèle hydropluviométrique conceptuel et son utilisation au Québec," Cahier ORSTOM, Serie Hydrologie, 7(2): 85-116.

[6] Girard, G., Charbonneau, R. et Morin, G., 1972. "Modèle hydrophysiographique," Symposium International sur les Techniques des Modèles Mathématiques Appliqués aux Systèmes de Ressource en eau, Proceedings, pp. 190-204, Environnement Canada.

[7] Kirchner, W. B., 1975. "An Examination of the Relationship between Drainage Bassin Morphology and the Export of Phosphorus," Limnol. Oceanog., 20(2): 267-270.

[8] MRN-INRS, 1974. "Planification de l'acquisition des données de qualité de l'eau au Québec," Tome 5: Présentation de la méthode, Min. des Richesses naturelles, Service qualité des equx, Q.E.10.

[9] Patalas, K. et Salki, A., 1973. "Crustacean plankton and the eutrophication of lakes in the Okanagan Valley, B.C.," J. Fish. Res. Board Can., 30: 519-542.

[10] Quebec Bureau de la Statistique, 1971. "Renseignements Statistiques 1971," Municipalités du Québec, Bureau de la Statistique du Québec, Service des Finances.

[11] Scott's, 1975. "Répertoire industriel du Québec," 7e édition, Penstock Publications Ltd.

[12] Sebert, L. M., 1972. "Chaque pouce carré - La projection universelle transverse de Mercator (système UTM)," Ministère de l'Energie, des Mines et des Ressources, Canada.

[13] Solomon, S. I., 1972. "Joint Mapping," Casebook of hydrological network design practice, WMO publication No. 342, 11-2.1.

[14] Statistique Canada, 1971. "Recensement du Canada - Population Québec," 1(1).

[15] Statistique Canada, 1971. "Recensement du Canada - Agriculture Québec," 4(2).

[16] Uttomark, P. D., Chapin, J. D. et Green, K. M., 1974, "Estimating nutrient loading of lakes from non-point sources," Report EPA 660/3-74-020, 112 p.

[17] Vollenweider, R. A., 1968. "Les bases scientifiques de l'eutrophisation des lacs et des eaux courantes sous l'aspect particulier du phosphore et de l'azote comme facteurs d'eutrophisation," OCDE, Paris, rapport DAS/CSI/68.27, pp. 95-148.

URBAN STORMWATER - A MAJOR CONTRIBUTOR TO WATER POLLUTION

By

Ian Cordery
School of Civil Engineering
The University of New South Wales
P.O. Box 1, Kensington, New South Wales, 2033
Australia

Abstract

Urban stormwater has been recognized for some time as a contributor of considerable pollution to waterways. However, little attention has been given to the prevention of pollution from this source. Most interest in urban stormwater has been directed towards flood effects and prevention. On the other hand, efforts are continually being made, at great cost, to improve the treatment of sewage effluent. It is the purpose of this paper to show that better treatment of sewage may not provide the greatest return for investment in water pollution control.

Stormwater quality data has been collected from three separately sewered, urban residential areas in Sydney, Australia, to attempt to characterize the movement of pollutants and nutrients. Over 100 water samples from 15 flood events were collected and each was analyzed for a minimum of 7 parameters, and in some cases, for up to 17 parameters. Each parameter varied in its own characteristic manner. The concentration of the pollution indicators, BOD and suspended solids, and of the nutrient ammonia decreased as floods progressed. However, the concentration of phosphate tended to remain constant throughout each flood. The average concentration of suspended solids during each flood was about the same as for raw sewage, but the other parameters, such as BOD, phosphate and ammonia were about 10% of typical raw sewage concentrations.

Some of the collected water samples were subjected to simple settling for periods of up to one hour. With only four minutes settling the suspended solids concentration was reduced by an average of 79%, BOD by 34%, phosphate by 57% and ammonia by 18%.

Estimates were made of the total pollution and nutrient loads leaving a 131 ha watershed in Sydney via both sewage and stormwater. The population of the area is about 5000 and the mean annual rainfall is 1100 mm. On an annual basis the pollution loads of the stormwater are about the same as those of the sewage, assuming the sewage has undergone secondary treatment. For the stormwater the load of suspended solids is about six times as high and of BOD about equal to the corresponding loads in secondary sewage effluent. This means that in absolute terms, simple settling would remove far more suspended solids, but slightly less BOD from the stormwater than tertiary treatment would remove from the sewage. A side benefit of settling the stormwater is the removal of 12% of the phosphate and 2% of the ammonia discharged in the combined wastewater, amounts which are at least equal to, and possibly greater than would be removed by normal tertiary treatment of the sewage without special treatment for nutrient removal. However, simple settling of stormwater for a few minutes, even to handle the spasmodic, short duration, high discharges which characterise urban stormwater runoff, may be cheaper than increasing the level of sewage treatment from secondary to tertiary.

Introduction

The change from rural to urban land-use has very large effects on streams. The quality of the water in the streams deteriorates to a considerable extent. This deterioration of water quality is caused by large volumes of sewage, accidentally or deliberately spilled industrial wastes and storm runoff which is usually of about the same quality as effluent from a secondary sewage treatment plant (Bryan, 1972).

Considerable interest has developed recently in the pollution of water resources by sewage. Local authorities are being required to produce higher quality effluents by a pollution conscious public. Hence a great deal of attention is currently being given to the quality of effluents from sewage treatment plants that are to be discharged into the environment, particularly inland waters. However, it is interesting to note that recent work has shown that the pollution load exerted on receiving water bodies by sewage and industrial effluents may not, in many cases, be the major cause of stream degradation. It has been shown that the pollution load exerted by urban stormwater may be as high, or even much higher than the load exerted by secondary sewage treatment plant effluent (Whipple et al., 1974).

These findings have very significant implications for the allocation of resources for water pollution control. It is possible that it would be more beneficial to attempt to provide simple treatment of urban stormwater than to provide higher levels of treatment of sewage effluent. However, to date the most common mode of reaction by local authorities to public pressure (in Australia, at least, and probably in many other places) concerning water pollution has been to attempt to improve the quality of sewage effluent.

The purpose of this paper is to present urban stormwater quality data collected over a period of 12 months at three sites in Sydney and to examine the possibility of obtaining worthwhile benefits from the treatment of urban stormwater.

Available stormwater quality data has enabled considerable understanding of quality aspects of urban runoff to be reached. However, very little insight has been gained of the processes involved in the contamination of rainwater once it has touched the ground surface. It is well known that early in a storm a flushing effect often occurs, when debris which has accumulated in dry times is washed into the drainage system. At the same time accumulations of material in the drains themselves are stirred up and swept along the pipe or channel system. It is also well known that the pollution load is much higher during flood flows than under low-flow conditions. However, the data obtained so far have not, in general, been sufficient to provide a clear understanding of the overall processes involved in the pollution of stormwater from which reliable, generally applicable predictive models can be derived.

The most readily observed and measured indicator of pollution in stormwater is the suspended solids concentration. A more definitive indicator of general pollution of water is the oxygen demand. Nutrient indicators such as phosphates and nitrates are also important in that they give some indication of the likelihood of eutrophication of water bodies. The other quality variables which appear to be of most significance for planning and water resources purposes, as well as for general pollution abatement are pathogen concentrations and the presence of toxic substances.

Several writers have shown that higher concentrations of some pollutants are found in urban stormwater than occur in raw domestic sewage. The general finding appears to be that urban stormwater is at least as heavily polluted as the effluent from secondary sewage treatment plants. Considerable amounts of lead from vehicle engines have also been observed in urban stormwater.

Stormwater Data for Sydney

Watersheds. Stormwater samples have recently been collected from three separately sewered watersheds in Sydney. Two of these watersheds, Musgrave Avenue Drain and Bunnerong Storm Water Channel (SWC), are in Sydney's Eastern Suburbs where the soils are predominantly sandy, underlain by considerable depths of sand with occasional outcrops of sandstone.

The third site is on Powells Creek at Strathfield where the soils are of the Wianamatta group--clays underlain by shale. All three watersheds are primarily residential areas, although Bunnerong SWC and Powells Creek contain minor commercial developments such as shopping centers. Over 20% of Musgrave Avenue watershed is parkland. A brief summary of the characteristics of the watersheds is given in Table 1.

Table 1. Characteristics of Drainage Areas

	Musgrave Avenue Drain	Powells Creek	Bunnerong SWC
Area (ha)	131	231	55
Maximum elevation difference (m)	70	40	20
Mean slope (%)	5.0	2.0	1.5
Mean annual rainfall (mm)	1150	990	1130
Land use	Residential—individual dwelling units set in about 0.04 ha, 10% of dwellings in multi unit blocks	Residential—primarily individual dwellings set in about 0.06 ha	Residential—individual units set in about 0.05 ha, some multi unit blocks, shopping center occupies 20% of area
Population (est.)	5000	6500	2000
Soil type	Sand	Clay	Sand
Area of parkland (ha)	27	3	0

Water samples were obtained adjacent to stream gaging stations on each of the watersheds. The three gaging stations are in brick or concrete lined open channels, just downstream of the emergence of water from pipes or enclosed concrete conduits.

Collection of Water Samples

Water samples were collected during low flow times, and during the passage of floods between June 1975 and May 1976. Twenty samples were obtained from low flows, and about 110 samples were obtained during the passage of 15 floods. Ten floods were sampled in Musgrave Avenue Drain, three in Powells Creek and two in Bunnerong SWC.

All samples were instantaneous, "grab" samples. No automatic sampling equipment was available. The samples were collected in large-mouthed plastic or glass containers of about two liters capacity, which, under flood flow conditions took less than one second to fill. In some cases four liter samples were obtained. The samples were "grabbed" so that they would be as representative as possible of the flow at the time. The depth and velocity of flood flows at the points of collection varied from a few centimeters and about 0.2 ms^{-1} to 0.6 meter and 2.5 ms^{-1}. Low flow samples were obtained at points where all or most of the flow could be caught, for example, at a step in a channel or at the outflow from a pipe into a channel.

Stormwater samples were tested for their constituent parameters using various standard tests which are listed by Cordery (1976).

Quality Parameters of Sampled Stormwater

Low flow data. A summary of the quality parameters for samples obtained under low flow conditions is given in Table 2. The mean values shown apply to the individual samples and are not flow weighted means. It can be seen that on some occasions the BOD and suspended solids loads are almost as high as would be expected for raw sewage. The highest dry weather BOD observed was 135 mg/l and the highest suspended solids concentration was 295 mg/l.

Flood flow data. Flood flow data are summarized in Table 3. Here, the mean concentration of each parameter shown is the flow weighted mean. For comparison purposes typical quality data for raw sewage and secondary effluent from the St. Marys Water Pollution Control Plant in Sydney are also shown in Table 3. The plant at St. Marys, which treats domestic and industrial wastes from a separately sewered area, comprises settling tanks and activated sludge treatment. The sewage is primarily drawn from residential areas with some light industry.

In Table 3 it can be seen that whilst the mean suspended solids concentration of the urban stormwater is the same as for raw sewage, individual values significantly higher than for raw sewage often occur, with a maximum value of 1400 mg/l being observed. The BOD of stormwater from the sandy soil watersheds was consistently higher than for secondary sewage-treatment effluent. The maximum BOD concentration observed for the stormwater was 145 mg/l. Fecal coliform counts are about one order of magnitude less than the total coliforms. During high flows the coliform counts tend to be slightly lower than under low flow conditions, but occasional very high values occur. Some parameters, such as sulphate, alkalinity, hardness, chloride and silicate were only measured from a few samples.

Table 2. Dry Weather Flow Quality Data

Parameter	Concentration (mg/1 except pH, coliforms and turbidity)								
	Musgrave Avenue Drain			Powells Creek			Bunnerong SWC		
	Number of Samples	Mean	Range	Number of Samples	Mean	Range	Number of Samples	Mean	Range
Suspended solids	10	59	0—295	4	85	10—240	6	19	8—115
Dissolved solids	8	250	210—330	4	690	600—840	5	184	90—220
Ammonia Nitrogen	10	1.7	0.3—4.9	4	1.14	0.4—2.1	6	0.88	0.32—2.6
Nitrate Nitrogen	9	1.5	1.0—2.5	2	0.98	—	6	0.39	0.12—1.05
Nitrite Nitrogen	8	0.21	0—0.76	2	0.14	0.13—0.16	5	0.02	0—0.03
Phosphate as P	9	1.6	0.8—3.2	4	2.2	0.7—4.9	5	1.35	0.8—2.0
BOD_5	10	31	4.2—135	4	7.6	3.3—12	6	19	5.9—63
pH	10	7.42	6.71—7.83	3	7.51	7.22—7.70	6	7.34	6.92—7.90
Total coliforms (MPN/100 ml)	8	5.3×10^6	0.5×10^6— 18×10^6	3	4×10^6	5×10^4— 11×10^6	5	3.3×10^6	0.2×10^6— 16×10^6
Hardness (total)	8	100	84—136	2	158	154—162	6	81	56—98
Alkalinity	4	50	32—62	2	86	85—88	4	58	40—76
Chloride	8	38	29—59	2	186	136—237	6	34	15—41
Sulphate	7	45	28—60	2	99	82—115	6	29	20—47
Silica	7	6.6	5.0—8.4	2	8.2	6.4—10.0	6	7.0	4.2—9.5
Turbidity (FTU)	10	66	9—250	2	132	100—165	6	27	8—52

Table 3. Comparison of Quality of Urban Flood Flows and Sewage Effluent

Parameter	Concentration (mg/l except pH and coliforms)										
	Musgrave Avenue Drain			Powells Creek			Bunnerong SWC			St. Marys Water Pollution Control Plant	
	Number of samples	Flow Weighted Mean	Range	Number of samples	Flow Weighted Mean	Range	Number of samples	Flow Weighted Mean	Range	Raw sewage effluent	Secondary sewage effluent
Suspended solids	69	276	12—940	31	187	30—1400	11	212	95—520	270	28
Dissolved solids	57	120	26—390	31	118	30—650	11	137	40—255	570	530
Ammonia Nitrogen	63	1.89	0.35—13.3	31	1.24	0.37—14.0	11	2.25	1.2—4.1	41	22
Nitrate Nitrogen	23	0.78	0.14—1.9	4	0.96	0.51—1.60	6	0.61	0.22—0.86	—	4
Phosphate as P	41	2.42	0.75—7.6	29	1.85	0.63—5.1	6	1.67	0.63—4.3	10.5	10
BOD_5	66	22.6	1.3—145	31	12.8	2.5—71	11	19.3	5.8—62	265	16
DO	22	5.7	0.3—7.9	6	3.1	0.5—5.0				0.6	3.5
pH	44	6.78	5.80—7.55	19	6.98	6.61—7.52	11	6.47	6.13—6.81	7.4	7.2
Fecal coliforms (MPN/100 ml)	12	51×10^4	0—120×10^4	7	56×10^4	9×10^4—93×10^4	2	6.5×10^4	4×10^4—9×10^4	20×10^6	1180
Total coliforms (MPN/100 ml)	15	3.3×10^6	40×10^4—23×10^6	7	4.4×10^6	23×10^4—9.3×10^6	3	2.0×10^6	23×10^4—4.6×10^6	—	—

Variation of Quality Parameters with Discharge

Discharge, concentration and load data from one flood in Musgrave Avenue Drain and one in Powells Creek are shown in Figs. 1 and 2. These data clearly show many quality characteristics of urban stormwater. For instance, Fig. 1 shows the "first flush" with high concentrations of pollutants resulting from two, early, light showers. However, it also indicates the significant increase in pollutant concentration whenever the discharge increases rapidly, as shown at 1725 hours and 1755 hours. The pollution concentration generally decreases with duration of rainfall, but as shown in Figs. 1b and 1c, the rate of transmission of pollutants (grams/s) by the system is much more dependent on the rate of flow than on the concentration. High concentrations at 1535 hours and

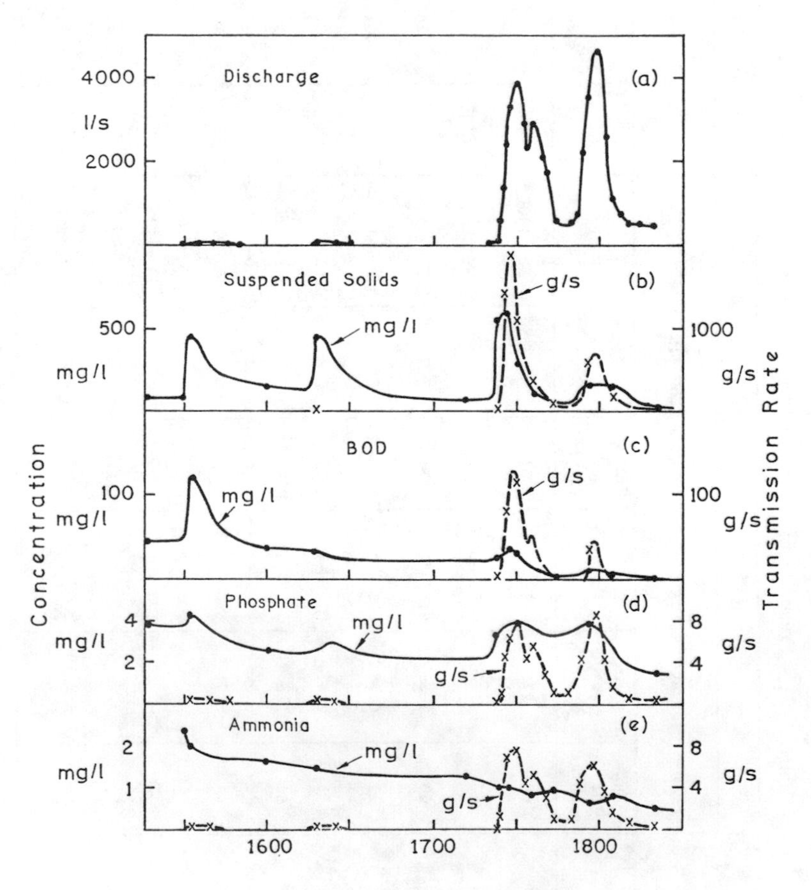

Fig. 1. Musgrave Drain October 20, 1975. Concentration (mg/l) and transmission rate (grams/s) of pollution and nutrient indicators

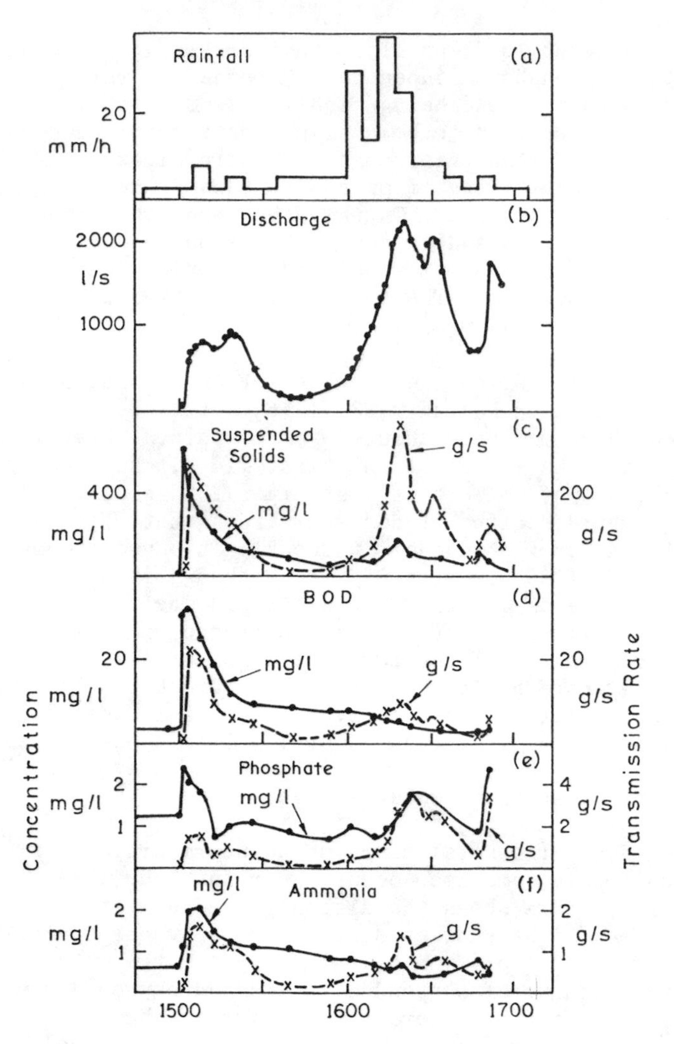

Fig. 2. Powell's Creek February 28, 1976. Concentration (mg/l) and transmission rate (grams/s) of pollution and nutrient indicators.

1610 hours associated with flows of less than 100 l/s resulted in the movement of a total of only 20 kg of suspended solids and 3 kg of BOD. However the higher flow with lower concentrations of pollutants between 1720 hours and 1810 hours resulted in the movement of 1150 kg of suspended solids and 100 kg of BOD past the gaging station. The total loads of pollutants from the 131 ha watershed in three hours on October 20, 1975, were just over 1.2 tons of suspended material and 100 kg of BOD.

The nutrient load was also considerable. As shown in Fig. 1d, the phosphate concentration tended to remain fairly constant, with the load being highly dependent on the rate of flow. This appears to be a significant result and has considerable importance for possible eutrophication of receiving waters. About 13 kg of phosphate passed the gaging station in this small flood. On the other hand the ammonia concentration fell steadily during the flood and about 11 kg of ammonia was carried. Most of the pollution load occurred in less than one hour. The total rainfall over the watershed for this storm was 13 mm.

Fig. 2 shows discharge, concentration and pollution load data for a storm on February 28, 1976, over Powells Creek watershed which produced about 20 mm of rain. This Figure clearly shows the effect of the first flush, which causes high concentrations of pollutants just after 1500 hours. It also shows the rapid decline of concentrations that often occurs before the discharge reaches its peak. This phenomenon is also shown at 1725 hours in Fig. 1, although it is not as clearly visible as in Fig. 2. This rapid decline of concentration of pollutants before the occurrence of peak discharge was observed in all of the larger floods sampled, and it was consistently more noticeable in Powells Creek than in Musgrave Avenue Drain. It was not observed in small floods which had peak discharges of less than about 1.5 liters/ha/s, presumably because in those cases there was more material available for a first flush than could be moved by the flood water.

When small freshets occur in the storm drains, very high pollution levels sometimes occur. For instance, Fig. 1 up to time 1700 hours shows the discharge and quality data which resulted from light showers over Musgrave Avenue Drain. The showers produced 1 or 2 mm of rainfall over the watershed. Whilst the amounts of suspended solids and oxygen demand resulting from these showers are not great, the high concentration of pollutants and nutrients could have a considerable effect on a receiving stream which had a predominantly rural watershed, which would be the usual situation for an inland city. Under these circumstances the receiving stream would be at a very low, dry weather discharge which would be unaffected by a shower yielding 1 mm of rain. The stream would probably have insufficient diluting power to prevent oxygen levels from dropping to a level which would probably cause the death of fish. The nutrient levels would be high enough to promote rapid growth of all forms of aquatic vegetation. Small showers such as this are a frequent occurrence in any area. Other examples are given by Cordery (1976). Because these small discharge, high concentration events occur frequently they could have a very significant effect on any predominantly rural stream which passes through, or close to, an urban development.

Observation of flows during floods has indicated that there are far more suspended solids than those measured in the samples. Considerable amounts of gross solids such as leaves, wood, cardboard, plastic and metal containers and sundry other large objects pass the measuring points during flood flows. These debris occur in large quantities at the same time as high concentrations of suspended solids are observed, i.e., during the "first flush" and when the discharge increases rapidly. The concentration of these large debris has not been estimated because suitable equipment for obtaining representative samples was not available.

Other parameters tend to vary in a characteristic way for all floods. As shown in Figs. 1 and 2, and mentioned earlier, the phosphate level remains approximately constant during the passage of each flood whilst the ammonia level tends to fall gradually with time. In an allegedly phosphate poor environment the amounts of phosphate carried are very significant. Large floods usually reduce the concentration of pollutants, but in the urban environment this does not appear to be the case for phosphates. Nitrate concentration varies very little during the passage of a flood except to decrease slightly during and after peak flows. Table 3 shows the small range of variation of nitrate in the stormwater.

The concentration of dissolved solids tends to fall very rapidly at the beginning of each increase in runoff and remain at a low level until the flood has passed. It is of interest to note from Table 2 that the concentration of dissolved solids under dry weather flow conditions is about 200 mg/l for the sandy soil basins but over three times as high for the clay soil, Powells Creek area. pH tends to fall early in the passage of a flood and to remain approximately constant. The data shown in Tables 2 and 3 indicate that flood water is slightly acidic whilst dry weather flow is slightly alkaline.

After prolonged rain, water in all three storm drains becomes relatively pollution free. Samples have been collected in Musgrave Avenue Drain and Powells Creek after several days of continuous rain during which about 150 mm was recorded. These samples indicated maximum suspended solids and BOD concentrations of 70 mg/l and 3 mg/l, respectively. All other concentrations were also low--as low as the lowest values observed during any other flood. This finding indicates that most of the pollution is washed from urban watersheds by the first 10-20 mm of rainfall, provided fairly high intensities occur, and that only minor amounts are removed by subsequent rain. This is indicated late in the events shown in both Fig. 1 and Fig. 2.

Oil slicks appear on the stormwater during all floods in the three drains sampled. Major difficulties were encountered in obtaining samples which could be assumed to indicate the concentration of oil in the water due to the fact that the oil is not distributed throughout the water but floats on the surface. Representative oil samples could only be obtained with highly sophisticated equipment, which was not available. One 2-liter sample collected from Musgrave Avenue Drain on November 17, 1975 after several millimeters of rain had washed the streets in the previous few hours contained 0.90 gram of oil. Another sample collected at the same point on February 1, 1976, contained 1.08 grams of oil in one liter of water.

Some authors (e.g., Bryan, 1972) have indicated significant lead concentrations in urban runoff, presumably originating from emissions from motor vehicle engines. Samples from the flood of October 20, 1975, discussed above and shown in Fig. 1 had maximum lead concentrations of 0.5 mg/l.

Load of Stormwater Pollutants

As discussed earlier, Figs. 1 and 2 show that the concentration of all constituents is highest early in each storm, but that the peak rate of movement of pollutants coincides not with the maximum concentration, but approximately with the maximum discharge. Other authors (e.g., McElroy and Bell, 1974) have found the same characteristics.

The storm of October 20, 1975 over Musgrave Avenue Drain watershed deposited 13 mm of rain in a period of 3 hours, as shown in Fig. 1. As discussed earlier the resulting flood carried 1200 kg of suspended solids, 100 kg of BOD, 13 kg of phosphate and 11 kg of ammonia in about 6500 m^3 of water. As shown elsewhere (Cordery, 1976) a storm which deposited 7.5 mm of rain over the watershed in about 10 minutes on November 23, 1975 removed 800 kg of suspended solids, 70 kg of BOD, 5 kg of phosphate and 4 kg of ammonia in 2100 m^3 of water. Storms of this size are fairly common in Sydney.

Analysis of records indicates that a daily rainfall of between 2.5 and 15 mm immediately following 3 or more dry days occurs about 19 times per year in Sydney. From data such as that shown in Figs. 1 and 2 it can be estimated that light showers of this type occurring 19 times per year over the 131 ha Musgrave Avenue Drain watershed probably cause the removal of up to 15 tons of suspended solids, 900 kg of BOD, 150 kg of phosphate and 120 kg of ammonia per year.

Samples obtained during major storms have shown that the concentration of most pollutants falls away to very low levels after the first 10-20 mm of rain has fallen. On the average about 10 major storms occur per year in Sydney. On an annual basis, it would seem that major storm events would remove considerable amounts of pollutants. The large volume, low concentration discharges from major storms could be expected to remove at least as much material as the smaller, more frequent events. Storms which do not fit into either of the categories discussed would also contribute to the pollution loads. Hence a very conservative estimate of the annual load of pollutants removed from the 131 ha watershed by stormwater would be about two and a half times the amounts removed by small storms, or 37.5 tons of suspended solids, 2.25 tons of BOD, 375 kg of phosphate and 300 kg of ammonia.

Dry weather flow in Musgrave Avenue Drain averages 3 ℓ/s, contributing about 80,000 m^3 per year. The product of this volume and the mean concentrations for low flows shown in Table 2 would give annual low-flow pollution loads of 4.8 tons of suspended solids, 2.4 tons of BOD, 130 kg of phosphate and 135 kg of ammonia. The estimated total annual load of pollutants carried in Musgrave Avenue Drain is shown in Table 4.

Table 4. Annual Pollutant Loads Carried in Component Discharges--Musgrave Avenue Drain

	BOD_5 (kg)	Suspended Solids (kg)	Phosphate as P (kg)	Ammonia Nitrogen (kg)
Floods resulting from 2.5 to 15 mm of rainfall	900	15 000	150	120
Major floods	1 350	22 500	225	180
Dry weather flow	2 400	4 800	130	135
Total	4 650	42 300	505	435
Total in kg/ha/yr	35.5	323	3.85	3.32

Treatment of Urban Stormwater

Treatment experience. The need for treatment of urban stormwater has been amply demonstrated above and should not be ignored in favour of treating all urban sewage at the tertiary level. The major difficulty in treatment of urban stormwater is the huge discharges which occur for very short periods of time with long periods between flows. Economically it would be quite unreasonable to provide a plant to treat stormwater at its unrestricted flow rate. Treatment of combined sewer effluent is becoming quite common. Where treatment is practiced, storage forms an integral part of almost all the treatment works (Field and Lager, 1975; Lager, 1974;

McPherson, 1974). The methods of treatment employed are, as one would expect for the combination of sewage and stormwater, similar to those used for sewage treatment, including screens, settling tanks, sand filters, trickling filters, oxidation ponds and aeration lagoons. Some systems have large concentrated storages whilst others have the storage located throughout the system. Examples are discussed by Lager (1974) and McPherson (1974).

Whilst considerable effort is now being made to treat combined sewage-stormwater flows there is little evidence of separate urban stormwater being treated, except as a byproduct of some other objective. For instance, stormwater held for several days in lakes and ponds which are part of the landscaping of an urban area undoubtedly undergoes considerable aeration and deposits most of its suspended load as it traverses the storage. However treatment of urban stormwater as a specific objective does not seem to be practiced anywhere, in spite of the fact that laboratory tests conducted almost 10 years ago (Evans et al., 1968) showed that significant improvements in stormwater quality could be achieved fairly simply.

Laboratory experiments. In an attempt to investigate the feasibility of simple, cheap treatment of urban stormwater to remove the significant pollutants, samples of freshly collected stormwater from a number of floods were settled in Imhoff cones and the supernatant liquid examined. The stormwater samples tested were grab samples from Musgrave Avenue Drain and Powells Creek. The samples were settled and tested within four hours of collection, wherever possible.

All samples were settled in 1 liter, 45 cm deep, Imhoff cones for 15 minutes and one hour. A few samples were settled for 4 minutes and 8 minutes and some others were settled for 2 hours and 24 hours. Supernatant liquid was drawn off at a depth of 25 cm to carry out tests for suspended solids, BOD, phosphate, ammonia and nitrate. In general it was found that settling had little or no effect on the nitrate concentration and only a small effect on ammonia concentration. However the other constituents were significantly reduced, even with as little as 4 minutes settling. A very wide range of percentage removal of each constituent was achieved. Table 5 and Fig. 3 summarise the test results. Initial BOD levels ranging from 2.3 mg/l to 108 mg/l were reduced after 15 minutes settling by from 1% to 72% with an average of 39%. The percentage reduction of BOD was quite unrelated to the initial BOD level. Large and small reductions were achieved for both high and low initial concentrations.

Table 5. Reduction of Concentration of Various Parameters by Simple Settling for Periods Shown

Constituent	Range of initial concentrations (mg/l)	Mean initial concentration (mg/l)	Mean reduction of concentration %			
			4 mins. settling	8 mins. settling	15 mins. settling	1 hour settling
BOD_5	2.3 — 108	49	34	36	39	41
Suspended Solids	50—1400	430	79	84	87	90
Phosphate	1.05— 7.6	4.15	57	61	62	69
Ammonia	1.5 —14.0	3.94	18	20	22	24

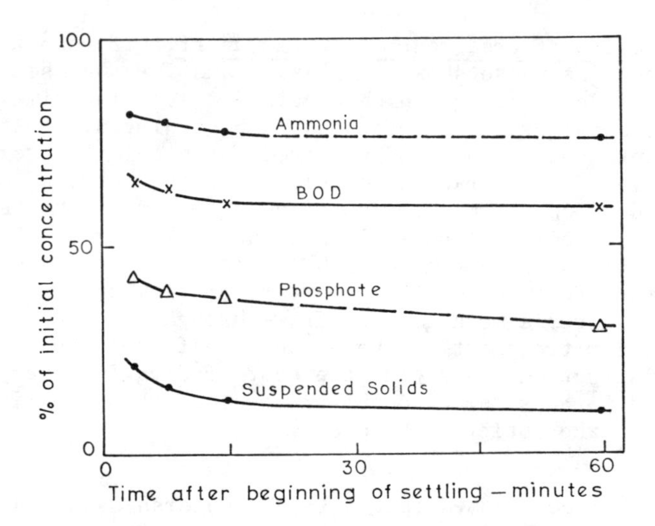

Fig. 3. Average reduction of concentration of pollutants and nutrients by setting stormwater

Removal of total phosphate after 15 minutes settling, with initial concentrations ranging from 1.05 mg/l to 7.6 mg/l, ranged from 47% to 82% with a mean value of 62%. Suspended solids removal was by far the most spectacular. It is probable that the reduction of the concentration of the other constituents was due, in part, to the removal of the suspended solids. After 4 minutes settling, the suspended solids concentrations, which ranged from 50 to 1400 mg/l, were reduced by an average of 79%. After 15 minutes settling, the suspended solids concentrations, which ranged from 50 to 1400 mg/l, were reduced by an average of 79%. After 15 minutes settling, solids were reduced by from 76% to 96%, with a mean removal of 87%.

The significant removal rates discussed above were observed at a depth of 25 cm. O'Connor and Eckenfelder (1958) have presented curves which clearly demonstrate an increase in settling velocity of suspended material with depth in the tank, due to increasing flocculation with depth. It would seem that since settling velocity increases with depth, the degree of removal of material observed at 25 cm depth after 4 minutes could be expected to be reproduced after 15 minutes of settling at a depth of more than 4 times the experimental depth. The precise depth at which the 4 minute, 25 cm result would be achieved after 15 minutes would depend on the flocculating nature of the stormwater, but it could be expected to be between 1.5 and 2 meters. This extrapolation of laboratory data would need to be tested in a pilot tank before it could be used for design purposes.

Benefits of treatment of storm runoff. If runoff from events such as those shown in Figs. 1 and 2 was passed through a 2 meters deep storage which provided a settling time of 15 minutes, considerable benefits could be achieved. A storage of perhaps 2000 m^3 would be required for a 131 ha watershed, with outlet works arranged to permit varying rates of outflow to provide about 15 minutes settling time, irrespective of the rate of inflow. The sedimentation tanks for settling urban runoff would not need to be as large as would be required for treatment of the same rate of flow of sewage since sewage requires a much greater flow-through time. Costs involved for treatment of the urban runoff would be the initial construction costs plus the cost of dredging every few years and perhaps three monthly removal of gross debris from screens at the outlet structure.

It will be assumed that a 131 ha watershed such as Musgrave Avenue Drain has a population of 5000 and that the constitution of the raw sewage from the area approximates that entering the St. Mary's Water Pollution Control Plant in Sydney. Typical concentrations of various constituents of the raw sewage and secondary effluent from the plant are shown in Table 3. The mean sewage discharge for Sydney is about 140 liters/person/day. Hence the annual loads of various pollutants in raw sewage and secondary sewage effluent from Musgrave Avenue Drain watershed would be as shown in Table 4. The stormwater loads are as shown in Table 4. Inspection of Table 6 indicates that, compared with secondary sewage treatment plant effluent, storm runoff carries about six times as much suspended solids, slightly more BOD, about 20% as much phosphate and 10% as much ammonia. Rows (7) and (8) of Table 6 indicate that, in terms of overall pollution abatement, benefits comparable to those obtainable from tertiary treatment of sewage can be achieved from a few minutes simple

Table 6. Estimated Annual Load of Various Constituents of Wastewater from Musgrave Avenue Drain Watershed

		BOD_5 (kg)	Suspended Solids (kg)	Phosphate as P (kg)	Ammonia Nitrogen (kg)
Raw Sewage	(1)	68 000	69 000	2 700	10 500
Secondary sewage effluent	(2)	4 100	7 200	2 500	5 600
Tertiary sewage effluent (est.)	(3)	2 050	3 600	2 500	5 600
Urban runoff	(4)	4 650	42 300	500	430
Urban runoff settled for 15 minutes in a 2 m deep tank	(5)	3 070	8 900	215	350
Secondary sewage effluent plus urban runoff	(6)	8 750	49 500	3 000	6 030
Tertiary sewage effluent plus urban runoff	(7)	6 700	45 900	3 000	6 030
Secondary sewage effluent plus settled urban runoff	(8)	7 170	16 100	2 715	5 950
Tertiary sewage effluent plus settled urban runoff	(9)	5 120	12 500	2 715	5 950

settling of storm runoff. This is based on the assumption that tertiary treatment would remove about 50% of the BOD and suspended solids from the secondary sewage effluent. Tertiary treatment of sewage would have a larger effect on the total BOD load than settling of stormwater but a much smaller effect on the suspended solids and little or no effect on the nutrient loads. Comparison of rows (6), (7) and (9) of Table 6 indicates that it is probably not worth providing tertiary treatment of sewage without treating storm runoff.

The results shown in Table 6 are based on the assumption that the population of the 131 ha watershed is 5000, or 38 per ha. If the population density were lower, the load of pollutants carried in sewage would be less but the load carried by the stormwater would only be slightly affected. Hence for a lower population density, treatment of urban runoff would provide greater advantages relative to tertiary treatment of sewage than those shown in Table 6.

The limited resources available for control of pollution and the results outlined above indicate that serious consideration should be given to the incorporation of simple treatment of storm runoff in the overall strategy of water pollution control in urban areas. Treatment of urban storm runoff may have little effect on the internal environment of a city but could make a considerable contribution to the quality of receiving waters downstream of the urban area.

Conclusion

Measurement of quality parameters of urban runoff in Sydney has shown that significant pollution and nutrient loads are carried in surface waters emanating from completely sewered areas. As observed in other urban areas the concentration of pollutants tends to increase rapidly at the beginning of flood flow and then to decrease as the flood progresses. During the early part of each flood the concentration of suspended solids is usually higher than for raw sewage and the BOD is usually between 20% and 50% of that of raw sewage. This early stormwater has significantly higher concentrations of pollution materials than secondary sewage effluent. During the later stages of urban floods the pollution concentrations usually fall to about the same level as for secondary sewage effluent.

The concentration of nutrients tends to remain constant or to fall slowly during the passage of a flood. Ammonia behaves similarly to BOD but phosphate concentration tends to remain approximately constant. This constant concentration of phosphate is superimposed by fairly high peak values which occur each time the discharge increases. The nutrient concentrations are usually in the range of 5-40% of those observed in secondary sewage effluent.

In the long term, the total load of pollutants carried in storm runoff is considerably higher than in secondary treatment plant effluent from the same area. The shock load of polluting material carried in stormwater can be even more important, especially for streams near inland cities. Floods which resulted from up to 15 mm rainfall carried very large loads of pollutants. Such storms over rural watersheds would often produce no significant runoff and hence the above loads would seriously degrade a receiving stream which would have little or no ability to dilute the urban runoff.

Simple laboratory tests were made to examine the possible benefits obtainable from treatment of urban runoff. The tests showed that for the Sydney watersheds simple settling of the urban runoff could reduce the pollution load contributed to receiving waters by an amount approximately equal to the reduction that could be obtained by tertiary treatment of the secondary effluent from the same area. It was also shown that tertiary treatment of secondary sewage effluent, without treatment of stormwater, would produce very little reduction in the overall pollution load carried in water flowing from an urban area.

The cost of treating storm runoff by simple settling would probably be much less than the cost of tertiary treatment of secondary sewage effluent from the same area.

Acknowledgments

The author is indebted to Mr. H. H. Rups for performing the oil and lead concentration tests and for other helpful advice concerning testing of water samples, and to the Secretary, Metropolitan Water, Sewerage and Drainage Board, Sydney for supplying sewage effluent data shown in Table 3.

References

[1] Bryan, E. H., 1972. "Quality of Stormwater Drainage From Urban Land," Water Resources Bulletin, Vol. 8, June, 578-588.

[2] Cordery, I., 1976. "Evaluation and Improvement of Quality Characteristics of Urban Stormwater," The University of New South Wales, Water Research Laboratory Report No. 147, October.

[3] Evans, F. L., Geldreich, E. E., Weibel, S. R. and Robeck, G. G., 1968. "Treatment of Urban Stormwater Runoff," Journal, Water Pollution Control Federation, Vol. 40, No. 5, May, R162-R170.

[4] Field, R. and Lager, J. A., 1975. "Urban Runoff Pollution Control--State-of-the-Art," Proceedings, ASCE, Journal of Environmental Engg Div., Vol. 101, No. EE1, Feb., 107-125.

[5] Lager, J. R., 1974. "Stormwater Treatment: Four Case Histories," Civil Engineering--ASCE, Vol. 44, No. 12, December, 40-44.

[6] McElroy, F. T. R. and Bell, J. M., 1974. "Stormwater Runoff Quality for Urban and Semi Urban/Rural Watersheds," Water Resources Research Center, Purdue Univ., Tech. Report No. 43, Feb.

[7] McPherson, M. B., 1974. "Innovation: A Case Study," ASCE Urban Water Resources Research Program, Tech. Memorandum No. 21, Feb.

[8] O'Connor, D. J. and Eckenfelder, W. W., 1958. "Evaluation of Laboratory Settling Data for Process Design," Biological Treatment of Sewage and Industrial Wastes, Vol. II, Eds. McCabe, Brother Joseph and Eckenfelder, W. W., Reinhold, New York.

[9] Whipple, W., Hunter, J. V. and Yu, S. L., 1974. "Unrecorded Pollution From Urban Runoff," Journal, Water Pollution Control Federation, Vol. 46, No. 5, May, 873-885.

PREDICTING THE EFFECT OF SHADE REMOVAL ON STREAM TEMPERATURE

By

David R. DeWalle
Associate Professor of Forest Hydrology
School of Forest Resources

Steven L. Taylor, Research Technologist
Institute for Research on Land and Water Resources

James A. Lynch
Assistant Professor of Forestry
School of Forest Resources
The Pennsylvania State University
University Park, PA 16802

Abstract

The temperature regime occurring after cool forest streams flow into unshaded channels was theoretically analyzed to determine the impact of riparian vegetation removal on stream temperatures. A model was developed to compute the temperature of water parcels as the parcels were routed through an unshaded reach. Heat exchange due to solar and longwave radiation, convection of latent and sensible heat and channel heat conduction were considered. Based upon the model, maximum daily water temperatures near the inlet in the unshaded reach increase almost linearly with distance. In this region absorbed solar radiation is the dominant heat exchange mechanism; the net sum of the other heat transfer mechanisms being negligible due to the relatively low water temperatures. Farther from the inlet, maximum water temperatures increase exponentially and approach an equilibrium condition as absorbed solar heat gains are balanced by heat losses. An analytical solution for the prediction of maximum daily temperatures increases in unshaded channels was developed by simplifying the heat exchange expressions.

Introduction

Thermal pollution of small forest streams by exposure to solar radiation may result from removal of riparian vegetation for construction of right-of-ways, urban development and timber harvesting. Many small forest streams support a delicate ecosystem which is intolerant of elevated temperatures likely to result from shade removal. A technique is needed to predict water temperatures within unshaded channels, especially maximum temperatures. The purpose of this paper is to present results of a theoretical analysis of stream temperatures as cool forest streams flow into unshaded channels. The theoretical analysis includes a computer simulation of the temperature distribution in unshaded channels and an approximate analytical solution for prediction of maximum daily water temperatures.

Computer Model

In the computer model the temperature of stream water parcels were computed as the parcels were routed through an unshaded channel reach. A hypothetical unshaded channel reach was divided into subsections of equal length separated by nodal points. At the inlet or first nodal point, new parcels of water were added at regular time intervals at an appropriate temperature. The temperature change of a water parcel with time as it flowed through the reach was computed from

$$\frac{dT}{dt} = \frac{Q}{\rho c d} \qquad (1)$$

in which T is water temperature, t is time, ρc is the volumetric heat capacity of water, d is stream depth, and Q is the net rate of heat exchange of the parcel which varies with T, t and x (the downstream distance). Stream velocity and depth were assumed constant in the reach. Heat exchange due to groundwater or tributary inflows was not considered.

The modified Euler method was used to numerically integrate Eq. (1) using a time step equal to the subsection length divided by stream velocity. With appropriate functions to describe the net rate of heat exchange, the model was cycled over diurnal time periods with varying meteorological conditions.

Heat exchange components. The net rate of heat exchange (Q) between the moving parcel and environment was the algebriac sum of rates of heat exchange due to absorbed solar radiation, net longwave radiation, evaporation/condensation, sensible heat exchange, and channel conduction. All but the latter term was calculated directly from given functions. Channel heat conduction was computed from a numerical solution to the Fourier heat conduction equation.

During the computer calculations the instantaneous flux density of downward solar irradiation (S_p) was allowed to vary with time of day according to potential solar beam irradiation equations given by Lee (1963) as:

$$S_p = (S_o/R^2)\ (\cos \phi \cos \delta)\ (\cos \omega t - \cos \omega t_s) \qquad (2)$$

in which S_o is the solar constant (1,394 W m^{-2}), R^2 is the square of the radius vector, ϕ is the latitude, δ is the solar declination, ωt is the hour angle at time t, and ωt_s is the hour angle at sunrise or sunset. The radius vector and solar declination are obtained from ephemeris tables for a specific data. Clear-sky conditions were assumed by calculating solar irradiation as 0.8 of the potential irradiation calculated above. The fraction of solar radiation reflected by the stream (r) was computed from an equation for clear sky given by Anderson (1954) as

$$r = 1.18\alpha^{-0.77} \tag{3}$$

in which α is the solar altitude or the arcsin of the product of the last two bracketed terms in Eq. (2).

The net longwave radiation exchange between the water and atmosphere was computed using the Stefan-Boltzmann equation (Munn, 1966). The water parcel was assumed to emit longwave radiation at water temperature with an emissivity of 0.97, while the atmosphere was assumed to emit at air temperature with an emissivity of 0.86. Air temperature was allowed to vary with time during the day as a sine function with an amplitude of 5°C which peaked three hours past solar noon.

Evaporation or condensation heat exchange was computed using an equation developed by Shulyakovskiy (1969) and modified by Ryan et al. (1974) which contains a virtual temperature difference term to account for the effects of free convection from a heated water surface. DeWalle (1976) found water temperature predictions for a thermally-loaded stream were significantly improved using this equation rather than an equation without such a correction. Latent heat exchange (LE) in W m^{-2} was calculated using this equation as

$$LE = \{3.14v_2 + 2.64(T_{vs} - T_{va2})^{1/3}\} (e_{a2} - e_s) \tag{4}$$

in which v_2 is relative wind velocity at 2 m height in m s^{-1}, T_{vs} is the virtual temperature of saturated air at the water surface temperature in °C, T_{va2} is the virtual temperature of the air at 2 m in °C, e_s is the saturation vapor pressure in mb at the water surface temperature, and e_{a2} is the atmospheric vapor pressure at 2 m in mb. Relative wind velocity is obtained by vector subtraction of wind and stream surface velocities. Water surface temperature was assumed equal to the parcel temperature. When the virtual temperature difference was negative, which occurred commonly in the initial subsections of the reach, this term was set to zero. Sensible heat exchange (H) was computed from LE using the Bowen ratio (β) as $\beta \cdot LE$ (Munn, 1966).

A constant relative wind velocity (v_2) was assumed throughout each 24-hr cycle period. Atmospheric vapor pressure (e_{a2}) was also assumed constant throughout the day and was computed as 0.6 of the saturation value at the maximum daily air temperature.

Channel heat conduction rates for each subsection in the unshaded reach were obtained by solving for the theoretical temperature distribution in the channel bottom. The bottom temperature distribution was computed using an implicit Crank-Nickelson numerical solution for the one-dimensional Fourier heat conduction equation. The channel bottom at each subsection was divided by nodal points with one centimeter spacing down to 30 cm depth. The deepest two nodal points were assumed to have the same temperature. Channel heat conduction rates were obtained by multiplying temperature changes with time in the channel bottom by the volumetric heat capacity of the bottom material at each depth and summing over the entire 30 cm deep layer. Two bottom materials were assumed: 1) a rock bottom with a thermal conductivity of 2.51 J cm^{-3} $°C^{-1}$, and 2) a mud bottom with a thermal conductivity of 0.84 W m^{-1} $°C^{-1}$ and heat capacity of 3.76 J cm^{-3} $°C^{-1}$.

The calculation procedure used to obtain a heat flux for a specific channel subsection and time interval was to initially use the computed channel heat flux from the preceding time interval in that subsection to predict a water temperature for the downstream nodal point using Eq. (1). Averaging this predicted water temperature with the upstream temperature, a new channel surface temperature boundary condition is obtained. Using this average temperature, a numerical solution for the temperature distribution with depth and heat flux for this time interval were again obtained. Using this channel heat flux, the downstream water temperature for this subsection was again estimated using Eq. (1). If this estimate of downstream water temperature differed by more than 0.1°C from the previous estimate using Eq. (1), the process was repeated beginning with a new average water temperature.

The initial conditions were established at six hours before solar noon with all bottom and stream nodal points at the inlet temperature. Initial conditions that would better represent the stream for maximum temperature calculations were obtained by cycling the program for consecutive 24 hour periods. Stable values for the initial conditions were obtained after three cycles.

Example stream temperature distributions. Temperatures of water parcels leaving the inlet at regular time intervals were calculated at various downstream distances in the unshaded reach. In Fig. 1 the typical distribution of maximum

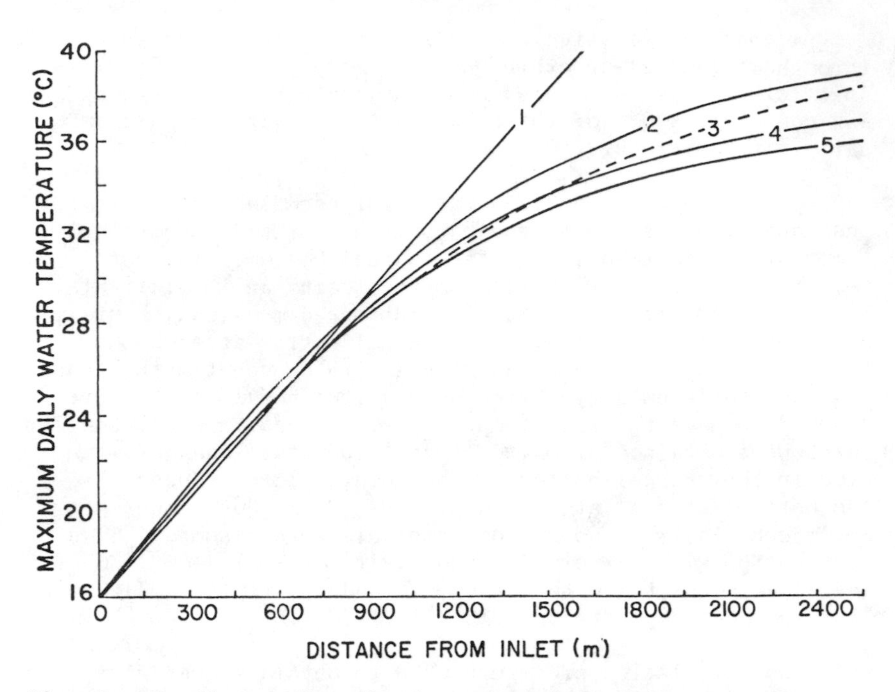

Fig. 1. Predicted distribution of maximum daily water temperatures in an unshaded channel with distance from the inlet using several head balance methods. A stream at 16°C flowing at 0.1 m depth and 0.15 m s^{-1} velocity into an unshaded reach with a 30°C maximum air temperature, 2 m s^{-1} wind velocity during a clear day on 1 August at 40° latitude is represented. The predicted temperature distribution using absorbed shortwave only is represented by line 1, line 2 shows temperatures for all heat exchange components but without channel conduction, line 3 is for all components with a rock channel without free convection, line 4 is for all components with a mud channel and, line 5 is for all components with a rock channel.

daily water temperatures with distance from the inlet is presented for a specific set of conditions and for various heat exchange calculation options. Calculations were made for a stream velocity of 0.15 m s^{-1}, a depth of 0.1 m, and a constant inlet water temperature of 16°C. Atmospheric conditions corresponded to a clear day on 1 August at 40° latitude with a maximum daily air temperature of 30°C and constant relative wind velocity of 2 m s^{-1}. Atmospheric vapor pressure was held constant at 60% relative humidity at the maximum daily air temperature in all calculations. Calculations were obtained for a stream where surface heat exchange was comprised of 1) absorbed solar only, 2) all heat exchange components except bottom conduction, 3) all components with a rock bottom but no free convection term in Eq. (4), 4) all components with a simulated mud bottom, and 5) all components with a simulated rock bottom.

Near the inlet, maximum daily water temperature increased almost linearly with distance, and all heat budget calculation techniques gave essentially the same results. Absorbed solar radiation alone accounted for nearly all temperature increases near the inlet, while the net total of longwave radiation, sensible heat, evaporation/condensation and channel conduction heat exchange was nearly zero due to low water temperatures. As the water heated, emitted longwave radiation, channel conduction and evaporation all contributed significantly to heat losses from the stream to balance absorbed solar heat gains. At this point in the unshaded reach temperature increases with distance from the inlet were reduced markedly and an equilibrium temperature was approached.

The importance of accounting for channel heat conduction can be determined by comparing the temperature distributions in Fig. 1 computed with and without a simulated rock channel. Accounting for channel heat conduction reduced the predicted maximum daily water temperatures up to 3°C at the maximum distances from the inlet. A rock bottom was more effective than mud in reducing the predicted water temperatures. Heat gained in the channel during mid-day was released the following night as the stream cooled.

When the free convection term in Eq. (4) was dropped in the calculation scheme, predicted water temperatures were affected very little. The free convection term in Eq. (4) with a 2 m s^{-1} wind velocity represents only about half of the calculated evaporative heat exchange for the virtual temperature difference of about 10°C which occurred at the end of the unshaded reach. The virtual temperature difference first became positive when the water temperature was about 28°C or at the 800 to 900 m distance.

The time of the maximum daily water temperature varied within the unshaded reach. Near the inlet, maximum water temperatures occurred at about solar noon. Farther from the inlet, maximum daily water temperatures occurred after solar noon by a time increment equal to about one-half of the distance from the inlet divided by the velocity. This indicates that parcels of water which achieve the maximum water temperature at distances far from the inlet have a travel time centered roughly on solar noon. At even greater distances downstream, the time of maximum daily temperatures becomes relatively constant.

Using the same meteorological and stream conditions used for Fig. 1 and a simulated mud bottom, a sensitivity analysis was performed to determine the most important variables affecting maximum daily water temperatures. The effects of a step change in each of the stream and meteorological variables on the predicted water temperatures at 600 and 1800 m are presented in Fig. 2a and 2b, respectively. At 600 m water temperature was about 25°C and still increasing almost linearly

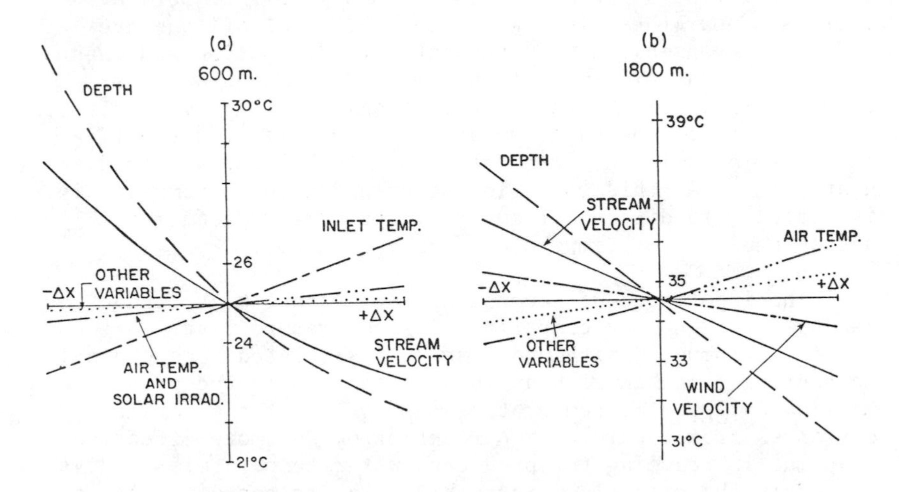

Fig. 2. Effect of a step-change in meteorological and stream variables at a) 600, and b) 1800 m from the inlet. Conditions are the same as used in Fig. 1 with a mud bottom. Step-change (Δx) for stream depth was 0.05 m; stream velocity, $\Delta x = 0.05$ s m^{-1}; inlet temperature, $\Delta x = 2$°C; wind velocity, $\Delta x = 1$ m s^{-1}; air temperature, $\Delta x = 5$°C; vapor pressure, $\Delta x = 5$ mb; and solar irradiation, $\Delta x = 50$ W m^{-1}.

with distance, while at 1800 m water temperature was about 34.5°C and the rate of increase with distance was declining. The step change imposed on each variable (Δx) was selected to approximate realistic variations. At 600 m (Fig. 2a) stream temperature predictions were most sensitive to depth, stream velocity and inlet temperature in that order. Meteorological variables were of minor importance. At 1800 m (Fig. 2b) stream depth and velocity were still important but meteorological conditions were of greater importance than at 600 m, especially air temperature and wind velocity. Inlet temperature was less important at the greater distance from the inlet.

Analytical Solution

Based upon the computer analysis results, an approximate analytical solution for maximum daily water temperatures in an unshaded reach was obtained. The general one-dimensional differential equation for the stream heat budget is

$$\delta T/\delta t + u\ \delta T/\delta x = Q/\rho cd \qquad (5)$$

where u is stream velocity and the other terms have been previously defined. An analytical solution to this equation is not possible without simplification of the net heat exchange rate term (Q).

In order to simplify Eq. (5), Q was considered to be composed of three terms: absorbed solar radiation (Q_s), channel heat conduction (Q_b) and the net sum of the other non-solar surface heat exchange components (Q_{ns}) including long-wave radiation, latent heat and sensible heat. Absorbed solar radiation is a function of time, the non-solar surface components vary with water temperature and time while channel heat conduction varies with water temperature, time and distance.

The time variation of solar irradiation was circumvented by using an average solar flux density over the travel time of water parcels producing maximum daily water temperatures. Based upon the observation that parcels producing maximum daily water temperatures have travel times roughly centered on solar noon, average absorbed solar radiation was computed by integrating the potential solar beam irradiation equation over travel time. Travel time was computed as distance from the inlet divided by stream velocity. Atmospheric transmission of 0.8 and a constant solar reflectivity of 0.05 were used in the calculations. Average absorbed solar radiation (Q_s) for maximum temperature parcels as a function of travel time and time of year are given in Fig. 3 for latitudes of 30°, 40° and 50°. An alternative procedure would be to calculate average absorbed solar for any date and latitude using

$$Q_s = Q_{max} - a\ x^2/u^2 \qquad (6)$$

in which Q_{max} is the maximum absorbed solar flux density, x is distance from the inlet, u is stream velocity, and a is computed as

$$a = 0.76\ S_o \cos\phi \cdot \cos\delta \cdot \omega^2/24R^2 \qquad (7)$$

in which $\omega^2 = 5.29 \times 10^{-9}\ rad^2\ s^{-2}$. Q_{max} can be calculated as 0.76 S_p with $\cos \omega t = 1$ in Eq. (2).

The results of the computer calculations indicated that an approximate linear relationship existed between water temperature and average channel heat conduction rate experienced

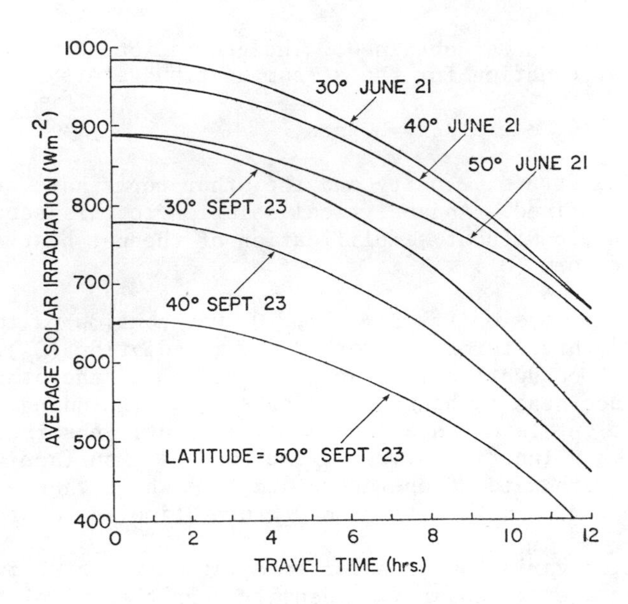

Fig. 3. Average absorbed shortwave radiation of maximum temperature water parcels as a function of travel time for latitudes of 30°, 40° and 50°C on 21 June and 23 September.

by the maximum temperature parcels. Thus channel heat conduction, Q_b in W m^{-2} was computed as

$$Q_b = A_b + B_b(T - T_i) \tag{8}$$

in which A_b is the channel conduction rate at the inlet at solar noon, B_b is the channel conduction coefficient and T_i is the inlet water temperature at solar noon. If inlet water temperature is held constant, $A_b = 0$; otherwise A_b can be calculated as

$$A_b = B_b(T_i - T_{ii}) \tag{9}$$

in which T_{ii} is the inlet temperature six hours before solar noon. A constant inlet temperature assumption can be used with little error, if the maximum expected daily forest stream temperature is used. The channel conduction coefficients (B_b) fit to the computer analysis results were 12.1 and 9.0 W m^{-2} °C^{-1} for rock and mud bottoms, respectively. The channel heat conduction rates for mud and rock bottoms are given in Fig. 4.

The non-solar surface heat exchange components were also approximated as a linear function of water temperature. In

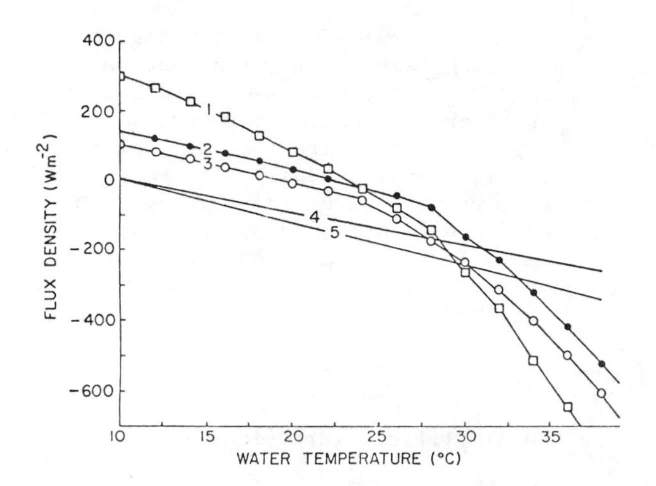

Fig. 4. Average channel heat conduction and non-shortwave surface heat exchange rates as a function of water temperature. Non-shortwave surface heat exchange rates (lines 1-3) given for: 1) 30°C air temperature, 1 m s^{-1} wind velocity and 25 mb vapor pressure; 2) 25°C, 1 m s^{-1}, 25 mb, and 3) 30°C, 3 m s^{-1}, 25 mb. Average channel heat conduction (lines 4 and 5) shown for: 4) mud bottom and 5) rock bottom.

Fig. 4 the non-solar heat exchange, computed as previously described for the computer analysis, is plotted as a function of water temperature for several combinations of wind velocity and air temperature. As indicated in Fig. 4, the slopes of the line segments of heat exchange coefficients vary markedly with wind velocity and somewhat less with air temperature. The non-solar heat exchange (Q_{ns}) must be approximated with two straight-line segments with different slopes and an intersection at the water temperature at which the free convection term in the evaporation equation becomes positive.

The equation for the first line segment before free convection begins is

$$Q_{ns} = A_{nfc} - B_{nfc}(T - T_i) \tag{10}$$

in which A_{nfc} is the non-solar heat exchange rate in W m^{-} at the inlet water temperature and B_{nfc} is the non-solar surface heat exchange coefficient before free convection begins in W m^{-2} °C^{-1}. After free convection the non-solar surface heat exchange becomes

$$Q_{ns} = A_{fc} - B_{fc}(T_i - T_{fc}) \tag{11}$$

where A_{fc} is the non-solar surface heat exchange rate in W m^{-2} at T_{fc}, T_{fc} is the water temperature at which free convection begins, and B_{fc} is the non-solar surface heat exchange coefficient after free convection begins in W m^{-2} °C^{-1}. Values for the slopes and constants for the line segments for various wind velocities, maximum daily air temperatures and inlet water temperatures are given in Table 1. Use of maximum daily air temperatures was justified by the relatively small effect of the time variation in air temperature on the heat exchange rate over the travel time of maximum temperature water parcels. The constant A_{fc} is calculated as

$$A_{fc} = A_{nfc} - B_{nfc}(T_{fc} - T_i) \tag{12}$$

Combining the simplified expressions for heat exchange gives

$$\begin{aligned} Q &= Q_s + Q_b + Q_{ns} \\ &= A + B(T - T_B) \end{aligned} \tag{13}$$

in which before free convection

$$A = Q_s + A_b + A_{nfc} \tag{14}$$

$$B = B_b + B_{nfc} \tag{15}$$

$$T_B = T_i \tag{16}$$

and after free convection

$$A = Q_s + A_b + A_{fc} \tag{17}$$

$$B = B_b + B_{fc} \tag{18}$$

$$T_B = T_{fc} \tag{19}$$

Since the time dependence of solar radiation exchange and wind velocity were eliminated by using averages and time dependence in air temperature was removed by using maximum daily air temperature, Eq. (5) could be rewritten as:

$$u\, dT/dx = Q/\rho cd \tag{20}$$

Substituting Eq. (13) for Q and letting $\Delta T = T - T_B$, Eq. (20) can be integrated to give

$$\Delta T = A/B(1 - \exp(Kx)) \tag{21}$$

Table 1. Heat Exchange Parameters for Analytical Solution to Maximum Daily Temperatures in Unshaded Channel Reach

	Wind Velocity							
	$1\ m\ s^{-1}$				$3\ m\ s^{-1}$			
Max. Air Temp. (°C)	25	30	35	40	25	30	35	40
Inlet Water Temp. (°C)	A_{ns} $(W\ m^{-2})$							
10	80	150	220	290	200	325	476	650
12	58	126	194	263	153	274	421	591
14	36	103	168	236	107	224	367	532
16	14	79	142	208	60	173	314	474
18	-8	56	116	181	14	123	260	415
20	-30	32	90	154	-33	72	206	356
B_{nfc} $(W\ m^{-2} {}^{\circ}C^{-1})$	11.0	11.8	13.0	13.6	23.3	25.3	26.9	29.4
B_{fc} $(W\ m^{-2} {}^{\circ}C^{-1})$	41	48	56	68	56	63	76	88
T_{fc} (°C)	23.8	28.5	33.0	37.6	23.8	28.5	33.0	37.6

where x is distance from the inlet and

$$K = B/\rho cud \tag{22}$$

If a rectangular channel is assumed

$$K \simeq Bw/\rho cq \tag{23}$$

where q is stream discharge and w is stream surface width which is more easily obtained than d. Solutions for ΔT are given in Fig. 5 for various combinations of A/B and Kx for

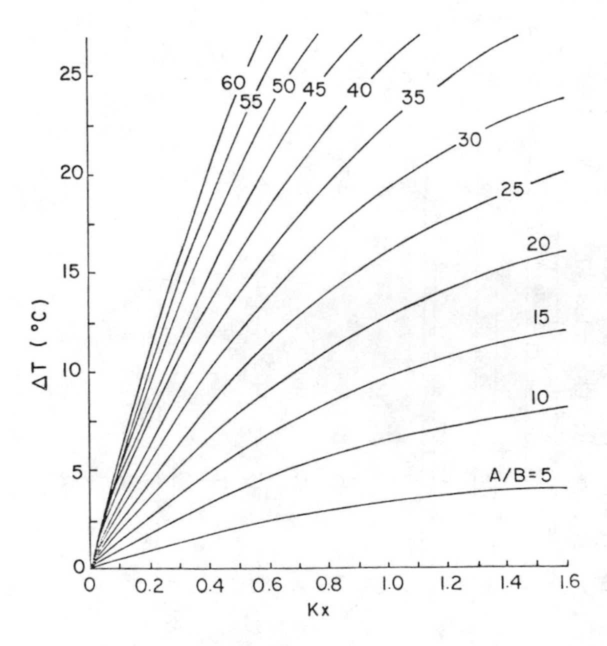

Fig. 5. Solution for maximum daily temperature increases in an unshaded channel near the inlet without free convection according to Eq. (21).

the region before free convection begins. In Fig. 6 similar solutions for ΔT are given for A/B and Kx for the region where free convection occurs. In this region, x is now the distance downstream from the point of onset of free convection (x_{fc}). In either region Q_s is obtained from Fig. 3 using the total distance from the inlet to compute travel time. Values of the water temperature at which free convection begins are given in Table 1 for various maximum air temperatures and an assumed 60% relative humidity.

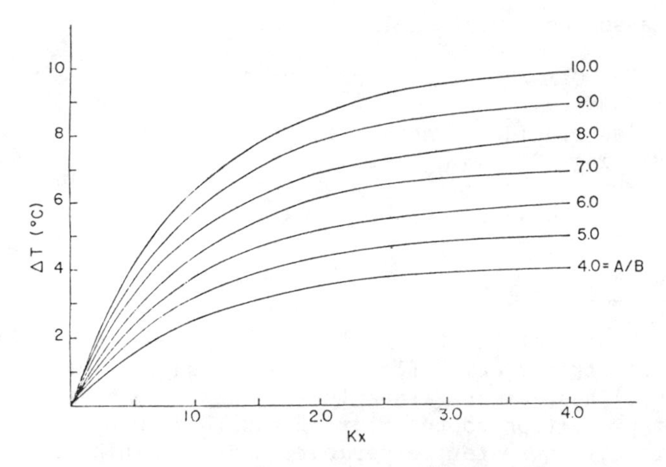

Fig. 6. Solution for maximum daily temperature increases in an unshaded channel beyond the point where free convection begins according to Eq. (21).

Use of Eq. (21) or Figs. 5 and 6 to predict maximum water temperature increases for a specified length of unshaded channel, requires an initial trial and error solution for x_{fc}; the distance from the inlet where free convection begins. A value of T_{fc} is selected from Table 1. A solution ΔT is obtained using Fig. 5 and an estimated x_{fc} until the distance from the inlet where $T = T_{fc}$ is found. Prediction of ΔT can then proceed for any proposed length of unshaded reach in either the region before or after free convection begins as needed.

The predicted maximum temperature increase (ΔT) from Eq. (21) will approach an equilibrium value at some point in the free convection region and decline thereafter. Equation (21) should only be used up to the point where

$$A - B(T - T_{fc}) \geq 0 \qquad (24)$$

For points farther downstream, where the left-hand side of Eq. (24) becomes negative, the equilibrium value should be used.

The foregoing procedure can be simplified for short channel lengths in the region before free convection begins. In this region temperature increases are primarily due to absorbed solar alone and $Q \simeq Q_s$. Thus Eq. (20) can be solved to give

$$\Delta T = Q_s x/\rho c u d \qquad (25)$$

or again assuming a rectangular channel

$$T = Q_s xw/\rho cq \tag{26}$$

Eq. (26) also provides a method to obtain a rough estimate of x_{fc} by setting $Q_s \simeq Q_{max}$ as

$$x_{fc} \simeq \rho cq(T_i - T_{fc})/Q_{max}w \tag{27}$$

Concluding Remarks

The results of the theoretical analysis indicate that cool forest streams which flow into unshaded reaches during mid-day will heat almost linearly with distance near the inlet in direct proportion to the flux of absorbed solar radiation. The relatively low water temperature near the inlet reduces the net flux of heat exchange due to channel conduction, long-wave radiation exchange, and atmospheric convection to negligible amounts. Condensation should commonly occur on the water surface near the inlet. However, as water temperature increases, heat losses due to these non-solar components will gradually increase. A point is reached downstream where the rate of temperature increase becomes exponential and stream temperatures in the reach gradually approach an equilibrium. In this region, channel heat conduction may be the same magnitude as the sum of the remaining non-solar heat exchange terms. Regardless, stream depth and velocity were found to be most important in determining maximum temperatures achieved in an unshaded reach.

The analytical solution for maximum stream temperatures in an unshaded reach reproduced the results of the computer runs very closely. The major difficulty is the trial and error solution for the distance from the inlet where free convection begins in the unshaded reach. This point will be closer to the inlet for high initial water temperatures in shallow slow moving streams. Estimates of x_{fc} using Eq. (27) for such conditions indicate free convection may begin within a few hundred meters of the inlet. In the channel region where $x < x_{fc}$, Eq. (26) which involves absorbed solar radiation only may be used to predict maximum water temperature increases in an unshaded reach. Brown (1970) proposed essentially the same equation for predicting the effect of clear-cutting on stream temperature. However, when $x > x_{fc}$, Eq. (21) which considers all major heat exchange components must be used.

Acknowledgment

Support for this study was provided by the School of Forest Resources, McIntire-Stennis Project 2224, "Forest Shade

Removal Effects on Stream Temperature: Response and Recovery Downstream" and the Institute for Research on Land and Water Resources at The Pennsylvania State University.

References

[1] Anderson, E. R., 1954. "Energy Budget Studies," U.S. Geol. Surv. Prof. Paper, 269: pp. 71-119.

[2] Brown, G. W., 1970. "Predicting the Effect of Clear-cutting on Stream Temperature," J. Soil and Water Conserv., 25(1): pp. 11-13.

[3] DeWalle, D. R., 1976. "Effect of Atmospheric Stability on Water Temperature Predictions for a Thermally-Loaded Stream," Water Resour. Res., 12(2): pp. 239-244.

[4] Lee, R., 1963. "Evaluation of Solar Beam Irradiation as a Climatic Parameter of Mountain Watersheds," Colo. State Univ., Hydrol. Papers, 2, 50 p.

[5] Munn, R. E., 1966. "Descriptive Micrometeorology," Academic Press, N.Y., 245 p.

[6] Ryan, P. J., Harleman, D. R., and Stolzenbach, K. D., 1974. "Surface Heat Loss from Cooling Ponds," Water Resour. Res., 10(5): pp. 930-938.

[7] Shulyakovskiy, L. G., 1969. "Formula for Computing Evaporation with Allowance for the Temperature of the Free Water Surface," Soviet Hydrol., Selected Papers, 6: pp. 566-573.

INTERPRETATION AND PREDICTION OF SPATIAL VARIATION IN TRACE METALS IN SMALL RIVERS BY CANONICAL AND DISCRIMINANT ANALYSIS

By

R. Herrmann, U. Bolz, W. Symader
Lehrstuhl für Hydrologie, Universitat Bayreuth

and

H. Rump
Institut für Naturschutz, Abt. Stadthygiene
Darmstadt, Germany

Abstract

The wide distribution of anthropogenic toxic trace metals has stimulated endeavours to develop reliable methods of water pollution detection and monitoring. *Fontinalis antipyretica* L. has been employed as an indicator of trace metals since this plant absorbs chemical elements directly from the water. An attempt was made to predict trace metal pollution on the basis of their concentrations in moss tissues by discriminant analysis and a classification procedure. Further the spatial variations in trace metals and their relations to other pollutants were interpreted by principal component analysis and canonical correlation.

Fontinalis antipyretica L. was collected from small rivers belonging to a wide range of natural regions with different anthropogenic influences: devonian mountains with quartzites, shales, greywacke and carbonaceous rocks, triassic hills with sandstones and limestones and loess plains. The grassland and forests decreases from the mountains towards the loess plains whereas the cropland, industrial activities and settlement area increases.

The organic moss material was digested with a mixture of HNO_3, H_2SO_4 and $HClO_4$, and then Cd, Cu, Pb and Zn were analysed by atomic absorption spectrophotometry.

For each metal, five or six groups of data were determined by a cluster analysis. Principal component analysis and an F-ratio of the predictor-variables were used to find a most favourable group of predictor variables by which an optimal separation between the different groups of data is possible. The discriminant functions, linear combinations of the predictors, together with the additional help of a classification procedure like Euklidic distances, may be used to assign an individual measurement to the group to which it best corresponds. The discriminant analysis shows that a linear combination of two or three of the 39 total predictors including runoff, namely a combination of 1) the percentage of forested area, 2) a weighted settlement index, and 3) the percentage of area covered with calcareons or quarternary rocks, have the best discriminant power. Tables were constructed that demonstrate the predicted versus actual group membership. Multivariate tests of significance were performed.

The interconnections between the trace metals, the chemical properties and the variables describing the natural and economic character of the river basins are shown for single variables by principle component analysis and for the above mentioned groups of variables by canonical correlation analysis. The latter reveals a higher correlation between the trace metal concentrations and the chemical properties than between the trace metal concentrations and the variables describing the natural and economic character of the river basin.

The Problem to be Solved

In Central Europe with its dense population and industrialization even small rivers are usually highly polluted. Regular control of the water quality in all of these river basins would cost a lot of time and money. Therefore, the aim of this investigation was to develop equations that allow the prediction of the spatial variance of some trace metals. Moreover, the relationship between some important pollutants and groups of pollutants should be found. Once these relations are known, future water quality control nets could be considerably simplified.

To begin with, the region where the investigation is carried out is described. Then the methods of research in the field and the laboratory are presented. The prediction is carried out by means of a multiple discriminant analysis and a classification procedure. The relation between the pollutants and groups of pollutants as well as some of the variables describing the natural and economic regions is examined by means of principle component analysis and canonical correlation analysis. These multivariate statistical methods are only explained as far as they contribute to the solution of water quality problems. Finally, the results are presented and discussed.

The Region of Investigation

The investigated river basins are part of the northern Eifel mountains and the Niederrheinische Bucht. They comprise the most important natural and economic regions of the German medium altitude mountains.

In the Eifel mountains there are rocks of palaeozoic age. The oldest rocks, namely quartzites, shales and greywacke, are found in the arch core; younger carboniferons rocks, e.g., carbonaceous in the troughs. On the northern fringe of the Eifel mountains parts of the palaezoic rocks are discordantly overlain by triassic rocks, above all sandstones and limestones. In the Niederrheinische Bucht, thick layers of tertiary and quarternary sediments, gravels, sands and loess, are deposited.

The river basins of the Eifel mountains with palaeozoic greywacke and shales are sparsely populated and are mostly used as grassland or forests. In regions with carbonaceous rocks, 50% to 75% of the land is used as cropland and the population density is rather high.

In the Niederrheinische Bucht, up to 85% of the loess soils are used as cropland; the population density is very high.

The pollution of the rivers depends on the population density, the land use and industry: iron and nonferrous industry in the Eifel valleys, food industry and paper mills in the Niederrheinische Bucht, and mainly chemical and engineering industry in the highly industrialized zone around Cologne.

The variables describing the natural and economical regions are explained in the section on 'methods of investigations'.

Methods of Investigation

Field work and chemical analysis. Within the different natural and economic regions of the northern Eifel mountains and the Niederrheinische Bucht, nineteen river basins were chosen in such a way that the distribution of the water quality data as well as the variables describing the natural and economic regions was as widely spread as possible.

As a measure of the mean trace metal concentration in the river water, its concentration in the water moss *Fontinalis antipyretica* L. was determined once at the end of a period of measurement. During this period, from 1973-74, all of the other variables with temporal variation had been measured once a week. Possible daily and weekly cycles had been taken into consideration. From these data the mean was computed.

The selected variables are explained below. The transformation into a Gaussian distribution was added. This was checked by means of a Kolmogorov-Smirnov test for the $\alpha_g = 0.05$ level.

1.	F(** 0.33)	: area of river basin (km^2)
2.	FD (SQRT)	: river density (km/km^{-2})
3.	REL (** 1.5)	: mean relief energy (m)
4.	WALD ((+ 1.)**0.33)	: area with forests (km^2)
5.	GR (ALOG 10)	: " " grassland (km^2)
6.	ACK ((+ 1.)**0.33)	: area with cropland (km^2)
7.	SIED (1./SQRT (+1))	: " " settlements (km^2)
8.	OED (1./SQRT (+1))	: " " fallow land (km^2)
9.	PCWALD ((+1)**0.66)	: percentage of area w/forests
10.	PCGR (ALOG 10)	: " " " w/grassland
11.	PCACK ((+1.)**0.66)	: " " " w/cropland
12.	PCSIED (+1.)	: " " " w/settlements
13.	PCOED (1./SQRT (+1.))	: " " " w/fallow land
14.	GEF (** 0.33)	: mean slope of the river basin (%o)
15.	FORM (ALOG 10)	: L /F (L = axial length, F: see (1))
16.	SIGEW (1./SQRT (+1.))	: weighted settlement area

$= \sum_{i=o}^{n} 0.8^{i} F_i$ The river basin is divided into rings of 5 km width aroung the gauge.
$i = o, \ldots, n$ number of ring
F_i= settlement area of the i-th ring (km^2)

17.	SAND (1./(+1.)	: area with sandstones (km^2)
18.	KALK (1./SQRT (+1)	: " " calcareous rocks (km^2)
19.	GG (ALOG 10 (+1)	: " " shales & greywacke (km^2)
20.	QUA (1./(+1))	: " " quarternary rocks (km^2)
21.	PCSAND (1./(+1.))	: percentage of area with sandstones
22.	PCKALK (1./SQRT (+1.)	: " " " " calcareous rocks
23.	PCGG (+1.)	: " " " " shales and greywacke
24.	PCQUA (1./SQRT (+1.)	: " " " " quarternary rocks
25.	BART ((+1.)**0.33)	: percentage of soil particles <0.01 mm
26.	CU (**0.33)	: content of Cu in *Fontinalis antipyretica* L. ($\mu g g^{-1}$)
27.	CD (1 /SQRT)	: content of Cd in *Fontinalis a.* L.
28.	PB (1./SQRT)	: " " Pb " " " "
29.	ZN (ALOG)	: " " Zn " " " "
30.	LF (**0.66)	: electrolytic conductivity ($\mu \int cm^{-1}$)
31.	OZ (**2)	: Oxygen content in percentage of volume of air saturation. For water saturation 21% was assumed.
32.	PH (**1.5)	: pH-value
33.	TR (SQRT)	: turbidity, measured as extinction (Ex 10^3) at 420 nm w/a photometre
34.	SE (ALOG 10)	: suspended matter ($mg\ l^{-1}$)
35.	NA (1./SQRT)	: Na was determined in water, which was not previously filtered, by a flame photometre ($mg\ l^{-1}$)

36. K (ACOG 10) : K was determined in the same way as number 35

37. CA (ALOG 10) : Ca was determined in the same way as number 35

38. CL (ALOG 10) : chloride was determined gravimetrically after precipitation by $AgHO_3$

39. PO4 (ALOG 10) : the water was filtered by a 0.45 µm membrane filter. Within 24 h, phosphate was determined by the colorimetric molybdenum blue method ($mg l^{-1}$)

40. NO3 (SQRT) : the water was filtered by a 0.45 µm membrane filter. Within 24 h, NO_3 was determined by the sodium salicylate method (Deutsche Einheitsverfahren (1972)), ($mg\ l^{-1}$)

41. NH4 (1./SQRT) : the water was filtered by a 0.45 µm membrane filter. Within 24 h, NH_4 was determined with Nessler's reagent (Deutsche Einheitsverfahren (1972)), ($mg\ l^{-1}$)

42. SO4 (1./SQRT) : Sulfates were titrated with EDTA solution against $BaCl_2$ ($mg\ l^{-1}$)

43. GHCH (ALOG 10) : the pesticide γ-benzene hexachloride (γ-BHC) was determined by GC/MS. Columns: alkali borate glas, 2.5 mm Ø, 2m

I. : 5% OV 1 on Chromosorb W-AW-DMCS (80/100 mesh)
II. : 3% OV 17 " " " "
III.: 5% OV 210" " " "
Temperatures: injector 240°C, cols. I and III 210°C; col. II 195°C, detector 260°C. Carrier gas: He 78 ml min^{-1}, purge gas Ar/methan 110 ml min^{-1}.

The above mentioned variables having been selected from a number of variables describing the hydrological system. The connections between single variables and groups of variables were explained and a prediction model of water quality was developed by means of principle component analysis, canonical correlation analysis, and multiple discriminant analysis.

Multivariate Statistical Methods

Principal component analysis. Principal component analysis was used to find groups of nonintercorrelated variables. X is a vector, formed from the n random variables of the matrix of variables. Each variable was converted to have the expectation zero and the variance 1. This vector X may be deduced from the equation:

$$X = AW \qquad (1)$$

Here W is a vector formed from $m < n$ mutually independent random variables which are characterized as principal components, each with expectation zero. A is an (nxm) matrix of coefficients. The variance-covariance matrix of X can be written as

$$C = E\,(XX') \tag{2}$$

where X' is the transpose of X. The matrix C has order n and rank $m < n$. A further matrix exists in the form

$$\lambda = A'A \tag{3}$$

where λ is an (mxn) diagonal matrix whose diagonal elements are nonzero eigenvalues of C.

The matrix W is given by

$$W = (1/\lambda)\ A'X \tag{4}$$

Here $(1/\lambda)$ is an (mxm) diagonal matrix, the diagonal elements of which are the reciprocals of the eigenvalues.

The following applies for $m=n$:

$$(1/\lambda)A' = A^{-1} \tag{5}$$

and for $m < n$:

$$(1/\lambda)A' = (A'A)^{-1}\ A' \tag{6}$$

where A^{-1} is the inverse of A. The variance-covariance matrix of W is then:

$$C_W = E\,(WW') = \lambda \tag{7}$$

The (i,i)th element of λ is the variance of the ith principal component of W with the following relation:

$$\lambda_{11} \geq \lambda_{22} \geq \cdots \geq \lambda_{mm}$$

From this, it follows that the first principal component W_1 explains the greater part of the variance of X, and less is explained by the following principal components W_i, $i > 1$. The principal components constitute the total variance of X. Thus:

$$\sum_{i=1}^{m} \lambda_{ii} = \sum_{j=1}^{n} c_{jj} \tag{8}$$

If the elements of X are the hydrological variables which together describe all the hydrological characteristics of a region, then each principal component, which is a weighted sum of the elements of X, represents a measure for some hydrological characteristics. If $\lambda_{ii} = 0$ for $i > 1$, the first principal component alone would describe all hydrological characteristics. That would imply that there is a high correlation between the ith and the jth variables. If there is no correlation, then the number of the principal components m would be equal to the number of variables n. The λ_{ii} then represent the variance of the ith principal component with the ith variable. The above description is based on Steele and Matalas (1971). A Gaussian distribution of the variable X_i is not necessary for the purpose of variable filtering (Anderson, 1958).

Cluster analysis. Further, cluster analysis helps to find groups of rivers with similar trace metal concentration. The algorithm of Ward (1963) was used.

Canonical correlation analysis. Suppose in a number of rivers conventional water quality data and the trace metal contents are measured. The goal of canonical correlation analysis is then to understand the relationship between these two sets of variables. For example, the conventional water quality data and the trace metal content may be weighted and written as two linear functions:

$$P = a_1 Cd + a_2 Cu + a_3 Pb + a_4 Zn \qquad (9)$$

$$Q = b_1 \text{el. conductivity} + b_2 pH + b_3 \text{ suspended load} + b_4 NO_3 \qquad (10)$$

The coefficients a_i and b_i are weights to be determined such that they yield an optimal simple correlation coefficient between P and Q, which are the canonical variables. This optimal correlation coefficient between the canonical variables is called canonical correlation coefficient and is a measure of the strength between the two sets of variables (in this example between the four trace metals and the four conventional water quality variables). The nature of the link may be indicated by the simple correlation coefficients between the original variables and the canonical variables.

Computational procedures. First, a transformation of the variables may be carried out to meet requirements of some statistical tests. Then a supermatrix of intercorrelations has to be computed, which expresses the relationship within and between two data sets.

This matrix can be partitioned so that

$$R = \left[\begin{array}{c|c} R_{11} & R_{12} \\ \hline R_{21} & R_{22} \end{array}\right]$$

where R_{11} is the matrix of intercorrelations among the p variables of the first data set,

R_{22} is the matrix of intercorrelations among the p variables of the second data set,

R_{12} is the matrix of intercorrelations of the first and second group, and

R_{21} is the transpose of R_{12}.

It is assumed that $p > q$.

In a third step the eigenvalues and eigenvectors of Eq. (11) are computed.

$$|R_{22}^{-1} R_{21} R_{11}^{-1} R_{12} - \Lambda^2 I| = 0 \tag{11}$$

A column matrix A_i which is the eigenvector of (11) contains the canonical coefficients of q variables of i-th canonical root.

Then the matrix, B_i, which contains the canonical coefficients of p variables of i-th canonical root is computed:

$$B_i = R_{11}^{-1} R_{12} A_i/\lambda_i \qquad (\lambda_i \neq 0) \tag{12}$$

Further: $B_i'B_i = 1$ and $A_i'A_i = 1$.

The canonical variables P and Q are shown by the equations

$$P_i = A_i'V_1, \quad \text{and} \quad Q_i = B_i'V_2 \tag{13}$$

V_1 and V_2 are the original variables of the first and second set of data.

Finally, chi-square tests are computed for each of the B canonical roots. The chance - probability for each root (= squared canonical correlation) serves to suggest the number of significant multivariable relationships between the two sets of data. The strengths of these relationships are shown by the canonical correlations. The nature of the link between the two sets of variables may be indicated by those original variables having larger simple correlation coefficients with the canonical variables.

Detailed discussion of canonical correlation analysis is given by Anderson (1958) and Lee (1969).

Discriminant analysis. Trace metal content in *Fontinalis antipyretica* L. growing in small rivers shall be estimated. Discriminant analysis with a classification procedure is proposed to solve this prediction task. The criterion variables (trace metals) and the predictor variables were measured in nineteen watersheds as mentioned previously.

Discriminant analysis is a multivariate technique that uses a group of variables to obtain maximum separation between different groups of data. By this analysis one or more discriminant functions are obtained which are linear combinations of the prediction variables. The discriminant scores are computed by these discriminant functions and then used within a classification procedure to find into which predetermined group of observations a new one fits best. To find the best combination of variables for the discriminant function, principal component analysis is used.

Computational procedures. The discriminant analysis was developed by Fisher (1936) and expanded to the multiple form by Bryan (1951). Miller (1962), Decoursey (1973) and Herrmann (1974) used the discriminant analysis for prediction tasks in meteorology, design review and hydrology.

The following short introduction into the mathematical derivation follows the excellent work of the last two authors. The total number of observations is:

$$N = \sum_{g=1}^{G} n_g \tag{14}$$

with n_g $(g=1,\ldots, G)$ the number of observations in each of G mutually exclusive groups.

Given P prediction variables, X_p $(p=1,\ldots, P)$ the linear discriminant function may be written as

$$T_j = V_{j1}X_1 + V_{j2}X_2 + \ldots + V_{jp}X_p \tag{15}$$

$$j = 1, \ldots, \min (G-1), P)$$

T_j is the discriminant score of the ith discriminant function and V_{ji} $(i = 1, \ldots, P)$ are the weights.

Discriminant analysis aims at the definition of the vector of weights V in order to obtain maximum separation between the G groups. This is done by maximizing

$$\lambda = \frac{V'BV}{V'WV} \stackrel{!}{=} \max \qquad (16)$$

where B = between-group deviation sums of squares and products matrix, and

W = within-group deviation sums of squares and products matrix.

The λ-values are maximized by means of the partial derivatives of λ in respect to V. Out of this, the following equation results:

$$(W^{-1}B - \lambda I)\ V = 0 \qquad (17)$$

and the determinantal equation:

$$/W^{-1}B - \lambda I/ = 0 \qquad (18)$$

in which W^{-1} is the inverse matrix of W and I the unit matrix. The vectors V_j are the solution of Eq. (17) with λ equal to λ_j and may be used for Eq. (15).

By Wilk's Λ (Cooley and Lohnes, 1966) it is shown that the P predictor variables are able to separate between the groups. There exists further an χ^2-test to evaluate the significance of the discriminant functions (Rao, 1952). The significance of the predictor variables will be found by an F-test.

Selection of variables. Wallis (1967) shows that selecting variables by means of principal component analysis is superior to the stepwise selection. If there is only one discriminant function Decoursey (1973) uses a principal component analysis with all prediction variables and the discriminant score. The principal component with the highest loading by the discriminant score contains the predictor variable with the highest discriminatory power and so on. Both procedures did not always lead to optimal results: it is possible that the second highest loading of a predictor variable in principal component may indicate the better discriminatory power. The objective is to find a combination of predictor variables which are not intercorrelated and which have the optimal discriminatory power. This power is checked by Wilk's Λ. The discriminatory power for a single predictor variable can be found by an F-test.

The independence of the steering variables may be seen from the principal component matrix. The normality was examined by a Kolmogorov-Smirnow test ($\alpha_g < 5\%$) and if necessary a transformation was performed.

Classification. Given a matrix C of the centroids:

$$C = V'\overline{X} \tag{19}$$

$\overline{X}$ = matrix of the G group means of the P steering variables. It is possible to compute the sum of deviation squares AQ_g between the centroids and the discriminant scores for each g = 1, ..., G group by:

$$AQ_g = \sum_{i=1}^{R} (T_i - C_{ig})^2 \tag{20}$$

R = number of discriminant functions. An individual observation is assigned to that group with the smallest sum of squared deviations (= Euclidic distance).

Results

Spatial distribution. The gauges of group 1 which are mostly to be found north of the Eifel mountains but not yet considerably influenced by industry show the smallest means for all metals (Table 1). They are followed by the gauges of group 2 with the exception of Pb and Cu the means of which are smaller in group 4. The gauges of group 2 show a higher

Table 1. Mean Trace Metal Concentrations ($\mu g g^{-1}$) in *Fontinalis antipyretica* L. for Each of the Five Predetermined Groups

group / metal	Cd	Pb	Zn	Cu
1	7	120	230	18
2	12	120	690	32
3	9	840	910	32
4	79	71	960	18
5	40	165	1800	66

inflow of trace metals from rocks as well as from pollution. In group 3, Cd has a smaller mean than in group 2 and Cu has the same mean as in group 2, whereas Zn shows a very high value and Pb the highest. The high Pb values are caused by ground water coming from sandstone aquifers with Pb-ores of the Mechernicher Trias Bucht.

The gauge of group 4 is noticeable for its high Cd and Zn load which may be explained as a result of industrial pollution. The gauge of group 5 is extremely loaded with Zn, Pb and Cu as a result of industrial pollution.

The data of the variables 1-25 and 30-43 may be found in Symader (1976) and Rump (1976) and the data of the trace metals in Bolz (1976). The gauges, marked by their abbreviated names, are shown in Fig. 1.

The gauges ARL (= 79 $\mu g g^{-1}$) and PLA (= 40 $\mu g g^{-1}$) show the highest Cd-values, which in the case of gauge ARL can be explained as industrial pollution whereas in the case of PLA as influence from the underground (Symader, 1977).

It is noticeable that river basins like REI (9.2 $\mu g g^{-1}$) and ERK (25 $\mu g g^{-1}$) with little influence by man still have high Cd-concentration in the water moss. As there are also high Zn-concentrations, natural deposits in the rocks may be assumed. Most of the rivers have a Cd-concentration of $2.5 < Cd\ (\mu g g^{-1}) < 8$, a range which may be explained as the natural background.

The very high Pb-concentrations of the gauges MEC (= 874 $\mu g g^{-1}$) and FRI (= 793 $\mu g g^{-1}$) result from the Pb-ores of the Mechernicher Trias Bucht and its Pb-industry. The Pb-concentrations of gauge KAL (= 267 $\mu g g^{-1}$) are lower and may depend on a non-iron-mill with melting processes. Most rivers are loaded with $100 < Pb\ (\mu g g^{-1}) < 170$ and only an analysis of sediment, suspended load and water may give criterions for natural or anthropogenic influence (Symader, 1977).

The water moss of gauge LAN and WEC show only 50 $\mu g g^{-1}$ Pb in spite of the considerable pollution of the river. Because of decomposition in the sludge and in the contact zone at negative redox potential the sulfide seems to be the stable phase, the metal thus being temporarily insoluble and not available for plants (the same happens with CD and Zn (Förstner and Müller, 1974)).

The highest Zn-concentrations are in the river basins above the gauges of PLA (= 1770 $\mu g g^{-1}$) and MEC (= 1390 $\mu g g^{-1}$). In the river basins of the gauge MEC there have been of old many brass and zinc works. The river basins with the lowest

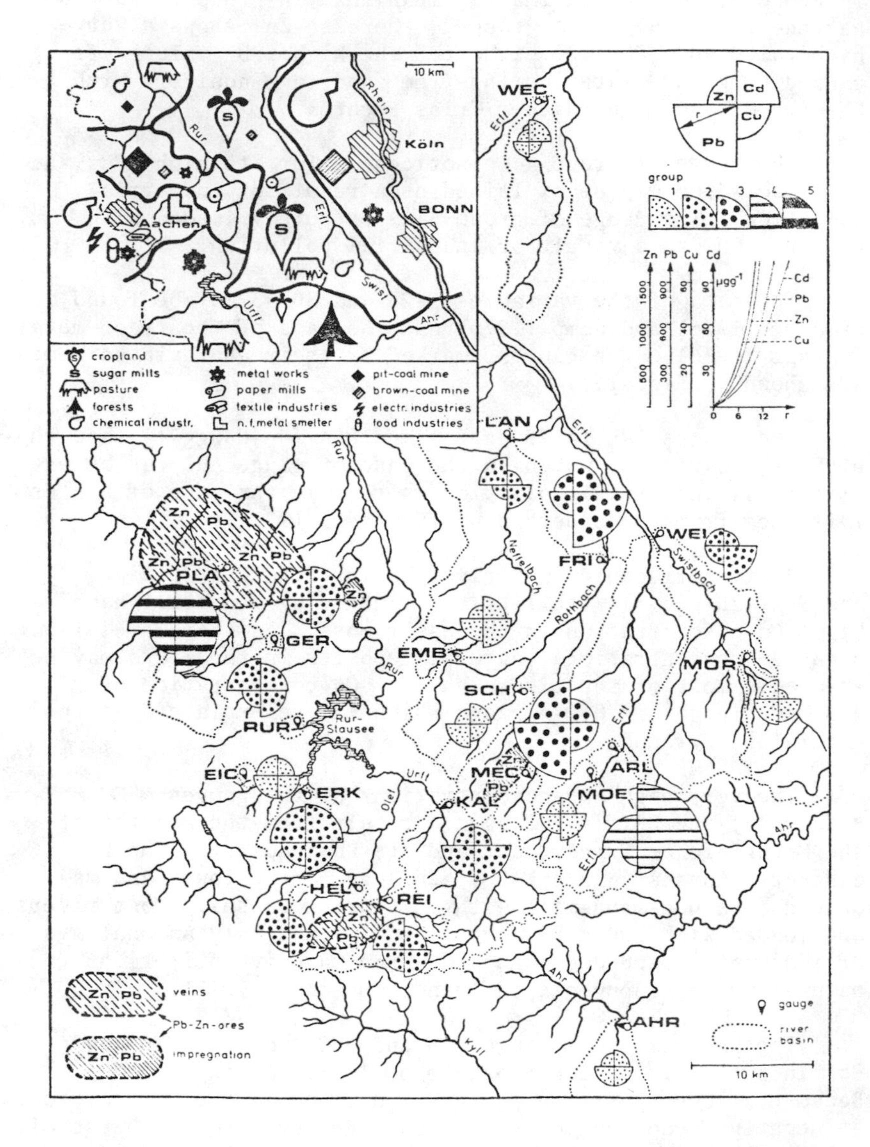

Fig. 1. Spatial distribution of trace metal concentrations in *Fontinalis antipyretica* L. (a water moss). The groups were found by cluster analysis.

Zn-concentrations (<250 μgg^{-1}) are generally those with predominantly agriculture, though often with high population density. The high population density does not influence the Zn concentration as highly as assumed (Hellmann, 1970).

The high Cu-concentrations (PLA (= 66 μgg^{-1})) and (KAL (= 52 μgg^{-1})) cannot be explained solely by industrial pollution as there are only small concentrations of Cu in the suspended load and in the water body whereas high concentrations occur in the sediments (Symader, 1977; Thomas, 1976).

Principal component analysis. The principal component analysis serves to reveal the inner relations between the predictor variables and the independent variables, so that groups of intercorrelated variables may be recognized (Table 2).

The first principal component is highly loaded by variables that describe settlement patterns and partly indirectly intensive land use (e.g., ACK, SIED, QUA). Contrary to this principal component, the highly loading variables of the second principal component describes forests and grassland as well as morphology and rocks of the basement. The following principal components are highly loaded by only a few variables or a single one, so that an interpretation is very easy.

The trace metals are not intercorrelated, so each of them loads one principal component. Pb, for example, loads highly the principal component 4, which is as well loaded by SAND. SAND shows the distribution of Pb-ores.

A more thorough interpretation of the principal component matrix is not attempted as its purpose is only to find groups of not intercorrelated variables. For this purpose, a Gaussian distribution is not necessary (Stammers, 1967).

Prediction models. For each metal the best prediction model will be presented. Parts of the model are the range of groups, the F-values of the predictor variables which show the discriminant power of each variable, the centroid matrix C and the coefficient matrix V of the significant discriminant functions (Tables 3, 4, 5 and 6).

Canonical correlation analysis. By means of the principal component analysis (Table 2), three groups of variables were thus put together so that the variables do not intercorrelate within a group. The variables of the first group: OED, SIGEW, KALK and GG describe the natural and economic character of the river basins. The trace metal concentration of the rivers is represented by the variables of the second group, namely CD, CU, PB and ZN. The variables of the third group show the chemical behavior of the rivers: LF, PH, SE and NO_3. A

Table 2. Principal Component Matrix (Varimax Rotated) for 43 Variables. All loadings >|0.6| are stated. 9 corresponds to loadings >0.85, other loadings are rounded to the next tenth.

Variable		Principle component 1	2	3	4	5	6	(10/13/16)
1	F	8						
2	FD	-6	6					
3	REL	-8						
4	WALD		9					
5	GR		7					
6	ACK	9						
7	SIE	9						
8	OED					-9		
9	PCWALD		7					
10	PCGR						8	
11	PACK	7						
12	PSIED	7						
13	POED					-9		
14	GEF							7
15	FORM							6
16	SIGEW	9						
17	SAND				9			
18	KALK			9				
19	GG		9					
20	QUA	9						
21	PCSAND							6
22	PCKALK			7				
23	PCGG	-6						
24	PCQUA	9						
25	B.ART			8				

Variable		Principle component 1	2	4	7	9/14
26	CU		6			
27	CD				8	
28	PB			6		
29	ZN					8
30	LF	9				
31	/2	-8				
32	PH					-8
33	TR			8		
34	SE			9		
35	NA	9				
36	K	9				
37	CA	6				
38	CL	9				
39	PO4	9				
40	NO3	9				9
41	NH4	8				
42	SO2	9				
43	GHCH	9				

Table 3. Prediction Model for Cd

	-0.367	2.115	-0.413
C =	0.310	1.730	-0.368
	0.565	1.596	-0.438
	1.572	2.214	-0.404

	-0.074	0.144	-0.986
V =	0.139	0.053	0.374
	-0.158	-0.009	+0.834

F-values of the variables

variable	F-value	P (%)
PCSIED	3.5	4.4
PCWALD	3.2	5.6
GR	1.3	30.1
PCKALK	0.97	56.3

Range of groups ($\mu g g^{-1}$)

group	range
1	2.8 - 6.0
2	6.0 - 8.4
3	8.4 -17.5
4	17.5 -40

Classification of observations

observed group / assigned group

	1	2	3	4
1	5	1	0	0
2	1	2	1	1
3	1	0	2	1
4	0	0	0	2

Table 4. Prediction Model for Pb

C =	-0.406	-0.419			
	-1.092	-0.203			
	-1.09	-0.517			
	-0.771	-0.307			

V =	-0.932	0.048	-0.026	+0.026	-0.357
	0.172	-0.061	-0.711	-0.499	+0.461

F-values of the variables

variable	F-value	P (%)
FD	7.96	0.2
PCSIED	3.68	3.6
PCQUA	3.59	3.9
KALK	1.44	27.0
OED	1.02	41.3

Range of groups ($\mu g g^{-1}$)

group	range
1	<66
2	66 - 119
3	119 - 530
4	530 - 874

Classification of observations

observed group	assigned group 1	2	3	4
1	2	0	0	0
2	0	4	1	1
3	0	4	4	1
4	0	0	0	2

Table 5. Prediction Model for Zn

	-2.434	8.363	2.133
	-2.469	7.402	2.854
C =	-0.308	7.162	1.999
	-0.873	7.995	2.614
	-0.090	9.756	2.578

	0.323	-0.670	-0.669
V =	0.405	0.857	-0.318
	0.075	-0.069	0.994

F-values of the variables

variable	F-value	P (%)
PCWALD	1.37	29
PCSIED	1.21	35
B.ART	0.68	52

Range of groups ($\mu g g^{-1}$)

group	range
1	150 - 290
2	290 - 510
3	510 - 840
4	840 -1210
5	1210 -1800

Classification of observations

observed group	assigned group				
	1	2	3	4	5
1	5	0	0	0	0
2	1	3	0	0	0
3	0	1	0	2	1
4	0	0	0	3	1
5	0	1	0	1	0

Table 6. Prediction Model for Cu

	0.906	-0.789	4.084
	0.592	-0.821	3.280
C =	0.852	-1.104	3.090
	0.782	-1.004	3.550
	0.678	-1.323	4.288

	0.981	-0.180	0.690
V =	-0.993	-0.108	0.051
	0.779	0.514	0.358

F-value of the variables

variable	F-value	P (%)
FD	1.81	18
WALD	1.30	32
PCSIED	1.05	42

Range of groups ($\mu g g^{-1}$)

group	range
1	6 - 15
2	15 - 20
3	20 - 30
4	30 - 46
5	> 46

Classification of observations

observed group	assigned group 1	2	3	4	5
1	2	0	1	0	0
2	0	3	0	0	0
3	2	0	3	1	0
4	1	1	0	2	1
5	0	1	0	0	1

modification of this group - LF was exchanged for GHCH - was additionally used for the canonical analysis.

For all canonical correlations only one dimension is significant, α_g being <0.005. Only when the trace metals and the natural and the economic character are correlated $\alpha_g = 0.26$.

The canonical correlation coefficients are shown in Table 7.

Table 7. Canonical Correlation Coefficients (-: not computed)

group of variables	1	2	3a	3b
1 natural and economic character	1	0.68	0.94	0.87
2 trace metals		1	0.92	0.87
3a chemical behaviour (with LF)			1	-
3b chemical behaviour (with GHCH)				1

The chemical behavior, expressed by the variables PH, SE, NO3 and LF (group 3a) shows the closest connection with the two other groups. LF being exchanged for GHCH (group 3b) the correlation is still high, though lower than in group 3a. It is thus clear that the chemical behavior may characterize the trace metal pollution and vice versa. On the contrary the natural and economic character of a river basin alone is not sufficient to explain the trace metal pollution - at least not by means of the variables chosen in this investigation.

In the following it is demonstrated, for each canonical analysis, how far the single variables of each group influence the relation between the various groups. For this purpose the original variables are correlated with the canonical variables.

In both groups (Table 8) only one variable is highly correlated with the canonical variable (NO3 or PB). Also the correlation matrix of the variables concerned shows that PB and NO3 correlate most (Table 9). The reason for this is that

Table 8. Correlation of the Variables of Group 2 (Trace Metals) and 3a (Chemical Behavior with LF) with Their Resp. Canonical Variables, thus Showing the Significance of Each Variable for the Canonical Correlation

2		3 a	
CU	0.35	LF	-0.30
CD	-0.21	PH	0.02
PB	-0.94	SE	0.07
ZN	0.06	NO3	0.70

Table 9. Correlation Coefficients Between the Variables of the Groups 2, 3a and 3b

	LF	PH	SE	NO3	GHCH
CU	-0.33	-0.26	-0.15	-0.17	-0.11
CD	0.49	0.36	0.26	0.07	0.45
PB	0.31	-0.04	-0.03	-0.61	0.25
ZN	-0.47	-0.44	-0.30	-0.11	-0.19

NO3 is easily washed out of the sandstones and their soils, and these sandstones on the other hand contain PB ores.

The significance of the correlation coefficients was not taken in consideration. Further, the interpretation of the signs must take into account the respective transformation of the variable into a Gaussian distribution.

Principally, there is no great difference between group 3a and 3b in their canonical correlation with group 2 (Table 10). The close connection between PB and NO3 was explained in Table 10.

The high correlations between the original and the canonical variables clearly reflect the complicated connections between the sets of data (Table 11). High concentrations of 8-BHC appear above all in river basins with high settlement density and thus resulting high detritus in the rivers. The

Table 10. Correlation of the Variables of Group 2 (Trace Metals) and 3b (Chemical Behavior with GHCH) with their Resp. Canonical Variables

	2		3 b
CU	0.04	PH	0.18
CD	-0.12	SE	0.18
PB	-0.,74	NO3	0.68
ZN	-0.27	GHCH	-0.42

Table 11. Correlation of the Variables of Group 1 (Natural and Economic Character) and Group 3a,b (Chemical Behavior) with their Resp. Canonical Variables

1		3 b	
OED	-0.23	PH	0.70
SIGEW	-0.85	SE	0.93
KALK	-0.70	NO3	0.18
GG	-0.29	GHCH	0.77
1		**3 a**	
OED	-0.08	LF	0.98
SIGEW	-0.87	PH	0.59
KALK	-0.58	SE	0.84
GG	-0.32	NO3	0.17

suspended load further depends on the erodability of the soils, especially soils on mesozoic sandstones and limestones as well as on devonic carbonaceous rocks. Besides waste water, it is first of all the groundwater from aquifers in limestones and older carboniferous rocks that influences the pH of the river water. When γ-BHC is replaced by the electrolytic conductivity (LF), there occurs small displacements in the association pattern because LF is a more general indicator for pollution than γ-BHC. Therefore LF shows a higher correlation with SIGEW than γ-BHC; similarly LF correlates higher with KALK than does γ-BHC (Table 12).

Table 12. Correlation between the Variables of the Groups 1, 3a and 3b

	LF	PH	SE	NO3	GHCH
OED	-0.00	-0.15	-0.31	0.18	-0.07
SIGEW	-0.83	-0.50	-0.69	0.10	-0.78
KALK	-0.46	-0.63	-0.52	-0.26	-0.22
GG	-0.27	0.63	-0.31	-0.37	-0.00

The comparatively bad results (Tables 3, 4, 5 and 6) of the prediction of trace metal concentrations by means of those variables describing the nature of the river basins correspond to the low canonical correlation coefficient of $R_c = 0.68$. Likewise, the pattern of the association of single variables shows a higher random part (Table 13). This part is caused

Table 13. Correlation of the Variables of Group 1 (Natural and Economic Character) and Group 2 (Trace Metals) with their Resp. Canonical Variables

1		2	
OED	-0.23	CU	0.47
SIGEW	-0.36	CD	0.07
KALK	0.50	PB	0.21
GG	0.64	ZN	0.52

by randomly arranged sites of sewage outlets. A corresponding behavior may be seen as well from the correlation coefficient matrix (Table 9).

Conclusions

By means of a principal component analysis, it is possible to find groups of intercorrelated variables. Consequently, the measuring program of the water quality may be reduced from 14 variables (without trace metals) to only 4 with hardly any loss of information. Further, prediction models based on discriminant analysis were built for 4 trace metals which show that variables like settlement density, river density, forests and area with quarternary rocks reveal the best predicting power. The canonical correlation between the chemical behavior of the rivers and the trace metal concentration in the water moss is comparatively high. Contrary to this there is only a low canonical correlation between the trace metal concentration and the variables describing the natural and economic character of the river basins. The latter variables on the contrary show a high canonical correlation with the chemical behavior of the rivers.

Acknowledgment

The authors gratefully acknowledge the receipt of a grant from the German Research Association.

References

[1] Anderson, T., 1958. "An Introduction to Multivariate Statistical Analysis," Wiley, New York, 374 p.

[2] Bolz, U., 1976. "Der Schwermetallgehalt in Wassermoosen in Fließgewässern der Bördenzone und Nordeifel als Ausdruck der Umweltverschmutzung," Unveröff, Staatsarbeit f. d. Lehramt an Gymnasien, Wiss. Prufungsamt d. Universität, Köln, p. 42.

[3] Bryan, J., 1951. "The Generalized Discriminant Function: Mathematical Foundation and Computation Routine," Harvard Educ. Rev., pp. 90-95.

[4] Cooley, W. and Lohnes, P., 1966. "Multivariate Procedures for the Behavioral Sciences," Wiley, New York, 211 p.

[5] Decoursey, D., 1973. "An Application of Discriminant Analysis in Design Review," Water Resources Research, 9: pp. 93-102.

[6] Deutsche Einheitsverfahren zue Wasser-, Abwasser-und Schlammuntersuchung, 1972. Verlag Chemie, Weinheim, loose leaves.

[7] Fisher, R. A., 1936. "The Use of Multiple Measurements in Taxonomic Problems," Ann. Eugenies, 7: pp. 179-188.

[8] Förstner, U. and Müller, G., 1974. "Schwermetalle in Flüssen und Seen als Ausdruck der Umweltverschmutzung," Springer, Heidelberg, 225 p.

[9] Hellmann, H., 1970. "Die Charakterisierung von Sedimenten auf Grund ihres Gehaltes an Spurenmetallen," Deutsche Gewässerkundl. Mitteilungen, 13: pp. 108-113.

[10] Herrmann, R., 1974. "Ein Anwendungsversuch der Mehrdimensionalen Diskriminanzanalyse auf die Abflubvorhersage," Catena, 1: pp. 367-385.

[11] Lee, P., 1969. "The Theory and Application of Canonical Trend Surfaces," J. Geology, 77: pp. 303-318.

[12] Miller, G., 1962. "Statistical Prediction by Discriminant Analysis," Meteorological Monographs, 4: 1-50.

[13] Rao, C., 1952. "Statistical Methods in Biometrical Research," Wiley, New York, 390 p.

[14] Rump, H., 1976. "Mathematische Vorhersagemodelle für Pestizide und Schadstoffe in Gewässern der Niederrheinishen Bucht und der Nordeifer," Kölner Geographische Arbeiten, 34: pp. 1-122.

[15] Stammers, W., 1967. "The Application of Multivariate Techniques in Hydrology," In: National Research Council of Canada, Subcommittee on Hydrology (ed.), Proceedings of Hydrology Symposium No. 5 held at McGill University 24. a. 25.th Febr. 1966, Queen's Printer, Ottowa: 255-270.

[16] Steele, T. and Matalas, N., 1971. "Principle Component Analysis of Streamflow Chemical Data," In: IASH (ed.), International Symposium on Mathematical Models in Hydrology, Warshaw, 1: pp. 663-682.

[17] Symader, W., 1976. "Multivariate Nährstoffuntersuchungen zu Vorhersagezwecken in Fließgewässern am Nordrand der Eifel," Kölner Geographische Arbeiten, 34: pp. 1-154.

[18] Symader, W., 1977. "Heavy Metals in Water, Suspended Matter and Sediment," In: IAHS (ed.), Symposium on the Effect of Urbanization and Industrialization on the Hydrological Regime and on Water Quality, Amsterdam, Oct. 2-7, 1977, accepted paper.

[19] Thomas, W., 1976. Personal communication.

[20] Wallis, J., 1967. "When is it Safe to Extend a Prediction Equation? - An Answer Based upon Factor and Discriminant Function Analysis," Water Resources Research, 3: pp. 375-384.

[21] Ward, J. H., 1963. "Hierarchiacal Grouping to Optimize an Objective Function," Americ. Statistical Assoc. J., 58: pp. 236-244.

MODELING OF WATER QUALITY SYSTEMS BY MULTIPLE FREQUENCY RESPONSE ANALYSIS

By

Gerhard Huthmann
Dip.-Geophys.
Federal Institute of Hydrology
Postfach 309
D-5400 Koblenz
Federal Republic of Germany

Abstract

Mathematical models have turned out to be a powerful tool in solving water quality management problems. The development of these mathematical models generally makes high demands on the data. These data may be obtained directly by collection of samples at some specified frequency and some specified sampling points for laboratory analysis to determine concentrations of inorganic constituents. As an operational alternative to provide this same information, a mathematical model is presented for the calculation of daily values by simulating water quality records that preserve the natural relationship between water quality and water quantity. This modeling technique allows cost savings in water quality network operation in terms of a decrease in the number of laboratory analyses by reducing the sampling frequency for several sampling points. In the mathematical model the water quality system is considered to be a multiple input-single output system. The relationship between these inputs and the output is determined by multiple input response functions which are calculated by multiple frequency response analysis using the equivalent representation of a time series in the frequency domain.

Following these preliminary theoretical considerations, the model is used to simulate daily values of a chloride record for a sampling site on the Rhine River by using the daily values of an upstream sampling site as well as the data of an important tributary sampling site.

With respect to the special sources of error, the results demonstrate the acceptable low level of errors inherent in this estimation process and the feasibility of reduced data-collection strategies without a great loss of information.

Introduction

As water pollution increases, the load of inorganic substances is of great importance. The increase of chloride (Cl^-) has been a especially grave problem for years now. The chloride load of the Rhine River, for example, has increased from 50 kg/s in 1875 to 350 kg/s in 1970. As river water becomes more and more important for meeting the growing demand for water, one cannot estimate the damage caused to a highly developed agriculture by irrigation with oversalted river water or the costs arising, for the water supplying communes which use bank filtrate or groundwater enriched with river water, from increasing corrosion of the pipes due to the increasing salinity.

Mathematical models, which make it possible to predict and control the water quality at a chosen point in the hydrologic system, are increasingly used for the optimal solution of such water management problems.

Arranging these mathematical models usually makes high demands on the existing data. Instead of yearly, monthly or weekly means, often daily values are needed to record the natural fluctuations of discharge and water quality. Frequently, the series of data from previous years are incomplete and have to be interpolated or extended to daily values.

Within the framework of the preliminary investigations for arranging a mathematical model for the prediction and optimal control of the chloride content of the Rhine River, a model has been developed which provides a possibility to close the lacunas in data stock while preserving the natural relationship between discharge and water quality.

With this model, the corresponding records for some stations within a system can be simulated which means that only control measurements are necessary. The sampling can therefore be reduced from daily to fortnightly measurements to reduce the costs of data collection. Without loss of generality, the mathematical model is based on chloride since this is one of the most important inorganic constituents for the description of water quality. Nevertheless, the mathematical model can be used in a similar way for all conservative water quality parameters.

Presentation of the Problem and Description of the Initial Data

The construction of a mathematical model for simulation of chloride records will be shown in a river reach in which the discharge is heavily influenced by a tributary (Rhine River from Karlsruhe to the mouth of the Main River).

Daily measurements from the gauges Q_1, Q_4 and Q_3 are available for the determination of discharge as shown in Fig. 1. The tributary's discharge is determined with the aid of measurements at gauge Q_2.

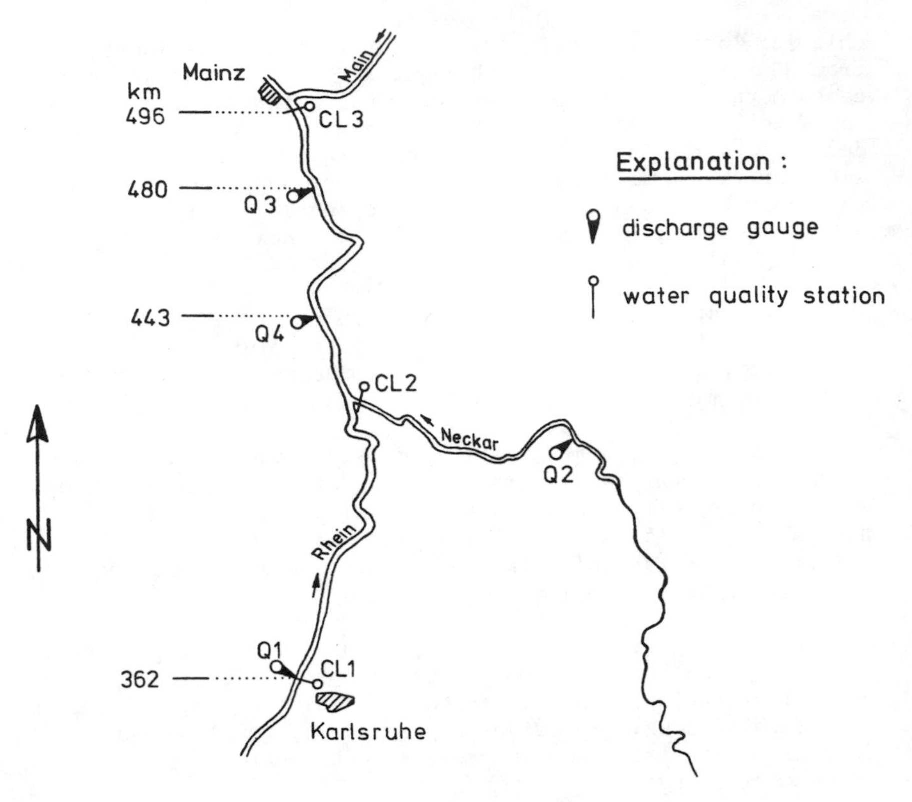

Fig. 1. Site plan showing the sub-basin of the Rhine River between Karlsruhe and Mainz.

The chloride sampling stations are CL1 (Maxau, km 363) and CL3 (Mainz, km 496, 200 m upstream of the mouth of the Main River) at the master stream and CL2 at the tributary about 0.3 km upstream of its inflow into the Main River. The aim of the investigations is the simulation of the chloride record for sampling station CL3 using the measurements of stations CL1 and CL2. This will save costs by reducing the sampling rate at station CL3 without a great loss of information. Only fortnightly checking measurements are necessary to control the reliability of the simulation.

One of the most important problems is the calculation of the chloride load for the corresponding stations. One can see from the site plan (Fig. 1) that direct calculation of the load from chloride content and discharge is possible only for station CL1. Since there are no direct discharge measurements for CL3 and CL2, discharge has to be computed with the aid of the data from gauges Q_3 and Q_2. The corresponding discharge hydrographs were calculated with consideration of an appropriate time lag and a gain factor determined from the relation of the size of the catchment areas. The analysis is based on records covering the years 1968 to 1970.

Description of the Water Chloride System

Due to the relationship between salt flow and water flow, a hydrological model of the river reach is a prerequisite condition for the development of the salinity flow system. The chloride concentration is characterized by the mass balance equation:

$$Cl_3^- = \frac{Q_1 \cdot Cl_1^- + Q_2 \cdot Cl_2^-}{Q_1 + Q_2} \tag{1}$$

Cl_i^- = concentration of chloride [mg/l]

Q_i = discharge [m^3/s]

The salinity flow system is superimposed on the hydrological system; measurements of the chloride concentrations in the water must be available for that purpose.

A simplified schematic diagram of the flow and salinity systems of a river reach is shown in Fig. 2. As is well known, the total inflow of a river reach is a combination of surface and subsurface inflow. The ultimate hydrologic flow input into a hydrologic system is precipitation in one form or another. However, when looking at a river reach, direct precipitation input to the model is very much overshadowed by river and tributary stream discharges into the system. For the salinity flow system, these inputs are even more important because of river transportation of salts into the river reach whereas precipitation is essentially free from salts. The municipal and industrial chloride inflow (i.e., potash mining) is important as well, since highly concentrated chloride solutions are pumped into the river system.

Fig. 2 shows the processes taking place in the water chloride system and indicates the spheres first influenced by increasing chloride concentration (Fig. 2).

As already mentioned in the introduction, agriculture is particularly affected by increasing salinity as irrigation with salted river water leads to losses. The same holds true for municipal and industrial water supply systems since corrosion in pipe networks increases badly with increasing salinity.

Relating Solute Concentrations to Stream Discharge

Since tributary discharge considerably influences the discharge conditions of the main river downstream of the mouth, it is necessary to respect the chloride of the tributary in the

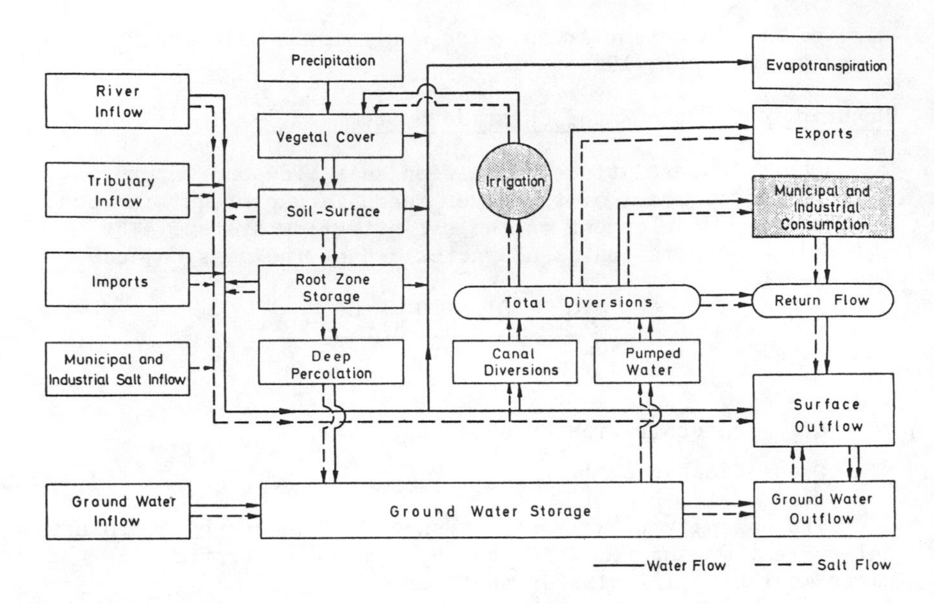

Fig. 2. Simplified flow diagram of hydrologic and salinity flow system.

calculation according to the mass balance equation. For this, the monthly measurements have to be interpolated to daily values. The daily discharge values can be used in the process. The major mineral constituents of natural water with negligibly artificial imports of chloride can be approximately represented by the equation

$$C(1) = k(1)/Q^{n(1)} \tag{2}$$

where C is solute concentration in mg/l and Q is stream discharge in m^3/s. K and n are regression parameters and 1 is the subscript for the corresponding ionic constituent.

Equation (2) has been well known for years and has been applied by others (for example, by Steele [1968] and Hall [1970]) using various modifications for many different problems with varying degrees of success.

The following three assumptions are implicit in the equation:

1) mixing is ideal (i.e., complete).
2) the dissolved constituents move in the flow system in the same fashion as does the water.

and 3) the influence of evapotranspiration or consumption is negligible.

To determine the regression coefficients k and n, one has to find the logarithm of the first equation and thus obtain a linear relation.

$$\log Cl^- = \log k - n \log Q \qquad (3)$$

K and n can be calculated analytically with the aid of least-squares fit analysis. Plotting the chloride content against discharge Q on logarithmic paper, it is possible to determine both coefficients graphically as well. On logarithmic paper, Relation (3) is a straight line.

Fig. 3 shows the chloride concentrations of the Neckar River as a function of actual discharge in logarithmic scale. The linear relation proves to be valid over the entire range of discharges.

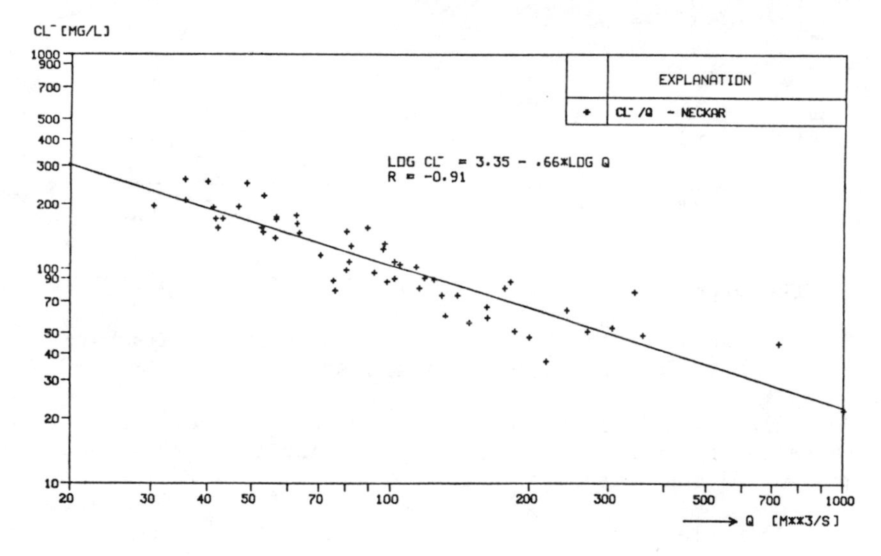

Fig. 3. Relation of Cl^--concentrations to stream discharge at station CL2 on the Neckar River.

The values for log k and n are 3.35 and 0.66, respectively. This leads to

$$Cl^- = 2238/Q^{0.66} \qquad (4)$$

The high correlation coefficient of $|r| = 0.91$ verifies the validity of the relation. The few isolated points above the linear trend obviously correspond to samples collected during the first heavy rainfalls after long periods of droughts.

It is impossible to determine the chloride concentration of station CL3 (Mainz) with the aid of equation (2) since the chloride concentration is strongly influenced by industrial waste waters. Fig. 4 describes the chloride measurements of station CL3 as a function of actual discharge in logarithmic scale.

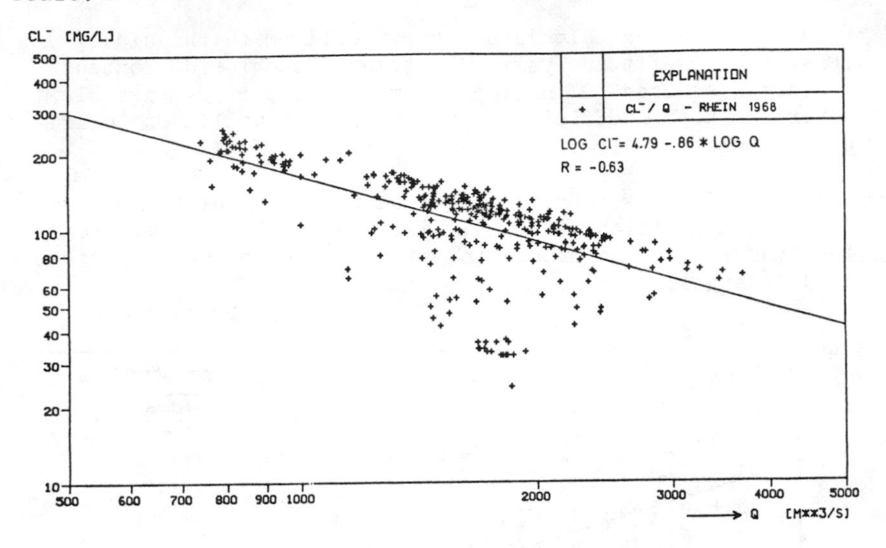

RELATION OF CL⁻ - CONCENTRATION TO STREAM DISCHARGE: STATION CL3

Fig. 4. Relation of Cl^--concentrations to stream discharge at station CL3 on the Rhine River.

The variations of the points around the linear trend indicates that relation (2) is not valid. The low coefficient of $|r| = 0.67$ confirms this.

In Table 1, the correlation coefficients are compared with the value of $|r|$ that was found for the Pescadero Creek (California) by Steele [1968]. The value of $|r| = 0.63$ for station CL3 is considerably below the reference values. Therefore, alternative methods have to be employed for computing the chloride concentration of station CL3. A method, which is based on multiple frequency response functions, has been used in the present paper.

Mathematical Basis

Introduction. For many hydrological problems, the river reach to be examined can be considered as a closed dynamic system in which the output $y(t)$ results from multiple inputs $x_I(t)$ (Fig. 5). To simplify the problem still further, the special case of one input and one output, as shown in Fig. 6, will be considered at this point.

Table 1 Regression Equations Relating Chloride Concentrations to Stream Discharge

Cl^- = chloride concentration [mg/l]
Q_f = discharge [cfs]
Q = discharge [m^3/s]

River/Station	Regression Equation	Correlation Coefficient	Curve linear Transformation
Pescadero Creek [Steele, 1968]	$\log Cl^- = 1.94-0.29 \log Q_f$	0.96	$Cl^- = 87.1/Q_f^{0.29}$
Neckar/CL2	$\log Cl^- = 3.35-0.66 \log Q$	0.91	$Cl^- = 2238/Q^{0.66}$
Rhein/CL3	$\log Cl^- = 4.79-0.86 \log Q$	0.63	$Cl^- = 61{,}569/Q^{0.69}$

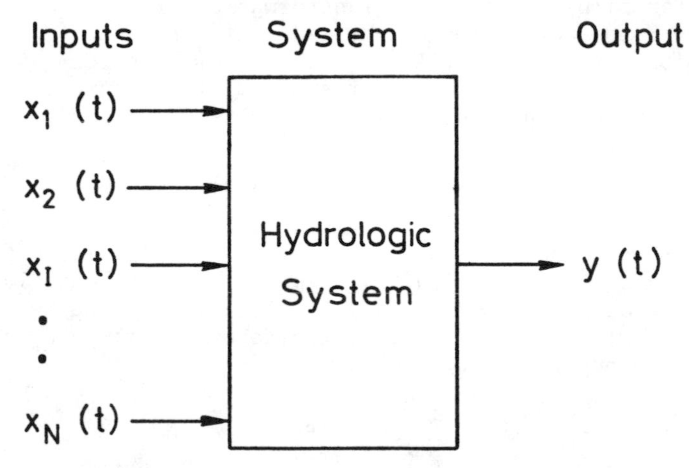

Fig. 5. Hydrologic system with multiple input and single output.

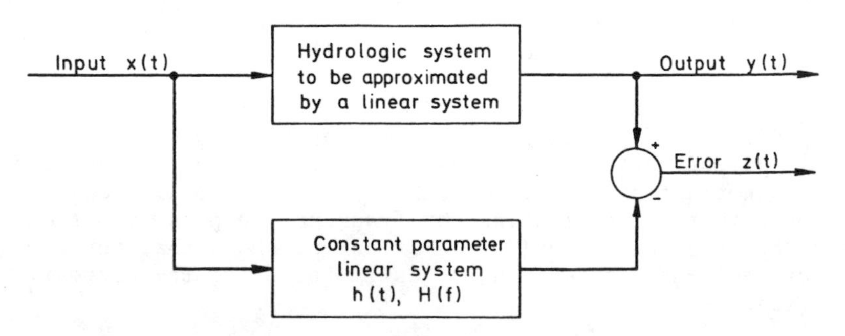

Fig. 6. Approximation of a hydrologic system by a single-input linear system.

When the deviations from the linear case are not too large, the output can be written in the form

$$y(t) = \int_{-\infty}^{+\infty} h(\tau)\ x\ (t-\tau)d\tau + z(t) \tag{5}$$

where $h(\tau)$ is the impulse response function and $z(t)$ is an error term which arises because the input and output variables may not be well controlled. $z(t)$ may also include quadratic and higher terms ommitted in the linear approximation.

Calculation of the Impulse Response Function of a Time-Invariant Single Input Linear System. The different methods of calculation of the impulse response function $h(\tau)$ have been discussed, for example, by Bendat and Piersol [1971], Jenkins and Watts [1969] and Blank et al. [1971).

For simplification, a transformation from the time domain to the frequency domain can be made with the aid of the Fourier transformation. If $X(f)$ is the Fourier transform of the input $x(t)$, and $Y(f)$ the Fourier transform of the output $y(t)$, the following important relation results:

$$Y(f) = H(f) \cdot X(f)\ . \tag{6}$$

In this way, the convolution integral (5) is replaced by a simple algebraic equation. This means that the convolution in the time domain changes to a multiplication in the frequency domain. $H(f)$ is the Fourier transform of $h(\tau)$ and defined as the frequency response function.

$$H(f) = \frac{Y(f)}{X(f)}\ . \tag{7}$$

However, it can be shown [Blank et al. 1971] that due to the noise in the input and output data, the Fourier transform method is no less immune to oscillations than the direct method, a numerical solution of the convolution integral equation for $h(\tau)$ in the time domain.

Hydrologic systems, in general, can be considered as highly damped systems and an oscillatory impulse response function is not anticipated from such a system. The difficulties can be removed by estimating the frequency response function using cross spectral analysis. It can be shown that the frequency response function can be estimated using the estimation equation:

$$\hat{H}(f) = \frac{\hat{G}_{xy}(f)}{\hat{G}_{xx}(f)} \tag{8}$$

In (8), $\hat{G}_{xy}(f)$ is the smoothed estimator of the cross spectrum between input and output and $\hat{G}_{xx}(f)$ is the smoothed estimator of the input power spectrum.

The main advantage of the spectral method is that the smoothing in the estimation procedure irons out erratic fluctuations and hence the estimator $h(\tau)$ obtained by taking the inverse Fourier transforms will be much smoother than that obtained by estimating $h(\tau)$ directly. Further, it is easily generalized to deal with multi-variate systems as will be shown in the next chapter.

The main advantage of the spectral method is that the smoothing in the estimation procedure irons out erratic fluctuations and hence the estimator $h(\tau)$ obtained by taking the inverse Fourier transform will be much smoother than that obtained by estimating $h(\tau)$ directly. Further, it is easily generalized to deal with multi-variate systems as will be shown in the next chapter.

<u>Calculation of the Partial Frequency Response Function of a Multiple Time-Invariant Linear System</u>. Various authors have applied the single input linear system as shown in Fig. 6 with varying success, for example Ribery [1964], Sauer [1973] and Wittenberg [1973].

This simple system can give accurate discharge calculations only for river reaches in which the discharge of the main river is not or insignificantly influenced by tributaries.

A look at the hydrologic map shows that this applies in very few cases. For modeling a river reach with one or more tributaries a procedure must, therefore, be found which will permit the treatment not of only one input, but of <u>multiple inputs</u> simultaneously. Fig. 7 is the schematic diagram of such a system based on multiple linear impulse response functions.

Mathematically, the linear system with multiple inputs can be expressed as:

$$y(t) = \sum_{I=1}^{N} \int_{-\infty}^{+\infty} h_I(\tau) \cdot x_I(t-\tau) d\tau + z(t) \tag{9}$$

where

N = number of inputs
$y(t)$ = output
$x_I(t)$ = inputs

$$y(t) = \sum_{I=1}^{N} y_I(t) + z(t) = \sum_{I=1}^{N} \int_{-\infty}^{\infty} h_I(\tau)\, x_I(t-\tau)\, d\tau + z(t)$$

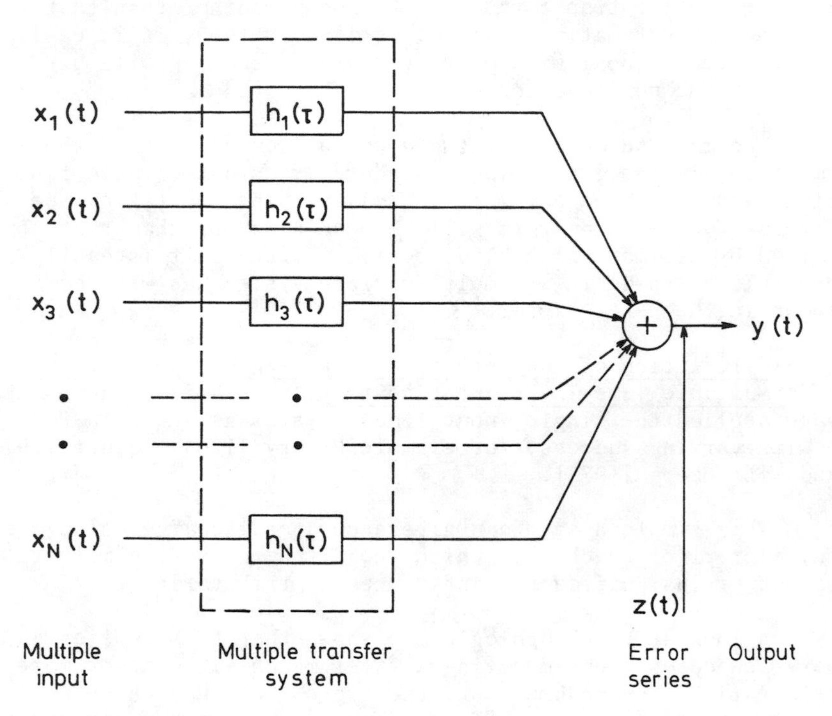

Fig. 7. Multiple linear system (MISO-multiple input/single output).

$h_I(t)$ = impulse response function connecting x_I and y, taking into consideration the $h_I(\tau)$ between all inputs x_J and $y (J \neq I)$

$z(t)$ = error as described above.

For the computation of the multiple frequency response functions, the following matrices and vectors are defined:

$\kappa(f) = [H_1(f), H_2(f), \ldots, H_N(f)]$ N-dimensional vector of the frequency response functions for one frequency f

$\xi_{xy}(f) = [G_{1y}(f), G_{2y}(f), \ldots, G_{Ny}(f)]$ — N-dimensional vector of the cross spectra of all inputs with the output, for one frequency f

$$\xi_{xx}(f) = \begin{bmatrix} G_{11}(f), & G_{12}(f), \ldots, & G_{1N}(f) \\ \cdot & & \cdot \\ \cdot & & \cdot \\ \cdot & & \cdot \\ G_{N1}(f), & G_{N2}(f), \ldots, & G_{NN}(f) \end{bmatrix}$$

N x N-matrix of the cross and power spectra of all inputs with each other, for one frequency f

Thus, the following term in matrix form is obtained for the response function:

$$\kappa'(f) = \xi_{xx}^{-1}(f) \cdot \xi'_{xy}(f) \tag{10}$$

where

ξ_{xx}^{-1} = inverse of ξ_{xx},

$\kappa'(f)$ = transposed vector of $\kappa(f)$.

Further calculations would exceed the framework of this paper. It should be pointed out, however, that the individual frequency response functions are not calculated independently, but that the influence of the remaining inputs is taken into account for every frequency response function.

Bendat and Piersol [1971] and Rodriguez-Iturbe [1967] examined the problem closely. Using a short-term forecasting of streamflow, Huthmann [1975] showed the high accuracy of hydrological parameter prediction which can be obtained with the above method. The mean standard error in the short term actual stage forecast up to one and two days ahead for the gauging station at Kaub on the Rhine was only 2.0 cm and 3.8 cm, respectively. This corresponds to 12.7 and 24.2 m^3/s. For a mean discharge of about 2000 m^3/s, this is an error of 0.6% to 1.2%.

Practical Aspects of Spectral Estimation

Following Bendat and Piersol [1971], the smoothed estimate of a true power spectrum is defined for arbitrary frequency f in the range $0 \le f \le f_c$

$$\hat{G}_{xx}(f) = 2\cdot\Delta t\ [\hat{R}_{xx}(0)+2\sum_{k=1}^{m-1} D(k)\cdot\hat{R}_{xx}(k)\cdot\cos(\frac{\pi\cdot k\cdot f}{f_c})] \tag{11}$$

where Δt is the time interval between samples, $\hat{R}_{xx}(k)$ is the estimate of the autocovariance function at lag k, m is the maximum lag number, $f_c = 1/2\,\Delta t$ is the Nyquist frequency and $D(k)$ is the lag weighting function.

Where the computation of frequency response functions and the associated coherence function is the main goal, the Parzen window has been found very useful because it maintains the sample coherence function between the theoretical values of 0 to 1.

The Parzen weighting function is given by the formula:

$$\begin{aligned} D(k) &= 1 - 6(k/m)^2 + 6(k/m)^3 && k = 0,1,2,\ldots,m/2 \\ &= 2\ (1 - k/m)^3 && k = m/2 + 1,\ldots,m \\ &= 0 && k > m \end{aligned} \tag{12}$$

The smoothed estimate of the true cross spectrum is a complex-valued quantity defined by

$$\hat{G}_{xy}(f) = \hat{C}_{xy}(f) - j\ \hat{Q}_{xy}(f) \tag{13}$$

with

$$\hat{C}_{xy}(f) = 2\cdot\Delta t\ [A_{xy}(0)+2\sum_{k=1}^{m-1} D(k)\cdot A_{xy}(k)\cdot\cos(\pi\cdot f\cdot k/f_c)] \tag{14}$$

and

$$\hat{Q}_{xy}(f) = 4\cdot\Delta t\ \sum_{k=1}^{m-1} D(k)\cdot B_{xy}(k)\cdot\sin(\pi\cdot f\cdot k/f_c) \tag{15}$$

where the $A_{xy}(k)$ and $B_{xy}(k)$ are the even and odd parts of the crosscovariance function, defined as

$$A_{xy}(k) = [\hat{R}_{xy}(k) + \hat{R}_{yx}(k)]\ /2 \tag{16}$$

$$B_{xy}(k) = [\hat{R}_{xy}(k) - \hat{R}_{yx}(k)]\ /2 \tag{17}$$

The general objective in any spectral analysis is to estimate the true spectrum as accurately as possible. This involves two requirements:

1) high fidelity

and 2) high stability.

Jenkins and Watts [1969] illustrate how these two requirements conflict and show that the important question in empirical spectral analysis is the choice of bandwidth of the spectral window determined by the maximum number of lags, m. In order to achieve high fidelity, the bandwidth of the spectral window must be of the same order as the width of the narrowest important detail in the spectrum. High fidelity is essential for computing frequency response functions. Therefore, the optimal choice of the bandwidth of the smoothing window has considerable influence on the calculation of frequency response functions.

In their investigation, Jenkins and Watts [1969] assume a constant bandwidth of the window for all frequency ranges. In practice, however, it has been found advisable to use different bandwidths for different frequency ranges so as to make the approximation to the theoretical spectrum as accurate and stable as possible. Fig. 8 shows the comparision of the theoretical spectral density functions Γxx of the AR-process $x(t) = x(t-1)-0.5\ x(t-2)+z(t)$ already used by Jenkins and Watts [1969] and the smoothed estimates $\hat{G}_{xx}$ of the spectral density function with constant and variable bandwidth. The graph shows a clear improvement of the estimate, especially in the range of high and low frequencies.

In the following investigation, the spectral functions are calculated with the aid of this method because practice has shown that often a few spectral ranges predominate while others are negligible.

<u>Calculation of the Impulse Response Function</u>

In order to compute $y(t)$ according Equation (9), it is necessary to transform the frequency response function into the time domain. Because of the stability conditions of linear systems in calculating practice, it is sufficient to define "memory lengths" n and p, which represent the upper and lower time limits of the impulse response function.

$$\hat{h}(k) = \Delta f \cdot \sum_{f=-f_c}^{f_c} \hat{H}(f) \cdot \exp(j \cdot 2\pi \cdot fk) \qquad -p \leq k \leq u \tag{18}$$

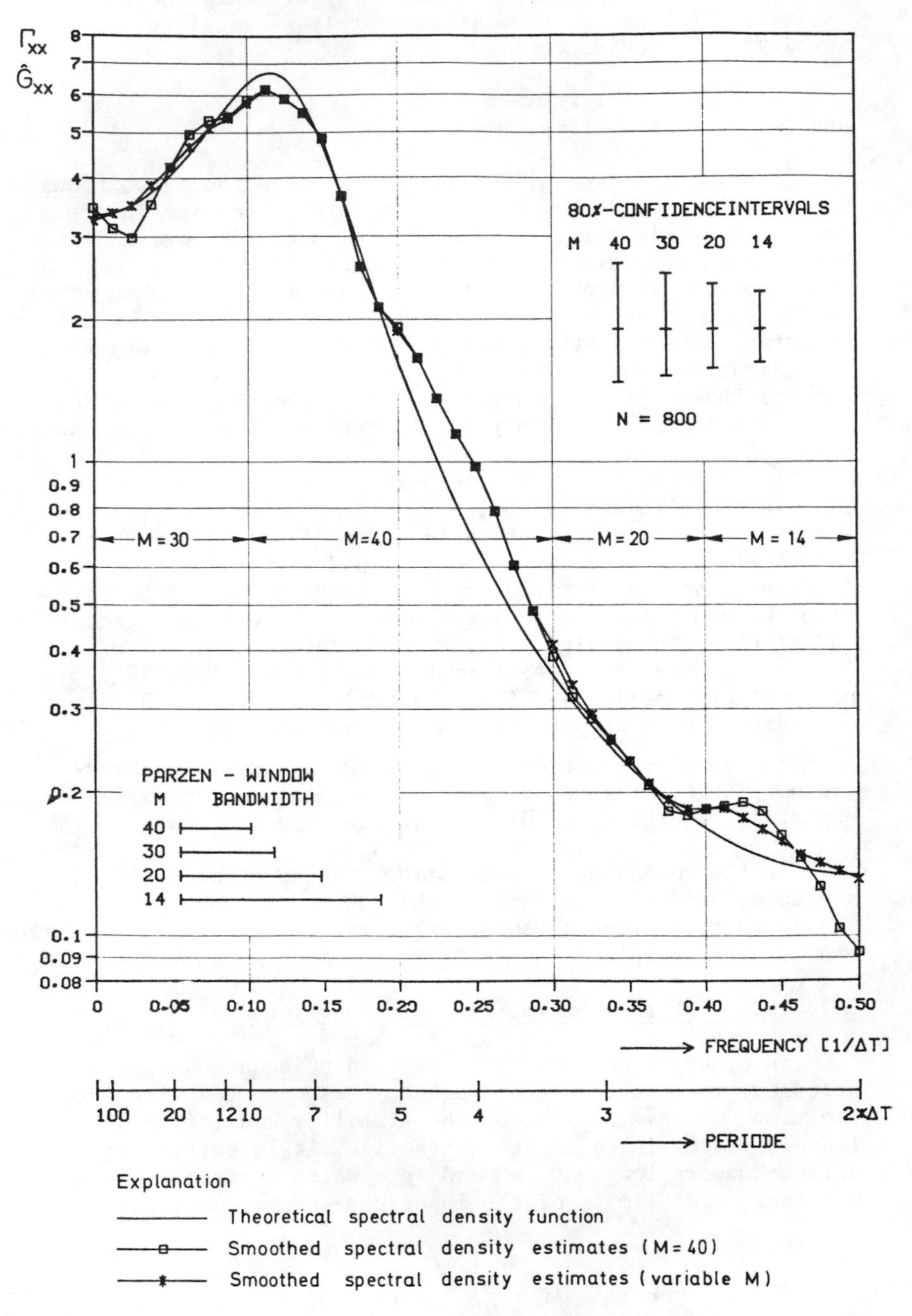

Fig. 8. Smoothed spectral density estimates for a second-order ar process (α_1 = 1.0, α_2 =-0.5; N = 800).

Note that in Equation (18), the impulse response function is calculated for $k < 0$ too, since the frequency response function does not show whether or not the system is physically realizable. The condition for a physically realizable system is:

$$h(k) = 0 \quad \text{for} \quad k < 0 \tag{19}$$

Unlike the optimization method in the time domain as applied by Wilke [1975] where $h(k) = 0$ for $k < 0$ is a constraint, the calculation in the frequency domain can lead to an impulse response function which shows values not equal to zero for $k < 0$. This leads to the discrete expression of Equation (9):

$$y_t = \Delta t \cdot \sum_{I=1}^{N} \sum_{k=-p}^{u} \hat{h}_{I,k} \quad x_{I,t-k} \tag{20}$$

$$1 \leq t \leq NGES\text{-}p$$

NGES = total number of data

Only if $p = 0$ can the simulation be carried out to NGES.

The impulse response functions of the chloride system obtained by inverse Fourier transform of the estimate of the frequency response functions are shown in Fig. 9. Due to the time lag of one day, the impulse response functions describe an approximately physically realizable system. Nevertheless, the calculation is performed with $p = 2$.

Model Verification and Statistical Evaluation

The aim of the investigations, as mentioned before, is to simulate the chloride hydrograph of station CL3 utilizing the chloride measurements of stations CL1 and CL2.

The term "simulation" is used here instead of "forecast" because chloride response curve calculation is not an actual forecast but the subsequent computation of a curve that has already been measured. This is, therefore, the simplest form of a multiple system, i.e., a system with two inputs and one output.

The large fluctuations of the chloride concentration in Rhine River water due to industrial waste water inflow can be seen clearly from Fig. 10 where the chloride concentration for the two input functions and the output function for the months January 1968 to April 1968 is indicated.

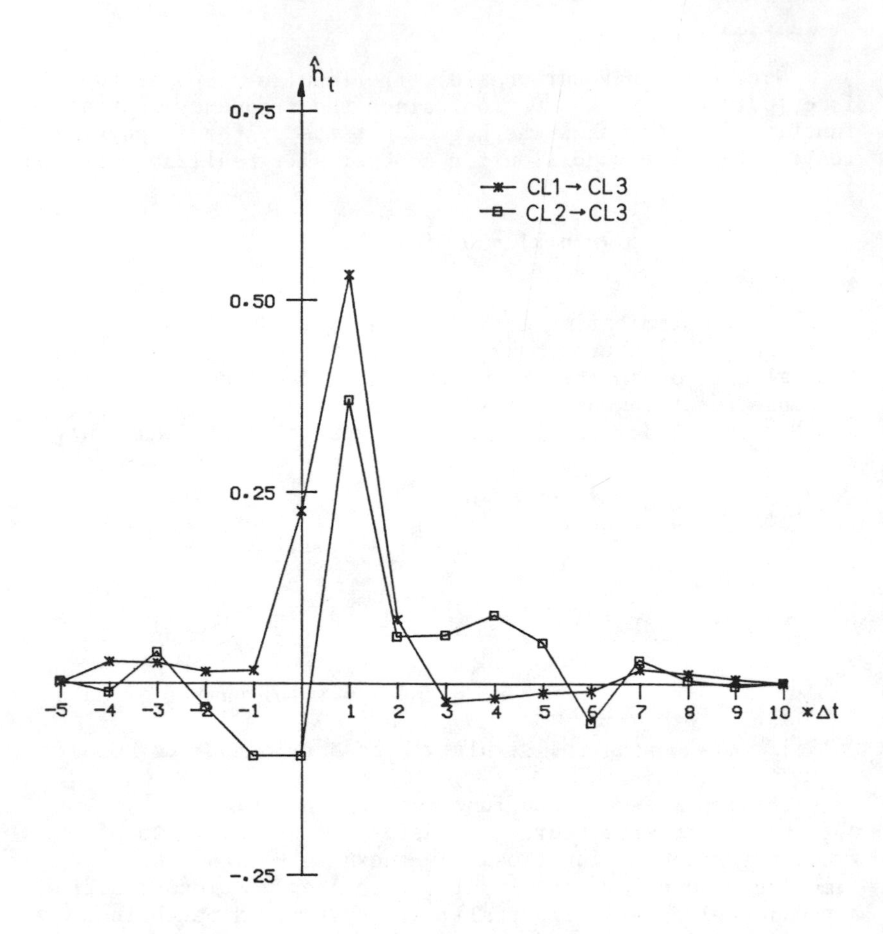

Fig. 9. Impulse response of the system obtained by inverse Fourier transform of the estimate of the frequency response function.

Provided that there is perfect mixing and that the dissolved constituents move in hand with the water, the frequency response functions can be determined from the chloride load curves. Smoothing windows with four different values of m have been applied in computing the estimates since the energy in the power spectra is not equally distributed over the whole of the examined frequency range.

Fig. 11 shows the result of the simulation after computing the chloride concentration. As had to be expected in view of the high variance of the chloride concentrations, the simulation does not reach the accuracy of the aforementioned discharge forecasts.

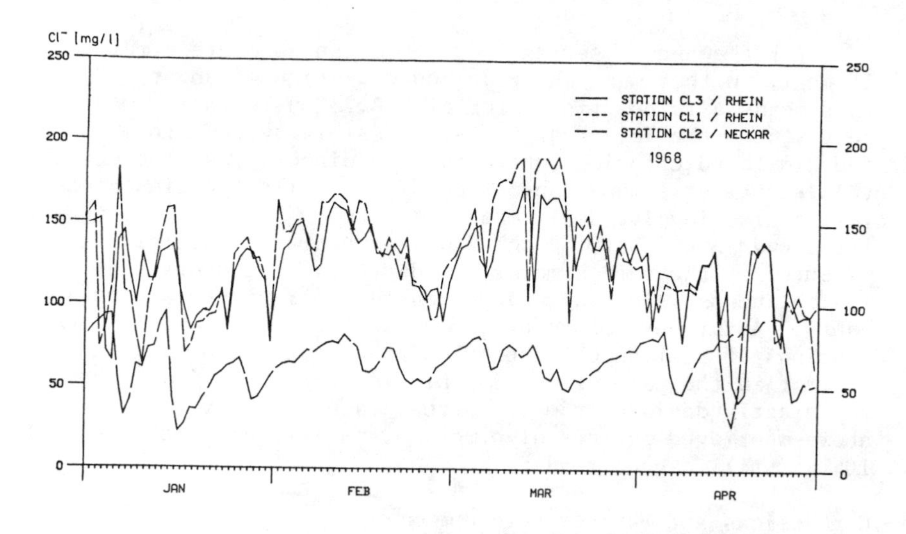

CHEMICAL HYDROGRAPHS FOR CHLORIDE CONCENTRATIONS

Fig. 10. Chemical hydrographs for chloride.

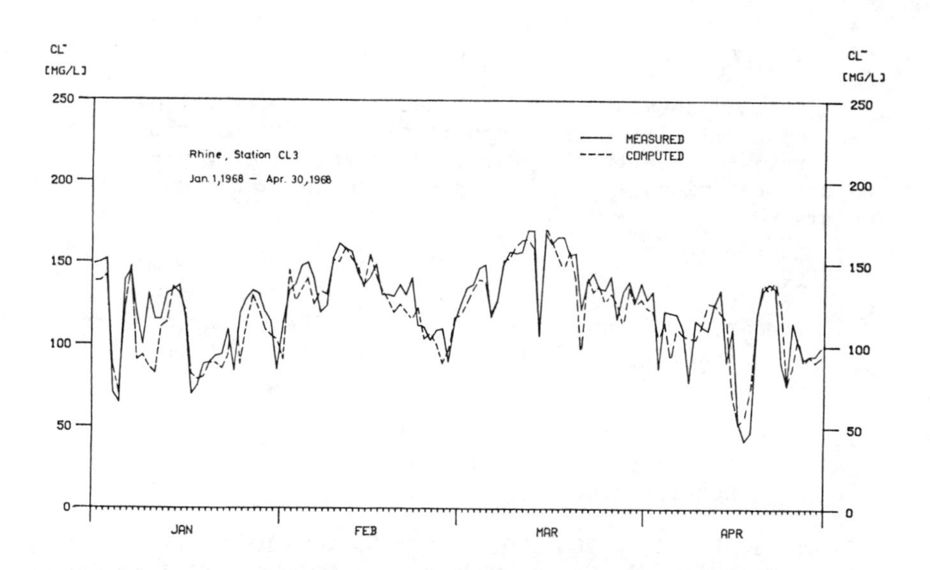

CHEMICAL HYDROGRAPH FOR CHLORIDE CONCENTRATION: STATION CL3

Fig. 11. Chemical hydrograph for chloride, comparing the measured and computed chloride concentrations at the station CL3 on the river Rhine.

For the annual series 1969, the mean absolute error is 15 mg/l. With a mean chloride concentration of about 150 mg/l, this means a forecasting error of 10%. Consideration must be given to the fact that it is impossible to compare measured and simulated chloride concentrations directly, as the measured values reflect a momentaneous condition while the simulated values also involve daily mean discharge. Furthermore, there is no guarantee that the chloride concentration was always measured at the same time of the day. Fig. 10 shows that the instantaneous value may change considerably within a couple of hours. A further source of error is that the chloride concentrations of station CL2 are calculated, not measured values, so that in the case of unfavorable weather conditions, they may greatly deviate from the actual values. In view of the above-mentioned sources of error, a statistical error of about 10% is still a good result.

Conclusions and Further Development

The investigations described above show that the method introduced here is not only well suited for the short term prediction of discharge hydrographs but also for filling gaps in hydrologic series and for the simulation of conservative water quality parameters. That means that with this method it is possible to simulate the effects of salt inputs on downstream sections of the river (with the tributaries taken into account).

Further development of the method to the level of a forecasting model will give an opportunity to control the salinity of streams optimally with due consideration being given to the natural relationships between discharge and water quality parameters.

References

[1] Bendat, J. S. and Piersol, A. G., 1971. "Random Data: Analysis and Measurement Procedures." John Wiley, New York.

[2] Blank, D., J. W. Delleur and A. Giogini, 1971. "Oscillatory Kernel Functions in Linear Hydrologic Models." Water Resources Research, 7: 1102-1117.

[3] Hall, F. R., 1970. "Dissolved Solids-Discharge Relationships, 1. Mixing Models." Water Resources Research, 6: 845-850.

[4] Huthmann, G., 1975. "Short-term Forecasting of Streamflow with the Aid of Multiple Frequency Response Functions." Proceedings of the Bratislava Symposium, September 1975. IAHS-AISH Publ. No. 115: 115-121.

[5] Jenkins, G. M., and D. W. Watts, 1969. "Spectral Analysis and Its Applications." Holden-Day, San Francisco.

[6] Rodriguez-Iturbe, I., 1967. "The Application of Cross-Spectral Analysis to Hydrologic Time Series." Colorado State University Hydrology Paper No. 24, Fort Collins, Colorado.

[7] Ribeny, F.M.J., 1964. "Flood Routing with A Unit Hydrograph Approach." J. Inst. Engrs. Australia, 36: 9-22.

[8] Sauer, V. B.,1973. "Unit Response Method of Open Channel Flow Routing." Journal of Hydraulics Div. ASCE, 99: 179-193.

[9] Steele, T. D., 1968. "Digital Computer Applications in Chemical-Quality Studies of Surface Water in a Small Watershed." IASH-UNESCO Symposium on Use of Analog and Digital Computers in Hydrology. Tucson, Arizona. IASH Publ. No. 80: 203-214.

[10] Wittenberg, H., 1973. "Die Bestimmung der Übertrangungsfunktion zwischen den Hochwasserbflüssen an zwei Pegeln der Brigach." Konferenz der Donauländer über hydrologische Vorhersagen. Varna, Bulgaria.

[11] Wilke, K., 1975. "Principles of Hydrological Forecasting by Multichannel Wiener Filtering." Proceedings of the Bratislava Symposium, September 1975. IAHS-AISH Publ. No. 115: 257-264.

APPLICATION OF RESIDUALS-MANAGEMENT TECHNIQUES FOR ASSESSING THE IMPACTS OF ALTERNATIVE COAL-DEVELOPMENT PLANS ON REGIONAL WATER RESOURCES

By

Ivan C. James, II
U.S. Geological Survey
410 National Center, Reston, Va. 22092

and

Timothy Doak Steele
U.S. Geological Survey
Mail Stop 415, Box 25046
Denver Federal Center, Lakewood, Colo. 80225

Abstract

The development of coal resources in a region will have a variety of effects on available water resources. Direct effects result from coal mining, processing, and transport and conversion techniques utilized in or proposed for the region. Indirect effects are related to the growth of population, public services, and commercial activities in the region. Both types of effects impact upon regional water resources in terms of quantity (water withdrawal and consumptive use) and quality (capacity to assimilate discharged residuals). Assessment of material and energy balances of the activities causing direct effects, coupled with an economic and technical evaluation of alternative methods of residuals modification, will provide the primary input data for regional environmental models. An analysis of residuals resulting from activities causing indirect effects will provide additional data for these models.

Residuals-management and environmental-modeling techniques are used to determine the potential effects of several plans for alternative levels of coal development and environmental standards in the Yampa River Basin, Colorado and Wyoming. The results will provide regional planners with a summary of the potential effects of coal development on the environment and economy upon which they can base decisions concerning the type and extent of development that would be authorized.

Introduction

The development of western coal resources will have a variety of effects and impacts on many of the people, institutions, and much of the environment, natural resources, and ecological resources of the region. There are a variety of ways in which the coal resources may be developed and utilized, each with different impacts. The type of mining and reclamation, type of transportation system, type of energy conversion process and rate of development are particularly determinative of the resulting impacts.

As Beckner (1976) has pointed out in discussing the geohydrologic data requirements for environmental impact statements, control of mining is exerted at amny political levels, from the county level through zoning to the Federal level through leasing policies. Policies on water-right transfer, relative pricing of energy and environmental-control requirements have definite impacts in an energy-developing region, even though they may be affected through indirect mechanisms.

The provision of information to the various levels of policy and decision making with the effects of their actions on the ambient social, economic, and environmental conditions in a region is a multi-step procedure with significant uncertainties related to the actions of other decision-making entities. In the first step, alternative policies must be converted into general developmental scenarios. The scenarios then are filled out to reflect the required level of infrastructure and supporting economic and natural-resource development. This complete specification of conditions is used to derive a series of models that estimate the impacts and environmental measures of interest. Finally, a mechanism for comparing the multidimensional set of impact measures must be applied to reduce the problem to manageable proportions.

Background. The U.S. Geological Survey is conducting studies of water use and residuals management in selected process industries, particularly the energy production and conversion industries. Initial results have been summarized in a draft report (James et al., 1977). At this stage of analysis, coal-mining and electric-power and coal-gasification plant processes have been evaluated for purposes of determining amounts and forms of generated residuals and alternatives for modifying these residuals through treatment. Objectives on ambient environmental quality can be met through residuals modification, but the type and degree of modification will depend upon other decisions on the location and amount of residuals released to the environmental media.

The Yampa River basin-assessment project is taking results from the residuals-management analysis, scaling the residuals loadings where necessary, and using these values to indicate the stresses on the environment of the basin (see Fig. 1). The modes of evaluation range from modeling

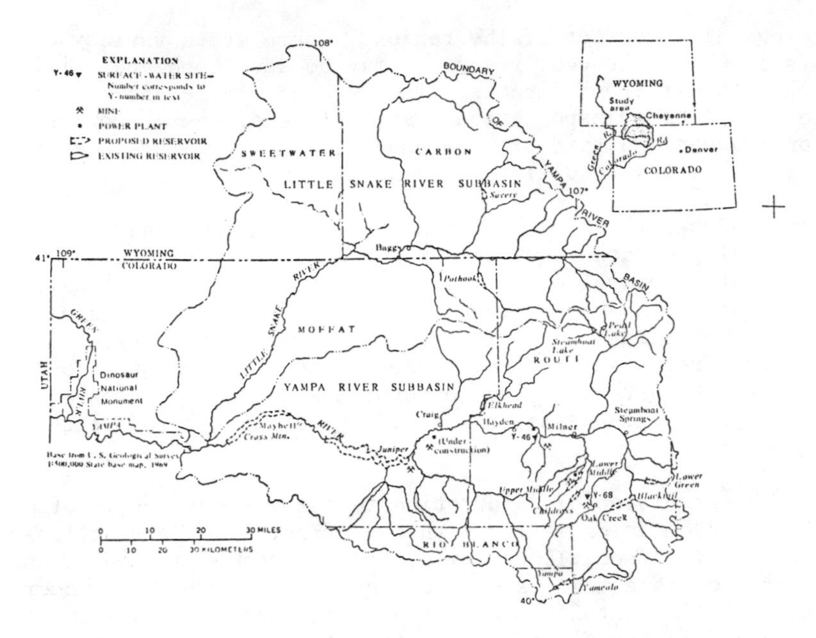

Fig. 1. Yampa River Basin, Colorado and Wyoming

applications, where such quantitative techniques are applicable to semi-quantitative or descriptive descriptions of projected environmental consequences. Coal-mining production in the basin is expected to increase from a 1975 level of 4.4 million tons (4.0 million tonnes) to 20 million tons (18 million tonnes) by 1990 (Steele, 1976). Details of the basin-assessment approaches and anticipated information products are outlined by Steele et al. (1976a; 1976b). The basin assessment serves as a real-case example for developing and utilizing methods of regional water-use evaluation, using techniques and information developed from the residuals-management analysis.

Objectives. The objectives of residuals-management analysis and the Yampa River basin-assessment projects are to develop and apply methods for investigating water use in a regional context. Individual water users respond to a variety of economic conditions and legal constraints which affect their water use. These conditions themselves are affected by the existence of other water uses and their opportunities for process modifications that lead to the most efficient production for a given technology set and prices of their inputs, including water. It is reasonable to believe that as competition drives up either the monetary or shadow price of water, shifts to less water-intensive processes would occur, with larger capital costs and probably lower operating efficiencies.

Thus the regional equilibrium for technology and water use must be studied as a whole in order to understand the marginal economics of production in the region.

Besides the adjustments to regional economics that affect water use, constraints on the form of residuals discharge have a substantial effect on water use. Perhaps the most significant and obvious constraint for the energy-conversion industry are requirements for cooling towers, which generally consume half again to twice as much water as is lost from stream evaporation in once-through cooling. Wet scrubbing of stack gases for control of SO_2 and particulates often requires stack-gas reheat, significantly decreasing plant output efficiency, and therefore increasing the cooling load for the same output. Thus, the conversion of the residual from a gaseous to a solid or liquid form has an impact on water use which may affect the amount of water rights purchased by industry. Moreover, eventual environmental damages from release of the residuals may be greater than anticipated.

Purpose and scope. The relative emphasis of this paper stresses the conceptual methodologies developed in the residuals-management analysis and applied to the studies comprising the Yampa River basin assessment. Specific results of the basin-assessment studies are being discussed in several reports in process. An attempt is made here to indicate those techniques and the sequences or patterns of problem identification and analysis that would have application in many of the regions undergoing accelerated coal-resource development. A common characteristic of these regions in the western United States is their limited water resource which is largely committed to existing economic sectors.

Approach

Planning and assessemnt of future conditions under uncertainty is a difficult task because of the multiplicative effect of the individual uncertainties. Not only are natural events uncertain, but political and economic realizations are so uncertain as to defy a rationally based probability assessment. An assessment directed towards a particular audience of decision makers would not even be able to specify the decision variables at the disposal of that audience because the variables vary with other political conditions. They are also often not well defined in the sense that any one decision-making entity has only partial control or just influence over a decision.

Under these conditions a direct linkage between an individual decision and the final outcomes is difficult to trace. It is possible, however, to make assessments of particular conditions. These conditions can be described for a future scenario and the assessment tied to that scenario. As to whether or not these future conditions will be met, or by what means, these questions are not addressed. A set of scenarios which in some manner spans the possible future conditions can be assessed without prejudice, and the desirability of the outcomes and means for obtaining them is left to the active participants in the decision-making process.

An outline of the flow of information and sequence of tasks for making an assessment is found in Fig. 2. A particular scenario for future energy development in combination with an economic input/output model is utilized for determining levels of economic output. Through various process models, residuals are generated and discharged to the environment through direct consequences of coal mining, processing, and conversion alternatives. Environmental impacts from population growth, related services, and non-coal economic sectors are termed secondary, and related residuals are generated empirically by coupling the basin's economic activity with residuals-generation coefficients. Results from these two concepts then feed into the various assessment methods, subject to constraints from assumed or proposed standards to implement specific environmental goals. Water-use implications with quality as well as quantity implications, interact with all of the above items and must be coupled with legal and institutional constraints on water-resource development as portrayed through State's water rights, basin compacts, and other applicable local, State and Federal regulations.

Brief descriptions of these concepts are outlined in the following sections. Applications of the methods imbedded in these concepts in both theoretical and real-case terms, the latter with respect to the Yampa River basin, then follow.

Economic input-output analysis. Over 30 sectors have been designated for describing economic activities in the Yampa River basin (Udis and Hess, 1976). Most represent single or aggregate codes from the U.S. Department of Commerce Standard Industrial Classification (SIC) scheme. Notable modifications for the Yampa River basin-assessment analysis include delineation of three agricultural sectors--(1) livestock, (2) irrigated, and (3) all other agriculture-- and breakdown of a "recreation" sector into winter-sports and other recreation (camping, fishing, etc.) sectors.

Regional total gross output figures for the economic sectors other than those related to coal mining and utilization are coupled with residuals coefficients (Udis et al., 1973; Howe et al., 1975) to estimate residuals loadings from these secondary sources. Total gross output values for Routt

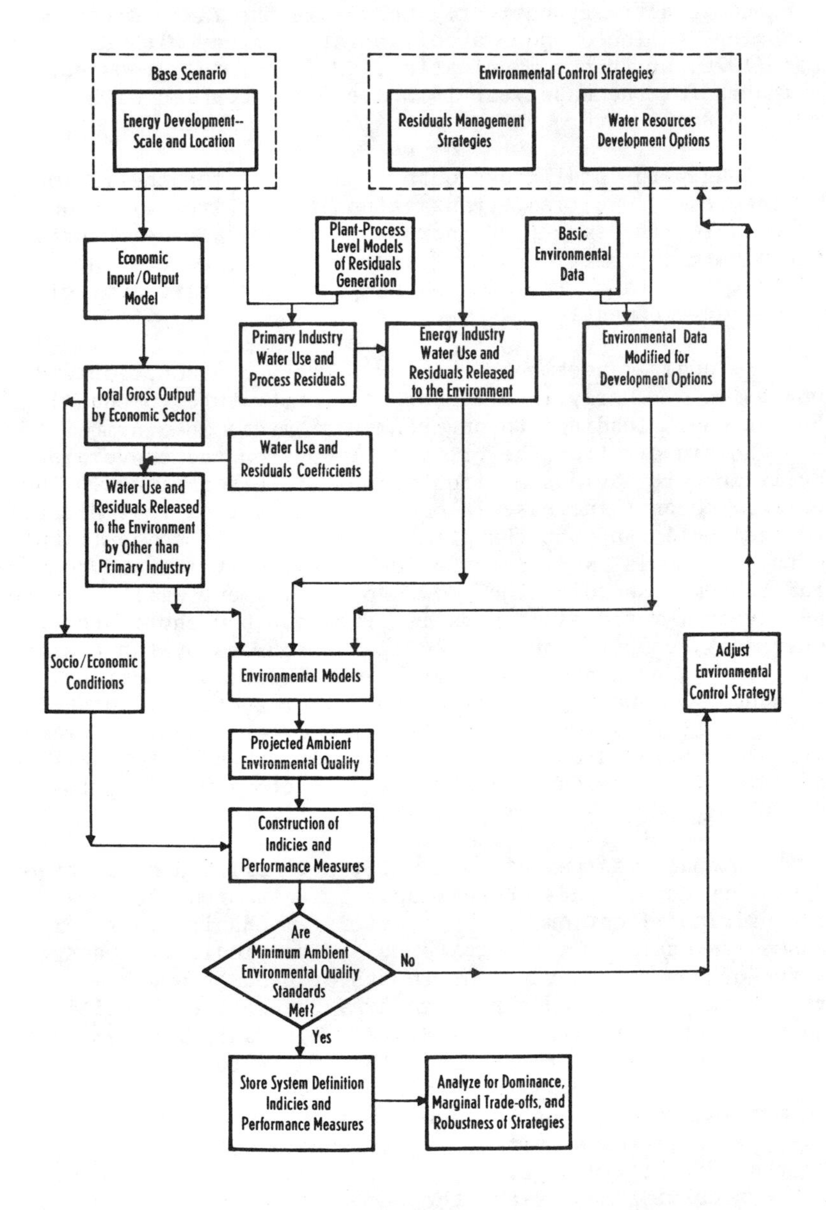

Fig. 2. Information flows and task sequence for basin assessment

and Moffat counties in northwestern Colorado, where the bulk of economic activity currently occurs in the Yampa River basin, are being developed under a collaborative study with Udis and Hess (1976; written communication, 1977). Total gross output estimates for the base year 1975 are being coupled with 5-year projections through the year 1990.

Along with total gross output estimates for the region for the years specified, transactions tables are being constructed for the same time increments; these are being used to evaluate interactions among sectors and to assess the possible impacts of introducing new or changing technologies of coal development.

Residuals management. New or increased production and conversion of energy in a region will contribute to changing the residuals loadings to environmental media in a number of ways. Residuals from the primary production and conversion facilities are obvious manifestations of this increase. The general regional increase in economic activities that directly and indirectly support the primary facilities also contributes to the residuals loadings. Besides the industrial and commercial economic sectors, the growth of employment will increase the level of residuals generation from the household sector. Conversely, competition for regional resources by the energy development and conversion sectors may reduce production, and hence residuals generation from other sectors. Notably, in the western United States, land and water resources transferred to energy uses are likely to bring changes to the types and amounts of residuals generated by sectors relating to agriculture.

Residuals management is usually accomplished by a set of regulatory constraints on residuals release or degree of residuals modification, or by a pricing mechanism on the released residuals. In the latter case, particularly, changes in the prices of other factor inputs or product and by-product outputs may, *ceteris paribus*, change the quantities and types of residuals released. For residuals management in a region, several desideratums might be set for evaluating the regional residuals-management program. These relate to the marginal conditions between costs and benefits, marginal costs of modifying a particular residual among industries, the transferability of residuals from one location to another in the receiving medium and the information and management costs of implementing and operating the residuals-management program. Whereas, a plant-process level model generally provides a determinate set of residuals outputs for a given specification of inputs, a regional residuals-management problem may require the estimation of equilibrium conditions which may not be a very easy task.

Environmental assessment. A range of environmental-assessment techniques can be applied to studying the various impacts of coal-resource and associated secondary development. Where relevant data are available and the nature of the impact justifies use of a quantitative approach, a specific process or statistical model might be applied. Where information or a quantitative modeling technique is deficient or lacking, or where the study does not justify use of a detailed analysis, an assessment of impacts may be semi-quantitative or descriptive (Hines et al., 1975). The following topic areas are being studied in the Yampa River basin assessment, where, in some cases, generated residuals loadings estimated from process models or secondary sectors are imposed as additional stresses of development on the environment (Bower and Basta, 1975): (1) surface-water reservoir development, (2) stream-salinity conditions, (3) regionalized stream-temperature analysis, (4) basin sediment yields, (5) areal variations for other selected water-quality variables, (6) regional ground-water aquifer systems with identification of stream losses or gains, (7) basinwide surface-water availability and seasonal fluctuations, (8) areal patterns of land use, vegetative cover, and snowpack, (9) regional ambient air-quality patterns, and (10) water use implied by various development alternatives. Results to date of studies encompassing several of the topic areas listed are discussed in this paper.

Environmental-control alternatives. Environmental standards for regional air quality, water quality, and disposal of solid residuals are set at both State and Federal levels. However, the States generally are given ultimate responsibility for enforcing these standards and for implementing new or revised standards proposed for achieving environmental goals.

Ramifications of environmental goals recently have been evaluated by the National Commission on Water Quality in the context of economic costs of achieving these goals, particularly with respect to uses of energy and water resources. Within a residuals-management analytical context (see above), an assessment of disposal alternatives of residuals loadings for a given process to environmental media is germane to the theme of this paper. For example, if air-quality standards for coal-conversion processes were to be particularly stringent in a region, technologies exist to transform gaseous residuals to liquid or solid forms. Such transformation might have highly undesireable consequences on regional water quality. In evaluating overall environmental impacts, one must be aware of such possible transformations and must recognize that residuals loadings cannot be eliminated entirely (Reiquam et al., 1975).

The Yampa River basin assessment seeks to evaluate alternative treatment processes for controls of residuals anticipated from coal mining and conversion as well as from growth in other economic sectors. Interactions between residuals controls and the various mixes of assumed environmental standards and other regulations should be evaluated carefully by both regional planners and resource managers to assure that societal goals are kept in perspective.

Water-use implications. The rate of water use in energy-developing regions will be the result of a complex interaction of supply and demand factors such as scale of development, technology, economics, politics, and local geohydrologic conditions. Supply conditions will be governed by the natural availability of water, possibilities for improving the dependability of natural supplies, import possibilities, and the legal system governing the transfer of water rights. Demand is affected by the scale of energy development, the amount of in-basin energy conversion, the technologies chosen for both the conversion processes and particularly the type of cooling system used for the conversion process, and the support service requirements of the chosen technologies.

One of the most significant water-use implications of energy development in the Yampa River basin will be the effect on agriculture of the purchase of water rights from this sector for transfer to energy-related economic sectors. Grazing is only somewhat dependent on irrigated pasture, but livestock raising in general depends upon irrigated hay crops for winter feeding. The sale of this irrigation water to other purposes would mean either a greater reliance on imported feeds, a conversion from year-around ranching to summer fattening, or a reduction in the scale of ranching activity. Another implication of energy development will be the increased use of the streams' assimilative capacity. Even if industry achieves a zero liquid-residuals discharge status, increases in urban population may contribute to higher stream loadings. Although mining may produce water in the early phases from de-watering of overburden and coal seams, the extensive utilization of sediment-control practices on reclaimed mine spoils may increase the loss of water through evapotranspiration.

The availability of water may become a constraint on the location of energy-conversion facilities in some of the energy-rich basins in the western United States. Social, environmental, and economic objectives will substantially affect the point in time when this constraint becomes effective.

Results

Residual generation. Estimates of residuals released from energy-development related economic activities are obtained from a variety of sources depending upon the relative importance of the sectors in contributing to the total residuals loads. Major potential industrial sources of residuals from new technologies such as coal gasification and liquefaction can best be assessed for residuals generation rates from detailed plant-level studies of process material and energy balances. These procedures are also useful for estimating residual generation rates in existing industries where economic or regulatory induced technical change may alter the quantity and composition of the residuals.

Estimates of residuals from economic sectors which contribute little to the total loads generally do not have to be made with such accuracy. For these estimates, coefficients of residuals discharge developed from national averages may suffice. Intermediate between these two extremes are a number of empirical procedures which utilize regional and site specific information either directly from measured residuals discharge or indirectly through estimates based on local values of econometric variables which are known to affect residuals generation.

The principal residuals associated with surface mining are water and wind-borne sediments. If no reclamation were to be practiced, these rates of sediment generation might average in the order of 600,000 tons per year and 30,000 tons per year respectively for a mining rate of 20 million tons per year. Delivery of the water-borne sediment to the main streams might be considerably reduced in quantity and delayed in time. If intensive reclamation practices are instituted, including irrigation for enhancement of vegetation and the construction of sediment detention facilities, water-borne sediment generation could probably be reduced to premining levels and wind-borne sediment to less than 10,000 tons per year. Current and proposed reclamation practices in the Yampa River basin do not require irrigation; therefore, actual yields may be somewhat higher than the estimates for the intensive reclamation effort.

If the 20 million tons of coal were converted into either gas or electricity, about 1.1 million tons of ash would be generated from the combustion of the coal or the chars from the gasification plant. In dry-bottom furnaces, 90 percent of this may be entrained in the stack gasses but electrostatic precipitation or the wet scrubbing necessary for SO_2 removal can remove up to 99 percent of this. Thus an air-pollution problem may be transformed to a solid residuals disposal problem.

If the coals utilized average 0.4 percent sulfur, then the 80,000 tons (73,000 t) per year of sulfur would end up as 3,200 tons (2,900 t) in the bottom ash and as 154,000 tons (140,000 t) of SO_2. For gasification plants, 96 percent of the gas would have to be treated at a 98-percent removal of sulfur to meet a 150-ppm limit on SO_2 assumed for a stack-emission standard. For electric-power plants, an estimated 57.5 percent of stack gases would have to be treated at a 90-percent removal level to comply with the same air-quality standard.

Complete sets of residuals coefficients are not presently available for estimating the increased residuals generation from the indirect economic expansions caused by increased energy production and conversion.

Estimates of water use. Projections of water use for coal mining in the Yampa River basin at the 20 million tons per year level range from 500 to 3,500 acre-ft per year depending upon the reclamation practices (James et al., 1977). The lower rate of water use corresponds to the requirements for sanitary uses and a minimal haul road dust control program. The higher rate of water use assumes a reclamation program including revegetation of mine spoils with three years of irrigation at diminishing rates of application. Rainfall in the mining areas of the Yampa River basin is generally adequate for revegetation without irrigation; however, rapid revegetation to dense stands of grasses and legumes for immediate sediment control could be enhanced by irrigation.

The projected mining rate of 20 million tons per year would be sufficient to support about 7,200 megawatts of electric power generation. If all of this coal were to be utilized for mine-mouth power generation in the basin, substantial quantities of water would be consumed in cooling--approximately 110,000 acre-ft per year if wet cooling towers were utilized for heat dissipation. Such a consumption represents about 10 percent of the average flow at the Maybell gaging station.

Using current input/output analyses (Udis and Hess, 1976) and water use coefficients from Gray and McKean (1975), it appears that increases in other economic activities spawned by this increase in energy production and conversion, including the household sector but excluding agriculture, would increase water consumption in the Yampa River basin by less than 2,000 acre-ft/year. In terms of the total water supply of the Yampa River basin, this is so small as to be inconsequential, but for the major trade centers it may require substantial increases to their existing supply system.

Environmental modeling and assessment. A range of modeling and semi-quantitative assessment techniques is being used to describe the environmental consequences of coal-resource development in the Yampa River basin. Relative emphasis is placed upon water-resource implications, particularly with regard to water quality. Presumedly, the techniques would be applicable to any region in the Rocky Mountain states which is undergoing such development. Some aspects of the following assessment techniques are described below: (1) reservoir modeling, (2) stream-salinity calculations, (3) selected water-quality implications, and (4) interactions between the basin's surface- and ground-water resources.

Reservoir development. The total reservoir storage capacity in the Yampa River basin is expected to increase from its current estimated level of 57,000 acre-ft (2.3 million m^3) (Steele et al., written comm., 1976) to a projected level of more than 1.9 million acre-ft (2.3 billion m^3), which exceeds by 21 percent the estimated long-term mean annual stream discharge from the basin (Steele, 1976, p. 12). Hydrologic impacts of selected major proposed reservoirs are being studied in the assessment. Selection was based upon several criteria including size, location, purpose, political sensitivity and probable success of implementation.

Historically for the 1922-75 water-year period, the combined annual flows from the Yampa and Little Snake Rivers (see Fig. 1) has ranged from 455,000 acre-ft (.056 billion m^3) to over 2,900,000 acre-ft (3.6 billion m^3). The long-term streamflow discharge from the basin based upon the same period of record, is about 1,500,000 acre-ft (1.9 billion m^3).

Within the 51.75 percent allocation to the State of Colorado of surface waters of the upper Colorado River basin after allocation of 50,000 acre-ft (62 million m^3) to Arizona (Upper Colorado River Basin Commission, 1965) a major stipulation in the Upper Colorado Basin Compact of 1948 is that a minimum of 5.0 million acre-ft (6.2 billion m^3) of water must pass the stream gage near the town of Maybell (Fig. 1) during any 10-year period. This is a minimum average flow of 500,000 acre-ft (0.62 billion m^3) annually, which has been met in all but one single year over the entire 1917-75 water-year period of record at that site.

A matrix of stream-discharge records has been generated at 36 flow points in the Yampa River basin for a 66-year period (1910-75 water years). Using interstation-correlation techniques (A.W. Burns, written communication, 1976), short-term records available at numerous sites were extended. Of the total of 2376 station years in the final streamflow matrix of annual and monthly values, 760 station years represented actual data collected at these sites. No attempt was made to adjust available streamflow records for upstream diversions. Parts of this matrix subsequently have been used to provide input to the reservoir-model and stream-salinity analyses.

Of major concern in the reservoir analysis would be hydrologic implications of construction of single major reservoirs and interactive effects of several reservoirs operating simultaneously in the basin's stream system. A modeling technique described by Adams (1975) adapted from earlier research by Markofsky and Harlemann (1971) is being used in the single-reservoir analysis. The multi-reservoir analysis will apply one of the HEC-series of models developed by the U.S. Army Corps of Engineers (1968, 1976) or the SIMYLD-II model developed by the Texas Water Development Board (1972). Reservoir effects to be studied include (1) downstream shifts in seasonal streamflow patterns, (2) evaporative losses, (3) in-reservoir water-quality characteristics, and (4) downstream water-quality impacts.

The following proposed reservoir projects in the Yampa River basin are being evaluated: (1) Juniper-Cross Mountain (Colorado River Water Conservation District, 1975), (2) Oak Creek Power (Blacktail, Main Green, Childress, and Middle Creek reservoirs) (Oak Creek Power Company, 1976), (3) Savery-Pothook (U.S. Department of Interior, 1976), and (4) Yamcolo (Western Engineers, Inc., 1974) (see Fig. 1). These represent a diversity of size, intended water uses, location in the stream system, and supporting agency. The study of reservoir impacts serves to demonstrate the assessment methodology and by no means will be exhaustive. Hydrologic implications will be interfaced with basinwide streamflow salt-loading calculations based upon ambient and projected conditions of reservoir development (see following section).

Ongoing work in the residuals-management analysis (R. M. Hirsch, written communications, 1976) is evaluating evaporative losses for alternative cooling systems for electric-power generation and coal-gasification plants. Critical variables include the magnitude of the thermal loading to a stream or pond, wind speed or direction, and receiving water temperatures. The range of evaporative heat loss under varying assumptions is from 25 to 86 percent. Overall impacts as to consumptive water uses are being tabulated as part of this analysis.

Stream-salinity calculations. With projected heavier demands being placed upon the Yampa River basin's water resources, one concern involves the coupled effects of water losses resulting from water-development plans to meet increasing demands on streamflow salinity. These effects are carried literally downstream to needed management decisions for the upper Colorado River basin and the Colorado River basin as a whole (Weatherford and Jacoby, 1975). Of concern here is that salinity changes resulting from increased water development and utilization in the Yampa River basin will be compatible with existing water laws, compact arrangements, and salinity-control policies for the entire Colorado River basin. Interactions between salinity and flow changes serve as an integral part of an analysis of impacts of alternative water-development plans.

In addition to the extensive impending reservoir development envisioned for the Yampa River basin, a shifting of water consumption from predominantly agricultural use to municipal and industrial (specifically, cooling and process waters for electric-power generation and coal-gasification plants) will have pronounced quality as well as quantity implications. A regional analysis of projected water development is being carried out with the assistance of a basin wide hydrologic and salt-load model developed by Ribbens (1975). Using mass-balance accounting procedures, water discharges and salinity concentrations and loads are calculated at designated flow points and routed through the basin's stream system.

Several input systems enable data of varying amounts and levels of detail to be utilized. The model uses a node-element system of streamflow-network depiction. Nodes are designated generally at points of streamflow discharge and quality information, and additional elements are added in stream reaches to be innundated by proposed reservoirs. Streamflow and salinity tributary inflows, imports, exports, and diversions must be specified. Preliminary model runs under pre-development conditions and considering all reservoir projects cited above except the Savery-Pothook project have been made.

Existing hydrologic and salinity conditions in the basin are being reproduced by this model with the aid of the generated streamflow matrix cited previously and historical water-quality data. Long-term records of specific conductance (beginning in the late 1950) are available at downstream flow-points on both the Yampa and Little Snake Rivers (Wentz and Steele, 1977) and historical (Iorns et al., 1965) and recent (Steele et al., 1976a) salinity data have been collected at numerous sites in the basin's streamflow system. Once this model has been calibrated, impacts of alternative coal-resource and regional economic development plans can be evaluated through projections of water demands of the various sectors, reservoir development, and generated water-borne residuals affecting salinity.

Other water-quality implications. As part of an assessment of current ambient stream quality in the Yampa River basin, a basinwide reconnaissance and quarterly sampling of selected sites were conducted (Steele and others, 1976a; U.S. Geological Survey, 1976). Certain anomalous conditions for selected trace-element concentrations have been noted, several of which can be attributed to (1) coal-mine drainage and (2) power-plant blowdown effluent (Wentz and Steele, 1976). Specifically, the Oak Creek drain (site Y-68, Fig. 1) appears to be an abandoned drainage tunnel and was sampled quarterly. Laboratory results indicate anomalously high concentrations of the following trace elements: total and dissolved Fe and Mn, total Cd, and dissolved Cu and Ni. In another example, lower Sage Creek (site Y-46, Fig. 1) during the reconnaissance sampling exhibited an apparent impact of discharge of cooling-tower blowdown effluent from the Hayden electric power-generation plant. Stream quality at this site was characterized by anomalously low pH (2.1 units) and high concentrations of sulfate, total and dissolved Cu, Fe, and Mn, and dissolved Pb and V (Wentz and Steele, 1976). However, these conditions for the latter station did not persist during subsequent sampling for a waste-assimilative study (Bauer et al., 1977; Giles and Brogden, 1976) and quarterly sampling through September 1976. This would suggest that such discharges occur intermittently at most and may have been associated solely with start-up operations in the expansion of the power plant (Robert Heard, personal communication, 1976).

As part of an assessment of stream-quality impacts of population growth, changing stream standards, and planned upgraded wastewater-treatment facilities for the Steamboat Springs area U.S. Environmental Protection Agency, 1976; (Wright-McLaughlin Engineers, 1975), an intensive data-collection effort and water-quality modeling analysis was carried out to evaluate the waste-assimilative capacity of the mainstem Yampa River between the towns of Steamboat Springs and Hayden (Fig. 1). Under varying assumptions of actual and projected loadings, treatment levels, and seasonal climatic and flow conditions, results of this analysis in general indicated that dissolved-oxygen levels would comply with stated goals but that high nutrient concentrations, particularly un-ionized ammonia, would violate proposed standards under certain conditions (Bauer and Steele, 1976; Bauer et al., 1977). By alerting regional water-resources managers to these potential problem areas, it is hoped that they can be resolved in a manner to be acceptable to the local citizenry at a reasonable cost so that aethestic appeal to the area for recreational uses can be maintained.

Besides the examples cited above, other basin-assessment studies by the U.S. Geological Survey are evaluating additional water-quality aspects of coal-resource development in the Yampa River basin (Steele et al., 1976b). Moreover, related projects funded by other local, State, and Federal agencies are contributing to assessment activities in this basin. Several of these are directed towards evaluating effects of mining and land-rehabilitation practices. For example, McWhorter, Skogerboe, and Skogerboe (1975) studied salt pick-up from waste spoils of the Edna strip mine southwest of Steamboat Springs (see Fig. 1), with 80 percent of the increase in salt load occurring during the period of snowmelt runoff (April - June). This study recently completed its field data-collection activities, and several other studies of a long-term nature are continuing (see for example, U.S. Geol. Survey, 1976; D. A. Woolhiser, written communication, 1976; R. G. Streeter, written communication, 1976).

Ground-water, surface-water interactions. As surface-water resources in the Yampa River basin become more competitive due to increasing demands by coal-resource and other economic development, ground-water resources in the basin will be sought after with increasing zeal as an alternative supply. Unfortunately, present knowledge of the hydraulic and water-quality characteristics of major geologic formations in the basin is limited (Brogden and Giles, 1976; Boettcher, 1972). However, utilizing available information, a preliminary appraisal of availability of ground-water resources has been completed for the basin (T. D. Steele et al., written communication, 1976) as well as for a specific area of interest south of the Yampa River between the towns of Craig and Steamboat Springs (see Fig. 1) (Warner and Brogden, 1976).

As might be expected, major aquifers in the basin are associated with those geologic structures containing the major coal seams (Speltz, 1974; McWhorter and Rowe, 1976). Recharge to these aquifers most likely occurs where the formations outcrop at the land surface, including areas crossed by stream channels. To date, most ground-water development has occurred in alluvial-filled valleys near stream channels. Flows from many of the wells drilled in the alluvium is induced stream recharge. Presumedly, this practice is monitored and controlled by the State Engineers Office of Colorado and Wyoming in their respective jurisdictions. As more development takes place, utilization of other aquifers is expected to increase. An evaluation of the hydraulics and quality characteristics of selected ground-water aquifers currently is in progress, and hopefully the potential of future expanded use of this supplemental resource can be defined in more detail.

Discussion

In applying residuals-management techniques to a regional environmental assessment, impacts of alternative environmental control strategies must be evaluated in terms of economic or societal benefits to indicate the tradeoffs of transforming any given residual to another form. Long-term implications from disposal of ultimate forms of residuals must also be considered.

The studies described in this paper are limited to just a few of the possible environmental impacts and to a time horizon over which forecasts regarding regional resource development have been made. "Post-audit" studies (W. B. Langbein, written communication, 1975) are strongly urged to measure the reliability of these assessments and to review the applicability of the methodologies developed and applied.

Potential benefits derived from more stringent environmental constraints and their effect on energy development and growth must be assessed in economic terms. Where causal relationships can be established between pollution levels and certain detrimental effects on society, development of damage functions (Kneese and Bower, 1968; Herschaft, 1976) is one means of quantifying the benefits so that they may be compared to the cost of alternative levels of pollution control. These tradeoffs must be depicted in an understandable and convincing way so that those using this information will consider it as objective and utilize it in a meaningful manner.

Two problems exist in assigning worth to technical-assessment information generated by studies such as those described in this paper. First, minimum and maximum levels of tolerable damages must be identified beyond which environmental costs are insensitive to changes in residuals modification efforts. Second, it should be recognized that regional resource management encompasses a diffuse and generally ill-defined mix of decision making at local, State, and Federal levels. As regional-resource management strategies become better understood as options for consideration in the interactive, multilevel decision-making process, the need for assessment studies will grow.

In summary, both the residuals-management analysis and the Yampa River basin-assessment studies are evaluating the water-use (quality as well as quantity) implications of alternative coal-resource development plans. The results of these studies will provide regional planners and resource managers with guidelines of the potential effects of coal-resource development on the environment and the economy of the region. Aided by the technical information derived from these and follow-up studies, they can better make or influence decisions concerning the type and extent of development that should be authorized.

References

[1] Adams, D. B., 1975. "Predicted and Observed Temperature and Water-Quality Changes of Lakes and Reservoirs," Intl. Assoc. of Hydrological Sciences Symposium, Tokyo, Japan, Publ. No. 117, pp. 873-882.

[2] Bauer, D. P. and Steele, T. D., 1976. "Waste Assimilative Capacity Analysis of a Mountain Stream," Am. Soc. of Civil Engr. Hydraulics Div. Specialty Conf., Applications of Mathematical and Physical Modeling in Hydraulic Engineering, Hydrology, and Water-Resources Engineering, Purdue Univ., W. Lafayette, Ind., August, abstract.

[3] Bauer, D. P., Steele, T. D., and Anderson, R. D., 1977. "Waste-Load Assimilative-Capacity Analysis of the Yampa River, Steamboat Springs to Hayden, Routt County, Colorado," U.S. Geol. Survey Water-Resources Inv. (draft), 87 p.

[4] Beckner, J. L., 1976. "Geohydrological Data Requirements for Environmental Impact Statements," Paper presented at Symposium on Methodologies for Environmental Assessment in Energy Developing Regions," Am. Geophys. Union Fall Annual Meeting, Dec., San Francisco, Calif., 11 p.

[5] Boettcher, A. J., 1972. "Ground-Water Occurrence in Northern and Central Parts of Western Colorado," Colo. Water Conservation Board, Water-Resources Circ. 15, 25 p.

[6] Bower, B. T., and Basta, D. J., 1973. "Residuals-Environmental Quality Management--Applying the Concept," Baltimore, Md., The Johns Hopkins Univ. Center for Metropolitan Plan. and Research, Oct., 88 p.

[7] Brogden, R. E. and Giles, T. F., 1977. "Reconnaissance of Ground-Water Resources in a Part of the Yampa River Basin between Craig and Steamboat Springs, Moffat and Routt Counties, Colorado," U.S. Geol. Survey Water-Resources Inv. (draft).

[8] Colorado River Water Conservation District, 1975. Federal Power Commission application for preliminary permit, Juniper - Cross Mountain Project, Water-development proposal and supporting material, 7 p. with appendices.

[9] Giles, T. F. and Brogden, R. E., 1977. "Basic Hydrologic Data, Yampa River Basin and Part of the White River Basin, Northwestern Colorado and South-Central Wyoming," U.S. Geol. Survey open-file report (draft), 11 p. and 3 tables.

[10] Hines, W. G., Rickert, D. A., McKenzie, S. W., and Bennett, J. P., 1975. "Formulation and Use of Practical Models for River-Quality Assessment," U.S. Geol. Survey Circ. 715-B, 13 p.

[11] Howe, C. W., Kreider, J.F., Udis, B., Hess, R.C., Orr, D.V., and Young, J.T., 1975. "Integrated Economic-Hydrosalinity-Air Quality for Oil Shale and Coal Development in Colorado," Boulder, Colorado Univ., Economics Dept., Rept. to Colorado Energy Research Inst.

[12] Iorns, W. V., Hembree, C. H., and Oakland, G. L., 1965. "Water Resources of the Upper Colorado River Basin--Technical Reports," U.S. Geol. Survey Prof. Paper 441, 370 p.

[13] James, I.C., II, Attanasi, E.D., Maddock, T.M., III, Chiang, S.H., and Matalas, N.C., 1977. "Water and Energy Use in Coal Conversion," draft open-file report, U.S. Geol. Survey, Reston, Va.

[14] Kneese, A. V. and Bower, B. T., 1968. "Managing Water Quality: Economics, Technology, Institutions," The John Hopkins Press, Baltimore, 328 p.

[15] Markofsky, M. and Harleman, D.R.F., 1971. "A Predictive Model for Thermal Stratification and Water Quality in Reservoirs," M.I.T., Hydrodynamics Lab Rept. No. 134, 283 p.

[16] McWhorter, D. B. and Rowe, J. W., 1976. "Inorganic Water Quality in a Surface Mined Watershed," Paper presented at Symposium on Methodologies for Environmental Assessments in Energy-Development Regions, Am. Geophys. Union 1976 Fall Annual Meeting, Dec., San Francisco, Calif., 27 p.

[17] McWhorter, D. B., Skogerboe, R. K. and Skogerboe, G. V., 1975. "Water Quality Control in Mine Spoils, Upper Colorado River Basin," U.S. Environmental Protection Agency, Environ. Protect., Tech. Series EPA-670/2-75-048, June, 99 p.

[18] Oak Creek Power Company, 1976. "Oak Creek Water and Power Project, Colorado," report by Van Sickle Associates, Inc., Consulting Engineers, Denver, Colo., Jan., 22 p.

[19] Reiquam, H., Dee, N., and Choi, P., 1975. "Assessing Cross-Media Impacts," Environmental Sci. Technology, Vol. 9, No. 2, Feb., pp. 118-120.

[20] Ribbens, R. W., 1975. "Program NW01, River Network Program, Users Manual," U.S. Bureau of Reclamation, Engrg. Res. Center, Div. of Planning and Coordination, Denver, Colo., 7 chapters, appendices.

[21] Speltz, C. N., 1974. "Strippable Coal Resource of Colorado--Location, Tonnage, and Characteristics," U.S. Dept. of Interior, Bureau of Mines, Prelim. Rept. 195, 63 p.

[22] Steele, T. D., 1976. "Coal-Resource Development Alternatives Residuals Management, and Impacts on the Water Resources of the Yampa River Basin, Colorado and Wyoming," Intl. Water Resources Assoc., Symposium on Water Resources and Fossil Fuel Production, Dusseldorf, Germany, Sept., 17 p.

[23] Steele, T. D., Bauer, D. P., Wentz, D. A. and Warner, J. W., 1976a. "An Environmental Assessment of Impacts of Coal Development on the Water Resources of the Yampa River Basin, Colorado and Wyoming, Phase-I Work Plan," U.S. Geol. Survey open-file report 76-367, Lakewood, Colo., 17 p.

[24] Steele, T. D., James, I. C., II, Bauer, D. P., and others, 1976b. "An Environmental Assessment of Impacts of Coal Development on the Water Resources of the Yampa River Basin, Colorado and Wyoming, Phase-II Work Plan," U.S. Geol. Survey open-file report 76-368, Lakewood, Colo., 33 p.

[25] Texas Water Development Board, 1972. "Economic Optimization and Simulation Techniques for Management of Regional Water Resource Systems," Texas Water Develop. Rept., 106 p.

[26] Udis, B. and Hess, R. C., 1976. "Input-Output Structure of the Economy of That Part of the Yampa River Basin in Colorado - 1975," draft report prepared for U.S. Geol. Survey, Dec., 127 p.

[27] Udis, B., Howe, C. W., and Kreider, J. F., 1973. "The Inter-Relationship of Economic Development and Environment Quality in the Upper Colorado River Basin--An Interindustry Analysis," Colorado Univ., Rept. to U.S. Dept. Comm., Econ. Devel. Admin., July, EDA Research Grant OER 351-G-71-8 (99-7-13215), 642 p.

[28] Upper Colorado River Commission, 1965. Selected legal references for the use of the Upper Colorado River Commission, Salt Lake City, Utah, 495 p.

[29] U.S. Army Corps of Engineers, 1968. "Hec-3, Reservoir Systems Analysis," Hydrol. Engr. Center Users Manual No. 238-53, 86 p.

[30] U.S. Army Corps of Engineers, 1976. "HEC-5C, Simulation of Flood Control and Conservation Systems, Users Manual," Hydrol. Engr. Center, Generalized Computer Program 723-500, 25 p. plus attachments.

[31] U.S. Department of the Interior, 1976. "Savery-Pothook Project, Colorado and Wyoming," draft Environmental Statement, Bureau of Reclamation, Upper Colorado Region, Salt Lake City, Utah, INT DES 76-37, Sept. 27, 1976, 9 chapters plus attachments.

[32] U.S. Environmental Protection Agency, 1976. "Draft Environmental Impact Statement," Steamboat Springs regional service authority, 201 wastewater facilities plan, prepared by Weiner and Assoc., Denver, Colo., 154 p. and appendices.

[33] U.S. Geological Survey, 1976. "Hydrologic Studies of the U.S. Geological Survey Related to Coal Development in Colorado," U.S. Geol. Survey open-file report 76-549, Aug., 22 p.

[34] Warner, J. W. and Brodgen, R. E., 1976. "Geohydrological Effects of Coal-Resource Development in the Yampa River Basin, Colorado and Wyoming," Geol. Soc. of Am., Hydrologeological Div., Sepcial Symposium on Hydrologic and Engr. Problems related to Coal Mining, Denver, Colo., Nov., abstract.

[35] Weatherford, G. D. and Jacoby, G. C., 1975. "Impact of Energy Development on the Law of the Colorado River," Nat. Resources Jour., Vol. 15, No. 1, pp. 171-213.

[36] Wentz, D. A. and Steele, T. D., 1976. "Surface-Water Quality in the Yampa River Basin, Colorado and Wyoming--An Area of Accelerated Coal Development," Paper presented at Engr. Foundation Conf. on Water for Energy and Development, Asilomar Conf. Center, Pacific Grove, Calif., Dec., 50 p.

[37] Western Engineers, Inc., 1975. "Yamcolo Reservoir Project, Feasibility Report," prepared for the Upper Yampa Water Conservancy District and the Colorado Water Conservation Board, Grand Junction, Colo., Nov., 8 chapters.

[38] Wright-McLaughlin Engineers, 1975. "Steamboat Springs Wastewater Management Report, Basic Information and Analyses," U.S. Environmental Protection Agency 201 Facilities Plan, Vol. I, Denver, Colorado.

THE ESTIMATE OF IONIC DISCHARGE DURING HIGH FLOWS IN SMALL TORRENT CATCHMENTS

By

Hans M. Keller
Swiss Federal Institute of Forestry Research
CH - 8903 Birmensdorf, Switzerland

Abstract

Ionic discharge as estimated from runoff and ion concentration during high flows in small torrent catchments is investigated for various levels of information intensity. Five models for ion loads are applied to ion concentration and runoff data from water samples taken at a gauging station with an automatic sampler during high flow events. Models, including ion concentration estimates from continuous records of electrical conductance, seem to be superior to those using direct relations between discharge and ion concentration. Also models taking into account the lag effect ion dilution during peak flows are to be recommended for application. The calcium loads for five storms during 1976 are presented.

Introduction

Information on water quality includes the estimation of ion or inorganic solute discharge in streamflow. In small mountainous catchments, the ionic output during high flows contribute a significant part of the annual loads. It seems therefore appropriate to perform such estimates derived from different sources of information and to consider their accuracy.

Steele (1976) presents a solution to the estimate of inorganic discharge in streamflow or groundwater. He is using daily records. In small mountainous torrent catchments however, most of the high flow events last up to 12 or 16 hours but rarely more than one day. Methods to estimate inorganic discharge are therefore investigated. Various types of mixing models are suggested, e.g., by Hall (1970, 1971) and by Johnson et al. (1969). They require only discharge and ion concentrations from instantaneous samples because rarely are continuous measurements of e.g., specific electrical conductance available to aid in the estimation of ion loads. Depending on the available information intensity, five different ways to estimate the total ion discharge during high flow events are therefore presented. Results of events in a small basin in the northern Prealps of Switzerland are given to illustrate the level of accuracy of the various approaches used.

Methods

The calculation of total ion discharge during a high flow event should make the best use of the information available. The following levels of increasing information intensity are considered:

- Continuous flow records and periodical analysis of instantaneous water samples for ion concentration; samples during the rising limb of the hydrograph are considered to be rare.
- Continuous flow records and additional high flow sampling for ion concentration.
- Continuous flow records and continuous monitoring of specific electrical conductance together with sampling for ion concentration during high flows.

According to this increasing intensity of available information the following five models to calculate ionic discharge are presented and discussed:

1. $$L = V \cdot c \qquad (1)$$

where L is the specific load of ions discharged during the high flow event in $kg.km^{-2}$, V is the streamflow volume of the event in $l.km^{-2}$, and c is the ion concentration in $mg.l^{-1}$. In this first model V is the total water volume passing the gauging station divided by the area of the catchment in km^2. From this volume the average discharge in $l\ sec^{-1}.km^{-2}$ is determined. The concentration c is expressed in terms of its relation to discharge. As shown in earlier reports (Keller, 1970a, 1970b) the relation between ion concentration and discharge may be expressed as

$$c = a + b.\ln q \qquad (2)$$

where q is the specific discharge in $l\ sec^{-1}\ km^{-2}$ and a and b are constants. The average discharge of the event is then used to estimate c from Eq. (2). The load L is now calculated from Eq. (1).

2. $$L = \Sigma(V_i \cdot c_i) \qquad (3)$$

where V_i and c_i are the volumes and concentrations of each interval of the event. The level of information is the same as in the first model, but the concentration c_i is determined for each interval from corresponding discharge values again using Eq. (2). Since this is a nonlinear relation, the estimates of L are expected to differ from those of Eq. (1). Since no distinction is made in either models, Eqs. (1) or (3), between rising and falling limb, Eq. (2) is used unaltered through the hydrograph.

3. $$L = V \cdot c_c \qquad (4)$$

where c_c is the mean ion concentration of the event derived from the relationship of the ion concentration to specific electrical conductance. This relation is assumed to be linear and of the form

$$c_c = a + b \cdot EC \qquad (5)$$

where EC is the specific electrical conductance in μmhos cm^{-1} at 20°C, a and b being constants. This approach makes use of the highest level of information mentioned above: continuous records of specific electrical conductance and sampling for ion concentration particularly during high flows. Through regression analysis the constants a and b for Eq. (5) are determined. Calculating the average value of electrical conductance EC over the time of the event, c_c is computed from Eq. (5) and used to solve for L in Eq. (4). This model assumes that the relation between electrical conductance and ion concentration remains unchanged during the entire event.

4. $$L = \Sigma(V_i \cdot c_{ci}) \qquad (6)$$

where c_{ci} is the ion concentration of each interval during the event as derived from the specific electrical conductance of the interval (Eq. 5). Again the same high level of information is needed as in the third model. Differences are due to the fact that this fourth model results in volume weighted concentrations versus the time weighted concentrations of the third approach, a fact which also has to be considered comparing models 1 and 2.

5. Many workers have shown significant differences in ion concentration-discharge relationships between rising and falling limb of the hydrograph, e.g., Hendrickson and Krieger (1960), Nakamura (1971), Glover and Johnson (1974). This fifth model is designed to account for these differences.

$$L = L_r + L_f \qquad (7)$$

where L_r is the ion load during the rising limb and L_f the load of the falling limb of the hydrograph, both determined as L in Eq. (4). The corresponding ion concentrations c_c are derived according to Nakamura (1971, p. 201) from linear relations between water discharges q (l sec^{-1} km^{-2}) and q c_c (mg sec^{-1} km^{-2}) of the form

$$q \cdot c_c = a + b \cdot q \qquad (8)$$

where a and b are constants for both rising and falling limb conditions. The concentrations c_c are then expressed as a hyperbolic relation:

$$c_c = a/q + b \tag{9}$$

Combining Eqs. (9) and (4) into Eq. (7), total ion load discharged during the high flow event is calculated.

Application

Information from a small (1.55 km^2) partly forested catchment in the Prealps of Switzerland was used to compare the five models to estimate calcium loads discharged during high flow events. The catchment Vogelbach 3 is described in Keller (1970a, 1974). Continuous records of streamflow and specific electrical conductance are available. Periodical manual sampling for ion concentrations (including calcium) during several years, as well as automatic samples taken during high flows in 1976 are the basic data for this comparison.

Five storms in 1976 have been chosen to estimate the total calcium load of each of the five events applying the five above mentioned procedures. A description of the storm conditions is given in Table 1.

Table 1. Storm Characteristics in Catchment Vogelbach 3, Switzerland

Date	Duration	Initial Disch'g	Total Precip.	Peak Disch.	Total Discharge		Mean Disch.	Runoff Coeffic.
	h	$l.sec^{-1}.km^{-2}$	mm	$l.sec^{-1}.km^{-2}$	$m^3.km^{-2}$	mm	$l.sec^{-1}.km^{-2}$	%
21.7.76	16.5	47.6	15.5	405	9730	9.7	163.8	62.5
28.7.76	13.5	112.6	11.5	325	8410	8.4	173.0	73.0
1.8.76	16.0	29.6	31.0	476	16538	16.5	287.1	53.2
10.8.76	11.5	23.2	26.0	508	8095	8.1	195.5	31.1
31.8.76	9.5	27.0	10.0	663	7210	7.2	210.9	72.0

The calcium concentrations in the test catchment showed relationships to discharge and electrical conductance with reasonable significance. Data for other ions are also available and could also be used. However, in some cases, additional parameters may be needed to make good estimates of the concentrations either from discharge (Keller, 1970b) or from electrical conductance.

The regression constants for Eqs. (2), (5), and (8) as well as the statistical parameters are given in Table 2. Earlier relations between calcium concentration and discharge (Keller, 1970a) differ slightly because of lacking high flow data. They should not be used for flow rates above 300 l sec^{-1} km^{-2}. In this study of high flow events automatic samples during peaks up to 660 l sec^{-1} km^{-2} are included and the corresponding regression constants as given in Table 2 (see Eq. 2) are used.

Table 2. The Statistics of the Regression Equations Relating Calcium Concentration c, c_c (mg l^{-1}) to Discharge q (l sec^{-1} km^{-2}) and Specific Electrical Conductance EC (μmhos cm^{-1} at 20°C), for Catchment Vogelbach 3.

Regression	Equation (2) $c = a + b.\ln q$	Equation (5) $c_c = a + b.EC$	Equation (9) rising $c_c = a/q + b$	Equation (9) falling $c_c = a/q + b$
a	56.05	-.66	265.0	620.5
b	- 6.11	.204	27.2	18.2
n	42	36	20	37
r^2	.57	.99	.98	.93

Results and Conclusions

The results of the calcium load calculations using the five different models are summarized in Table 3. The load estimates yield standard deviations of less than 15 kg Ca km^{-2} or about 8% of the mean. Even using the approach with the lowest level of information, acceptable results for any of the chosen storms are obtained. However, if we consider that in these areas of the Swiss Prealps high flows above 150 l sec^{-1} km^{-2} yield about 22% of the annual flow during only 5% of the time (Keller, 1970a), significant systematic errors in computing ion loads from small mountainous catchments become important. The following observations seem therefore appropriate:

- The second model, (Eq. 3), seems to yield consistently lower calcium loads than the first, Eq. (1). The

Table 3. The Calcium Loads of Five Storm Events in the Vogelbach 3 Catchment Estimated from Five Different Models.

Date	Average Load		1 $L = V \cdot c$		2 $L = \Sigma(V_i \cdot c_i)$		3 $L = V \cdot c_c$		4 $L = \Sigma(V_i \cdot c_{ci})$		5 $L = L_r + L_f$		
	$\bar{x}$	s	L	c	L	$\bar{c}_i$ 1)	L	c_c	L	$\bar{c}_{ci}$ 1)	L	L_r	L_f
	$kg.km^{-2}$		$kg.km^{-2}$	$mg.l^{-1}$	$kg.km^{-2}$	$mg.l^{-1}$	$kg.km^{-2}$	$mg.l^{-1}$	$kg.km^{-2}$	$mg.l^{-1}$	$kg.km^{-2}$		
21.7.76	230	9.9	242	24.9	232	23.9	236	24.3	217	22.3	224	63	61
%			112		107		109		100		103		
28.7.76	207	8.7	202	24.7	204	24.9	214	26.2	218	26.7	197	54	143
%			93		94		98		100		90		
1.8.76	354	13.3	355	21.5	345	22.4	376	22.8	349	21.1	343	35	308
%			102		99		108		100		98		
10.8.76	191	14.4	193	23.8	178	22.0	213	26.3	194	24.0	178	25	153
%			99		92		110		100		92		
31.8.76	156	7.7	168	23.4	154	21.4	153	21.2	147	20.4	156	13	143
%			114		105		104		100		106		

1) $\bar{c}_i$ and $\bar{c}_{ci}$ are calculated as L/V.

logarithmic dilution gives less weight to large volumes during peak flows than the straight linear approach. The variation between storms regarding the differences between model 1 and 2 seems to be due to the rather low coefficient of determination (r^2 = .57, see Table 2).

- Similarly, model 4, Eq. (6), yields almost consistently calcium loads about 6% lower than the third model (Eq. 4). Both take into consideration the lag effect of calcium concentration versus peak flow by using continuous records of electrical conductance. However, only model 4 gives the correct weight to the volumes of water being discharged. Due to the good prediction equation used in model 3 (coefficient of determination r^2 = 0.99), the variation between storms is small. If the level of information is sufficient, this fourth approach is considered to yield the most accurate estimates of calcium loads during high flow events and is therefore used as a basis of comparison (100%).

- The fifth approach yields results closely related to model 2. Instead of a logarithmic, a hyperbolic dilution is used and particular consideration is given to the separation of rising and falling limb of the hydrograph and hence to the earlier mentioned lag effect. However, since the hydrograph is divided into only two parts, many transitions are neglected and the result is a less accurate estimate.

In conclusion, model 4 should be given preference if the level of information is sufficient. In addition to flow records the continuous records of specific electrical conductance are most important. If only irregular records of electrical conductance are available, preference should be given to model 5. In the case of periodical or irregular sampling for ion concentrations only model 2 should be used. The use of models 1 and 3 may result in systematic errors which tend to overestimate ion discharge during high flow events.

Acknowledgment

The help of Mr. A. Storrer for construction and maintenance of the automatic sampler for high flow events and for the compilation of the data is gratefully acknowledged.

References

[1] Glover, B. J. and Johnson, P., 1974. "Variation in the Natural Chemical Concentration of River Water during Flood Flows, and the Lag Effect," Journal of Hydrology, 22: pp. 303-316.

[2] Hall, F. R., 1970. "Dissolved Solids-Discharge Relationships. 1. Mixing Models," Water Resources Research, 6(3), pp. 845-850.

[3] Hall, F. R., 1971. "Dissolved Solids-Discharge Relationships. 2. Applications to Field Data," Water Resources Research, 7(3): pp. 591-601.

[4] Hendrickson, G. E. and Krieger, R. A., 1960. "Relationship of Chemical Quality of Water to Stream Discharge in Kentucky," International Geological Congress, 21st, Copenhagen, Rept. Pt. 1: pp. 66-75.

[5] Johnson, N. M., Likens, G. E., Bormann, F. H., Fisher, D. W., and Pierce, R. S., 1969. "A Working Model for the Variation in Stream Water Chemistry at the Hubbard Brook Experimental Forest, New Hampshire," Water Resources Research, 5(6): pp. 1353-1363.

[6] Keller, H. M., 1970a. "Der Chemismus Kleiner Bäche in Teilweise Bewaldeten Einzugsgebieten in der Flyschzone eines Voralpentales," Mitt. eidg. Anst. forstl. Versuchswesen 46: pp. 113-155.

[7] Keller, H. M., 1970b. "Factors Affecting Water Quality of Small Mountain Catchments," Journal of Hydrology, New Zealand, 9(2): pp. 113-141.

[8] Keller, H. M., 1974. "Ueber den Chemismus kleiner Bäche in den Flyschvoralpen der Schweiz," Mitteilungen Arbeitskreis Wald und Wasser, Essen, Nr. 6: pp. 29-42.

[9] Nakamura, R., 1971. "Runoff Analysis by Electrical Conductance of Water," Journal of Hydrology, 14: 197-212.

[10] Steele, T. D., 1976. "A Bivariate-Regression Model for Estimating Chemical Composition of Streamflow or Ground-Water," Hydrological Sciences Bulletin, 21: 149-161.

MODELLING OF RIVER POLLUTION BY FINAL EFFLUENTS FROM WASTE TREATMENT PLANTS

By

G. Römberg
DFVLR - INSTITUT für Strömungsmechanik
Bunsenstrabe 10
34 Göttingen
West Germany

Abstract

In modern environmental planning, the river takes the role of an additional stage of waste treatment: residual pollution of the final effluents, which the river is able to cope with, is deliberately tolerated in order to reduce the operation costs and construction costs of waste treatment plants. A physical model adapted for computational evaluation is presented. With respect to modern policies of environmental protection, the model predicts the river pollution due to each plant separately.

The river pollution is modelled as the dispersion of clouds of polluted fluid elements originating from the nearfield mixing regions of the jets of final effluents. These fluid elements are considered as once charged tank reactors, in which the biological self-purification takes place. The resulting relations for the various dependent variables are modified balance equations in integral form with the exception of the relations for the oxygen deficit. The evaluation of these resulting relations requires a comparatively small amount of numerical work. If the water temperature is constant and the composition of each final effluent is independent of time, the modified balance equations prove to be special versions of the π-theorem.

The determination of the parameters of the model (model identification) can be divided into a + 1 independent steps:

1. Identifying the model of biological self-purification assigned to the polluted water elements from the αth nearfield mixing region ($\alpha = 1,\ldots,a$; a = number of waste treatment plants).

The αth identification ($\alpha = 1,\ldots,a$) is carried out using laboratory experiments with two water samples from the αth nearfield mixing region.

2. Providing the required information about the physical parameters with the aid of experiments.

The dependent variables of the model are cross-sectional means of the following concentrations due to each plant: concentration of the degradable organic matter, concentration of the nondegradable organic matter and oxygen deficit.

The parameters of the model are: volume rate of flow, cross-section, emission rate of the degradable organic matter of each plant, emission rate of the nondegradable organic matter of each plant, absorption coefficient and various dimensionless biological parameters occurring in the model of self-purification.

Introduction

In modern environmental planning, the role of a final stage of waste treatment is assigned to a river polluted by final effluent inputs from waste treatment plants. A residual pollution of the final effluents, which the river is able to cope with, is deliberately tolerated to reduce the operation costs and construction costs of waste treatment plants. Adopting this policy, the waste treatment plant constructor has to start from a complex system which consists of the river and the waste treatment plants joined to the river. The river (mainstream) may have contributary branches. The mainstream pollution is assumed to be due to organic materials in the final effluents. Let us define the clean mainstream by certain minimum requirements for the water quality. Recommendations for such standards have been discussed recently in connection with environmental protection (Knöpp, Schott, 1971, p. K13; Böhnke, 1975).

In the present paper we consider only a part of the complex system mentioned above: the mainstream without the stream fauna and the stream flora apart from the micro-organisms bringing about the biological self-purification. In what follows "mainstream" means the part of the complex system just described.

In slightly polluted well-oxygenated stream water, certain micro-organisms which require free oxygen (aerobic micro-organisms) bring about the elimination of the bio-degradable organic materials. Such materials are utilized as food, partially converted to cell substance and partially broken down to harmless end-products such as water and carbon dioxide with the aid of dissolved oxygen. This important process is called biological self-purification (Klein, 1972).

Adopting the policy outlined above the following fluid mechanical problem arises with regard to the mainstream: modelling of the final stage of waste treatment in continuous operation. "Continuous operation" implies time-independence of relevant time-averaged quantities. The treatment of this problem is divided into two steps:

1. Mapping the original problem onto a suitable mathematical form
2. Model identification.

The first step is to reduce the original problem to the determination of unknown parameters. This implies finding relevant variables and mathematical relations among them. The second step is to determine the unknown parameters with the aid of pertinent experiments.

Modelling of river pollution has been studied by a number of investigators (reviews are included in the lecture of Preissmann (1971) and the report of Stehfest (1973) with 97 references, the major portion referring to biological self-purification). These studies are based on the early work by Streeter and Phelps (Preissmann, 1971). The original problem is mapped onto a boundary value problem associated with a system of coupled differential equations. The systems used by the various investigators differ widely by the accuracy with which they cover the biological self-purification.

Stehfest (1973) recently proposed a system of differential equations for rivers, whose benthos is ngeligible. The parameters in this system are assumed constant. Their determination is based on experimental values of dependent variables and is achieved by applying Bellman's quasilinearization technique and the method of least squares. Stehfest assumes among other things: the mainstream temperature is constant at cross-sections moving downstream at the mean velocity, defined as volume rate of flow divided by cross-sectional area, and "complete homogeneity" exists in the river cross-sections. (Stehfest, 1973). Another shortcoming of the model is that it cannot distinguish the pollutions due to the various polluters. This, however, is desirable with regard to modern policies of environmental protection.

The present paper gives a physical model of pollution by final effluents from waste treatment plants. The pollution is modelled as the dispersion of clouds of polluted fluid elements originating from the nearfield mixing regions of the jets of final effluent in the mainstream. We consider the polluted fluid elements as once charged tank reactors, in which the biological self-purification takes place. The model of purification used is based on a preparatory investigation (Romberg, 1976). Based on the idea just outlined, the original problem is mapped onto balance equations in integral form by using maximum properties of certain probabilities. These analytical expressions are comparatively easy to survey and their practical application requires a comparatively small amount of computational work. They correspond to the solution of the boundary value problem associated with the models based on the early work by Streeter and Phelps. The identification of the present physical model can be separated into two independent steps:

1. Providing the necessary informations on the fluid mechanical parameters (volume rate of flow of the mainstream, mainstream cross-section, emission rate of the degradable and nondegradable organic materials, respectively, of each waste treatment plant) by pertinent experiments.

2. Identifying the model of purification for the biological self-purification in the polluted fluid elements from the nearfield mixing region of the αth jet of final effluent $(\alpha = 1,\ldots,a)$.

Step 2 is achieved using laboratory experiments with two water samples from the αth nearfield mixing region $(\alpha = 1, \ldots,a)$ (Romberg, 1976).

The present model avoids the above-mentioned shortcomings of Stehfest's model. With respect to modern policies of environmental protection the present model predicts the mainstream pollution due to each plant separately.

The dependent variables of the present model are weighted cross-sectional means of concentrations of pollution.

For an infinitely long mainstream of constant volume rate of flow downstream of the final effluent inputs it is shown that the weighted means agree with the appropriate unweighted downstream of the final effluent inputs.

A generalization of the described model in order to cover the oxygen deficit due to each polluter is in progress. The section on oxygen balance gives a provisional account of the generalization.

Basic Considerations and Assumptions

The mainstream exhibits three operations which enable it to take the role of a final stage of waste treatment: transport, dilution and chemical conversion.

Transport. The pure mainstream water is considered as a carrier medium which can be charged with organic materials from the waste treatment plants. The pollution load can be subdivided into various phases which come into three categories: true solutions (molecular phases), colloidal suspensions (colloidal phases) and coarse suspensions (coarse phases)

(Soo, 1967, Franke, 1970, and Kelker et al., 1971). Carrier medium and the various phases are considered as true continua occupying the same space. A single velocity ν (called barycentric velocity) is assigned to the carrier medium, the molecular and colloidal phases, whereas each coarse phase has its own velocity. The flow of the carrier medium is turbulent. Its irregular motion in a perpendicular direction causes forces at the surface of the coarse particles, which prevent sedimentation by gravity. A similar mechanism renders possible the pneumatic and hydraulic transport of suspended granular material in horizontal pipes.

In modelling the (convective) transport of organic materials from final effluents, only one velocity, the barycentric velocity (flow velocity), is needed, except for the so-called nearfield mixing regions explained later.

Dilution. This operation covers the following phenomena: molecular and turbulent diffusion.

To explain the molecular diffusion, we consider an areal element vector dF moving at flow velocity in the mainstream. The flow of phase i (molecular or colloidal) through the surface of dF in its direction is $J_i \cdot dF$, where the point means scalar product and J_i is the molecular diffusion flux of phase i.

To explain the turbulent diffusion, we start from the relations

$$\nu = \overline{\nu} + \nu', \quad \rho_{A\alpha} = \overline{\rho_{A\alpha}} + \rho'_{A\alpha}, \quad \rho_{U\alpha} = \overline{\rho_{U\alpha}} + \rho'_{U\alpha} \tag{1}$$

where the overbar denotes the ensemble mean and the turbulent fluctuations are denoted by primes. $\rho_{A\alpha}$ and $\rho_{U\alpha}$ are, respectively, the density of the degradable organic pollution and the density of the nondegradable organic pollution due to the αth waste treatment plant $(\alpha = 1,\ldots,a)$.

Let us now focus our attention on an areal element vector moving at the velocity $\overline{\nu}$. The flux of the degradable organic matter due to the αth waste treatment plant through the surface of this areal element vector is (in the absence of molecular diffusion and outside the nearfield mixing regions) $\rho_{A\alpha}\nu'$.

$$\overline{\rho_{A\alpha}\nu'} = \overline{\rho'_{A\alpha}\nu'} = J_{A\alpha} \tag{2}$$

is the turbulent diffusion flux of the degradable organic matter due to the αth waste treatment plant (outside the nearfield mixing regions). For the nondegradable organic matter due to the αth waste treatment plant, we have (outside the nearfield mixing regions)

$$\overline{\rho_{U\alpha}\nu'} = \overline{\rho_{U\alpha}\nu'} = J_{U\alpha} \quad . \tag{3}$$

The dilution of the organic materials in the mainstream is achieved in the main by turbulent diffusion. The contribution of molecular diffusion is negligible apart from the nearfield mixing regions (Romberg, 1976).

The mechanism of transfer of pollutants from the αth waste treatment plant $(\alpha = 1,\ldots,a)$ to the mainstream consists in mixing of the αth turbulent jet of final effluent $(\alpha = 1,\ldots,a)$ with the turbulent mainstream. Density and temperature differences between the core of the αth jet $(\alpha = 1,\ldots,a)$ and the surrounding mainstream water are considered negligible. The region of the jet between the inlet cross-section and the cross-section of the jet, where the core ends, is termed nearfield.

The nearfield mixing region is the part of the nearfield surrounding the core. Molecular diffusion and differences between the barycentric velocity and the velocities of the coarse phases cause charging of pure water elements with organic pollutants and changes of the pollution loadings of effluent elements. We can consider the nearfield mixing regions as sources of fluid elements charged with organic pollutants.

Chemical conversion is assumed to be achieved only by the biological self-purification. The influence of chemical conversion on the flow velocity is considered negligible. Furthermore, the benthos is assumed negligible in modelling of the biological self-purification in the polluted fluid elements originating from the nearfield mixing regions.

We divide the mainstream, which may have tributary branches, up into L reaches, so that the ensemble-averaged mainstream temperature can be considered as constant in each reach in the absence of heat pollution and each nearfield lies inside a reach.

We assume continuous operation of the total system composed of the mainstream and adjoining waste treatment plants. This implies that the distribution functions of the following quantities are independent of time: ν, $\rho_{U\alpha}$, $e_{U\alpha}$ and $e_{A\alpha}$, where the last two symbols stand for the emission rate of nondegradable and degradable organic pollution, respectively, from the αth waste treatment plant $(\alpha = 1,\ldots,a)$.

The time-averaged outflow of degradable organic pollution due to the αth waste treatment plant through inlet cross-sections of tributary branches is considered negligible and similarly for the nondegradable organic pollution. These inlet cross-sections serve as partial bounds of the mainstream.

Transport of organic pollution through the banks and the bottom of the mainstream is excluded.

Mainstream Pollution by Nondegradable Organic Materials

$\overline{\rho_{U\alpha}}$ and $\overline{\rho_{A\alpha}}$ $(\alpha = 1,\ldots,a)$ or suitable averages of these quantities are relevant variables for the modelling of the final stage of waste treatment.

We introduce a fixed volume, bounded by the mainstream bottom, the mainstream banks, the water surface and two mainstream cross-sections F_1 and F outside the nearfields. The cross-section F is located downstream of the αth nearfield and the cross-section F_1 is located upstream of the αth nearfield. We set down the balance equation for the nondegradable organic pollution due to the αth waste treatment plant with regard to the fixed volume. Under the assumptions made as jet, this balance equation can be written as:

$$\int_{F_1} \rho_{U\alpha}\, \nu \cdot d\, F_1 + \int_{F_1} J_{U\alpha} \cdot d\, F_1 + \overline{e_U} = \int_F \overline{\rho_{U\alpha}}\, \overline{\nu} \cdot dF + \int_F J_{U\alpha} \cdot dF \tag{4}$$

where $d\,F_1$ and $d\,F$ are areal element vectors (in downstream direction) of F_1 and F, respectively.

The diffusion term on the left side of Equation (4) is small compared with the first term and similar for the right side of Equation (4) (see Fischer, 1973). If we take these facts into consideration and assume

$$v_N > 0 \quad \text{within } F_1 \tag{5}$$

the balance Equation (4) can be simplified to give

$$\overline{e_{U\alpha}} = \int_F \overline{\rho_{U\alpha}}\ \overline{v_N}\, dF \quad . \tag{6}$$

dF denotes an area element of a mainstream cross-section, $\overline{v_N}$ is the component of $\overline{\nu}$ parallel to $d\,F_1$ and dF respectively.

Using the volume rate of flow

$$\dot{V} = \int_F \overline{v_N}\, dF \tag{7}$$

and the averages

$$\hat{\rho}_{U\alpha} = \int_F \overline{\rho_{U\alpha}}\ \overline{v_N}\, dF \Big/ \int_F \overline{v_N}\, dF \tag{8}$$

Statement (6) can be rewritten as

$$\hat{\rho}_{U\alpha}\ \dot{V} = \overline{e_{U\alpha}} \quad . \tag{9}$$

Let us assume

$$\dot{V} > 0 \ \text{(for any mainstream cross-section)},\ \ \overline{v_N} > 0 \ \text{within inside } F. \tag{10}$$

$\hat{\rho}_{U\alpha}$ then means a value attained by the variable $\overline{\rho_{U\alpha}}$ at a certain point of F. This value represents a measure for the pollution of the mainstream cross-section F by nondegradable organic materials due to the αth waste treatment plant. The product $\hat{\rho}_{U\alpha}\ \dot{V}\ \Delta\ t$ gives approximately the average amount of degradable organic pollution due to the αth waste treatment plant in a disk. The disk is bounded in the upstream direction by the cross-section F. The symbol Δt denotes a sufficiently small positive time difference.

The volume rate of flow $\dot{V}$ and the average emission rate $\overline{e_{U\alpha}}$ are measurable quantities.

The ensemble-mean values in Equations (4) to (10) can be considered as time-average values (ergodic theorem).

The assumptions underlying Equations (9) and (10) do not exclude phenomena such as secondary flow in the mainstream, motion of granular material at the mainstream bottom, percolation and evaporation (of pure water) and time-dependence of the structure of each final effluent. The volume rate of flow may depend on the position of the cross-section because of evaporation and percolation of pure water.

The previous considerations in this section apply to the mainstream pollution by degradable organic materials in the case of frozen biological self-purification.

Mainstream Pollution by Degradable Organic Materials

Dispersion of Polluted Fluid Elements from the Nearfield Mixing Region of the αth Jet of Final Effluent in the Mainstream.

Let N_α be the number of fluid elements which leave the nearfield mixing region of the αth jet of final effluent forever during the finite time interval $t_a \leq \sigma \leq t_b$. These fluid elements are loaded with degradable organic pollution. They can be considered to agree in volume and shape when they leave the nearfield mixing region of the αth jet of final effluent forever. Having left the nearfield mixing region forever, each of these polluted fluid elements will join various lumps in course of time, but will not lose its identity. Percolation and evaporation with these elements are considered negligible.

We focus our attention on a marked element among the N_α elements which is at the position x_α at time t_c, $t_a \leq t_c \leq t_b$. We consider this position to lie, for a large number of realizations, in the nearfield mixing region of the αth jet of final effluent.

Consider at time $t > t_b$ a position x in the mainstream lying outside the nearfield of each jet of final effluent and inside a suitably chosen closed surface bounding a domain of volume V_H. Let us define

$$P_\alpha(x, t, x_\alpha, t_c) = V_H^{-1}, \qquad (11)$$

if the marked element, which was at position x_α at time t_c, lies inside the surface under consideration at time t, otherwise

$$p_\alpha(x, t, x_\alpha, t_c) = 0 \quad . \qquad (12)$$

We can think of the ensemble mean $\overline{P_\alpha(x, t, x_\alpha, t_c)}$ as the probability density of finding the marked element at position x at time t.

Let $\mu_\alpha(t)$ denote the amount of degradable organic materials due to the αth waste treatment plant in the marked fluid element at time t. The ensemble-averaged density of the degradable organic materials from the αth waste treatment plant at position x at time t caused by the individual element under consideration can then be written in the form:

$$\overline{\mu_\alpha(t)\ P_\alpha(x,\ t,\ x_\alpha,\ t_c)}\ . \tag{13}$$

With the aid of the relations

$$\mu_\alpha(t) = \overline{\mu_\alpha(t)} + \mu'_\alpha(t),\quad P_\alpha(x,\ t,\ x_\alpha,\ t_c) = \overline{P_\alpha(x,\ t,\ x_\alpha,\ t_c)} + P'_\alpha(x,\ t,\ x_\alpha,\ t_c) \tag{14}$$

Expression (13) can be rewritten to give:

$$\overline{\mu_\alpha(t)\ P_\alpha(x,\ t,\ x_\alpha,\ t_c)} = \overline{\mu_\alpha(t)}\ \overline{P_\alpha(x,\ t,\ x_\alpha,\ t_c)} + \overline{\mu'_\alpha(t)\ P_\alpha(x,\ t,\ x_\alpha,\ t_c)}\ . \tag{15}$$

Strictly speaking the N_α elements leave the nearfield mixing region of the αth jet of final effluent forever at different positions at different times. Suppose these differences are negligible in calculating the ensemble-averaged density of the degradable organic materials from the αth waste treatment plant at position x at time t, caused by the N_α polluted fluid elements, we then obtain for this ensemble average

$$\overline{M_\alpha(t)\ P_\alpha(x,\ t,\ x_\alpha,\ t_c)} = \overline{M_\alpha(t)}\ \overline{P_\alpha(x,\ t,\ x_\alpha,\ t_c)} + \overline{M'_\alpha(t)\ \ P'_\alpha(x,\ t,\ x_\alpha,\ t_c)} \tag{16}$$

where $M_\alpha(t)$ is the total amount of the degradable organic materials due to the αth waste treatment plant contained in the N elements at time t. We have

$$M_\alpha(t) = \overline{M_\alpha(t)} + M'_\alpha(t)\ . \tag{17}$$

Consider now the case which differs from the case underlying Equation (16) only by the fact that no biological self-purification occurs for $\sigma \geq t_b$. The statement corresponding to Equation (16) is then

$$\overline{M_\alpha(t_b)P_\alpha(x, t, x_\alpha, t_c)} = \overline{M_\alpha(t_b)}\ \overline{P_\alpha(x, t, x_\alpha, t_c)} + \\ + \overline{M'_\alpha(t_b)P'_\alpha(x, t, x_\alpha, t_c)} \tag{18}$$

under the assumption that penetration of a polluted fluid element into a nearfield mixing region after having left another nearfield mixing region forever can be neglected.

The assumption introduced between Equation (15) and Equation (16) requires "fading memory" of the fluid elements dispersion process under consideration of its history and the following constraints for x and t in Equation (16) and Equation (18):

1) $t - t_b >> t_b - t_a$
2) The distance between the cross-section containing x and the cross-section containing x_α must be large compared with the characteristic length of the αth nearfield mixing region. Here it is understood that the distance is measured along the mainstream axis.

The second term on the right side of Equation (16) can be estimated as follows:

$$\overline{M'_\alpha(t)P'_\alpha(x, t, x_\alpha, t_c)} = \overline{M'_\alpha(t)\{P_\alpha(x, t, x_\alpha, t_c) - \overline{P_\alpha(x, t, x_\alpha, t_c)}\}}$$

$$\leq \overline{|M'_\alpha(t)|P_\alpha(x, t, x_\alpha, t_c)} << \overline{M_\alpha(t)}\ \overline{P_\alpha(x, t, x_\alpha, t_c)} \quad \text{for}$$

$$\overline{P_\alpha(x, t, x_\alpha, t_c)} \neq 0 \tag{19}$$

under the assumption

$$\frac{|M'_\alpha(t)|}{\overline{M(t)}} << \text{ for all realizations where } P_\alpha(x, t, x_\alpha, t_c) = 0. \tag{20}$$

To establish the estimate

$$\overline{M'_\alpha(t_b)P'(x, t, x_\alpha, t_c)} << \overline{M_\alpha(t_b)}\ \overline{P_\alpha(x, t, x_\alpha, t_c)} \quad \text{for}$$

$$P_\alpha(x, t, x_\alpha, t_c) \neq 0 \tag{21}$$

for the second term on the right side of Equation (18), the assumption

$$\frac{|M'_\alpha(t_b)|}{\overline{M_\alpha(t_b)}} \ll 1 \text{ for all realizations where } P_\alpha(x, t, x_\alpha, t_c) \neq 0 \tag{22}$$

is enough.

The biological self-purification in the nearfield G_α of the αth jet of final effluent and in the N_α fluid elements is considered negligible during the time interval $t_a \leq \sigma \leq t_b$. The following balance equation then holds

$$\int_{t_a}^{t_b} e_{A\alpha}\, dt + \left(\int_{G_\alpha} \rho_{A\alpha}\, d\,V\right)_{t_a} - \left(\int_{G_\alpha} \rho_{A\alpha}\, d\,V\right)_{t_b} = M_\alpha(t_b) \tag{23}$$

where $d\,V$ stands for a volume element of G_α. The continuous operation of the final stage of waste treatment implies

$$\overline{\left(\int_{G_\alpha} \rho_{A\alpha}\, d\,V\right)_{t_a} - \left(\int_{G_\alpha} \rho_{A\alpha}\, d\,V\right)_{t_b}} = 0 \ . \tag{24}$$

Hence, the ensemble-averaged balance Equation (23) can be written as

$$\overline{M_\alpha(t_b)} = \overline{e_{A\alpha}}(t_b - t_a) \ . \tag{25}$$

The quotient $\overline{M_\alpha(t)}/\overline{M_\alpha(t_b)}$ depends at most on t, t_b, and $t_b - t_a$. Truncating the Taylor expansion of this quotient with respect to $t_b - t_a$ after the first term, we obtain

$$\frac{\overline{M_\alpha(t)}}{\overline{M_\alpha(t_b)}} = Q_\alpha(t - t_b, \ t_b) \ . \tag{26}$$

The first term depends, at most, on $t - t_b$ and t_b.

The continuous operation of the final stage of waste treatment implies that the mainstream turbulence is steady in time. Hence, the probability density in the first term on the right side of Equation (16) cannot depend on t and t_c separately but only on the combination $t - t_c$

$$\overline{P_\alpha(x, t, x_\alpha, t_c)} = \overline{P_\alpha}(x, x_\alpha, t - t_c) \quad . \tag{27}$$

Divide the interval $-\infty < \sigma < t$ up into equal subintervals. The interval $t_a \le \sigma \le t_b$ can be considered to be any of these subintervals. The ensemble - averaged density of the degradable organic pollution due to the αth waste treatment plant at position x at time t can be represented as a sum of contributions which are of the type of the first term on the right side of Equation (16). If the increment time $t_b - t_a$ is sufficiently small, the sum of contributions under consideration can be represented as an integral

$$\overline{\rho_{A\alpha}(x,t)} = \int_0^\infty Q_\alpha(\tau, t-\tau)\ \overline{e_{A\alpha}}\ \overline{P_\alpha}(x, x_\alpha, \tau)\ d\tau \tag{28}$$

where τ is an integration variable. $\overline{\rho_{A\alpha}}$ will, in general, depend on time of continuous operation of the final stage of waste treatment if the structure of the final effluent from the αth waste treatment plant depends on time. In the case of frozen biological self-purification, one has to replace $Q_\alpha(\tau, t-\tau)$ by 1 in Equation (28), thus eliminating t on the right side of Equation (28).

Multiplying Equation (28) by $\overline{v_N}$ and then averaging over the cross-section yields

$$\overline{\overline{\rho_{A\alpha}}\ \overline{v_N}} = \overline{e_{A\alpha}} \int_0^\infty Q_\alpha(\tau, t-\tau)\ \overline{\overline{P_\alpha}(x, x_\alpha, \tau)\ \overline{v_N}}\ d\tau \tag{29}$$

where it is understood that the position x lies in F.

The broken line means cross-sectional averaging. In the case of frozen biological self-purification, Equation (29) must agree with Equation (6) if we replace the subscript U by A in Equation (6).

We recognize

$$\int_0^\infty F\ \overline{\overline{P_\alpha}(x, x_\alpha, \tau)\ \overline{v_N}}\ d\tau = 1 \tag{30}$$

where F is the area of the mainstream cross-section F. With the aid of Equation (30) and a mean value theorem, Equation (29) can be rewritten to give:

$$\hat{\rho}_{A\alpha}\ \dot{V} = \overline{e_{A\alpha}}\ Q_{\alpha}(\tau_{\alpha m}, t-\tau_{\alpha m}), \quad \hat{\rho}_{A\alpha} = \int_{F} \overline{\rho_{A\alpha}}\ \overline{v_N}\ dF / \int_{F} \overline{v_N}\ d\ F \qquad (31)$$

where $\tau_{\alpha m}$ denotes a suitable τ-value such that $0 \leq \tau_{\alpha m} \leq \infty$. The τ-value $\tau_{\alpha m}$ depends on x but not on t. The product $FP_{\alpha}(x, x_{\alpha}, \tau)\ \overline{v_N}\ \Delta t$ means the probability of finding the marked fluid element, which was at position x_{α} at time $t - \tau$, inside a disk at time t. The disk is bounded in the upstream direction by the mainstream cross-section F and its volume is $\dot{V}\Delta t$. The integrand of Equation (30) is expected to attain a sharp maximum so that the τ-value of this maximum can be set equal to $\tau_{\alpha m}$ in a good approximation. $Q_{\alpha}(\tau_{\alpha m}, t-\tau_{\alpha m})$ can be considered as the ratio formed of two "cross-sectional pollutions". These are due to the αth waste treatment plant in the normal case exhibiting biological self-purification and the case of frozen biological self-purification, respectively.

To determine $\tau_{\alpha m}$ we start from the dispersion of polluted fluid elements considered in the beginning of section α. These fluid elements leave the αth nearfield mixing region forever during the time interval $t_a \leq \sigma \leq t_b$. Their pollution is due to the αth waste treatment plant. Let us set down a balance equation for the mainstream pollution by this cloud of polluted fluid elements assuming frozen biological self-purification. With regard to a fixed mainstream section of infinitesimal thickness bounded in the upstream direction by F, the balance equation reads

$$\frac{\partial\,(F\overline{\rho_{A\alpha}})}{\partial\sigma} + \frac{\partial}{\partial x}\,\{F\overline{\rho_{A\alpha}}\ \overline{v_N}\} = 0 \quad . \qquad (32)$$

In Equation (32) and in the subsequent part of this section, the symbol $\overline{\rho_{A\alpha}}$ stands for the ensemble-averaged density of organic pollution due to the cloud of polluted fluid elements under consideration. x denotes a length coordinate measured along the mainstream axis. x increases in downstream direction. The mainstream axis is perpendicular to the mainstream cross-sections.

Separate the average mass flux $\overline{\overline{\rho_{A\alpha}}}\;\overline{\overline{v_N}}$ into a "convective" part and a "dispersion" part

$$\overline{\overline{\rho_{A\alpha}}\;\overline{v_N}} = \overline{\overline{\rho_{A\alpha}}}\;\overline{\overline{v_N}} + \overline{(\overline{\rho_{A\alpha}} - \overline{\overline{\rho_{A\alpha}}})(\overline{v_N} - \overline{\overline{v_N}})} \quad . \tag{33}$$

Substituting Equation (33) into Equation (32) and neglecting dispersion (see Verboom, 1976, pp. 299-300) yields

$$\{\frac{\partial}{\partial \sigma} + \overline{\overline{v_N}}\,\frac{\partial}{\partial x}\} \quad (F\;\overline{\overline{\rho_{A\alpha}}\;\overline{v_N}}) = 0 \; . \tag{34}$$

According to the preceding considerations, the average density $\overline{\rho_{A\alpha}}$ can be expressed as follows

$$\overline{\rho_{A\alpha}} = \overline{M_\alpha(t_b)}\;\overline{P_\alpha}(x,\, x_\alpha,\, \sigma - t_c) \quad . \tag{35}$$

Thus

$$\overline{\overline{\rho_{A\alpha}}\;\overline{v_N}} = \overline{M_\alpha(t_b)}\;\overline{\overline{P_\alpha}(x,\, x_\alpha,\, \sigma - t_c)\;\overline{v_N}} \quad . \tag{36}$$

With the aid of the relations (34) and (36), we obtain:

$$\{\frac{\partial}{\partial \sigma} + \overline{\overline{v_N}}\,\frac{\partial}{\partial x}\} \quad (F\;\overline{\overline{P_\alpha}(x,\, x_\alpha,\, \sigma - t_c)\;\overline{v_N}}) = 0 \tag{37}$$

This differential equation holds for mainstream cross-sections lying downstream of the αth nearfield and outside of the remaining nearfields. Furthermore, σ and x in Equation (37) have to satisfy the conditions A1 and A2 (formulated between Equation (18) and Equation (19)). The considerations leading to the differential Equation (37) refer to the cloud of the N_α fluid elements. Similar considerations can be made for the individual marked fluid element which was at position x_α at time t_c. Taking $\mu_\alpha(\sigma)$ as constant, the latter considerations demonstrate that Equation (37) holds for an arbitrary cross-section $\mathcal{F}(x)$ downstream of x_α for a finite time interval. This interval includes the time at which $F\;\overline{\overline{P_\alpha}(x,\, x_\alpha,\, \sigma - t_c)\;\overline{v_N}}$ attains its maximum at the fixed position x. We now recognize the sharp maximum of the integrand of Equation (30) moves downstream along the stream axis with velocity $\overline{\overline{v_N}}$. Thus

$$\tau_{\alpha m} = \int_{x_\alpha}^{x} \frac{d\xi}{\overline{v_N}(\xi)}$$

where ξ is an integration variable and x_α is the x-value of the mainstream cross-section through x_α.

<u>Biological Self-Purification in the Polluted Fluid Elements from the Nearfield Mixing Region of the αth Jet of Final Effluent in the Mainstream</u>. Let the nearfield mixing region of the αth jet of final effluent be inside the λth reach given by the x-interval $x_\lambda \leq x \leq x_{\lambda+1}$. The x-value x_λ is assigned to the mainstream cross-section bounding the λth reach in the upstream direction.

The marked fluid element considered in the previous section will stay in the λth reach during a time interval $t_\lambda \leq \sigma \leq t_{\lambda+1}$. For the lower bound t_λ and the residence time $t_{\lambda+1} - t_\lambda$, we put

$$t_\lambda = t_b, \quad t_{\lambda+1} - t_\lambda = \int_{x_\alpha}^{x_{\lambda+1}} \frac{d\xi}{\overline{v_N}(\xi)} \quad . \tag{39}$$

The marked fluid element will stay in the $(\lambda + \mu)^{th}$ reach during the time interval $t_{\lambda+\mu} \leq \sigma \leq t_{\lambda+\mu+1} (1 \leq \mu \leq L - \lambda)$. The residence time $t_{\lambda+\mu+1} - t_{\lambda+\mu}$ is supposed to be

$$t_{\lambda+\mu+1} - t_{\lambda+\mu} = \int_{x_{\lambda+\mu}}^{x_{\lambda+\mu+1}} \frac{d\xi}{\overline{v_N}(\xi)} \; (1 \leq \mu \leq L - \lambda) \tag{40}$$

Equation (40) is inapplicable for $L = \lambda$.

We adopt the well-stirred reactor model for the nearfield mixing region of the αth jet of final effluent with regard to the following: the average density of degradable organic pollution, the average density of the active bio-mass of the aerobic micro-organisms, the average pH-value, the average temperature, the structure of the aerobic micro-flora and the structure of the degradable organic pollution.

The turbulent fluctuations about the above-mentioned ensemble averages are considered small in comparison with the respective averages.

From the viewpoint of chemical engineering, the marked fluid element can be considered for $\sigma \geq t_b$ as a once charged aerated tank reactor (Romberg, 1976).

In an investigation while preparing the present paper, a model of purification (including model identification) is developed (Romberg 1976). The application of this model to the biological self-purification in the marked fluid element during $t_b \leq \sigma \leq t_{\lambda+1}$ is straight-forward.

The cited model of purification needs a slight modification to be able to model the biological self-purification in the marked fluid element during its stay in the $(\lambda + \mu)^{th}$ reach. This modification is straight-forward and does not alter the form of Equations (32), (49), (50), (55), (56), (69), (73), to (76) in the preparatory investigation if the following interpretations of symbols occurring in the preparatory investigations are used:

q	ratio of the concentration of the degradable organic materials in the marked fluid element at time σ to the concentration of the degradable organic materials in the marked fluid element at time $t_{\lambda+\mu}$
t_b	$t_{\lambda+\mu}$
t_c	$t_{\lambda+\mu+1} - t_{\lambda+\mu}$
$m_{1I}(t)$	change of the total amount of oxygen in sample bottle I during $0 \leq \sigma - t_{\lambda+\mu} \leq t \leq$ $\leq t_{\lambda+\mu+1} - t_{\lambda+\mu}$
$\dfrac{\rho(t_b)\, V_{FI}}{f}$	oxygen consumption in sample bottle I from $\sigma = t_{\lambda+\mu}$ until the biological self-purification is finished.

The temperature of the contents of sample bottle I has to agree sufficiently well with the ensemble-averaged temperature of the $(\lambda+\mu)^{th}$ reach in the vicinity of the marked fluid element under consideration. The temperature of the contents of sample bottle I has to agree with the ensemble-averaged mainstream temperature at $x = x_{L+1}$ for $\sigma \geq t_{L+1}$.

Determination of $Q_\alpha(\tau_{\alpha m}, t - \tau_{\alpha m})$. The following relations hold for the marked fluid element under consideration during its stay in the $(\lambda+\mu)^{th}$ reach $(\mu = 0, \ldots, L - \lambda)$

$$q_{\alpha} = A_{\alpha\mu} \cdot B_{\alpha\mu} \tag{41}$$

where

$$A_{\alpha\mu} = \frac{\kappa_{\alpha\mu}(t_b) + 1}{\kappa_{\alpha\mu}(t_b) + \exp[D_{\alpha\mu}(t_b) \frac{-t_{\lambda+\mu}+1}{t^*}]} \tag{42}$$

and

$$B = \prod_{\nu=0}^{\mu-1} \frac{\kappa_{\alpha\nu}(t_b) + 1}{\kappa_{\alpha\nu}(t_b)+\exp[D_{\alpha\nu}(t_b) \frac{-t_{\lambda+\nu} + t_{\lambda+\nu+1}}{t^*}]}, \mu \geq 1; B_0 = 1 . \tag{43}$$

Comments on various time intervals

$$t_{\lambda+\nu+1} - t_{\lambda+\nu} = \int_{x_{\lambda+\nu}}^{x_{\lambda+\nu+1}} \frac{d\xi}{\bar{\bar{\bar{v}}}_N(\xi)} \text{ for } \nu \geq 1, \tag{44}$$

$$t_{\lambda+1+\nu} - t_{\lambda+\nu} = \int_{x_{\alpha}}^{x_{\lambda+1}} \frac{d\xi}{\bar{\bar{\bar{v}}}_N(\xi)} \text{ for } \nu = 0 , \tag{45}$$

$$t - t_{\lambda+\mu} = \int_{x_{\alpha}}^{x} \frac{d\xi}{\bar{\bar{\bar{v}}}_N(\xi)} - \int_{x_{\alpha}}^{x_{\lambda+\mu}} \frac{d\xi}{\bar{\bar{\bar{v}}}_N(\xi)} \text{ for } \mu \geq 1, \tag{46}$$

$$t - t_{\lambda+\mu} = \int_{x_{\alpha}}^{x} \frac{d\xi}{\bar{\bar{\bar{v}}}_N(\xi)} \text{ for } \mu = 0 . \tag{47}$$

Comments on various denotations

$D_{\alpha\mu}(t_b, \kappa_{\alpha\mu}(t_b)$ — D-value and κ-value obtained by applying the above-mentioned slightly modified model of purification to the marked fluid element under consideration during its stay in the region $x_{\lambda+\mu} \leq x < x_{\lambda+\mu+1}$.

q_α — ratio of the concentration of the degradable organic materials in the marked fluid element at time t, when the element is at cross-section x, to the concentration of the degradable organic materials in the marked element at time $t_b = t - \tau_{\alpha m}$ where $\tau_{\alpha m} = \int_{x_\alpha}^{x} d\xi / \overline{v_N}(\xi)$

t^* — characteristic time

The preceding considerations lead to

$$q_\alpha = \frac{M_\alpha(t)}{M_\alpha(t_b)} = \frac{\overline{M_\alpha(t)}}{\overline{M_\alpha(t_b)}} = Q_\alpha(t - t_b,\ t_b) \text{ for } t = t_b + \tau_{\alpha m}. \tag{48}$$

Summary

In the previous sections, the final stage of waste treatment is modelled. This physical model is adapted for computational evaluation.

The main assumptions underlying the model are:

1. Continuous operation.
2. The mainstream pollution is due to organic materials in the final effluents, not heat pollution.
3. Average outflow of organic matter through the banks and the bottom is ngeligible.
4. No reverse flow outside of the nearfield mixing regions in the main motion described by the velocity $\underline{v}$.
5. Penetration of a polluted fluid element into a nearfield mixing region, after having left the nearfield mixing region forever from which it originated, can be neglected. Percolation and evaporation with such a fluid element is negligible.
6. We adopt the well-stirred reactor model for the nearfield mixing regions with regard to the following: the average density of the degradable organic pollution and of the active bio-mass of the aerobic micro-organisms, respectively, and the average pH-value, the average temperature, the structure of the aerobic micro-flora and of the degradable organic pollution, respectively.

7. The model of purification developed in the preparatory investigation is applicable inside each reach.
8. The benthos can be neglected in modelling the biological self-purification in the polluted fluid elements originating from the nearfield mixing regions.

Input

Discharge location, division into reaches: x_α, and $x_{\lambda+\nu}(\nu = 1,\ldots,L + 1-\lambda)$. The emission rates of the αth waste treatment plant: $\overline{e_{A\alpha}}$, and $\overline{e_{U\alpha}}$. For an arbitrary mainstream cross-section F outside of the nearfields and downstream of the nearfield of the αth jet of final effluent; the mean velocity $\overline{v_N}$ and the cross-sectional area F.

To determine the capacity of self-purification from degradable organic matter due to the αth plant, described by $Q_\alpha(\tau_{\alpha m},t-\tau_{\alpha m})$: the mean velocity $\overline{v(\xi)}$ for $x_\alpha \leq \xi \leq x$, the biological parameters $D_{\alpha\nu}(t_b)$ and $\kappa_{\alpha\nu}(t_b)$ $(\nu = 0,\ldots,\mu)$ which relate to the biological self-purification in the polluted fluid elements leaving the th nearfield mixing region forever during $t_a \leq \sigma \leq t_b$.

Output

Cross-sectional pollution at F by organic materials from the αth plant, described by

$$\hat{\rho}_{A\alpha} = \int_F \overline{\rho_{A\alpha}}\ \overline{v_N}\ dF / \int_F \overline{v_N}\ dF, \quad \hat{\rho}_{U\alpha} = \int_F \overline{\rho U_\alpha}\ \overline{v_N}\ dF / \int_F \overline{v_N}\ dF \tag{49}$$

as a function of $\overline{e_{A\alpha}}$, $\overline{e_{U\alpha}}$ and $\overline{v_N}$, F, $Q_\alpha(\tau_{\alpha m},t-\tau_{\alpha m})$ at F :

$$\begin{pmatrix} \rho_{A\alpha} \\ \rho_{U\alpha} \end{pmatrix} = (F\ \overline{v_N})^{-1} \begin{pmatrix} \overline{e_{A\alpha}}\ Q_\alpha(\tau_{\alpha m},t-\tau_{\alpha m}) \\ \overline{e_{U\alpha}} \end{pmatrix} \quad (\alpha = 1,\ldots,a) \tag{50}$$

In the section entitled, "Mainstream Pollution by Nondegradable Organic Materials," the capacity of self-purification Q_α is represented as a function of input data.

If F does not satisfy the condition formulated between Equation (18) and Statement (19), we can, in general, put $Q_{\alpha}(\tau_{\alpha m}, t-\tau_{\alpha m}) = 1$ at F.

Mainstream of Constant Temperature

Figure 1 illustrates the cross-sectional pollution due to the degradable organic matter form the αth plant depending on the volume rate of flow, emission rate and biological self-purification after

$$\frac{\hat{\rho}_{A\alpha}\dot{V}}{e_{A\alpha}} = \frac{\kappa_{\alpha 0}(t_b) + 1}{\kappa_{\alpha 0}(t_b) + \exp\{D_{\alpha 0}(t_b)\frac{\tau_{\alpha m}}{t^*}\}} \quad (\alpha = 1,\ldots,a) \tag{51}$$

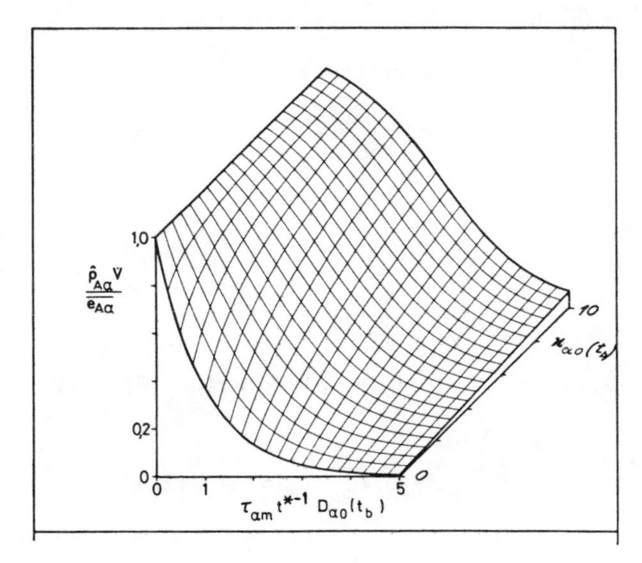

Fig. 1. Cross-sectional pollution of the mainstream by degradable organic matter from the αth waste treatment plant, after formula (51).

If the structure of the final effluent from the αth plant does not depend on time, the biological parameters $x_{\alpha 0}(t_b)$ and $D_{\alpha 0}(t_b)$ are constants. Formula (51) then proves to be a special version of the π-theorem where the flow time $\tau_{\alpha m}$ serves as an independent variable.

Remarks on Dispersion

The following relation holds for an arbitrary mainstream cross-section

$$\overline{\rho_{X\alpha} v_N} = \overline{\rho_{X\alpha}}\;\overline{v_N} + \Delta_{X\alpha} \quad \text{for} \quad X = A \quad \text{and} \quad X = U \tag{52}$$

where

$$\overline{(\rho_{X\alpha} - \overline{\rho_{X\alpha}})(v_N - \overline{v_N})} = \Delta_{X\alpha} \quad \text{for} \quad X = A \quad \text{and} \quad X = U \tag{53}$$

is the dispersion of the degradable and nondegradable organic pollution from the αth waste treatment plant, respectively. Assuming

$$\dot{V} > 0 \tag{54}$$

Statement (52) can be rewritten as

$$\hat{\rho}_{X\alpha} = \overline{\rho_{X\alpha}} + \overline{v_N}^{-1} \Delta_{X\alpha} \quad \text{for} \quad X = A \quad \text{and} \quad X = U \quad . \tag{55}$$

Hence

$$\hat{\rho}_{X\alpha} = \overline{\rho_{X\alpha}} \quad \text{for} \quad X = A \quad \text{and} \quad X = U \;, \tag{56}$$

if the dispersion $\Delta_{X\alpha}$ is negligible.

Cross-sectional averaging of Formula (28) gives

$$\overline{\rho_{A\alpha}} = \int_0^\infty Q_\alpha(\tau, t-\tau)\; \overline{e_{A\alpha}}\; \overline{P_\alpha(x, x_\alpha, \tau)}\, d\tau \;. \tag{57}$$

$\overline{P_\alpha(x, x_\alpha, \tau)}$ can be expected to attain a sharp maximum at the position $\tau = \tau_{\alpha m}$. Let us denote the case of frozen biological self-purification by the subscript f. With the aid of the Equations (31) and (57) we obtain

$$\frac{\overline{\rho_{A\alpha}}}{(\overline{\rho_{A\alpha}})_f} = \frac{\hat{\rho}_{A\alpha}}{(\hat{\rho}_{A\alpha})_f} \;. \tag{58}$$

Using the Relations (58) and (55) we find

$$\Delta_{A\alpha} = Q_{\alpha}(\tau_{\alpha m}, t-\tau_{\alpha m})(\Delta_{A\alpha})_f \tag{59}$$

Statements (57), (58) and (59) hold for mainstream cross-sections F, the x-coordinate of which satisfies the Condition A2 formulated between Equation (18) and Statement (19).

Substituting the well-known relation

$$(\Delta_{A\alpha})_f = - D_{A\alpha} \frac{d(\overline{\overline{\rho_{A\alpha}}})_f}{dx} \tag{60}$$

where $D_{A\alpha} > 0$ is the dispersion coefficient, and the balance equation

$$(\hat{\rho}_{A\alpha})_f \dot{V} = \overline{e_{A\alpha}} \tag{61}$$

into the relation

$$(\hat{\rho}_{A\alpha})_f = (\overline{\overline{\rho_{A\alpha}}})_f + \overline{\overline{v_N}}^{-1} (\Delta_{A\alpha})_f \quad , \tag{62}$$

one obtains the differential equation

$$- D_{A\alpha} F \frac{d\Phi_{A\alpha}}{dx} + \dot{V} \Phi_{A\alpha} = 1 \tag{63}$$

where we have for brevity:

$$\Phi_{A\alpha} = \frac{(\overline{\overline{\rho_{A\alpha}}})_f}{\dot{V}(\hat{\rho}_{A\alpha})_f} [1 - \frac{F}{\overline{e_{A\alpha}}} (\Delta_{A\alpha})_f] \frac{1}{\dot{V}} \quad . \tag{64}$$

Relations (61) to (64) hold downstream of the αth nearfield for mainstream cross-sections, where $\dot{V} > 0$.

Let us introduce a new independent variable

$$\zeta = \int_0^x \frac{\dot{V}(\xi)\, d\xi}{D_{A\alpha}(\xi)F(\xi)} \tag{65}$$

where

$$0 < K_U < D_{A\alpha}, \; F, \; \dot{V} < K_0 < \infty \quad \text{for} \quad x \geq 0 \quad . \tag{66}$$

K_U and K_0 denote bounds. Transforming the differential equation (63) to the new independent variable yields

$$- \frac{d\Phi_{A\alpha}}{d\zeta} + \Phi_{A\alpha} = \dot{V}^{-1} \qquad (67)$$

The general solution of this differential equation reads

$$\Phi_{A\alpha} = \{C - \int_0^{\zeta} \frac{d\xi}{V(x(\xi))e^{\xi}} \; e^{\zeta} \quad \text{for} \quad \zeta \geq 0 \qquad (68)$$

where C is an arbitrary constant. $x = 0$ is assigned to a mainstream cross-section arbitrarily chosen downstream of the αth nearfield. $x(\zeta)$ is the inverse transformation (65).

With the aid of Statements (61), (64) and (66) we find that $\Phi_{A\alpha}$ is bounded for $\zeta \geq 0$. For an infinitely long mainstream of constant volume rate of flow for $x \geq 0$, it follows

$$(\overline{\rho_{A\alpha}})_f = (\hat{\rho}_{A\alpha})_f \; , \quad (\Delta_{A\alpha})_f = 0 \quad \text{for} \quad x \geq 0 \; . \qquad (69)$$

Assuming

$$(\Delta_{A\alpha})_f = 0 \quad \text{for} \quad 0 \leq x \leq x_e \qquad (70)$$

we conclude, with the aid of Equation (64) and Equation (67):

$$\dot{V} = \text{const} \quad \text{for} \quad 0 \leq x \leq x_e \quad . \qquad (71)$$

x_e denotes the upper bound of a closed interval. Thus, we recognize that vanishing dispersion is incompatible with a volume rate of flow depending on position.

The considerations leading to the results (69) and (71) apply also to the nondegradable organic pollution from the αth waste treatment plant.

Let us now assign $x = 0$ to a mainstream cross-section, which is located downstream of the αth nearfield and satisfies the condition A2 formulated between Equation (18) and Statement (19). Suppose only one waste treatment plant is jointed to the mainstream (a = 1). Equation (58) and Equation (69) then hold for $x \geq 0$. For an infinitely long mainstream of constant volume rate of flow for $x \geq 0$, it then follows with the aid of Equation (55) and Equation (59)

$$\hat{\rho}_{A\alpha} = \overline{\rho_{A\alpha}}, \quad \Delta_{A\alpha} = 0 \quad \text{for} \quad x \geq 0 . \tag{72}$$

Assuming

$$Q_\alpha \neq 0, \quad \Delta_{A\alpha} = 0 \quad \text{for} \quad 0 \leq x \leq x_e \tag{73}$$

we obtain with the aid of the Relations (59), (64), and (67):

$$\dot{V} = \text{const} \quad \text{for} \quad 0 \leq x \leq x_e \tag{74}$$

Statement (74) is in accord with Equation (71).

For a position - dependent volume rate of flow in the form

$$\dot{V} = \frac{1}{A + B\, e^{-K\zeta}} \quad \text{for} \quad \zeta \geq 0 \tag{75}$$

where the parameters A, B and K satisfy the conditions

$$A,\ K > 0, \quad |B| < A \quad , \tag{76}$$

the general solution takes the form

$$\Phi_{A\alpha} = A + \frac{B}{K+1}\, e^{-K\zeta} \quad \text{for} \quad \zeta \geq 0 \tag{77}$$

Figure 2 illustrates the formula

$$\dot{V}\, \Phi_{A\alpha} = \frac{1 + \frac{1}{K+1}\frac{B}{A}\, e^{-K\zeta}}{1 + \frac{B}{A}\, e^{-K\zeta}} \tag{78}$$

In the last formula it is understood that the following constraints are imposed on A, B, K and ζ,

$$0 \leq \frac{B}{A}\, e^{-K\zeta} < 1, \quad 0 < (K+1)^{-1} < 1 . \tag{79}$$

The curves $\dot{V}\,\Phi_{A\alpha} = \text{const}$ $(1/2 < \text{const} < 1)$ represent hyperbolic arcs.

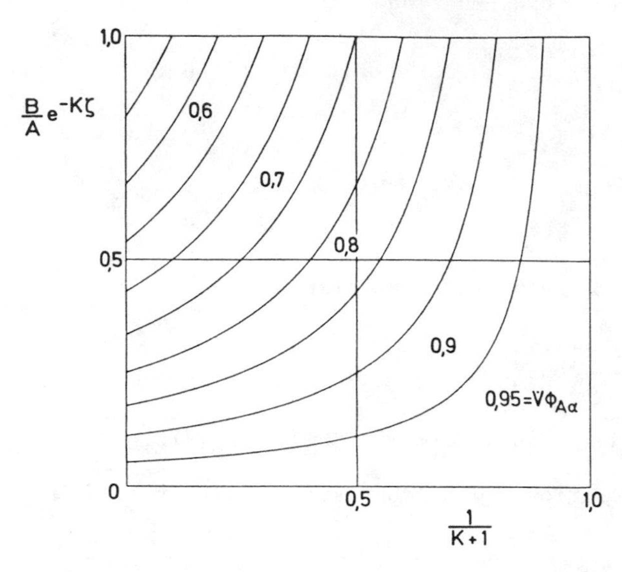

Fig. 2. Dispersion in a mainstream with position-dependent volume rate of flow, after formula (78).

Oxygen Balance

Consider an element of space moving with the mean velocity field $\overline{\overline{v_N}}$. The element has the form of a sufficiently thin disk and is bounded by two mainstream cross-sections, the banks, the bottom and the water surface. We set down the oxygen balance assigned to the αth waste treatment plant, in the frame of reference just introduced. Neglecting the ensemble mean of the outflow of oxygen through the surface of the disk apart from the water surface and neglecting production of oxygen due to the photosynthetic activity of plants, the ensemble - averaged oxygen balance under consideration can be written in the form,

$$\frac{D\overline{\overline{S}}}{D\sigma} = C(\overline{\overline{S_g}} - \overline{\overline{S}}) + \Gamma_\alpha \tag{80}$$

where the operator $D/D\sigma$ is defined as follows

$$\frac{D}{D\sigma} = \frac{\partial}{\partial\sigma} + \overline{\overline{v_N}}\,\frac{\partial}{\partial x} \; . \tag{81}$$

S and S_g denote, respectively, the ensemble mean of the density of dissolved oxygen and of the saturation density. Many factors are found to affect the absorption coefficient C (Preissman, 1971; Klein, 1972, pp. 232-237). Pollution by well-purified detergent-free final effluents has a relatively small effect on the absorption coefficient. Anionic surface-active agents derived from detergents in final effluents, however, can produce a considerable reduction in the absorption coefficient. We can assign a biochemical oxygen demand to the degradable organic matter due to the αth waste treatment plant in the moving disk. Γ_α is the rate of change of the ensemble mean of the biochemical oxygen demand just introduced. Utilizing considerations of section D we find

$$\Gamma_\alpha = f_\alpha^{-1} \frac{D\overline{\overline{\overline{\rho}}}_\alpha}{D\sigma} \tag{82}$$

at any mainstream cross-section satisfying the following conditions: the cross-section lies downstream of the αth nearfield and outside of the remaining $a-1$ nearfields, the x-coordinate of the cross-section satisfies the Condition A2 (formulated between Equation (18) and Statement (19) and

$$\int_{x_\alpha}^{x} d\xi \, / \, \overline{\overline{\overline{v}}}(\xi) >> t_b - t_a \quad . \tag{83}$$

If the structure of the αth final effluent is independent of time, the coefficient f_α is a constant. f_α^{-1} can be explained as the oxygen consumption in the marked fluid element (considered in the section entitled, "Mainstream Pollution by Degradable Organic Materials") per unit mass of degradable organic pollution in the marked fluid element. Using Equations (31), (38), and (56) the average density $\overline{\overline{\overline{\rho}}}_{A\alpha}$ in Equation (82) can be expressed as follows,

$$\overline{\overline{\overline{\overline{\rho}}}}_{A\alpha} = \overline{e}_{A\alpha} \, \dot{V}^{-1} \, Q_\alpha(\tau_{\alpha m}), \quad \tau_{\alpha m} = \int_{x_\alpha}^{x} d\xi / \overline{\overline{\overline{v}}}(\xi) \tag{84}$$

where it is understood that x signifies the position of the disk.

Consider now a mainstream of constant temperature where $a = 1$. The structure of the final effluent is independent of time and $\overline{\overline{S}}_g$, $\dot{V}$ and C, are positive constants downstream of the discharge location. In this case, Equation (84) becomes,

$$\overline{\rho_{A\alpha}} = \overline{e_{A\alpha}} \; \dot{V}^{-1} \; \frac{\kappa + 1}{\kappa + \exp(D\sigma)} , \tag{85}$$

if we introduce the notations $\kappa = \kappa_{\alpha 0}(t_b)$, $D = t^{*-1} \; D_{\alpha 0}(t_b)$ and prescribe that the disk lies at $x = x_\alpha$ at $\sigma = 0$. To simplify the balance Equation (80) in the special case under consideration, dimensionless quantities are introduced,

$$Y = \frac{f \; (\overline{S_g} - \overline{S}) \; \dot{V}}{\overline{e_A} \; (\kappa + 1) \; K} , \quad T = C\sigma, \quad K = \frac{D}{C} . \tag{86}$$

The balance Equation (80) then becomes

$$\frac{DY}{DT} + Y = \frac{\exp\;(K\,T)}{[\kappa + \exp(K\,T)]^2} \tag{87}$$

Requiring

$$Y = 0 \quad \text{at} \quad \sigma = 0 , \tag{88}$$

the formal solution of the differential Equation (84) becomes:

$$Y = \{\int_0^T \frac{\exp\;[K + 1)\;\tau]}{[\kappa + \exp\;(K\;\tau)]^2} \; d\tau\} \; e^{-T} . \tag{89}$$

The condition, that the aerobic micro-organisms must not be inhibited by shortage of dissolved oxygen requires the restriction of the validity of Formula (89) for a given $\kappa > 0$ and $K > 0$ to an initial T-range, where the ratio

$$\frac{Y \; e_{A\alpha} (\kappa + 1) \; K}{f_\alpha \; \overline{S_g} \; \dot{V}}$$

is sufficiently smaller than one.

Figure 3 illustrates Formula (89). The curves $K = \text{const} > 0$ are oxygen sag curves. The coordinate of the peak of the oxygen sag curve is plotted against K in Figure 4. The parameter of the graphs in Figure 4 is κ.

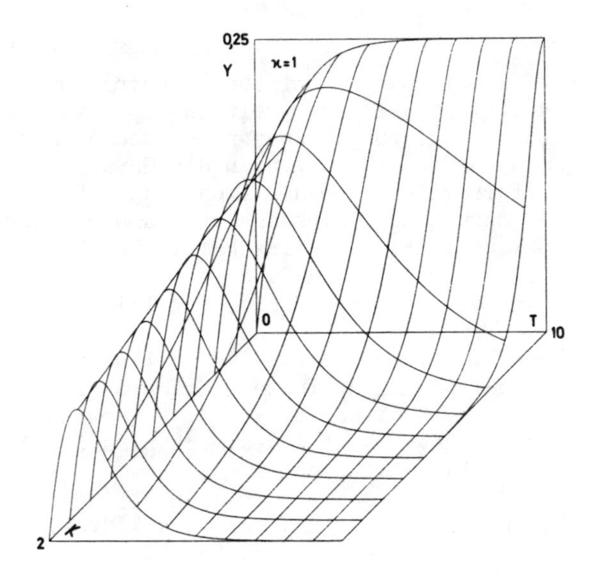

Fig. 3. Relief map illustrating the oxygen deficit after formula (89).

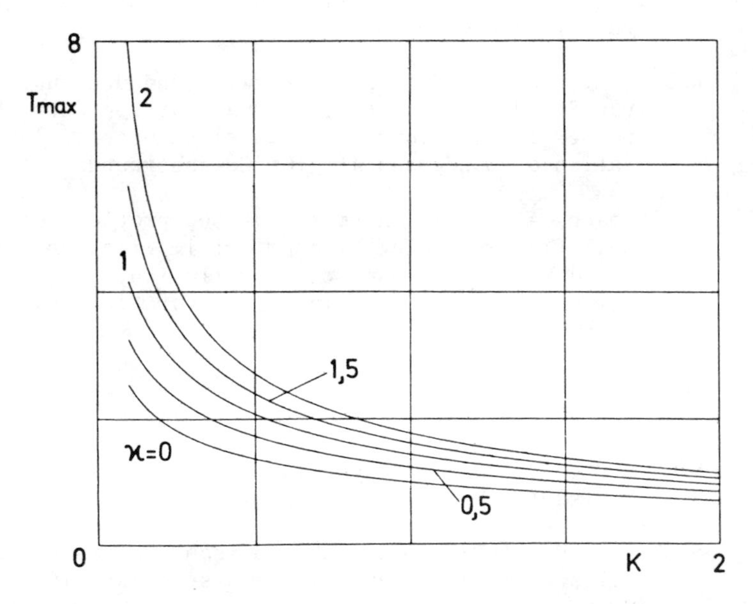

Fig. 4. Coordinate of the peak of the oxygen sag curve as a function of the dimensionless parameters K and x.

References

[1] Böhnke, B., 1975. Anzustrebende Reiningungsleistung mit Rücksicht auf das Umweltschutzprogramm der Bundesregierung, paper presented at meeting No. 343-75 "Anforderungen an die Gawässergüte und die erforderliche Abbauleistung von Kläranlagen zur Sicherung der Trinkwassergewinnung unter besonderer Berücksichtigung wirtschaftlicher Möglichkeiten der Leistungssteigerung überlasteter Kläranlagen" in Haus der Technik, Essen, April 1975.

[2] Fischer, H. B., 1973. Longitudinal Dispersion and Turbulent Mixing in Open-Channel Flow. In: M. Van Dyke, W. G. Vincenti and J. V. Wehausen (Editors), Annual Reviews of Fluid Mechanics 5. Annual Reviews Inc., Palo Alto, pp. 59-78.

[3] Franke, H., 1970. Lexikon der Physik, Vol. 5. Deutscher Taschenbuch Verlag, München, 320 pp.

[4] Kelker, H., Klages, F., Schwarz, R. and Wannagat, U., 1971. Chemielexikon. Fischerbücherei, Frankfurt, 400 pp.

[5] Klein, L., 1972. River Pollution 2: Causes and Effects. Butterworths, London, 456 pp.

[6] Knöpp, H., Schott, N., 1971. Entwicklung und Anwendung von Standardwerten für die Qualität von Oberflächenwasser als Instrument des Umweltschutzes. Investigation of Bundesanstalt für Gewässerkunde.

[7] Preissmann, A., 1971. A review of the problems of river pollution including biochemical implications. In: Fluid Mechanical Aspects of Pollution, Lecture Series 35, von Karman Institute for Fluid Dynamics, 19 pp.

[8] Romberg, G., 1976. Biologische Selbstreinigung verschmutzter Flüsse, Wärme-und Stoffübertrangung, 9: 226-246.

[9] Soo, S. L., 1967. Fluid Dynamics of Multiphase Systems. Blaisdell, Waltham, 524 pp.

[10] Stehfest, H., 1973. Modelltheoretische Untersuchungen zur Selbstreiningung von Fliebgewässern. KFK 1654 UF, 90 pp.

[11] Verboom, G. K., 1976. The advection - dispersion equation for an anisotropic medium solved by fractional-step methods. In: C.A. Brebbia (Editor), Mathematical Models for Environmental Problems. Pentech, London, pp. 299-312.

STREAMFLOW QUALITY AND QUANTITY RELATIONSHIPS ON A FOREST CATCHMENT IN ALBERTA, CANADA

By

T. Singh
Northern Forest Research Centre
Canadian Forestry Service, Environment Canada
5320 122 Street, Edmonton, Alberta
T6H 3S5 Canada

Abstract

The composition of main constituents in the natural waters of remote catchments is usually difficult to determine. A data base, however, is required to assess impact of intended land use practices on upland areas which are the main source of prime quality water. Relationships between water quality and the readily available data on streamflow can provide the required baselines for comparative purposes.

Dilution effects in the stream waters of Marmot Experimental Watershed were examined to investigate water quality and quantity relationships. A number of mathematical models incorporating linear and nonlinear functional forms were hypothesized. The model parameters were determined from 31 samples collected and analyzed during 1971 and 1972. Calcium, Mg, Na, K, HCO_3, SO_4, Cl, and SiO_2 were the main constituents of these waters. Of the 28 postulated models, the regression models incorporating current flow, specific conductance, and a lagged variable of streamflow gave the best fit for most of the analyzed constituents. Other models based on the relationships of the above mentioned constituents to the components of streamflow (base flow and storm flow) are presently being developed and tested.

The coefficients for all models were derived from the collected data by the method of least squares. The models' prediction capabilities were tested by comparison with actual quality data acquired subsequently. The models are reassessed on this basis and ranked according to the magnitude of the sum of the squared deviations of observed values from those predicted by each model.

Introduction

The study of the composition and concentration of dissolved chemical constituents in natural waters is important in assessing the environmental impacts of land use practices. The laboratory analyses required for such determinations are costly and time-consuming. A simple scheme is needed for remote catchments, where feasible, to predict the concentration of commonly occurring constituents from easily determined hydrologic variables.

An obvious variable is streamflow. Johnson et al. (1969), Pinder and Jones (1969), Hall (1970, 1971), Pionke, Nicks and Schoof (1972), and Steele (1973, 1976) have shown how the relationship between stream discharge and water chemistry can be used as a predictive tool. The present study explores the predictive feasibility of using the relationships of streamflow and its components with the inorganic constituents in the streams originating from upland mountain forest watersheds.

Water is the chief agent releasing and transporting nutrients in an ecosystem (Hewlett and Nutter, 1969). Relationships of chemical constituents with streamflow are thus important for an understanding of the ecosystems in relation to forestry practices causing changes in nutrient exports.

As Hem (1970) states, the streams having the most consistent relationship between water discharge and dissolved-solids concentrations ought to be the streams that receive a large part of their mineral load from a relatively constant source upstream. The stream used in this study fulfills this criterion, and thus further provides a basis for testing this hypothesis.

Study area. The study was conducted in Marmot Creek experimental watershed, situated on the eastern slopes of the Rocky Mountains, about 80 km west of Calgary. It ranges in elevation from about 1500 m to 2750 m (mean elevation 2113 m) and represents spruce-fir [*Picea engelmanni* Parry, *P. glauca* (Moench) Voss, and *Abies lasiocarpa* (Hook.) Nutt.] vegetation typical of the Saskatchewan River headwaters. The watershed was selected in 1962 as the first major research effort under the Alberta Watershed Research Program. In 1974 a treatment was applied on the Cabin subbasin to determine the effects of commercial logging on water yield, quality, and regime. The treatment consisted of clear-cutting six blocks totalling 40 ha.

The geology, soils, and vegetation of the area are described by Stevensen (1967), Beke (1969), and Kirby and Ogilvie (1969), respectively. Jeffrey (1965), Golding (1970), Singh and Kalra (1972), Singh (1976), and Telang et al. (1976) have described the hydrological and water quality aspects of the research program in the watershed.

Methods

Collection of data. The water quality samples collected from the permanent gaging site were analyzed at the Water Quality Laboratory of Environment Canada, Calgary. Stream discharge at the time of sampling was also determined from the hydrograph. Samples were collected at least monthly when the streamflow was nearly constant, as in the winter months, and more frequently in other months when it showed greater variation. Only samples collected from the Main Marmot Creek prior to commercial logging were used in this study.

Modelling data. The 31 samples collected during the 2-year period (1971-1972) were used for model building. These samples provided concentration data (mg/ℓ) for Ca, Mg, Na, K, HCO_3, SO_4, SiO_2, Cl and total dissolved solids which were treated as the variables to be predicted. The streamflow data (ℓ/sec) and specific conductance (μS/cm) were the predictor variables.

Test data. The data collected during 1973 were used for testing or validating the models. The same input data were used for all models. As no data were available for SO_4, the models for this constituent could not be tested for the present. Further tests and updating of the models are planned for future years when more data become available.

Modelling procedures. Stream discharge at the time of sampling (X_1) was used to find linear and nonlinear relationships with the concentration of each constituent. Another variable, specific conductance (X_2), was included later to improve the goodness of fit. In order to incorporate the lagged effect, one more variable for the mean daily streamflow on the previous day (X_3) was also included as a predictor variable to further improve the fit.

All models were fitted to the data by the method of least squares. In addition to the estimation of model parameters, the correlation coefficient and the standard error of estimate were also computed.

For validation, the estimated values from each model were compared with the actual data. Each residual was listed and used for determining sum of residuals, sum of the squared deviations, mean square error, and standard error of estimate.

A computer program (Bathlahmy, 1972) was used on the 1973 data for determining the groundwater, interflow, and rapid flow components of stream discharge. A model based on three predictor variables (base flow, storm flow, and specific conductance) was derived for each modelled constituent. The model parameters, the correlation coefficient, and the standard error of estimate were also computed for this model by the method of least squares.

Results

The summary of modelling data is provided in Table 1. Table 2 gives a summary of the test data for comparative purposes.

Twenty-eight models were tried in all (Table 3). Of these, 12 had only one predictor variable, 12 had two predictor variables, and 4 had three predictor variables included.

Although outputs of model parameters and related statistics were obtained for all models, only results from I, XIII and XXV are listed in Tables 4, 5, and 6 and presented here. These tables summarize comparative statistics for the individual constituents Ca, Mg, Na, K, HCO_3, Cl, and SiO_2 and also for the total dissolved solids.

Table 7 gives the related information on model parameters and other statistics when two of the three predictor variables of model XXV are replaced by the variables containing information on current base flow (or groundwater flow) and storm flow.

Discussion

The dependence of the concentration of chemical constituents of stream waters on discharge is evident from the generally highly significant relationships listed in Table 4. Total dissolved solids and Ca^{2+} show the most significant relationship, whereas Cl^- relationship is the least, and statistically nonsignificant. The negative sign of the correlation coefficient shows dilution effects, i.e., the concentration of constituents become lower if streamflow increases.

The slope (b) representing change in concentration with unit change in streamflow is the highest for total dissolved solids and the lowest for K. From Table 4, the individual constituents can be ranked in order of the absolute value of b as $HCO_3^- > Ca^{2+} > Mg^{2+} > SiO_2 > Na^+ > K^+ > Cl^-$.

The relationships improve considerably when specific conductance is included as an additional variable in the prediction model. The use of specific conductance for estimating the concentration of total dissolved solutes has been tested earlier (Singh and Kalra, 1975). In this paper its use has been extended to other dissolved constituents. The results (Table 5) show improvements in all the tested constituents.

Table 1. Summary of Modelling Data, Marmot Creek

Variable	Mean	Standard Deviation	Minimum	Maximum	Coefficient of Variation
Ca^{2+} (mg/ℓ)	44.2	10.4	25.2	56.6	23.61
Mg^{2+} "	11.9	3.1	6.8	17.2	25.87
Na^{+} "	1.1	0.6	0.4	2.5	51.90
K^{+} "	0.5	0.1	0.3	0.7	21.09
HCO_3^- "	173.6	40.7	101.7	232.8	23.44
CL^- "	0.2	0.1	0.1	0.3	41.45
SiO_2 "	4.5	0.8	2.9	5.5	18.48
Total dissolved solids "	164.4	42.1	83.0	226.5	25.58
Current streamflow (ℓ/sec)	279.7	388.2	11.6	1441.3	138.79
Specific conductance (μS/cm)	302.9	66.8	182.0	406.0	22.06
Streamflow previous day (ℓ/sec)	284.0	390.7	11.6	1452.6	137.54

Table 2. Summary of Test Data, Marmot Creek

Variable	Mean	Standard Deviation	Minimum	Maximum	Coefficient of Variation
Ca^{2+} (mg/ℓ)	41.8	10.1	28.2	62.6	24.09
Mg^{2+} "	10.5	2.6	3.4	16.7	24.72
Na^{+} "	1.1	0.4	0.6	2.0	38.01
K^{+} "	0.5	0.1	0.3	0.8	19.25
HCO_3^- "	161.9	34.0	115.0	230.0	21.01
Cl^- "	0.4	0.2	0.1	1.0	58.01
SiO_2 "	4.7	0.6	3.6	6.0	12.18
Total dissolved solids "	152.1	33.9	107.0	218.0	22.29
Current streamflow (ℓ/sec)	369.9	289.3	15.3	996.7	78.20
Specific conductance (μS/cm)	282.6	62.6	204.0	452.0	22.15
Streamflow previous day (ℓ/sec)	354.4	261.6	15.6	985.4	73.82

Table 3. Mathematical Models for Estimating Concentration (mg/ℓ) of Dissolved Constituents (Y); X_1 is Current Streamflow (ℓ/sec), X_2 is Specific Conductance (µS/cm), and X_3 is Mean Streamflow (ℓ/sec) on Previous Day

Code	Model
One predictor variable:	
I	$Y = a + bX_1$
II	$Y = a + bX_2$
III	$Y = a + bX_3$
IV	$Y = a + b(\ln X_1)$
V	$Y = a + b(\ln X_2)$
VI	$Y = a + b(\ln X_3)$
VII	$\ln Y = a + bX_1$
VIII	$\ln Y = a + bX_2$
IX	$\ln Y = a + bX_3$
X	$\ln Y = a + b(\ln X_1)$
XI	$\ln Y = a + b(\ln X_2)$
XII	$\ln Y = a + b(\ln X_3)$
Two predictor variables:	
XIII	$Y = a + bX_1 + cX_2$
XIV	$Y = a + bX_1 + cX_3$
XV	$Y = a + bX_2 + cX_3$
XVI	$Y = a + b(\ln X_1) + c(\ln X_2)$
XVII	$Y = a + b(\ln X_1) + c(\ln X_3)$
XVIII	$Y = a + b(\ln X_2) + c(\ln X_3)$
XIX	$\ln Y = a + bX_1 + cX_2$
XX	$\ln Y = a + bX_1 + cX_3$
XXI	$\ln Y = a + bX_2 + cX_3$
XXII	$\ln Y = a + b(\ln X_1) + c(\ln X_2)$
XXIII	$\ln Y = a + b(\ln X_1) + c(\ln X_3)$
XXIV	$\ln Y = a + b(\ln X_2) + c(\ln X_3)$
Three predictor variables:	
XXV	$Y = a + bX_1 + cX_2 + dX_3$
XXVI	$Y = a + b(\ln X_1) + c(\ln X_2) + d(\ln X_3)$
XXVII	$\ln Y = a + bX_1 + cX_2 + dX_3$
XXVIII	$\ln Y = a + b(\ln X_1) + c(\ln X_2) + d(\ln X_3)$

Table 4. Model Parameters and Related Statistics for Estimating Concentration (mg/ℓ) of Dissolved Constituents (Y) Using Current Streamflow (X_1, ℓ/sec) as Predictor Variable (Model: $Y = a + bX_1$)

Constituent (mg/ℓ)	Model Parameters		r*	S_E*		Sum of prediction residuals
	a	b		Model	Predicted	
Ca^{2+}	50.6317	-0.0229	-0.85	5.56	6.86	-10.23
Mg^{2+}	13.4855	-0.0057	-0.72	2.18	2.01	-27.70
Na^+	1.3937	-0.0016	-0.69	0.42	0.42	8.03
K^+	0.4913	-0.0001	-0.23	0.10	0.13	2.41
HCO_3^-	197.565	-0.0858	-0.82	23.74	22.08	-138.92
Cl^-	0.1864	-0.00003	-0.09	0.08	0.35	7.09
SiO_2	5.0612	-0.0027	-0.77	0.54	0.90	19.77
Total dissolved solids	190.179	-0.0920	-0.85	22.58	20.41	-122.03

* r is correlation coefficient and S_E is standard error of estimate.

Table 5. Model Parameters and Related Statistics for Estimating Concentration (mg/ℓ) of Dissolved Constituents (Y) Using Current Streamflow (X_1, ℓ/sec) and Specific Conductance (X_2, μS/cm) as Predictor Variables (Model: $Y = a + bX_1 + cX_2$)

Constituent (mg/ℓ)	Model Parameters			R*	S_E*		Sum of prediction residuals
	a	b	c		Model data	Predicted data	
Ca^{2+}	3.7706	-0.0035	0.1368	0.984	1.94	3.46	20.09
Mg^{2+}	-3.4576	0.0013	0.0494	0.94	1.10	2.20	-15.59
Na^+	-1.5724	0.0002	0.0084	0.85	0.31	0.28	6.20
K^+	0.5393	-0.0001	-0.0001	0.24	0.10	0.13	2.39
HCO_3^-	-5.2898	-0.0016	0.5920	0.985	7.33	8.23	5.72
Cl^-	0.1035	0.00002	0.0002	0.14	0.08	0.37	6.91
SiO_2	1.9047	-0.0008	0.0089	0.85	0.46	0.84	17.26
Total dissolved solids	4.8911	-0.0151	0.5407	0.977	9.27	7.72	2.14

* R is multiple correlation coefficient and S_E is standard error of estimate.

Table 6. Model Parameters and Related Statistics for Estimating Concentration (mg/ℓ) of Dissolved Constituents (Y) Using Current Streamflow (X_1, ℓ/sec), Specific Conductance (X_2, μS/cm), and Mean Streamflow on the Previous Day (X_3, ℓ/sec) as Predictor Variables (Model: $Y = a + bX_1 + cX_2 + dX_3$)

Constituent (mg/ℓ)	Model parameters				R*	S_E*
	a	b	c	d		
Ca^{2+}	-6.0787	-0.00003	0.00014	0.1739	0.95	2.93
Mg^{2+}	3.5253	0.00001	-0.00009	0.0271	0.78	1.54
Na^{+}	-0.9534	-0.000002	-0.000004	0.0071	0.93	0.13
K^{+}	0.1487	-0.0000007	-0.000002	0.0014	0.53	0.10
HCO_3^-	4.7719	-0.00007	-0.00002	0.5680	0.986	4.67
Cl^-	0.2865	-0.000006	0.000002	0.0013	0.68	0.19
SiO_2	3.8966	-0.000007	-0.000003	0.0039	0.62	0.46
Total dissolved solids	-0.6559	-0.00015	0.00007	0.5635	0.990	4.19

* R is multiple correlation coefficient and S_E is standard error of estimate.

Table 7. Model Parameters and Related Statistics for Estimating Dissolved Constituents Using Components of Hydrograph. X_1 is Base Flow (ℓ/sec), X_2 is Storm Flow (ℓ/sec), and X_3 is Specific Conductance (μS/cm) (Model: $Y = a + bX_1 + cX_2 + dX_3$)

Constituent (mg/ℓ)	Model parameters				R*	S_E*		Sum of prediction residuals
	a	b	c	d		Model data	Predicted data	
Ca^{2+}	3.4044	-0.0090	0.1378	0.0056	0.984	1.97	3.71	23.62
Mg^{2+}	-3.3459	0.0030	0.0491	-0.0017	0.94	1.12	2.29	-16.67
Na^{+}	-1.7233	-0.0009	0.0088	0.0012	0.85	0.31	0.32	6.68
K^{+}	0.5626	0.00004	-0.0002	-0.0002	0.25	0.10	0.14	2.31
HCO_3^-	-5.0626	0.0018	0.5914	-0.0035	0.985	7.46	8.41	4.56
Cl^-	0.1904	0.0005	0.000004	-0.0005	0.26	0.08	0.37	6.77
SiO_2	1.9188	-0.0007	0.0089	-0.0001	0.85	0.47	0.85	17.22
Total dissolved solids	4.5247	-0.0206	0.5417	0.0056	0.977	9.44	8.01	5.67

* R is multiple correlation coefficient and S_E is standard error of estimate.

The inclusion of mean streamflow on the previous day as yet another variable produced a slight improvement in the fit, as indicated by the R and S_E values in Table 6. These improvements occurred primarily in K^+ and Cl^-, as shown by the R values and the sum of prediction residuals for these constituents.

The models incorporating logarithmic transformation showed slightly higher R values in some cases, but were not preferred over those presented here. The slightly higher R values may be due to the slightly lesser deviations in the transformed data, rather than actual data (Brownlee, 1953). It was therefore decided to use models incorporating untransformed data only.

The use of the components of streamflow in deriving model equations for the 1973 data shows the best overall fit for all constituents. The R values for the modelled constituents range from 0.53 to 0.990 (Table 7). The R values for the model XXV in Table 6 had a range of 0.25 to 0.985.

The models based on the components of streamflow and specific conductance can therefore provide the most accurate estimates. Further work along these lines is in progress and consists of testing nonlinear predictive models for constituents which have multiple correlation coefficients lower than 0.90.

Acknowledgments

The assistance of Z. Fisera in collection of samples and of W. Chow in running computer analyses is gratefully acknowledged. Streamflow data were provided by Water Survey of Canada.

References

[1] Bathlahmy, N., 1972. "Hydrograph Analysis: A Computerized Separation Technique," Intermountain Forest and Range Experiment Station, Ogden, Utah, USDA Forest Service Paper INT-122, 19 p.

[2] Beke, G. J., 1969. "Soils of Three Experimental Watersheds in Alberta and Their Hydrologic Significance," Ph.D. Thesis, Univ. of Alberta, Edmonton, 456 p.

[3] Brownlee, K. A., 1953. "Industrial Experimentation," Chemical Publishing Co., Inc., New York, N.Y., 194 p.

[4] Golding, D. L., 1970. "Research Results from Marmot Creek Experimental Watershed, Alberta, Canada," In: Symposium on the Results of Research on Representative and Experimental Basins, International Association of Scientific Hydrology, Publ. 96:397-404.

[5] Hall, F. R., 1970. "Dissolved Solids-Discharge Relationships. 1. Mixing Models," Water Resources Research, 6:845-850.

[6] Hall, F. R., 1971. "Dissolved Solids-Discharge Relationships. 2. Applications to Field Data," Water Resources Research, 7:591-601.

[7] Hem, J. D., 1970. "Study and Interpretation of the Chemical Characteristics of Natural Water," U.S. Dept. of the Interior, USGS Water Supply Paper 1473, Washington, D.C., 363 p.

[8] Hewlett, J. D. and Nutter, W. L., 1969. "An Outline of Forest Hydrology," University of Georgia Press, Athens, 132 p.

[9] Jeffrey, W. W., 1965. "Experimental Watersheds in the Rocky Mountains, Alberta, Canada," In: Representative and Experimental Areas, International Association of Scientific Hydrology, Publ. 66:502-521.

[10] Johnson, N. M., Likens, G. E., Bormann, F. H., Fisher, D. W. and Pierce, R. S., 1969. "A Working Model for the Variation in Stream Water Chemistry at the Hubbard Brook Experimental Forest," Water Resources Research, 5:1353-1363.

[11] Kirby, C. L. and Ogilvie, R. T., 1969. "The Forests of Marmot Creek Watershed Research Basin," Environment Canada, Canadian Forestry Service, Publ. 1259, 37 p.

[12] Pinder, G. F. and Jones, J. F., 1969. "Determination of the Groundwater Component of Peak Discharge from the Chemistry of Total Runoff," Water Resources Research, 6:1353-1363.

[13] Pionke, H. B., Nicks, A. D. and Schooff, R. R., 1972. "Estimating Salinity of Streams in the Southwestern United States," Water Resources Research, 8:1597-1604.

[14] Singh, T., 1976. "Yields of Dissolved Solids from Aspen-Grassland and Spruce-Fir Watersheds in Southwestern Alberta," Journal of Range Management, 29:401-405.

[15] Singh, T. and Kalra, Y. P., 1972. "Water Quality of an Experimental Watershed during the Calibration Period," Paper presented at the 19th Pacific Northwest Regional Meeting, American Geophysical Union, Vancouver, B.C., Abstract in Transactions, 54:139.

[16] Singh, T. and Kalra, Y. P., 1975. "Specific Conductance Method for In Situ Estimation of Total Dissolved Solids," Journal of American Water Works Association, 67:99-100.

[17] Steele, T. D., 1973. "Simulation of Major Inorganic Chemical Concentrations and Loads in Streamflow," USGS Computer Contribution PB-222 556, Washington, D.C.

[18] Steele, T. D., 1976. "A Bivariate-Regression Model for Estimating Chemical Composition of Streamflow or Ground-Water," Hydrological Sciences Bulletin, 21:149-161.

[19] Stevensen, D. R., 1967. "Geological and Groundwater Investigations in the Marmot Creek Experimental Basin of Southwestern Alberta, Canada," M.Sc. Thesis, Univ. of Alberta, Edmonton, 106 p.

[20] Telang, S. A., Baker, B. L. and Hodgson, G. W., 1976. "Water Quality and Forest Management: Chemical and Biological Processes in a Forest-Stream Ecosystem of the Marmot Creek Drainage Basin," Environmental Sciences Centre (Kananaskis), University of Calgary, Calgary, Alberta.

EFFECTS OF THE HYDROMETEOROLOGICAL FACTORS ON SUSPENDED SEDIMENT OUTPUT OF A CATCHMENT IN NEW ENGLAND

By

A. S. Zakaria
Department of Geography
National University of Malaysia
Kuala Lumpur, Malaysia

Abstract

The suspended sediment output of streams is the product of the interaction of a complex set of interrelated environmental factors such as hydrometeorological conditions, topography, vegetation, lithology and land use management. This study endeavours to observe the relationship between the suspended sediment output from a drainage basin and one of the environmental variables, namely the hydrometeorological factors. The experimental catchment selected is wholly underlain by granitic bedrock. Observations were made once a week and additional samples were collected before, during and after rain events. The study shows that the hydrometeorological factors by themselves can only explain less than 40 percent of the total variance observed in the suspended sediment output over the study period which indicates that other environmental factors are also important. In terms of these factors, the study shows that runoff variables are better related to the sediment concentration than the rainfall variables. This is inevitable since runoff parameters condense all the sediment producing factors which include rainfall variables, while the rainfall variables express the quantity of moisture and energy input without gauging their effectiveness in relation to changing catchment cover with time. This time-dependent catchment condition is well accommodated, though not fully, in the runoff variables. Among the runoff variables, peak discharge prior to observation is prominent. Instantaneous discharge, often found to be the most significant independent variable in other studies, is not as important in the present situation. Rainfall intensity which indicates the energy input, is the most important rainfall variable. Both principal component and multiple regression analysis show that in term of the hydrometeorological variables, the main direction of variance operating in the suspended sediment model can be explained by the antecedent flood condition, energy input and season of the year.

Introduction

The study of fluvial geomorphology has, in recent years, demonstrated two new trends after several phases of growth over the past century. Firstly, there is a marked increase in the study of contemporary processes in contrast to the earlier over-concentration on studies of denudation chronology and secondly, there is a growing interest in the use of small experimental catchments as the basic study unit. These process-orientated studies are devoted mainly to the examination of sediment (including solute) yields and morphometry, generally, in relation to the rainfall and runoff characteristics. They are accompanied by various complementary techniques such as the use of the whole catchment (e.g., Imeson, 1970 and Edwards, 1973), channel reaches (Bevan, 1970 and Loughran, 1973) or reservoirs (e.g., Glymp, 1954; Roehl, 1962). The study of sediment yields and attempts to relate these to a number of hydrometeorological and watershed characteristics have taken two main directions namely, short-term and long-term investigations. The former may range from one event to a year period while the latter may extend over several years. Apparently, general observations on the variability of sediment yield have appeared frequently in literature while detailed studies are few, for the principle reason that most large experimental catchments are heterogeneous in terms of their physical characteristics. This paper endeavours to describe the variability of sediment yield from a small catchment in relation to changes in hydrometeorological conditions. In terms of the general geomorphological studies, this work is contemporary process-oriented which involves the whole catchment based on a short-term temporal observation extending over one-year period.

Theoretical Basis of the Study

The sediment output from a drainage basin is the net result of the interaction of a complex set of interrelated environmental processes within the basin ecosystem which tends to represent an energy balance and to include series of quasi-equilibrium states. Guy (1964) divided these factors into active and passive groups. The active forces are such climatic elements as the nature of precipitation, runoff, infiltration, temperature and wind. The passive forces are basin characteristics such as rock type, soil, topography, vegetation and land use. To these elements, human activities must be added.

Past works on the subject have indicated that the relative importance of the individual element in influencing the variations in sediment outputs of streams depends upon the scale of investigation. In space, on the global scale, when the widest possible range of variations are taken into consideration, climatic variables have been shown to explain most of the long term differences in mineral outputs. Fournier (quoted by Douglas, 1969) relates the mineral outputs, on a world scale, to rainfall and relief while Strahler (1957), Schumm (1965) and Strakhov (1967) have inferred that mineral outputs vary in a single direction with climate. Langbein and Schumm (1958) express the sediment yields of streams as the function of the effective mean annual precipitation. Studies made in areas with constant climatic factors have shown that most of the spatial variations in sediment outputs are related to the geological and geomorphological differences (Schumm, 1954; Maner, 1958; Schumm and Hadley, 1963; and Lustig, 1965). Reducing the scale to areas of similar geology and lithology, vegetation and land use, climate seems to explain most of the variations (Anderson, 1954, 1957; Williams, 1964; Mustonen, 1967). In time, over millions of years, the changing of soil and climatic conditions, topography and surface cover are likely to be the important elements. By altering the scale of investigation in terms of time or space or both, the influence of one or more elements can be considered as constant. The sediment outputs from a drainage basin, therefore, reflect the variations in the relationship between active and passive forces and the nature of and variations in the interrelationships between all the variables that exist within the system. Ideally, the study of sediment outputs from a catchment should include investigations of the relationships between the environmental factors and sediment outputs and the effect of changes in the interrelationship of the independent variables on sediment output variations. However, with the present limited and scattered knowledge on the processes operating within and between the environmental systems, this is not practical. It is necessary to select one or more elements from the total complex. In making this selection it is, however, impossible to isolate the effects of a single factor and hence considerations are made to the other closely related factors.

The approach of this study is based on the premise that the suspended sediment outputs of streams are essentially a function of a wide range of hydrometeorological conditions, topography, vegetation, lithology and landuse management. In this context, the sediment outputs measured at the gauging site are taken as the dependent variables. The others above are equated as independent variables.

Physiographic Background of the Experimental Catchment

Sandy catchment is a subcatchment of the Dumaresq basin located north of Armidale in New South Wales (Fig. 1). The catchment is developed on adamellite with grey-brown podzolic soil association. In an attempt to stress the significance of the interrelationships between the physiographic elements

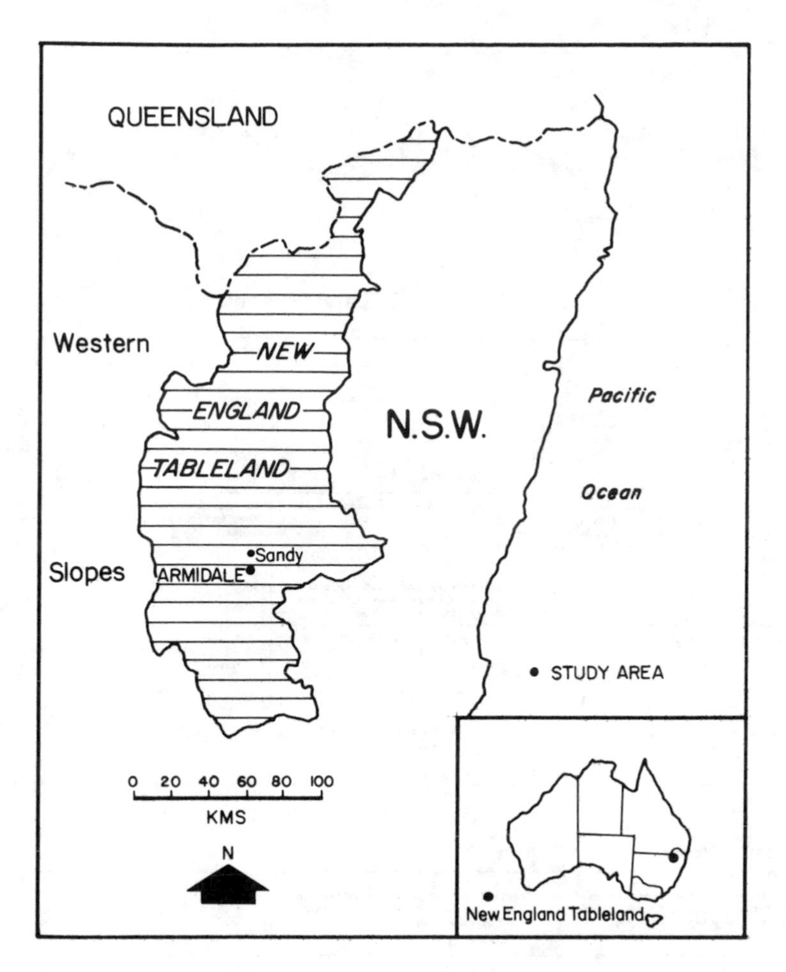

Fig. 1. The location of the Sandy experimental catchment

in influencing the sediment outputs, the integrated land system approach (Christian and Stewart, 1953) was employed (Tables 1 & 2, Fig. 2). The related quantitative indices of the catchment were calculated (Table 3) and channel characteristics were summarized in Table 4.

Land systems attempt to integrate the characteristics of the physical environment. It is well-known that these characteristics individually will influence the hydrological processes and the sediment generation in a particular catchment. Thus, the rock type underlying a catchment will, through its permeability characteristics, affect the proportion of baseflow and quickflow runoff components. Furthermore, the nature of the rock type will affect the load characteristics of the discharge. However, the bedrock may not be exposed extensively

Table 1. Duval Land System

LAND UNITS	SOIL[1]	VEGETAL COVER[2]
1. Granitic dome: Dome-like mass with granitic outcrops consisting of large whalebacks, tors and pavements. Short convex slopes up to 2° and approximately 100 m long with good surface and sub-surface drainage.	Clay loam or sandy and gritty skeletal soil with incipient B horizon.	Dry sclerophyll woodland without the mesomorphic shrub layer. E. viminalis and E. caligirosa are dominant with the mean height of about 20 m. The secondary layer of about 2-6 m composed of almost entirely of Casuarina, Acacia and Banksia spp. The ground is fully covered with native grass in between granitic outcrops
2. Hillslopes: Rugged slopes ranging from 15° to 30°. Their length varies from 250 m to 500 m. The slopes are dominated by granitic outcrops consisting of whalebacks, tors and pavements. Good surface and sub-surface drainage.	Dominated by grey-brown podzolics with small areas of skeletal soils on the very steep slopes. The grey-brown podzolics are with loamy sand or sandy loams and gritty A horizon and sandy and gritty clay B horizon. The A horizon is very compacted and indurated and once it is exposed rill and gully erosions become prominant.	Similar to unit 1 with the thinning of trees and secondary layer down the lope slopes.
3. Footslopes and Plain: The footslopes are smooth and gentle averaging 6° with very sparse and isolated outcrops of small granitic boulders and tors. Surface and subsurface drainage are good with piping and seepages evidently illustrated. The accumulation of debris is evident on the lower footslopes. The plains are waterlogged with impeded vertical movement of water. The slope varies from 0° to 3° and the width from 500 m to 600 m.	The higher slopes are dominated by grey-brown podzolics similar to the hillslopes type and the lower slopes are with yellow podzolics. The yellow podzolics are characterised by loamy sand A horizon and clay loam slightly waterlogged B horizon. A horizon is compacted and indurated with slightly waterlogged character.	Thick grass cover with sparse and isolated pockets of E. viminalis and E. coligirosa.
4. Drainage tracks: Up to 10 m wide with shallow trench-like channels and grassy depressions up to 2 m wide. The beds are characterised by pools and riffles and grass covered in the lower reaches. Bank erosion is prominent in some places where the process of bankslip is not impeded by any vegetation cover.	Yellow podzolics underlain all the drainage tracks except the main creek within the plain land unit and the swampy areas where alluvial soils are found. The alluvial soils are characterised by clay loam subangular blocky earthy top soil structure and loamy sand granular subsoil structure.	Grass covered except where bank erosion and gullying are active.

1. See Gibbons and Hallsworth (1953); Hallsworth et.al. (1952); Premier's Dept. N.S.W. (1953); Warner (1963) Jessup (1965) and McConnell et.al. (1966).

2. See Williams (1963); Anderson (1956) and Plummer (1965).

Table 2. The Brothers Land System

LAND UNITS	SOIL[1]	VEGETAION COVER[2]
L. Interfluves: The unit consists of convex crest and rounded hills. The hills vary in height from 110 m to 150 m while the convex crests are berween 275 m and 500 m wide. They are higher and broader on the basaltic parent materials. Generally the surface and subsurface drainage are good.	Two main types of soil were observed: the chocolate on the basaltic parent material and grey-brown podzolics on the granitic and sedimentary parent material. The chocolate soils are characterised by clay loam or light clay A horizon with small to moderate numbers of basaltic pebbles and stone on the surface and in the horizon, and clay B horizon with moderate to strong grade blocky structure and contain small to large numbers of basaltic pebbles and stones. The grey-brown podzolic on granitic parent materials are with loamy sand or sandy loam A horizon and sandy and gritty clay loam or sandy and gritty clay B horizon. The grey-brown podzolics on the sedimentary parent material are with sandy loam to clay loam A horizon and clay B horizon.	Scattered pockets of timber (E. viminalis and E. coligirosa) with fairly thick grass covered ground surface. Small areas in the upper and lower parts of the catchment are utilised for fodder cultivation.
2. Footslopes: characterised by gentle slopes varying from 1° on the granitic parent rocks to about 7° on the sedimentary rocks. They vary in length from 250 m to 750 m. The valleys are narrower and steeper on the sedimentary rocks. The drainage are generally good.	On the higher basaltic parent materials transitional soils developed with trace of basaltic pebbles. The A horizon is loam or clay loam and the B horizon is of clay. The lower slopes are underlain by 4 soil types - chocolate, grey-brown and yellow podzolics and alluvial. The chocolate soils are similar to those in unit 1, the grey-brown podzolics on the granite are similar to those found in the Duval land system and those on the sedimentary parent rocks are similar to those found in the interfluve land unit above but with deeper profile. The yellow podzolics are generally similar to those found in Duval land system with numerous angular fragments of weathered parent materials scattered throughout the profile.	Scattered pockets of timber (E. viminalis and E. coligirosa) with native grass, semi improved and improved pastures coverage.
3. Drainage tracks: up to 30 m wide with shallow trench-like water channels up to 2 m wide. Bank erosions are widespread along the main channel. The channels beds are characterised by grass covered riffles and gravelly pools in the lower reaches.	Yellow podzolics underly most of the drainage tracks except along the lower reaches of the main stream where alluvial soils are developed.	Fairly thick grass cover except along the reaches where bank erosion is active.

1. See Table 1.

2. See Table 1.

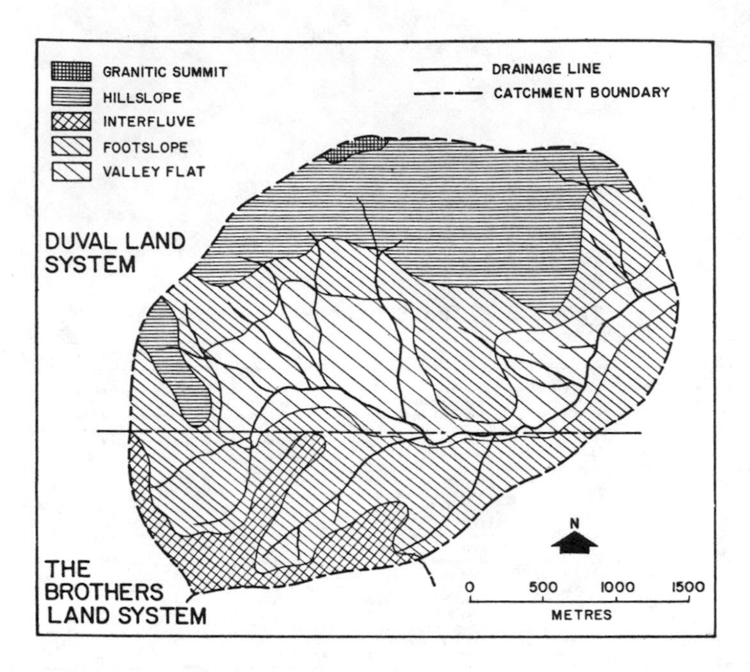

Fig. 2. The land systems of Sandy Catchment

Table 3. Morphometric Indices

Indices	Sandy
Form Factor (Horton)	0.61
Compactness Coefficient	0.37
Elongation Ratio	0.85
Size (km^2)	7.64
Relief Ratio	0.10
Total Stream Length (km)	17.25
Drainage Density (km/km^2)	2.26

Table 4. Channel Characteristics

Features	Sandy
Unprotected well defined channels (km)	4.00
Protected channel mainly with trees along the banks (km)	2.50
Channels with only grass on their banks (km)	0.75
Grassy depressions (km)	10.00
Total Stream Length	

so that its influence is expressed indirectly through the character of soils and superficial deposits. Where exposed, for example on slopes, these can influence sediment production and their infiltration characteristics can affect the relative significance of throughflow, interflow and overland flow contributing to the runoff pattern of the catchment. These infiltration characteristics will of course be greatly influenced by the vegetation or the land use of the surface not only because of the direct influence of vegetation, for example in the interception of precipitation, but also because of the way in which vegetation affects the infiltration properties of the soil through the development of root networks. In any basin with specific characteristics of rock type, soil and vegetation, the slope of the basin will also exercise an influence simply as a result of the potential energy which is afforded by high relief and which influences stream velocity and therefore sediment production. Although relief and slope are both important topographic characteristics, the most significant topographic feature of the drainage basin is perhaps the stream channel network. This is simply because channel flow is more rapid and capable of producing more mechanical sediment than is sheet flow, interflow or groundwater flow. Therefore, the extent of the stream network and the character of the channels in that network will affect the pattern of runoff and suspended sediment production.

The relatively easily-eroded, granitic-derived, podzolic soils of Sandy with the good internal and external drainage and steep slopes should be susceptible to a relatively high rate of erosion and to the production of high peak runoff.

This is, however, very much influenced by the nature and amount of precipitation. Taking the amount of precipitation as constantly high, the processes of slope erosion are significantly impeded by the high percentage of woodland (Fig. 3) and grass cover on the granitic dome and hillslopes and by the dense grass cover on the lower footslopes and the plain.

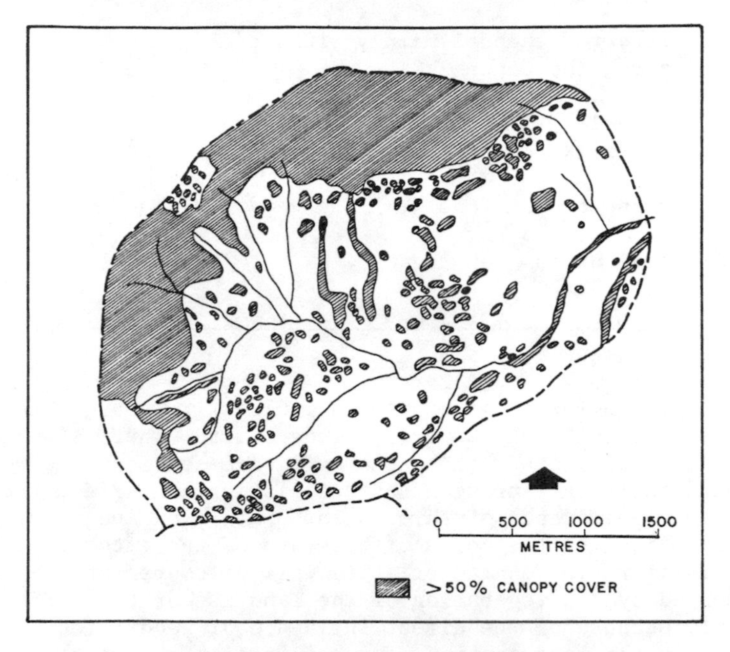

Fig. 3. Canopy cover of Sandy

The cover on the interfluve and slopes will increase the lag time period of runoff concentration. The cover on the lower footslopes and plain will increase the probability of entrapment of the particulate materials besides increasing the lag time period. While the quantity of particulate sediments are expected to be relatively low, the runoff amount is expected to be high with a gradual rise in the hydrograph reflecting the effect of the high canopy and high percentage of vegetal cover.

In addition to the above physiographic variables, the morphometric factors are also important in the sediment production and transportation processes. They are also important in elucidating some aspects of the rainfall-runoff relationships. Table 3 provides the morphometric values for the catchment.

The form factor (Horton, 1932) provides some indication of the runoff peak. High values are associated with high runoff while low values are related to low runoff. The compactness coefficient (Gravelius, 1914) reflects the rate of peak discharge from the basin. Low values suggest rapid rate of discharge and vice versa. However, the situation can be significantly altered by the surface conditions (soil permeability, slope and vegetal cover) of the catchment and the stream network. The elongation ratio (Schumm, 1956) which expresses the dimensionless ratio of the diameter of a circle having the same area as the basin and the maximum length of the basin parallel to the principal drainage line, reflects the time of concentration of the floodflow. Values close to unity are related to short concentration.

The size of a drainage area has a considerable effect on sediment yield. Total yields are sometimes found to be inversely related to the size of a basin (Gottschalk, 1964), reflecting the probability of entrapment and lodgement of particles with an increase in the drainage size. Branson and Owen (1970), Maner (1958) and Schumm (1954) found that the annual sediment yields per unit drainage basin area is a positive function of the relief ratio which is the dimensionless ratio between basin relief and its maximum length. High values are associated with high sediment discharge and vice versa. The drainage density which Horton (1932) expressed as the ratio of total stream length to catchment area together with the nature of the stream channels are indicative of the efficiency of the sediment production and runoff generation processes. High drainage density with well-defined channels suggests greater runoff (Gottschalk, 1964) and sediment yields, depicting not only the importance of the morphometric variables but also channel characteristics. Figure 4 and Table 4 illustrate the channel characteristics of Sandy catchments.

Climatic Elements

The climatic pattern of the study areas is very much influenced by their elevation, inland location and the prevailing pressure systems and their associated airflow types (Thompson, 1970, 1973a, 1973b). Being on the eastern periphery of the Australian continent they lie within the zone of thermal contrasts, the juxtaposition of warm moist Pacific air to the east and the dry continental air to the west with its significant seasonal temperature differences. Their altitude (over 1,000 m) and inland location (c. 132 km from the coast) have significantly influenced the maximum and minimum temperatures, the amount of rainfall and the intensity of windspeed. The atmospheric dynamics caused considerable variations in the temperature pattern and pronounced irregularities in rainfall distribution.

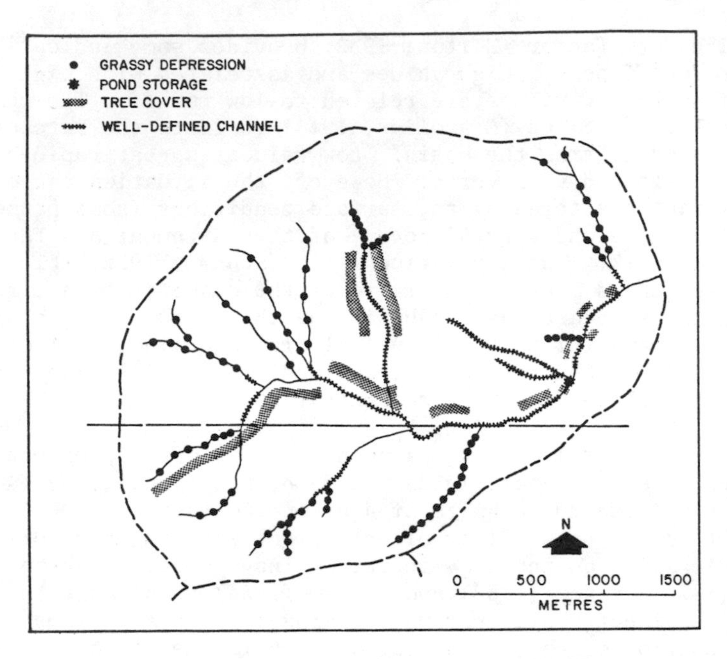

Fig. 4. Channel characteristics of Sandy

Variation in the summer temperatures are significantly associated with the types and location of the pressure systems. The preponderance of anticyclogenesis during this season brings calm and clear nights facilitating considerable amounts of terrestrial radiation and nocturnal chilling which occasionally resulted in the extreme diurnal temperature range of up to 20°C. The mean minimum temperatures in January and February are close to 14°C and occasionally fall to 8°C. When the anticyclonic ridges migrate eastward, onshore moist maritime airstreams (anticyclonic maritime) are generated. The resultant orographic turbulence caused cloudy wet weather with moderate temperatures. Temperature and rainfall data for the study area are summarized in Tables 5, 6 and 7.

Since the operation of other factors in the sediment production and runoff processes is dependent on the climatic elements, it is therefore expected that the sediment yield and runoff pattern will vary in accordance with variations in the rainfall and temperature patterns. The winter/spring rains are expected to encourage more runoff since the rate of evapotranspiration is low during these periods. Spring rainfall is critical for the growth of vegetation which will afford the ground surface more protection against erosion. The effectiveness of the high summer rain is reduced quite considerably by the evapotranspiration losses (Thompson, 1970) which are related to the high summer temperatures. The irregularity and

Table 5. Mean Monthly Rainfall, Armidale (1900-1966)

Month	Amount	% of total
January	92.40	12.20
February	85.45	11.28
March	65.75	8.68
April	43.00	5.68
May	39.25	5.18
June	55.70	7.35
July	50.75	6.70
August	50.43	6.66
September	49.68	6.56
October	64.48	8.51
November	76.75	10.13
December	83.50	11.02

Table 6. Distribution of Rainfall and Associated Pressure Systems, Armidale (Data Source: Thompson, 1970)

Pressure system types	% of total rainfall	Intensity (mm/day)	% of total raindays
Anticyclonic Ridges	33.60	6.5	41.1
Tropical cyclones	0.50	3.0	1.4
Tropical troughs	14.60	12.0	10.4
Depression	7.90	11.9	5.8
Wave Depression	5.80	15.0	3.2
Troughs from depressions and wave depressions	15.50	7.0	17.2
Troughs from subpolar wave depression	22.10	8.3	20.9

Table 7. Mean Temperature of Armidale (°C)

	Mean daily maximum	Mean daily minimum	Main daily	Mean daily range
January	27.1	13.6	20.4	13.5
February	26.4	13.2	19.8	13.2
March	24.1	11.2	17.6	12.9
April	20.2	7.6	13.9	12.6
May	16.2	4.0	10.1	12.2
June	13.0	1.6	7.3	11.4
July	12.2	1.0	6.6	12.8
August	14.0	1.2	7.7	12.8
September	17.7	3.8	10.8	13.9
October	21.4	7.3	14.3	14.1
November	24.5	10.2	17.3	14.3
December	26.3	12.4	19.3	13.9

unreliability of any rainfall type render it difficult to establish any seasonal rainfall-runoff and sediment yield patterns. Any specific rainfall type, save those associated with the tropical lows, can be expected at any time of the year.

Method of Analysis

1) Rainfall quantity. The basic instrument used for this measurement is Wilh Lambrecht KG-Gottingen type 1509 automatic monthly recording rain-gauge. It was located at the center of the catchment.

2) Stream discharge. Stream discharge measurements were made using OTT Current Meter 14701 Type X and Bristol Water Level recorder. A rating curve describing the river discharge and river stage relationship was established.

3) Suspended sediment. Weekly samples were obtained with the use of US DH-48 depth integrated hand sampler. Additional samples were collected during and after every rain event.

The amount of suspended load was determined by filtration as described in Aerospace Recommended Practice ARP 785 (Society of Automotive Engineers, 1963) and Appendix II of ASTM Method D2276-65T (American Society for Testing and Materials 1968). The filter equipment used are those illustrated by Harmon (1965). The total suspended sediment load was estimated based on the combination of computed (using rating-curve) and observed suspended sediment concentration.

4) Statistical analysis. The suspended sediment concentration/independent variable relationships will be described by employing the stepwise multiple regression and principal component analytical approach.

Throughout this chapter, suspended sediment concentration will be utilized as the dependent variable. It is felt that the relationship between suspended sediment and water discharge is more meaningfully expressed by using suspended sediment concentration rather than load because the suspended sediment load will inevitably correlate significantly with water discharge (Walling, 1974a). Guy (1964) had shown that a series of samples of similar suspended sediment concentration and a wide range of water discharges are found to give a perfect and significant correlation when suspended sediment load values were related to water discharge values. When the suspended sediment concentrations and loads for Sandy were related to the water discharge (Figs. 5 and 6), a good correlation was obtained between water discharge and suspended sediment load. The independent variables are listed in Table 8.

Relationships between Hydrometeorological Parameters and Suspended Sediment Concentration

The total observation. Data collected during the study period were analysed at two levels. The first analysis involved the data for the total one-year observation which included both the weekly and the additional information gathered during rain events. The main objective of the analysis was to look at the general relationship between the dependent and independent variables, if any. The second level of analysis involved those data collected only immediately before and after and during rain events. The aim of the second analysis was to see the significance of the rainfall variables in the suspended sediment model.

During the period under study, the concentrations observed in Sandy were generally low, even during storm events. The values ranged from 0.2 mg ℓ^{-1} to 104.4 mg ℓ^{-1}.

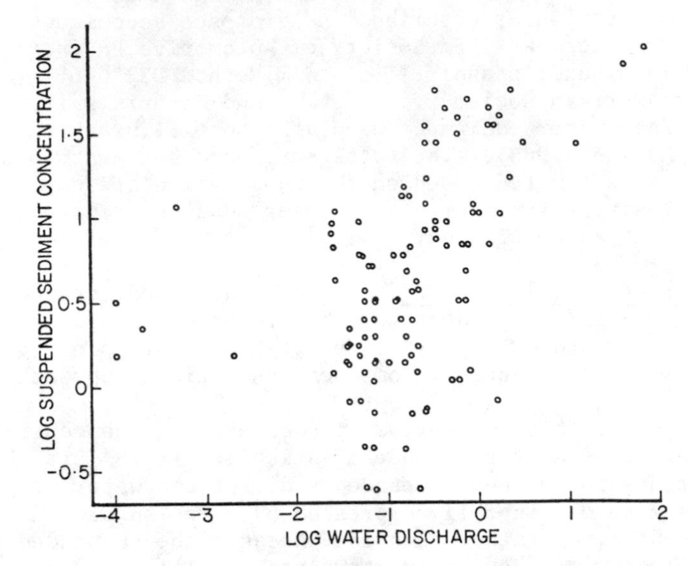

Fig. 5. Relationship between suspended sediment concentration and water discharge

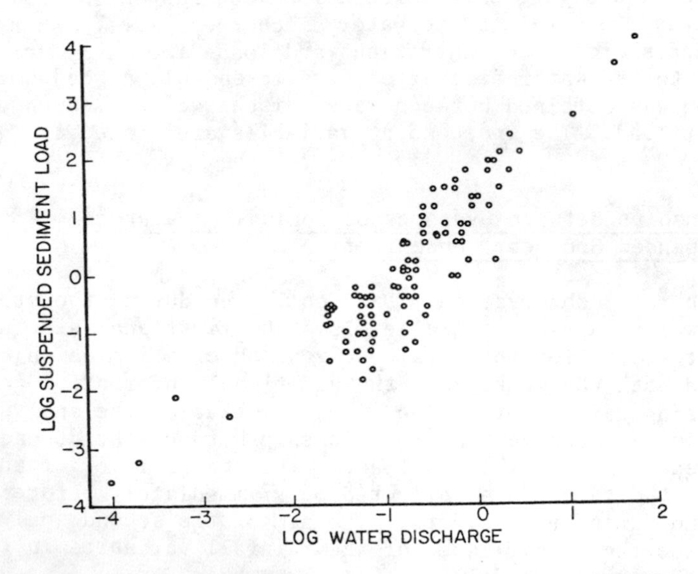

Fig. 6. Relationship between suspended sediment discharge and water discharge

Table 8. The Independent Variables Selected

Rainfall variables

- Total rainfall 48 hours before sampling (TP3)
- Maximum 15-min. rain intensity 48 hours before sampling (INT7)
- Maximum 30-min. rain intensity 48 hours before sampling (INT8)
- Maximum 60-min. rain intensity 48 hours before sampling (INT9)
- Duration of rain 48 hours before sampling (DUR3)
- Antecedent precipitation index (API)
- Product of KE (Kinetic energy of the storm 48 hours before sampling (KE)) and amount of max 30-min rain 48 hours before sampling (KEA)

Runoff variables

- Flood intensity of the hydrograph rise before the sampled hydrograph (FINT)
- Ratio of FINT and period of time between the passage of the hydrograph peak and sampling (FIT)
- Minimum discharge before sampling (MQBS)
- Absolute discharge (Q)
- Maximum discharge before sampling (PQBS)
- Flood intensity of the sampled hydrograph (FISH)
- Maximum discharge of sampled hydrograph (PQSH)

Temperature variable

- Water temperature (TEMP)

Concentration of over 100 mg ℓ^{-1} was recorded only once after a long intense spring storm (30.10.74 to 2.11.74). The suspended sediment concentrations exceeded 50 mg ℓ^{-1} only on four occasions and, on 86 other occasions of the total 116 observations, the values were less than 10 mg ℓ^{-1}. Concentrations were generally low in autumn and winter and increased with the coming of spring when most of the high concentration values were recorded towards the end of the season and early summer. Both the maximum and the minimum concentrations were much smaller than those reported by Burkhardt (1967) and Bevan (1970) for small rural catchments around Armidale. These differences can be explained by the contrast in the basin characteristics, hydrological regime and sampling and analytical techniques adopted. The different basins not only react differently to the agents of denudation but also to different combinations of these agents. Dissimilarity in hydrological regime results in different sediment producing processes. Hence contrast in suspended sediment concentrations is inevitable. The sampling procedure and experimental design adopted and the equipment used in the field by Burkhardt (1967), Bevan (1970) and the author rendered more frequent sampling impractical. Hence, the maximum observed values may not be the 'true' maximum values although presumably the minimum approximate 'true' minimum values. Based on the trend of concentration and water discharge hydrographs rise (Fig. 7) there is a high probability that maximum concentration falls

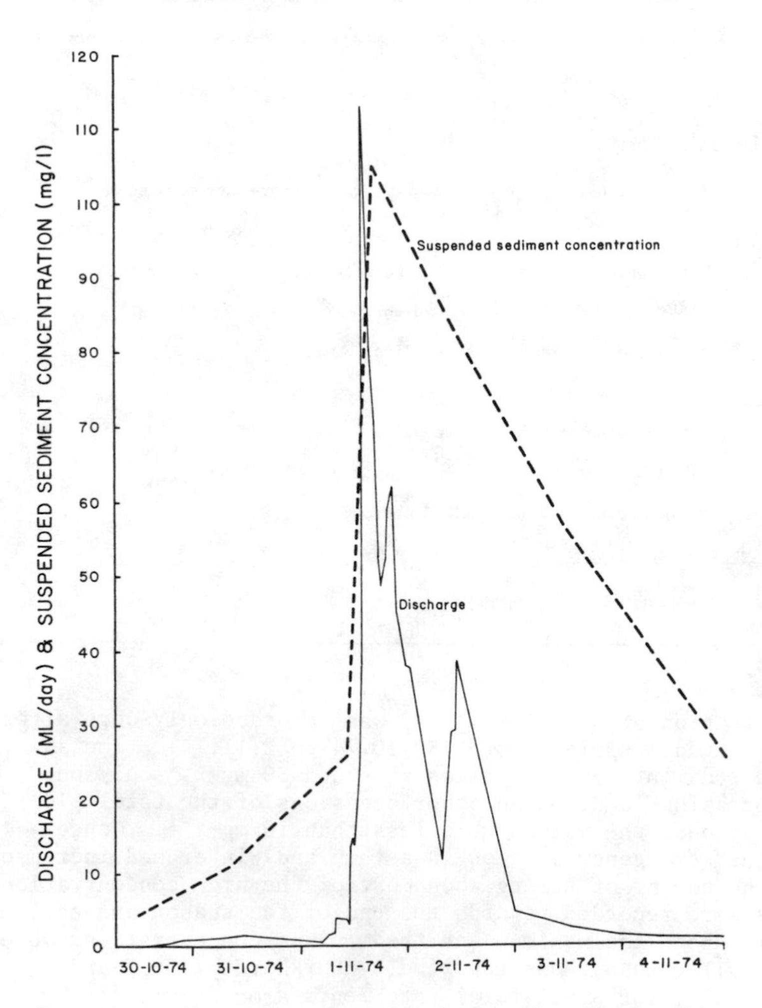

Fig. 7. Sedimentograph and hydrograph relationships during storm events in Sandy

between the graphed sediment concentration peak and the following plotted point. Therefore, it is possible that the maximum suspended sediment concentration is somewhere near, if not exceedint 200 mg ℓ^{-1}. Of equal importance is the range of water discharge from which observations were made. A set of suspended sediment concentration data based on water discharge ranging from baseflow to maximum discharge is expected to differ from another set based on just from medium to high flows or any flow range that differs from the former, even though the observations were made during a similar period. Similarly, differences are expected for sets of data collected within two different periods or lengths of time although the water discharge range is similar. Different analytical methods adopted and different types of apparatus used undoubtedly cause some degree of variance in the result obtained. With low sediment concentration values, any small difference in the value obtained is sufficient to double or triple or halve the real values. Unless the observations were made under identical conditions, comparisons of the suspended sediment concentration values are misleading and often meaningless.

In an attempt to account for the variations in the suspended sediment concentration in relation to water discharge, a scatter was plotted (Fig. 5), using the conventional log-log coordinates. A marked scatter is evident. The correlation coefficient of 0.40 though significant at the 0.1 percent level is small and hence precludes any possibility of obtaining a highly significant straightline relationship between these two variables. The wide scatter partly emphasis the severe limitations of the use of the general rating-curve (Guy, 1964) and also the multivariate controls upon the suspended sediment concentration fluctuations.

A proportion of the scatter can be explained in terms of the seasonal effect. There is a clear indication of higher suspended sediment concentrations in spring (Table 9) which are associated with the greater incidence of intense rainfall (Table 10). All the spring (Sept.-Nov.) and summer (Dec.-Feb.) rises were associated with intense rainfall with KEA values more than 10. Only three such occasions were observed in autumn (March-May) and winter (June-Aug.). The dry and dusty summer conditions provide a greater availability of sediment for transportation. Summer low flows carry more sediment than low winter flows as the result of the variable amounts of algae transported by these flows. The frequency of 'chance' erosion increases with the greater mobility of cattle and sheep within the catchment. Hence at any given water discharge value, spring/summer concentrations are higher than those of autumn/winter. A similar pattern was observed by Guy (1964) in the United States and Hall (1967), Imeson (1970) Walling and Teed (1971) and Walling (1947a,b) in Great Britain.

TABLE 9: SEASONAL DISTRIBUTION OF MEAN OBSERVED SUSPENDED SEDIMENT CONCENTRATION-SANDY

Season	Mean Suspended Sediment Concentration (mg ℓ^{-1})	Number of Observations
Autumn	2.77	36
Winter	9.62	28
Spring	22.58	28
Summer	8.44	24

TABLE 10: NUMBER OF FLOODS AND KEA VALUES (Kinetic energy x amount of maximum 30-mins rainfall intensity within 48 hours before the observation) BY SEASONS.

	Spring/Summer	Autumn/Winter
Flood rise	22	13
KEA > 20	8	3
KEA > 10	15	3

The relationships between the trends of the sediment concentration and water discharge rises and falls accommodate some of the scatter. Figures 7 and 8 illustrate considerable lag periods between the sediment concentration and water discharge peaks. The concentration peak occurs several hours after the water discharge peak so that sometimes, between the peaks, there is an inverse relationship between the parameters of discharge and concentration. The severity of this effect on the scatter is dependent on the duration of the lag period. In the case of a short lag period between the water discharge and sediment hydrograph, the effect is not so apparent. However, if the lag period extends for days as observed by Lewis (1912) and reported by Gregory and Walling (1973), the effect

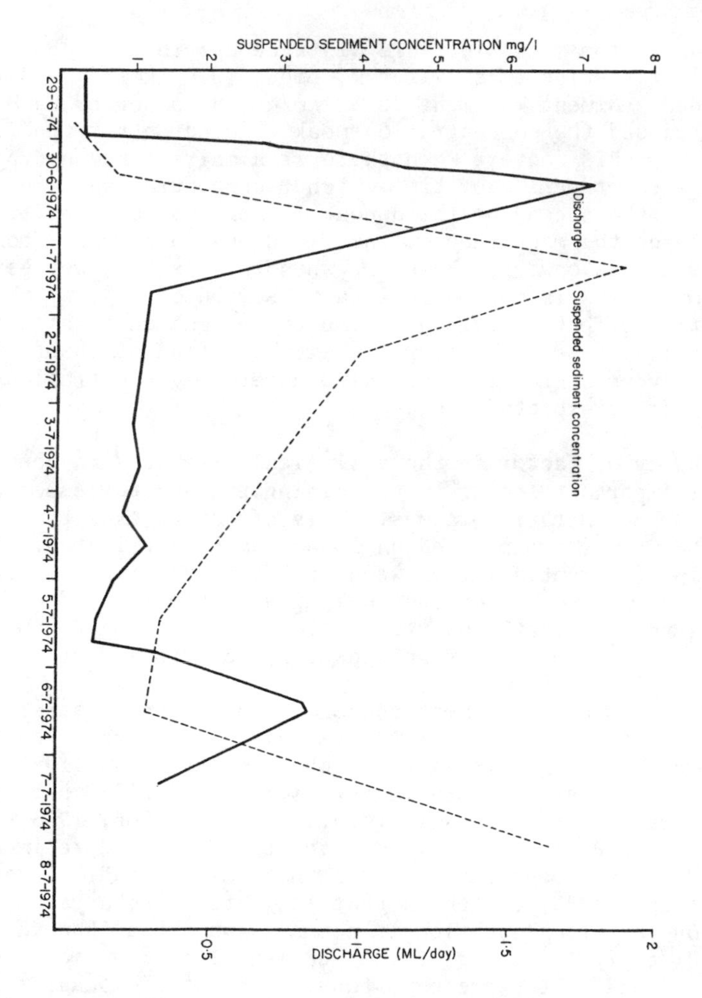

Fig. 8. Sedimentograph and hydrograph relationships during storm events in Sandy

invalidates a straight line rating relationship between discharge and suspended sediment concentration. Consequently, the hysteretic effects will be significant. A similar effect is expected, though to a lesser degree, for all hydrograph rises in the present study of the Sandy catchment. The trend of the sediment concentration rise during a storm hydrograph (Figs. 7 and 8) shows that the concentration increases at a rate similar to stream flow on the rising stage but decreases less rapidly than the stream flow on the falling stage. Hence, the concentration for a given discharge value during storm events in Sandy is higher on the falling than on the rising stage. Such a situation indicates that the suspended sediment originates from a considerable distance upstream. The differences in the velocity of water movement (0.61 km/hr) and the suspended sediment movement (0.32 km/hr) as observed by Heidel (1956) caused the concentration peak to occur after the flood crest. In this context when there is a multipeaked hydrograph within a short period of time, higher concentrations are expected on the second or third peak rather than on the first even though the magnitude of the flood peak decreases from the first peak onward. However, when the distance between the flood peaks is considerable as observed in Fig. 8, the exhaustion effect is evident. The concentration is highest on the first rise. Therefore, over the period of observation, at any given water discharge value numerous concentration values can be expected.

The error factor in the analytical processes may be another important factor in explaining the scatter especially with low concentration values. This effect is exaggerated when the values are plotted on logarithmic coordinates. The error in differentiating between such a narrow range of concentration, ±1 mg ℓ^{-1} on the analytical scale is sufficient to inflate or deflate low values of less than 1 mg ℓ^{-1} by 50-100 percent to cause significant hysteretic effects.

The suspended sediment concentration exponent obtained from the simple concentration/water discharge regression is 0.288 which is higher than those obtained by Bevan (1970) for the two small catchments adjacent to Sandy (0.2425 and 0.1549). Other works within Australia (e.g., Douglas, 1966, 1973; Loughran, 1968, 1969, 1973) reported the suspended sediment/water discharge regression values in terms of load. Gregory and Walling (1973) contended that the water discharge exponent of a concentration equation is normally one less than the water discharge exponent of the sediment load equation and this is found in the present study. The water discharge exponent of 0.288 on the concentration equation is comparable to those obtained by Douglas (1973) for small catchments in Queensland and New South Wales. The value is, however, low when compared with those evaluated for larger Australian streams (Loughran, 1973) and other streams of the world (Table 11). Although the correlation coefficients are useful in as much as they describe the bivariate relationships existing between the variables involved, they are difficult to interpret. The reason is that the factors influencing the suspended sediment concentration are closely interrelated and do not exist in terms of simple bivariate relationships. To overcome this limitation the multivariate analytical technique was investigated.

Table 11. Sediment Load Exponents

RIVER	BASIN AREA km^2	EXPONENT	SOURCE
Rio Galisteo US)	-	1.25	Leopold & Miller
Rio Peuco (US)	-	1.58	" (1956)
Brandywine (US)	813	2.37	Wolman (1955)
Hodge Beck (UK)	18.9	3.11	Imeson (1970)
Catchwater (UK)	15.4	1.35	"
Glomma St. 1 (Norwa)	8842	3.2	Nordseth (1973)
Glomma St.2 (Norway)	8842	4.2	"
Rapaalven (Sweden)	684	3.37	Axelsson (1967)
Gombak (Malaysia):			
i) Field centre	26.5	2.339	Douglas (1968b)
ii) 12½ milestone	44.86	2.772	"
iii) Jn. Pekeliling	140.00	1.885	"
iv) Sg. Pasir	3.26	2.685	"
Dumaresq:			
i) Wright College	64.8	1.21	Bevan (1970)
ii) Western Tributary	8.0	1.26	"
iii) Eastern Tributary	8.4	1.20	"
Dumaresq:			
i) Wright College	88.0	1.31	Burkhardt (1967)
ii) Faulkner Street	115.0	1.03	"
iii) Tombs Farm	138.0	1.09	"
Way Way Creek		0.90	Fletcher (1972)
Starlight Creek		1.67	"
Stoney Creek		0.95	"
Chandler at:			
Fassifern	16.0	1.48	Loughran (1973)
Maryburn	58.0	1.54	"
Camperdown	37.0	1.51	"
Lyndhurst	59.0	1.78	"
Spring Creek at Lynoch	38.0	1.53	"

Multivariate Analysis

The multiple regression model that best explains the suspended sediment concentration variation is log SSC = -0.1696 + 0.2893 log PQBS + 0.4619 + 1.2880 Sin TEMP. The prominence of the PQBS (maximum discharge before sampling) occurs because it summarizes the sediment producing processes and reflects the slow rate of sediment movements in relation to water movements. The API indicates the importance of the antecedent moisture status, while the TEMP parameter shows the importance of season. The proportion of the variance explained by this model is 35.64 percent which is rather low. The low value indicates the inefficiency of the parameters to measure changes occurring within the catchment which significantly cause infrequent erosion.

The fluctuation in suspended sediment concentration must be viewed in the context of the low concentrations that prevail. The lowest concentration recorded was 0.2 mg ℓ^{-1}, about 55 percent of the observations gave less than 5 mg ℓ^{-1} and only 25 percent had concentrations higher than 10 mg ℓ^{-1}. The concentrations are so low that any small addition of sediment resulting from small scale erosional events, such as caused by human or animal activities or any action of climatic elements, might cause considerable deviation in the relationships. An addition of about 1 mg ℓ^{-1} to those values less than 5 mg ℓ^{-1} could change the relationship by at least 20 percent and in the case of the lowest concentration the deviation is as much as 500 percent. This 'chance' erosional effect and the associated processes are not adequately accommodated in any of the variables within the multiple regression model. The effects of such small scale erosional events might not be noticeable in larger streams or in streams having a discharge partly derived from surface runoff with high concentration values.

The wide range of processes that could lead to a very small increase in sediment concentration is such that it would be impossible to measure or record each event. Rainfall characteristics can be adequately recorded, but the changes in the surface cover and the collapse of bank materials, particularly on very small scale, in all cases, cannot be detected easily. Large changes to the sediment concentrations can possibly be related to particular erosional events. For example, the high concentrations recorded from October 6, 1974 to October 12, 1974 were associated with the rain when a 4-meter deep well was dug to install the automatic float recorder. The highest concentration of 104.4 mg ℓ^{-1} was associated with the long intense spring storm, the largest storm recorded.

The rain-event observation. Generally, rain-event data which refers to the data collected during or after rain-events is less variable than the total set and it provides higher correlation coefficients between the hydrometeorological parameters and suspended sediment conecntrations.

The multiple regression model found to best describe the fluctuations in sediment concentration is log SSC = 0.4029 + 0.3312 log PQBS + 0.4716 log INT7. The model, explaining just over 45.16% of the total variance, is better than that developed for the total set of data but the number of observations is smaller. The inclusion of INT7 demonstrates the importance of energy supply to the catchment. PQBS is again prominent for reasons advanced earlier. The low level of explanation given by the multiple regression model again reflects the similar problems seen in the overall observations. Hence, the explanations of this still hinge upon two main factors. Firstly, the 'chance' and 'minor' erosional events caused by wide range of processes effectively upset the relationships. The influence is significant since the concentration values, in most cases, are very small. Secondly, the varied and random nature of the above influences in time and space within the catchment made them harder to be expressed by any set of hydrometeorological parameters.

The above analyses selected three parameters to explain the variation in sediment concentration for the overall observations and two parameters for the rain-event observations. To identify the interrelationships between the variables involved in each regression model, correlation matrices for overall data and rain-event data were developed (Tables 12 and 13). Apparently, the variables selected in any one model

Table 12. Correlation Matrix of Sandy
(Total observation N = 116)

	TEMP	Q	API	FINT	FIT	MQBS	PQBS	FISH	PQSH	SSC
TEMP	1.00									
Q	-0.03	1.00								
API	0.13	0.19	1.00							
FINT	0.02	0.78	0.42	1.00						
FIT	0.05	0.83	0.10	0.64	1.00					
MQBS	0.08	0.44	0.43	0.85	0.25	1.00				
PQBS	0.08	0.85	0.36	0.90	0.77	0.73	1.00			
FISH	0.01	0.76	0.37	0.86	0.55	0.78	0.77	1.00		
PQSH	0.01	0.79	0.25	0.70	0.62	0.53	0.69	0.92	1.00	
SSC	-0.06	0.70	0.39	0.72	0.54	0.55	0.76	0.67	0.66	1.00

Table 13. Correlation Matrix of Sandy-Rain-Event Samples (N = 85)

	Q	TEMP	TP3	INT7	INT8	INT9	DUR3	API	FINT	FIT	MQBS	PQBS	FISH	PQSH	SSC
Q	1.00														
TEMP	-0.02	1.00													
TP3	0.46	0.19	1.00												
INT7	0.36	0.33	0.57	1.00											
INT8	0.40	0.30	0.63	0.98	1.00										
INT9	0.37	0.32	0.69	0.94	0.97	1.00									
DUR3	0.09	0.05	0.56	0.03	0.01	0.04	1.00								
API	0.18	0.14	0.54	0.26	0.26	0.26	0.39	1.00							
FINT	0.78	0.01	0.53	0.44	0.48	0.44	0.14	0.43	1.00						
FIT	0.83	0.05	0.29	0.26	0.28	0.24	0.02	0.09	0.63	1.00					
MQBS	0.44	0.07	0.41	0.36	0.40	0.36	0.11	0.44	0.85	0.24	1.00				
PQBS	0.85	0.07	0.46	0.38	0.42	0.38	0.10	0.35	0.90	0.77	0.72	1.00			
FISH	0.76	0.03	0.54	0.50	0.53	0.49	0.12	0.37	0.86	0.78	0.78	0.77	1.00		
PQSH	0.79	0.02	0.48	0.46	0.49	0.47	0.10	0.23	0.70	0.52	0.52	0.68	0.92	1.00	
SSC	0.73	-0.02	0.48	0.51	0.54	0.49	0.07	0.37	0.76	0.57	0.57	0.79	0.70	0.68	1.00

are not significantly interrelated. In terms of the hydrometeorological factors, the basic denudation system for the overall model over the length of time considered can therefore be envisaged as being influenced by PQBS and API which in turn are subjected to the season of the year, expressed as TEMP. The rain-event denudation system is dependent upon the antecedent catchment conditions and sediment producing processes expressed by PQBS and the supply of energy for erosive and transporting works expressed by INT7.

From the principal component analysis, six eigenvectors were obtained (Table 14) for all of the selected independent variables (Table 8). The degree of association between a variable and a component is indicated by the factor loading (Table 15). Component one is the most significant. The variables which are significantly associated with this component are flood intensity (FINT) and peak discharge before sampling (PQBS). It condenses the antecedent wetness condition. Component two reflects the energy supply as indicated by the significant association of INT7 (max. 15-min. intensity), INT8 (max. 3-min intensity), INT9 (max. 60-min intensity) and KEA (product of kinetic energy and amount of rain at INT8) with this component. Component three describes the amount of moisture available within the catchment. TP3 (total 48-hour rain), API (antecedent precipitation index) and DUR3 (Duration of 48-hour rain) are the three significant variables in this component. Components four to six which collectively explain only 12.23 percent of the total variance can be considered

Table 14. Principal Component Analysis

Component	Eigenvalue	Accumulated % of Total Variance	Percent Explained
1	7.3616	57.59	57.59
2	2.3248	75.78	18.19
3	1.3649	86.45	10.67
4	0.7443	92.27	5.82
5	0.5318	96.43	2.16
6	0.2873	98.68	2.25

Table 15. Factor Loadings

Eigenvector	Component 1	Component 2	Component 3	Component 4	Component 5	Component 6
Variance	8.0036	2.3695	1.3694	0.7453	0.5321	0.3097
SSC	0.2855	0.1314	0.0550	0.0367	0.0233	0.2763
Q	0.2753	0.2566	0.1432	-0.3197	0.0037	0.0224
TEMP	0.0501	-0.3245	-0.0661	-0.3434	-0.8431	-0.1287
API	0.1572	-0.0105	-0.5213	0.1977	-0.2343	0.5092
TP3	0.2514	-0.1469	-0.3796	-0.1917	0.2251	-0.0346
INT7	0.2620	-0.3694	0.1387	0.0503	0.0808	0.0679
INT8	0.2753	-0.3538	0.1181	0.0421	0.1258	0.0323
INT9	0.2650	-0.3723	0.0750	0.0040	0.1455	0.0068
DUR3	0.0584	0.0241	-0.6502	-0.3478	0.2643	-0.2331
FINT	0.3039	0.2154	-0.0397	0.1596	-0.1382	0.0231
FIT	0.2177	0.2713	-0.2173	-0.4559	0.0130	0.3040
MQBS	0.2451	0.1536	-0.1365	0.5583	-0.1476	-0.1980
PQBS	0.2910	0.2701	0.0376	0.0061	-0.0693	0.2205
FISH	0.3062	0.1610	-0.0020	0.1265	-0.1013	-0.4264
PQSH	0.2821	0.1577	0.0971	-0.1188	-0.0022	-0.4711
KEA	0.2763	-0.3489	0.1277	0.0478	0.1163	0.0359
SUM OF SQUARES	1.0000	1.0000	1.0000	1.0000	1.0000	1.0000

as insignificant due to eigenvalue less than unity (Cooley and Lohnes, 1962).

The principal component analysis has effectively reduced the 15 hydrometeorological parameters to 6 underlying dimensions. The largest components, one and two, are represented graphically in Fig. 9. Apparently all of the runoff variables

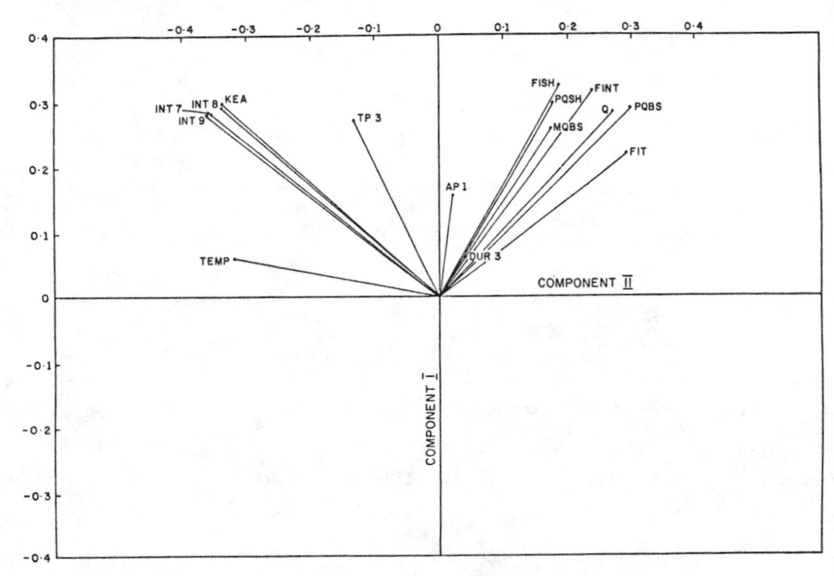

Fig. 9. Plot of loadings on Component I against loadings on Component II for suspended sediment concentration of Sandy

are in the first quadrant and the rainfall variables in the fourth quadrant. Components one and two measure two trends in the hydrometeorological parameters which were earlier observed in the multiple regression model. These trends, an increase in sediment concentration with increased rate and magnitude of antecedent flood and the increase in sediment concentration with increased energy input, are condensed into two unrelated new variables. Although principal component analysis gives a coordinated expression of the relationship between all of the variables, no single factor or component can be pinpointed as the sole cause of the sediment concentration variation. The cause is still the product of a complex interplay of a number of factors which are sorted out into the six dimensions or components. Only the relative importance of each component can be elaborated.

The above analysis demonstrates that over the length of time being considered, the magnitude and rate of the preceding flood (component one) is responsible for the greatest concentration variation, the effect of which is subjected to the energy input (component two).

Suspended Sediment Discharge

The suspended sediment discharge was calculated from the rating curve and observed values. Since the frequency of sampling during my particular storm event is small, at most four, and the number of storm events is only three, detail statistical analysis to determine the importance of any particular hydrometeorological parameters in determining the amount of load discharged, as employed by Walling (1974), is not possible.

Over the study period, a total of about 2.493 tons of suspended sediment were discharged. Figure 10 shows the

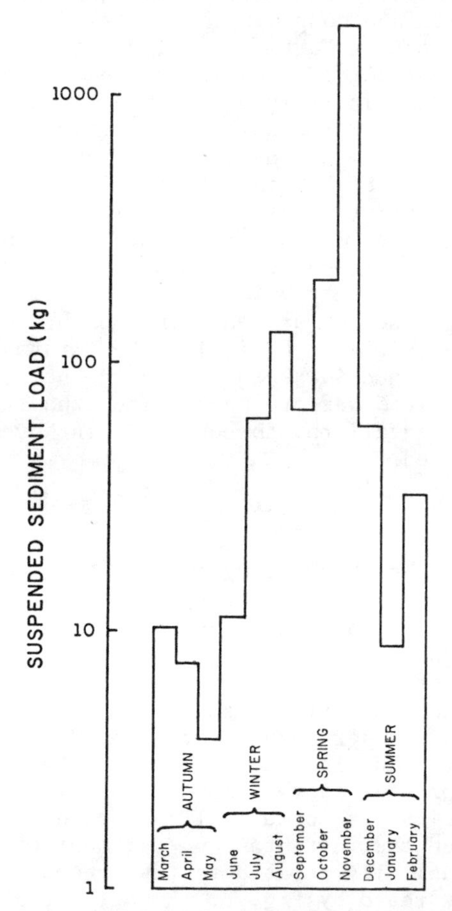

Fig. 10. Monthly sediment discharge from Sandy

monthly distribution. There appears to be an obvious seasonal pattern which is the reflection of the effectiveness of the denudation processes. Spring and summer (September-February), which are associated with the high frequency of storm events and the dry, dusty catchment conditions, account for about 90 percent of the total load discharged. Most of the load was transported out by infrequent flood flows following storm events. Low flows are very insignificant. About 91 percent of the load over the year was removed by this infrequent phenomena with the greatest flood responsible for more than 75 percent of the year's total. On a time scale, about 35 percent of the total load was removed in only 0.27 percent of the period of observation and 67.3 percent was removed in just 0.60 percent of the observed period.

The effectiveness of any individual flood is influenced by the antecedent conditions which are in accordance with those situations observed in the multivariate analysis (multiple regression and component analysis). The storm which caused the flood on November 1, 1974 was less effective than that of December 12, 1974 as indicated by the KI values. However, the intense rain is spread over a longer period and the total rainfall (TP) is much more than that for flood 4. In addition, the moisture status of the catchment (API) prior to November 1, 1974 is higher than that prior to December 11, 1974, therefore, relatively less rainwater is retained in the soil during the November 1, 1974 to November 4, 1974 period to replenish the moisture deficit. Looking at the antecedent conditions, it is shown that prior to November 1, 1974 there was no significant rise for at least one month. Hence the flood of November 1, 1974 is basically the first catchment flushing process. Flood 4 was the fifth in the month and therefore not much load was produced. The exhaustion effect is very apparent when all the three floods in November and December are compared.

Conclusion

The above analyses have brought to light numerous features regarding the relationships of the hydrometeorological factors and the suspended sediments on Sandy Creek. Instantaneous discharge, often found to be the most significant independent variable in other studies, is not as important in the present situation. Instead, peak discharge before observation is prominent. Like any other runoff variable, PQBS embraces a series of sediment producing factors. This variable is prominent in the present context due to the slow rate of sediment movement and its peak comes well after the water discharge peak. Hence the instantaneous sediment concentrations or discharge is basically related to the preceding discharge. This is well illustrated by Fig. 8. Runoff variables are better related to the sediment concentrations than the rainfall variables. This is inevitable since runoff parameters condense all of the sediment producing factors which include rainfall variables, while rainfall variables express the quantity or moisture and energy input without gauging their effectiveness in relation to the changing catchment cover with time. However, the time-dependent catchment condition is well accommodated into the runoff variables. There is the tendency for a seasonal variation to occur in suspended sediment concentration. This is partly due to the differences in the rainfall-runoff relationship with season, the intensity and frequency of rain events and to some degree the influence of algae growth in the streams in warm seasons.

The low concentration values observed in this catchment render it sensitive to any minor erosional events. This phenomena and the infrequent erosional events become significant in explaining the variations of sediment concentration. Their effects are very eminent with low concentrations resulting in high variability.

The principal component and the multiple regression analyses both show that the main direction of variance operating in the suspended sediment model can be explained in terms of the antecedent flood condition, energy input and season of the year.

Most of the sediment outputs from the catchment are transported by the infrequent event which occurs once or twice a year. Low flows are insignificant in transporting the load out of the catchment.

The low level of explanation obtained in the multiple regression equations suggests that other factors are as important, if not moreso, in influencing the suspended sediment output. The hydrometeorological factors alone cannot fully explain all of the variations.

References

[1] American Society for Testing and Materials, 1968. "Tentative Methods of Tests for Particulate Contaminant in Aviation Turbine Fuels," Book of ASTM Standards, 17, pp. 837-850.

[2] Anderson, H. W., 1954. "Suspended Sediment Discharge as Related to Streamflow, Topography, Soil and Landuse," Trans. Amer. Geophy. Union, 35(2), pp. 268-281.

[3] Anderson, H. W., 1957. "Relating Sediment Yield to Watershed Variables," Trans. Amer. Geophys. Union, 38(6), pp. 921-924.

[4] Anderson, R. H., 1956. "The Trees of New South Wales," 3rd edition, Government Printer, Sydney.

[5] Axelsson, V., 1967. "The Laitaure Delta," Geografiska Annaler, 49A(1), pp. 1-127.

[6] Bevan, J. R., 1970. "The Dumaresq Creek Catchment Above Armidale - A Water Balance and Sediment Transport Study," Unpubl. M.Sc. Thesis, U.N.E.

[7] Branson, F. E. and Owen, J. R., 1970. "Plant Cover, Runoff and Sediment Relationships on Mancos Shale in Western Colorado," Water Resources Research, 6(3), pp. 783-790.

[8] Burkhardt, J., 1967. "Sediment Transport in the Dumaresq Creek with Special Reference to Siltation in the City of Armidale," Unpub. Hons. Thesis, Univ. New England, Armidale.

[9] Christian, C. S. and Stewart, G. A., 1953. "General Report on Survey of Katherine-Darwin Region, 1946," CSIRO Aust. Land Res. Ser., No. 1.

[10] Douglas, I., 1966. "Denudation Rates and Water Chemistry of Selected Catchments in Eastern Australia and Their Significance for Tropical Geomorphology," Unpub. Ph.D. Thesis, Aust. National University, Canberra.

[11] Douglas, I., 1968b. "Erosion in the Sungei Gombak Catchment, Selangor, Malaysia," Jour. of Tropical Geography, 26, pp. 1-16.

[12] Douglas, I., 1969. "The Efficiency of Humid Tropical Denudation Systems," Trans. Inst. Br. Geogr., 46, 1-16.

[13] Douglas, I., 1973. "Rates of Denudation in Selected Small Catchments in Eastern Australia," University of Hull, Occasional Papers in Geography, No. 21.

[14] Edwards, A.M.C., 1973a. "Dissolved Load and Tentative Solute Budgets of Some Norfolk Catchments," Jour. Hydrol. 18, pp. 201-217.

[15] Edwards, A.M.C., 1973b. "The Variation of Dissolved Constituents with Discharge in Some Norfolk Rivers," Jour. Hydrol., 18, pp. 219-242.

[16] Fletcher, G. F., 1972. "Fluvial Erosion of a Selected Area of the Mid-North Coast, New South Wales," Unpub. B.A. (Hons.) thesis, Univ. New England, Armidale.

[17] Gibbons, F. R. and Hallsworth, E. G., 1953. "The Soils of the New England Region," In: The New England Region-A Preliminary Survey of Resources, Govt. Printer, Sydney.

[18] Gottschalk, L. C., 1964. "Reservoir Sedimentation," In: V.T. Chow (ed.), Handbook of Applied Hydrology, 17-1-17-34.

[19] Gravelius, 1914. "Flusshunde," Berlin and Leipzig.

[20] Gregory, K. J. and Walling, D. E., 1973. "Drainage Basin Form and Process," Edward Arnold, London.

[21] Guy, H. P., 1964. "An Analysis of Some Storm-Period Variables Affecting Stream Sediment Transport," U.S. Geol. Sur. Prof. Paper, 462-E.

[22] Hall, D. G., 1967. "The Pattern of Sediment Movement in the River Tyne," Internat. Assoc. Sci. Hydr. Pub. 75, pp. 117-142.

[23] Hallsworth, E. G., Costin, A. B., and Gibbons, F. R., 1953. "Studies in Pedogenesis VI on the Classification of Soils Showing Features of Podzol Morphology," Jour. Soil Sci., 4, pp. 241-256.

[24] Harmon, N. R., 1965. "Filtering Water Samples Using Vacuum from Automobile Intake Manifold," U.S. Geol. Surv. Water Supply Paper, 1822, pp. 61-62.

[25] Heidel, S. G., 1956. "The Progressive Lag of Sediment Concentration with Flood Waves," Trans. Am. Geophys. Union, 37(1), pp. 56-66.

[26] Horton, R. E., 1932. "Drainage Basin Characteristics," Trans. Amer. Geophys. Union, 13, pp. 350-361.

[27] Imeson, A. C., 1970. "Erosion in Three East Yorkshire Catchments and Variations in Dissolved, Suspended and Bedload," Unpub. Ph.D. Thesis, University of Hull.

[28] Imeson, A. C., 1971. "Hydrological Factors Influencing Sediment Concentration Fluctuations in Small Drainage Basins," Earth Science Jour., 5(2), pp. 71-78.

[29] Imeson, A. C., 1974. "The Origin of Sediment in a Moorland Catchment with Particular Reference to the Role of Vegetation," In Gregory and Walling (eds.) Fluvial Processes in Instrumented Watersheds, Inst. Br. Geogr. Sp. Publ. No. 6, pp. 59-72.

[30] Jessup, R. W., 1965. "The Soils of the Central Portion of the New England Region, N.S.W.," CSIRO Soil Publication No. 21.

[31] Langbein, W. B. and Schumm, S. A., 1958. "Yield of Sediment in Relation to Mean Annual Precipitation," Trans. Amer. Geophys. Union, 39, pp. 1076-1084.

[32] Leopold, L. B. and Miller, J. P., 1956. "Ephemeral Streams - Hydraulic Factors and Their Relation to the Drainage Net," U.S. Geol. Sur. Prof. Paper 282-A.

[33] Loughran, R. J., 1968. "Fluvial Erosion on the New England Tableland, N.S.W.," Unpub. M.Sc. Thesis, Univ. New England, Armidale, N.S.W.

[34] Loughran, R. J., 1969. "Fluvial Erosion in Five Small Catchments Near Armidale, N.S.W.," Research Series in Physical Geography, No. 1, Univ. New England, Armidale.

[35] Loughran, R. J., 1973. "The Downstream Variation of Suspended Sediment and Total Solute Transport in Two N.S.W. River Basins," Unpub. Ph.D. Thesis, Univ. New England, Armidale.

[36] Lustig, L. K., 1965. "Sediment Yield of the Castaic Watershed, Western Los Angeles County, California. A Quantitative Geomorphic Approach," U.S. Geol. Sur. Prof. Paper 422-F.

[37] Maner, S. B., 1958. "Factors Affecting Sediment Delivery Rates in the Red Hills Physiographic Area," Trans. Amer. Geophys. Union, 39, pp. 669-675.

[38] Mustonen, S. E., 1967. "Effects of Climatologic and Basin Characteristics on Annual Runoff," Water Resour. Res., 3(1), pp. 123-130.

[39] Nordseth, N., 1973. "Fluvial Processes and Adjustments on a Braided River. The Islands of Koppangsoyene on the River Glomma," Norsk Geografisk Tidsskrift, 27(2), pp. 77-108.

[40] Plummer, B.A.G., 1965. "Forests and Woodlands," In: Warner, R. F. (ed.) A Preliminary Report on the Geography of Dumaresq Shire, Council of Dumaresq Shire, Armidale, N.S.W., IX/1-IX/10.

[41] Premier's Department, 1953. "The New England Region, A Preliminary Survey of Resources," Sydney, pp. 21-26.

[42] Schumm, S. A., 1954. "The Relation of Drainage Basin Relief to Sediment Loss," Internat. Assoc. Sci. Hydrol. Publ. 36, pp. 216-219.

[43] Schumm, S. A., 1956. "The Evolution of Drainage Systems and Slopes in Badlands at Perth Amboy, New Jersey," Geol. Soc. Amer. Bull., 67, pp. 597-646.

[44] Schumm, S. A., 1965. "Quaternary Palaeohydrology," In: H. E. Wright and D. G. Frey (eds.) The Quaternary of the United States, Princeton, pp. 783-794.

[45] Schumm, S. A. and Hadley, R. F., 1963. "Progress in the Application of Landform Analysis in Studies of Semi-Arid Erosion," U.S. Geol. Sur. Prof. Paper 598.

[46] Society of Automotive Engineers, 1963. "Procedure for the Determination of Particulate Contamination in Hydraulic Fluids by the Control Filter Gravimetric Procedure, Aerospace Recommended Practice," ARP, 785.

[47] Strahler, A. N., 1957. "Quantitative Analysis of Watershed Geomorphology," Trans. Amer. Geophys. Union, 38, pp. 913-920.

[48] Strakhov, N. M., 1967. "Principles of Lithogenesis. 1," London, Oliver and Boyd.

[49] Thompson, R. D., 1970. "The Dynamic Climatology of Northeast New South Wales," Ph.D. Thesis, Univ. New England, Armidale.

[50] Thompson, R. D., 1973a. "The Contribution of Airflow Circulation to Local Temperatures and Rainfall in the New England Area, N.S.W. Australia," Arch. Meteorol. Geophys. Biochem., Ser. B. 21, pp. 175-188.

[51] Thompson, R. D., 1973b. "Some Aspects of the Synoptic Mesoclimatology of the Armidale District, N.S.W. Australia," Jour. Appl. Meteorol., 12, pp. 578-588.

[52] Walling, D. E., 1974a. "Suspended Sediment and Solute Yields from a Small Catchment Prior to Urbanization," In: Gregory and Walling (eds.) Fluvial Processes in Instrumented Watersheds, Inst. Br. Geogr. Spec. Pub. No. 6, pp. 169-192.

[53] Walling, D. E. and Teed, A., 1971. "A Simple Pumping Sampler for Research into Suspended Sediment Transport in Small Catchments," Jour. Hydrol. 13, pp. 325-337.

[54] Warner, R. F., 1963. "New England Essays," Univ. New England, Armidale, N.S.W.

[55] Williams, J. B., 1963. "The Vegetation of Northern N.S.W. from the Eastern Scarp to the Western Slopes - A General Transect," In: Warner, R. F. (ed.) New England Essays, Univ. New England, Armidale, N.S.W., pp. 41-52.

[56] Williams, R. C., 1964. "Sedimentation in Three Small Drainage Basins in the Alsea River Basin, Oregon," U.S. Geol. Surv. Circ. 490.

[57] Wolman, M. G., 1955. "The Natural Channel of Brandywine Creek, Pennsylvania," U.S. Geol. Sur. Prof. Paper, 271.

DISCUSSIONS - SECTION V
Surface Water Quality

T. Singh*

Water quality analyses are expensive, yet baseline data under varying input conditions are required to assess landuse practices. A simple procedure is needed to predict ionic concentrations with reasonable accuracy. Streamflow measurements were used as the main basis for such predictions because streamflow is a routinely measured variable on all gauged watersheds.

The paper presents models which can be used to predict concentrations of eight constituents from streamflow and specific conductance measurements. An improvement in prediction is further relieved if streamflow is divided into two main components: stormflow (rapid flow plus interflow) and baseflow. These components can be readily obtained from streamflow measurements by running a computer program.

The approach suggested can be used on any mountain stream. Data encompassing full range of variability are needed to determine model parameters. Once this has been done, only occasional updating is required for checking the accuracy of the predictions by calculating the errors of estimate as shown in the paper.

The need for this study arose from the fact that appreciable changes in many of the studied nutrients were noticed in the Hinton (Alberta) clearcuts as a result of pulpwood harvesting operations. Techniques like the one presented here can be used to provide a measure of the change in the desired constituents as a result of such operations by using streamflow components and specific conductance.

Dale Huff**

On the issue of treating urban storm water by simple settling: Have you also considered the approach that would attempt to eliminate the P and suspended solids at the source by a comprehensive street sweeping program? In Madison, Wisconsin, the Lake Wingra studies of the International Biological Program showed that there are certain

*Canadian Forestry Service, Northern Forest Research Centre, 5320 - 122 Street, Edmonton, Alberta, Canada.

**Civil Engineering Department, Stanford University, Stanford, California.

critical periods during the year when sweeping could be very effective and the City has instituted an education program together with street sweeping to reduce nutrient loadings to the lakes around the city. I think it is an option worth considering.

<u>Ian Cordery*</u>

Discussion was of paper - nothing other than contained in submitted paper was stated. Answer to questions:

Question - Large amounts of floating materials. These were not measured but were visually observed to be highly correlated with suspended solids and would need to be removed by simple screening of stormwater, either at the inlet or outlet of settling ponds.

Question - Was treatment of catchment such as sweeping considered: This was not considered as a treatment alternative but could reduce wasteload carried in stormwater considerably. Public education would seem to be the major requirement - to reduce dumping of trash on the urban catchment surface (grass cuttings, paper, cans, etc.) and to have the public require municipal authorities to clean streets.

<u>Daniel Cluis**</u>

(A nutrient transport model based on exports by land-uses.) The relatively important contribution in nutrients by livestock (more than 50%) was shown, and the trend to mass-breeding from traditional breeding may in the future permit the extraction of highly concentrated nutrients (point-sources) in priority to the construction of new municipal treatment plants.

<u>Jacque Bernier***</u>

Stochastic Point of View in Oxygen Balance Models in Streams. The author summarizes his paper in English.

*University of NSW, Australia.

**INRS-Eau, Universite du Quebec, C.P. 7500, Quebec 10, Quebec, Canada.

***Laboratoire National d'Hydraulique, 6, quai Watier, 78400 - Chatou, France.

Michael C. Quick*

In Cordery's slides, a major problem in storm runoff appeared to be the floatables which are certainly not removed by settling. Can you comment on this aspect which is primarily an aesthetic problem.

A. Nir**

Question to the coauthor of Dr. Woolhiser. Did you observe in your experiments any time lag between the arrival of the water and the concentration of pollutants and is there any estimate for their velocity ratio?

A. S. Zakaria***

1. Correct the title of paper. Should read: Effects of the hydrometeorological factors on suspended sediment output of a catchment in New England, Australia.

2. Emphasizing the importance of factors other than hydrometeorological in the suspended sediment output model. Hydrometeorological factors only account for 45% of the total variance.

3. In terms of the hydrometeorological factors, antecedent runoff and rainfall and temperature variables are important in suspended sediment output.

4. Peak water discharges/storm discharges are the important transporting media of suspended sediment load. Floods account for more than 75% of the load.

5. In terms of time, 67.3% were transported in just 0.60 % of the total study period.

Thomas G. Sanders***

1. Question concerned type or W.Q. parameter: I said it was a conservative, dissolved tracer Rhodamine WT. I mentioned that we controlled most hydrological inputs, rainfall

*Address unknown

**Isotope Dept., The Weizmann Institute of Science, Rehovot, Israel.

***Dept. of Geography, National University of Malaysia, Kuala Lumpur, Malaysia.

****Dept. of Civil Engineering, Colorado State University, Fort Collins, Colorado.

intensity, duration, watershed shape, slope, roughness, to determine effect on water quality hydrograph.

2. Question concerning velocity of wave and particles. I did not have data at hand but data proved that wave moved faster than the particle.

David R. DeWalle*

The problem we were addressing in this study was the effect of shade removal associated with forest clearing on maximum stream temperatures. Currently a method developed by Brown is being used to predict clearing effects. His technique shows that maximum stream temperatures achieved in an unshaded reach increase linearly with distance from the inlet. Our calculations verify his technique for positions close to the inlet, but at points farther from the inlet maximum temperatures curvilinearly approach an equilibrium where absorbed solar heat gains are balanced by losses due to bottom conduction, longwave radiation, evaporation and sensible heat convection.

Stephen J. Burges**

The general report leaves the impression that little has been done with respect to determining model parameters from "noisy" data. Much work has been done on parameter identification and an updating parameter estimates. Appropriate material can be found in the following references:

Gelb, A., (Editor), Applied Optimal Estimation, MIT Press, Cambridge, Mass., 1974.

Box, G.E.P. and G. M. Jenkins, Time Series Analysis, Forecasting and Control, Holden Day, San Francisco, 1970.

Proceedings IIASA/WMO Workshop on "Recent Developments in Real-Time Forecasting/Control of Water Resource Systems," Oct. 18-20, 1976, Luxemburg, Austria.

---

*Forest Hydrology, Pennsylvania State University, University Park, Pennsylvania 16802.

**Civil Engineering Department, University of Washington, Seattle, Washington.

Gerhard Huthmann*

In order to complete the general report, I would like to show you some slides.

First of all I have to explain what the problem was. The first slide (Fig. 1 in the paper) shows the subbasin of the River Fhine between Karlsruhe and Mainz. The problem is to compute the chloride concentration at station C13 by measuring the chloride concentration and discharges at stations C11 and C12. For a natural water, the concentration of chloride can be represented approximately by the equation

$$C = k/Q^n$$

where C is the concentration in mg/l and Q is stream discharge in m^3/s. k and n are regression parameters. On logarithmic paper, this relation is a straight line.

$$\log C = \log k - n \log Q$$

The second slide (Fig. 4) shows the chloride concentration of station C13 against the actual discharge. It indicates that the above relation is not valid, because the chloride concentration is strongly influenced by industrial waste waters. The low correlation coefficient of 0.67 confirms this.

Therefore, alternative methods have to be employed for predicting the chloride concentration of station C13.

In this paper, I have used a method called Mutliple Frequency Response Analysis. It depends on the equivalent representation of a time series in the frequency domain. Time is too short to go into details. But let me point out the main features.

In this method, the hydrological system is regarded as a dynamic linear system with multiple input and single output (MISO).

The next slide (Fig. 7) shows the scheme of such a system based on multiple linear impulse response functions. The system I have used in the paper describes the simplest form of a multiple system, i.e., a system with two inputs and one output. The impulse response functions for this system are shown in the next slide (Fig. 9). Please notice, that h_t is also computed for t less than zero. This is very

---

*Federal Institute of Hydrology, Postfach 309, D-5400 Koblenz, Federal Republic of Germany.

unconvenient but due to the fact that my time is short, I can not go into details. If you are interested in this feature, please don't hesitate to ask me during the discussion.

However, I would like ot show you the result. The next slide (Fig. 10) shows the chemical hydrographs of the chloride concentration at input stations C11 and C12 and output station C13. The very high variation is a result of industrial waste waters.

The result of the simulation is shown in the next slide (Fig. 11). The mean error is about 10%, which, because of data error, is not a breath-taking but still a very useful result. I may point out that the prediction of chloride is only *one* example of Multiple Frequency Response Analysis. This technique can be used in the same way for all conservative water quality parameters.

Timothy D. Steele*

The Yampa River Basin Assessment reflects one end of an analytical spectrum in promoting a regional overview to screen out the several water-quality and water-quantity impacts of coal-resource development in the basin. In this semi-arid area, quality interactions with water availability are quite important. The stream system presently is largely unregulated; however, numerous major reservoirs and transbasin diversions have been proposed. Institutional and legal aspects of water, undoubtedly, will play a major role in constraining future demands and changes in water use. Primary impacts of coal mining, transport, and conversion are being evaluated using mass-and-energy balance-process models for residuals discharge and water use. Secondary impacts of increased population growth and development of industries, commerce, and services are being evaluated using regional economic input-output growth factors coupled with residuals and water-use coefficients. Interactions of discharged residuals with current or proposed environmental controls are important in determining the types and forms of residuals that are to be assimilated into the environment.

---

*U.S. Geological Survey, Federal Center, Lakewood, Colorado.